AF353146

Marine Compounds and Cancer 2020

Marine Compounds and Cancer 2020

Editors

Friedemann Honecker
Sergey A. Dyshlovoy

MDPI • Basel • Beijing • Wuhan • Barcelona • Belgrade • Manchester • Tokyo • Cluj • Tianjin

Editors
Friedemann Honecker
Tumor and Breast Center ZeTuP
St. Gallen
Switzerland

Sergey A. Dyshlovoy
University Medical Center
Hamburg-Eppendorf, Germany;
A.V. Zhirmunsky National
Scientific Center of Marine
Biology, Russia

Editorial Office
MDPI
St. Alban-Anlage 66
4052 Basel, Switzerland

This is a reprint of articles from the Topical Collection published online in the open access journal *Marine Drugs* (ISSN 1660-3397) (available at: https://www.mdpi.com/journal/marinedrugs/special_issues/marine-compounds-cancer).

For citation purposes, cite each article independently as indicated on the article page online and as indicated below:

LastName, A.A.; LastName, B.B.; LastName, C.C. Article Title. *Journal Name* **Year**, *Volume Number*, Page Range.

ISBN 978-3-0365-0630-2 (Hbk)
ISBN 978-3-0365-0631-9 (PDF)

Contents

About the Editors

Friedemann Honecker, MD, PhD, studied medicine in Germany and the UK between 1991 and 1999. In 1999, he started his medical and scientific career in internal medicine with a focus on oncology and hematology at the University Medical Center Tübingen, Germany. From 2002 to 2004, he earned his PhD in experimental pathology at Rotterdam Erasmus MC in Rotterdam, NL, focusing on germ cell tumor development and mechanisms of drug resistance. From 2005 to 2013, he worked in the Department of Oncology and Hematology at Hamburg University Hospital, both as a senior consultant in Oncology and group leader of the "Laboratory of Experimental Oncology" of the University Cancer Center Hamburg (UCCH), a comprehensive cancer center. Since 2013, he has worked as oncologist and researcher at the Tumor and Breast Center ZeTuP in St. Gallen, Switzerland. His main research interests are breast cancer and genitourinary cancers, the development of new anticancer substances, and the treatment of elderly cancer patients. He has published over 100 original Pub-Med listed articles and reviews and is author of numerous book chapters.

Sergey A. Dyshlovoy, PhD, trained in chemistry and biochemistry in the Russian Federation and in Germany. He started his scientific career in 2006. Between 2009 and 2012, he earned his PhD in chemistry, focusing on the elucidation of structure and mode of action of novel small-molecule marine bioactive compounds with anticancer activity in the G.B. Elyakov Pacific Institute of Bioorganic Chemistry (Vladivostok, Russian Federation). Since 2012, he has been a senior researcher in both the aforementioned institute and the Far Eastern Federal University (Vladivostok). Since 2013, he has worked as a postdoctoral researcher at the University Medical Center Hamburg-Eppendorf (Hamburg, Germany) in the Department of Oncology. Since 2019, he has been a senior researcher in the A.V. Zhirmunsky National Scientific Center of Marine Biology (Vladivostok, Russian Federation). Since 2014, he has acted as a member of the Commission of Experts of the Russian Scientific Foundation. His main research interest is the investigation of the mechanisms of anticancer action of novel marine compounds. To date, he has published over 80 research articles as well as several books and patents. He is a member of the Editorial board and reviewer in many scientific journals.

Preface to "Marine Compounds and Cancer 2020"

In 2019, the scientific and medical community celebrated the 50th anniversary of the introduction of the very first marine drug, Cytarabine (aka Ara-C, Cytosar-U®), in clinical use. In 1969, it was approved by the Food and Drug Administration (FDA) for the treatment of leukemia. At the beginning of 2021, the list of approved marine-derived anticancer drugs consists of nine substances, five of which received approval within the last two years, demonstrating the rapid evolution of the field. Many more are in different phases of clinical testing, including phase III trials, and a plethora of substances have already been examined regarding activity in vitro and in vivo. Technical innovation and increasingly precise research tools allow the dissection of the molecular mode of action of these cytotoxic substances, often uncovering the specific drug targets in cancer cells. This development will blur the boundaries between targeted and untargeted therapy, and will hopefully lead to a more direct use of cancer medicine based on a molecular rationale of activity. The Topical Collection "Marine Compounds and Cancer" of the journal *Marine Drugs* covers this whole field, from agents with cancer-preventive activity, to novel and previously characterized compounds with anticancer activity, both in vitro and in vivo, as well as reports on the latest status of the development of drugs under clinical investigation. In addition, special focus is placed on the current shortfalls and possible strategies to overcome obstacles in the area of marine anticancer drug development.

Friedemann Honecker, Sergey A. Dyshlovoy
Editors

Editorial

Marine Compounds and Cancer: Updates 2020

Sergey A. Dyshlovoy [1,2,3,*] and Friedemann Honecker [2,4]

1 Laboratory of Pharmacology, A.V. Zhirmunsky National Scientific Center of Marine Biology, Far Eastern Branch, Russian Academy of Sciences, 690041 Vladivostok, Russia
2 Laboratory of Experimental Oncology, Department of Oncology, Hematology and Bone Marrow Transplantation with Section Pneumology, Hubertus Wald-Tumorzentrum, University Medical Center Hamburg-Eppendorf, 20251 Hamburg, Germany; Friedemann.Honecker@zetup.ch
3 Martini-Klinik, Prostate Cancer Center, University Hospital Hamburg-Eppendorf, 20251 Hamburg, Germany
4 Tumor and Breast Center ZeTuP St. Gallen, 9000 St. Gallen, Switzerland
* Correspondence: dyshlovoy@gmail.com

Received: 1 December 2020; Accepted: 6 December 2020; Published: 15 December 2020

By the end of the year 2020, there are nine marine-derived anticancer drugs available on the market, and the field is currently growing exponentially. This process is stipulated by improvements in the development of biomedical sciences in general and recent approval of new and exciting anticancer medications in particular, which were developed based on small molecules of marine origin.

Looking back, it is noteworthy that at the very beginning of 2018, when we published an article on updates in the field of marine anticancer agents, there were only four marine-derived drugs approved for the treatment of cancer and cancer-related conditions [1]. Those were **cytarabine** (Cytosar-U®, the very first marine-derived drug [2] approved in 1969 produced by Pfizer [3]), **trabectedin** (Yondelis®, produced by PharmaMar), **eribulin mesylate** (Halaven®, produced by Eisai Inc.), and the antibody–drug conjugate (ADC) **brentuximab vedotin** (Adcetris®, produced by Seattle Genetics) [1]. Within only three years since 2018, five (!) new drugs have been approved for the treatment of different cancer types all over the world; two of them have been approved only recently in 2020. Thus, to the four "marine" pharmaceuticals listed above, the following medications were added:

- **Plitidepsin** (Aplidin®, produced by PharmaMar), dehydrodidemnin B, is an ascidian depsipeptide binding to eEF1A2 and inducing an oxidative stress in cancer cells; the drug was first approved in 2018 in Australia for the treatment of multiple myeloma, leukemia, and lymphoma [4].

- **Polatuzumab vedotin** (Polivy™, produced by Genentech, Roche) is an ADC that consists of MMAE (monomethyl auristatin E, an analogue of dolastatin 10, which is a peptide toxin of symbiotic marine cyanobacteria) conjugated with the CD76b-specific monoclonal antibody polatuzumab. The antibody provides a specific delivery of MMAE to cancer cells, where following the proteolytic ADC cleavage and release of the "warhead" molecule (MMAE), an inhibition of tubulin polymerization leading to the cancer cells' death can be achieved. The drug was approved by the FDA in 2019 for the treatment of B-cell lymphomas, non-Hodgkin lymphomas, and chronic lymphocytic leukemia [5].

- **Enfortumab vedotin** (PADCEV™, produced by Astellas Pharma and Seattle Genetics) is another ADC consisting of MMAE (see above) and an antibody specific to nectin-4. It was approved in 2019 for the treatment of metastatic urothelial cancer [6].

- **Belantamab mafodotin** (Blenrep™, produced by GlaxoSmithKline) is yet another ADC consisting of MMAF (monomethyl auristatin F, one of the MMAE derivatives) as the warhead, bound to an antibody targeting BCMA (B-cell maturation antigen). Similar to MMAE, MMAF targets tubulin polymerization. The drug was approved in 2020 for the treatment of relapsed and refractory multiple myeloma [7].

- **Lurbinectedin** (Zepzelca^{TM}, produced by PharmaMar) is a synthetic derivative of trabectedin (see above) that binds to the minor groove of DNA and exerts its anticancer action via inhibition and degradation of RNA polymerase II. The drug was approved in 2020 for the treatment of metastatic small cell lung cancer [8].

According to the Marine Pharmacology web page provided by Prof. Alejandro M. S. Mayer and colleagues (https://www.midwestern.edu/departments/marinepharmacology.xml), there are currently another 23 "marine" molecules in different stages of clinical development in various cancer entities [9]. The vast majority of these drug candidates (i.e., 19 out of 23 (83%)) are being tested as anticancer drugs. It should, however, be noted that most of the molecules (70%) either are ADC derivatives of MMAE or MMAF or are already approved drugs undergoing trials in entities where they have not yet been approved (e.g., lurbinectedin in ovarian, breast, and small cell lung cancer) [9].

To keep track in this dynamic area and also to offer a platform on research dealing with new and promising marine compounds possessing anticancer activity, we started a topical collection, "Marine Compounds and Cancer" (http://www.mdpi.com/journal/marinedrugs/special_issues/marine-compounds-cancer), in 2015 [10]. This topical collection covers the whole scope of agents showing in vitro and in vivo anticancer properties, which are able to prevent cancer development or can kill existing cancer cells. We publish data on both novel and previously characterized compounds, which either just have started to come to the attention of biomedical scientists or already have become established drugs [1,10,11].

A number of articles have been published in the topical collection before 2018 [1,10]. Since then, many new high-quality manuscripts have been submitted and subsequently published. In total, 6 review and 24 original research articles have been accepted in this topical collection.

In the following, we will briefly review the data presented in those publications. Li and colleagues reviewed the recent data on chemopreventive, antineoplastic, chemosensitive, procoagulant, and anticoagulant activities of **sepia ink polysaccharide** [12]. Ćetković et al. summarized findings on **cancer-related genes and proteins** found in marine sponges and provided insight into sponge genome and proteome [13]. Fan et al. reviewed **marine-derived compounds that have been described to be active in human prostate cancer models** both in vitro and in vivo. Molecules with activity in this entity belong to different molecular classes, such as nucleotides, amides, quinones, polyethers, and peptides, and possess different anticancer-related activities, such as antioxidant, antiangiogenetic, antiproliferative, and apoptosis-inducing activities [14]. A nice concise review by Martínez Andrade and coauthors covers the topic of **marine microalgae and their unique molecules** that have shown anticancer properties [15]. Van Andel et al. outlined different **chromatographic bioanalytical methods that are used for the quantitative determination of marine-derived molecules**, which have shown anticancer properties [16]. Ha et al. compiled recent findings on the design, synthesis, and biological activity of so-called **hybrid (chimera) molecules, which are based on marine natural compounds** and their derivatives [17].

In the 24 research articles published in the topical collection since the beginning of 2018, data from a number of both new and previously known marine-derived compounds have been reported. Guo et al. synthesized a novel bromophenol derivative (**BOS-102**), which is active in a human lung cancer cell model. The underlying mechanism of action could be identified as an induction of apoptosis and cell cycle arrest via ROS-mediated PI3K/Akt and MAPK signaling [18]. Aldairi and colleagues described **glycosaminoglycan-like polysaccharides** isolated from the marine mollusk *Cerastoderma edule*. This compound has exhibited antiproliferative activity in chronic myeloid leukemia as well as in relapsed acute lymphoblastic leukemia models [19]. Manh Hung et al. studied the effect of **gliotoxin** in combination with adriamycin on non-small cell lung cancer cells resistant to Adriamycin. The authors showed that gliotoxin can induce an intrinsic apoptotic response in cancer cells and activate the p53 protein. Additionally, gliotoxin enhanced the cytotoxic effect of Adriamycin [20]. Loret et al. isolated and characterized the small protein **BDS-5** from the sea anemone *Anemonia viridis*, which shows similarities to Kunitz-type inhibitors. BDS-5 has shown to possess antiangiogenic activity, which may be exploited in anticancer therapy [21]. Using a pheochromocytoma model, Bechmann and coauthors showed anticarcinogenic and antimetastatic activities of **aeroplysinin-1 and isofistularin-3**, compounds that were previously isolated from the marine sponge *Aplysina aerophoba*. Additionally, aeroplysinin-1 downregulated integrin β1 [22]. Hao and colleagues studied the anticancer activity of **phycocyanin** in non-small cell lung cancer cells. The authors reported that phycocyanin can regulate NF-κB signaling and induce apoptosis and cell cycle arrest, and suppresses cell migration, proliferation, and colony formation [23]. An article by Lin et al. describes an anticancer effect of the sponge-derived pentacyclic alkaloid **manzamine A** in colorectal cancer cells. In their research, using a microarray-based gene expression analysis, the authors revealed an effect of the treatment on caspase-dependent intrinsic apoptosis, DNA repair executed via an inhibition of CDKs p53/p21/p27 cell cycle arrest, and mRNA metabolism [24]. Rath et al. reported an anticancer effect of the well-known spongian alkaloid **fascaplysin** in lung cancer cells and circulating tumor cells from lung cancer. Fascaplysin induced ATM signaling, initiated by treatment-induced DNA damage, and increased the anticancer action of cisplatin [25]. Zhu and colleagues showed the antiangiogenic activity of **rLm-cystatin F**, a homologue of cystatin F, which was isolated from the buccal glands of *Lampetra morii*. rLm-cystatin F can suppress the migration, invasion, adhesion, and tube formation of HUVEC cells [26]. Ting and Chen demonstrated the anticancer activity of the antimicrobial peptide **tilapia piscidin 4** (TP4) derived from the fish species *Nile tilapia* in non-small cell lung cancer cells. TP4 induced necrotic death of cancer cells and increased the effect of the EGFR inhibitors erlotinib and gefitinib in EGFR-mutated non-small cell lung cancer cells [27]. Liang and colleagues utilized a fragment-based drug design in order to optimize the structure of the spongian brominated tyrosine **itampolin A**, which has previously been reported to be a potent p38α inhibitor. The authors synthesized and selected the most potent derivative, which showed activity in non-small cell lung cancer cells [28]. Qiao et al. investigated the anticancer activity of **tetracenomycin X** in human lung cancer cells in vitro and in vivo. They showed that the compound induces cell cycle arrest via both direct induction of cyclin D1 degradation by the proteasomal system and indirect downregulation of cyclin D1 due to the activation of p38 and c-JUN [29]. Groult and colleagues showed that algal **λ-carrageenan oligosaccharides** inhibit the migration of MDA-MB-231 breast cancer cells [30]. Xu et al. reported new and previously known natural dimeric naphthopyrones that are cytotoxic in several human cancer cells lines and could demonstrate that this cytotoxicity is executed via ROS-mediated apoptosis. Additionally, the authors showed that the PI3K/Akt pathway also plays a role in inducing this cytotoxic effect [31]. Zhou et al. reported on the activity of four new ansamycins, namely, **divergolides T–W**, and two previously known individual compounds isolated from the culture of mangrove-derived actinomycete *Streptomyces* sp. KFD18. Some of these compounds exhibited potent cytotoxicity in several human cancer cell lines, executed via the apoptotic pathway [32]. Lin and colleagues reported a suppressive effect of **actinomycin V** on the migration and invasion of human breast cancer cells. This effect can be explained by the inhibition of the Snail/Slug-mediated EMT (epithelial–mesenchymal transition) under drug treatment [33]. Teruya et al. analyzed the activity of a low **molecular weight fucoidan**. They identified that it can suppress the growth of fibrosarcoma cancer cells without having an effect on the proliferation of noncancer TIG-1 cells. The underlying mechanism of this activity has been identified as a specific inhibition of the PD-L1/PD-L2 expression in cancer cells [34]. An analysis of the **αO-conotoxin GeXIVA** activity in a breast cancer cell model performed by Sun and colleagues revealed the antiproliferative activity of this conotoxin. This effect can be explained by the downregulating α9-nAChR (α9 nicotine acetylcholine receptor), ultimately leading to cell cycle arrest [35]. Kapustina et al. reported the isolation of four new humulane sesquiterpenoids, **leptogorgins A–C** and a new dihydroxyketosteroid, **leptogorgoid A**. Some of these compounds exhibited cytotoxicity and selectivity in human drug-resistant prostate cancer cells in vitro [36]. Zhou et al. described the proapoptotic activity of the previously known spongian scalarane sesterterpenoid **12-deacetyl-12-epi-scalaradial** in human cancer HeLa cells. The authors postulate that this cytotoxic effect is executed via the MAPK/ERK pathway, as well as by the activation of the Nur77 nuclear receptor activity [37]. Capasso et al. showed an antiproliferative activity and selectivity of **mycalin**

A and its derivatives, synthesized by the same group, in melanoma and cervical cancer models [38]. Shubina and colleagues reported the discovery of a new structural group of spongian monosulfated polyoxygenated steroids named gracilosulfates. In their work, they isolated seven **gracilosulfates, A–G**. These molecules can inhibit the expression of PSA in human prostate cancer cells, thereby inducing an anticancer effect [39]. Finally, Delgado-Roche et al. reported that metabolites of *Thalassia testudinum*, in particular **thalassiolin B**, exhibit chemopreventive and antigenotoxic activity, which can be of use in anticancer therapy. This effect was at least partially mediated by the inhibition of the CYP1A1-mediated benzo[a]pyrene-induced transformation (i.e., by suppressing the effects of oxidative and mutagenic stress) [40].

The scientific and medical community embraces new biologically active compounds, which will hopefully be further developed into clinically useful drugs. Putting marine-derived molecules in the focus of research on natural products and medical chemistry has already resulted in the development of several effective drugs that have saved thousands of lives. Therefore, we thank all the authors who have contributed to this important field and have added to the topical collection "Marine Compounds and Cancer" of *Marine Drugs*!

Funding: This research received no external funding.

Conflicts of Interest: The authors declare no conflict of interest.

References

1. Dyshlovoy, S.A.; Honecker, F. Marine Compounds and Cancer: 2017 Updates. *Mar Drugs* **2018**, *16*, 41. [CrossRef] [PubMed]
2. Bergmann, W.; Feeney, R.J. Contributions to the study of marine products. XXXII. The nucleosides of spongies. I. *J. Org. Chem.* **1951**, *16*, 981–987. [CrossRef]
3. Stonik, V. Marine natural products: A way to new drugs. *Acta Nat.* **2009**, *2*, 15–25. [CrossRef]
4. Brönstrup, M.; Sasse, F. Natural products targeting the elongation phase of eukaryotic protein biosynthesis. *Nat. Prod. Rep.* **2020**, *37*, 752–762. [CrossRef]
5. Deeks, E.D. Polatuzumab Vedotin: First Global Approval. *Drugs* **2019**, *79*, 1467–1475. [CrossRef]
6. Hanna, K.S. Enfortumab vedotin to treat urothelial carcinoma. *Drugs Today* **2020**, *56*, 329–335. [CrossRef] [PubMed]
7. Markham, A. Belantamab Mafodotin: First Approval. *Drugs* **2020**, *80*, 1607–1613. [CrossRef] [PubMed]
8. Markham, A. Lurbinectedin: First Approval. *Drugs* **2020**, *80*, 1345–1353. [CrossRef] [PubMed]
9. Mayer, A. Marine Pharmaceutical: The Clinical Pipeline. Available online: https://www.midwestern.edu/departments/marinepharmacology/clinical-pipeline.xml (accessed on 27 November 2020).
10. Dyshlovoy, S.A.; Honecker, F. Marine Compounds and Cancer: Where Do We Stand? *Mar. Drugs* **2015**, *13*, 5657–5665. [CrossRef] [PubMed]
11. Dyshlovoy, S.A.; Honecker, F. Marine Drugs Acting as Autophagy Modulators. *Mar. Drugs* **2020**, *18*, 53. [CrossRef] [PubMed]
12. Li, F.; Luo, P.; Liu, H. A Potential Adjuvant Agent of Chemotherapy: Sepia Ink Polysaccharides. *Mar. Drugs* **2018**, *16*, 106. [CrossRef]
13. Ćetković, H.; Halasz, M.; Herak Bosnar, M. Sponges: A Reservoir of Genes Implicated in Human Cancer. *Mar. Drugs* **2018**, *16*, 20. [CrossRef] [PubMed]
14. Fan, M.; Nath, A.K.; Tang, Y.; Choi, Y.-J.; Debnath, T.; Choi, E.-J.; Kim, E.-K. Investigation of the Anti-Prostate Cancer Properties of Marine-Derived Compounds. *Mar. Drugs* **2018**, *16*, 160. [CrossRef] [PubMed]
15. Martínez Andrade, K.A.; Lauritano, C.; Romano, G.; Ianora, A. Marine Microalgae with Anti-Cancer Properties. *Mar. Drugs* **2018**, *16*, 165. [CrossRef] [PubMed]
16. Van Andel, L.; Rosing, H.; Schellens, J.H.; Beijnen, J.H. Review of Chromatographic Bioanalytical Assays for the Quantitative Determination of Marine-Derived Drugs for Cancer Treatment. *Mar. Drugs* **2018**, *16*, 246. [CrossRef] [PubMed]
17. Ha, M.W.; Song, B.R.; Chung, H.J.; Paek, S.-M. Design and Synthesis of Anti-Cancer Chimera Molecules Based on Marine Natural Products. *Mar. Drugs* **2019**, *17*, 500. [CrossRef] [PubMed]

18. Guo, C.-L.; Wang, L.-J.; Zhao, Y.; Liu, H.; Li, X.-Q.; Jiang, B.; Luo, J.; Guo, S.-J.; Wu, N.; Shi, D.-Y. A Novel Bromophenol Derivative BOS-102 Induces Cell Cycle Arrest and Apoptosis in Human A549 Lung Cancer Cells via ROS-Mediated PI3K/Akt and the MAPK Signaling Pathway. *Mar. Drugs* **2018**, *16*, 43. [CrossRef] [PubMed]

19. Aldairi, A.F.; Ogundipe, O.D.; Pye, D.A. Antiproliferative Activity of Glycosaminoglycan-Like Polysaccharides Derived from Marine Molluscs. *Mar. Drugs* **2018**, *16*, 63. [CrossRef]

20. Manh Hung, L.V.; Song, Y.W.; Cho, S.K. Effects of the Combination of Gliotoxin and Adriamycin on the Adriamycin-Resistant Non-Small-Cell Lung Cancer A549 Cell Line. *Mar. Drugs* **2018**, *16*, 105. [CrossRef]

21. Loret, E.P.; Luis, J.; Nuccio, C.; Villard, C.; Mansuelle, P.; Lebrun, R.; Villard, P.H. A Low Molecular Weight Protein from the Sea Anemone Anemonia viridis with an Anti-Angiogenic Activity. *Mar. Drugs* **2018**, *16*, 134. [CrossRef]

22. Bechmann, N.; Ehrlich, H.; Eisenhofer, G.; Ehrlich, A.; Meschke, S.; Ziegler, C.G.; Bornstein, S.R. Anti-Tumorigenic and Anti-Metastatic Activity of the Sponge-Derived Marine Drugs Aeroplysinin-1 and Isofistularin-3 against Pheochromocytoma In Vitro. *Mar. Drugs* **2018**, *16*, 172. [CrossRef] [PubMed]

23. Hao, S.; Yan, Y.; Li, S.; Zhao, L.; Zhang, C.; Liu, L.; Wang, C. The In Vitro Anti-Tumor Activity of Phycocyanin against Non-Small Cell Lung Cancer Cells. *Mar. Drugs* **2018**, *16*, 178. [CrossRef] [PubMed]

24. Lin, L.-C.; Kuo, T.-T.; Chang, H.-Y.; Liu, W.-S.; Hsia, S.-M.; Huang, T.-C. Manzamine A Exerts Anticancer Activity against Human Colorectal Cancer Cells. *Mar. Drugs* **2018**, *16*, 252. [CrossRef] [PubMed]

25. Rath, B.; Hochmair, M.; Plangger, A.; Hamilton, G. Anticancer Activity of Fascaplysin against Lung Cancer Cell and Small Cell Lung Cancer Circulating Tumor Cell Lines. *Mar. Drugs* **2018**, *16*, 383. [CrossRef] [PubMed]

26. Zhu, M.; Li, B.; Wang, J.; Xiao, R. The Anti-Angiogenic Activity of a Cystatin F Homologue from the Buccal Glands of Lampetra morii. *Mar. Drugs* **2018**, *16*, 477. [CrossRef] [PubMed]

27. Ting, C.-H.; Chen, J.-Y. Nile Tilapia Derived TP4 Shows Broad Cytotoxicity toward to Non-Small-Cell Lung Cancer Cells. *Mar. Drugs* **2018**, *16*, 506. [CrossRef]

28. Liang, J.-W.; Wang, M.-Y.; Wang, S.; Li, X.-Y.; Meng, F.-H. Fragment-Based Structural Optimization of a Natural Product Itampolin A as a p38α Inhibitor for Lung Cancer. *Mar. Drugs* **2019**, *17*, 53. [CrossRef]

29. Qiao, X.; Gan, M.; Wang, C.; Liu, B.; Shang, Y.; Li, Y.; Chen, S. Tetracenomycin X Exerts Antitumour Activity in Lung Cancer Cells through the Downregulation of Cyclin D1. *Mar. Drugs* **2019**, *17*, 63. [CrossRef]

30. Groult, H.; Cousin, R.; Chot-Plassot, C.; Maura, M.; Bridiau, N.; Piot, J.-M.; Maugard, T.; Fruitier-Arnaudin, I. λ-Carrageenan Oligosaccharides of Distinct Anti-Heparanase and Anticoagulant Activities Inhibit MDA-MB-231 Breast Cancer Cell Migration. *Mar. Drugs* **2019**, *17*, 140. [CrossRef]

31. Xu, K.; Guo, C.; Meng, J.; Tian, H.; Guo, S.; Shi, D. Discovery of Natural Dimeric Naphthopyrones as Potential Cytotoxic Agents through ROS-Mediated Apoptotic Pathway. *Mar. Drugs* **2019**, *17*, 207. [CrossRef]

32. Zhou, L.-M.; Kong, F.-D.; Xie, Q.-Y.; Ma, Q.-Y.; Hu, Z.; Zhao, Y.-X.; Luo, D.-Q. Divergolides T–W with Apoptosis-Inducing Activity from the Mangrove-Derived Actinomycete Streptomyces sp. KFD18. *Mar. Drugs* **2019**, *17*, 219. [CrossRef] [PubMed]

33. Lin, S.; Zhang, C.; Liu, F.; Ma, J.; Jia, F.; Han, Z.; Xie, W.; Li, X. Actinomycin V Inhibits Migration and Invasion via Suppressing Snail/Slug-Mediated Epithelial-Mesenchymal Transition Progression in Human Breast Cancer MDA-MB-231 Cells In Vitro. *Mar. Drugs* **2019**, *17*, 305. [CrossRef] [PubMed]

34. Teruya, K.; Kusumoto, Y.; Eto, H.; Nakamichi, N.; Shirahata, S. Selective Suppression of Cell Growth and Programmed Cell Death-Ligand 1 Expression in HT1080 Fibrosarcoma Cells by Low Molecular Weight Fucoidan Extract. *Mar. Drugs* **2019**, *17*, 421. [CrossRef] [PubMed]

35. Sun, Z.; Bao, J.; Zhangsun, M.; Dong, S.; Zhangsun, D.; Luo, S. αO-Conotoxin GeXIVA Inhibits the Growth of Breast Cancer Cells via Interaction with α9 Nicotine Acetylcholine Receptors. *Mar. Drugs* **2020**, *18*, 195. [CrossRef]

36. Kapustina, I.I.; Makarieva, T.N.; Guzii, A.G.; Kalinovsky, A.I.; Popov, R.S.; Dyshlovoy, S.A.; Grebnev, B.B.; von Amsberg, G.; Stonik, V.A. Leptogorgins A–C, Humulane Sesquiterpenoids from the Vietnamese Gorgonian Leptogorgia sp. *Mar. Drugs* **2020**, *18*, 310. [CrossRef] [PubMed]

37. Zhou, M.; Peng, B.-R.; Tian, W.; Su, J.-H.; Wang, G.; Lin, T.; Zeng, D.; Sheu, J.-H.; Chen, H. 12-Deacetyl-12-epi-Scalaradial, a Scalarane Sesterterpenoid from a Marine Sponge Hippospongia sp., Induces HeLa Cells Apoptosis via MAPK/ERK Pathway and Modulates Nuclear Receptor Nur77. *Mar. Drugs* **2020**, *18*, 375. [CrossRef] [PubMed]

38. Capasso, D.; Borbone, N.; Terracciano, M.; Di Gaetano, S.; Piccialli, V. Antiproliferative Activity of Mycalin A and Its Analogues on Human Skin Melanoma and Human Cervical Cancer Cells. *Mar. Drugs* **2020**, *18*, 402. [CrossRef]
39. Shubina, L.K.; Makarieva, T.N.; Denisenko, V.A.; Popov, R.S.; Dyshlovoy, S.A.; Grebnev, B.B.; Dmitrenok, P.S.; von Amsberg, G.; Stonik, V.A. Gracilosulfates A–G, Monosulfated Polyoxygenated Steroids from the Marine Sponge Haliclona gracilis. *Mar. Drugs* **2020**, *18*, 454. [CrossRef]
40. Delgado-Roche, L.; Santes-Palacios, R.; Herrera, J.A.; Hernández, S.L.; Riera, M.; Fernández, M.D.; Mesta, F.; Garrido, G.; Rodeiro, I.; Espinosa-Aguirre, J.J. Interaction of Thalassia testudinum Metabolites with Cytochrome P450 Enzymes and Its Effects on Benzo(a)pyrene-Induced Mutagenicity. *Mar. Drugs* **2020**, *18*, 566. [CrossRef]

Publisher's Note: MDPI stays neutral with regard to jurisdictional claims in published maps and institutional affiliations.

Article

Efficacy and Mechanism of Action of Marine Alkaloid 3,10-Dibromofascaplysin in Drug-Resistant Prostate Cancer Cells

Sergey A. Dyshlovoy [1,2,3,4,*], Moritz Kaune [1], Jessica Hauschild [1], Malte Kriegs [5,6], Konstantin Hoffer [5,6], Tobias Busenbender [1], Polina A. Smirnova [4], Maxim E. Zhidkov [4], Ekaterina V. Poverennaya [7], Su Jung Oh-Hohenhorst [3,8], Pavel V. Spirin [9], Vladimir S. Prassolov [9], Derya Tilki [3,10], Carsten Bokemeyer [1], Markus Graefen [3] and Gunhild von Amsberg [1,3]

[1] Laboratory of Experimental Oncology, Department of Oncology, Hematology and Bone Marrow Transplantation with Section Pneumology, Hubertus Wald-Tumorzentrum, University Medical Center Hamburg-Eppendorf, Martinistrasse 52, 20251 Hamburg, Germany; moritz.kaune@stud.uke.uni-hamburg.de (M.K.); j.hauschild@uke.de (J.H.); tobias.busenbender@gmx.de (T.B.); c.bokemeyer@uke.de (C.B.); g.von-amsberg@uke.de (G.v.A.)

[2] Laboratory of Pharmacology, A.V. Zhirmunsky National Scientific Center of Marine Biology, Far Eastern Branch, Russian Academy of Sciences, Palchevskogo str. 17, 690041 Vladivostok, Russian

[3] Martini-Klinik, Prostate Cancer Center, University Hospital Hamburg-Eppendorf, Martinistrasse 52, 20251 Hamburg, Germany; s.oh-hohenhorst@uke.de (S.J.O.-H.); d.tilki@uke.de (D.T.); graefen@martini-klinik.de (M.G.)

[4] School of Natural Sciences, Far Eastern Federal University, FEFU Campus, Ajax Bay 10, Russky Island, 690922 Vladivostok, Russian; pollianna_95@mail.ru (P.A.S.); zhidkov.me@dvfu.ru (M.E.Z.)

[5] Department of Radiotherapy & Radiation Oncology, Hubertus Wald Tumorzentrum–University Cancer Center Hamburg (UCCH), University Medical Center Hamburg-Eppendorf, Martinistrasse 52, 20251 Hamburg, Germany; m.kriegs@uke.de (M.K.); k.hoffer@uke.de (K.H.)

[6] UCCH Kinomics Core Facility, Hubertus Wald Tumorzentrum–University Cancer Center Hamburg (UCCH), University Medical Center Hamburg-Eppendorf, Martinistrasse 52, 20251 Hamburg, Germany

[7] Laboratory of Proteoform Interactomics, Institute of Biomedical Chemistry, Pogodinskaya str. 10/8, 119121 Moscow, Russian; k.poverennaya@gmail.com

[8] Institute of Anatomy and Experimental Morphology, University Medical Center Hamburg-Eppendorf, Martinistrasse 52, 20246 Hamburg, Germany

[9] Engelhardt Institute of Molecular Biology, Russian Academy of Sciences, Vavilova 32, 119991 Moscow, Russian; spirin.pvl@gmail.com (P.V.S.); prassolov45@mail.ru (V.S.P.)

[10] Department of Urology, University Hospital Hamburg-Eppendorf, Martinistrasse 52, 20251 Hamburg, Germany

* Correspondence: dyshlovoy@gmail.com

Received: 5 November 2020; Accepted: 27 November 2020; Published: 1 December 2020

Abstract: Efficacy and mechanism of action of marine alkaloid 3,10-dibromofascaplysin (DBF) were investigated in human prostate cancer (PCa) cells harboring different levels of drug resistance. Anticancer activity was observed across all cell lines examined without signs of cross-resistance to androgen receptor targeting agents (ARTA) or taxane based chemotherapy. Kinome analysis followed by functional investigation identified JNK1/2 to be one of the molecular targets of DBF in 22Rv1 cells. In contrast, no activation of p38 and ERK1/2 MAPKs was observed. Inhibition of the drug-induced JNK1/2 activation or of the basal p38 activity resulted in increased cytotoxicity of DBF, whereas an active ERK1/2 was identified to be important for anticancer activity of the alkaloid. Synergistic effects of DBF were observed in combination with PARP-inhibitor olaparib most likely due to the induction of ROS production by the marine alkaloid. In addition, DBF intensified effects of platinum-based drugs cisplatin and carboplatin, and taxane derivatives docetaxel and cabazitaxel. Finally, DBF inhibited AR-signaling and resensitized AR-V7-positive 22Rv1 prostate cancer cells to enzalutamide, presumably due to AR-V7 down-regulation. These findings propose DBF to be a promising novel drug candidate for the treatment of human PCa regardless of resistance to standard therapy.

Keywords: fascaplysin; prostate cancer; JNK1/2; natural products; synergism

---

1. Introduction

In the past decades, treatment of metastatic prostate cancer (mPCa) tremendously improved resulting in increased overall survival (OS) and better quality of life of the patients. However, despite the high efficacy of androgen receptor targeting agents (ARTA) and docetaxel in the hormone naïve stage of disease, loss of efficacy of standard therapies is eventually observed in the course of treatment reflected by reduced PSA-responses and a decreased OS with each additional treatment line [1–3]. To date, different mechanisms of resistance have been identified. Thus, modification of the androgen receptor (AR) including amplification, mutation, and alternative splice variants as well as PTEN loss were found to be associated with decreased sensitivity to ARTAs, while up-regulation of p-glycoprotein and the induction of life-prolonging autophagy mediate resistance to standard chemotherapy [2]. To date, treatment strategies to overcome drug resistance in mPCa are limited and novel agents effectively targeting drug-resistant prostate cancer are urgently needed.

Marine invertebrates are a rich source of new molecules having a unique structure and promising biological activity [4,5]. A significant proportion of these compounds are cytotoxic to cancer cells making them promising candidates for further development. By the end of 2020, there will be 10 clinically-approved anticancer drugs created on the bases of small molecules isolated from marine organisms [6,7]. In addition, many more are in different stages of clinical and preclinical development and around 1000 new structures are reported every year [8,9].

Fascaplysin is a red bioactive pigment initially isolated from the marine sponge *Fascaplysinopsis* sp. [10]. This alkaloid possesses a 12*H*-pyrido[1,2-*a*:3,4-*b'*]diindole core [11] and reveals a broad spectrum of biological activities, including antifungal, antibacterial, antiviral, antimalarial, and antitumor effects [11–13]. Its antitumor activity was evaluated in different human cancer cells lines including lung, prostate, breast, and colorectal cancer, melanoma and leukemia [13–20]. Fascaplysin and similar alkaloids seem to mediate their activity by inhibition of cyclin-dependent kinase 4 (CDK4) and intercalation with dsDNA, which leads to a G1-phase cell cycle arrest and ultimately to the apoptotic cancer cell death [11,12,21,22]. However, recently it was reported that fascaplysin can kill cancer cells independently from the presence and activity of CDK4, suggesting other molecular targets to be involved in its mechanism of action [17]. Indeed, several other molecular targets and cellular processes, both pro-apoptotic and pro-survival, have also been reported to be affected by this marine alkaloid. Thus, fascaplysin was capable to inhibit VEGFR2, TRKA, surviving, and HIF-1α [17]. In lung and colorectal cancer cells, synergistic effects with AKT and AMPK inhibitors were observed, emphasizing its possible implication in combinational therapy [23]. Additionally, an induction of cytoprotective autophagy was found in human breast cancer cells as well as human vascular endothelial cells treated with fascaplysin and its derivatives [18,19]. This event was linked to the PI3K/AKT/mTOR signaling inhibition in the treated cells [18,24]. Autophagy is often induced in cancer cells exposed to chemotherapeutics as one of the basic survival mechanisms, helping the cells to overcome stress conditions [25,26].

Several halogenated fascaplysin derivatives were found to exhibit antitumor activity as well. Thus, 3-chlorofascaplysin suppressed angiogenesis and tumor growth of breast cancer cells [18]. The brominated derivatives, i.e., 3- and 10-bromofascaplysins induced a caspase-dependent apoptosis in human leukemia cells at nanomolar concentrations [13]. Additionally, these compounds were active in rat C6 glioma cells model [27]. Of note, in this experiment the brominated derivatives were more active in comparison with "mother" fascaplysin molecule [27].

3,10-Dibromofascaplysin (DBF) is a novel halogenated fascaplysin alkaloid initially isolated from the marine sponge *Fascaplysinopsis reticulata* [28] and later synthesized by our group [20]. Recently, we identified DBF to be active in human prostate cancer cells during a small-scale screening of semi-synthetic fascaplysin derivatives. In contrast to the other synthesized derivatives, DBF revealed a smooth cytotoxicity profile, suggesting a wide therapeutic window [20]. In addition, DBF was found to affect cellular metabolism, which further leads to cancer cell death [20]. In the current research we evaluated the activity of DBF in human prostate cancer cell lines harboring different levels of drug resistance to currently available standard therapies. Mechanism of action and molecular targets were examined by a kinome profiling approach.

2. Results and Discussion

2.1. 3,10-Dibromofascaplysin (DBF) Induces Apoptotic Cell Death of Drug-Resistant Prostate Cancer Cells

Overcoming drug resistance is a major challenge in the treatment of advanced prostate cancer. 3,10-Dibromofascaplysin (DBF, Figure 1A)—a new halogenated fascaplysin—showed promising activity in previous screening experiments [20]. Therefore, we evaluated cytotoxicity of this marine alkaloid in different human drug-resistant prostate cancer cell lines in vitro.

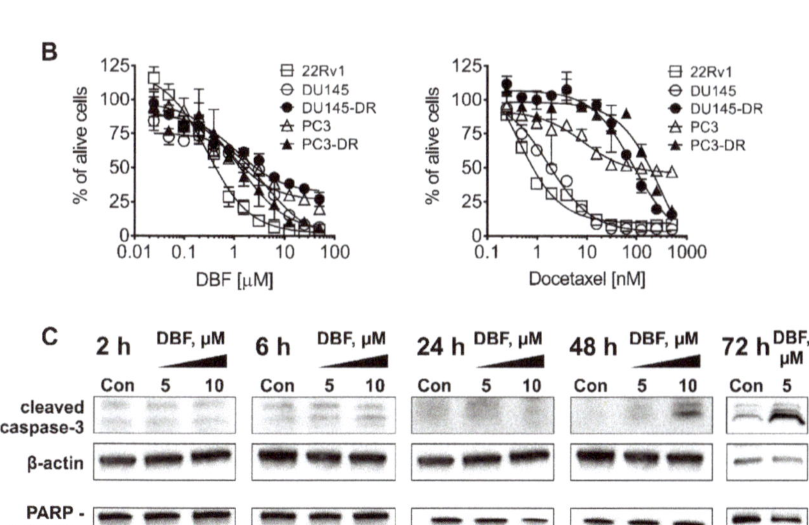

Figure 1. Cytotoxicity and selectivity of DBF. (A), The structure of DBF. **(B)**, Cytotoxicity profiles of DBF in human prostate cancer cell lines resistant to hormone therapy or docetaxel. Cell viability was measured using MTT assay following 72 h of incubation. **(C)**, Western blotting analysis of the protein expression in 22Rv1 cells treated with DBF for indicated time. β-actin was used as a loading control.

22Rv1, PC3 and DU145 cells reveal resistance to AR-targeting therapies e.g., abiraterone and enzalutamide. In 22Rv1 cells, resistance is mediated by the expression of AR splice variant

7 (AR-V7) [29], which lacks an androgen binding site and induces permanent auto-activation of the ARs [30]. PC3 and DU145 cells lack AR expression and thus do not require androgens for growth and proliferation [29]. DBF was found to be cytotoxic in all cell lines investigated at micro- and nanomolar concentrations with the highest activity in 22Rv1 cells (Table 1). The docetaxel-resistant PC3 and DU145 sublines (PC3-DR and DU145-DR) were generated using continuous incubation of PC3 and DU145 with increasing concentrations of docetaxel until reaching a concentration of 12.5 nM as previously described [31]. Notably, the PC3-DR and DU145-DR cells are ~50-fold less sensitive to docetaxel compared to their parental cell lines (Figure 1B). Remarkably, IC_{50} of DBF in PC3-DR cells was only 2-fold higher compared to PC3 cells, and DU145-DR cells were even more sensitive to DBF than DU145 cells suggesting no cross-resistance between docetaxel and DBF (Figure 1B, Table 1).

Table 1. Cytotoxicity of DBF in different prostate cancer cells. Cells were incubated with the drug for 72 h. Docetaxel was used as a reference compound.

Cell Line	IC_{50}, 72 h	
	DBF [μM]	Docetaxel [nM]
22Rv1	0.29 ± 0.04	0.38 ± 0.08
PC3	0.79 ± 0.17	8.55 ± 3.09
PC3-DR	1.51 ± 0.35	355.8 ± 148.7
DU145	4.19 ± 0.81	1.72 ± 0.19
DU145-DR	1.25 ± 0.27	89.4 ± 13.2

In a next step, we evaluated the mechanism of action of DBF in prostate cancer. 22Rv1 cells were chosen as they revealed the highest sensitivity to the marine compound. First, apoptotic markers including poly(ADP-ribose)polymerase (PARP) and caspase-3 were examined to determine the character of cell death mediated by DBF (Figure 1C). In fact, cleavage of both proteins first appeared 48 h after treatment of 22Rv1 cells at concentrations of 5 and 10 μM. Hence, these conditions were chosen for further experiments.

2.2. DBF Induces Alterations of Protein Tyrosine Kinases Activity

Protein kinases are important target molecules for various anticancer drugs [32]. They catalyze the phosphorylation of target proteins thereby modulating their activity. Serine/threonine kinases (STK) belong to a group of protein kinases, involved in critical processes related to both, cellular death and survival [33]. In fact, modulation of STKs activity has been identified as a mechanism of action of a number of clinically-approved drugs, such as cobimetinib, palbociclib, axitinib, sunitinib, and others [33]. Hence, we assessed the effect of DBF on STKs activity by functional kinomics assay using the PamTechnology® (http://www.pamgene.com, Figure 2A–E) [34]. This method allows to determine specific STK activity changes in living cells. Short-term treatment for 2 h was chosen in order to minimize the detection of unspecific effects secondary to cell death related events. Results are shown for the control group vs. the treated group, as a log2 of signal intensity per peptide (Figure 2A–D). No significant changes of the overall STKs activity were observed between the groups (Figure 2C). However, further analyses of specific STK potentially affected by DBF predicted an activation of JNK1 kinase (Figure 2E), belonging to the group of the mitogen activated protein kinases (MAPKs). Additionally, other MAPKs, such as p38 and ERK1, were also predicted to be activated by DBF, however, with a lower specificity score (Figure 2E). The above mentioned MAPKs are associated with different cancer-related processes, however, with ambiguous impact on cancer cell elimination and growth inhibition [32,35]. In prostate cancer, MAPKs were reported to have an important impact on tumor growth [35–37]. Additionally, a crosstalk with AR-signaling has been demonstrated, especially for JNK1/2 [38–40]. Thus, the changes of JNK1/2 activity and other MAPKs were further examined.

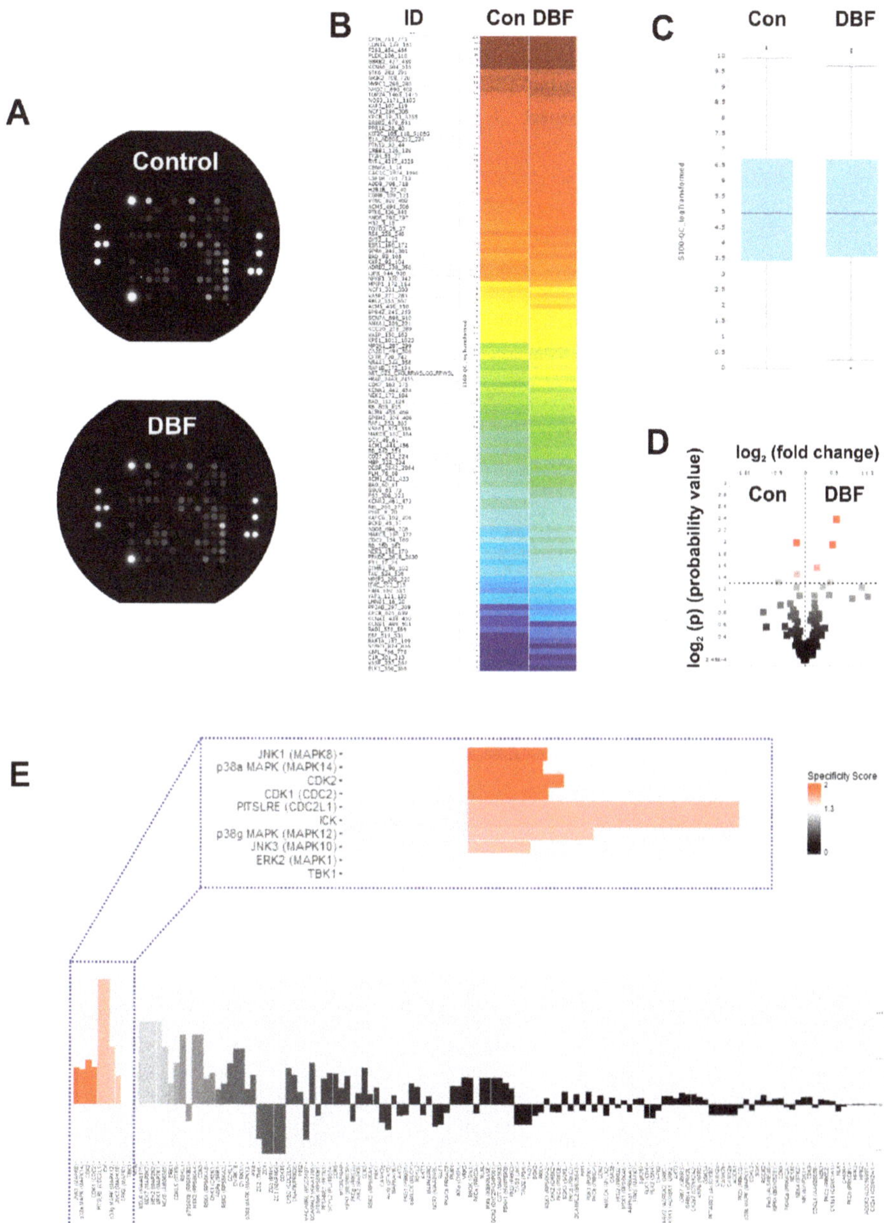

Figure 2. Functional kinome profiling of serine/threonine kinases (STKs). 22Rv1 cells were treated with DBF for 2 h; proteins were extracted and analyzed using STK-PamChip® (sequence-specific peptide phosphorylation assay) and anti-phospho-STK antibodies. (**A**), Microphotographs of STK-PamChip®. The generated data is represented as heat-map plot (**B**), box plots (**C**), or volcano-plot (**D**). Red dots indicate peptide substrates having significantly increased phosphorylation in comparison to control samples ($\log_2(p) > 1.3$, dotted line, (**D**)). (**E**), Upstream analysis of the treatment-affected kinases in 22Rv1 cells. Normalized kinase statistic > 0 indicates higher kinase activity in DBF-treated cells; specificity score > 1.3 indicates statistically significant changes.

2.3. Validation of Kinome Analysis Data

JNK1 was the top-ranked kinase predicted to be activated by DBF in prostate cancer 22Rv1 cells (Figure 2E). Hence, time-dependent Western blotting-based analyses of JNK1/2 activation were performed for further validation (Figure 3A). Indeed, a phosphorylation of JNK1/2 was only observed in PCa cells exposed to short-term DBF treatment (2 h), whereas no activation, or even phospho-JNK1/2 degradation was found after 6 h to 48 h (Figure 3A). More detailed examinations revealed a pronounced JNK1/2 phosphorylation already after 15 min of DBF treatment. Thus, JNK1/2 phosphorylation is one of the very first cellular events following drug exposure (Figure 3D), long before first apoptotic signs appear (48 h, Figure 1C). Due to a potential crosstalk of different MAPKs, we examined the effect of DBF on p38 and ERK1/2 MAPKs, which were also predicted to be affected by kinomic analysis (Figure 3B,C,E,F). Of note, no significant alterations of p38 and ERK1/2 were observed at any time point ranging from 15 min to 48 h of treatment (Figure 3B,C,E,F).

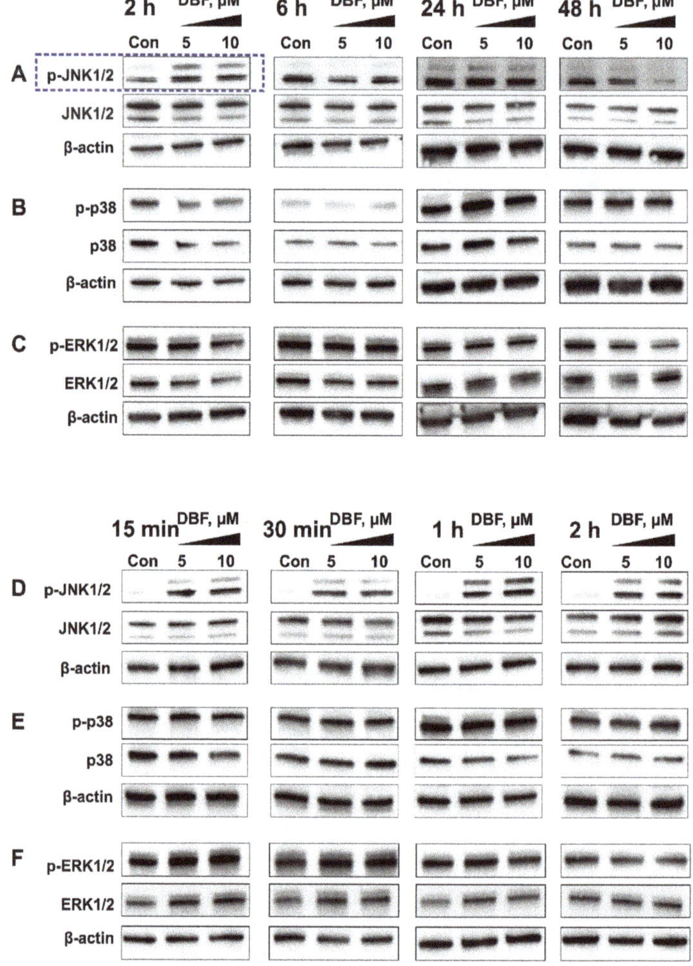

Figure 3. Validation of kinome analysis data using Western blotting. Western blotting analyses of the phosphorylated and total JNK1/2 (**A,D**), p38 (**B,E**), and ERK1/2 (**C,F**) kinases in 22Rv1 cells treated with DBF for 2–48 h (**A–C**) or 15 min–2 h (**D–F**). β-actin was used as a loading control.

2.4. Role of JNK1/2 and Other MAPKs in Cytotoxic Effect of DBF

In malignant conditions, the impact of JNK1/2, p38, and ERK1/2 MAPKs depend on the cellular context as well as the nature of stimuli ranging from pro-survival to pro-apoptotic effects [41]. In order to determine the role of JNK1/2 and the other MAPKs in execution of the biological effects of DBF, we applied a co-treatment with specific inhibitors. Therefore, selective JNK1/2 inhibitor SP600125 (Figure 4A), p38 inhibitor SB203580 (Figure 4B), MEK1 inhibitor PD98039 (Figure 4C), as well as ERK1/2 inhibitor FR180204 (Figure 4D) were tested. Note, MEK1/2 kinase exclusively and directly activates ERK1/2 [42] indicating that an inactivation of MEK1/2 leads to an inhibition ERK1/2 [42]. Due to the cytotoxic nature of most of the inhibitors mentioned above, we used combinations of several active concentrations to determine a synergistic or antagonistic effect of the inhibitors on cytotoxic activity of DBF. The data were generated using MTT assay and were further analyzed using SynergyFinder 2.0 software and a Zero interaction potency (ZIP) reference model [43] (Figure 4A–D). We generated heat-maps for the effects of DBF and the MAPK inhibitors alone, and for their respective combinations (Figure 4A–D). Our analyses revealed pronounced synergistic effects of DBF with JNKi (SP600125) and p38i (SB203580) suggesting a pro-survival role of both kinases in the cellular response to DBF treatment (Figure 4A,B). In contrast, the MEK/ERKi (PD98039/FR180204) antagonized cytotoxic effects of DBF (Figure 4C,D) indicating ERK1/2 to exert a cytotoxic function in DBF-treated 22Rv1 cells. Additionally, to confirm these data we applied the lower non-cytotoxic concentrations of the inhibitors in order to avoid unspecific cytotoxicity-related effects (Figure 4E–H). Thus, in line with the above described results (Figure 4A–D), co-treatment with JNKi and p38i increased the cytotoxic effects of DBF, while in contrast combination with MEKi and ERKi reduced DBF mediated cytotoxicity (Figure 4E–H).

Interestingly, in our experiments we observed a transient (temporal) activation of JNK1/2, which takes place within first two hours, and then decreases to the basal level by the time point of 6 h (Figure 3A,D). A number of previous studies report that activation of MAPKs may have either transient or sustained character in the same model, depending on stimulus nature (reviewed in [44]). Moreover, the time course of MAPK activation may be critical for the specific outcome of this event as well as for the cellular fate [44–47]. In particular, it has been shown that sustained activation of JNK normally leads to the apoptotic program activation, whereas temporal short-term activation of this MAPK stimulates the cellular survival [45]. In line with this, the observed short-term activation of JNK1/2 in DBF-stimulated cells (Figure 3A,D) which was identified as a pro-survival event (Figure 4A,E)

In summary, DBF treatment is accompanied by JNK1/2 activation. The inhibition of this process could synergistically increase an anticancer effect of the marine alkaloid. Although, no regulation of p38 or ERK1/2 was observed in our experiments, the generated results suggest an involvement of both kinases in the maintenance of cellular processes, which are important for survival and death of the drug-treated cancer cells, correspondingly (Figure 4C,D).

2.5. Effect of DBF in Combination with Platinum and Taxane Agents

Taxanes are routinely applied in advanced prostate cancer, while platine-based therapy shows pronounced activity in patients harboring DNA repair defects or aggressive variants of the disease. In order to determine the clinical relevance of DBF in potential combinational therapies, co-treatment with cisplatin and carboplatin (platinum drugs, DNA-binding agents inducing cross-links) (Figure 5A,B), as well as docetaxel and cabazitaxel (taxane derivates, which mediate a microtubuline stabilization) was performed (Figure 5C,D). The activity of the drugs in combinations was analyzed by SynergyFinder 2.0 software using a ZIP reference model.

Remarkably, for DBF combinations with all afore mentioned drugs pronounced synergistic effects were detected (Figure 5A–D). Of note, synergism was observed for the whole range of carboplatin and cabazitaxel concentrations (Figure 5B,D), whereas for combinations with cisplatin and docetaxel synergistic effects were observed only for low doses of the drugs (Figure 5A,C).

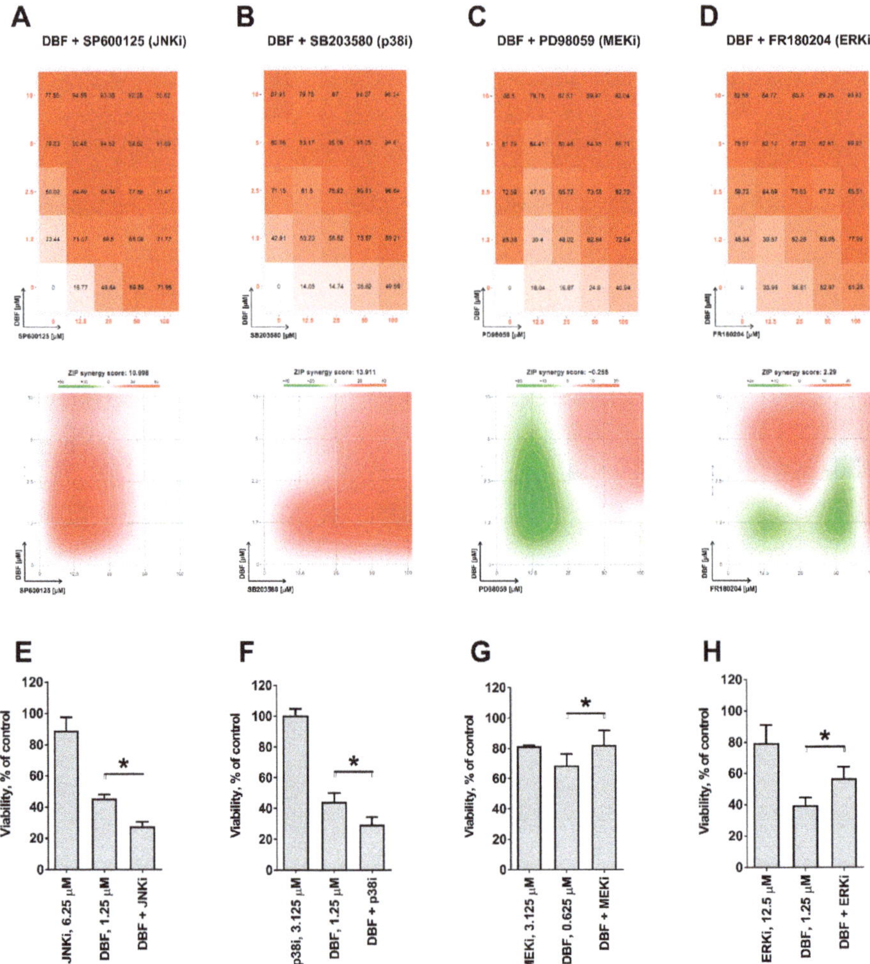

Figure 4. Analysis of the effect on DBF in combination with MAPKi. 22Rv1 cells were co-treated with DBF in combination with JNK1/2 inhibitor SP600125 (**A,E**), p38 inhibitor SB203580 (**B,F**), MEK1/2 inhibitor PD98059 (**C,G**), or ERK1/2 inhibitors FR180204 (**D,H**) for 48 h. The viability was measured using the MTT assay and the effect of the drug combination (synergism / additive effect / antagonism) was calculated and visualized using SynergyFinder 2.0 software and a ZIP reference model (**A–D**); or presented as of cell viability in % of control (**E–H**). Red regions indicate synergism; white—additive effect; green—antagonism (**A–D**). * $p < 0.05$, one-way ANOVA test.

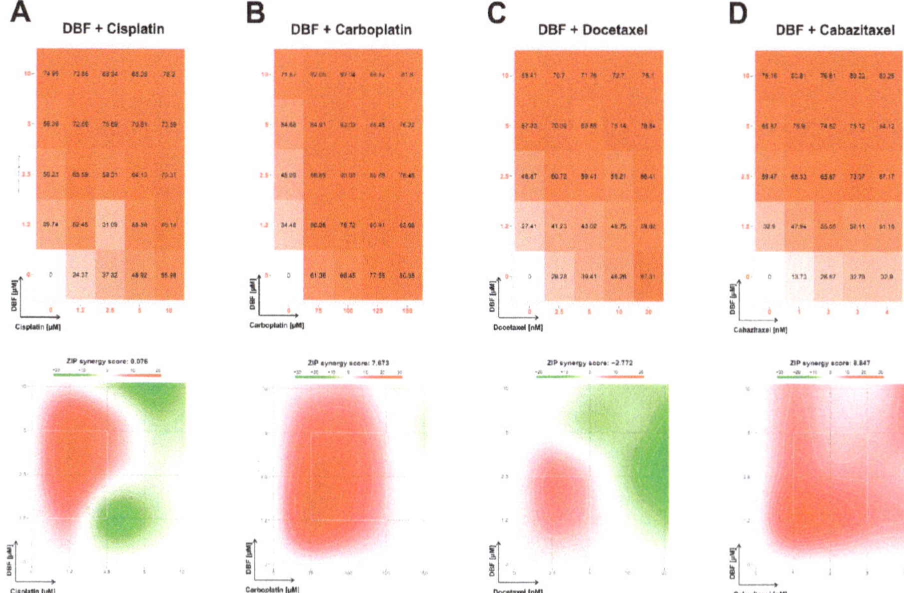

Figure 5. Analysis of the effect on DBF in combination with platinum and taxane drugs. 22Rv1 cells were co-treated with DBF in combination with cisplatin (**A**), carboplatin (**B**), docetaxel (**C**), or cabazitaxel (**D**) for 48 h. The viability was measured using the MTT assay and the effect of the drug combination (synergism/additive effect/antagonism) was calculated and visualized using SynergyFinder 2.0 software and a ZIP reference model. Red regions indicate synergism; white—additive effect; green—antagonism.

2.6. Effect of DBF in Combination with Olaparib

We further examined the effects of DBF in combination with PARP inhibitor olaparib. Olaparib has recently been approved by the FDA for the treatment of patients with deleterious germline or somatic homologous recombination repair (HRR) gene-mutated metastatic castration-resistant PCs. PARP holds major functions on DNA repair of single strand breaks (ssDNA), induced by different stress factors [48]. Inhibition of ssDNA reparation by PARP inhibitors results in DNA double strand (dsDNA) breaks eventually leading to synthetic cell death in cells carrying HRR defects [49–51]. 22Rv1 cells are known to bear a BRCA2 defect, the most frequent HRR alteration in PCa, revealing high sensitivity to olaparib [52]. Remarkably, DBF exhibited a strong synergistic effect when combined with olaparib at the whole range of concentrations of both drugs (Figure 6A). We hypothesized that this may be due to ssDNA breaks following treatment with DBF, potentially mediated by increased production of reactive oxygen species (ROS) [53]. Hence, we have examined the induction of ROS in 22Rv1 cells treated with DBF. Notably, a significant increase of the ROS level was detected already after 1 h of treatment with 1 μM of DBF (Figure 6B,C), whereas the DNA damage could be detected only after longer time of treatment with the same concentration of the drug (Figure 6D). Thus, the ROS induction is a primary effect of the alkaloid in cancer cells. Moreover, the previously described JNK1/2 activation (Figure 2A,B), may also result from oxidative stress [54].

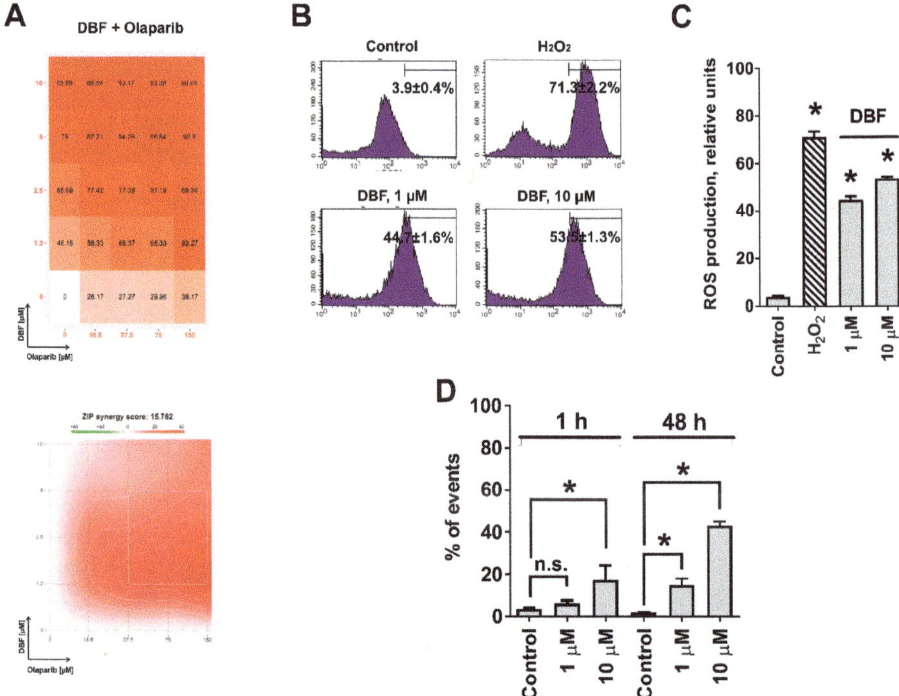

Figure 6. Analysis of the effect on DBF on ROS production and DNA damage. (**A**), 22Rv1 cells were co-treated with DBF in combination with olaparib for 48 h. The viability was measured using the MTT assay and the effect of the drug combination (synergism / additive effect / antagonism) was calculated and visualized using SynergyFinder 2.0 software and a ZIP reference model. Red regions indicate synergism; white—additive effect; green—antagonism. (**B,C**), Effect of DBF on ROS production in 22Rv1 cells following 1 h of treatment. The analysis was performed using CM-H$_2$DCFDA staining and flow cytometry technique (**B**). The ROS level was quantified with Cell Quest Pro software (**C**). H$_2$O$_2$ (200 μM) was used as a positive control. (**D**), Analysis of DNA break in 22Rv1 cells treated with DBF for indicated time. n.s.—non-significant ($p > 0.05$), * $p < 0.05$, one-way ANOVA test.

2.7. Effect of DBF on AR Signaling

Hormone therapy is a key component in the treatment of advanced prostate cancer. AR targeting agents (ARTA), e.g., enzalutamide, competitively inhibit the AR resulting in suppression of AR-signaling which in turn leads to inhibition of prostate cancer cell growth and viability [52,55]. Resistance to ARTA is mediated by AR amplification, mutation, or alternative splicing. AR splice variant V7 (AR-V7) lacks a C-terminal androgen binding domain and therefore cannot bind androgens or antiandrogens. An auto-activated AR signaling results in cell proliferation and survival [30]. Along with an AR-full length (AR-FL) expression, 22Rv1 cells are also known to express AR-V7, and therefore are resistant to enzalutamide [29]. Interestingly, DBF is able to resensitize 22Rv1 cells to enzalutamide showing synergistic effects with the ARTA (Figure 7A). In order to explain this phenomenon we examined the effect of DBF on the expressional levels of both AR-FL and AR-V7. Notably, we observed a down-regulation of AR-V7 and other AR splice variants (AR-Vs) in the treated 22Rv1 cells (Figure 7B). Additionally, the expression of AR-FL was also suppressed, suggesting an inhibition of AR signaling (Figure 7B). In line with this, a down-regulation of PSA expression was detected (Figure 7B). Note, PSA is a down-stream target of AR pathway reflecting its activity. Hence, we speculate that DBF is capable of AR-FL/V7-dependent signaling inhibition caused by a down-regulation of both AR-FL and AR-V7 expression. Moreover, due to the suppression of AR-V7 the investigated alkaloid recalls sensitivity of 22Rv1 cells to enzalutamide, making DBF a promising drug candidate for the treatment of castration-resistant PCa.

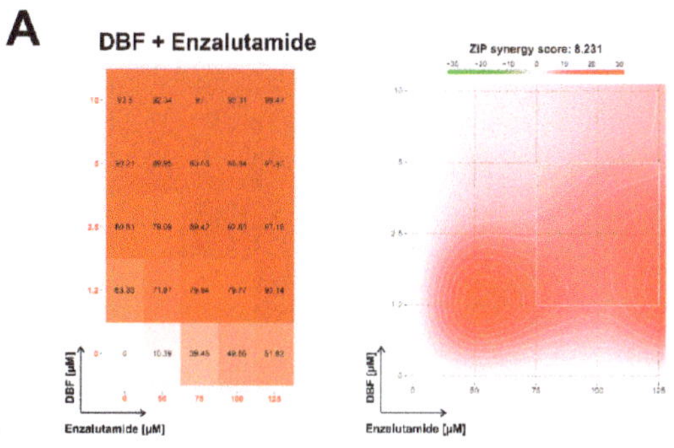

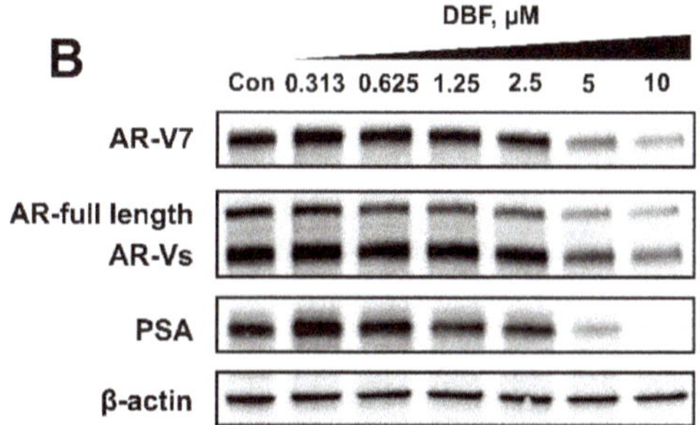

Figure 7. Analysis of the effect on DBF on AR signaling. (**A**), 22Rv1 cells were co-treated with DBF in combination with enzalutamide for 48 h. The viability was measured using the MTT assay and the effect of the drug combination (synergism/additive effect/antagonism) was calculated and visualized using SynergyFinder 2.0 software and a ZIP reference model. Red regions indicate synergism; white—additive effect; green—antagonism. (**B**), Effect of DBF on the expression of several proteins involved in the androgen receptor signaling following 72 h of treatment. Protein expression was accessed using Western blotting. β-actin was used as a loading control.

3. Materials and Methods

3.1. 3,10-Dibromofascaplysin

The marine alkaloid 3,10-dibromofascaplysin (DBF) was synthesized and purified as previously reported [20]. The purity of the compounds has been confirmed using H^1 NMR and high-resolution mass spectrometry.

3.2. Reagents and Antibodies

CM-H_2DCFDA was purchased from Molecular probes (Invitrogen, Eugene, OR, USA); MTT (3-(4,5-dimethylthiazol-2-yl)-2,5-diphenyltetrazolium bromide) and propidium iodide (PI)—from Sigma (Taufkirchen, Germany). cOmplete™ *EASY*packs protease inhibitors cocktail and PhosSTOP™ *EASY*packs phosphotase inhibitors cocktail—from Roche (Mannheim, Germany). Anisomycin—from NeoCorp (Weilheim, Germany). N-acetylcystein—from MedChemExpress (Monmouth Junction, NJ, USA). Docetaxel, cabazitaxel, cisplatin and carboplatin—from a Pharmacy of the University Hospital Hamburg-Eppendorf (Hamburg, Germany). RNase—from Carl Roth (Karlsruhe, Germany). Primary and secondary antibodies are listed in Table 2.

Table 2. List of antibodies used.

Antibodies	Clonality	Source	Cat.-No.	Dilution	Manufacturer
anti-AR	pAb	rabbit	sc-816	1:200	Santa Cruz
anti-AR-V7	mAb	rabbit	198394	1:1000	abcam
anti-cleaved Caspase-3	mAb	rabbit	#9664	1:1000	Cell Signaling
anti-ERK1/2	mAb	mouse	#9107	1:2000	Cell Signaling
anti-JNK1/2	mAb	rabbit	#9258	1:1000	Cell Signaling
anti-mouse IgG-HRP		sheep	NXA931	1:10,000	GE Healthcare
anti-p38	mAb	rabbit	#9212	1:1000	Cell Signaling
anti-PARP	pAb	rabbit	#9542	1:1000	Cell Signaling
anti-phospho-ERK1/2	mAb	rabbit	#4377	1:1000	Cell Signaling
anti-phospho-JNK1/2	mAb	rabbit	#4668	1:1000	Cell Signaling
anti-phospho-p38	mAb	rabbit	#4511	1:1000	Cell Signaling
anti-PSA/KLK3	mAb	rabbit	#5365	1:1000	Cell Signaling
anti-rabbit IgG-HRP		goat	#7074	1:5000	Cell Signaling
anti-α-Tubulin	mAb	mouse	T5168	1:5000	Sigma-Aldrich
anti-β-Actin-HRP	pAb	goat	sc-1616	1:10,000	Santa Cruz

3.3. Cell Lines and Culture Conditions

The human prostate cancer cell lines DU145, PC-3 and 22Rv1 were purchased from ATCC (Manassas, VA, USA). The docetaxel-resistant DU145-DR and PC3-DR cell lines, generated as described previously [31], were kindly provided by Prof. Z. Culig, Innsbruck Medical University, Austria. The cell lines used had a passage No. $\leq$ 50, and were continuously kept in culture for a max. of 3 months. All cell lines used were recently authenticated by commercial service (Multiplexion GmbH, Heidelberg, Germany). Cells were cultured as monolayers in a humidified atmosphere with 5% CO_2 at 37 °C. For cell culture, the RPMI medium supplemented with Glutamax™-I (gibco® Life technologies™, Paisley, UK) containing and 1% penicillin/streptomycin (Invitrogen) and 10% fetal bovine serum (FBS, gibco® Life technologies™) was used. Cells were regularly inspected for stable phenotype and mycoplasma infection.

3.4. MTT Assay

Cell viability was evaluated using MTT assay as previously reported [56]. Cells were seeded in 96-well plates (6000 cells/well in 100 µL/well, unless otherwise stated), incubated overnight and the medium was replaced with fresh medium containing the investigated drugs at indicated concentrations. Cells were incubated for the indicated time and MTT reagent was added. After 2 to 4 h of incubation the media was aspirated, the plates were dried (1 h at RT), and 50 µL/well of DMSO were added to each well. The absorbance was measured using Infinite F200PRO reader (TECAN, Männedorf, Switzerland). Cell viability and IC_{50}s were calculated using The GraphPad Prism software v. 7.05 was used (GraphPad Prism software Inc., La Jolla, CA, USA)

3.5. Kinase Activity Profiling

Kinase activity profiling was performed as previously reported [57,58]. The PamStation®12 machine (PamGene International, 's-Hertogenbosch, The Netherlands) and STK-PamChip® arrays were used for profiling of the affected serine/threonine kinases.

In brief, the cell were lysed using M-PER Mammalian Extraction Buffer (Pierce, Waltham, MA, USA) containing Protease and Phosphatase Inhibitor Cocktails (Pierce). Then, 1 µg of total extracted protein mixture with ATP were applied per each array. The phosphorylation of the specific peptide was detected using primary anti-phospho-Ser/Thr antibodies and secondary polyclonal swine anti-rabbit

Immunoglobulin-FITC antibody and CCD camera The signals were further analyzed using the BioNavigator software v. 6.0 (BN6, PamGene International).

3.6. Western Blotting

The assay was performed as previously reported [59]. 22Rv1 cells were seeded in Petri dishes (ø 6 cm, 10^6 cells/well in 5 mL/dish), incubated overnight and treated with the compounds in fresh culture media (5 mL/dish) for indicated time. Then, cells were harvested and the proteins were extracted and separated using PAGE in gradient ready-made gels. The proteins were then transferred onto ø 0.2 µm pore PVDF membrane, followed by blocking and treated with primary and secondary antibodies. The signals were detected using the ECL chemiluminescence system (Thermo Scientific, Rockford, IL, USA). The antibodies used are listed in Table 2.

3.7. Determination of Drug Combination Effects

The effects (synergistic versus antagonistic) of MAPK inhibitors (SP600125, SB203580, PD98059, and FR180204), clinically used cytotoxic chemotherapeutics (cisplatin, carboplatin, docetaxel, and cabazitaxel), PARP inhibitor olaparib, or AR inhibitor enzalutamide on the cytotoxic activity of DBF was determined using the Zero interaction potency (ZIP) reference model [43] and the online-based SynergyFinder 2.0 software (https://synergyfinder.fimm.fi, [60]). The experiments were performed as described previously [61]. The 22Rv1 cells (12,000 cells/well of 96-well plate) were co-treated with the defined concentrations of inhibitors and DBF for 48 h in 100 µL/well of 10% FBS/RPMI media. The cytotoxic effects of the individual drugs and its combinations were measured using MTT assay and the data was analyzed using SynergyFinder 2.0 software. Deviations between observed and expected responses with positive (red areas) and negative δ-values (green areas) indicate synergy and antagonism, respectively.

3.8. Analysis of Intracellular ROS Level

Intracellular ROS levels were accessed using the flow cytometry technique and CM-H$_2$DCFDA staining (Cat. No. C6827, Molecular probes, Invitrogen, Eugene, OR, USA) following the previously reported protocol [62,63]. In brief, 10^5 of 22Rv1 cells were seeded in 12-well plates and incubated overnight. The media was exchanged with 4 µM CM-H$_2$DCFDA solution in pre-warmed PBS (0.5 mL/sample) and the cells were incubated for 30 min in the dark (37 °C, 5% CO$_2$). Then the staining solution was replaced with pre-warmed PBS containing investigated compounds at the defined concentrations and the plates were incubated for another 2 h. The cells were harvested and immediately analyzed by flow cytometry.

3.9. Analysis of DNA Damage

The effect of the treatment on DNA damage was analyzed using flow cytometry technique and PI staining. The cells were incubated overnight in 6-well plates (0.2×10^6 cells/well in 2 mL/well), then the media was exchanged with fresh drug-containing media and the cells were incubated for another 48 h. After that, the cells were harvested by trypsinization, twice washed with PBS, resuspended in 70% EtOH/H$_2$O and incubated overnight at −20 °C. The cells were then pelleted, air-dried, and incubated in 0.2 mL/sample the staining solution containing RNase (0.2 mg/mL) and PI (0.02 mg/mL) in PBS for 30 min in the dark at RT. Then additional 0.2 mL/sample of PBS was added to each sample and the cells were immediately analyzed by FACS Calibur (BD Bioscience, Bedford, MA, USA) instrument. The results were analysed using the BD Bioscience Cell Quest Pro v.5.2.1. software (BD Bioscience, San Jose, CA, USA). The cells containing damaged DNA were identified as sub-G1 population.

3.10. Data and Statistical Analysis

For statistical analyses the GraphPad Prism software v. 7.05 was used (GraphPad Prism software Inc., La Jolla, CA, USA). Data are presented as mean ± SD. The unpaired Student's *t*-test (for comparison of two groups) or one-way ANOVA in combination with Dunnett's post-hoc test (for comparison of multiple groups) were used to compare the treated groups with control group. All the experiments were performed in triplicates, unless otherwise stated. Statistically significant differences (*) was assumed if $p < 0.05$.

4. Conclusions

Here, we investigated the mechanism of action of the marine natural alkaloid 3,10-dibromofascaplysin (DBF), which has been previously identified by us as a promising compound in screening experiments. DBF induced apoptosis in human prostate cancer cells, including hormone- and docetaxel-resistant lines, at low micro- and nanomolar concentrations. No cross-resistance to currently available standard therapies was observed. The kinome analysis followed by the validation and functional experiments identified JNK1/2 as one of the molecular targets of DBF in the cells. Inhibition of JNK1/2 and p38 by specific inhibitors led to the increase of cytotoxic activity of DBF, whereas active ERK1/2 was identified to be important for the cytotoxic activity of the alkaloid. Remarkably, DBF strongly synergized with olaparib due to the induction of ROS production, as well as with other clinically approved platinum and taxane agents. Additionally, the compound inhibited AR-signaling and resensitized AR-V7-positive prostate cancer cells to enzalutamide, presumably due to AR-V7 inhibition. A combination of these properties makes this marine alkaloid a promising drug candidate for the treatment of human castration-resistant PCa.

Author Contributions: Conception and design: S.A.D. and G.v.A.; development of methodology: S.A.D.; acquisition of data: S.A.D., J.H., M.K. (Malte Kriegs), K.H., T.B.; data analysis: S.A.D., M.K. (Moritz Kaune), M.K. (Malte Kriegs), K.H., P.A.S., M.E.Z., E.V.P., G.v.A.; data interpretation: All authors; drug-resistant cell lines production: S.J.O.-H.; compound synthesis and purification: P.A.S., M.E.Z.; writing of the manuscript—original draft preparation: S.A.D. and G.v.A.; writing of the manuscript—review and editing: All authors; review and/or revision of the final version of the manuscript: All authors; artwork: S.A.D.; fundraising: M.E.Z. and G.v.A.; study supervision: S.A.D. and G.v.A. All authors have read and agreed to the published version of the manuscript.

Funding: This study was supported by RFBR grant number 20-33-70075 (synthesis of the compound examination of the mechanism of action) and by RFBR grant number 20-04-00089 (investigation of activity in drug-resistant cell lines).

Acknowledgments: The authors are thankful to Zoran Culig (Innsbruck Medical University, Austria) for providing the drug-resistant PC3-DR and DU145-DR cells lines, and to Tina Rohlfing (University Medical Center Hamburg-Eppendorf) for the support in performing of the biological experiments.

Conflicts of Interest: The authors declare no conflict of interest.

References

1. Beyer, J.; Albers, P.; Altena, R.; Aparicio, J.; Bokemeyer, C.; Busch, J.; Cathomas, R.; Cavallin-Stahl, E.; Clarke, N.W.; Claßen, J.; et al. Maintaining success, reducing treatment burden, focusing on survivorship: Highlights from the third European consensus conference on diagnosis and treatment of germ-cell cancer. *Ann. Oncol.* **2013**, *24*, 878–888. [CrossRef]

2. Caffo, O.; De Giorgi, U.; Fratino, L.; Alesini, D.; Zagonel, V.; Facchini, G.; Gasparro, D.; Ortega, C.; Tucci, M.; Verderame, F.; et al. Clinical outcomes of castration-resistant prostate cancer treatments administered as third or fourth line following failure of docetaxel and other second-line treatment: Results of an Italian multicentre study. *Eur. Urol.* **2015**, *68*, 147–153. [PubMed]

3. Armstrong, C.M.; Gao, A.C. Drug resistance in castration resistant prostate cancer: Resistance mechanisms and emerging treatment strategies. *Am. J. Clin. Exp. Urol.* **2015**, *3*, 64–76. [PubMed]

4. Stonik, V.A. Marine natural products: A way to new drugs. *Acta Nat.* **2009**, *2*, 15–25.

5. Molinski, T.F.; Dalisay, D.S.; Lievens, S.L.; Saludes, J.P. Drug development from marine natural products. *Nat. Rev. Drug Discov.* **2009**, *8*, 69–85. [CrossRef]

6. Dyshlovoy, S.A.; Honecker, P.D.D.F. Marine Compounds and Cancer: The First Two Decades of XXI Century. *Mar. Drugs* **2019**, *18*, 20. [CrossRef]

7. Mayer, A. Marine Pharmaceutical: The Clinical Pipeline. 2020. Available online: https://www.midwestern.edu/departments/marinepharmacology/clinical-pipeline.xml (accessed on 2 November 2020).

8. Dyshlovoy, S.A.; Honecker, P.D.D.F. Marine Compounds and Cancer: Where Do We Stand? *Mar. Drugs* **2015**, *13*, 5657–5665. [CrossRef]

9. Dyshlovoy, S.A.; Honecker, F. Marine Compounds and Cancer: 2017 Updates. *Mar. Drugs* **2018**, *16*, 41. [CrossRef]

10. Roll, D.M.; Ireland, C.M.; Lu, H.S.M.; Clardy, J. Fascaplysin, an unusual antimicrobial pigment from the marine sponge *Fascaplysinopsis* sp. *J. Org. Chem.* **1988**, *53*, 3276–3278. [CrossRef]

11. Bharate, S.S.; Manda, S.; Mupparapu, N.; Battini, N.; Vishwakarma, R.A. Chemistry and Biology of Fascaplysin, a Potent Marine-Derived CDK-4 Inhibitor. *MiniRev. Med. Chem.* **2012**, *12*, 650–664. [CrossRef]

12. Lin, J.; Yan, X.-J.; Chen, H. Fascaplysin, a selective CDK4 inhibitor, exhibit anti-angiogenic activity in vitro and in vivo. *Cancer Chemother. Pharmacol.* **2006**, *59*, 439–445. [CrossRef] [PubMed]

13. Kuzmich, A.S.; Fedorov, S.N.; Shastina, V.V.; Shubina, L.K.; Radchenko, O.S.; Balaneva, N.N.; Zhidkov, M.E.; Park, J.I.; Kwak, J.Y.; Stonik, V.A. The anticancer activity of 3-and 10-bromofascaplysins is mediated by caspase-8,-9,-3-dependent apoptosis. *Biorg. Med. Chem.* **2010**, *18*, 3834–3840. [CrossRef] [PubMed]

14. Popov, A.M.; A Stonik, V. Physiological activity of fascaplisine—An unusual pigment from tropical sea fishes. *Antibiot. Khimioterapiia Antibiot. Chemoterapy* **1991**, *36*, 12–14.

15. Dembitsky, V.M.; Gloriozova, T.; Poroikov, V.V. Novel Antitumor Agents: Marine Sponge Alkaloids, their Synthetic Analogs and Derivatives. *MiniRev. Med. Chem.* **2005**, *5*, 319–336. [CrossRef]

16. Hamilton, G. Cytotoxic Effects of Fascaplysin against Small Cell Lung Cancer Cell Lines. *Mar. Drugs* **2014**, *12*, 1377–1389. [CrossRef]

17. Oh, T.-I.; Lee, Y.-M.; Nam, T.-J.; Ko, Y.-S.; Mah, S.; Kim, J.; Kim, Y.; Reddy, R.H.; Kim, Y.J.; Hong, S.; et al. Fascaplysin Exerts Anti-Cancer Effects through the Downregulation of Survivin and HIF-1α and Inhibition of VEGFR2 and TRKA. *Int. J. Mol. Sci.* **2017**, *18*, 2074. [CrossRef]

18. Sharma, S.; Guru, S.K.; Manda, S.; Kumar, A.; Mintoo, M.J.; Prasad, V.D.; Sharma, P.R.; Mondhe, D.M.; Bharate, S.B.; Bhushan, S. A marine sponge alkaloid derivative 4-chloro fascaplysin inhibits tumor growth and VEGF mediated angiogenesis by disrupting PI3K/Akt/mTOR signaling cascade. *Chem. Interact.* **2017**, *275*, 47–60. [CrossRef]

19. Meng, N.; Mu, X.; Lv, X.; Wang, L.; Li, N.; Gong, Y. Autophagy represses fascaplysin-induced apoptosis and angiogenesis inhibition via ROS and p8 in vascular endothelia cells. *Biomed. Pharmacother.* **2019**, *114*, 108866. [CrossRef]

20. Zhidkov, M.E.; Smirnova, P.A.; Tryapkin, O.A.; Kantemirov, A.V.; Khudyakova, Y.V.; Malyarenko, O.S.; Ermakova, S.; Grigorchuk, V.P.; Kaune, M.; Von Amsberg, G.; et al. Total Syntheses and Preliminary Biological Evaluation of Brominated Fascaplysin and Reticulatine Alkaloids and Their Analogues. *Mar. Drugs* **2019**, *17*, 496. [CrossRef]

21. Soni, R.; Müller, L.; Furet, P.; Schoepfer, J.; Stephan, C.; Zumstein-Mecker, S.; Fretz, H.; Chaudhuri, B. Inhibition of Cyclin-Dependent Kinase 4 (Cdk4) by Fascaplysin, a Marine Natural Product. *Biochem. Biophys. Res. Commun.* **2000**, *275*, 877–884. [CrossRef]

22. Hörmann, A.; Chaudhuri, B.; Fretz, H. DNA binding properties of the marine sponge pigment fascaplysin. *Bioorg. Med. Chem.* **2001**, *9*, 917–921. [CrossRef]

23. Oh, T.-I.; Lee, J.H.; Kim, S.; Nam, T.-J.; Kim, Y.-S.; Kim, B.M.; Yim, W.J.; Lim, J.-H. Fascaplysin Sensitizes Anti-Cancer Effects of Drugs Targeting AKT and AMPK. *Molecules* **2017**, *23*, 42. [CrossRef] [PubMed]

24. Kumar, S.; Guru, S.K.; Pathania, A.S.; Manda, S.; Kumar, A.; Bharate, S.B.; Vishwakarma, R.A.; Malik, F.; Bhushan, S. Fascaplysin Induces Caspase Mediated Crosstalk Between Apoptosis and Autophagy Through the Inhibition of PI3K/AKT/mTOR Signaling Cascade in Human Leukemia HL-60 Cells. *J. Cell. Biochem.* **2015**, *116*, 985–997. [CrossRef] [PubMed]

25. Mathew, R.; Karantza-Wadsworth, V.; White, E. Role of autophagy in cancer. *Nat. Rev. Cancer* **2007**, *7*, 961–967. [CrossRef]

26. Kung, H.-J.; Changou, C.; Nguyen, H.G.; Yang, J.C.; Evans, C.P.; Bold, R.J.; Chuang, F. Autophagy and Prostate Cancer Therapeutics. In *Advanced Structural Safety Studies*; Springer Science and Business Media LLC: Berlin/Heidelberg, Germany, 2013; pp. 497–518.

27.	Lyakhova, I.A.; Bryukhovetsky, I.S.; Kudryavtsev, I.V.; Khotimchenko, Y.S.; Zhidkov, M.E.; Kantemirov, A.V. Antitumor Activity of Fascaplysin Derivatives on Glioblastoma Model In Vitro. *Bull. Exp. Biol. Med.* **2018**, *164*, 666–672. [CrossRef]

28.	Segraves, N.L.; Robinson, S.J.; Garcia, D.; Said, S.A.; Fu, X.; Schmitz, F.J.; Pietraszkiewicz, H.; Valeriote, A.F.A.; Crews, P. Comparison of Fascaplysin and Related Alkaloids: A Study of Structures, Cytotoxicities, and Sources. *J. Nat. Prod.* **2004**, *67*, 783–792. [CrossRef]

29.	Sampson, N.; Neuwirt, H.; Puhr, M.; Klocker, H.; Eder, I.E. In vitro model systems to study androgen receptor signaling in prostate cancer. *Endocr. Relat. Cancer* **2013**, *20*, R49–R64. [CrossRef]

30.	Nelson, P.S. Targeting the androgen receptor in prostate cancer—A resilient foe. *N. Engl. J. Med.* **2014**, *371*, 1067–1069. [CrossRef]

31.	Puhr, M.; Hoefer, J.; Schäfer, G.; Erb, H.H.; Oh, S.J.; Klocker, H.; Heidegger, I.; Neuwirt, H.; Culig, Z. Epithelial-to-Mesenchymal Transition Leads to Docetaxel Resistance in Prostate Cancer and Is Mediated by Reduced Expression of miR-200c and miR-205. *Am. J. Pathol.* **2012**, *181*, 2188–2201. [CrossRef]

32.	Dhillon, A.S.; Hagan, S.; Rath, O.; Kolch, W. MAP kinase signalling pathways in cancer. *Oncogene* **2007**, *26*, 3279–3290. [CrossRef]

33.	Bhullar, K.S.; Lagarón, N.O.; McGowan, E.M.; Parmar, I.; Jha, A.; Hubbard, B.P.; Rupasinghe, H.P.V. Kinase-targeted cancer therapies: Progress, challenges and future directions. *Mol. Cancer* **2018**, *17*, 48. [CrossRef] [PubMed]

34.	Struve, N.; Binder, Z.A.; Stead, L.F.; Brend, T.; Bagley, S.J.; Faulkner, C.; Ott, L.; Müller-Goebel, J.; Weik, A.-S.; Hoffer, K.; et al. EGFRvIII upregulates DNA mismatch repair resulting in increased temozolomide sensitivity of MGMT promoter methylated glioblastoma. *Oncogene* **2020**, *39*, 3041–3055. [CrossRef] [PubMed]

35.	Rodríguez-Berriguete, G.; Fraile, B.; Martínez-Onsurbe, P.; Olmedilla, G.; Paniagua, R.; Royuela, M. MAP Kinases and Prostate Cancer. *J. Signal Transduct.* **2012**, *2012*, 169170. [CrossRef]

36.	Xu, R.; Hu, J. The role of JNK in prostate cancer progression and therapeutic strategies. *Biomed. Pharmacother.* **2020**, *121*, 109679. [CrossRef] [PubMed]

37.	Yang, Y.-M.; Bost, F.; Charbono, W.; Dean, N.; McKay, R.; Rhim, J.S.; Depatie, C.; Mercola, D. C-Jun NH(2)-terminal kinase mediates proliferation and tumor growth of human prostate carcinoma. *Clin. Cancer Res.* **2003**, *9*, 391. [PubMed]

38.	Hu, J.; Wang, G.; Sun, T. Dissecting the roles of the androgen receptor in prostate cancer from molecular perspectives. *Tumor Biol.* **2017**, *39*, 1010428317692259. [CrossRef]

39.	Liu, P.-Y.; Lin, S.-Z.; Sheu, J.J.-C.; Lin, C.-T.; Lin, P.-C.; Chou, Y.-W.; Liu, C.-A.; Chiou, T.-W.; Harn, H.-J. Regulation of androgen receptor expression byZ-isochaihulactone mediated by the JNK signaling pathway and might be related to cytotoxicity in prostate cancer. *Prostate* **2012**, *73*, 531–541. [CrossRef]

40.	Tang, F.; Kokontis, J.; Lin, Y.; Liao, S.; Lin, A.; Xiang, J. Androgen via p21 Inhibits Tumor Necrosis Factor α-induced JNK Activation and Apoptosis. *J. Biol. Chem.* **2009**, *284*, 32353–32358. [CrossRef]

41.	Liu, J.; Lin, A. Role of JNK activation in apoptosis: A double-edged sword. *Cell Res.* **2005**, *15*, 36–42. [CrossRef]

42.	Samatar, A.A.; Poulikakos, P.I. Targeting RAS–ERK signalling in cancer: Promises and challenges. *Nat. Rev. Drug Discov.* **2014**, *13*, 928–942. [CrossRef]

43.	Yadav, B.; Wennerberg, K.; Aittokallio, T.; Tang, J. Searching for Drug Synergy in Complex Dose–Response Landscapes Using an Interaction Potency Model. *Comput. Struct. Biotechnol. J.* **2015**, *13*, 504–513. [CrossRef] [PubMed]

44.	Tomida, T. Visualization of the spatial and temporal dynamics of MAPK signaling using fluorescence imaging techniques. *J. Physiol. Sci.* **2014**, *65*, 37–49. [CrossRef] [PubMed]

45.	Ventura, J.J.; Hübner, A.; Zhang, C.; Flavell, R.A.; Shokat, K.M.; Davis, R.J. Chemical genetic analysis of the time course of signal transduction by JNK. *Mol. Cell* **2006**, *21*, 701–710. [CrossRef] [PubMed]

46.	Qiu, M.-S.; Green, S.H. PC12 cell neuronal differentiation is associated with prolonged p21ras activity and consequent prolonged ERK activity. *Neuron* **1992**, *9*, 705–717. [CrossRef]

47.	Traverse, S.; Gomez, N.; Paterson, H.; Marshall, C.; Cohen, P. Sustained activation of the mitogen-activated protein (MAP) kinase cascade may be required for differentiation of PC12 cells. Comparison of the effects of nerve growth factor and epidermal growth factor. *Biochem. J.* **1992**, *288*, 351–355. [CrossRef]

48.	Sachdev, E.; Tabatabai, R.; Roy, V.; Rimel, B.J.; Mita, M. PARP Inhibition in Cancer: An Update on Clinical Development. *Target. Oncol.* **2019**, *14*, 657–679. [CrossRef]

49.	Bochum, S.; Berger, S.; Martens, U.M. Olaparib. *Recent Results Cancer Res.* **2018**, *211*, 217–233.

50. Lord, C.J.; Ashworth, A. PARP inhibitors: Synthetic lethality in the clinic. *Science* **2017**, *355*, 1152–1158. [CrossRef]

51. Li, L.; Karanika, S.; Yang, G.; Wang, J.; Park, S.; Broom, B.M.; Manyam, G.C.; Wu, W.; Luo, Y.; Basourakos, S.; et al. Androgen receptor inhibitor–induced "BRCAness” and PARP inhibition are synthetically lethal for castration-resistant prostate cancer. *Sci. Signal.* **2017**, *10*, eaam7479. [CrossRef]

52. Feiersinger, G.E.; Trattnig, K.; Leitner, P.D.; Guggenberger, F.; Oberhuber, A.; Peer, S.; Hermann, M.; Skvortsova, I.; Vrbková, J.; Bouchal, J.; et al. Olaparib is effective in combination with, and as maintenance therapy after, first-line endocrine therapy in prostate cancer cells. *Mol. Oncol.* **2018**, *12*, 561–576. [CrossRef]

53. Lee, J.; Kim, Y.; Lim, S.; Jo, K. Single-molecule visualization of ROS-induced DNA damage in large DNA molecules. *Analyst* **2016**, *141*, 847–852. [CrossRef] [PubMed]

54. Zhang, S.; Lin, Y.; Kim, Y.-S.; Asharani, P.V.; Liu, Z.-G.; Shen, H.-M. c-Jun N-terminal kinase mediates hydrogen peroxide-induced cell death via sustained poly(ADP-ribose) polymerase-1 activation. *Cell Death Differ.* **2007**, *14*, 1001–1010. [CrossRef] [PubMed]

55. Shafi, A.A.; Yen, A.E.; Weigel, N.L. Androgen receptors in hormone-dependent and castration-resistant prostate cancer. *Pharmacol. Ther.* **2013**, *140*, 223–238. [CrossRef] [PubMed]

56. Dyshlovoy, S.A.; Madanchi, R.; Hauschild, J.; Otte, K.; Alsdorf, W.H.; Schumacher, U.; Kalinin, V.I.; Silchenko, A.S.; Avilov, S.A.; Honecker, F.; et al. The marine triterpene glycoside frondoside A induces p53-independent apoptosis and inhibits autophagy in urothelial carcinoma cells. *BMC Cancer* **2017**, *17*, 1–10. [CrossRef] [PubMed]

57. Arni, S.; Le, T.H.N.; De Wijn, R.; García-Villegas, R.; Dankers, M.; Weder, W.; Hillinger, S. Ex vivo multiplex profiling of protein tyrosine kinase activities in early stages of human lung adenocarcinoma. *Oncotarget* **2017**, *8*, 68599–68613. [CrossRef] [PubMed]

58. Dyshlovoy, S.A.; Kaune, M.; Kriegs, M.; Hauschild, J.; Busenbender, T.; Shubina, L.K.; Makarieva, T.N.; Hoffer, K.; Bokemeyer, C.; Graefen, M.; et al. Marine alkaloid monanchoxymycalin C: A new specific activator of JNK1/2 kinase with anticancer properties. *Sci. Rep.* **2020**, *10*, 13178. [CrossRef] [PubMed]

59. Dyshlovoy, S.A.; Pelageev, D.N.; Hauschild, J.; Sabutskii, Y.E.; Khmelevskaya, E.A.; Krisp, C.; Kaune, M.; Venz, S.; Borisova, K.L.; Busenbender, T.; et al. Inspired by Sea Urchins: Warburg Effect Mediated Selectivity of Novel Synthetic Non-Glycoside 1,4-Naphthoquinone-6S-Glucose Conjugates in Prostate Cancer. *Mar. Drugs* **2020**, *18*, 251. [CrossRef] [PubMed]

60. Ianevski, A.; Giri, A.K.; Aittokallio, T. SynergyFinder 2.0: Visual analytics of multi-drug combination synergies. *Nucleic Acids Res.* **2020**, *48*, W488–W493. [CrossRef]

61. Dyshlovoy, S.A.; Tabakmakher, K.M.; Hauschild, J.; Shchekaleva, R.K.; Otte, K.; Guzii, A.G.; Makarieva, T.N.; Kudryashova, E.K.; Fedorov, S.N.; Shubina, L.K.; et al. Guanidine Alkaloids from the Marine Sponge Monanchora pulchra Show Cytotoxic Properties and Prevent EGF-Induced Neoplastic Transformation In Vitro. *Mar. Drugs* **2016**, *14*, 133. [CrossRef]

62. Dyshlovoy, A.S.; Pelageev, N.D.; Hauschild, J.; Borisova, L.K.; Kaune, M.; Krisp, C.; Venz, S.; Sabutskii, E.Y.; Khmelevskaya, A.E.; Busenbender, T.; et al. Successful Targeting of the Warburg Effect in Prostate Cancer by Glucose-Conjugated 1,4-Naphthoquinones. *Cancers* **2019**, *11*, 1690. [CrossRef]

63. Dyshlovoy, S.A.; Rast, S.; Hauschild, J.; Otte, K.; Alsdorf, W.H.; Madanchi, R.; Kalinin, V.I.; Silchenko, A.S.; Avilov, S.A.; Dierlamm, J.; et al. Frondoside A induces AIF-associated caspase-independent apoptosis in Burkitt lymphoma cells. *Leuk. Lymphoma* **2017**, *58*, 2905–2915. [CrossRef] [PubMed]

Publisher's Note: MDPI stays neutral with regard to jurisdictional claims in published maps and institutional affiliations.

Article

Interaction of *Thalassia testudinum* Metabolites with Cytochrome P450 Enzymes and Its Effects on Benzo(a)pyrene-Induced Mutagenicity

Livan Delgado-Roche [1,2,†], Rebeca Santes-Palacios [3,†], José A. Herrera [4], Sandra L. Hernández [5], Mario Riera [1], Miguel D. Fernández [1], Fernando Mesta [6], Gabino Garrido [7], Idania Rodeiro [1,*] and Jesús Javier Espinosa-Aguirre [5,*]

[1]   Departamento de Farmacología, Instituto de Ciencias del Mar (ICIMAR), Loma 14, Alturas del Vedado, Plaza de la Revolución, La Habana 10600, Cuba; ldelgado@liomont.com.mx (L.D.-R.); mario@icimar.cu (M.R.); migueldavid@cebimar.cu (M.D.F.)
[2]   Dirección Médica, Laboratorios Liomont S.A. de C.V., Carretera México-Toluca 5420, Ciudad de México 05320, Mexico
[3]   Laboratorio de Toxicología Genética, Instituto Nacional de Pediatría, Insurgentes Sur 3700, Insurgentes Cuicuilco, Ciudad de México 04530, Mexico; rebeca.santes@inp.mx
[4]   Instituto de Ciencia y Tecnología de Materiales (IMRE), Universidad de La Habana, Zapata y G, Vedado, Plaza de la Revolución, La Habana 10400, Cuba; jose@imre.oc.uh.cu
[5]   Departamento de Medicina Genómica y Toxicología Ambiental, Instituto de Investigaciones Biomédicas, Universidad Nacional Autónoma de México (UNAM), Ciudad Universitaria, Ciudad de México 04510, Mexico; slhernandez@biomedicas.unam.mx
[6]   Escuela Nacional de Medicina y Homeopatía, Instituto Politécnico Nacional, Guillermo Massieu Helguera 239, Ciudad de México 07320, Mexico; lmestac1800@alumno.ipn.mx
[7]   Departamento de Ciencias Farmacéuticas, Facultad de Ciencias, Universidad Católica del Norte, Angamos 0610, Antofagasta 1240000, Chile; gabino.garrido@ucn.cl
*   Correspondence: idania.rodeiro@infomed.sld.cu (I.R.); jjea@iibiomedicas.unam.mx (J.J.E.-A.); Tel.: +53-7-8819812 (I.R.); +52-556229214 (J.J.E.-A.)
†   Both authors equally contributed to this work.

Received: 31 October 2020; Accepted: 17 November 2020; Published: 19 November 2020

Abstract: The aim of the present work was to evaluate the effects of *Thalassia testudinum* hydroethanolic extract, its polyphenolic fraction and thalassiolin B on the activity of phase I metabolizing enzymes as well as their antimutagenic effects. Spectrofluorometric techniques were used to evaluate the effect of tested products on rat and human CYP1A and CYP2B activity. The antimutagenic effect of tested products was evaluated in benzo[a]pyrene (BP)-induced mutagenicity assay by an Ames test. Finally, the antimutagenic effect of *Thalassia testudinum* (100 mg/kg) was assessed in BP-induced mutagenesis in mice. The tested products significantly ($p < 0.05$) inhibit rat CYP1A1 activity, acting as mixed-type inhibitors of rat CYP1A1 (Ki = 54.16 ± 9.09 µg/mL, 5.96 ± 1.55 µg/mL and 3.05 ± 0.89 µg/mL, respectively). Inhibition of human CYP1A1 was also observed (Ki = 197.1 ± 63.40 µg/mL and 203.10 ± 17.29 µg/mL for the polyphenolic fraction and for thalassiolin B, respectively). In addition, the evaluated products significantly inhibit ($p < 0.05$) BP-induced mutagenicity in vitro. Furthermore, oral doses of *Thalassia testudinum* (100 mg/kg) significantly reduced ($p < 0.05$) the BP-induced micronuclei and oxidative damage, together with an increase of reduced glutathione, in mice. In summary, *Thalassia testudinum* metabolites exhibit antigenotoxic activity mediated, at least, by the inhibition of CYP1A1-mediated BP biotransformation, arresting the oxidative and mutagenic damage. Thus, the metabolites of *T. testudinum* may represent a potential source of chemopreventive compounds for the adjuvant therapy of cancer.

Keywords: *Thalassia testudinum*; thalassiolin B; polyphenols; CYP1A1; benzo[a]pyrene; chemoprevention

---

1. Introduction

Air pollution-related diseases represent a major environmental problem affecting health worldwide. Outdoor and indoor pollution has been tightly associated with the incidence of respiratory diseases, including lung cancer. In accordance with WHO estimates, lung cancer causes about 6% of premature deaths related to outdoor air pollution, as well as tobacco consumption [1]. Polycyclic aromatic hydrocarbons (PAHs) are pollutants widely distributed in the environment as a result of organic matter incomplete combustion. Furthermore, PAHs are present in commercial products consumed by humans (e.g., tobacco) [2]. Benzo[a]pyrene (BP), one of the main PAH air pollutants, is metabolized by CYP1A1 and CYP1B1 enzymes and it is biotransformed into carcinogenic (±)-B[a]P-r-7, t-8-dihydrodiol-t-9,10 (BPDE) epoxide species [2,3]. BPDE reacts with DNA to produce BPDE-N2-deoxyguanosine adducts, promoting DNA mutations and carcinogenesis [4–7]. The presence of BPDE-DNA adducts in human lung cells has been well documented to be related to the initiation of pulmonary cancer [5,8].

Advances in the pharmaceutical industry improve cancer treatment; however, the discovery of new effective chemopreventive agents is still necessary. Cancer chemoprevention by natural or synthetic agents capable of avoiding, reversing or suppressing carcinogenic progression has become a plausible strategy to arrest cancer mortality [9]. There are several classes of cancer chemopreventive agents including blocking agents, which act at the initiation stage of carcinogenesis by inhibiting pro-carcinogen activating enzymes by inducing carcinogen-detoxifying enzymes, by enhancing antioxidant activity or by inducing DNA repair enzymes [10]. Cytochrome P450 (CYP450) enzymes are a superfamily of hemoproteins that catalyze the biotransformation of not only a wide array of drugs and endogenous substances, but also the bioactivation of many pro-carcinogens [11]. Consequently, specific CYP enzymes have been identified as potential targets for cancer chemoprevention [12].

Polyphenols exert a wide range of beneficial effects beyond their antioxidant and anti-inflammatory properties [13–15]. Data evidence from in vitro [16–20] and in vivo studies [21–24] suggest the chemoprotective role of polyphenols against lung carcinogenesis probably resulting from three main mechanisms: antioxidant activity, regulation of phase I and II enzymes and regulation of cell survival pathways [24].

Marine plants are a potential source of secondary metabolites with beneficial properties [25–32]. *Thalassia testudinum* seagrass grows abundantly in the Caribbean Sea, particularly in the Cuban coasts. A previous study reports sulfated glycoside flavone thalassiolin B (TB) (chrysoeriol7-β-D-glucopyranosyl-2″-sulphate, Figure 1) as the most abundant bioactive component within the *T. testudinum* crude hydroethanolic extract (Th) [33]. Other phenolic compounds have been identified in the extract, including apigenin-7-O- β-D-glucopyranosyl-2″-sulfate (thalassiolin C), chrysoeriol-7-O-β-D-glucopyranoside, apigenin-7-O- β-D-glucopyranoside, dihydroxy-3′,4′-dimethoxyflavone 7-O-β-D-glucopyranoside, luteolin-3′-sulphate, chrysoeriol and apigenin [34]. Th shows in vitro scavenger activity for $^{\bullet}OH$, $RO_2^{\bullet}$, $O_2^{\bullet-}$ and $DPPH^{\bullet}$ free radicals and in vivo antioxidant effects against brain and liver induced-lipid peroxidation in mice [34,35]. In addition, Th shows acute anti-inflammatory effects in mice [36] and it displays selective anti-proliferative activity against cancer cells compared to normal cells [37]. Besides, the extract also inhibits drug efflux by ABCG2/breast cancer resistance protein (BCRP) and ABCB1/P-glycoprotein (MDR1 gene), increasing intracellular accumulation of anticancer agents [38,39]. Thus, the marine angiosperm *T. testudinum* has been considered a natural source of potential antitumor agents.

On the other hand, Th modulates the activity of different isoforms of P450 system, including CYP1A and 2B families [38,40]; however, these interactions are not well characterized yet. As CYP1A and 2B subfamilies are involved in the metabolism of several mutagens and carcinogens, the enzymatic inhibition could be associated with decreased carcinogenic risk. Thus, the aim of the present work was to further characterize the effects of *T. testudinum* extract and its polyphenolic components (polyphenolic fraction of the hydroethanolic extract, PF) on CYP1A and CYP2B enzymatic activity, an also to evaluate the effects of Th on BP-induced mutagenicity.

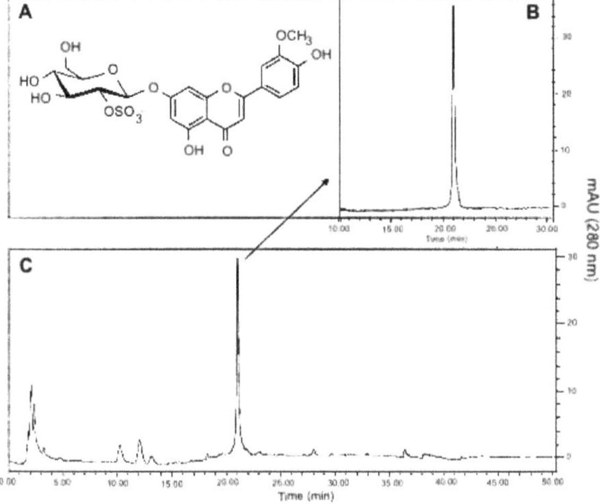

Figure 1. Chromatographic profile of thalassiolin B isolated from *T. testudinum* hydroethanolic extract. (**A**) Chemical structure of thalassiolin B (chrysoeriol 7-β-D-glucopyranosyl-2″-sulphate), the main component of *T. testudinum* extract. (**B**) HPLC of thalassiolin B standard. (**C**) HPLC profile of *T. testudinum* hydroethanolic extract. The authors have the right to use this figure.

2. Results

2.1. Tested Compounds Modulate Rat CYP1A But Not CYPB2 Activity

The enzymatic activity of CYP1A1/2 CYP2B1/2 was measured in rat liver microsomes in the presence, or not, of Th, PF or TB. Test products shown no interference with the fluorescence of resorufin even at the highest concentration tested. No appreciable changes in the activity of both CYP2B isoforms were observed (data not shown). In contrast, Th, PF and TB modulated the rat CYP1A activities as shown in Table 1. The enzymatic activity of both CYP1A1 and CYP1A2 was modulated by the test natural products; however, CYP1A1 was more sensitive than CYP1A2. The PF and TB showed a significant ($p < 0.05$) higher inhibitory effect than the crude extract (Th) on CYP1A isoforms; meanwhile, the Th showed no significant inhibition for CYP1A2.

Table 1. Rat CYP1A1/2 activity modulation by *T. testudinum* extract and its components.

CYP	Product	Test Product Concentrations (µg/mL)				
		2.5	5.0	12.5	25.0	50.0
rCYP1A1	Th	74.88 ±2.92 [a]	63.11 ±8.22 [a,*]	49.68 ±8.04 [a,**]	ND	ND
	PF	46.61 ±3.10 [b,**]	43.31 ±0.94 [b,**]	31.76 ±6.64 [b,**]	27.07 ±1.95 [a,**]	24.00 ±2.97 [a,**]
	TB	53.83 ±1.08 [c,*]	42.97 ±4.66 [b,**]	42.35 ±0.84 [a,**]	29.99 ±3.03 [a,**]	35.15 ±1.92 [b,**]
		2.5	5.0	12.5	25.0	50.0
rCYP1A2	Th	100.47 ±10.66 [a]	99.88 ±5.25 [a]	94.65 ±14.80 [a]	ND	ND
	PF	83.93 ±2.55 [a]	71.87 ±2.85 [b]	62.05 ±4.92 [b,*]	55.02 ±2.84 [a,*]	56.02 ±2.04 [a,*]
	TB	81.01 ±8.06 [a]	75.04 ±2.78 [b]	57.08 ±4.70 [b,*]	63.36 ±4.90 [a,*]	57.03 ±4.07 [a,*]

Test products were added at 2.5–50.0 µg/mL to the incubation mixture containing rat liver microsomes (80 µg) and 7-ethoxyresorufin (1 µM) for CYP1A1 or 7-methoxyresorufin (5 µM) for CYP1A2. The values represented the means ± SD of CYP activities (% respect control) from three independent experiments. Each sample was running by triplicate. The enzymatic activity in absence of test products was taken as 100%. rCYP1A1, rCYP1A2: rat CYPs; Th: *T. testudinum* extract; PF: polyphenolic fraction; TB: thalassiolin B. Different letters ([a,b,c]) represent statistical differences ($p < 0.05$) between test products; * $p < 0.05$, ** $p < 0.01$ when compared with control (100% enzyme activity).

2.2. *T. testudinum Extract, Polyphenolic Fraction and Thalassiolin B Are CYP1A1 Mixed-Type Inhibitors*

Once CYP1A1 was identified as the most sensitive enzyme, kinetics experiments were performed in order to elucidate the type of inhibition induced by PF and TB. Demethylation of EROD in the presence of rat liver microsomes showed typical Michaelis–Menten kinetics for evaluated products (Figure 2A–C). Using non-linear regression and a Lineweaver–Burk plot, it was determined that Th, PF and TB are mixed-type inhibitors for rat CYP1A1 (Figure 2D–F) with constants of inhibition (Ki) of 54.16 ± 9.09 µg/mL, 5.96 ± 1.55 µg/mL and 3.05 ± 0.89 µg/mL, respectively (Table 2). In a mixed-type inhibition model, affinity changes of the enzyme for the substrate in presence of an inhibitor is determined by the parameter alpha (α), which was 8.66 ± 2.82 for Th, 370.60 ± 56.86 for PF and 3.65 ± 0.86 for TB (Table 2).

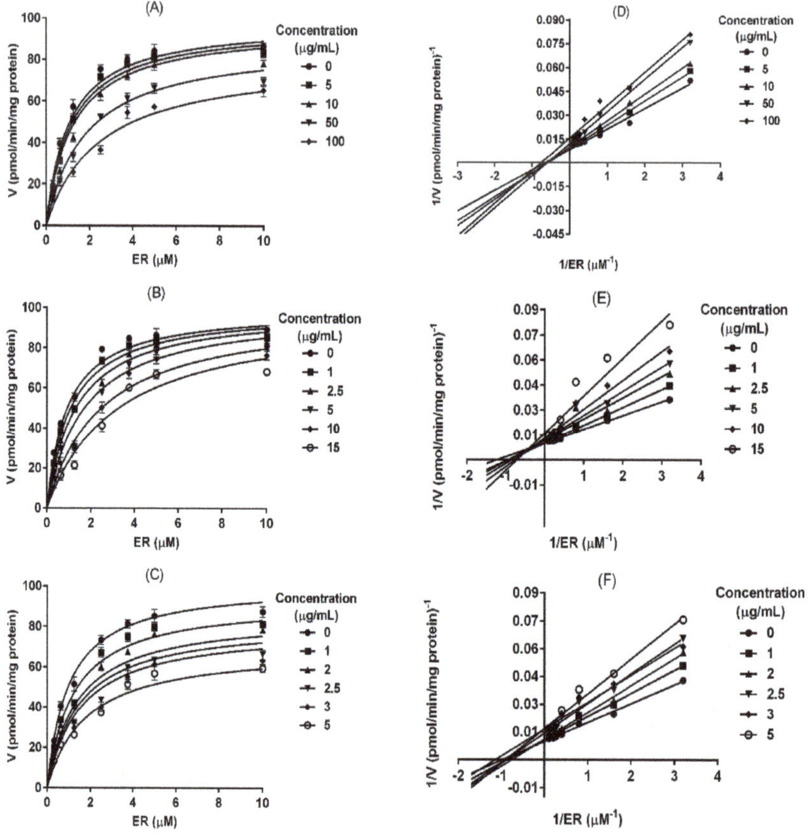

Figure 2. Inhibition kinetics of rat CYP1A1 by *T. testudinum* extract, polyphenolic fraction and Thalassiolin B. The fluorescence was recorded every 15 s during 15 min; reactions consisted in 80 µg protein, 0.32–10 µM 7-ethoxyresorufin, and 50 mM NADPH. For inhibition assays, the test products were added at different concentrations to the reaction mixture. Each point in (**A–C**) represents the mean ± SD from three independent experiments. (**D–F**) Lineweaver-Burk plot analyses were done to obtain the kinetic parameters. (**A,D**) *T. testudinum* extract (Th); (**B,E**) polyphenolic fraction (PF); (**C,F**) thalassiolin B (TB).

Table 2. Kinetics parameters for rat and human CYP1A1 inhibition.

Inhibitor	Parameter	Rat CYP1A1	Human CYP1A1
ER	Vmax (pmol/min/mgPr)	2396.00 ± 116.20	95.20 ± 8.14
	Km (µM)	0.42 ± 0.05	0.34 ± 0.02
Th	Type of inhibition	Mixed	-
	Ki (µg/mL)	54.16 ± 9.09	-
	α	8.66 ± 2.82	-
PF	Type of inhibition	Mixed	Mixed
	Ki (µg/mL)	5.96 ± 1.55	197.10 ± 63.40
	α	370.60 ± 56.86	7.14 ± 5.67
TB	Type of inhibition	Mixed	Non-competitive
	Ki (µg/mL)	3.05 ± 0.89	203.10 ± 17.29
	α	3.65 ± 0.86	-

Data represent the mean ± SD of three independent experiments. Kinetic parameters were obtained by a nonlinear regression analysis of experimental data fitted to Michaelis-Menten equation. EROD (7-ethoxyresorufin) was used as substrate, Vmax: maximum velocity, Km: Michaelis-Menten constant. Th: *T. testudinum* hydroethanolic extract; PF: polyphenolic fraction; TB: thalassiolin B. Human recombinant CYP1A1 was obtained from *E. coli*.

2.3. Polyphenolic Fraction and Thalassiolin B Modulate the Human CYP1A1 Activity

Taking into account that rat CYP1A1 was more sensitive than the CYP1A2 isoform, the effect of tested products on human recombinant CYP1A1 activity was evaluated. The results showed a significant ($p < 0.05$) inhibition of human recombinant CYP1A1 by PF and TB, while Th only exhibited a slight inhibitory effect on enzyme activity at the highest concentration (Figure 3A). The biochemical characterization of human CYP1A1 inhibition resulted in a mixed-type for PF and non-competitive inhibition for TB (Figure 3B–E). A moderate inhibitory potential was found for PF and TB on human CYP1A1 with Ki of 197.10 ± 63.40 µg/mL and 203.10 ± 17.29 µg/mL, respectively (Table 2). These results showed a potential inhibitory effect of phase I carcinogen-metabolizing enzymes CYP1A1 by *T. testudinum* metabolites.

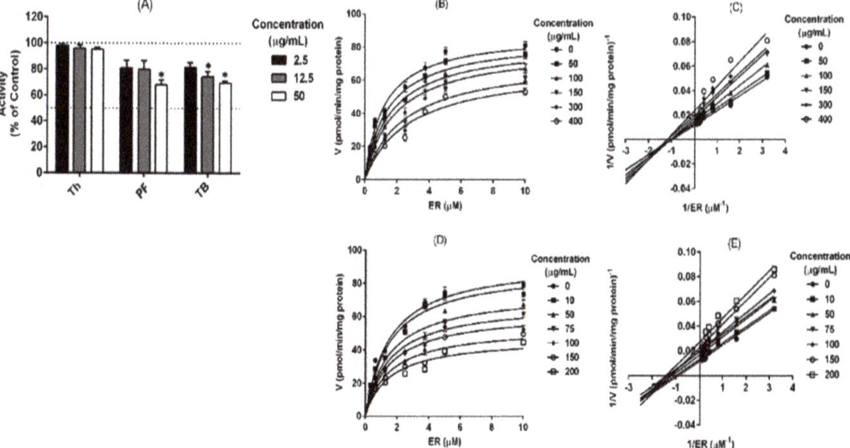

Figure 3. Inhibition of human recombinant CYP1A1 activity and kinetics by the tested products. (**A**) The fluorescence was recorded every 15 s during 15 min; reactions consisted in 40 µg protein, 0.32–10 µM 7-ethoxyresorufin, and 50 mM NADPH. For inhibition assays, the products were added at different concentrations to the reaction mixture. Each point in (**A**) represents the mean ± SD from three independent experiments. (**B–E**) Lineweaver–Burk plot analysis was done to obtain the kinetic parameters. (**B,C**) polyphenolic extract; (**D,E**) thalassiolin B. Th, *Thalassia testudinum* hydroethanolic extract; PF, polyphenolic fraction; TB, thalassiolin B. * Statistical differences ($p < 0.05$).

2.4. Antimutagenic Effect of T. testudinum Extract, Polyphenolic Fraction and Thalassiolin B against Benzo[a]pyrene-Induced Mutagenicity in S. typhimurium

The bioactivation of BP by CYP1A1 leads to carcinogenic effects [2,5]. Since Th, PF and TB exert inhibitory effects on both rat and human CYP1A1 activity, we explored the potential anti-mutagenic effect of these compounds by using the Ames test. First, we evaluate the cytotoxicity of test products (up to 1000 µg/mL) on *S. typhimurium*. The frequency of spontaneous reversion in controls (S9 plus vehicle) did not differ from the historically controls of our laboratory for *S. typhimurium*. Therefore, the test products are not cytotoxic for *S. typhimurium*, whereas BP induced a significantly ($p < 0.05$) increase of revertant frequency. Interestingly, the number of revertant colonies decreased in a dose-dependent fashion in presence of test natural products. The inhibition percentage of BP-induced mutagenicity in presence of S_9 activation mixture achieved 27% for the highest tested dose of the extract (1000 µg/mL), 34% for PF (500 µg/mL) and 33% for TB (400 µg/mL) (Table 3). These results suggested that Th, PF and TB possess antimutagenic effects under these experimental conditions.

Table 3. Effects of *T. testudinum* extract and its components on benzo[a]pyrene-induced mutagenicity in *Salmonella typhimurium*.

Treatments	His$^+$ Revertants/Plate (% Inhibition)
S_9-Control	23.1 ± 2.0
S_9-Control vehicle + BP	746.2 ± 32.3
S_9-Th (10 µg/mL) + BP	714.4 ± 8.5
S_9-Th (100 µg/mL) + BP	678.0 ± 11.2 ** (10%)
S_9-Th (1000 µg/mL) + BP	547.4 ± 5.9 *** (27%)
S_9-PF (5 µg/mL) + BP	713.4 ± 8.8
S_9-PF (50 µg/mL) + BP	665.5 ± 10.1 ** (11%)
S_9-PF (500 µg/mL) + BP	497.3 ± 10.7 *** (34%)
S_9-TB (400 µg/mL) + BP	512.1 ± 28.4 *** (32%)

Data represent mean ± SD of histidine revertant colonies number in TA98 *S. typhimurium* strain of two independent experiments by triplicate. Incubations were in presence of rat liver microsomal mix (S9). Th: *T. testudinum* extract; PF: polyphenolic fraction; TB: thalassiolin B, BP: benzo[a]pyrene. ** $p < 0.01$, *** $p < 0.001$, ANOVA followed by Tukey test, compared to control vehicle + BP. %inhibition: percentage of inhibition number revertants/plate in regards to "S9-Control (vehicle) + BP" group, as (1-(colonies/plates with BP + product)/(colonies/plate with just BP)) × 100%.

2.5. T. testudinum Extract Reduces Oxidative Damage and Micronuclei Formation in BP-Exposed Mice

The capacity of Th to reduce BP-induced oxidative damage in mice was evaluated by measuring the serum levels of malondialdehyde (MDA), advanced oxidation protein products (AOPP) and reduced glutathione (GSH). As we expect, in Th pre-treated mice, the levels of MDA and AOPP were significantly lower ($p < 0.05$) than in BP control animals. In accordance with the reduction of oxidative damage, GSH level was significantly increased ($p < 0.05$) in Th pre-treated animals (Table 4). Additionally, BP induced a significant micronuclei formation in bone marrow of BP-exposed animals compared to cells from control animals (vehicle). Meanwhile, in Th pre-treated mice it was found a significant reduction ($p < 0.05$) of micronuclei formation (Table 5). These results indicate a protective effect by the extract obtained from *T. testudinum* leaves against BP-induced DNA damage together with a modulation of systemic oxidative stress in mice.

Table 4. Effects of *T. testudinum* extract pre-treatment on oxidative stress biomarkers in mice after exposure to benzo(a)pyrene.

Treatment (mg/kg)	MDA (μM/mgPr)	AOPP (μM chloramines/mgPr)	GSH (μM/mgPr)
Control (vehicle)	3.17 ± 0.5	7.41 ± 1.3	495.1 ± 67.8
BP	7.31 ± 0.2 [a]	14.67 ± 1.2 [a]	149.7 ± 63.5 [a]
Th + BP	5.55 ± 0.6 [b]	11.64 ± 1.5 [b]	352.8 ± 35.1 [b]

Values are expressed as mean $\pm$ SD (concentration *per* mg protein). [a] significant difference regarding control, [b] significant difference regarding to BP group. Control: animals received 7 daily oral doses of distilled water and one dose of oil (BP vehicle). Th: *T. testudinum* extract, BP: animals received 250 mg/kg benzo(a)pyrene, BP + *T. testudinum*: 7 days oral pre-treatment with 100 mg/kg *T. testudinum* extract before receiving BP dose. ANOVA-Dunnett post hoc-test, $p < 0.05$. MDA: malondialdehyde, AOPP: advanced oxidation protein products, GSH: reduced glutathione.

Table 5. Effects of *T. testudinum* extract on benzo(a)pyrene-induced micronucleus in mice bone marrow.

Treatment (mg/kg)	PCE/NCE	MN/PCE
Control (vehicle)	1.8 ± 0.26 [b]	4.0 ± 0.7 [b]
BP	3.4 ± 0.78 [a]	17.0 ± 1.7 [a]
Th + BP	2.0 ± 0.21 [b]	7.0 ± 1.0 [a,b]

Values are expressed as mean $\pm$ SD. MN: micronucleus, PCE: polychromatic erythrocytes, NCE: normochromatic erythrocytes (2000 cells/animal), Control: Animals received 7 daily oral doses of distilled water and one dose of oil (BP vehicle). Th: *T. testudinum* extract, BP: animals received 250 mg/kg of benzo(a)pyrene, BP + *T. testudinum*: 7 days' oral pre-treatment with 100 mg/kg *T. testudinum* extract before receiving BP dose, [a] significant difference regarding control, [b] significant difference regarding to BP group. ANOVA–Dunnett post hoc-test, $p < 0.05$.

3. Discussion

There is evidence on the inhibitory effects of plant-derived phenols in PAH-induced mutagenesis and carcinogenesis [41,42]. The antimutagenic and chemopreventive properties of polyphenols have been associated with modulation of CYP450-mediated metabolism of mutagens, as well as to the interaction with active mutagenic metabolites [43]. It has been suggested that flavones which contain free 5- and 7-hydroxyls are potent inhibitors of cytochrome CYP1Al/2 [44]. Therefore, they may be useful as chemopreventive agents against PAH-induced carcinogenesis. The extract obtained from the leaves of *T. testudinum* marine plant is rich in flavonoids and other polyphenols (29.5% $\pm$ 1.2% total polyphenols, proanthocyanidins 21.0% $\pm$ 2.3%, total flavonoids 4.6% $\pm$ 0.2%, expressed as g per 100 g of the dry extract, % *w/w*) suggesting a potential inhibitory capacity on CYPs enzymatic activity. Derivate of *T. testudinum* are under preclinical investigation as new nutraceutical with promising active pharmacological effects [45–47]. The *T. testudinum* extract constitutes a potential source of chemopreventive agents because of its effectiveness as anti-inflammatory and antioxidant, which has been demonstrated previously [33–35,45–47].

In an earlier report, we describe the inhibitory effects of the Th and TB on the CYP1A1 activity in human hepatocytes [38]. The current work was aimed to further understanding of the effects of Th, PF and TB on CYP1A activity. This work let us evaluate the role of polyphenolic constituents of the extract, as substances responsible of its bioactive properties, in particular on phase I metabolizing enzymes.

Reduced O-Dealkylation of 7-ethoxyresorufin and 7-methoxyresorufin demonstrated the capacity of tested products to inhibit rat and human CYP1A1/2 activities. The polyphenolic-rich fraction isolated from the extract and TB revealed to be more active than Th, which in part supported our early hypothesis. All the natural products showed a mixed-type inhibition on rat CYP1A1, describing typical Michaelis–Menten kinetics, meanwhile exhibiting a differential response on human CYP1A1. Th did not inhibit the human CYP1A1 isoform. However, the PF and TB showed relevant inhibitory effects with differences in their kinetics and inhibition mechanism. PF acted as a mixed-type inhibitor, while TB behaved as non-competitive inhibition kinetics. These results were in agreement with previous reports on interspecies differences regarding the inhibitory potency of natural compounds [47,48] and also provided additional evidence on the interactions of polyphenols with CYP450 system.

Other naturally occurring polyphenolic compounds, such as galangin (3,5,7-trihydroxyflavone) and apigenin (5,7,4-trihydroxyflavone) also inhibit CYP1A1 and CYP1A2 activities [49]. Reported values of apparent Ki for the inhibition of CYP1A1 by galangin is 0.015 µM [50] and 0.32 µM apigenin [51]. On the other hand, there are fewer reports regarding the inhibition of CYP1B1 catalytic activity by naturally occurring compounds than CYP1A1 and CYP1A2. In fact, the extract and its polyphenolic components did not modulate CYP2B family activity under our experimental conditions.

The capacity of polyphenols and flavonoids to inhibit CYP1A1/2 isoforms has been proposed as the main mechanism supporting the antimutagenic and chemopreventive activities of these natural compounds [52]. Metabolic activation from BP to BPDE is believed to be essential for the mutagenic and carcinogenic properties of this pollutant [53,54]. CYP1A1 plays an important role in the biotransformation/activation of BP [55,56]. Thus, we assessed the antimutagenic properties of the extract and its derivate in *S. typhimurium* TA98 by the mutagenesis assay of Maron and Ames (1983) [57]. The aim of this assay was to explore the biological consequences of the CYP1A modulation by the tested products. BP bioactivation was achieved by phenobarbital-induced rat liver S_9 fraction. It is widely accepted that the Ames assay is useful for correlating in vitro mutagenesis and in vivo carcinogenicity in animals and humans. Th, PF and TB exerted a protective effect against BP-induced mutagenicity in *S. typhimurium*, evidenced here by the reduction of His$^+$ revertant colonies per plate, which suggested a potential antimutagenic activity of this marine organism.

Interestingly, in a previous report we observed a significant increase in BP mutagenicity after incubation with S_9 fractions obtained from rats, orally treated during 10 days with doses of 200 and 400 mg/kg of Th; however, mutant colonies were reduced at low evaluated doses (20 mg/kg) [40]. We also reported a differential response of CYP1A1, 1.5-fold increase after treatment with 200 mg/kg of Th, a discrete increase at 400 mg/kg dose and no significant changes at the low dose of *T. testudinum* extract. Differences in CYP activity suggest that polyphenol concentration could be a critical factor mediating the interaction between *T. testudinum* extract and CYP1A1 activity or BP mutagenicity. The present work may support this hypothesis, since BP-induced DNA damage, micronucleus frequency and oxidative damage diminished in orally pre-treated mice with 100 mg/kg of Th. These results are in agreement with the in vitro Ames test results and contrast with previous reports of our group in which a high dose of Th was assessed.

Antioxidant and anti-inflammatory properties of *T. testudinum* extract have been also observed under similar experimental conditions [46,58]. Accordingly, *T. testudinum* extract could protect DNA by different mechanisms, where antioxidant effects and the modulation of BP bioactivation should be conjugated. The capacity of *T. testudinum* components to inhibit CYP1A enzymatic activity could certainly be involved in the observed chemoprotective effect, but the alternative hypothesis which considers that Th components may act as scavengers that bind reactive metabolites of BP must not be discarded (assessed in ongoing studies). Previously, we demonstrated that the extract protects rat hepatocytes from terbutil-hidroperoxide-induced GSH depletion, together with higher catalase and superoxide dismutase activities compared with controls [35]. Th also showed protective effects against ethanol-, carbon tetrachloride- and lipopolysaccharide-induced cytotoxicity in rat hepatocytes [35]. Therefore, it seems that a combination of different mechanisms might be involved in the complex interaction of *T. testudinum* extract and the ADME cellular process, which in turn has influence its bioactive properties.

DNA oxidative damage is critical for BP in vivo toxicity [59–61]. It has been also reported that BP induces genetic lesions such as DNA single-strand breaks, DNA–protein cross-links and chromosomal aberrations [60]. The protective effects of Th against acute exposure to BP were corroborated here by mean the micronucleus assay. As we expected, Th reduced not only the BP-oxidative damage but also the micronucleus formation, meanwhile an increase of antioxidant defenses, such as GSH levels, was observed.

These results altogether support new findings on the modulation of CYP system and particularly CYP1A1 by *T. testudinum* polyphenols. As mentioned before, CYP1A1/2 activity plays a significant role in the activation/detoxification balance of pro-carcinogens in hepatic and extrahepatic tissues. Therefore, the inhibitory effects of polyphenols present in *T. testudinum* marine plant on CYP1A activity may explain, in part, the antimutagenic and/or chemopreventive properties of these natural products.

4. Materials and Methods

4.1. Materials and Reagents

Beta-naphtoflavone (β-NF), 7-ethoxyresorufin (EROD), methoxyresorufin (MROD), benzyloxiresorufin (BROD), penthoxyresorufin (PROD), resorufin and benzo[a]pyrene (BP) were purchased from Sigma (St. Louis MO, USA). Culture media were obtained from BD Difco (Trenton, NJ, USA). The *Escherichia coli* DH5α recombinant and *Salmonella typhimurium* strains were kindly provided by Dr. Peter Guengerich (Vanderbilt University, Nashville, TN, USA).

4.2. Vegetal Material

Thalassia testudinum (Banks and Soland ex. Koenig) was collected during March 2017 in "Guanabo" beach (22°05′45″ N, 82°27′15″ W). The specimen was identified by Dr. J.A. Areces (Institute of Oceanology, Havana, Cuba) and a voucher sample (No. IdO40) is deposited in the herbarium of the Cuban National Aquarium. The leaves were washed with water to eliminate sediments and the excess of salt; then, the plant material was dried at room temperature. Whole dry and ground *T. testudinum* leaves (840 g) were continuously extracted with ethanol-H_2O (50:50, *v/v*) during 24 h at room temperature. The extract was filtered and concentrated under reduced pressure and temperature (40 °C) to yield 54 g of crude extract (Th).

Th was used as a starting material to obtain a polyphenol-enriched fraction (PF), where non-polar components were excluded by chloroform extraction (1:10, w/v). The resultant PF was filtered and dried at room temperature. Later, TB (1-chrysoeriol 7-β-D-glucopyranosyl-2″-sulphate) was isolated by electrospray ionization mass spectrometry (ESIMS) with an [M − H]− ion at m/z = 541. The structure was confirmed by spectroscopic analysis (^{1}H and ^{13}C nuclear magnetic resonance) as shown in Figure 1, and compared to reported data [38].

4.3. Rat Liver S9 and Microsomal Fraction Obtaining

The S9 fraction was obtained as previously described [57]. To obtain the microsomal fraction, S_9 was split into 1 mL aliquots and centrifuged at 100,000× *g* and 4 °C per 60 min. The pellet was resuspended in 0.1 M phosphate buffer (pH 7.4) plus 0.25 M sucrose and it was centrifuged again at 100,000× *g* at 4 °C per 60 min. The pellet (microsomal fraction) was resuspended in 0.1 M phosphate buffer (pH 7.4), 1 mM EDTA, 0.1 mM dithiothreitol and 20% *v/v* glycerol and stored at −80 °C until use.

4.4. Bacterial Membrane Fraction Obtaining

The isolation of membrane fractions from *Escherichia coli* expressing a recombinant human CYP1A1 was performed as previously described [62,63]. Briefly, overnight cultured *E. coli* DH5α were diluted 1:100 in Terrific Broth/ampicillin (100 μg/mL) medium containing 1 mM isopropyl b-D thiogalactoside, 0.5 mM aminolevulinic acid, 1 mM thiamine and trace salts. Bacteria cultures were grown during 24 h at 30 °C and 150 rpm shaking. Thereafter, *E. coli* membrane fractions were isolated from the whole pellets by serial d ultracentrifugation steps.

4.5. Enzymatic Activity Assays

4.5.1. CYP1A1 and CYP1A2 Activities

CYP enzymatic activities were evaluated in rat liver microsomes and in bacterial membrane fraction by spectrofluorometric assay as previously described [48,64], with slight modifications. The reaction mixtures contained: rat liver microsomes (80 µg) or bacterial membrane fraction (40 µg), EROD (1 µM) or MROD (5 µM) substrate and 2.5 to 50 µg/mL of test products (Th, PF or TB). Then, buffer solution (Tris-HCl (50 mM) and MgCl2 (25 mM), pH = 7.6) was added to reaction mixtures and incubated 3 min at 37 °C. The reaction started by the addition of NADPH (0.5 mM). The fluorescence units were registered at 20 s intervals during 15 min in a hybrid multi-mode microplate reader (Synergy H4, Biotech). Finally, CYPs activities were calculated from a resorufin standard curve (5–50 pmol/mL). Interference of tested products with the fluorescence of resorufin was also carried out.

4.5.2. CYP2B1 and CYP2B2 Activities

The activities of CYP2B1-related penthoxyresorufin O-dealkylase (PROD) and CYP2B2-related benzyloxyresorufin O-dealkylase (BROD) were determined in rat liver microsomes by spectrofluorometric techniques as previously described [64], with minor modifications. The assays were conducted under similar conditions previously described for CYP1A activity.

4.6. Kinetic Analysis of Enzyme Inhibition

Different concentrations of substrates (0.32–10 µM of EROD) mixed with rat liver microsomes (80 µg) or *E. coli* membrane fraction (40 µg) were used for enzymatic kinetics assays in presence or absence of the tested products. Kinetic constants were obtained by a nonlinear regression analysis of experimental data fitted to the Michaelis–Menten equation with competitive, non-competitive and mixed-type inhibition models (GraphPad Prism version 6 software). Kinetic analysis was also shown by using the Lineweaver–Burk plot.

4.7. Ames Test

The effects of the tested products against BP-induced mutagenicity was assessed according to Maron and Ames [57], in *Salmonella typhimurium* strain TA98 and rat liver S_9 fraction. The reaction mixture contained 0.6% agar, 0.5% NaCl, 0.5 mM biotin and 0.05 mM L-histidine. A solution of the tested products (100 µL) at different concentrations (Th: 10, 100, 1000 µg/mL; PF: 5, 50, 500 µg/mL; TB: 400 µg/mL), overnight cultured *S. typhimurium* (10^8 cells), S9 (500 µL) and benzo(a)pyrene (10 µg/plate) were added to the reaction mixture (2 mL). Phosphate buffer (500 µL) was used instead of the S_9 fraction as negative control. Afterward, the plates were incubated at 37 °C for 48 h and the number of revertant colonies (His+) was quantified. The inhibition percentage of BP-induced mutagenicity was calculated as (1-(number colonies/plates with mutagen plus the tested products)/(number colonies/plate with just mutagen)) × 100%.

4.8. Effects of the T. testudinum Extract against BP-Induced DNA Damage in Mice

Balb/c male mice (20–25 g) were obtained from Centro para la Producción de Animales de Laboratorio (CENPALAB, Havana, Cuba). The animals were adapted to standard conditions (temperature: 20 ± 2 °C, humidity: 40–60%, 12 h light/dark cycle) during a week. Mice were fed with a standard diet and water ad libitum. Experimental procedures were carried out in accordance with European regulations on animal protection (Directive 86/609), and the Guide for the Care and Use of Laboratory Animals as adopted and promulgated by the US National Institute of Health (NIH Publication № 85–23, revised 1996). The experimental protocol was approved by the Institutional Animal Care and Ethical Committee from the Institute of Marine Sciences (ICIMAR), Havana, Cuba (Protocol number 1805, Date of approval: 28 March 2018). Four groups were included in the study (five animals per group). The first group was animals received seven daily oral doses of distilled water and one dose of oil (BP vehicle). The second group only received one oral dose of benzo(a)pyrene (250 mg/kg), the last day of the experiment. The third group was orally treated with Th aqueous solution (100 mg/kg) for seven days and one hour after the last administration of the extract, animals received an oral dose of BP (250 mg/kg). After 24 h, all the animals were sacrificed, and blood samples were obtained by cardiac puncture and centrifuged at $3000 \times g$ for 10 min, at 4 °C. Serum was collected and stored at −80 °C until use.

4.8.1. Oxidative Stress Biomarkers Determination

Serum markers of oxidative stress were determined in mice. Malondialdehyde (MDA) content was determined as previously described [65]. Reduced glutathione (GSH) levels were measured at 412 nm after precipitation of thiol proteins by using the Ellman's reagent (5,5′dithiobis-2-nitrobenzoic acid, Sigma, Burbank, CA, USA) in accordance with Sedlak and Lindsay [66]. The advanced oxidation protein products (AOPP) were quantified with potassium iodide (1.16 M) followed by the addition of acetic acid. The absorbance was immediately read at 340 nm. AOPP concentration was expressed as μM of chloramine-T. Concentration of oxidative stress biomarkers was expressed per mg of protein.

4.8.2. Micronuclei Formation Determination

After mice euthanasia, the femurs were removed, and its proximal end was shortened until the marrow canal became visible. One milliliter of serum was introduced into the bone canal and the marrow was aspirated and flushed several times. Cells were centrifuged at 1000 rpm for 5 min at 4 °C. Afterward, they were fixed in methanol and stained with Giemsa 5% (*v/v*) for 12 min. The presence of micronuclei was determined in a sample of 2000 polychromatic erythrocytes (PCE). Normochromatic erythrocytes (NCE) were also scored in 200 erythrocytes samples to determine the PCE/NCE ratio [67].

4.9. Statistical Analysis

Statistical analyses were performed with GraphPad Prism 5.0 (GraphPad, La Jolla, CA, USA). Revertants/plate, enzymatic activities, MDA, GST, AOPP levels, and micronucleus data were expressed as mean ± SD values. For multiple mean comparisons was used a one-way ANOVA followed by Dunnett or Tukey non-parametric tests. The level of statistical significance was set to * $p < 0.05$, ** $p < 0.01$, or *** $p < 0.001$.

5. Conclusions

In summary, our results suggest that in vitro and in vivo antimutagenicity effects of *Thalassia testudinum* extract and its polyphenolic components may be mediated, at least, by the inhibitory effect on phase I metabolizing enzymes, in particular the CYP1A family. As consequence, reduced oxidative stress and mutagenic effects were observed. The results altogether contribute to support the potential of *T. testudinum* as a source of chemoprotective compounds against air pollution-mediated carcinogenesis.

Author Contributions: J.J.E.-A., I.R., R.S.-P. and L.D.-R. were responsible for the general study concept, conducting experimental analysis, interpreting and discussing of results, and drafting the manuscript. L.D.-R., F.M., J.A.H., S.L.H., M.R., M.D.F. and G.G. contributed, analyzed and interpreted the chemical and biological experiments. J.J.E.-A., I.R., R.S.-P., L.D.-R. were responsible of discussing of results. F.M., R.S.-P., L.D.-R. carried out statistical data analysis. Funding acquisition: I.R., G.G., J.J.E.-A. and L.D.-R. All authors have read and agreed to the published version of the manuscript.

Funding: This research was funded by: CITMA P211LH005-019 (Cuba); the program *Becas de Excelencia del Gobierno de México para Extranjeros* 2016 (México); FONDECYT 1130601 (Chile), as well as *Programas Institucionales del Instituto de Investigaciones Biomédicas, UNAM* (México).

Acknowledgments: The authors acknowledge the technical support given by Cindel Cuellar.

Conflicts of Interest: The authors declare no conflict of interest. The funders had no role in the design of the study; in the collection, analyses or interpretation of data; in the writing of the manuscript or in the decision to publish the results.

References

1. World Health Organization. 7 Million Premature Deaths Annually Linked to Air Pollution. Available online: https://www.who.int/mediacentre/news/releases/2014/air-pollution/en/#:~{}:text=25%20March%202014%20%7C%20Geneva%20%2D%20In,result%20of%20air%20pollution%20exposure (accessed on 24 March 2020).
2. Kim, J.H.; Stansbury, K.H.; Trush, M.A.; Strickland, P.T.; Sutter, T.R. Metabolism of benzo[a]pyrene and benzo[a]pyrene-7,8-diol by human cytochrome P450 1B1. *Carcinogenesis* **1998**, *19*, 1847–1853. [CrossRef] [PubMed]
3. Schwarz, D.; Kisselev, P.; Cascorbi, I.; Schunck, W.H.; Roots, I. Differential metabolism of benzo[a]pyrene and benzo[a]pyrene-7,8-dihydrodiol by human CYP1A1 variants. *Carcinogenesis* **2001**, *22*, 453–459. [CrossRef] [PubMed]
4. Osborne, M.R.; Brookes, P.; Beland, F.A.; Harvey, R.G. The reaction of (±)-7α, 8β-dihydroxy-9β, 10β-epoxy-7,8,9,10-tetrahydrobenzo(a)pyrene with dna. *Int. J. Cancer* **1976**, *18*, 362–368. [CrossRef] [PubMed]
5. Boysen, G.; Hecht, S.S. Analysis of DNA and protein adducts of benzo[a]pyrene in human tissues using structure-specific methods. *Mutat. Res.* **2003**, *543*, 17–30. [CrossRef]
6. Rojas, M.; Marie, B.; Vignaud, J.M.; Martinet, N.; Siat, J.; Grosdidier, G.; Cascorbi, I.; Alexandrov, K. High DNA damage by benzo[a]pyrene 7,8-diol-9,10-epoxide in bronchial epithelial cells from patients with lung cancer: Comparison with lung parenchyma. *Cancer Lett.* **2004**, *207*, 157–163. [CrossRef]
7. Alexandrov, K.; Rojas, M.; Rolando, C. DNA damage by benzo(a)pyrene in human cells is increased by cigarette smoke and decreased by a filter containing rosemary extract, which lowers free radicals. *Cancer Res.* **2006**, *66*, 11938–11945. [CrossRef]
8. Pfeifer, G.; Yoon, J.H.; Liu, L.; Tommasi, S.; Wilczynski, S.P.; Dammann, R. Methylation of the RASSF1A gene in human cancers. *Biol. Chem.* **2002**, *383*, 907–914. [CrossRef]
9. Sporn, M.B.; Suh, N. Chemoprevention of cancer. *Carcinogenesis* **2000**, *21*, 525–530. [CrossRef]
10. Stoner, G.D.; Morse, M.A.; Kelloff, G.J. Perspectives in cancer chemoprevention. *Environ. Health Perspect.* **1997**, *105*, 945–954.
11. Guengerich, F.P.; Shimada, T. Activation of procarcinogens by human cytochrome P450 enzymes. *Mutat. Res.* **1998**, *400*, 201–213. [CrossRef]
12. Yang, C.S.; Smith, T.J.; Hong, J.Y. Cytochrome P-450 Enzymes as Targets for Chemoprevention against Chemical Carcinogenesis and Toxicity: Opportunities and Limitations. *Cancer Res.* **1994**, *54*, 1982s–1986s. [PubMed]
13. Scalbert, A.; Johnson, I.T.; Saltmarsh, M. Polyphenols: Antioxidants and beyond. *Am. J. Clin. Nutr.* **2005**, *81*, 215S–217S. [CrossRef] [PubMed]
14. Pandey, K.B.; Rizvi, S.I. Plant polyphenols as dietary antioxidants in human health and disease. *Oxid. Med. Cell Longev.* **2009**, *2*, 270–278. [CrossRef] [PubMed]
15. Zhang, H.; Tsao, R. Dietary polyphenols, oxidative stress and antioxidant and anti-inflammatory effects. *Curr. Opin. Food Sci.* **2016**, *8*, 33–42. [CrossRef]
16. Khan, N.; Mukhtar, H. Tea polyphenols for health promotion. *Life Sci.* **2007**, *81*, 519–533. [CrossRef] [PubMed]
17. Hessien, M.; El-Gendy, S.; Donia, T.; Sikkena, M.A. Growth inhibition of human non-small lung cancer cells h460 by green tea and ginger polyphenols. *Anticancer Agents Med. Chem.* **2012**, *12*, 383–390. [CrossRef] [PubMed]
18. Zhu, X.; Wetta, H. Genetics and epigenetics in tumorigenesis: Acting separately or linked. *Austin J. Clin. Med.* **2014**, *1*, 1016.
19. Barron, C.C.; Moore, J.; Tsakiridis, T.; Pickering, G.; Tsiani, E. Inhibition of human lung cancer cell proliferation and survival by wine. *Cancer Cell. Int.* **2014**, *14*, 6. [CrossRef]
20. Bauer, D.; Pimentel de Abreu, J.; Silva, H.S.; Goes-Neto, A.; Bello, M.G.; Teodoro, A.J. Antioxidant Activity and Cytotoxicity Effect of Cocoa Beans Subjected to Different Processing Conditions in Human Lung Carcinoma Cells. *Oxid. Med. Cell. Longev.* **2016**, *2016*, 7428515. [CrossRef]

21. Rajendran, P.; Ekambaram, G.; Sakthisekaran, D. Cytoprotective Effect of Mangiferin on Benzo(a)pyrene-Induced Lung Carcinogenesis in Swiss Albino Mice. *Basic Clin. Pharmacol. Toxicol.* **2008**, *103*, 137–142. [CrossRef]

22. Fantini, M.; Benvenuto, M.; Masuelli, L.; Frajese, G.V.; Tresoldi, I.; Modesti, A.; Bei, R. In vitro and in vivo antitumoral effects of combinations of polyphenols, or polyphenols and anticancer drugs: Perspectives on cancer treatment. *Int. J. Mol. Sci.* **2015**, *16*, 9236–9282. [CrossRef] [PubMed]

23. Kou, X.; Han, L.; Li, X.; Xue, Z.; Zhou, F. Antioxidant and antitumor effects and immunomodulatory activities of crude and purified polyphenol extract from blueberries. *Front. Chem. Sci. Eng.* **2016**, *10*, 108–119. [CrossRef]

24. Amararathna, M.; Johnston, M.R.; Vasantha, H.P. Plant Polyphenols as Chemopreventive Agents for Lung Cancer. *Int. J. Mol. Sci.* **2016**, *17*, 1352. [CrossRef] [PubMed]

25. Simmons, T.; Andrianasolo, E.; McPhail, K.; Flatt, P.; Gerwick, W.H. Marine natural products as anticancer drugs. *Mol. Cancer Ther.* **2005**, *4*, 333–342. [PubMed]

26. Folmer, F.; Jaspars, M.; Dicato, M.; Diederich, M. Photosynthetic marine organisms as a source of anticancer compounds. *Phytochem. Rev.* **2010**, *9*, 557–579. [CrossRef]

27. Malve, H. Exploring the ocean for new drug developments: Marine pharmacology. *J. Pharm. Bioallied. Sci.* **2016**, *8*, 83–91. [CrossRef]

28. Rowley, D.C.; Hansen, M.S.; Rhodes, D.; Sotriffer, C.A.; Ni, H.; McCammon, J.A.; Bushman, F.D.; Fenical, W. Thalassiolins A-C: New marine-derived inhibitors of HIV cDNA integrase. *Bioorg. Med. Chem.* **2003**, *10*, 3619–3625. [CrossRef]

29. Kontiza, I.; Stavri, M.; Zloh, M.; Vagias, C.; Gibbons, S.; Roussis, V. New metabolites with antibacterial activity from the marine angiosperm *Cymodocea nodosa*. *Tetrahedron* **2008**, *64*, 1696–1702. [CrossRef]

30. Achamlale, S.; Rezzonico, B.; Grignon-Dubois, M. Evaluation of *Zostera detritus* as a potential new source of zosteric acid. *J. Appl. Phycol.* **2009**, *21*, 347–352. [CrossRef]

31. Carbone, M.; Gavagnin, M.; Mollo, E.; Bidello, M.; Roussis, V.; Cimino, G. Further syphonosides from the sea hare *Syphonota geographica* and the sea-grass *Halophila stipulacea*. *Tetrahedron* **2008**, *64*, 191–196. [CrossRef]

32. Hamdy, A.H.; Mettwally, W.S.; Abou El Fotouh, M.; Rodriguez, B.; El-Dewany, A.I.; El-Toumy, S.A.; Hussein, A.A. Bioactive Phenolic Compounds from the Egyptian Red Sea Seagrass *Thalassodendron ciliatum*. *Z. Für Nat. C* **2012**, *67*, 291–296. [CrossRef] [PubMed]

33. Regalado, E.L.; Rodríguez, M.; Menéndez, R.; Concepción, Á.A.; Nogueiras, C.; Laguna, A.; Rodríguez, A.A.; Williams, D.E.; Lorenzo-Luaces, P.; Valdés, O.; et al. Repair of UVB-damaged skin by the antioxidant sulphated flavone glycoside thalassiolin B isolated from the marine plant *Thalassia testudinum* Banks ex Konig. *Mar Biotechnol.* **2009**, *11*, 74–80. [CrossRef] [PubMed]

34. Regalado, E.L.; Menendez, R.; Valdés, O.; Morales, R.A.; Laguna, A.; Thomas, O.P.; Hernandez, Y.; Nogueiras, C.; Kijjoa, A. Phytochemical Analysis and Antioxidant Capacity of BM-21, a Bioactive Extract Rich in Polyphenolic Metabolites from the Sea Grass *Thalassia testudinum*. *Nat. Prod. Commun.* **2012**, *7*, 47–50. [CrossRef]

35. Rodeiro, I.; Donato, M.; Martinez, I.; Hernandez, I.; Garrido, G.; González-Lavaut, J.; Menendez, R.; Laguna, A.; Castell, J.V.; Gómez-Lechón, M.J. Potential hepatoprotective effects of new Cuban natural products in rat hepatocytes culture. *Toxicol. In Vitro* **2008**, *22*, 1242–1249. [CrossRef]

36. Garateix, A.; Salceda, E.; Menéndez, R.; Regalado, E.L.; López, O.; García, T.; Morales, R.A.; Laguna, A.; Thomas, O.P.; Soto, E. Antinociception produced by *Thalassia testudinum* extract BM-21 is mediated by the inhibition of acid sensing ionic channels by the phenolic compound thalassiolin B. *Mol. Pain* **2011**, *7*, 10. [CrossRef] [PubMed]

37. Rodeiro, I.; Hernández, I.; Herrera, J.A.; Riera, M.; Donato, M.T.; Tolosa, L.; González, K.; Ansoar, Y.; Gómez-Lechón, M.J.; Vanden Berghe, W.; et al. Assessment of the cytotoxic potential of an aqueous-ethanolic extract from Thalassia testudinum angiosperm marine grown in the Caribbean Sea. *J. Pharm. Pharmacol.* **2018**, *70*, 1553–1560. [CrossRef] [PubMed]

38. Rodeiro, I.; Gomez-Lechon, M.J.; Tolosa, L.; Pérez, G.; Hernández, I.; Menéndez, R.; Regalado, E.L.; Castell, J.V.; Donato, M.T. Modulation of biotransformation and elimination systems by BM-21, an aqueous ethanolic extract from *Thalassia testudinum*, and thalassiolin B on human hepatocytes. *J. Funct. Foods* **2012**, *4*, 167–176. [CrossRef]

39. Miguel, V.; Otero, J.A.; Barrera, B.; Rodeiro, I.; Prieto, J.G.; Merino, G.; Álvarez, A.I. ABCG2/BCRP interaction with the sea grass *Thalassia testudinum*. *Drug Metab. Pers. Ther.* **2015**, *30*, 251–256. [CrossRef]

40. Rodeiro-Guerra, I.; Hernández-Ojeda, S.L.; Herrera-Isidrón, J.A.; Hernández-Balmaseda, I.; Padrón-Yaquis, S.; del Rosario Olguín-Reyes, S.; Alejo-Rodríguez, P.L.; Ronquillo-Sánchez, M.D.; Camacho-Carranza, R.; del Valle, R.M.; et al. Study of the interaction of an extract obtained from the marine plant *Thalassia testudinum* with pahse I metabolism in rats. *Rev. Int. Contam. Ambient.* **2017**, *33*, 547–557. [CrossRef]

41. Clementino, M.; Shi, X.; Zhang, Z. Prevention of Polyphenols against Carcinogenesis Induced by Environmental Carcinogens. *J. Environ. Pathol. Toxicol. Oncol.* **2017**, *36*, 87–98. [CrossRef]

42. Omidian, K.; Rafiei, H.; Bandy, B. Polyphenol inhibition of benzo[a]pyrene-induced oxidative stress and neoplastic transformation in an in vitro model of carcinogenesis. *Food Chem. Toxicol.* **2017**, *106*, 165–174. [CrossRef] [PubMed]

43. Shimada, T. Inhibition of Carcinogen-Activating Cytochrome P450 Enzymes by Xenobiotic Chemicals in Relation to Antimutagenicity and Anticarcinogenicity. *Toxicol. Res.* **2017**, *33*, 79–96. [CrossRef] [PubMed]

44. Chae, Y.H.; Marcus, C.E.; Ho, D.K.; Cassady, J.M.; Baird, W.M. Effects of synthetic and naturally occurring flavonoids on benzo[a]pyrene metabolism by hepatic microsomes prepared from rats treated with cytochrome P-450 inducers. *Cancer Lett.* **1991**, *60*, 15–24. [CrossRef]

45. De la Torre, E.; Rodeiro, I.; Menéndez, R.; Pérez, C.D. *Thalassia testudinum*, a sea plant with great therapeutical potentialities. *Rev. Cubana Plant. Med.* **2012**, *17*, 288–296.

46. Menéndez, R.; García, T.; Garateix, A.; Morales, R.A.; Regalado, E.L.; Laguna, A.; Valdés, O.; Fernández, M.D. Neuroprotective and antioxidant effects of *Thalassia testudinum* extract BM-21, against acrylamide-induced neurotoxicity in mice. *J. Pharm. Pharmacogn. Res.* **2014**, *2*, 53–62.

47. Arora, S. In vivo prediction of CYP-mediated metabolic interaction potential of formononetin and biochanin a using in vitro human and rat CYP450 inhibition data. *Toxicol. Lett.* **2015**, *239*, 1–8. [CrossRef]

48. Delgado-Roche, L.; Santes-Palacios, R.; Herrera, J.A.; Hernández, S.L.; Riera, M.; Fernández, M.D.; Mesta, F.; Garrido, G.; Rodeiro, I.; Espinosa-Aguirre, J.J. Regulation of Human Cytochrome P4501A1 (hCYP1A1): A Plausible Target for Chemoprevention? *Biomed. Res. Int.* **2016**, *2016*, 5341081.

49. Santes-Palacios, R.; Marroquín-Pérez, A.; Hernández-Ojeda, S.L.; Camacho-Carranza, R.; Govozensky, T.; Espinosa-Aguirre, J.J. Human CYP1A1 inhibition by flavonoids. *Toxicol. In Vitro* **2020**, *62*, 104681. [CrossRef]

50. Zhai, S.; Dai, R.; Friedman, F.K.; Vestal, R.E. Comparative inhibition of human cytochromes P450 1A1 and 1A2 by flavonoids. *Drug Metab. Dispos.* **1998**, *26*, 989–992.

51. Pastrakuljic, A.; Tang, B.K.; Roberts, E.A.; Kalow, W. Distinction of CYP1A1 and CYP1A2 activity by selective inhibition using fluvoxamine and isosafrole. *Biochem. Pharmacol.* **1997**, *53*, 531–538. [CrossRef]

52. Schwarz, D.; Roots, I. In vitro assessment of inhibition by natural polyphenols of metabolic activation of procarcinogens by human CYP1A1. *Biochem. Biophys. Res. Commun.* **2003**, *303*, 902–907. [CrossRef]

53. Sundberg, K.; Dreij, K.; Seidel, A.; Jernström, B. Glutathione conjugation and DNA adduct formation of dibenzo[a,l]pyrene and benzo[a]pyrene diol epoxides in V79 cells stably expressing different human glutathione transferases. *Chem. Res. Toxicol.* **2002**, *15*, 170–179. [CrossRef] [PubMed]

54. Alexandrov, K.; Cascorbi, I.; Rojas, M.; Bouvier, G.; Kriek, E.; Bartsch, H. CYP1A1 and GSTM1 genotypes affect benzo[a]pyrene DNA adducts in smokers' lung: Comparison with aromatic/hydrophobic adduct formation. *Carcinogenesis* **2002**, *23*, 1969–1977. [CrossRef] [PubMed]

55. Baird, W.; Mahadevan, B. Carcinogenic polycyclic aromatic hydrocarbon-DNA adducts and mechanism of action. *Environ. Mol. Mutagen.* **2005**, *45*, 106–114. [CrossRef]

56. Hodek, P.; Koblihová, J.; Kizek, R.; Frei, E.; Arlt, V.M.; Stivorobá, M. The relationship between DNA adduct formation by benzo[a]pyrene and expression of its activation enzyme cytochrome P450 1A1 in rat. *Environ. Toxicol. Pharmacol.* **2013**, *36*, 989–996. [CrossRef]

57. Maron, D.M.; Ames, B.N. Revised methods for the Salmonella mutagenicity test. *Mutat. Res.* **1983**, *113*, 173–215. [CrossRef]

58. Fernández, M.D.; Llanio, M.; Arteaga, F.; Dajas, F.; Echeverri, C.; Ferreira, M.; Hernández, I.; Cabrera, B.; Rodríguez, M.; Aneiros, A. Propiedades anti-inflamatoria-analgésica y antioxidante de una planta marina. *Avicennia* **2003**, *16*, 31–35.

59. Kamaraj, S.; Vinodhkumar, R.; Anandakumar, P.; Jagan, S.; Ramakrishnan, G.; Devaki, T. The effects of quercetin on antioxidant status and tumor markers in the lung and serum of mice treated with benzo(a)pyrene. *Biol. Pharm. Bull.* **2007**, *30*, 2268–2273. [CrossRef]

60.   Singh, H.P.; Batish, D.; Kohli, R.; Arora, K. Arsenic-induced root growth inhibition in mung bean (Phaseolus aureus Roxb.) is due to oxidative stress resulting from enhanced lipid peroxidation. *Plant. Growth Regul.* **2007**, *53*, 65–73. [CrossRef]
61.   Nebert, D.W.; Shi, Z.; Gálvez-Peralta, M.; Uno, S.; Dragin, N. Oral benzo[a]pyrene: Understanding pharmacokinetics, detoxication, and consequences—Cyp1 knockout mouse lines as a paradigm. *Mol. Pharmacol.* **2013**, *84*, 304–313. [CrossRef]
62.   Guo, Z.; Gillam, E.M.; Ohmori, S.; Tukey, R.H.; Guengerich, F.P. Expression of modified human cytochrome P450 1A1 in Escherichia coli: Effects of 5′ substitution, stabilization, purification, spectral characterization, and catalytic properties. *Arch. Biochem. Biophys.* **1994**, *312*, 436–446. [CrossRef] [PubMed]
63.   Guengerich, F.P.; Martin, M.V. Purification of cytochromes P450: Products of bacterial recombinant expression systems. *Methods Mol. Biol.* **2006**, *320*, 31–37. [PubMed]
64.   Burke, M.D.; Thompson, S.; Weaver, R.J.; Wolf, C.R.; Mayer, R.T. Cytochrome P450 specificities of alkoxyresorufin O-dealkylation in human and rat liver. *Biochem. Pharmacol.* **1994**, *48*, 923–936. [CrossRef]
65.   Ohkawa, H.; Ohishi, N.; Yagi, K. Assay for lipid peroxides in animal tissues by thiobarbituric acid reaction. *Anal. Biochem.* **1979**, *95*, 351–358. [CrossRef]
66.   Sedlak, J.; Lindsay, R.H. Estimation of total, protein-bound, and nonprotein sulfhydryl groups in tissue with Ellman's reagent. *Anal. Biochem.* **1968**, *25*, 192–205. [CrossRef]
67.   Hayashi, M.; Tice, R.R.; MacGregor, J.T.; Anderson, D.; Blakey, D.H.; Kirsh-Volders, M.; Oleson, F.B.; Pacchierotti, F.; Romagna, F.; Shimada, H.; et al. In vivo rodent erythrocyte micronucleus assay. *Mutat. Res.* **1994**, *312*, 293–304. [CrossRef]

Publisher's Note: MDPI stays neutral with regard to jurisdictional claims in published maps and institutional affiliations.

Article

Gracilosulfates A–G, Monosulfated Polyoxygenated Steroids from the Marine Sponge *Haliclona gracilis*

Larisa K. Shubina [1], Tatyana N. Makarieva [1,*], Vladimir A. Denisenko [1], Roman S. Popov [1], Sergey A. Dyshlovoy [1,2,3], Boris B. Grebnev [1], Pavel S. Dmitrenok [1], Gunhild von Amsberg [2,3] and Valentin A. Stonik [1]

[1]   G.B. Elyakov Pacific Institute of Bioorganic Chemistry, Far Eastern Branch of the Russian Academy of Sciences, Pr. 100-let Vladivostoku 159, 690022 Vladivostok, Russia; shubina@piboc.dvo.ru (L.K.S.); vladenis@piboc.dvo.ru (V.A.D.); prs_90@mail.ru (R.S.P.); dyshlovoy@gmail.com (S.A.D.); grebnev_bor@mail.ru (B.B.G.); paveldmt@piboc.dvo.ru (P.S.D.); stonik@piboc.dvo.ru (V.A.S.)

[2]   Department of Oncology, Hematology and Bone Marrow Transplantation with Section Pneumology, Hubertus Wald-Tumorzentrum, University Medical Center Hamburg-Eppendorf, 20251 Hamburg, Germany; g.von-amsberg@uke.de

[3]   Martini-Klinik, Prostate Cancer Center, University Hospital Hamburg-Eppendorf, 20251 Hamburg, Germany

*   Correspondence: makarieva@piboc.dvo.ru; Tel.: +7-950-295-66-25

Received: 21 July 2020; Accepted: 27 August 2020; Published: 30 August 2020

Abstract: Seven new polyoxygenated steroids belonging to a new structural group of sponge steroids, gracilosulfates A–G (**1–7**), possessing 3β-*O*-sulfonato, 5β,6β epoxy (or 5(6)-dehydro), and 4β,23-dihydroxy substitution patterns as a common structural motif, were isolated from the marine sponge *Haliclona gracilis*. Their structures were determined by NMR and MS methods. The compounds **1**, **2**, **4**, **6**, and **7** inhibited the expression of prostate-specific antigen (PSA) in 22Rv1 tumor cells.

Keywords: polyoxygenated steroids; sponge; *Haliclona gracilis*; anticancer activity

---

1. Introduction

Marine organisms are known as a rich source of unique bioactive sulfate-containing metabolites [1]. Sulfated derivatives of different chemical classes (aliphatic compounds, steroids, terpenoids, carotenoids, aromatic compounds, alkaloids, carbohydrates, etc.) have been identified from them [1–4]. Marine invertebrates such as starfishes, ophiuroids, and ascidians contain mainly mono- and disulfated polyoxygenated steroids [1], which are almost exclusively marine secondary metabolites [1], while terrestrial sulfated polyoxygenated steroids are relatively rare. In fact, there is only one report concerning isolation of sulfated polyoxygenated steroid from plants [5].

Sulfated steroids represent one of the most numerous classes of sponge metabolites [1–3] Marine sponges provide a great structural diversity of bioactive sulfated polyoxygenated steroids, including nitrogen-containing [6], halogenated [7,8], monosulfated [9–17], disulfated [18], and trisulfated [7,8] steroids, as well as tetra- [19] or pentasulfated dimeric steroid derivatives [20]. The monosulfated polyoxygenated steroids account for only a part of these metabolites. Some of them show antimicrobial [13] and/or antifungal [10,13,15] and cytotoxic [10] activities or enhance glucose uptake via the AMPK signaling pathway [9].

During the search for bioactive compounds from the Northwestern Pacific deep-water marine invertebrates [21,22], we collected the pale orange sponge *Haliclona gracilis* near Shikotan Island, Russia, whose extract exhibited hemolytic and antifungal activities.

The genus *Haliclona* (order Haplosclerida, family Halinidae) is represented by more than 600 species [23]. Marine sponges of *Haliclona* genus have been extensively examined, and more than 200 various bioactive metabolites including steroids, alkaloids, macrolides, polyketides, cyclic peptides, long-chain sphingoid bases, merohexaprenoids, and cyclic bis-1,3-dialkylpyridinium salts have been isolated, and different activities, including cytotoxic and antitumor effects, have been reported [1].

Sponges of *Haliclona* genus have provided very few sulfated steroids [1]. Thus, only two trisulfated steroids have been isolated in one Indo-Pacific *Haliclona* sponge [17], while monosulfated polyoxygenated steroids have never been isolated from this genus. Moreover, thus far, the sponge *H. gracilis* has not been chemically investigated.

The ^{1}H NMR analysis of the fractions obtained after diverse chromatographic separations suggested the presence of polar metabolites, inspiring our extensive investigation. Here, we report the details of the isolation and structure determination of compounds **1–7**, belonging to a new group of naturally occurring monosulfated polyoxygenated steroids with a 3β-*O*-sulfonato, 5β,6β-epoxy (or 5(6)-dehydro), or 4β,23-dihydroxy substitution pattern as a common structural motif. Additionally, anticancer activities of **1, 2, 4, 6**, and **7** were evaluated.

2. Results and Discussion

The concentrated EtOH extract of the sponge was partitioned between *n*-BuOH and H$_2$O. The organic extract was concentrated and the obtained residue was fractionated by flash chromatography on a YMC gel column. Further separation using reversed-phase HPLC resulted in the isolation of seven new steroids, gracilosulfates A–G (**1–7**, Figure 1).

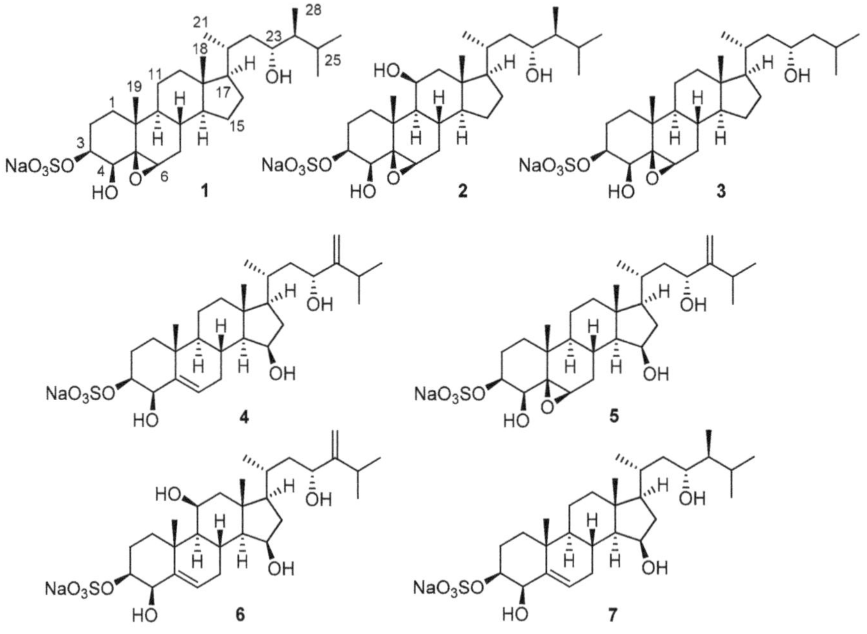

Figure 1. The structures of **1–7**.

Compound **1** was isolated as a white, amorphous solid. The molecular formula of **1** was determined to be $C_{28}H_{47}NaO_7S$ from the [M − Na]$^-$ ion peak at *m/z* 527.3045 in the (−)HRESIMS spectrum. The fragment ion peak at *m/z* 97.9606 in the (−)HRESIMS/MS spectrum and absorption band at 1213 cm^{-1} in the IR spectrum revealed the presence of a sulfate group in **1**.

The ^{1}H NMR spectrum of **1** (Table 1) showed signals attributable to six methyl groups at δ$_H$ 1.16 (s), 0.94 (d), 0.91 (d), 0.82 (d), 0.74 (d), and 0.70 (s); four oxygen-bearing methine protons at δ$_H$ 4.27 (d),

3.55 (br.d), 3.53 (br.d), and 3.17 (br.d); and a series of other methine and methylene multiplets. The ^{13}C NMR (Table 2) and DEPT spectra of **1** revealed the presence of 28 signals, corresponding to 6 methyls, 8 methylenes, 11 methines, and 3 nonprotonated carbons (one bearing oxygen atom). These data evidenced a C-28 steroidal skeleton. Structure determination of **1** began with HMBC correlations from CH$_3$-19 to C-1, C-5, C-9, and C-10. The COSY correlations (Figure 2) delineated the spin system H$_2$-1 to H-4, which included protons of oxygenated methines at C-3 and C-4 based on their characteristic chemical shifts. The sequences of protons from H-6 to H-8, H-8 to H$_2$-12, H-8 to H$_3$-27, and H-24 to H$_3$-28 were also established from COSY correlations and indicated the third oxymethine group at C-23. The cross peaks H-4/O<u>H</u> and H-23/O<u>H</u> in the COSY spectrum recorded in DMSO$-d_6$ (Figure S11) and the ^{13}C chemical shifts for C-4 (δ_C 77.7) and C-23 (δ_C 71.7) implied OH substitution, while the chemical shift for C-3 at δ_C 80.1 was more consistent with a sulfate half-ester O(SO$_3$)Na [14]. The ^{13}C NMR signals at δ_C 64.2 (CH) and 66.4 (C) and ^{1}H NMR signal at δ_H 3.17 indicated the presence of trisubstituted epoxy ring [24]. The epoxy group was placed at C-5 and C-6 on the basis of HMBC correlations from H$_2$-1, H-4, and H-19 to C-5 and from H-6 to C-4, C-8, and COSY correlations H-6/H$_2$-7.

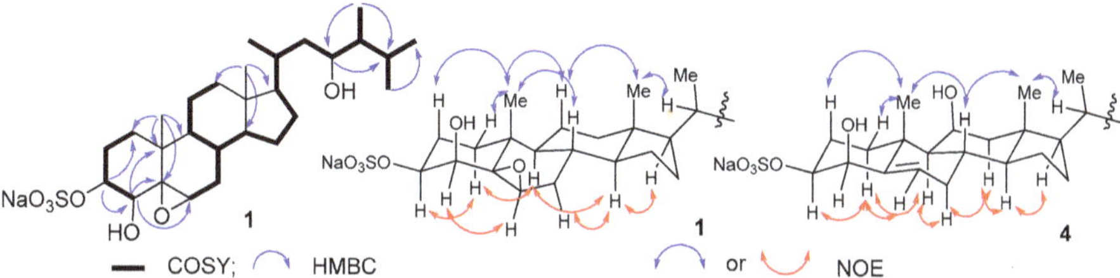

Figure 2. Key COSY and HMBC correlations for **1**, and NOESY correlations for **1** and **4**.

The large coupling constant of H-3 with H-2ax (J = 11.5 Hz) and small coupling constant of H-4 with H-3 (J = 3.0 Hz) pointed to β orientations for the 3-O(SO$_3$)Na and 4-OH groups. The configuration of the 5β,6β-epoxy group was established by the NOESY correlations H-6/H-4. The same evidence was earlier used for β-orientation of epoxide group in a steroid from the soft coral *Dendronephthya gigantea* [25]. The key NOESY correlations H$_3$-19/H-1β, H-2β, H-8, H-11β; H$_3$-18/H-20, H-8, H-11β; H-9/H-1α, H-14; H-1α H-3α; H-4α/H-6; and H-17/H-14α, H$_3$-21 confirmed the 3β,4β,5β,6β configurations of the oxygenated carbons and H-8β, H-9α, H-14α, and H-17α configurations of the ring portion in **1**. The 20R configuration was demonstrated by the NOESY cross-peak H$_3$-18/H-20 and chemical shift value of CH$_3$-21 at δ_H 0.94 [26].

The absolute configuration at C-23 was assigned by application of the Mosher's method. Esterification of **1** with (R)- and (S)-α-methoxy-α-(trifluoromethyl)-phenylacetyl chlorides (MTPACl) yielded the 23-MTPA adducts **1S** and **1R**, respectively, while C-4 hydroxy group was not modified. Interpretation of ^{1}H NMR chemical shift differences $\Delta\delta$ between **1S** and **1R** (Figure 3) revealed that the absolute configuration of C-23 is R. The $J_{H23/H24}$ coupling constant was 7.3 Hz, which indicated *anti* relationship of the H-23 and H-24 protons [27] (Figure 4). The NOESY cross peak for H$_2$-22/H$_3$-28 suggested the *gauche* relationship between the C-22 methylene and C-28 methyl groups, as shown in Figure 3. These data allowed us to determine the 24S absolute configuration.

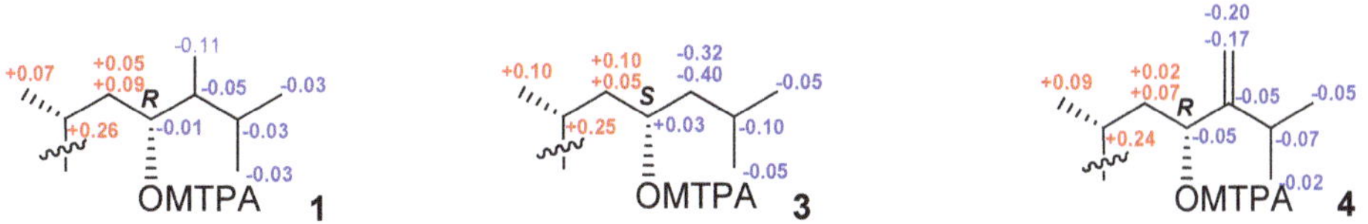

Figure 3. $\Delta\delta$ values ($\delta_S - \delta_R$) for 23(S)- and 23(R)-MTPA esters of compounds **1**, **3**, and **4**.

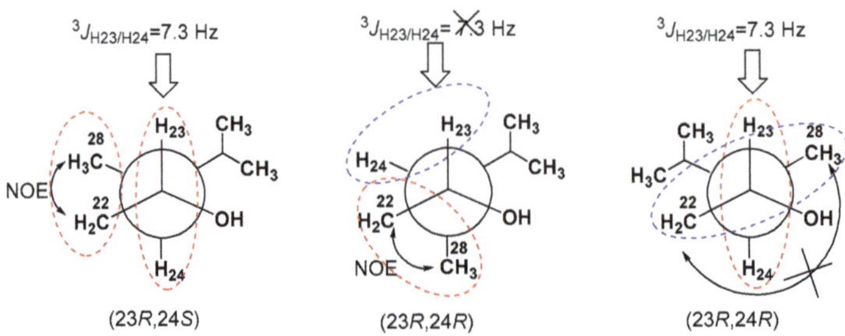

Figure 4. *J*-based configuration analysis and NOESY data of compound **1**.

Thus, the structure of **1** was defined as (20*R*,23*R*,24*S*)-4*β*,23-dihydroxy-5*β*,6*β*-epoxy-24-methylcholest-3*β* yl sulfate and was named gracilosulfate A.

The molecular formula of the second isolated compound, gracilosulfate B (**2**), was determined as $C_{28}H_{47}NaO_8S$ (*m/z* 543.3002 [M − Na]⁻) on the basis of the negative ion HRESIMS analysis. The ^1H NMR data of **2** resembled those of **1**, except for the presence of an oxymethine group in **2** (δ_H 4.14) instead of a methylene group for **1**. Analysis of 1D and 2D NMR (COSY, HSQC, and HMBC) spectra allowed us to assign all the observed ^1H and ^{13}C signals for **2** (Tables 1 and 2). The localization of the additional hydroxy group at C-11 followed from the HMBC correlations between H-11, C-8, and C-13 (Figure S16) and COSY data. The equatorial disposition of H-11 was evident from the small $^3J_{HH}$ vicinal coupling to H-9 and H_2-12 (Table 1) and confirmed by the relatively low field shift of H-8 (Table 1) caused by the 1,3-diaxial relationship of this proton to the hydroxy group at C-11. The configurations of other stereogenic centers of the ring portion were assigned using similar principles used for **1**. The similarity of the NMR data of the side chains of steroids **2** and **1** suggested the same (20*R*,23*R*,24*S*) configuration. Thus, gracilosulfate B was defined as (20*R*,23*R*,24*S*)-4*β*,11*β*,23-trihydroxy-5*β*,6*β*-epoxy-24-methylcholest-3*β* yl sulfate.

The molecular formula of gracilosulfate C (**3**), determined as $C_{27}H_{45}NaO_7S$ from HRESIMS data (*m/z* 513.2891[M–Na]⁻), was one methylene unit less than that of **1**. The spectroscopic properties of **3** were similar to those of **1** and differed only by the signals of steroid side chain (Tables 1 and 2). A combination of 2D NMR data showed the lack of a C-28 methyl group, while the remaining portion of the molecule was intact in **1**. The configuration of the ring moiety of **3** was assumed to be the same as that of **1** on the basis of the complete overlapping of proton and carbon resonances in NMR spectra. The configuration of the stereogenic center at C-23 was determined by the MTPA method as 23*S* (Figure 3). Thus, the gracilosulfate C was defined to be 24-demethyl derivative of gracilosulfate A (**1**), namely, (20*R*,23*S*)-4*β*,23-dihydroxy-5*β*,6*β*-epoxycholest-3*β* yl sulfate.

Gracilosulfate D (**4**) with a molecular formula $C_{28}H_{45}NaO_7S$, confirmed by HRESIMS, was isolated as an optically active white amorphous solid. In addition to the signals relative to 3*β*-*O*-sulfonato-4*β*,23-dihydroxy structure, the ^1H and ^{13}C NMR spectra of **4** (Tables 1 and 2) revealed signals of trisubstituted double bond (δ_H 5.70 and δ_C 144, 130.0), oxygenated methine group (δ_H 4.15 and δ_C 70.9), and terminal methylene group (δ_H 4.84, 5.03, and δ_C 106.8). The HMBC correlations between H_3-19 and C-5 (δ_C 130.0), and between olefinic proton at δ_H 5.70 and C-4 (δ_C 77.6), were consistent with a double bond at C-5/C-6 position. The HMBC correlations from the oxymethine proton at δ_H 4.15 to C-13 and C-17 (Figure S32), in addition to COSY data (Figure S30), allowed placement of a hydroxy group at C-15 position, whereas the HMBC correlations between H_2-28 and C-23, C-24, and C-25 confirmed the position of terminal methylene group at C-24. The coupling pattern associated with H-15 (ddd, *J* = 7.9, 5.8, 2.2 Hz) indicated that the hydroxy group at C-15 is *β*-positioned [28]. The configurations of other stereocenters of the steroid nucleus were assigned by NOESY (Figure 2) and coupling constants data (Table 1).

Table 1. ^{1}H NMR data for compounds **1–7** in CD$_3$OD.

Position	1 δ_H (J in Hz)	2 δ_H (J in Hz)	3 δ_H (J in Hz)	4 δ_H (J in Hz)	5 δ_H (J in Hz)	6 δ_H (J in Hz)	7 δ_H (J in Hz)
1α	1.39, m	1.37, m	1.37, m	1.15, m	1.37, m	1.31, m	1.15, m
1β	1.05, m	2.10, m	2.06, m	1.90, m	2.07, m	2.12, dt (13.3, 3.3)	1.89, dt (13.3, 3.3)
2α	1.82, m	1.84, m	1.82, m	1.81, m	1.83, m	1.84, m	1.82, m
2β	2.10, m	2.12, m	2.10, m	2.11, m	2.11, m	2.21, m	2.12, m
3	4.27, ddd (11.5, 4.1, 3.0)	4.29, ddd (11.5, 4.1, 3.0)	4.27, ddd (11.5, 4.1, 3.0)	4.18, ddd (12.2, 4.3, 3.3)	4.28, ddd (11.5, 4.1, 3.0)	4.19, ddd (12.0, 3.9, 3.1)	4.18, ddd (11.7, 4.0, 3.1)
4	3.55, br d (3.0)	3.58, br d (3.0)	3.56, br d (3.0)	4.42, dd (3.3, 1.3)	3.57, br d (3.0)	4.40, dd (3.3, 1.3)	4.42, dd (3.3, 1.3)
5							
6	3.17, br d (2.7)	3.12, br d (2.7)	3.17, br d (2.5)	5.70, dd (5.0, 2.4)	3.19, br d (2.5)	5.59, dd (4.2, 3.0)	5.70, dd (5.0, 2.4)
7α	1.29, m	1.35, m	1.27, m	1.68, ddd (18.2, 10.3, 2.3)	1.32, m	1.80, ddd (18.0, 9.8, 2.6)	1.68, ddd (18.0, 10.3, 2.3)
				2.40, m		2.53, ddd (18.8, 6.6, 4.3)	2.39, m
7β	2.08, m	2.19, m	2.08, m		2.39, m		
8	1.43, m	1.79, m	1.42, m	1.99, dd (10.8, 5.7)	1.85, m	2.42, m	1.99, m
9	0.68, dd (12.0, 4.5)	0.78 dd, (11.5, 3.0)	0.68, dd, (11.6, 4.7)	0.98, m	0.74, m	1.01, m	0.98, m
11α	1.40, m	4.14, br q, (3.0)	1.40, m	1.50, m	1.38, m	4.29, br q, (3.4)	1.49, m
11 β	1.43, m		1.44, m	1.52, m	1.43, m		1.51, m
12α	1.14, m	1.30, m	1.12, m	1.18, m	1.10, m	1.36, m	1.18, m
12 β	2.02, m	2.23, dd, (13.3, 3.0)	2.02, m	2.03, dt (12.6, 3.6)	1.97, dt (12.3, 3.7)	2.22, m	2.03, dt (12.5, 3.3)
13							
14	0.95, m	0.94, m	0.94, m	0.90, m	0.78, dd (11.3, 5.7)	0.89, m	0.90, m
15	1.64, m	1.65, m	1.64, m	4.15, ddd (7.9, 5.8, 2.2)	4.15, ddd (8.1, 5.7, 2.3)	4.18, ddd (7.8, 5.7, 2.2)	4.15, ddd (7.7, 5.6, 2.0)
	1.05, m	1.15, m	1.06, m				
16	1.86, m	1.83, m	1.86, m	2.40, m	2.36, m	2.37, m	2.40, m
	1.36, m	1.39, m	1.35, m	1.41, ddd (14.3, 10.4, 2.3)	1.37, m	1.41, m	1.39, m
17	1.08, m	1.03, m	1.08, m	1.05, m	1.02, m	1.00, m	1.07, m
18	0.70, s	0.93, s	0.69, s	1.00, s	0.94, s	1.20, s	0.99, s
19	1.16, s	1.43, s	1.16, s	1.24, s	1.19, s	1.49, s	1.24, s
20	1.73, m	1.72, m	1.72, m	1.88, m	1.86, m	1.88, m	1.87, m
21	0.94, d (6.7)	0.96, d (6.7)	0.95, d (6.7)	1.02, d (6.7)	0.99, d (6.7)	1.04, d (6.7)	0.97, d (6.7)
22	1.41, m	1.39, m	1.48, m	1.59, ddd (13.7, 10.3, 2.3)	1.57, ddd (14.1, 10.5, 2.7)	1.58, ddd (14.1, 10.5, 2.5)	1.43,
	1.04, m	1.03, m	0.98, m	1.11, m	1.10, m	1.11, m	1.07, m
23	3.53, ddd (9.1, 7.3, 2.0)	3.53, ddd (9.4, 7.3, 2.0)	3.70, m	4.13, br d (10.5)	4.11, br d (10.5)	4.13, br d (10.5)	3.55, ddd (9.3, 7.1, 2.0)
24	1.29, m	1.28, m	1.38, m				1.31, m
			1.14, m				
25	1.91, m	1.91, m	1.75, m	2.26, septet (6,7)	2.24, septet (6,7)	2.25, septet (6,7)	1.93, m
26	0.82, d (6.6)	0.82, d (6.6)	0.90, d (6.8)	1.06, d (6.9)	1.05, d (6.8)	1.06, d (6.8)	0.83, d (6.9)
27	0.91, d (6.6)	0.91, d (6.6)	0.91, d (6.8)	1.08, d (6.9)	1.07, d (6.8)	1.08, d (6.8)	0.92, d (6.9)
28	0.74, d (6.8)	0.74, d (6.8)		5.03, t (1.2)	5.03, t (1.2)	5.03, t (1.2)	0.75, d
				4.84, br s	4.84, br s	4.84, br s	

The absolute configuration at C-23 was deduced by theMTPA method. Treatment of **4** with (*R*)- and (*S*)- MTPACl yielded the corresponding 4,23-bis-MTPA adducts **4S** and **4R**, respectively. The $\Delta\delta$ values around the C-23 stereocenter between the adducts **4S** and **4R** (Figure 3) indicated the 23*S* configuration and, therefore, the structure of **4** was assigned as (20*R*, 23*S*)-4β,15β,23-trihydroxy-24-methylenecholest-5(6)-en-3βyl sulfate.

The molecular formula of $C_{28}H_{45}NaO_8S$ was assigned by HRESIMS to gracilosulfate E (**5**). The 1D (Tables 1 and 2) and 2D NMR analysis showed that gracilosulfate E (**5**) differs from **4** in the 5,6-epoxy group, replacing trisubstituted double bond. The configurations of the ring moiety were assigned on the basis of the analyses of proton–proton coupling constants (Table 1) and NOESY data. The absolute configuration of the side chain of **5** was determined to be the same as in **4** by comparison of ^{1}H and ^{13}C chemical shifts. Thus, gracilosulfate E (**5**) was determined to be (20*R*,23*R*)-4β,15β,23-trihydroxy-5β,6β-epoxy-24-methylenecholest-3β yl sulfate.

Gracilosulfate F (**6**) of molecular formula $C_{28}H_{45}NaO_8S$ was a close analogue of gracilosulfate D (**4**) showing only an additional oxygen atom. Inspection of 1D (Tables 1 and 2) and 2D NMR data allowed placement of an additional hydroxy group at C-11. The configuration at C-11 was deduced from NOESY correlation of H-11 to axial proton H-1 and small vicinal coupling constant of H-11 (Table 1), which is consistent with an equatorial disposition for this proton, thereby placing the hydroxy group in an axial position. The configurations of remaining stereogenic centers of the ring portion were the same as those of **4**, as established on the basis of analyses of proton–proton coupling constants (Tables 1 and 2) and NOESY data. The absolute configuration of the side chain was determined to be the same as that of 4 by comparison of ^{1}H and ^{13}C chemical shifts, and finally the structure of **6** was established as (20*R*, 23*R*)-4β,11β,15β,23-tetrahydroxy-24-methylenecholest-5(6)-en-3βyl sulfate.

Gracilosulfate G (**7**) showed the molecular formula $C_{28}H_{47}NaO_7S$ as determined by HRESIMS. On the basis of the results of the 1D NMR spectra, we were able to assign a trisubstituted double bond and four oxygen-bearing methine groups. The same steroid core constitution and configurations as in gracilosulfate D (**4**) were inferred from 1D (Tables 1 and 2) and 2D NMR analysis. The proton and carbon resonances attributable to the side chain of **7** were coincident with those of **1** and **2** (Tables 1 and 2). Thus, gracilosulfate G was defined as (20*R*, 23*R*, 24*S*)-4β,15β,23-trihydroxy-24-methylcholest-5(6)-en-3βyl sulfate.

Next, antitumor activity of compounds **1**, **2**, **4**, **6**, and **7** were determined in human prostate cancer cells 22Rv1. Of note, this cell line reveals resistance to androgen receptor (AR)-targeted therapy due to the expression of AR-V7 (AR transcript variant V7), which lacks the androgen-binding site [29,30]. The compounds exhibited moderate cytotoxic activity in the cancer cells after 48 h of treatment. Thus, compound **7** exhibited $IC_{50} = 64.4 \pm 14.9$ µM, while the other tested compounds had $IC_{50} > 100$ µM (docetaxel was used as a positive control and exhibited $IC_{50} = 17.3 \pm 6.3$ nM). However, all compounds were able to effectively inhibit the expression of PSA (prostate-specific antigen) in 22Rv1 cells (Figure 5). Earlier, only two monosulfated polyoxygenated steroids have been shown to exert cytotoxic activity on human cancer cell lines [10]. On the other hand, non-sulfated polyoxygenated steroid aragusterol with potent antitumor activities was isolated from a sponge of the genus *Xestospongia* [31]. Interestingly, for compounds **6** and **7**, this effect was already detected at a concentration of 10 µM. PSA is a well-known downstream target of AR signaling. Thus, suppression of PSA expression may indicate an inhibition of this pathway. AR signaling is essential for the growth and survival of prostate cancer cells, with its targeting playing a central role in the modern therapy of advanced prostate cancer. The ability of the isolated compounds to suppress AR signaling can be explained by the similarity of their structures to androgen ligands, which may result in a binding to androgen receptors and therefore blocking of AR-mediated signaling in prostate cancer cells.

Table 2. ^{13}C NMR data[a] for compounds 1–7 in CD$_3$OD.

Position	1	2	3	4	5	6	7
	δ_C, Type	δ_C, Type	δ_C, Type	δ_C, Type	δ_C, Type	δ_C, Type	δ_C, Type
1	39.2, CH$_2$	40.2, CH$_2$	39.2, CH$_2$	39.2, CH$_2$	39.2, CH$_2$	38.2, CH$_2$	39.4, CH$_2$
2	23.7, CH$_2$	24.1, CH$_2$	23.7, CH$_2$	24.6, CH$_2$	23.7, CH$_2$	24.0, CH$_2$	24.4, CH$_2$
3	80.1, CH	79.9, CH	80.1, CH	82.5, CH	80.1, CH	82.0, CH	82.1, CH
4	77.7, CH	78.0, CH	77.7, CH	77.6, CH	77.7, CH	76.9, CH	77.6, CH
5	66.4, C	66.2, C	66.4, C	144.1, C	66.3, C	145.4, C	144.1, C
6	64.2, CH	63.3, CH	64.2, CH	130.0, CH	64.2, CH	129.5, CH	130.0, CH
7	34.2, CH$_2$	33.6, CH$_2$	34.2, CH$_2$	33.0, CH$_2$	33.6, CH$_2$	33.2, CH$_2$	33.0, CH$_2$
8	31.7, CH	28.8, CH	31.7, CH	29.4, CH	27.4, CH	26.4, CH	29.5, CH
9	53.9, CH	57.8, CH	53.9, CH	52.9, CH	54.3, CH	56.2, CH	59.2, CH
10	36.9, C	37.9, C	36.9, C	38.0, C	36.9, C	38.6, C	38.1, C
11	23.1, CH$_2$	69.5, CH	23.1, CH$_2$	22.1, CH	23.1, CH$_2$	69.5, CH	22.2, CH
12	41.8, CH$_2$	51.1, CH$_2$	41.8, CH$_2$	43.0, CH$_2$	43.1, CH$_2$	52.3, CH$_2$	43.0, CH$_2$
13	44.1, C	43.7, C	44.1, C	44.0, C	43.9, C	43.4, C	44.0, C
14	58.1, CH	61.2, CH	58.1, CH	63.5, CH	62.8, CH	65.6, CH	63.6, CH
15	25.9, CH$_2$	25.7, CH$_2$	25.9, CH$_2$	71.3, CH	70.9, CH	71.3, CH	71.2, CH
16	29.8, CH$_2$	29.7, CH$_2$	29.8, CH$_2$	42.7, CH$_2$	42.6, CH$_2$	42.3, CH$_2$	42.6, CH$_2$
17	59.0, CH	59.8, CH	58.9, CH	59.1, CH	59.0, CH	59.9, CH	59.4, CH
18	12.8, CH$_3$	16.1, CH$_3$	12.8, CH$_3$	15.7, CH$_3$	15.4, CH$_3$	18.1, CH$_3$	15.6, CH$_3$
19	19.1, CH$_3$	21.8, CH$_3$	19.1, CH$_3$	22.0, CH$_3$	19.0, CH$_3$	25.3, CH$_3$	21.9, CH$_3$
20	34.1, CH	34.2, CH	34.1, CH	34.2, CH	34.2, CH	34.3, CH	33.9, CH
21	19.5, CH$_3$	19.5, CH$_3$	19.7, CH$_3$	19.6, CH$_3$	19.5, CH$_3$	19.5, CH$_3$	19.7, CH$_3$
22	41.7, CH$_2$	41.8, CH$_2$	46.2, CH$_2$	45.0, CH$_2$	45.1, CH$_2$	45.1, CH$_2$	41.9, CH$_2$
23	71.7, CH	71.8, CH	68.1, CH	72.5, CH	72.5, CH	72.6, CH	71.8, CH
24	47.4, CH	47.4, CH	49.5, CH	162.4, C	162.3, C	162.4, C	47.4, CH
25	29.6, CH	29.6, CH	26.4, CH	32.2, CH	32.1, CH	32.2, CH	29.6, CH
26	22.5, CH$_3$	22.5, CH$_3$	24.4, CH$_3$	24.4, CH$_3$	24.4, CH$_3$	24.4, CH$_3$	22.5, CH$_3$
27	18.3, CH$_3$	18.4, CH$_3$	23.3, CH$_3$	23.6, CH$_3$	23.7, CH$_3$	23.7, CH$_3$	18.4, CH$_3$
28	11.3, CH$_3$	11.3, CH$_3$		106.8, CH$_2$	106.8, CH$_2$	106.8, CH$_2$	11.4, CH$_3$

[a] Assignments were confirmed by HSQC and HMBC (8Hz) data.

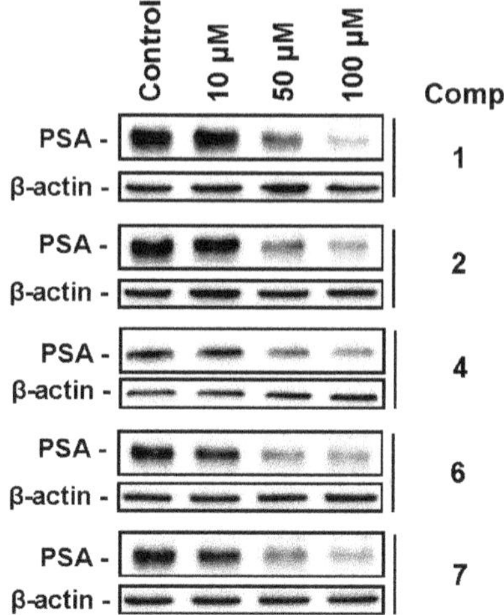

Figure 5. Effects of the compounds on PSA expression in 22Rv1 cells. The cells were treated with the compounds for 24 h, then the proteins were extracted and examined using Western blotting. β-actin was used as a loading control.

3. Materials and Methods

3.1. General Procedures

Optical rotations were measured using a PerkinElmer 343 polarimeter (Waltham, MA, USA). IR spectra were recorded using spectrophotometer Equinox 55 (Bruker, Ettlingen, Germany). The ^{1}H and ^{13}C NMR spectra were obtained using Bruker Avance III-700 and Bruker Avance III HD-500 spectrometers (Bruker, Ettlingen, Germany). Chemical shifts were referenced with Me$_4$Si as an internal standard. ESI mass spectra (including HRESIMS) were measured using Bruker maXis Impact II mass spectrometer (Bruker Daltonics, Bremen, Germany). Low-pressure column liquid chromatography was performed using YMC Gel ODS-A (YMC Co., Ltd., Kyoto, Japan). HPLC was performed using Shimadzu Instrument equipped with RID-10A refractive index detector (Shimadzu Corporation, Kyoto, Japan) and YMC-Pack ODS-A (250 × 10 mm) column (YMC Co., Ltd., Kyoto, Japan).

3.2. Animal Material

Specimens of *Haliclona gracilis* were collected off the coast of Shikotan Island (43°28′0 N; 146°48′9 E) by dredging at 145 m depth on June 2017, and identified by Grebnev B. B. using the morphology of skeleton and spicules. Comparison the data of #050-078 with the corresponding characteristics of *Haliclona gracilis* and their complete coincidence supported the sponge identification as *Haliclona gracilis* [32]. A voucher specimen is deposited under registration number 050-078 in the collection of marine invertebrates of the Pacific Institute of Bioorganic Chemistry (Vladivostok, Russia).

3.3. Extraction and Isolation

The freshly collected specimens were immediately frozen and stored at −18 °C until use. Animal material (dry weight 20 g) were crushed and extracted with EtOH (2 × 1 L). The EtOH extract after evaporation in vacuo was partitioned between H$_2$O and *n*-BuOH. The *n*-BuOH-soluble materials were partitioned with aqueous EtOH and *n*-hexane. The EtOH-soluble layer was fractioned by flash column chromatography on YMC gel ODS-A (75 μm), eluting with a step gradient of H$_2$O – EtOH (100:0 − 20:80) with monitoring by HPLC. The fractions that eluted with 40% EtOH were further purified by repeated reversed-phase HPLC (YMC ODS-A column (250 × 10 mm), 1.5 mL/min, H$_2$O-EtOH, 40:60 +1% AcONH$_4$) to afford, in order of elution, compounds **6** (2 mg), **2** (3 mg), **4** (6 mg), **5** (1 mg), **3** (1 mg), **7** (4 mg), and **1** (8 mg) with retention times (t_R) of 14.0, 17.5, 18.4, 22.5, 26.2, 32.5, and 36.1 min, respectively.

3.4. Compound Characterization Data

*Gracilosulfate A (**1**)*: white, amorphous solid; $[\alpha]_D^{20}$ +6 (*c* 0.2, EtOH); IR (KBr) ν_{max} 3467, 2957, 1457, 1242, 1002, 939 cm^{-1}; ^{1}H, ^{13}C NMR, Tables 1 and 2; HRESIMS *m/z* 527.3045 [M−Na]$^-$ (calcd for C$_{28}$H$_{47}$O$_7$S, 527.3048).

*Gracilosulfate B (**2**)*: white, amorphous solid; $[\alpha]_D^{20}$ +22 (*c* 0.2, EtOH); IR (KBr) ν_{max} 3446, 2947, 1457, 1242, 937 cm^{-1}; ^{1}H, ^{13}C NMR, Tables 1 and 2; HRESIMS *m/z* 543.3002 [M−Na]$^-$ (calcd for C$_{28}$H$_{47}$O$_8$S, 543.2997).

*Gracilosulfate C (**3**)*: white, amorphous solid; $[\alpha]_D^{20}$ ≈0 (*c* 0.1, EtOH); IR (KBr) ν_{max} 3465, 2960, 1450, 1240 cm^{-1}; ^{1}H, ^{13}C NMR, Tables 1 and 2; HRESIMS *m/z* 513.2891 [M−Na]$^-$ (calcd for C$_{27}$H$_{45}$O$_7$S, 513.2891).

*Gracilosulfate D (**4**)*: white, amorphous solid; $[\alpha]_D^{20}$ −40 (*c* 0.2, EtOH); IR (KBr) ν_{max} 3436, 2956, 1457, 1242, 1065, 998 cm^{-1}; ^{1}H, ^{13}C NMR, Tables 1 and 2; HRESIMS *m/z* 525.2890 [M−Na]$^-$ (calcd for C$_{28}$H$_{45}$O$_7$S, 525.2891).

Gracilosulfate E (**5**): white, amorphous solid; $[\alpha]_D^{20}$ ~0 (*c* 0.1, EtOH); IR (KBr) ν_{max} 3440, 2938, 1457, 1241, 936 cm^{-1}; ^{1}H, ^{13}C NMR, Tables 1 and 2; HRESIMS *m/z* 541.2845 [M−Na]$^-$ (calcd for C$_{28}$H$_{45}$O$_8$S, 541.2841).

Gracilosulfate F (**6**): white, amorphous solid; $[\alpha]_D^{20}$ −17 (*c* 0.2, EtOH); IR (KBr) ν_{max} 3456, 2942, 1457, 1242, 998 cm^{-1}; ^{1}H, ^{13}C NMR, Tables 1 and 2; HRESIMS *m/z* 541.2840 [M−Na]$^-$ (calcd for C$_{28}$H$_{45}$O$_8$S, 541.2841).

Gracilosulfate G (**7**): white, amorphous solid; $[\alpha]_D^{20}$ −32 (*c* 0.1, EtOH); IR (KBr) ν_{max} 3440, 2932, 1653, 1457, 1240 cm^{-1}; ^{1}H, ^{13}C NMR, Tables 1 and 2; HRESIMS *m/z* 527.3057 [M − Na]$^-$ (calcd for C$_{28}$H$_{47}$O$_7$S, 527.3048).

Preparation of MTPA esters of compounds **1**, **3**, and **4**

To duplicate solutions of compound **1** (2 mg each) in 100 μL of anhydrous pyridine, we added (*R*)- or (*S*)-MTPACl (10 μL). After stirring for 30 min at rt, the reaction mixtures were concentrated under reduced pressure and separated by HPLC (YMC ODS-A column (250 × 10 mm), H$_2$O-EtOH, 24:76 + 1% AcONH$_4$) to afford the (*S*)- or (*R*)-MTPA esters of **1**. The (*S*)- or (*R*)-MTPA derivatives of **3** and **4** were also prepared in a similar manner.

(S)-MTPA ester of 1 (1S): white, amorphous solid; ^{1}H NMR (CD$_3$OD, 500 MHz) δ_H 5.35 (1H, dd, *J* = 11.2, 4.7 Hz, H-23), 1.76 (1H, m, H-22), 1.52 (1H, m, H-24), 1.47 (1H, m, H-25), 1.39 (1H, m, H-20), 1.13 (1H, m, H-22), 0.99 (3H, d, *J* = 6.7 Hz, H-21), 0.94 (3H, d, *J* = 6.6 Hz, H-27), 0.86 (3H, d, *J* = 6.6 Hz, H-26), 0.76 (3H, d, *J* = 6.7 Hz, H-28), 0.62 (3H, s, H-18). HRESIMS *m/z* 779.3210 [M + Cl]$^-$ (calcd for C$_{38}$H$_{55}$ClF$_3$O$_9$S, 779.3213).

(R)-MTPA ester of 1 (1R): white, amorphous solid; ^{1}H NMR (CD$_3$OD, 500 MHz) δ_H 5.36 (1H, dd, *J* = 11.2, 4.7 Hz, H-23), 1.69 (1H, m, H-22), 1.57 (1H, m, H-24), 1.50 (1H, m, H-25), 1.13 (1H, m, H-20), 1.04 (1H, m, H-22), 0.97 (3H, d, *J* = 6.6 Hz, H-27), 0.92 (3H, d, *J* = 6.7 Hz, H-21), 0.89 (3H, d, *J* = 6.6 Hz, H-26), 0.87 (3H, d, *J* = 6.7 Hz, H-28), 0.45 (3H, s, H-18). HRESIMS *m/z* 779.3210 [M + Cl]$^-$ (calcd for C$_{38}$H$_{55}$ClF$_3$O$_9$S, 779.3213).

(S)-MTPA ester of 3 (3S): white, amorphous solid; ^{1}H NMR (CD$_3$OD, 500 MHz) δ_H 5.33 (1H, m, H-23), 1.78 (1H, m, H-22), 1.49 (1H, septet, *J* = 6.6 Hz,, H-25), 1.41 (1H, m, H-20), 1.30 (1H, m, H-24), 1.19 (1H, m, H-22), 0.99 (3H, d, *J* = 6.5 Hz, H-21), 0.98 (1H, m, H-24), 0.90 (3H, d, *J* = 6.6 Hz, H-27), 0.87 (3H, d, *J* = 6.6 Hz, H-26), 0.63 (3H, s, H-18). HRESIMS *m/z* 765.3060 [M + Cl]$^-$ (calcd for C$_{37}$H$_{53}$ClF$_3$O$_9$S, 765.3056).

(R)-MTPA ester of 3 (3R): white, amorphous solid; ^{1}H NMR (CD$_3$OD, 500 MHz) δ_H 5.30 (1H, m, H-23), 1.62 (1H, m, H-24), 1.68 (1H, m, H-22), 1.38 (1H, m, H-24), 1.59 (1H, septet, *J* = 6.6 Hz,, H-25), 1.16 (1H, m, H-20), 1.14 (1H, m, H-22), 0.89 (3H, d, *J* = 6.5 Hz, H-21), 0.95 (3H, d, *J* = 6.6 Hz, H-27), 0.92 (3H, d, *J* = 6.6 Hz, H-26), 0.42 (3H, s, H-18). HRESIMS *m/z* 765.3060 [M + Cl]$^-$ (calcd for C$_{37}$H$_{53}$ClF$_3$O$_9$S, 765.3056).

Bis(S)-MTPA ester of 4 (4S): white, amorphous solid; ^{1}H NMR (CD$_3$OD, 500 MHz) δ_H 6.02 (1H, dd, *J* = 3.3, 1.1 Hz, H-4), 5.47 (1H, brd, *J* = 11.1 Hz, H-23), 4.87 (1H, t, *J* = 1.2 Hz, H-28), 4.84 (1H, brs, H-28), 1.91 (1H, m, H-22), 2.25 (1H, septet, *J* = 6.6 Hz, H-25), 1.67 (1H, m, H-20), 1.28 (1H, m, H-22), 1.10 (3H, d, *J* = 6.6 Hz, H-27), 1.03 (6H, d, *J* = 6.6 Hz, H-21, 26), 0.93 (3H, s, H-18). HRESIMS *m/z* 957.3675 [M − Na]$^-$ (calcd for C$_{48}$H$_{59}$F$_6$O$_{11}$S, 957.3688).

Bis(R)-MTPA ester of 4 (4R): white, amorphous solid; ^{1}H NMR (CD$_3$OD, 500 MHz) δ_H 5.91 (1H, dd, *J* = 3.3, 1.1 Hz, H-4), 5.52 (1H, brd, *J* = 11.1 Hz, H-23), 5.07 (1H, t, *J* = 1.2 Hz, H-28), 5.01 (1H, brs, H-28), 2.32 (1H, septet, *J* = 6.6 Hz, H-25), 1.89 (1H, m, H-22), 1.43 (1H, m, H-20), 1.21 (1H, m, H-22), 0.94 (3H, d, *J* = 6.5 Hz, H-21), 1.12 (3H, d, *J* = 6.6 Hz, H-27), 1.08 (3H, d, *J* = 6.6 Hz, H-26), 0.55 (3H, s, H-18). HRESIMS *m/z* 957.3675 [M − Na]$^-$ (calcd for C$_{48}$H$_{59}$F$_6$O$_{11}$S, 957.3688).

3.5. Bioactivity Assay

3.5.1. Reagents

The MTT reagent (thiazolyl blue tetrazolium bromide) was purchased from Sigma (Taufkirchen, Germany).

3.5.2. Cell Lines and Culture Conditions

The human prostate cancer cell line 22Rv1 was purchased from ATCC (Manassas, VA, USA). Cells were cultured according to the manufacturer's instructions in RPMI media containing 10% FBS (Invitrogen, Carlsbad, USA). Cells were continuously kept in culture for a maximum of 3 months, and were routinely examined for stable phenotype and mycoplasma contamination.

3.5.3. In Vitro MTT-Based Drug Sensitivity Assay

The in vitro cytotoxicity of individual substances was evaluated using a MTT-based assay, which was performed as previously described [33]. Treatment time was 48 h.

3.5.4. Western Blotting

Preparation of protein extracts and Western blotting were performed as described previously [34]. For the detection of PSA, expression the anti-PSA/KLK3 antibodies was used (Cell Signaling, #5365, 1:1000). Treatment time was of 24 h.

4. Conclusions

In summary, we isolated gracilosulfates A-G, new steroids from the marine sponge *H. gracilis*, possessing a rare 3β-O-sulfonato, 4β-hydroxy moiety [1]. To date, only one pregnane steroid [35] and two polyhydroxy steroids [36] with such a fragment have been isolated from the sponge *Stylopus australis* and the starfish *Coscinasterias tenuispina*, respectively. In addition, the 5β,6β epoxy fragment is unprecedented in sulfated steroids [1]. Finally, the combination of 3β-O-sulfonato, 5β,6β-epoxy (or 5(6)-dehydro), and 4β,23-dihydroxy moieties is unprecedented, taking into account structures of all previously known natural sulfated steroids. Interestingly, these compounds are able to inhibit PSA expression in human hormone-independent prostate cancer cells, suggesting inhibition of AR signaling, a central target for the treatment of advanced prostate cancer.

Supplementary Materials: The following are available online at http://www.mdpi.com/1660-3397/18/9/454/s1, Copies of HRESIMS, and 1D- and 2D-NMR spectra of **1–7**, and photo of the marine sponge *Haliclona gracilis* (#050-078).

Author Contributions: L.K.S. isolated the metabolites; T.N.M. and L.K.S. elucidated structures; S.A.D. performed the bioactivity assays; V.A.D. performed the NMR spectra; R.S.P. and P.S.D. performed the mass spectra; B.B.G. performed species identification of the sponge; G.v.A. and V.A.S. assisted the results discussion; T.N.M., L.K.S., and V.A.S. wrote the paper, which was revised and approved by all the authors. All authors have read and agreed to the published version of the manuscript.

Funding: Isolation and establishment of chemical structures were partially supported by the RSF grant #20-14-00040 (Russian Science Foundation). Search for bioactive compounds from the Northwestern Pacific deep-water marine invertebrates was partially supported by the Grant of the Ministry of Science and Higher Education of the Russian Federation, grant #2020-1902-01-006.

Acknowledgments: The authors are thankful to Ms. Jessica Hauschild (University Medical Center Hamburg-Eppendorf) for assistance in the biological experiments. The study was carried out on the equipment of the Collective Facilities Center "The Far Eastern Center for Structural Molecular Research (NMR/MS) PIBOC FEB RAS."

Conflicts of Interest: The authors declare no conflict of interest.

References

1. Carroll, A.R.; Copp, B.R.; Davis, R.A.; Keyzers, R.A.; Prinsep, M.R. Marine Natural Products. *Nat. Prod. Rep.* **2020**, *37*, 175–223. [CrossRef]
2. Carvalhal, F.; Correia-da-Silva, M.; Sousa, E.; Pinto, M.; Kijjoa, A. Sources and biological activities of marine sulfated steroids. *J. Mol. Endocrinol.* **2018**, *61*, T211–T231. [CrossRef] [PubMed]
3. Kornprobst, J.M.; Sallenave, C.; Barnathan, G. Sulfated compounds from marine organisms. *Comp. Biochem. Physiol.* **1998**, *119B*, 1–51.
4. Kellner Filho, L.C.; Picao, B.W.; Silva Marcio, L.A.; Cunha, W.R.; Pauletti, P.M.; Dias, G.M.; Copp, B.R.; Bertanha, C.S.; Januario, A.H. Bioactive aliphatic sulfates from marine invertebrates. *Mar. Drugs* **2019**, *17*, 527. [CrossRef] [PubMed]
5. Xu, Y.-M.; Marron, M.T.; Seddon, E.; McLaughlin, S.P.; Ray, D.T.; Whitesell, L.; Gunatilaka, A.A.L. 2,3-Dihydrowithaferin A-3β-O-sulfate, a new potential prodrug of withaferin A from aeroponically grown Withania somnifera. *Bioorg. Med. Chem.* **2009**, *17*, 2210–2214. [CrossRef]
6. Morinaka, B.I.; Pawlik, J.R.; Molinski, T.F. Amaranzoles B-F, imidazole-2-carboxy steroids from the marine sponge *Phorbasam aranthus*. C24-N- and C24-O-analogues from a divergent oxidative biosynthesis. *J. Org. Chem.* **2010**, *75*, 2453–2460. [CrossRef]
7. Guzii, A.G.; Makarieva, T.N.; Denisenko, V.A.; Dmitrenok, P.S.; Burtseva, Y.V.; Krasokhin, V.B.; Stonik, V.A. Topsentiasterol sulfates with novel iodinated and chlorinated side chains from a marine sponge Topsentia sp. *Tetrahedron Lett.* **2008**, *49*, 7191–7193. [CrossRef]
8. Tabakmakher, K.M.; Makarieva, T.N.; Denisenko, V.A.; Popov, R.S.; Dmitrenok, P.S.; Dyshlovoy, S.A.; Grebnev, B.B.; Bokemeyer, C.; von Amsberg, G.; Cuong, N.X. New trisulfated steroids from the vietnamese marine sponge *Halichondria vansoesti* and their PSA expression and glucose uptake inhibitory activities. *Mar. Drugs* **2019**, *17*, 445. [CrossRef]
9. Woo, J.-K.; Ha, T.K.Q.; Oh, D.-C.; Oh, W.-K.; Oh, K.-B.; Shin, J. Polyoxygenated steroids from the sponge *Clathria gombawuiensis*. *J. Nat. Prod.* **2017**, *80*, 3224–3233. [CrossRef]
10. Boonlarppradab, C.; Faulkner, D.J. Eurysterols A and B, cytotoxic and antifungal steroidal sulfates from a marine sponge of the genus *Euryspongia*. *J. Nat. Prod.* **2007**, *70*, 846–848. [CrossRef]
11. Zhang, H.J.; Sun, J.B.; Lin, H.W.; Wang, Z.L.; Tang, H.; Cheng, P.; Chen, W.S.; Yi, Y.H. A new cytotoxic cholesterol sulfate from marine sponge *Halichondria rugosa*. *Nat. Prod. Res.* **2007**, *21*, 953–958. [CrossRef] [PubMed]
12. Makarieva, T.N.; Stonik, V.A.; D'yachuk, O.G.; Dmitrenok, A.S. Annasterol sulfate, a novel marine sulfated steroid, inhibitor of glucanase activity from the deep-water sponge *Poecillastra laminaris*. *Tetrahedron Lett.* **1995**, *36*, 129–132. [CrossRef]
13. Tsukamoto, S.; Matsunaga, S.; Fusetani, N.; Van Soest, R.W.M. Acanthosterol sulfates A–J: Ten new antifungal steroidal sulfates from a marine sponge *Acanthodendrilla* sp. *J. Nat. Prod.* **1998**, *61*, 1374–1378. [CrossRef] [PubMed]
14. Gabant, M.; Schmitz-Afonso, I.; Gallard, J.-F.; Menou, J.-L.; Laurent, D.; Debitus, C.; Al-Mourabit, A. Sulfated steroids: Ptilosteroids A–C and ptilosaponosides A and B from the Solomon Islands marine sponge *Ptilocaulis spiculifer*. *J. Nat. Prod.* **2009**, *72*, 760–763. [CrossRef]
15. DiGirolamo, J.A.; Li, X.-C.; Jacob, M.R.; Clark, A.M.; Ferreira, D. Reversal of fluconazole resistance by sulfated sterols from the marine sponge *Topsentia* sp. *J. Nat. Prod.* **2009**, *72*, 1524–1528. [CrossRef]
16. Radwan, M.M.; Manly, S.P.; Ross, S.A. Two new sulfated sterols from the marine sponge *Lendenfeldia dendyi*. *Nat. Prod. Commun.* **2007**, *2*, 901–904. [CrossRef]
17. Sperry, S.; Crews, P. Haliclostanone sulfate and halistanol sulfate from an Indo-Pacific *Haliclona* sponge. *J. Nat. Prod.* **1997**, *60*, 29–32. [CrossRef]
18. Fusetani, N.; Matsunaga, S.; Konosu, S. Bioactive marine metabolites II. Halistanol sulfate, an antimicrobial novel steroid sulfate from the marine sponge *halichondria* cf. *moorei bergquist*. *Tetrahedron Lett.* **1981**, *22*, 1985–1988. [CrossRef]
19. Cheng, J.-F.; Lee, J.-S.; Sun, F.; Jares-Erijman, E.A.; Cross, S.; Rinehart, K.L. Hamigerols A and B, unprecedented polysulfate sterol dimers from the mediterranean sponge *Hamigera hamigera*. *J. Nat. Prod.* **2007**, *70*, 1195–1199. [CrossRef]

20. Ushiyama, S.; Umaoka, H.; Kato, H.; Suwa, Y.; Morioka, H.; Rotinsulu, H.; Losung, F.; Mangindaan, R.E.P.; de Voogd, N.J.; Yokosawa, H.; et al. Manadosterols A and B, sulfonated sterol dimers inhibiting the ubc13-uev1a interaction, isolated from the marine sponge *Lissodendryx fibrosa. Nat. Prod.* **2012**, *75*, 1495–1499. [CrossRef]

21. Guzii, A.G.; Makarieva, T.N.; Denisenko, V.A.; Gerasimenko, A.V.; Udovenko, A.A.; Popov, R.S.; Dmitrenok, P.S.; Golotin, V.A.; Fedorov, S.N.; Grebnev, B.B.; et al. Guitarrins A-E and Aluminumguitarrin A: 5-azaindoles from the northwestern pacific marine sponge *Guitarra fimbriata. J. Nat. Prod.* **2019**, *82*, 1704–1709. [CrossRef] [PubMed]

22. Shubina, L.K.; Makarieva, T.N.; von Amsberg, G.; Denisenko, V.A.; Popov, R.S.; Dmitrenok, P.S.; Dyshlovoy, S.A. Monanchoxymycalin C with anticancer properties, new analogue of crambescidin 800 from the marine sponge *Monanchora pulchra. Nat. Prod. Res.* **2019**, *33*, 1415–1422. [CrossRef] [PubMed]

23. Appeltans, W.; Boucher, P.; Boxshall, G.A.; De Broyer, C.; de Voogd, N.J.; Gordon, D.P.; Hoeksema, B.W.; Horton, T.; Kennedy, M.; Mees, J.; et al. (Eds.) *World Register of Marine Species.* 2012. Available online: http://www.marinespecies.org (accessed on 20 November 2019).

24. Santafe´, G.; Paz, V.; Rodrıguez, J.; Jimenez, C. Novel cytotoxic oxygenated C29 sterols from the Colombian marine sponge *Polymastia tenax. J. Nat. Prod.* **2002**, *65*, 1161–1164. [CrossRef] [PubMed]

25. Wu, J.; Xi, Y.; Huang, L.; Li, G.; Mao, Q.; Fang, C.; Shan, T.; Jiang, W.; Zhao, M.; He, W.; et al. A Steroid-type antioxidant targeting the Keap1/Nrf2/ARE signaling pathway from the soft coral *Dendronephthya gigantea. J. Nat. Prod.* **2018**, *81*, 2567–2575. [CrossRef]

26. Vanderah, D.J.; Djerassi, C. Marine natural products. Synthesis of four naturally occurring 20β-H cholanic acid derivatives. *J. Org. Chem.* **1978**, *43*, 1442–1448. [CrossRef]

27. Matsumori, N.; Kaneno, D.; Murata, M.; Nakamura, H.; Tachibana, K. Stereochemical determination of acyclic structures based on carbon-proton spin-coupling constants. A method of configuration analysis for natural products. *J. Org. Chem.* **1999**, *64*, 866–876. [CrossRef]

28. Ivanchina, N.V.; Kicha, A.A.; Kalinovsky, A.I.; Dmitrenok, P.S.; Dmitrenok, A.S.; Chaikina, E.L.; Stonik, V.A.; Gavagnin, M.; Cimino, G. Polar steroidal compounds from the far eastern starfish *Henricia Leviuscula. J. Nat. Prod.* **2006**, *69*, 224–228. [CrossRef]

29. Sampson, N.; Neuwirt, H.; Puhr, M.; Klocker, H.; Eder, I.E. In vitro model systems to study androgen receptor signaling in prostate cancer. *Endocr. Relat. Cancer* **2013**, *20*, R49–R64. [CrossRef]

30. Nelson, P.S. Targeting the androgen receptor in prostate cancer—A resilient foe. *N. Engl. J. Med.* **2014**, *371*, 1067–1069. [CrossRef]

31. Iguchi, K.; Fujita, M.; Nagaoka, H.; Mitome, H.; Yamada, Y. Aragusterol Aa potent antitumor marine steroid from the Okinawan sponge of the genus *Xestospongia. Tetrahedron Lett.* **1993**, *34*, 6277–6280. [CrossRef]

32. Koltun, V.M. *Demospongian Sponges of the Northern and Far-Eastern Seas*; Publisher of USSR Academy of Sciences: Moscow, Russia; Leningrad, Russia, 1959; pp. 1–236. (In Russian)

33. Dyshlovoy, S.A.; Venz, S.; Hauschild, J.; Tabakmakher, K.M.; Otte, K.; Madanchi, R.; Walther, R.; Guzii, A.G.; Makarieva, T.N.; Shubina, L.K.; et al. Antimigrating activity of marine alkaloid monanchocidin A, proteome-based discovery and confirmation. *Proteomics* **2016**, *16*, 1590–1603. [CrossRef] [PubMed]

34. Dyshlovoy, S.A.; Tabakmakher, K.M.; Hauschild, J.; Shchekaleva, R.K.; Otte, K.M.; Guzii, A.G.; Makarieva, T.N.; Kudryashova, E.K.; Shubina, L.K.; Bokemeyer, C.; et al. Anticancer activity of eight rare guanidine alkaloids isolated from marine sponge *Monanchora pulchra. Mar. Drugs* **2016**, *14*, 133. [CrossRef] [PubMed]

35. Prinsep, M.R.; Blunt, J.W.; Munro, M.H.G. A new sterol sulfate from the marine sponge *Stylopus australis. J. Nat. Prod.* **1989**, *52*, 657–659. [CrossRef]

36. Riccio, R.; Iorizzi, M.; Minale, L. Starfish saponins XXX. Isolation of sixteen steroidal glycosides and three polyhydroxysteroids from the mediterranean starfish *Coscinasterias tenuispina. Bull. Soc. Chim. Belg.* **1986**, *95*, 869–893. [CrossRef]

Article

Antiproliferative Activity of Mycalin A and Its Analogues on Human Skin Melanoma and Human Cervical Cancer Cells

Domenica Capasso [1], Nicola Borbone [2], Monica Terracciano [2], Sonia Di Gaetano [3,* and Vincenzo Piccialli [4,*

[1] CESTEV, University of Naples Federico II, 80145 Naples, Italy; domenica.capasso@unina.it
[2] Department of Pharmacy, University of Naples Federico II, 80131 Naples, Italy; nicola.borbone@unina.it (N.B.); monica.terracciano@unina.it (M.T.)
[3] Institute of Biostructures and Bioimaging, CNR, 80134 Naples, Italy
[4] Department of Chemical Sciences, University of Naples Federico II, 80126 Naples, Italy
* Correspondence: digaetan@unina.it (S.D.G.); vinpicci@unina.it (V.P.)

Received: 21 June 2020; Accepted: 27 July 2020; Published: 29 July 2020

Abstract: Mycalin A, a polybrominated C_{15} acetogenin isolated from the encrusting sponge *Mycale rotalis*, displays an antiproliferative activity on human melanoma (A375) and cervical adenocarcinoma (HeLa) cells and induces cell death by an apoptotic mechanism. Various analogues and degraded derivatives of the natural substance have been prepared. A modification of the left-hand part of the molecule generates the most active substances. A structurally simplified lactone derivative of mycalin A, lacking the C1–C3 side chain, is the most active among the synthesized compounds exhibiting a strong cytotoxicity on both A375 and HeLa cells but not but not on human dermal fibroblast (HDF) used as healthy cells. Further evidence on a recently discovered chlorochromateperiodate-catalyzed process, used to oxidise mycalin A, have been collected.

Keywords: Mycalin A; C_{15} acetogenins; synthetic analogues; antiproliferative activity; A375 and HeLa cell lines

---

1. Introduction

C_{15} acetogenins are typical metabolites from red algae belonging to the genus *Laurencia*. They are usually halogenated substances including one or more cyclic ethers with different ring sizes and an enyne or a bromoallene terminal unit [1]. Linear C_{15} acetogenins are less common [1,2], and some of them, such as laurencenyne and *trans*-laurencenyne, have been hypothesized to be precursors of their cyclic congeners [3]. The investigation of specimens of algae *Laurencia* genus collected in different geographical areas continues to provide new C_{15} acetogenins displaying a large variety of substitution patterns and structural complexity, and even rearranged skeletons [4].

In 1990 three new members of this class of substance (mycalin A–C, **1–3**, Figure 1) (†) were isolated by our group from the encrusting sponge *Mycale rotalis* [5,6] collected in Italy along the Sicily coast. This finding was rather unusual and was explained by hypothesizing that the sponge could preferentially grow on an alga of the *Laurencia* species engulfing it entirely. This agrees well with the findings by Rinehart et al. [7] who had previously shown that metabolites peculiar to *Laurencia* species could be detected from other marine organisms.

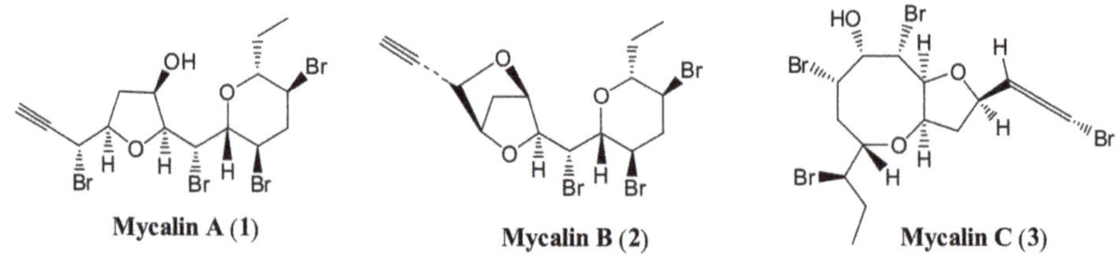

Figure 1. C_{15} acetogenins isolated from the sponge *Mycale rotalis*.

Mycalin A (**1**) is the most abundant metabolite isolated from the natural source. It possesses an unusual terminal bromopropargylic side chain, *cis*-THF (tetrahydrofuran) and *cis*-THP (tetrahydropyran) rings and four bromine atoms, a unique structural feature among this kind of metabolite [1]. Mycalin A has subsequently been isolated from two algae of *Laurencia* genus namely *Laurencia paniculata* [8] (‡), collected at Çeşmealt near Izmir (Turkey) and *Laurencia obtusa* [9] (§), collected in the Greek Ionic sea. The former organism also contained mycalin B [8] (‡) (**2**, Figure 1), a substance closely related to mycalin A. These findings support the hypothesis about the algal origin of the mycalins. The structural determination of compounds **1–3** was accomplished through spectral evidence and chemical correlation and derivatization work, and their relative and absolute stereochemistry was firmly established by X-ray analyses.

C_{15} acetogenins have exhibited a broad range of biological properties including anti-inflammatory [10] and anti-tumor activities, among others [1]. In this paper we report that mycalin A (**1**) and some of its synthetic analogues (**4–11**) possess a strong antiproliferative activity on human melanoma (A375) and human cervical adenocarcinoma (HeLa) cells. A degraded C4 lactone derivative of mycalin A (**11**) has displayed a remarkable selective cytotoxicity towards tumor cells in comparison with healthy control cells.

2. Results and Discussion

2.1. Biological Assays on Mycalin A

The search for new biologically active compounds is an important objective in medicinal chemistry and historically the marine environment has furnished a great number of natural products some of which have served as lead compounds for drug development [11]. As a part of our ongoing interest on the search for new anti-tumor substances, either of synthetic or natural origin [12–14], the effect of mycalin A (**1**) has been evaluated on the proliferation of human tumor cell lines of different histological origin. In particular, human malignant melanoma (A375), human cervical adenocarcinoma (HeLa) and human breast cancer (MCF-7) cells have been investigated. Human dermal fibroblasts (HDF) were used as healthy cells to evaluate the selectivity of action of the examined compound. All the cells were incubated with mycalin A at a 10 μM concentration for 48 h. As shown in Figure 2A, mycalin A (**1**) induced a strong inhibition on the proliferation of A375 and HeLa cells of about 80% and 70%, respectively. Interestingly, a low inhibition was observed on MCF-7 cells, which indicates the selectivity of **1** towards specific tumor cell lines, an attractive characteristic in terms of cancer therapy. Nonetheless, a very strong toxicity was observed against the healthy HDF cells.

As a next step, the ability of mycalin A to induce cell death through an apoptotic mechanism was tested on the A375 cell line that had shown the best response in terms of cytotoxicity among all the studied cells. The cell sample was treated with compound **1** at a 10 μM concentration for 48 h and then the apoptosis analysis was performed with the annexin V-FITC (V-fluorescein isothiocyanate)/propidium iodine (PI) double staining method using flow cytometry analysis. In this experiment, FITC-labelled annexin V, that specifically identifies apoptotic cells, was used in combination with PI, that is able to stain positively late stage apoptotic and necrotic cells. Therefore, the necrotic cells stain with only PI, advanced apoptotic cells with PI and FITC, and early apoptotic cells with only

FITC. The results indicated that the entire population of the cells treated with mycalin A was in the late apoptotic stage (Figure 2B). This result is in agreement with the high anti-proliferative activity shown by **1** on the A375 cells under the same conditions (Figure 2A).

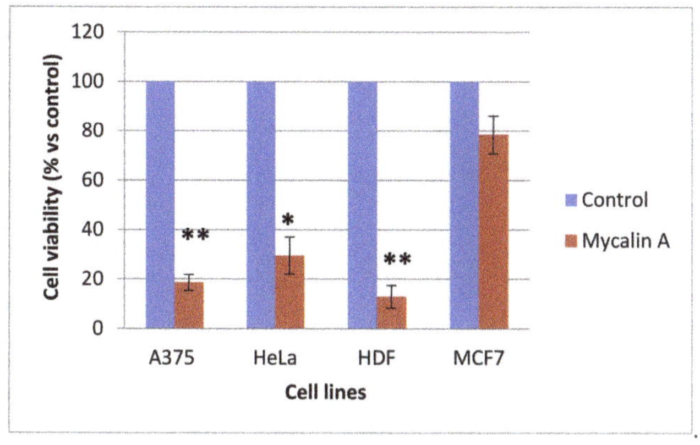

A

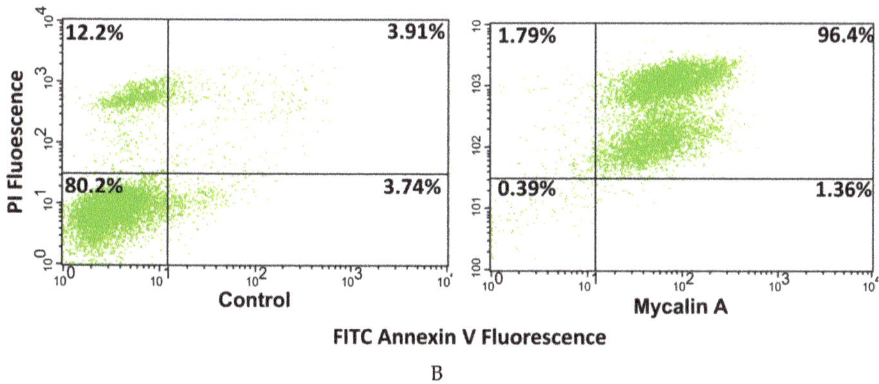

B

Figure 2. (**A**) Effect of mycalin A on tumor and healthy cell proliferation. The cells were incubated in the presence of 10 µM mycalin A, at 37 °C. Blue bars: control (vehicle treated cells); red bars: treated cells. Error bars represent ± SE (standard error), ** $p < 0.01$, * $p < 0.05$. (**B**) Apoptosis analysis with the annexin V-FITC (V-fluorescein isothiocyanate)/PI (propidium iodine) double staining method on A375 cells. The cells were treated with mycalin A (**1**), at 37 °C, at 10 µM concentration, for 48 h. The control is the vehicle treated sample. Lower left quadrant: viable cells; upper left: necrotic cells; upper right: advanced apoptotic cells; lower right: early apoptotic cells. This picture is representative of two independent experiments.

Furthermore, the dose–response curves (Figure 3) and the corresponding IC$_{50}$ values, calculated on responsive cells (A375, HeLa and HDF, Table 1), confirm the high activity of mycalin A and its lack of selectivity towards tumor cells.

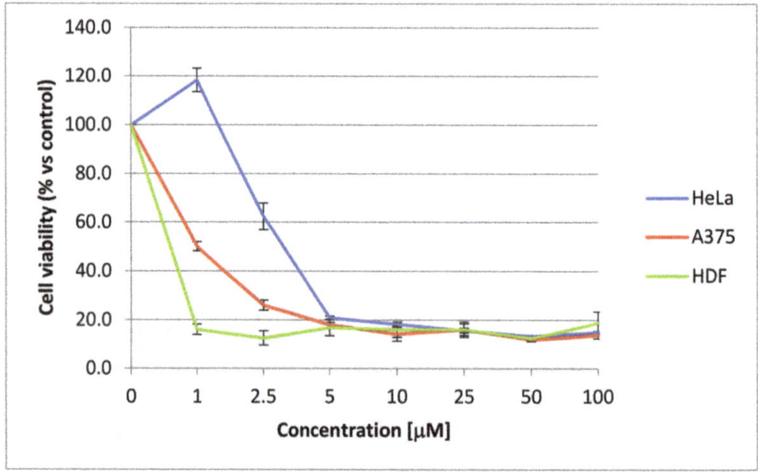

Figure 3. Dose-response curves obtained using the indicated concentrations of mycalin A (**1**) on responsive cells. The proliferation was determined by MTT assays. The results are presented as the percentage of proliferating cells compared to the control and are expressed as means ± SE of at least three independent experiments performed in triplicate. Statistical analyses: on HeLa $p < 0.05$ at 2.5 μM and $p < 0.01$ from 5 to 100 μM, on A375 $p < 0.01$ from 1 to 100 μM and on HDF $p < 0.01$ from 1 to 100 μM.

Table 1. IC$_{50}$ values of mycalin A (**1**) obtained on the responsive treated cells, after 48h incubation.

Cell Line	IC$_{50}$ (μM)
A375	1.1 ± 0.29
HeLa	3.9 ± 1.88
HDF	0.32 ± 0.22

2.2. Synthesis of Analogues of Mycalin A

The above interesting results on the activity of mycalin A prompted us to design simplified analogues of mycalin A which could retain the antitumor activity of the parent compound while exhibiting a limited toxicity towards healthy cells. To achieve this objective, mycalin A was subjected to a modification of its functional groups and a structural simplification, including degradation. An overview of the prepared derivatives **2** and **4–11** is shown in Figure 4. A detailed explanation of the chemistry follows.

Acetylation is a simple way to decrease the polarity of a substance and, potentially, to facilitate its penetration into the cell membrane. Thus, mycalin A was acetylated under standard conditions with Ac$_2$O/pyridine at 50 °C for 3 h, to give acetylated mycalin A **4** (Figure 4). Indeed, this simple structural modification further increased the antitumor potency of this substance (see later Figure 5A,B). Other potential sites of reactivity were then taken into consideration, such as the triple bond as well as the C-Br and ethereal C-O bonds. In this respect, the left-hand moiety of mycalin A proved to be the most easily modifiable. It is well known that ethereal C-O and C-Br bonds can be reductively cleaved by catalytic hydrogenation under standard conditions. Based on this consideration and literature precedents [15], we were confident that compound **1** could be structurally simplified when subjected to hydrogenation, possibly even giving open chain mycalin A analogues. Indeed, when **1** was subjected to catalytic hydrogenation with Pd(OH)$_2$/C (10%) in EtOH, a mixture of compounds **5** and **7** was obtained in the approximate ratio of 1:2, in favor of **7** (Scheme 1). Prolonged reaction times only resulted in higher amounts of **7**, suggesting that **5** at least partly converts into **7**. Accordingly, compound **5** could be transformed into **7** in high yields under the same reaction conditions (Scheme 1). The 4-propyl analogue **5** derived from the alkyne saturation and the C3-Br reductive cleavage. Further cleavage of the C4-O bond of the THF ring of **1** gave the open-chain diol **7**. This favorable selectivity concerning the C-Br and ethereal C-O bond cleavage allowed us to obtain compound **7** with the intact THP ring and a functionalization that paved the way for the preparation of other THP derivatives, thus permitting us to access mycalin A right-hand model compounds. In particular, the presence of a 7,8-halohydrin subunit in **7** allowed us the preparation of the corresponding 7,8-epoxide analogue (Scheme 2), thus increasing the functional group diversity and, taking into account the biological activity exhibited by a variety of natural epoxides, the probability of accessing a cytotoxic substance.

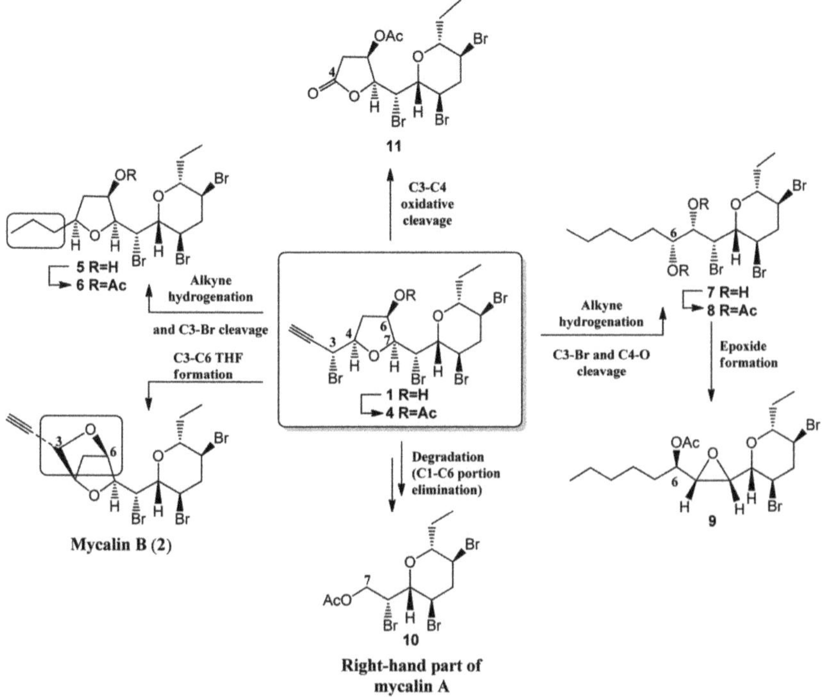

Figure 4. Overview of mycalin A derivatives.

Scheme 1. Hydrogenation of mycalin A (**1**). The H-4/H-7 nOe (nuclear Overhauser effect) correlation in **6** is highlighted.

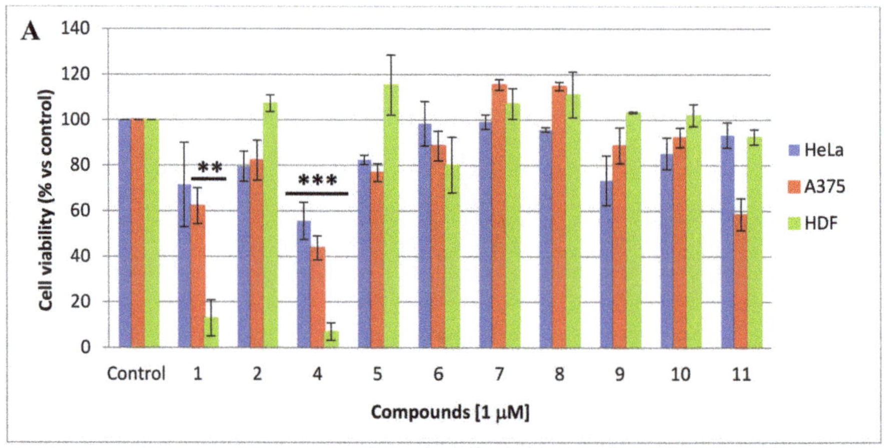

Scheme 2. Preparation of epoxide **9**. The H-6/H-9 nOe correlation is highlighted.

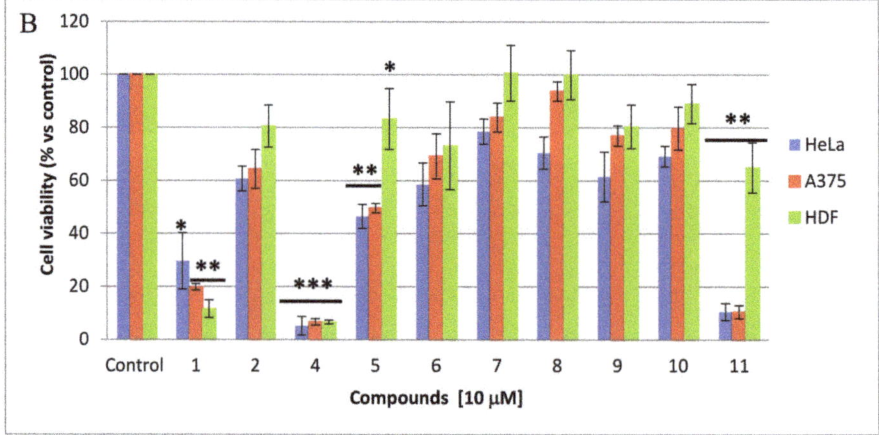

Figure 5. *Cont.*

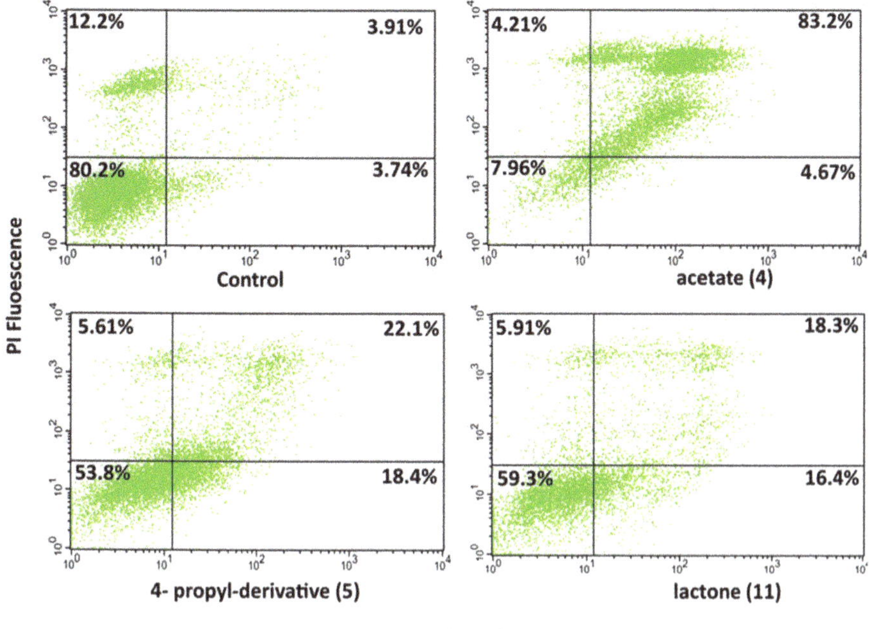

Figure 5. Effect of mycalin A and its derivatives on tumor and normal cell proliferation. The A375, HeLa and HDF cells were incubated in the presence of compounds **1**, **2** and **4–11**, solubilized in dimethyl sulfoxide DMSO (dimethyl sulfoxide), at 1 μM (**A**) or 10 μM (**B**) concentrations, at 37 °C for 48 h. Error bars represent ± SE; *** $p < 0.001$ ** $p < 0.01$ * $p < 0.05$ (**C**) Apoptosis analysis with the annexin V-FITC/PI double staining method on A375 cells. The cells were treated with the indicated compounds, at 37 °C, at 10 μM concentration, for 48 h. The control is the vehicle treated sample. Lower left quadrant: viable cells; upper left: necrotic cells; upper right: advanced apoptotic cells; lower right: early apoptotic cells. This picture is representative of two independent experiments.

The structures of **5** and **7** were confirmed by extensive 2D NMR (two-dimensional nuclear magnetic resonance) experiments on the corresponding acetate derivatives **6** and **8**, respectively, which were obtained under usual acetylation conditions (Scheme 1). The complete proton and carbon assignments of **6** and **8** are reported in Table S1 and S2, respectively. In particular, the presence of a strong nOe correlation between the H-4 and H-7 protons in **6** (Scheme 1 and Figure S12) confirmed that the *cis* configuration of the THF ring had been preserved under the hydrogenation conditions. On the other hand, the replacement of the methine proton signal resonating at δ 3.96 (H-4) in **6** (Scheme 1) with the methylene signal resonating at δ 1.26 (H$_2$-4) in **8** (Scheme 1), confirmed the cleavage of the THF ring at the C4-O bond in **1**. Based on the retention of the *S* configuration at C7 on reductive opening of the THF ring, the *7S,8R* configuration could be formulated for the 7,8-halohydrin system in **7**.

Epoxide **9** was obtained on treatment of diacetate **8** with catalytic amounts of K_2CO_3 in MeOH, followed by acetylation (Scheme 2). The presence of an nOe correlation between H-6 and H-9 protons in **9** (Scheme 2; see Table S3 for the complete NMR assignment), pointed to the *cis* configuration of the epoxide ring and, consequently, to the all *syn* arrangement of the C6–C8 segment in **8**. Therefore, we could assign the 6*R*,7*S*,8*S* configuration at the C6–C8 stereocentres in **9**.

The presence of the C6-C7 diol system in **7** allowed us to achieve the synthesis of the structurally simplified THP derivative **10** (Scheme 3), lacking the THF-containing left-hand (C1–C6) portion of **1**. To this end, first the oxidative cleavage of **7** was attempted by treatment with $NaIO_4$ in CH_2Cl_2/H_2O (1:1). However, a slow process and low yields of the expected aldehyde **12** (Scheme 3) resulted. Similar results were obtained by using acetone/water as solvent. Likewise, silica supported $NaIO_4$ also gave unsatisfactory results. Eventually, the treatment of **7** with lead tetraacetate (LTA) in acetic acid quickly gave aldehyde **12** as the sole product whose immediate reduction with $NaBH_4$ in EtOH, followed by acetylation, afforded the degraded product **10** in 45% yield, over three steps (Scheme 3; Table S4).

Mycalin B was then prepared (Scheme 4). This substance, possessing a further THF ring including part of the C4 side chain and the C6 OH group, introduces a structural perturbation in the left-hand moiety of the mycalin A. Therefore, it was seen as a compound able to provide information on the possible involvement of the left-hand part of mycalin A functional groups in its pharmacophoric portion. Thus, compound **1** was treated with KOH in MeOH at r.t. The internal nucleophilic substitution was very fast, and compound **2** was obtained in a 70% yield. This substance showed spectroscopic (Figures S3 and S4) and chromatographic properties identical to those exhibited by an authentic sample of mycalin B [5].

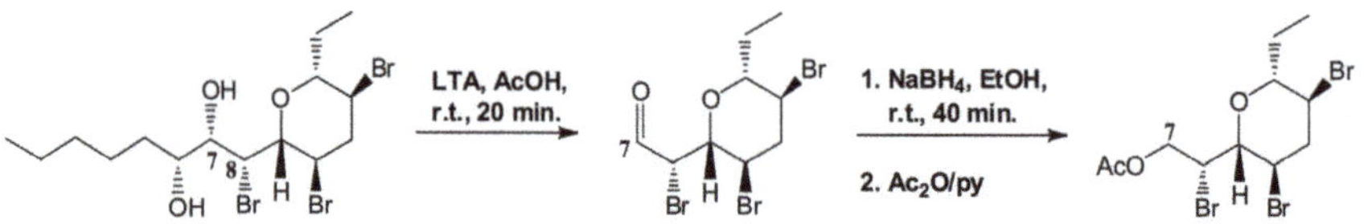

Scheme 3. Synthesis of the degraded derivative **10**.

Scheme 4. Synthesis of mycalin B **2**.

Lastly, lactone **11** was prepared by using a PCC (pyridinium chlorochromate)-catalyzed process (PCC/H_5IO_6 system) recently discovered in our laboratory [16], starting from mycalin A acetate (Scheme 5). Okamura et al. [17] have postulated that the actual oxidizing agent, under the above conditions, is chlorochromateperiodate (CCP), an oxidizing species more powerful than PCC, derived from the condensation of PCC and H_5IO_6 with the elimination of water. As we observed for the preparation of the corresponding benzoate [12], this process failed to go to completion under previously used conditions. However, in this instance we were able to force the process up to the complete consumption of the starting material by means of three further additions of freshly prepared CCP. In this way, lactone **11** was obtained in a 55% isolated yield. The loss of the C1-C3 carbon chain and the formation of the lactone ring was confirmed by ^{13}C-NMR (Figure S34) and HMBC (Figure S38) spectra. An nOe correlation between Ha-5 and H-7 protons in **11** further confirmed the configuration at C-7 (Scheme 5 right and Figure S36). Due to the limited amount of mycalin A, no further optimization of the process could be carried out at this stage. However, the observed behavior suggests a quick destruction of the active species and this process is certainly worth further investigation. Considering the synthetic importance of γ-lactones and the biological activity displayed by a variety of natural substances belonging to this class of compound, further studies are currently in progress in our laboratories.

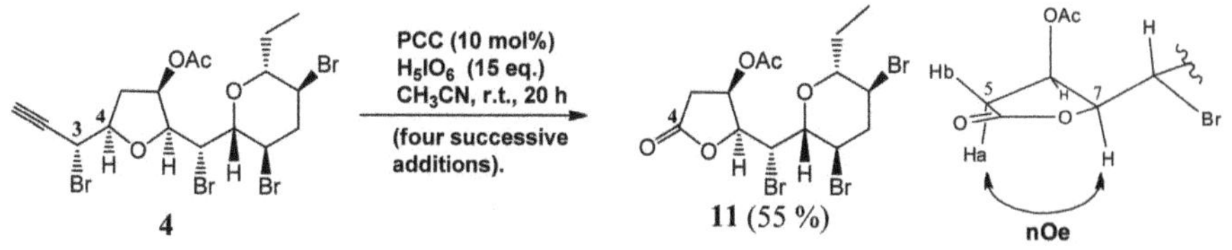

Scheme 5. Synthesis of lactone **11**. The 3D lactone ring part structure and the Ha-5/H-7 nOe correlation are highlighted on the right.

A summary of the synthesized mycalin A derivatives is shown in Table 2.

Table 2. Summary of the synthesized derivatives of mycalin A.

Mycalin A (1)	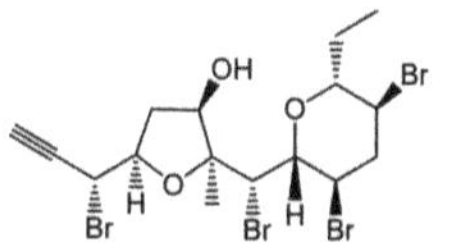	Mycalin B (2)	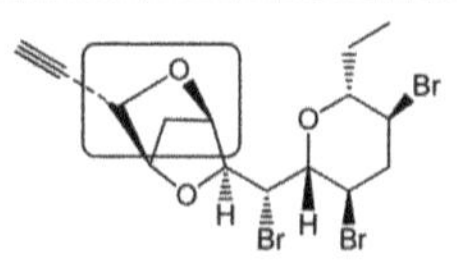
Mycalin A acetate (4)	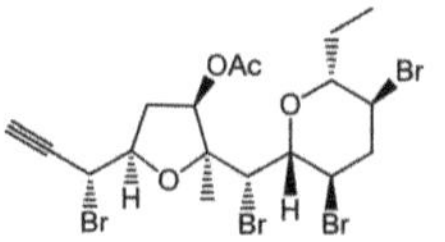	5	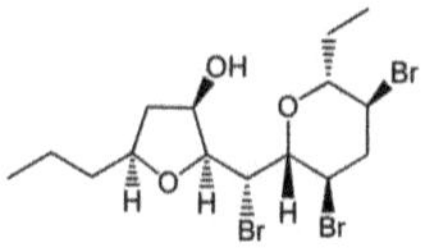
6	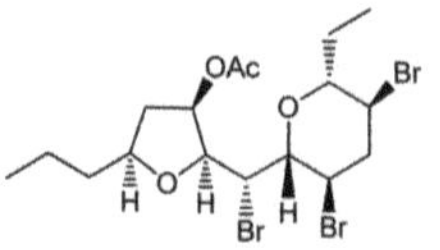	7	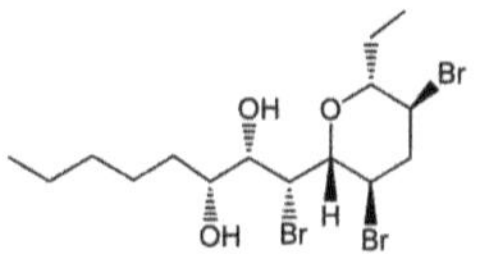
8	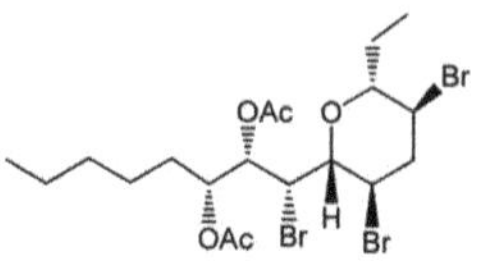	9	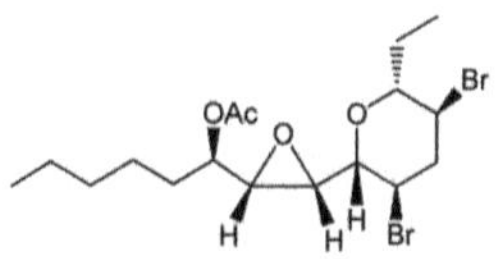
10	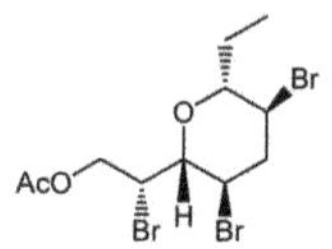	11	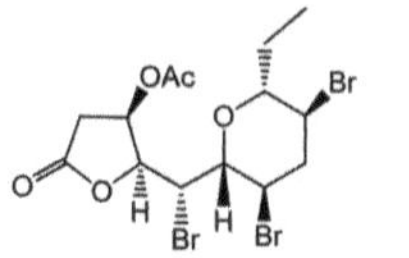

2.3. Biological Assays

To evaluate the cytotoxic activity of compounds **2** and **4–11** on the A375, HeLa and HDF cells, an initial screening was carried out at 1 µM and 10 µM concentrations for 48 h (Figure 5A,B). The cell viability was measured by means of the MTT assay. Among all the tested compounds, a remarkable cytotoxic activity was exhibited by acetate **4** and lactone **11** on the A375 and HeLa cells. Therefore, the acetylation of **4** increased the cytotoxicity of mycalin A towards the A375 and HeLa cells even at 1 µM, reaching a prominent effect at 10 µM. It is to be said that compound **4** also showed a toxic effect towards healthy cells at both the used concentrations. On the other hand, lactone **11** caused a drastic decrease in the cell viability up to 10%, compared to the control, at a dose of 10 µM on both cell lines. In addition, differently from acetate **4**, lactone **11** showed a considerable selectivity towards tumor cells, with a cytotoxicity of only about 35% on the HDF cells. Among the remaining compounds, 4-propyl-derivative **5** showed a good cytotoxicity on both the tumor cells and only a slight toxicity on the healthy cells, at a 10 µM concentration.

Then, FITC-AnnexinV assays were performed on the A375 cell line to evaluate the ability of **4**, **5**, and **11** to induce cell death through apoptosis. The cell sample was treated with such compounds at a 10 µM concentration for 48 h. The results indicated that all the tested compounds induced apoptotic effects (Figure 5C). In particular, **4** showed the entire population in late apoptosis, as observed for mycalin A. Differently, **5** and **11** displayed a lower level of apoptosis, at the same concentration.

These data indicate that the THF-containing substances are the most active and that the absence of the THF portion strongly reduces the activity, as shown by the THP derivatives **7–10** which display the lowest cytotoxic activity. This suggests that the THF-containing portion of mycalin A is essential for the activity. However, this does not mean that it alone is sufficient for the activity. The synthesis of further simplified analogues, only containing the left-hand part of the molecule, may help to clarify this point. The fact that mycalin B (**2**), where the C4 side-chain and the C6-OH group are engaged in a further THF ring, exhibits a reduced cytotoxicity compared with mycalin A, further supports the importance of the THF-containing, or THF-mimicking, portion of the substance and also suggests that the side-chain and/or the hydroxyl group could play a role in the cytotoxic effect of mycalin A.

Prompted by the results obtained by the anti-proliferative assays, the bioactivity profile of the most interesting compounds **4**, **5**, and **11** was evaluated using a concentration ranging from 0.25 µM to 25 µM for **4** and from 1 µM to 100 µM for **5** and **11**. The results are depicted in Figure 6A–C. From these dose–response analyses, the IC_{50} for each substance could be evaluated (Table 3) and some observations could be made. Acetate **4** showed a high cytotoxicity upon all the tested cells, including the healthy cells and its IC_{50} values confirm the increased activity compared to mycalin A. Compound **5** showed comparable IC_{50} values on the tumor cells as well as on the healthy cells and consequently lacking selectivity. Finally, compound **11** proves to be a very interesting substance, showing an IC_{50} of about 1 µM towards the melanoma (A375) cell line and an IC_{50} on HDF cells that is an order of magnitude higher. Remarkably, **11** retained the same activity of **1** on tumor cells, further increasing its selectivity.

Table 3. IC_{50} values obtained from the dose–response curves of acetate **4**, 4-propyl-derivative **5** and lactone **11** on all the tested cell lines, after 48h incubation.

Cell Line	4	5	11
A375	0.62 ± 0.20	14.11 ± 3. 5	1.1 ± 0.22
HeLa	0.92 ± 0.23	18.5 ± 5.0	6.7 ± 2.3
HDF	0.11 ± 0.004	23.2 ± 7.8	11.2 ± 6.3

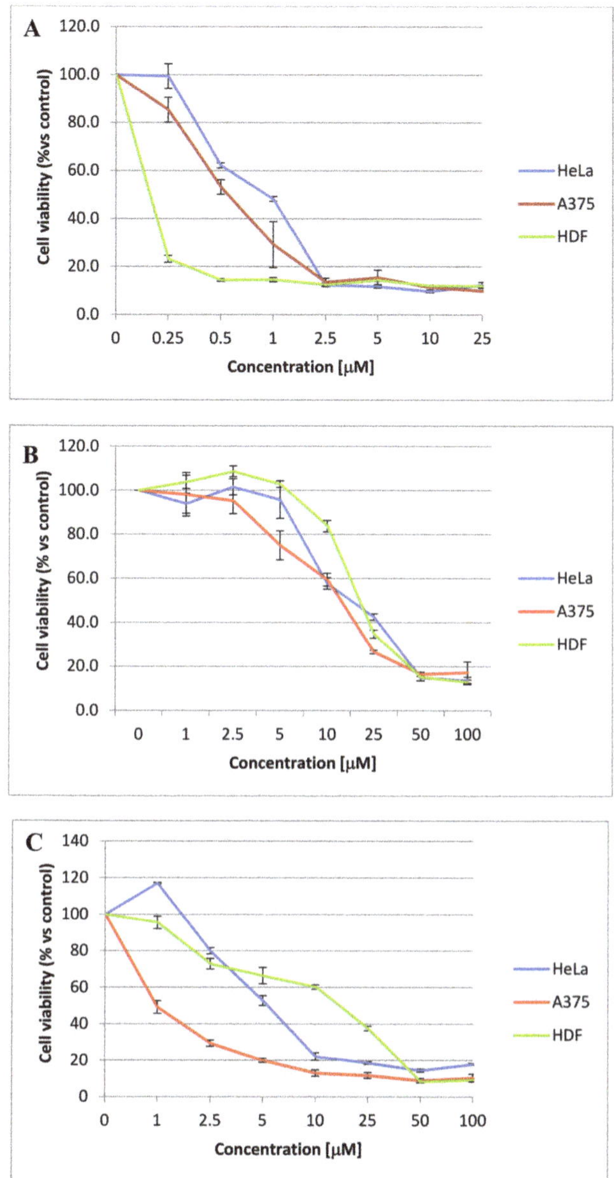

Figure 6. Dose-response curves of **4** (**A**), **5** (**B**), and **11** (**C**). The compounds were tested at different concentrations on the indicated cell lines. The proliferation was evaluated using the MTT assay. The results are presented as the percentage of proliferating cells with respect to the control (vehicle treated cells) and are expressed as means ± SE of at least three independent experiments performed in triplicate. Statistical analysis **4**: on HeLa $p < 0.001$ from 0.5 to 25 μM, on A375 $p < 0.01$ from 0.5 to 25 μM, on HDF $p < 0.001$ from 0.25 to 25 μM. **5**: on HeLa $p < 0.01$ from 10 to 100 μM, on A375 $p < 0.01$ at 10 and 25 μM, $p < 0.05$ at 50 and 100 μM, on HDF $p < 0.05$ at 10 μM and $p < 0.001$ from 25 to 100 μM. **11**: on HeLa $p < 0.01$ from 2.5 to 100 μM, on A375 $p < 0.05$ at 2.5 μM, $p < 0.01$ from 5 to 100 μM, on HDF $p < 0.01$ at 10 μM and $p < 0.05$ from 25 to 100 μM.

3. Materials and Methods

3.1. General Experimental Procedures

All reagents were purchased (Aldrich and Fluka) at the highest commercial quality and used without further purification. The reactions were monitored using thin layer chromatography (TLC) carried out on precoated silica gel plates (Merck 60, F254, 0.25 mm thick). Merck silica gel (Kieselgel 40, particle size 0.063–0.200 mm) was used for the column chromatography. All the HPLC (high performance liquid chromatography) separations and purifications were performed on a Phenomenex Luna column (25 cm × 4.6 mm, 5 μm) using *n*-hexane/EtOAc mixtures as eluent, at a flow rate of 1 mL/min. Na_2SO_4 was used as a drying agent for aqueous work-up. Nuclear magnetic resonance (NMR) experiments were performed using either a Bruker 700 MHz AvanceNeo spectrometer (Billerica, MA, US) equipped with a triple resonance cryoprobe or a Bruker 400 MHz Avance spectrometer (Billerica, MA, US) in $CDCl_3$. Proton chemical shifts were referenced to the residual $CHCl_3$ signal (7.26 ppm). ^{13}C-NMR chemical shifts were referenced to the solvent (77.0 ppm). Coupling constants (J) are given in Hertz. Abbreviations for signal coupling are as follows: s = singlet, d = doublet, t = triplet, q = quartet, m = multiplet, b = broad. The optical rotations were measured using a JASCO P-2000 polarimeter (JASCO, Oklahoma City, OK, US) at the sodium D line. The high-resolution mass spectra were recorded by infusion on a Thermo Linear Trap Quadrupole (LTQ) Orbitrap XL mass spectrometer (Thermo Fisher Scientific, Waltham, MA, US) equipped with an electrospray source in the positive mode using MeOH as the solvent.

3.2. Mycalin A 1

Mycalin A (**1**) was isolated from the sponge *Mycale rotalis* collected in the Stagnone di Marsala lagoon (Sicily) in the spring of 1989 [5,6]. For details of the isolation of mycalin A see ref. 6. A pure sample of mycalin A for the biological assays was obtained by filtration on a silica gel pad (eluent *n*-hexane-EtOAc, 4:6) followed by HPLC purification using *n*-hexane-EtOAc (75:25) as the eluent. Copy of the NMR spectra are reported in the Supplementary Materials.

3.3. Mycalin A Acetate 4

Excess acetic anhydride (0.5 mL) was added to a solution of alcohol **1** (18.2 mg, 0.032 mmol) in pyridine (0.5 mL) and the mixture stirred at 50 °C. After 3 h the reaction mixture was evaporated under reduced pressure to give acetyl derivative **4** (19.5 mg) as a single spot by TLC (*n*-hexane-EtOAc, 7:3, Rf = 0.56) The crude was subjected to HPLC purification using *n*-hexane-EtOAc, 8:2 as the eluent to give **4** (18.0 mg, 93%) as a colourless oil. **4**: $[\alpha]_D^{20}$ = + 41.1 (c = 0.20, $CHCl_3$). IR (infrared) (neat) λ_{max} 2121, 1737, 1374, 1241, cm^{-1}. ^1H-NMR (400 MHz, $CDCl_3$): δ 5.27 (1H, bdd, J = 5.4, 3.1), 4.65 (1H, dd, J = 9.5, 1.5), 4.59 (1H, dd, J = 6.7, 2.3), 4.34 (1H, ddd, 8.8, 6.6, 4.8), 4.24 (1H, dd, J = 9.6, 3.2), 4.17 (1H, ddd, J = 12.3, 9.5, 4.5), 3.72 (1H, ddd, J = 11.9, 10.2, 4.4), 3.39 (1H, dt, J = 9.5, 2.3), 3.21 (1H, dd, J = 9.6, 1.5), 2.98 (1H, ddd, J = 12.9, 4.4, 4.4), 2.68–2.59 (2H, overlapped m's), 2.39 (1H, ddd, J = 12.6, 12.6, 12.6), 2.29 (1H, dd, J = 15.6, 5.0), 2.12 (3H, s, acetate), 2.02 (1H, m), 1.54 (1H, m), 0.93 (3H, t, J = 7.3). ^{13}C-NMR: (100 MHz, $CDCl_3$): δ 170.1, 83.7, 83.6, 80.2, 79.7, 79.3, 76.5, 72.1, 51.5, 47.1, 46.6, 45.1, 37.4, 37.0, 25.7, 21.2, 9.5. HRESIMS (high resolution electrospray ionization mass spectrometry) *m/z*: calcd for $C_{17}H_{22}^{79}Br_4NaO_4$ 628.8149 [M + Na]$^+$, found: 628.8140.

3.4. Propyl Derivative 5

$Pd(OH)_2/C$ (20% *w/w*, 8.5 mg) was added to mycalin A (**1**, 42.3 mg, 0.074 mmol) in EtOH (2 mL) and the mixture stirred under hydrogen atmosphere. A vacuum-fill technique was used using a hydrogen balloon and a three-way vacuum adapter. After 2 h the reaction mixture was filtered over celite, and the filtrate was dried under reduced pressure to give a mixture of the propyl-derivative **5** and the diol **7** (40.5 mg) as a colorless oil. Further filtration of the crude on a silica gel pad, eluting with *n*-hexane-EtOAc (7:3), afforded an oily material (29.8 mg), essentially constituted by a mixture of **5** and

7 (TLC: *n*-hexane-EtOAc, 7:3. **5**: Rf = 0.60; **7**: Rf = 0.67). HPLC separation with *n*-hexane-EtOAc (7:3) as the eluent gave pure compounds **5** (8.2 mg, 23%) and **7** (15.9 mg, 43%) as colourless oils. **5**: $[\alpha]_D^{20}$ = +4.2 (c = 0.17, CHCl$_3$). IR (neat) λ_{max} 3446 (broad) cm^{-1}. ^{1}H-NMR: (400 MHz, CDCl$_3$): δ 4.64 (1H, dd, *J* = 9.2, 1.8), 4.24–4.14 (2H, overlapped m's), 3.96–3.87 (2H, overlapped m's), 3.81 (1H, dd, *J* = 9.8, 1.9), 3.76 (1H, ddd, *J* = 12.1, 10.0, 4.3), 3.44 (1H, dt, *J* = 9.1, 2.3), 3.00 (1H, ddd, *J* = 12.9, 4.3, 4.3), 2.48 (2H, overlapped m's), 2.04 (1H, m), 1.77 (1H, m), 1.68–1.33 (5H, overlapped m's), 0.96 (3H, t, *J* = 7.2), 0.94 (3H, t, *J* = 7.2). ^{13}C-NMR: (100 MHz, CDCl$_3$): δ 84.8, 83.8, 80.0, 71.9, 53.3, 47.7, 46.8, 45.3, 42.2, 38.9, 25.8, 19.2, 14.0, 9.7. HRESIMS *m/z*: calcd for C$_{15}$H$_{25}$^{79}Br$_3$NaO$_3$ 512.9252 [M + Na]$^+$, found: 512.9240.

3.5. Propyl Acetate 6

Acetic anhydride (0.4 mL) was added to a stirred solution of **5** (8.2 mg, 0.017 mmol) in pyridine (0.4 mL), at room temperature. After 16 h the reaction mixture was evaporated under reduced pressure to give acetyl derivative **6** (9.0 mg) as a colourless oil. HPLC purification (eluent: *n*-hexane-EtOAc, 7:3) gave the pure compound **6** (8.8 mg, 98%) suitable for the biological assays. **6**: $[\alpha]_D^{20}$ = +15.3 (c = 0.71, CHCl$_3$). IR (neat) λ_{max} 1737, 1375, 1239 cm^{-1}. ^{1}H-NMR: (700 MHz, CDCl$_3$): δ 5.19 (ddd, *J* = 6.3, 3.1, 1.1, 1H), 4.69 (dd, *J* = 9.7, 1.6, 1H), 4.17 (ddd, *J* = 12.1, 9.6, 4.6, 1H), 4.06 (dd, *J* = 9.7, 3.1, 1H), 3.96 (dq, *J* = 8.4, 6.4, 1H), 3.73 (ddd, *J* = 12.0, 10.0, 4.3, 1H), 3.38 (ddd, *J* = 10.0, 8.8, 2.4, 1H), 3.22 (dd, *J* = 9.6, 1.6, 1H), 2.98 (ddd, *J* = 13.0, 4.3, 4.3, 1H), 2.55 (ddd, *J* = 14.6, 8.4, 6.3, 1H), 2.40 (1H, ddd, *J* = 13.0, 12.1, 12.0), 2.12 (s, 3H, acetate), 2.06–1.98 (m, 1H), 1.79–1.71 (m, 1H), 1.66 (ddd, *J* = 14.5, 6.4, 1.1, 1H), 1.56–1.46 (m, 2H), 1.41 (dddd, *J* = 12.9, 10.3, 7.4, 5.5, 1H), 1.35 (dddd, *J* = 13.0, 10.6, 7.4, 5.4, 1H), 0.93 (t, *J* = 7.4, 6H, 2×Me). ^{13}C-NMR: (176 MHz, CDCl$_3$): δ 170.6, 83.6, 82.3, 80.2, 77.5, 73.2, 52.1, 47.4, 46.7, 45.0, 39.7, 38.2, 25.7, 21.4, 19.2, 14.1, 9.7. HRESIMS *m/z*: calcd for C$_{17}$H$_{27}$^{79}Br$_3$NaO$_4$ 554.9357 [M + Na]$^+$, found: 554.9368.

3.6. THP Diol 7

Compound **7** was synthesized as described above (see propyl derivative **5**). **7**: Amorphous solid. $[\alpha]_D^{20}$ = +9.9 (c = 0.57, CHCl$_3$). IR (neat) λ_{max} 3414 (broad) cm^{-1}. ^{1}H-NMR: (400 MHz, CDCl$_3$): δ 4.67 (1H, dd, *J* = 6.3, 1.7), 4.14 (1H, ddd, *J* = 12.2, 12.2, 4.5), 3.81–3.67 (4H, overlapped m's), 3.48 (1H, ddd, *J* = 8.6, 8.6, 2.3), 3.01 (1H, ddd, *J* = 12.9, 4.5, 4.5), 2.81 (2H, bs, 2xOH), 2.47 (1H, ddd, *J* = 12.3, 12.3, 12.3), 2.04 (1H, m), 1.66–1.26 (9H, overlapped m's), 0.95 (3H, t, *J* = 7.3), 0.90 (3H, t, *J* = 6.9). ^{13}C-NMR: (100 MHz, CDCl$_3$): δ 83.8, 81.4, 75.7, 71.3, 59.2, 47.1, 46.9, 45.1, 33.8, 31.7, 25.8, 25.3, 14.0, 9.5. HRESIMS *m/z*: calcd for C$_{15}$H$_{27}$^{79}Br$_3$NaO$_3$ 514.9408 [M + Na]$^+$, found: 514.9426.

3.7. THP Diol Acetate 8

Acetic anhydride (0.3 mL) was added to a stirred solution of diol **7** (7.0 mg, 0.014 mmol) in pyridine (0.3 mL), and the mixture stirred at 50 °C. After 3 h the reaction mixture was evaporated under reduced pressure to give diacetate **8** (8.3 mg) as a colorless oil (TLC: *n*-hexane-EtOAc, 85:15, Rf = 0.54). Pure **8** (7.9 mg, 96%), suitable for the biological assays, was obtained by HPLC purification (eluent: *n*-hexane-EtOAc, 85:15). **8**: $[\alpha]_D^{20}$ = +35.8 (c = 0.12, CHCl$_3$). IR (neat) λ_{max} 1742, 1371, 1217, cm^{-1}. ^{1}H-NMR (700 MHz, CDCl$_3$) δ 5.49 (dd, *J* = 9.5, 2.0, 1H), 5.09 (ddd, *J* = 7.7, 6.0, 2.0, 1H), 4.55 (dd, *J* = 9.4, 1.9, 1H), 4.13 (ddd, *J* = 12.1, 9.5, 4.5, 1H), 3.72 (ddd, *J* = 12.1, 10.0, 4.3, 1H), 3.47 (ddd, *J* = 10.0, 8.6, 2.4, 1H), 3.43 (dd, *J* = 9.5, 1.9, 1H), 2.96 (ddd, *J* = 12.9, 4.5, 4.3, 1H), 2.44 (ddd, *J* = 12.9, 12.1, 12.0, 1H), 2.17 (s, 3H, acetate), 2.13 (s, 3H, acetate), 2.04 (m, 1H), 1.55 (m, 1H), 1.50 (m, 1H), 1.34–1.21 (m, 8H), 0.99 (t, *J* = 7.4, 3H), 0.86 (t, *J* = 7.0, 3H). ^{13}C-NMR: (176 MHz, CDCl$_3$): δ 170.7, 170.0, 83.7, 79.4, 74.3, 71.6, 54.8, 47.5, 47.3, 45.0, 31.5 (2C), 25.8, 24.7, 22.4, 21.1, 20.8, 13.9, 9.7. HRESIMS *m/z*: calcd for C$_{19}$H$_{31}$^{79}Br$_3$NaO$_5$ 598.9619 [M + Na]$^+$, found: 598.9627.

3.8. THP Epoxyde 9

K$_2$CO$_3$ (3 eq, 2.2 mg, 0.016 mmol), was added to a stirred solution of diacetate **8** (3.0 mg, 0.005 mmol) in MeOH (1 mL), at room temperature. After 48 h AcOH was added up to pH = 6 and the reaction mixture was taken to dryness. The solid residue was partitioned between water and EtOAc and the organic phase was dried and evaporated to give an oil (2.9 mg). TLC analysis revealed still the presence of a trace amount of the starting diol. HPLC separation (eluent: *n*-hexane-EtOAc, 85:15) gave the pure 7-hydroxy epoxide (1.7 mg, 80%). Acetylation under usual conditions afforded acetate **9** (1.9 mg, 100%) as a colourless oil. **9**: $[\alpha]_D^{20}$ = −7.7 (c = 0.21, CHCl$_3$). IR (neat) λ_{max} 1742, 1463, 1371, 1239 cm^{-1}. ^{1}H-NMR (700 MHz, CDCl$_3$) δ 4.89 (ddd, *J* = 8.7, 8.7, 3.9, 1H), 3.84 (ddd, *J* = 12.4, 10.0, 4.4, 1H), 3.73 (ddd, *J* = 12.4, 10.1, 4.4, 1H), 3.51 (dd, J = 10.0, 7.1, 1H), 3.45 (ddd, J = 10.1, 7.3, 2.7, 1H), 3.05 (dd, *J* = 8.7, 4.1, 1H), 3.02 (dd, *J* = 7.1, 4.1, 1H), 2.98 (dt, *J* = 12.4, 4.4, 1H), 2.43 (ddd, *J* = 12.2, 12.2, 12.0, 1H), 2.11 (s, 3H, acetate), 1.98 (ddq, *J* = 14.9, 7.4, 2.7, 1H), 1.75–1.52 (m, 3H overlapped), 1.36 (m, 1H), 1.34–1.20 (m, 5H overlapped), 0.95 (t, *J* = 7.4, 3H), 0.88 (t, *J* = 7.0, 3H). ^{13}C-NMR: (176 MHz, CDCl$_3$): δ 170.4, 83.0, 80.5, 71.9, 57.7, 56.6, 47.1, 46.1, 45.6, 32.1, 31.8, 25.8, 24.6, 22.5, 21.1, 14.0, 9.0. HRESIMS *m/z*: calcd for C$_{17}$H$_{28}$^{79}Br$_2$NaO$_4$ 477.0252 [M + Na]$^+$, found: 477.0238.

3.9. Degraded THP Derivative 10

Crystalline lead tetraacetate (2 eq., 12.5 mg, 0.028 mmol) was added to a solution of diol **7** (7.0 mg, 0.014 mmol) in AcOH (1.0 mL). After 20 min., TLC analysis (*n*-hexane-EtOAc, 75:25) revealed the disappearance of the starting material and the formation of a slightly less polar substance. After a further 10 min., two drops of ethylene glycol were added and the mixture stirred for 10 min., diluted with water and extracted with CHCl$_3$. The organic layer was washed with a sat. aqueous NaHCO$_3$ solution, dried and evaporated to give 5.3 mg of aldehyde **12** as a smelling oil. NaBH$_4$ (a tip of spatula, excess) was added to a solution of the crude aldehyde **12** (4.0 mg, 0.010 mmol) in EtOH (1 mL), and the suspension stirred at room temperature for 40 min. Then, the excess reagent was destroyed by the addition of a few drops of AcOH, water was added, and the mixture extracted with CHCl$_3$. The organic phase was washed with a sat. aqueous NaHCO$_3$ solution, dried and evaporated to give 4.8 mg of the crude alcohol. HPLC purification of the latter gave 3.8 mg of the pure C7 alcohol.

Acetylation of the above alcohol with Ac$_2$O/pyridine (0.5 mL/0.5 mL) overnight afforded 4.0 mg of the acetyl derivative **10**. Filtration of this material on a short pad of silica gel (eluent: *n*-hexane-EtOAc, 9:1), gave pure acetate **10** (2.8 mg, 45% over three steps) as a colorless oil. **10**: $[\alpha]_D^{20}$ = +18.2 (c = 0.19, CHCl$_3$). IR (neat) λ_{max} 1746, 1227 cm^{-1}. ^{1}H-NMR: (700 MHz, CDCl$_3$): δ 4.60 (ddd, *J* = 8.3, 6.7, 1.8, 1H), 4.43 (dd, *J* = 11.3, 6.7, 1H), 4.40 (dd, *J* = 11.3, 8.3, 1H), 4.10 (ddd, *J* = 12.1, 9.7, 4.5, 1H), 3.75 (ddd, *J* = 12.0, 10.0, 4.3, 1H), 3.51 (dd, *J* = 9.7, 1.8, 1H), 3.43 (ddd, *J* = 10.0, 8.5, 2.5, 1H), 3.01 (ddd, *J* = 12.1, 4.5, 4.3, 1H), 2.46 (ddd, *J* = 12.9, 12.1, 12.0, 1H), 2.11 (s, 3H, acetate), 2.03 (ddq, 14.5, 7.4, 2.5 1H), 1.55 (ddq, *J* = 14.5, 8.5, 7.4, 1H), 0.96 (t, *J* = 7.4, 3H). ^{13}C-NMR (176 MHz, CDCl$_3$): δ 170.2, 83.7, 79.3, 64.7, 49.9, 47.5, 46.6, 45.3, 25.8, 20.7, 9.4. HRESIMS *m/z*: calcd for C$_{11}$H$_{17}$^{79}Br$_3$NaO$_3$ 456.8626 [M + Na]$^+$, found: 456.8629.

3.10. Lactone 11

PCC (10 mol%, 150 µL of a 0.01 M stock solution in acetonitrile) was added to a vigorously stirred suspension of H$_5$IO$_6$ (15 eq., 0.022 mmol, 49.8 mg) in acetonitrile (100 µL) at room temperature. After 5 min., compound **4** (9.0 mg, 0.015 mmol) dissolved in acetonitrile (100 µL + 2 × 100 µL rinse) was added to give a yellow cloudy mixture. After 16 h TLC analysis (*n*-hexane-EtOAc, 8:2) revealed the presence of a product at Rf = 0.3. No further progress of the process was observed after 2 h. Therefore, freshly prepared CCP (12 mol%, 180 µL of a 0.01 M stock solution in acetonitrile) was added to the reaction mixture. After 15 min., TLC analysis revealed the further progress of the process, but successive TLC analyses indicated no further consumption of the starting material. Two successive additions of the same amounts of CCP were required to drive the process to completion (overall further 2 h). Then, EtOH (1 mL) was added and the mixture was taken to dryness. Filtration of this material on a short pad of silica gel (eluent: CHCl$_3$-MeOH, 95:5) followed by HPLC purification (*n*-hexane-EtOAc, 7:3) gave pure acetate **11** (4.1 mg, 55%) as a colourless oil. $[\alpha]_D^{20}$ = +27.7 (c = 0.13, CHCl$_3$). IR (neat) λ_{max} 1794, 1745, 1225 cm^{-1}. ^{1}H-NMR: (700 MHz, CDCl$_3$): δ 5.46 (1H, dd, *J* = 5.1, 3.3), 4.85 (1H, dd, *J* = 9.8, 3.3,), 4.75 (1H, dd, *J* = 9.8, 1.7), 4.19 (1H, ddd, *J* = 12.1, 9.5, 4.6), 3.75 (1H, ddd, *J* = 12.0, 10.5, 4.3), 3.42 (1H, ddd, *J* = 10.5, 8.4, 2.5), 3.26 (1H, dd, *J* = 9.5, 1.7), 3.01 (1H, ddd, *J* = 12.9, 4.6, 4.3), 2.95 (1H, dd,

J = 18.1, 5.1), 2.69 (1H, d, J = 18.1), 2.41 (1H, ddd, J = 12.9, 12.1, 12.0), 2.17 (3H, s, acetate), 2.04 (1H, m), 1.59 (1H, m), 0.96 (3H, t, J = 7.4). ^{13}C-NMR: (176 MHz, CDCl$_3$): δ 171.9, 169.8, 83.7, 83.0, 79.3, 69.5, 49.8, 46.7, 46.1, 44.9, 37.9, 25.7, 21.0, 9.5. HRESIMS *m/z*: calcd for C$_{14}$H$_{19}$^{79}Br$_3$NaO$_5$ 526.8680 [M + Na]$^+$, found: 525.8671.

3.11. Mycalin B 2

KOH (excess) was added to a stirred solution of **1** (2.5 mg, 0.044 mmol) in MeOH (0.5 mL). After 10 min AcOH was added up to neutrality. The mixture was taken to dryness, the residue was partitioned between water and EtOAc and the organic phase was dried and evaporated. HPLC separation of the crude (eluent: *n*-hexane-EtOAc, 8:2) gave pure **2** (1.5 mg, 70%) as a colourless oil. **2**: For the $[\alpha]_D^{20}$, IR and MS (mass spectrometry) data see refs. 5 and 8. ^{1}H-NMR: (400 MHz, CDCl$_3$): δ 4.60 (1H, bd, J = 2.4), 4.55 (1H, bd, J = 2.1), 4.51 (1H, bd, J = 2.2), 4.39 (1H, dd, J = 8.9, 2.0), 4.24 (1H, bd, J = 8.9), 4.13 (1H, ddd, J = 12.7, 9.6, 4.4), 3.73 (1H, ddd, J = 12.2, 9.9, 4.5), 3.58 (1H, dd, J = 9.5, 2.0), 3.38 (1H, ddd, J = 9.5, 9.5, 2.3), 3.00 (1H, ddd, J = 12.8, 4.3, 4.3), 2.59 (1H, d, J = 2.1), 2.44 (1H, ddd, J = 12.3, 12.3, 12.3), 2.38 (1H, dd, J = 10.6, 2.9), 2.10 (1H, dd, J = 10.6, 2.3), 2.04 (1H, m), 1.52 (1H, m), 0.92 (3H, t, J = 7.3). ^{13}C-NMR: (100 MHz, CDCl$_3$): δ 86.0, 83.9, 81.0, 80.7, 79.5, 76.5, 76.1, 73.6, 56.1, 47.5, 46.5, 45.1, 36.6, 25.8, 9.8.

3.12. Cell Lines and Culture Conditions

The human cervical adenocarcinoma (HeLa), human breast adenocarcinoma (MCF-7) and human malignant melanoma (A375) cell lines were from ATCC (U.S.). The normal human dermal fibroblasts (HDF) were kindly provided by Dr. Annalisa Tito (Arterra, Biosciences). The cells were grown in DMEM (Dulbecco's Modified Eagle Medium) supplemented with 10% fetal bovine serum (FBS), 1% glutamine, 100 U/mL penicillin and 100 µg/mL streptomycin (Euroclone, Milan, Italy) and maintained in humidified air containing 5% CO$_2$, at 37 °C.

3.13. Cell Proliferation Assay

The cells were plated at a density of 1200 cells/well for A375, 1000 cells/well for HeLa and 2000 cells/well for HDF and MCF-7, in 96-well microplates (Corning). After 24 h incubation, the cells were treated with different concentrations of each of the compounds **1**, **2** and **4–11**, previously solubilized in DMSO (vehicle). The cell proliferation was determined after 48 h of treatment by using MTT (Sigma Aldrich, Milan, Italy) as previously reported. The plates were then analyzed by using a microplate reader (Enspire, Perkin Elmer, Cambridge, MA, USA) at 570 nm. The mean value ± SE of the adherent cells for each treatment was expressed as the relative percentage of the cell number with respect to the cells treated with the vehicle (control). Statistical differences were determined by the Student's test, paired, two-sided. All the experiments were performed at least in triplicate and repeated at least 3 times; a *p* value less than 0.05 was considered to be significant. The IC$_{50}$ values were obtained by the Prism 6.01 software (GraphPad San Diego, CA, U.S.) by extrapolating them from the dose–response curves data [18].

3.14. Apoptosis Experiments

The apoptosis assays were performed on the A375 cells seeded at a density of 1x10^5 cells/well in 6 well plates and incubated with compounds **1, 4, 5, and 11** at 37 °C and at 10 µM concentration for 48 h. The apoptosis was then analysed by staining with annexin V/FITC and propidium iodine (PI) (eBioscience). Briefly, after incubation, the cells were detached with accutase solution (eBioscience), harvested and washed with cold PBS [19]. Subsequently, the cells were treated following the manufacturer's instructions. The percentage of cell undergoing apoptosis or necrosis was quantified using a flow cytometer (Becton Dickinson, San Diego, CA, U.S.) equipped with the Cell Quest software version 3.5.1

4. Conclusions

In conclusion, in this study we have shown that mycalin A, a polybrominated C_{15} acetogenin of marine origin, is a strongly cytotoxic and pro-apoptotic substance on A375 tumor cells. The synthesis of some simplified analogues of this substance has been accomplished. It has been observed that the left-hand portion of mycalin A is essential for the biological activity since its perturbation, or its absence, strongly affects the antiproliferative activity, as the reduced cytotoxicity of mycalin B and the THP derivatives show. A simplified C4-lactone analogue of the mycalin A acetate has displayed a high cytotoxicity towards A375 malignant cells mediated by an apoptotic process, a result which appears very important considering that the tested A375 cells derive from a tumor with a negative prognosis. In addition, this substance has shown a reasonable selectivity with respect to healthy cells. A less pronounced activity on tumor cells is displayed by the 4-propyl-analogue of mycalin A acetate that did not show any selectivity. Studies are in progress to synthesize further simplified analogues of mycalin A and of its lactone derivative and to identify their biological target(s). Due to its high anti-proliferative and apoptotic activity, the acetate derivative of mycalin A (**4**), will also be the subject of further structural modifications with the aim of increasing its selectivity of action.

Supplementary Materials: The following are available online at http://www.mdpi.com/1660-3397/18/8/402/s1, Figure S1: ^{1}H NMR spectrum of compound **1** (CDCl$_3$, 400 MHz), Figure S2: ^{13}C NMR spectrum of compound **1** (CDCl$_3$, 100 MHz), Figure S3: ^{1}H NMR spectrum of compound **2** (CDCl$_3$, 400 MHz), Figure S4: ^{13}C NMR spectrum of compound **2** (CDCl$_3$, 100 MHz), Figure S5: ^{1}H NMR spectrum of compound **4** (CDCl$_3$, 400 MHz), Figure S6: ^{13}C NMR spectrum of compound **4** (CDCl$_3$, 100 MHz), Figure S7: ^{1}H NMR spectrum of compound **5** (CDCl$_3$, 400 MHz), Figure S8: ^{13}C NMR spectrum of compound **5** (CDCl$_3$, 100 MHz), Figure S9: ^{1}H NMR spectrum of compound **6** (CDCl$_3$, 400 MHz)., Figure S10: ^{13}C NMR spectrum of compound **6** (CDCl$_3$, 100 MHz), Figure S11: 2D COSY spectrum of compound **6** (CDCl$_3$, 700 MHz), Figure S12: 2D NOESY spectrum of compound **6** (CDCl$_3$, 700 MHz), Figure S13: 2D HSQC spectrum of compound **6** (CDCl$_3$, 700 MHz), Figure S14: 2D HMBC spectrum of compound **6** (CDCl$_3$, 700 MHz), Figure S15: ^{1}H NMR spectrum of compound **7** (CDCl$_3$, 400 MHz), Figure S16: ^{13}C NMR spectrum of compound **7** (CDCl$_3$, 100 MHz), Figure S17: ^{1}H NMR spectrum of compound **8** (CDCl$_3$, 400 MHz), Figure S18: ^{13}C NMR spectrum of compound **8** (CDCl$_3$, 100 MHz), Figure S19: 2D COSY spectrum of compound **8** (CDCl$_3$, 700 MHz), Figure S20: 2D NOESY spectrum of compound **8** (CDCl$_3$, 700 MHz), Figure S21: ^{1}H NMR spectrum of compound **9** (CDCl$_3$, 400 MHz), Figure S22: ^{13}C NMR spectrum of compound **9** (CDCl$_3$, 100 MHz), Figure S23: 2D COSY spectrum of compound **9** (CDCl$_3$, 700 MHz), Figure S24: 2D NOESY spectrum of compound **9** (CDCl$_3$, 700 MHz), Figure S25: 2D HSQC spectrum of compound **9** (CDCl$_3$, 700 MHz), Figure S26: 2D HMBC spectrum of compound **9** (CDCl$_3$, 700 MHz), Figure S27: ^{1}H NMR spectrum of compound **10** (CDCl$_3$, 400 MHz), Figure S28: ^{13}C NMR spectrum of compound **10** (CDCl$_3$, 100 MHz), Figure S29: 2D COSY spectrum of compound **10** (CDCl$_3$, 400 MHz), Figure S30: 2D NOESY spectrum of compound **10** (CDCl$_3$, 400 MHz), Figure S31: 2D HSQC spectrum of compound **10** (CDCl$_3$, 700 MHz), Figure S32: 2D HMBC spectrum of compound **10** (CDCl$_3$, 700 MHz), Figure S33: ^{1}H NMR spectrum of compound **11** (CDCl$_3$, 400 MHz), Figure S34: ^{13}C NMR spectrum of compound **11** (CDCl$_3$, 100 MHz), Figure S35: 2D COSY spectrum of compound **11** (CDCl$_3$, 700 MHz), Figure S36: 2D NOESY spectrum of compound **11** (CDCl$_3$, 700 MHz), Figure S37: 2D HSQC spectrum of compound **11** (CDCl$_3$, 700 MHz), Figure S38: 2D HMBC spectrum of compound **11** (CDCl$_3$, 700 MHz), Table S1: ^{1}H and ^{13}C chemical shift data for compound **6** (700 MHz, CDCl$_3$), Table S2: ^{1}H and ^{13}C chemical shift data for compound **8** (700 MHz, CDCl$_3$), Table S3: ^{1}H and ^{13}C chemical shift data for compound **9** (700 MHz, CDCl$_3$), Table S4: ^{1}H and ^{13}C chemical shift data for compound **10** (700 MHz, CDCl$_3$), Table S5: ^{1}H and ^{13}C chemical shift data for compound **11** (700 MHz, CDCl$_3$).

Author Contributions: Conceptualization, V.P., D.C. and S.D.G.; Investigation, V.P., D.C., S.D.G., N.B. and M.T.; Data curation, S.D.G., N.B. and M.T.; Writing—original draft preparation, V.P. and D.C.; Writing—review and editing, V.P., N.B. and D.C.; Supervision, V.P.; Funding acquisition, V.P., D.C. and S.D.G. All authors have read and agreed to the published version of the manuscript.

Funding: This research was funded by the Department of Chemical Sciences, University of Naples Federico II and by MIUR, Programma Operativo Nazionale Ricerca e Competitività 2007-2013, PON 01/02388.

Acknowledgments: The authors are grateful to Annalisa Tito (Arterra Biosciences, Naples, Italy) for providing the normal human dermal fibroblasts (HDF) and to Roberto Centore of the University of Naples Federico II for the helpful discussion on the crystallographic data reported for mycalin A.

Conflicts of Interest: The authors declare no conflict of interest.

Appendix A

† Since in the original manuscripts (see refs. 5 and 6) no common names were given to these compounds we propose here the common names mycalin A, B and C for compounds 1, 2 and 3, respectively.

‡ Comparison of the X-ray molecular structure and crystallographic data reported by Imre et al. for their compound 2 with those exhibited by mycalin A unambiguously shows that they are identical compounds. In particular, the molecular formula and the spatial group are the same, and the elemental cell parameters differ within 0.2%. However, their drawing incorrectly shows the relative stereochemistry of this compound and does not match the X-ray image they report. Similarly, the configuration of mycalin B (their compound 1) is incorrectly reported.

§ The configuration of mycalin A isolated from this alga is incorrectly reported. In particular, the configuration of the stereogenic centres C-3, C-9, C-10, C-12 and C-13 is inverted.

References

1. Wanke, T.; Philippus, A.C.; Zatelli, G.A.; Vieira, L.F.O.; Lhullier, C.; Falkenberg, M. C_{15} acetogenins from the *Laurencia* complex: 50 years of research-An overview. *Rev. Bras. De Farmacogn.* **2015**, *25*, 569–587. [CrossRef]
2. Zhou, Z.F.; Menna, M.; Cai, Y.S.; Guo, Y.W. Polyacetylenes of marine origin: Chemistry and bioactivity. *Chem. Rev.* **2015**, *115*, 1543–1596. [CrossRef] [PubMed]
3. Kigoshi, H.; Shizuri, Y.; Niwa, H.; Yamada, K. Four new C_{15} acetylenic polyenes of biogenetic significance from the red alga *Laurencia okamurai*: Structures and synthesis. *Tetrahedron* **1986**, *42*, 3781–3787. [CrossRef]
4. Perdikaris, S.; Mangoni, A.; Grauso, L.; Papazafiri, P.; Roussis, V.; Ioannou, E. Vagiallene, a rearranged C_{15} acetogenin from *Laurencia obtusa*. *Org. Lett* **2019**, *21*, 3183–3186. [CrossRef] [PubMed]
5. Giordano, F.; Mayol, L.; Notaro, G.; Piccialli, V.; Sica, D. Structure and absolute configuration of two new polybrominated C_{15} acetogenins from the sponge *Mycale rotalis*. *J. Chem. Soc. Chem. Commun.* **1990**, 1559–1561. [CrossRef]
6. Notaro, G.; Piccialli, V.; Sica, D.; Mayol, L.; Giordano, F. A Further C_{15} nonterpenoid polybromoether from the encrusting sponge *Mycale rotalis*. *J. Nat. Prod.* **1992**, *55*, 626–632. [CrossRef]
7. Rinehart, K.L.; Johnson, R.D.; Paul, I.C.; McMillan, J.A.; Siuda, J.F.; Krejcarek, G.E. Identification of compounds in selected marine organisms by gas chromatography-mass spectrometry, field desorption mass spectrometry, and other physical methods. In *Food-drugs from the sea*; Webber, H.H., Ruggieri, G.D., Eds.; Marine Technol. Soc.: Washington, DC, USA, 1976; pp. 434–442.
8. Imre, S.; Aydoğmuş, Z.; Güner, H.; Lotter, H.; Wagner, H. Polybrominated nonterpenoid C_{15} compounds from *Laurencia paniculata* and *Laurencia obtusa*. *Z. Für Naturforsch.* **1995**, *50C*, 743–747. [CrossRef]
9. Mihopoulos, N.; Vagias, C.; Scoullos, M.; Roussis, V. Laurencienyne B, a new acetylenic cyclic ether from the red alga *Laurencia obtusa*. *Nat. Prod. Lett.* **1999**, *13*, 151–156. [CrossRef]
10. Koutsaviti, A.; Daskalaki, M.G.; Agusti, S.; Kampranis, S.C.; Tsatsanis, C.; Duarte, C.M.; Roussis, V.; Ioannou, E. Thuwalallenes A–E and thuwalenynes A–C: New C_{15} acetogenins with anti-inflammatory activity from a Saudi Arabian red sea *Laurencia* sp. *Mar. Drugs* **2019**, *17*, 644. [CrossRef] [PubMed]
11. Huffman, B.J.; Shenvi, R.A. Natural products in the "marketplace": Interfacing synthesis and biology. *J. Am. Chem. Soc.* **2019**, *141*, 3332–3346. [CrossRef] [PubMed]
12. Caso, A.; Laurenzana, I.; Lamorte, D.; Trino, S.; Esposito, G.; Piccialli, V.; Costantino, V. Smenamide A analogues. Synthesis and biological activity on multiple myeloma cells. *Mar. Drugs* **2018**, *16*, 206. [CrossRef] [PubMed]
13. Caso, A.; Mangoni, A.; Piccialli, G.; Costantino, V.; Piccialli, V. Studies toward the synthesis of smenamide A, an antiproliferative metabolite from *Smenospongia aurea*: Total synthesis of *ent*-smenamide A and 16-*epi*-smenamide A. *Acs Omega* **2017**, *2*, 1477–1488. [CrossRef] [PubMed]
14. Piccialli, V.; Zaccaria, S.; Borbone, N.; Oliviero, G.; S.; Hemminki, A.; Cerullo, V.; Romano, V.; Tuzi, A.; Centore, R. Discovery of a novel one-step RuO_4-catalysed tandem oxidative polycyclization/double spiroketalization process. Access to a new type of polyether bis-spiroketal compound displaying antitumor activity. *Tetrahedron* **2010**, *66*, 9370–9378. [CrossRef]

15. Suzuki, M.; Koizumi, K.; Kikuchi, H.; Suzuki, T.; Kurosawa, E. Epilaurallene, a new nonterpenoid C_{15}-bromoallene from the red alga *Laurencia nipponica* Yamada. *Bull. Chem. Soc. Jpn.* **1983**, *56*, 715–718. [CrossRef]

16. Piccialli, V. A novel PCC-catalyzed process involving the oxidative cleavage of an α-bromomethyl-tetrahydrofuran bond. Synthesis of (2S,3R)-2-{(R)-bromo[(2R,3R,5S,6R)-3,5-dibromo-6-ethyltetrahydro-2H-pyran-2-yl]methyl}-5-oxotetrahydrofuran-3-yl benzoate. *Molbank* **2017**, M969. [CrossRef]

17. Okumura, A.; Kitani, M.; Murata, M. Kinetic studies of the catalytic oxygen exchange of chromate ions with water by periodate ions. *Bull. Chem. Soc. Jpn.* **1994**, *67*, 1522–1530. [CrossRef]

18. Comegna, D.; Zannetti, A.; Del Gatto, A.; De Paola, I.; Russo, L.; Di Gaetano, S.; Liguoro, A.; Capasso, D.; Saviano, M.; Zaccaro, L. Chemical modification for proteolytic stabilization of the selective $\alpha_v \beta_3$ integrin RGDechi peptide: In vitro and in vivo activities on malignant melanoma cells. *J. Med. Chem.* **2017**, *60*, 9874–9884. [CrossRef] [PubMed]

19. Capasso, D.; Di Gaetano, S.; Celentano, V.; Diana, D.; Festa, L.; Di Stasi, R.; De Rosa, L.; Fattorusso, R.; D'Andrea, L.D. Unveiling a VEGF-mimetic peptide sequence in the IQGAP1 protein. *Mol. Biosyst.* **2017**, *13*, 1619–1629. [CrossRef] [PubMed]

 *marine drugs* 

Article

12-Deacetyl-12-epi-Scalaradial, a Scalarane Sesterterpenoid from a Marine Sponge Hippospongia sp., Induces HeLa Cells Apoptosis via MAPK/ERK Pathway and Modulates Nuclear Receptor Nur77

Mi Zhou [1,†], Bo-Rong Peng [2,3,†], Wenjing Tian [1], Jui-Hsin Su [2,4], Guanghui Wang [1], Ting Lin [1], Dequan Zeng [1], Jyh-Horng Sheu [3,5,6,*] and Haifeng Chen [1,*]

[1] Fujian Provincial Key Laboratory of Innovative Drug Target, School of Pharmaceutical Sciences, Xiamen University, Xiamen 361005, China.; zmxmuer@163.com (M.Z.); tianwj@xmu.edu.cn (W.T.); guanghui@xmu.edu.cn (G.W.); linting@xmu.edu.cn (T.L.); xmu-zdq@126.com (D.Z.)

[2] National Museum of Marine Biology and Aquarium, Pingtung 944, Taiwan; pengpojung@gmail.com (B.-R.P.); x2219@nmmba.gov.tw (J.-H.S.)

[3] Department of marine Biotechnology and Resources, National Sun Yat-sen University, Kaohsiung 804, Taiwan

[4] Graduate Institute of Marine Biology, National Dong Hwa University, Pingtung 944, Taiwan

[5] Department of Medical Research, China Medical University Hospital, China Medical University, Taichung 404, Taiwan

[6] Institute of Natural Products, Kaohsiung Medical University, Kaohsiung 807, Taiwan

[*] Correspondence: sheu@mail.nsysu.edu.tw (J.-H.S); haifeng@xmu.edu.cn (H.C.)

[†] Equal contributions, first author.

Received: 5 June 2020; Accepted: 18 July 2020; Published: 21 July 2020

Abstract: 12-Deacetyl-12-*epi*-scalaradial, a scalarane sesterterpenoid from a marine sponge *Hippospongia sp*, has been reported to possess cytotoxic activity on HepG2, MCF-7, and HCT-116 cells. However, there is no research to indicate that 12-deacetyl-12-*epi*-scalaradial exhibited anticancer effect on cervical cancer HeLa cells. The aim of this study was to investigate the anticancer activity of 12-deacetyl-12-*epi*-scalaradial against HeLa cells and to explore the mechanism. The results from a methylthiazolyldiphenyl-tetrazolium (MTT) assay suggested that 12-deacetyl-12-*epi*-scalaradial suppressed the proliferation of HeLa cells and flow cytometry analysis showed 12-deacetyl-12-*epi*-scalaradial could induce the apoptosis of HeLa cells in dose- and time-dependent manner. Western blotting analysis demonstrated that 12-deacetyl-12-*epi*-scalaradial triggered apoptosis via mediating the extrinsic pathway and was found to suppress MAPK/ERK pathway which was associate with cancer cell death. Nur77, a critical number of orphan nuclear receptors, plays diverse roles in tumor development as a transcription factor and has been considered as a promising anticancer drug target. The dual-luciferase reporter assays suggested that 12-deacetyl-12-*epi*-scalaradial could selectively enhance the trans-activation activity of Nur77. Furthermore, Western blotting analysis and fluorescence quenching showed that 12-deacetyl-12-*epi*-scalaradial could induce the phosphorylation of Nur77 and interact with the ligand-binding domain (LBD) of Nur77. Our research confirmed 12-deacetyl-12-*epi*-scalaradial as a potential agent for cervical cancer therapy and provided a view that 12-deacetyl-12-*epi*-scalaradial may be a modulator of Nur77.

Keywords: 12-deacetyl-12-*epi*-scalaradial; HeLa cells; apoptosis; Nur77; MAPK/ERK pathway

1. Introduction

Cancer, a hyper-proliferative disorder, ranks as the primary cause of mortality worldwide. Cervical cancer, with an increasing annual incidence and mortality rate, is one of the most frequently diagnosed cancers among women globally [1]. Even though the combination of chemotherapy and radiotherapy has had a certain effect on treating cervical cancer, it was accompanied with many side effects [2]. Therefore, it is worth persistently identifying small candidate molecules for the treatment of cervical cancer.

Nur77 (also named TR3, NR4A1, or NGFI-B), an orphan member of the nuclear receptor superfamily, participates in numerous cellular process such as growth, survival, and apoptosis. The biological effects of Nur77 are regulated by many factors, including its modulators [3,4]. Many reports showed that the natural modulators of Nur77 could lead to apoptosis in certain cancer cells. Cytosporone B, a polyketone from the marine fungi *Dothiorella sp.*, was the first natural agonist of Nur77and induced apoptosis via modulating the transactivation of Nur77 [5]. In addition, several other compounds targeting Nur77 derived from nature or synthesis were also found to interact with Nur77 and mediate the apoptosis of cancer cells [6–10]. Thus, Nur77 can be considered as a promising anticancer drug target. It is also of great significance to continually discover modulators of Nur77 from natural products which contribute to the development of Nur77-targeting anticancer drugs.

Marine natural products represent an extraordinarily rich source of chemical and biological diversity with unique chemical skeletons and outstanding functionalities [11,12]. A large number of terpenoids possessing novel carbon skeletons and a wide variety of biological activities were found in marine organisms, which was different from those present in terrestrial species. Scalarane sesterterpenoids, one of the unique types of terpenoid, are exclusively derived from sponges and shell less molluscs. Many of these compounds have been reported to show significant antitumor activity in vitro or in vivo [13–18], while most of their mechanisms of action remain unclear. Moreover, 12-deacetyl-12-*epi*-scalaradial, a scalarane sesterterpenoid, has been reported to show growth inhibitory activity against hepatocellular carcinoma, breast adenocarcinoma cells, and colorectal carcinoma [19]. However, whether 12-deacetyl-12-*epi*-scalaradial exhibited cytotoxicity against cervical cancer cells were still unreported so far. Therefore, in this work, we evaluated the antitumor effect and mechanism of 12-deacetyl-12-*epi*-scalaradial in HeLa cells, and further examined whether 12-deacetyl-12-*epi*-scalaradial modulated the transactivation activity and phosphorylation of Nur77. Further, we investigated the interaction between 12-deacetyl-12-*epi*-scalaradial and the ligand-binding domain (LBD) of Nur77.

2. Results

2.1. 12-Deacetyl-12-epi-scalaradial Inhibits HeLa Cell Proliferation

Briefly, 12-deacetyl-12-*epi*-scalaradial was isolated from marine sponge *Hippospongia sp* and its chemical structure was identified by a comparison with the literature [19] (Figure 1A). An MTT assay was used to investigate whether 12-deacetyl-12-*epi*-scalaradial showed cytotoxic effects against human cervical cancer cells. As shown Figure 1B, the HeLa cell viability was dose-dependently inhibited with an increasing concentration of 12-deacetyl-12-*epi*-scalaradial. Meanwhile, 12-deacetyl-12-*epi*-scalaradial exhibited significant inhibitory effects on HeLa cells under the concentration of 30 µM. Moreover, the IC_{50} value of 12-deacetyl-12-epi-scalaradial against HeLa cells was 13.74 µM, which is lower than the IC_{50} values of other cell lines in the reports with 36 µM for MCF-7, 23.4 µM for HepG2, and 27.1 µM for HCT-1116 [19]. The above data suggested that 12-deacetyl-12-*epi*-scalaradial possessed cytotoxic activities against HeLa cells, while HeLa cells seem to be more sensitive to 12-deacetyl-12-*epi*-scalaradial.

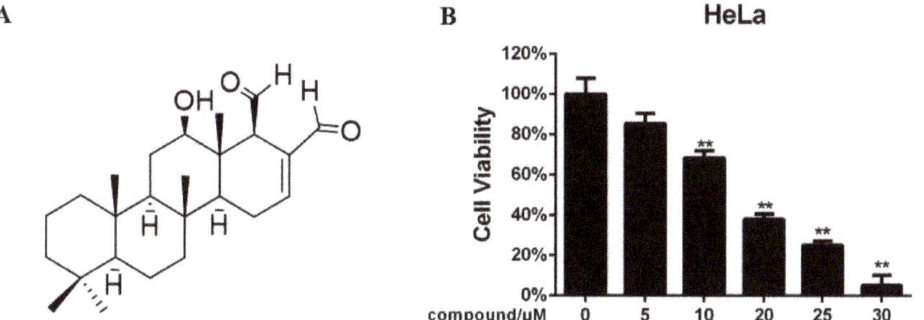

Figure 1. 12-Deacetyl-12-*epi*-scalaradial inhibits HeLa cell proliferation. (**A**) The structure of 12-deacetyl-12-*epi*-scalaradial. (**B**) HeLa cells were treated with various concentration of 12-deacetyl-12-*epi*-scalaradial (0, 5, 10, 20, 25, 30 μM) for 48 h. After incubation, cell viability of HeLa cells was measured by MTT assay. The values are the mean ± SD for four independent replicates. (** $p < 0.01$, compared with untreated cells).

2.2. 12-Deacetyl-12-epi-scalaradial Induces HeLa Cells Apoptosis

Programmed cell death, also named apoptosis, is a critical process for erasing unneeded or unhealthy cells [20]. In many cancers, apoptosis was blocked by the abnormal activation of pro-apoptotic proteins or inhibition of anti-apoptotic pathways. Therefore, apoptosis induction is associated with the antiproliferation of tumour cells [21]. To investigate the ability of 12-deacetyl-12-*epi*-scalaradial to induce HeLa cells apoptosis, the cells were pretreated with compound (30 μM), and apoptosis was measured by flow cytometry. Indeed, 12-deacetyl-12-*epi*-scalaradial displayed a robust inducement of apoptotic activity in HeLa cells. Figure 2A shows the percentage of Annexin V-stained HeLa cells was 5.71% for the control, but the apoptosis percentages increased to 40.3% and 51.6% after treatment of compound (30 μM) for 12 and 24 h, respectively. Furthermore, the percentage of apoptotic cells was increased from 4.80% to 61.4% with a concentration of 12-deacetyl-12-*epi*-scalaradial in the range of 0 to 30 μM (Figure 2B). Thus, 12-deacetyl-12-*epi*-scalaradial significantly induced HeLa cells apoptosis in a time- and dose-dependent manner.

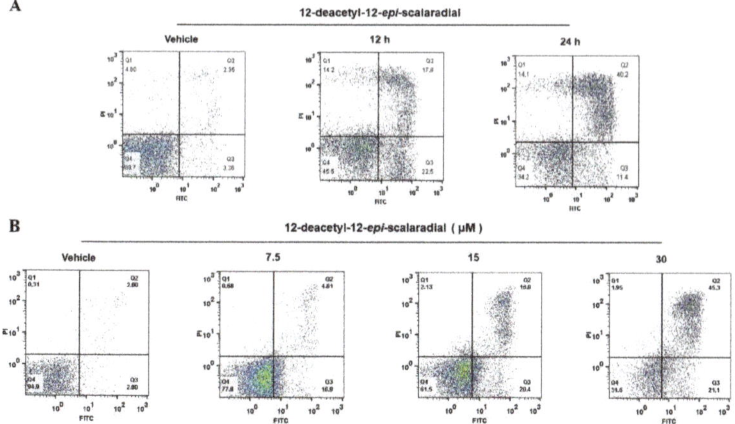

Figure 2. 12-Deacetyl-12-*epi*-scalaradial induces HeLa cells apoptosis. (**A**) Flow Cytometry analysis via Annexin V/PI staining was used to investigate apoptosis induced by 12-deacetyl-12-*epi*-scalaradial. HeLa cells were treated with 30 μM compound for 12 or 24 h; (**B**) HeLa cells were treated with different concentrations of 12-deacetyl-12-*epi*-scalaradial (0, 7.5, 15, 30 μM) for 24 h.

2.3. 12-Deacetyl-12-epi-scalaradial Induces PARP Cleavage and Activates Caspase Pathway in HeLa Cells

The presence of cleaved PARP is considered as one of the biomarkers for the detection of apoptosis in many cell lines [22–24]. In this study, Western blotting was used to identify PARP cleavage. As shown in Figure 3A, PARP cleavage was observed after HeLa cells were exposed to 30 µM compound for 6 h, which was more obvious when treated with compound for longer time. Dose response study showed that the PARP cleavage was stronger when treated with higher concentration of 12-deacetyl-12-*epi*-scalaradial for the same time (Figure 3B). These results were consistent with the flow cytometry analysis.

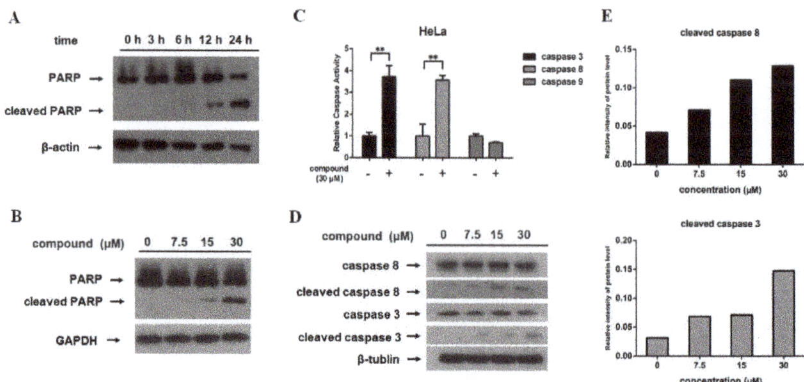

Figure 3. 12-Deacetyl-12-*epi*-scalaradial induces PARP cleavage and activates caspase pathway in HeLa cells. (**A**) HeLa cells treated with 12-deacetyl-12-*epi*-scalaradial (30 µM) for 3, 6, 12, 24 h were analyzed by western blotting. (**B**) HeLa cells were incubated with different concentrations of 12-deacetyl-12-*epi*-scalaradial (0, 7.5, 15, 30 µM) for 24 h. The expression level of PARP and cleaved PARP were detected through western blotting (**C**) After treated with 12-deacetyl-12-*epi*-scalaradial (30 µM) for 12 h, caspases activities were measured by Caspase-Glo assay kit (Promega). The values are the mean + SD for three independent replicates. (**D**) Expression levels of caspase 3, caspase 8, cleaved caspase 3, cleaved caspase 8 and β-tublin were measured using western blotting after treated with 12-deacetyl-12-*epi*-scalaradial. (**E**) Quantification of western blotting in (**D**).

Caspases are crucial mediators of cell apoptosis. In mammals, while apoptosis occurs, caspases are activated in a protease cascade that leads to the activation or disablement of key structural proteins, signal pathways, and homeostatic and repair enzymes [25]. In order to investigate whether the caspases pathway contributed to the apoptotic effects of 12-deacetyl-12-*epi*-scalaradial, the Caspase-Glo assay kit was used to measure the caspases activity in HeLa cells with compound treatment. Figure 3C showed that the activities of caspase 3 and caspase 8 were increased after HeLa cells was exposed to 12-deacetyl-12-*epi*-scalaradial, while caspase 9 was not activated. As is known to all, activated caspase 8 is not only cleaved but also induce the cleavage of downstream effector caspase 3 [26]. Western blotting and relative intensity analysis further confirmed that the protein level of cleaved caspase 8 was increased dose dependently after treatment with 12-deacetyl-12-*epi*-scalaradial in HeLa cells, and cleaved caspase 3 was observed when HeLa cells were exposed to the compound in 30 µM (Figure 3D,E). These results were consistent with the analysis for cleaved PARP. The above data indicated that 12-deacetyl-12-*epi*-scalaradial induced PARP cleavage in a time- and concentration-dependent manner and it may induce apoptosis via the activation of caspase in human cervical cancer HeLa cells.

2.4. 12-Deacetyl-12-epi-scalaradial Suppresses MAPK/ERK Pathway

The mitogen-activated protein kinase (MAPK) pathway plays crucial roles in various cellular functions, including growth, differentiation, and metastasis. MAPKs consists of three subfamilies including the extracellular signal-regulated kinase (ERK), c-Jun N-terminal kinase (JNK), and p38 kinase. The MAPK signaling pathway was closely involved in tumor cell apoptosis [27,28]. PI3K/Akt pathway is also associated with physiological process including survival and apoptosis in cancer cells [29]. In this work, the effects of 12-deacetyl-12-*epi*-scalaradial on proteins involved in the MAPK and PI3K/Akt pathway were investigated through western blotting assay. As shown in Figure 4A,B, the expression level of phosphorylation of ERK was reduced after treatment with 12-deacetyl-12-*epi*-scalaradial in a dose-dependent manner, while the expression level of p-JNK, p-p38, and p-Akt showed no difference. The above data demonstrated that 12-deacetyl-12-*epi*-scalaradial may suppress the MAPK/ERK pathway in HeLa cells.

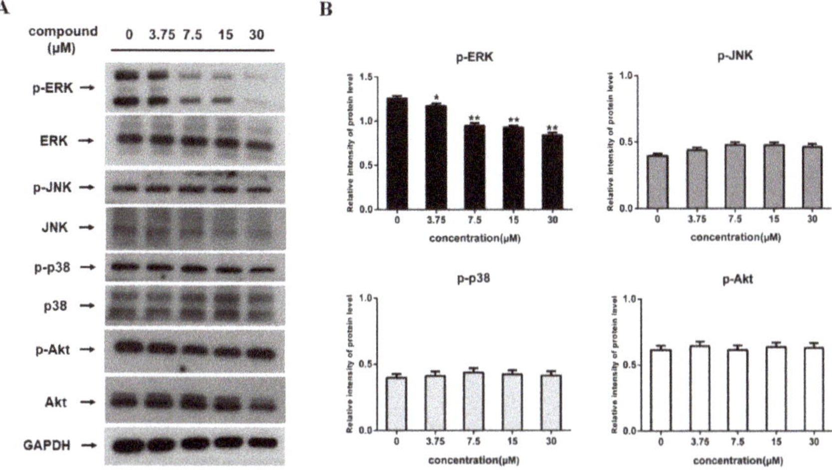

Figure 4. 12-Deacetyl-12-*epi*-scalaradial suppresses MAPK/ERK pathway. (**A**) The expression levels of ERK, p-ERK, JNK, p-JNK, p38, p-p38, AKT, p-AKT and GAPDH were analyzed by western blotting. (**B**) Quantification of p-ERK, p-JNK, p-p38 and p-AKT in (A) (* $p < 0.05$, ** $p < 0.01$, compared with untreated cells).

2.5. 12-Deacetyl-12-epi-scalaradial Modulates Trans-Activation Activity and Phosphorylation of Nur77, and Interacts with Nur77-LBD

Nur77, an important orphan member of the nuclear receptor superfamily, mediates survival or apoptosis in cancer cells. Some reports had shown that several signaling pathways, including the MAPK/ERK pathway, contributed to cell apoptosis through regulating the trans-activation function of Nur77 [30,31]. Meanwhile, a study showed that natural terpenoids exhibited antitumor activities via modulating the function of Nur77 [32]. Therefore, a dual-luciferase reporter gene assay, a method to identify small modulators of Nur77 [5,33–35], was used to evaluate the effect of 12-deacetyl-12-*epi*-scalaradial on trans-activation of Nur77. As shown in Figure 5A, the ascension of relative Nur77 luciferase reporter-gene activity was observed with increasing concentration of 12-deacetyl-12-*epi*-scalaradial. Our previous studies as well used dual-luciferase reporter gene assay to identify small modulators of Retinoid X Receptor-alpha (RXRα) [36–39], which is also an important nuclear receptor. Thus, to explore whether 12-deacetyl-12-*epi*-scalaradial selectively affected trans-activation activity of Nur77, the effect of compound on trans-activation activity of RXRα was measured. As a result, 12-deacetyl-12-*epi*-scalaradial has no influence on the trans-activation activity of RXRα with the same concentration (Figure 5B). Furthermore, the deactivation of MAPK/ERK pathway was reported to contribute to the phosphorylation of Nur77-mediated apoptosis [30,31]. Therefore, Western blotting was used to analyze the level of phosphorylation of Nur77 after treatment with

12-deacetyl-12-*epi*-scalaradial. Figure 5C exhibited that the expression level of phosphorylation of Nur77 increased clearly when exposed to high concentration of compound.

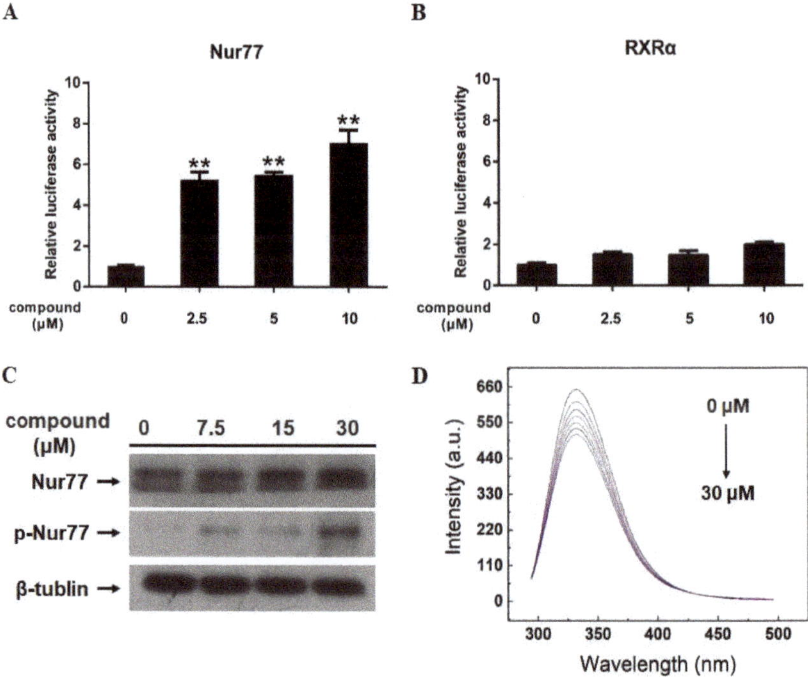

Figure 5. 12-Deacetyl-12-*epi*-scalaradial stimulates trans-activation activity of Nur77 and interacts with Nur77-LBD. (**A**) After treated with different concentrations of 12-deacetyl-12-*epi*-scalaradial (0, 2.5, 5, 10 μM) for 12 h, the relative luciferase reporter-gene activity for Nur77 and RXRα were measured by dual-luciferase reporter gene assay. (**B**) After treated with different concentrations of 12-deacetyl-12-epi-scalaradial (0, 7.5, 15, 30 μM) for 12 h, the expression levels of Nur77, p-Nur77 and GAPDH were analysed by western blotting. (**C**) The expression levels of Nur77, p-Nur77 and β-tublin were analyzed by western blotting. (**D**) After treated with 12-deacetyl-12-*epi*-scalaradial (0–30 μM), the interacting affinity of compound toward Nur77-LBD was measured via fluorescence quenching technology at 298K.

The phenyl alanine, tyrosine, and tryptophan in proteins are the donors of the fluorescence and the interaction between ligands and proteins will lead to the quenching of fluorescence. Thus, fluorescence quenching technology has been an effective way to study the binding of ligand to the target protein [40–42] and has been successfully used to investigate the interactions between natural modulators and the Nur77 [5,43]. In this study, the Nur77-LBD (ligand-binding domain) protein was purified and its interaction with 12-deacetyl-12-*epi*-scalaradial was measured by fluorescence quenching technology. As shown in Figure 5D, the fluorescence of Nur77-LBD protein was quenched in a dose-dependent manner after 12-deacetyl-12-*epi*-scalaradial treatment. The above results suggested that 12-deacetyl-12-*epi*-scalaradial could modulate trans-activation activity and phosphorylation of Nur77, and interact with Nur77-LBD.

3. Discussion

Cervical cancer is a growing worldwide health problem for women. There are several therapies for its treatment, such as surgery, radiotherapy and chemotherapy, but side effects remain a rigorous problem [1,2]. Therefore, it is urgent to develop effective drug for the treatment of cervical cancer. The marine natural products have shown extensive antitumor activity, and continuous attention has been drawn to discover drug candidates from marine organisms [44]. Scalarane sesterterpenoids, a unique type of terpenoids deriving from marine organisms, exhibit significant potential in the inhibition of cancer cells proliferation. Hyatelactam, a scalarane sesterterpenoid isolated from *Hyatella intestinalis,* showed inhibitory activity against human colon cancer HT-29 cells [16]. Another scalarane sesterterpenoid, hippospongide B, exhibited cytotoxicities against HCT-116, T-47D, K562, and DLD-1 tumor cells [17]. Heteronemin, also a scalarane sesterterpenoid from *Hyrtios sp.,* could induce the apoptosis of cancer cells via modulating the NF-κB and MAPKs signal pathways [4,45]. Even though scalarane sesterterpenoids are considered as a rich source of cancer therapeutic agents, their anticancer mechanism still needs to be investigated.

Briefly, 12-deacetyl-12-*epi*-scalaradial is a scalarane sesterterpenoid isolated from a marine sponge *Hippspongia sp,* which has been reported to inhibit the proliferation of several cancer cell lines containing HepG2, MCF-7 and HCT-116 [19]. However, the effects of 12-deacetyl-12-*epi*-scalaradial towards cervical cancer HeLa cells and its possible mechanism of antitumor activity remain largely unknown. Our results confirmed 12-deacetyl-12-*epi*-scalaradial inhibited the proliferation of HeLa cells (Figure 1B) with the IC_{50} value of 13.74 μM, which is lower than that of the other cell lines. Furthermore, we demonstrated that 12-deacetyl-12-*epi*-scalaradial could induce the apoptosis of HeLa cells (Figures 2 and 3A,B). This further certificated that scalarane sesterterpenoids possessed broad antitumor activity, and showed promise as a leading cervical cancer therapeutic compound.

As is well known, the caspase pathway plays an essential role in cell apoptosis. Caspase enzymes family contains two major types of members, one type of apoptotic caspases is initiator caspases, which is activated through recruiting to signaling complexes as well as providing a link between cell signal pathway and apoptotic execution, such as caspase 8 and caspase 9. Caspase-8 is mediator of extrinsic apoptosis pathway, and the intrinsic apoptosis pathway leads to activation of caspase-9 [46]. Another type of caspases, named effector caspases, are activated by initiator caspases and most of the cellular substrates are cleaved by them. Among effector caspases, caspase-3 is the most frequently activated death protease in mammals, and poly (ADP-ribose) polymerase (PARP) is one of the downstream effectors of caspase 3 [47]. A CaspACE Assay System and Western blotting suggested that 12-deacetyl-12-*epi*-scalaradial simultaneously activated caspase 8 and caspase 3 enzymes (Figure 3C–E) to induce apoptosis of HeLa cells through extrinsic pathway. The MAPK and PI3K/Akt pathways are also important pathways associated with cancer cells apoptosis. The extracellular signal-regulated kinase (ERK), p38, and Jun N-terminal kinase (JNK) are three major subfamilies of the MAPK pathway. Among them, ERK and Akt signaling mediate the pro-survival effects of cancer cells, whereas the activation of JNK kinase and p38 kinase are involved in pro-apoptosis progress in cancer cells [48–50]. The data from this work showed that 12-deacetyl-12-*epi*-scalaradial induced the suppression of phosphorylation of ERK, contributed to the inhibitory effect on proliferation of HeLa cells (Figure 4A,B), and further confirmed that MAPK/ERK and caspase pathways played crucial roles in tumour cell apoptosis. Our result enriched the mechanism information concerning how scalarane sesterterpenoids exert antitumor effects.

The extrinsic pathway causes apoptosis through transmembrane receptor-mediated interactions including FasL and TRAIL. The interaction between FasL and TRAIL recruits cytoplasmic adapter proteins and caspase 8, resulting in the oligomerization and activation of caspase 8 that leads to apoptosis [51]. Several studies showed that FasL and TRAIL were promoted to enhance cancer cells apoptosis via Nur77 activated by DIM-C-pPhOCH$_3$ compound [34,52], that indicated Nur77 could modulate extrinsic pathway to induce tumor cells apoptosis. Moreover, as a critical member of nuclear receptor superfamily, Nur77 is linked to various biological signaling pathways including MAPK and PI3K/Akt pathways in cancer cells. ERK pathway has been reported to depress Nur77-mediated apoptosis, and ERK inhibitor enhanced Nur77 to increase caspase 3 activity and induce apoptosis in HepG2 cells [31]. Our research indicated that 12-deacetyl-12-*epi*-scalaradial could selectively stimulate the trans-activation of Nur77 (Figure 5A,B). Furthermore, the phosphorylation of Nur77 is associated with MAPK/ERK pathway [30,31]. Our results showed that 12-deacetyl-12-*epi*-scalaradial suppressed the activation of MAPK/ERK pathway (Figure 4A,B) and induced the phosphorylation of Nur77 (Figure 5C). Interestingly, p38 and JNK pathways are also important signaling pathways for Nur77-modulated apoptosis. The phosphorylation of p38 and JNK enhance expression and translocation of Nur77, that lead to occurrence of cancer cells apoptosis [8,9,53]. The activation of PI3K/Akt pathway inhibits Nur77-induced apoptosis in cancer cells, respectively [54,55]. However, in this study, 12-deacetyl-12-*epi*-scalaradial was found to have no effects on JNK, p38, and Akt pathways. Previous reports and this work made us speculate that 12-deacetyl-12-*epi*-scalaradial may induce HeLa cell apoptosis by modulating Nur77 to mediate caspase and MAPK/ERK pathways. We also used fluorescence quenching technology to investigate interaction between 12-deacetyl-12-*epi*-scalaradial and Nur77-LBD (Figure 5D). However, regrettably, due to the insufficient amounts of compounds, we couldn't further investigate the specificity of 12-deacetyl-12-*epi*-scalaradial binding to Nur77. These data showed the possibility of scalarane sesterterpenoids being a small modulator of Nur77 and provided a new direction for discovering natural modulators of Nur77.

In conclusion, our results demonstrated that 12-deacetyl-12-*epi*-scalaradial, which is a scalarane sesterterpenoid, possessed anticancer activity against human cervical cancer HeLa cells via inducing apoptosis. Additionally, our data suggested that 12-deacetyl-12-*epi*-scalaradial may induce apoptosis through extrinsic pathway and MAPK/ERK pathway. Furthermore, 12-deacetyl-12-*epi*-scalaradial modulated the trans-activation activity and phosphorylation of Nur77 and interacted with Nur77-LBD. Our finding provides evidence about 12-deacetyl-12-*epi*-scalaradial as a potential agent of cervical cancer therapy and a small modulator of Nur77 for the first time.

4. Materials and Methods

4.1. Chemicals and Reagents

HeLa cell line was purchased from the China Cell Bank of the Institute of Biochemistry and Cell Biology in Shanghai, China. Methylthiazolyldiphenyl-tetrazolium bromide (MTT) was purchased from Aladdin, China. Annexin V-FITC Apoptosis Detection Kits was purchased from Vazyme, China. Caspase-Glo assay kit was purchased from Promega, USA.

4.2. Isolation of Natural Products

Briefly, 12-deacetyl-12-*epi*-scalaradial was isolated from *Hippospongia sp.* The fresh sponge (1.2 kg wet weight) was extracted with EtOAc. The crude extract fraction (15.3 g) was separated on silica gel and eluted with using the mixtures of n-hexane/ EtOAc to obtain fractions 1-13. Purification from fraction 9 (5.4551 g) through repeated preparative TLC with n-Hexane–Acetone (10:1) afforded 12-deacetyl-12-*epi*-scalaradial.

4.3. Cell Cultures

HeLa cells were cultured in Eagle's Minimum Essential Medium (MEM) supplemented with 10% FBS in a humidified atmosphere containing 5% CO_2 at 37 °C.

4.4. Cytotoxic Activity

Anti-proliferative activity of compound was tested by MTT assay. HeLa cells were seeded into 96-well-plate at a density of 5×10^3 cells per well and allowed to settle 12 h. Then, the cells were treated with varied concentrations of 12-deacetyl-12-*epi*-scalaradial. After 48 h, 15 µL of MTT reagent and 60 µL MEM were added and incubated for 4 h. After removing the supernatant, the transformed crystals were dissolved in DMSO (100 µL) and measured at 490 nm using microplate reader (Thermo

Multiskan MK3, Thermo Scientific, Helsinki, Finland). The cell proliferation-inhibition rate (%) was calculated as follows:

$$\text{Growth Rate (\%)} = [(\text{ODsample-ODblank})/(\text{ODcontrol-ODblank})] \times 100\%$$

4.5. Flow Cytometry Analysis

The apoptotic rate of HeLa cells was assessed by flow cytometry using the Annexin-V/FITC Apoptosis Detection Kit according to manufacturer's protocol. Cells were seeded into 6-well plates at a density of 1×10^5 cells per well for 24 h. Cells were digested by trypsin with no EDTA and collected by centrifugation and resuspended in 100 µL of Binding buffer (1×) after 12-deacetyl-12-*epi*-scalaradial treatment. Finally, HeLa cells were stained with Annexin-V/FITC and PI for 10 min at room temperature in the dark. The stained cells were diluted with 1× binding buffer and analyzed by flow cytometry. Data were analyzed with FlowJo 10.

4.6. Western Blot Analysis

After treated with 12-deacetyl-12-*epi*-scalaradial, HeLa cells were lysed in RIPA buffer. Equal amounts of protein lysates were separated by SDS-PAGE and transferred onto polyvinylidene difluoride membranes. The membranes were blocked with 5% no-fat milk in TBST for 1 h, then incubated with primary antibodies at 4 °C overnight. After washing three times with TBST, the membranes were detected with secondary antibodies for 1 h under room temperature. The final immunoreactive products were detected by using an enhanced chemiluminescence system and examined by densitometric analysis.

4.7. Caspase Activation Activity

Caspase-3, -8, and -9 activities were measured using a Caspase-Glo assay kit (Promega) according to manufacturer's protocol. HeLa cells were seeded in 96 well-culture dish and treated with 12-deacetyl-12-*epi*-scalaradial (30 µM). After 12 h, 100 µl Caspase-Glo Reagent were added to each well 12 h later. After 1 h incubation, the cells were measure the luminescence of each sample in a plate-reading luminometer as directed by the luminometer manufacturer.

4.8. Dual-Luciferase Reporter Assay

The 293T cells (293T) were cultured at 37 °C in DMEM with 10% FBS for 24 h and then seeded at a density of 1×10^4 cells per well in 96-well plates. Three plasmids, namely pGL5-NURE luciferase reporter vector (15 ng/mL), myc-Nur77 vector (15 ng/mL), and pC DNA-Renila (15 ng/mL), were co-transfected by Liposome 2000. As to RXRα transcriptional activities, there are two plasmids, pGL5 luciferase reporter vector (30 ng/mL) and pGAL4-RXRα-LBD vector (15 ng/mL). After 12 h, cells were incubated with the compounds at different concentrations for 12 h. The activities of firefly luciferase and renilla luciferase were measured using the Dual-Luciferase Assay System Kit (Promega).

4.9. Nur77-LBD Protein Purified

The ligand-binding domain (LBD) of human Nur77 (genes 367–598) was cloned as N-terminal histidine-tagged fusion protein in pET15b vector and transformed to E. coli BL21(DE3). Cells were grown at 37 °C in LB medium until $\text{OD}_{600} = 0.6$–0.8 and then protein expression was induced by 1 mM IPTG. After sustained at 25 °C for 16 h, cells were harvested and sonicated. The target protein was purified using Ni^{2+}-NTA agarose column at low temperatures.

4.10. Fluorescence Quench Analysis

The fluorescence experiment was performed using a Cary Eclipse Fluorescence Spectrophotometer (Varian). Fluorescence spectra (300–500 nm) were conducted via excitation wavelength of 289 nm and slit widths to 10/10 nm at 289T.

4.11. Statistical Analysis

The values are presented as mean ± SD and were analyzed statistically with student's t-tests for simple comparisons between groups, one-way analyses of variance (ANOVA) for more than two groups, and Tukey's multiple comparisons test using GraphPad Prism 6.0. Differences were considered statistically significant at $p < 0.05$.

Author Contributions: Conceptualization, M.Z.; Data curation, M.Z.; Funding acquisition, J.-H.S. (Jyh-Horng Sheu) and H.C.; Investigation, M.Z. and B.-R.P.; Project administration, H.C.; Resources, J.-H.S. (Jui-Hsin Su), T.L., D.Z., and G.W.; Supervision, H.C.; Validtion, M.Z.; Visualization, M.Z.; Writing—original draft, M.Z.; Writing—review & editing, W.T. and H.C. All authors have read and agreed to the published version of the manuscript.

Funding: This research was funded by Fundamental Research Funds for the Central University (20720160117 and 20720190079), National Natural Science Foundation of China (No. U1605221).

Acknowledgments: The authors would like to thank the Core Facility of Biomedical of Xiamen University for letting us use the instruments and equipment.

Conflicts of Interest: The authors declare no conflict of interest.

References

1. Small, W.; Bacon, M.A.; Bajaj, A.; Chuang, L.T.; Fisher, B.J.; Harkenrider, M.M.; Jhingran, A.; Kitchener, H.C.; Mileshkin, L.R.; Viswanathan, A.N.; et al. Cervical cancer: A global health crisis. *Cancer* **2017**, *123*, 2404–2412. [CrossRef] [PubMed]

2. Bahrami, A.; Hasanzadeh, M.; Hassanian, S.M.; ShahidSales, S.; Ghayour-Mobarhan, M.; Ferns, G.A.; Avan, A. The potential value of the PI3K/Akt/mTOR signaling pathway for assessing prognosis in cervical cancer and as a target for therapy. *J. Cell. Biochem.* **2017**, *118*, 4163–4169. [CrossRef] [PubMed]

3. Mohan, H.M.; Aherne, C.M.; Rogers, A.C.; Baird, A.W.; Winter, D.C.; Murphy, E.P. Molecular pathways: The role of NR4A orphan nuclear receptors in cancer. *Clin. Cancer Res.* **2012**, *18*, 3223–3228. [CrossRef] [PubMed]

4. Moll, U.M.; Marchenko, N.; Zhang, X. p53 and Nur77/TR3—transcription factors that directly target mitochondria for cell death induction. *Oncogene* **2006**, *25*, 4725–4743. [CrossRef]

5. Zhan, Y.; Du, X.; Chen, H.; Liu, J.; Zhao, B.; Huang, D.; Li, G.; Xu, Q.; Zhang, M.; Weimer, B.C.; et al. Cytosporone B is an agonist for nuclear orphan receptor Nur77. *Nat. Chem. Biol.* **2008**, *4*, 548–556. [CrossRef]

6. Kolluri, S.K.; Bruey-Sedano, N.; Cao, X.; Lin, B.; Lin, F.; Han, Y.-H.; Dawson, M.I.; Zhang, X.-K. Mitogenic effect of orphan receptor TR3 and its Regulation by MEKK1 in lung cancer cells. *Mol. Cell. Biol.* **2003**, *23*, 8651–8667. [CrossRef]

7. Li, H. Cytochrome c release and apoptosis induced by mitochondrial targeting of nuclear orphan receptor TR3. *Science* **2000**, *289*, 1159–1164. [CrossRef]

8. Thigpen, T. The role of chemotherapy in the management of carcinoma of the cervix. *Cancer J.* **2003**, *9*, 425–432. [CrossRef]

9. Zhou, Y.; Zhao, W.; Xie, G.; Huang, M.; Hu, M.; Jiang, X.; Zeng, D.; Liu, J.; Zhou, H.; Chen, H.; et al. Induction of Nur77-dependent apoptotic pathway by a coumarin derivative through activation of JNK and p38 MAPK. *Carcinogenesis* **2014**, *35*, 2660–2669. [CrossRef]

10. Jeong, J.H.; Park, J.-S.; Moon, B.; Kim, M.C.; Kim, J.-K.; Lee, S.; Suh, H.; Kim, N.D.; Kim, J.-M.; Park, Y.C.; et al. Orphan nuclear receptor Nur77 translocates to mitochondria in the early phase of apoptosis induced by synthetic chenodeoxycholic acid derivatives in human stomach cancer cell line SNU-1. *Ann. N. Y. Acad. Sci.* **2003**, *1010*, 171–177. [CrossRef]

11. Tarhouni-Jabberi, S.; Zakraoui, O.; Ioannou, E.; Riahi-Chebbi, I.; Haoues, M.; Roussis, V.; Kharrat, R.; Essafi-Benkhadir, K. Mertensene, a halogenated monoterpene, induces G2/M cell cycle arrest and caspase dependent apoptosis of human colon adenocarcinoma HT29 cell line through the modulation of ERK-1/-2, AKT and NF-κB signaling. *Mar. Drugs* **2017**, *15*, 221. [CrossRef]

12. Molinski, T.F.; Dalisay, D.S.; Lievens, S.L.; Saludes, J.P. Drug development from marine natural products. *Nat. Rev. Drug Discov.* **2009**, *8*, 69–85. [CrossRef]

13. Roy, M.C.; Tanaka, J.; de Voogd, N.; Higa, T. New scalarane class sesterterpenes from an indonesian sponge, phyllospongia sp. *J. Nat. Prod.* **2002**, *65*, 1838–1842. [CrossRef]

14. Chang, Y.-C.; Tseng, S.-W.; Liu, L.-L.; Chou, Y.; Ho, Y.-S.; Lu, M.-C.; Su, J.-H. Cytotoxic sesterterpenoids from a sponge hippospongia sp. *Mar. Drugs* **2012**, *10*, 987. [CrossRef] [PubMed]

15. Hahn, D.; Won, D.H.; Mun, B.; Kim, H.; Han, C.; Wang, W.; Chun, T.; Park, S.; Yoon, D.; Choi, H.; et al. Cytotoxic scalarane sesterterpenes from a Korean marine sponge psammocinia sp. *Bioorg. Med. Chem. Lett.* **2013**, *23*, 2336–2339. [CrossRef] [PubMed]

16. Renner, M.K.; Jensen, P.R.; Fenical, W. ChemInform abstract: Mangicols: Structures and biosynthesis of a new class of sesterterpene polyols from a marine fungus of the genus fusarium. *ChemInform* **2000**, *31*. [CrossRef]

17. Cassiano, C.; Esposito, R.; Tosco, A.; Zampella, A.; D'Auria, M.V.; Riccio, R.; Casapullo, A.; Monti, M.C. Heteronemin, a marine sponge terpenoid, targets TDP-43, a key factor in several neurodegenerative disorders. *Chem. Commun.* **2014**, *50*, 406–408. [CrossRef]

18. Aoki, S.; Higuchi, K.; Isozumi, N.; Matsui, K.; Miyamoto, Y.; Itoh, N.; Tanaka, K.; Kobayashi, M. Differentiation in chronic myelogenous leukemia cell K562 by spongean sesterterpene. *Biochem. Biophys. Res. Commun.* **2001**, *282*, 426–431. [CrossRef]

19. Elhady, S.; El-Halawany, A.; Alahdal, A.; Hassanean, H.; Ahmed, S. A new bioactive metabolite isolated from the red sea marine sponge hyrtios erectus. *Molecules* **2016**, *21*, 82. [CrossRef]

20. Danial, N.N.; Korsmeyer, S.J. Cell death: Critical control points. *Cell* **2004**, *16*, 205–219. [CrossRef]

21. Fesik, S.W. Promoting apoptosis as a strategy for cancer drug discovery. *Nat. Rev. Cancer.* **2005**, *5*, 876–885. [CrossRef] [PubMed]

22. Duriez, P.J.; Shah, G.M. Cleavage of poly (ADP-ribose) polymerase: A sensitive parameter to study cell death. *Biochem. Cell Biol.* **1997**, *75*, 337–349. [CrossRef] [PubMed]

23. Oliver, F.J.; de la Rubia, G.; Rolli, V.; Ruiz-Ruiz, M.C.; de Murcia, G.; Murcia, J.M. Importance of Poly (ADP-ribose) polymerase and its cleavage in apoptosis. *J. Biol. Chem.* **1998**, *273*, 33533–33539. [CrossRef] [PubMed]

24. Scovassi, C.S.A.I. Poly (adp-ribose) polymerase-1 cleavage during apoptosis. *Apoptosis Int. J. Program. Cell Death.* **2002**, *7*, 321. [CrossRef]

25. Porter, A.G.; Jänicke, R.U. Emerging roles of caspase-3 in apoptosis. *Cell Death Differ.* **1999**, *6*, 99–104. [CrossRef] [PubMed]

26. Weinlich, R.; Oberst, A.; Beere, H.M.; Green, D.R. Necroptosis in development, inflammation and disease. *Nat. Rev. Mol. Cell Biol.* **2017**, *18*, 127–136. [CrossRef]

27. Seger, R.; Wexler, S. The MAPK signaling cascades. *Faseb J.* **1995**, *9*, 726–735. [CrossRef]

28. Fang, J.Y.; Richardson, B.C. The MAPK signalling pathways and colorectal cancer. *Lancet Oncol.* **2005**, *6*, 322–327. [CrossRef]

29. Fresno, V.; Casado, E.; de Castro, J.; Cejas, P.; Belda, C.; Gonzalez, M. PI3K/Akt signalling pathway and cancer. *Cancer Treat. Rev.* **2004**, *30*, 193–204. [CrossRef]

30. Wang, A.; Rud, J.; Olson, C.M.; Anguita, J.; Osborne, B.A. Phosphorylation of Nur77 by the MEK-ERK-RSK cascade induces mitochondrial translocation and apoptosis in T Cells. *J. Immunol.* **2009**, *183*, 3268–3277. [CrossRef]

31. Yang, H.; Nie, Y.; Li, Y.; Wan, Y.-J.Y. ERK1/2 deactivation enhances cytoplasmic Nur77 expression level and improves the apoptotic effect of fenretinide in human liver cancer cells. *Biochem. Pharmacol.* **2011**, *81*, 910–916. [CrossRef] [PubMed]

32. Hu, M.; Luo, Q.; Alitongbieke, G.; Chong, S.; Xu, C.; Xie, L.; Chen, X.; Zhang, D.; Zhou, Y.; Wang, Z.; et al. Celastrol-Induced Nur77 interaction with TRAF2 alleviates inflammation by promoting mitochondrial ubiquitination and autophagy. *Mol. Cell.* **2017**, *66*, 141–153. [CrossRef]

33. Liu, J.; Zeng, H.; Zhang, L.; Zhan, Y.; Chen, Y.; Wang, Y.; Wang, J.; Xiang, S.; Liu, W.; Wang, W.; et al. A unique pharmacophore for activation of the nuclear orphan receptor Nur77 in vivo and in vitro. *Cancer Res.* **2010**, *70*, 3628–3637. [CrossRef] [PubMed]

34. Qiu, D.; Zhou, M.; Lin, T.; Chen, J.; Wang, G.; Huang, Y.; Jiang, X.; Tian, W.; Chen, H. Cytotoxic components from hypericum elodeoides targeting RXRα and inducing HeLa cell apoptosis through Caspase-8 activation and PARP cleavage. *J. Nat. Prod.* **2019**, *82*, 1072–1080. [CrossRef] [PubMed]

35. Lakshmi, S.P.; Reddy, A.T.; Banno, A.; Reddy, R.C. Molecular, chemical, and structural characterization of prostaglandin A2 as a novel agonist for Nur77. *Biochem. J.* **2019**, *496*, 2757–2767. [CrossRef] [PubMed]

36. Tian, W.-J.; Qiu, Y.-Q.; Yao, X.-J.; Chen, H.-F.; Dai, Y.; Zhang, X.-K.; Yao, X.-S. Dioxasampsones A and B, two polycyclic polyprenylated acylphloroglucinols with unusual epoxy-ring-fused skeleton from Hypericum sampsonii. *Org. Lett.* **2014**, *16*, 6346–6349. [CrossRef]

37. Tian, W.-J.; Yu, Y.; Yao, X.-J.; Chen, H.-F.; Dai, Y.; Zhang, X.-K.; Yao, X.-S. Norsampsones A–D, four new decarbonyl polycyclic polyprenylated acylphloroglucinols from Hypericum sampsonii. *Org. Lett.* **2014**, *16*, 3448–3451. [CrossRef]

38. Qiu, D.; Zhou, M.; Chen, J.; Wang, G.; Lin, T.; Huang, Y.; Yu, F.; Ding, R.; Sun, C.; Tian, W.; et al. Hyperelodiones A-C, monoterpenoid polyprenylated acylphloroglucinols from Hypericum elodeoides, induce cancer cells apoptosis by targeting RXRα. *Phytochemistry* **2020**, *170*, 112216. [CrossRef]

39. Wu, H.; Zeng, W.; Chen, L.; Yu, B.; Guo, Y.; Chen, G.; Liang, Z. Integrated multi-spectroscopic and molecular docking techniques to probe the interaction mechanism between maltase and 1-deoxynojirimycin, an α-glucosidase inhibitor. *Int. J. Biol. Macromol.* **2018**, *114*, 1194–1202. [CrossRef]

40. Bozoğlan, B.K.; Tunç, S.; Duman, O. Investigation of neohesperidin dihydrochalcone binding to human serum albumin by spectroscopic methods. *J. Lumin.* **2014**, *155*, 198–204. [CrossRef]

41. Ajmal, M.R.; Abdelhameed, A.S.; Alam, P.; Khan, R.H. Interaction of new kinase inhibitors cabozantinib and tofacitinib with human serum alpha-1 acid glycoprotein. A comprehensive spectroscopic and molecular Docking approach, spectrochim. *Acta Part A Mol. Biomol. Spectrosc.* **2016**, *159*, 199–208. [CrossRef] [PubMed]

42. Zhang, D.; Chen, Z.; Hu, C.; Yan, S.; Li, Z.; Lian, B.; Xu, Y.; Ding, R.; Zeng, Z.; Zhang, X.; et al. Celastrol binds to its target protein via specific noncovalent interactions and reversible covalent bonds. *Chem. Commun.* **2018**, *54*, 12871–12874. [CrossRef] [PubMed]

43. von Schwarzenberg, K.; Vollmar, A.M. Targeting apoptosis pathways by natural compounds in cancer: Marine compounds as lead structures and chemical tools for cancer therapy. *Cancer Lett.* **2013**, *332*, 295–303. [CrossRef] [PubMed]

44. Schumacher, M.; Cerella, C.; Eifes, S.; Chateauvieux, S.; Morceau, F.; Jaspars, M.; Dicato, M.; Diederich, M. Heteronemin, a spongean sesterterpene, inhibits TNFα-induced NF-κB activation through proteasome inhibition and induces apoptotic cell death. *Biochem. Pharmacol.* **2010**, *79*, 610–622. [CrossRef] [PubMed]

45. Zahedifard, M.; Faraj, F.L.; Paydar, M.; Looi, C.Y.; Hajrezaei, M.; Hasanpourghadi, M.; Kamalidehghan, B.; Majid, N.A.; Ali, H.M.; Abdulla, M.A. Synthesis, characterization and apoptotic activity of quinazolinone Schiff base derivatives toward MCF-7 cells via intrinsic and extrinsic apoptosis pathways. *Sci. Rep.* **2015**, *5*, 11544. [CrossRef] [PubMed]

46. Kumar, S. Caspase function in programmed cell death. *Cell Death Differ.* **2007**, *14*, 32–43. [CrossRef]

47. Bryant, K.L.; Stalnecker, C.A.; Zeitouni, D.; Klomp, J.E.; Peng, S.; Tikunov, A.P.; Gunda, V.; Pierobon, M.; Waters, A.M.; George, S.D.; et al. Combination of ERK and autophagy inhibition as a treatment approach for pancreatic cancer. *Nat. Med.* **2019**, *25*, 628–640. [CrossRef]

48. Wagner, E.F.; Nebreda, Á.R. Signal integration by JNK and p38 MAPK pathways in cancer development. *Nat. Rev. Cancer.* **2009**, *9*, 537–549. [CrossRef]

49. Janku, F.; Yap, T.A.; Meric-Bernstam, F. Targeting the PI3K pathway in cancer: Are we making headway? *Nat. Rev. Clin. Oncol.* **2018**, *15*, 273–291. [CrossRef]

50. Fritsch, M.; Günther, S.D.; Schwarzer, R.; Albert, M.-C.; Schorn, F.; Werthenbach, J.P.; Schiffmann, L.M.; Stair, N.; Stocks, H.; Seeger, J.M.; et al. Caspase-8 is the molecular switch for apoptosis, necroptosis and pyroptosis. *Nature* **2019**, *575*, 683–687. [CrossRef]

51. Chintharlapalli, S.; Burghardt, R.; Papineni, S.; Ramaiah, S.; Yoon, K.; Safe, S. Activation of Nur77 by selected 1,1-Bis(3'-indolyl)-1-(p-substituted phenyl)methanes induces apoptosis through nuclear pathways. *J. Biol. Chem.* **2005**, *280*, 24903–24914. [CrossRef] [PubMed]

52. Yoon, K.; Lee, S.-O.; Cho, S.-D.; Kim, K.; Khan, S.; Safe, S. Activation of nuclear TR3 (NR4A1) by a diindolylmethane analog induces apoptosis and proapoptotic genes in pancreatic cancer cells and tumors. *Carcinogenesis* **2011**, *32*, 836–842. [CrossRef]

53. Kim, H.-J.; Kim, J.-Y.; Lee, S.J.; Kim, H.-J.; Oh, C.J.; Choi, Y.-K.; Lee, H.-J.; Do, J.-Y.; Kim, S.-Y.; Kwon, T.-K.; et al. α-Lipoic acid prevents neointimal Hyperplasia via induction of p38 mitogen-activated protein Kinase/Nur77-Mediated apoptosis of vascular smooth muscle cells and accelerates postinjury reendothelialization. *Arterioscler. Thromb. Vasc. Biol.* **2010**, *30*, 2164–2172. [CrossRef]

54. Han, Y.-H.; Cao, X.; Lin, B.; Lin, F.; Kolluri, S.K.; Stebbins, J.; Reed, J.C.; Dawson, M.I.; Zhang, X.-K. Regulation of Nur77 nuclear export by c-Jun N-terminal kinase and Akt. *Oncogene* **2006**, *25*, 2974–2986. [CrossRef] [PubMed]

55. Lee, K.-W.; Cobb, L.J.; Paharkova-Vatchkova, V.; Liu, B.; Milbrandt, J.; Cohen, P. Contribution of the orphan nuclear receptor Nur77 to the apoptotic action of IGFBP-3. *Carcinogenesis* **2007**, *28*, 1653–1658. [CrossRef] [PubMed]

Article

Leptogorgins A–C, Humulane Sesquiterpenoids from the Vietnamese Gorgonian *Leptogorgia* sp.

Irina I. Kapustina [1], Tatyana N. Makarieva [1,*], Alla G. Guzii [1], Anatoly I. Kalinovsky [1], Roman S. Popov [1], Sergey A. Dyshlovoy [1,2,3], Boris B. Grebnev [1], Gunhild von Amsberg [2,3] and Valentin A. Stonik [1]

[1] G.B. Elyakov Pacific Institute of Bioorganic Chemistry, Far Eastern Branch of the Russian Academy of Sciences, Pr. 100-let Vladivostoku 159, 690022 Vladivostok, Russia; ikapust@rambler.ru (I.I.K.); gagry@rambler.ru (A.G.G.); kaaniv@piboc.dvo.ru (A.I.K.); prs_90@mail.ru (R.S.P.); dyshlovoy@gmail.com (S.A.D.); grebnev_bor@mail.ru (B.B.G.); stonik@piboc.dvo.ru (V.A.S.)

[2] Department of Oncology, Hematology and Bone Marrow Transplantation with Section Pneumology, Hubertus Wald-Tumorzentrum, University Medical Center Hamburg-Eppendorf, 20251 Hamburg, Germany; g.von-amsberg@uke.de

[3] Martini-Klinik, Prostate Cancer Center, University Hospital Hamburg-Eppendorf, 20251 Hamburg, Germany

* Correspondence: makarieva@piboc.dvo.ru; Tel.: +7-950-295-66-25

Received: 26 May 2020; Accepted: 10 June 2020; Published: 13 June 2020

Abstract: Leptogorgins A–C (**1**–**3**), new humulane sesquiterpenoids, and leptogorgoid A (**4**), a new dihydroxyketosteroid, were isolated from the gorgonian *Leptogorgia* sp. collected from the South China Sea. The structures were established using MS and NMR data. The absolute configuration of **1** was confirmed by a modification of Mosher's method. Configurations of double bonds followed from NMR data, including NOE correlations. This is the first report of humulane-type sesquiterpenoids from marine invertebrates. Sesquiterpenoids leptogorgins A (**1**) and B (**2**) exhibited a moderate cytotoxicity and some selectivity against human drug-resistant prostate cancer cells 22Rv1.

Keywords: gorgonian; *Leptogorgia*; humulane sesquiterpenoids; anticancer activity

---

1. Introduction

Marine gorgonian corals have been reported to be a rich source of isoprenoids with unprecedented chemical structures and biological activities [1]. Species of the genus *Leptogorgia* (Gorgoniidae) have been shown to produce cembranoids [2–7], polyoxygenated steroids [8–12], alkaloids [13], fatty acids [14], homarine [15], thyroxine, and vitamin D [16]. To date, different humulane-type sesquiterpenoids have been found in plants [17–19], liverworts [20], and fungi [21–23]. However, until recently they were not found in marine invertebrates, including gorgonians. Interestingly, two new norhumulene were isolated from the soft coral *Sinularia hirta* [24]. In addition, one more norhumulene was found in a formazan soft coral *Sinularia gibberosa* [25]. Humulanes from the peeled stems of *Syringa pinnatifida* inhibit NO production in LPS-induced RAW264.7 macrophage cells and decrease the TNF-α and IL-6 levels in RAW264.7 cells [26]. Additionally, plant cytochrome P450 was reported to catalyse the conversion of α-humulene into 8-hydroxy-α-humulene [27].

For some humulenes, an antitumor activity was reported. Thus, zurumbone (2,6,9-humulatriene-8-one), as an active component of the *Zingiber aromaticum* extract, was shown to be active in human cancer HT-29, CaCO-2, and NCF-7 cell lines. Remarkably, it was more active than curcumin, which was used as a reference compound [28]. Herein, we report the structures and biological activities of three new humulane sesquiterpenoids, leptogorgins A–C (**1**–**3**), and a new steroid, leptogorgoid A (**4**), from the gorgonian *Leptogorgia* sp. (Figure 1).

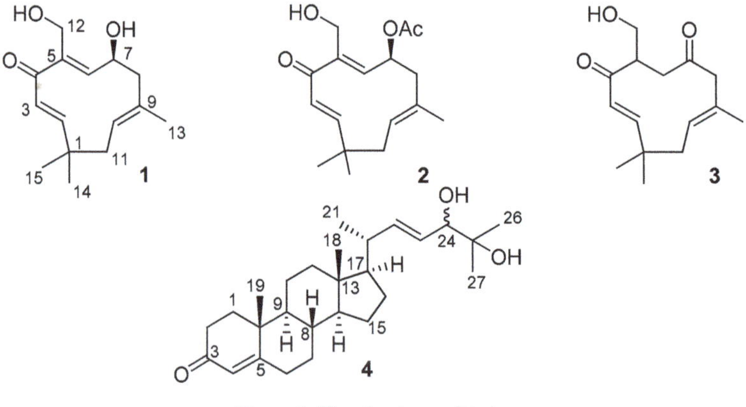

Figure 1. The structures of **1–4**.

2. Results and Discussion

The EtOH extract of the gorgonian *Leptogorgia* sp. (registration number O38-011) was concentrated and partitioned between aqueous EtOH and *n*-hexane. The EtOH-soluble materials were separated by silica gel flash chromatography, followed by Sephadex LH-20 column chromatography and normal and reversed-phase HPLC to give leptogorgins A–C (**1–3**, 2.5, 0.8, and 1.0 mg, respectively) and leptogorgoid A (**4**, 0.6 mg).

Compound **1** was isolated as a colourless oil. The HRESIMS of **1** showed an [M + Na]$^+$ ion peak at *m/z* 273.1459 and an [M − H]$^-$ ion peak at *m/z* 249.1498, which indicated a molecular formula of $C_{15}H_{22}O_3$. The ^{13}C NMR spectrum displayed 15 signals, which could be assigned to a sesquiterpene substructure. Analysis of the ^{1}H, ^{13}C, and HSQC NMR spectra (Table 1) revealed signals indicative of one ketocarbonyl (δ_C 200.8, C-4), one oxymethine (δ_H 4.21/δ_C 71.7, C-7), one oxymethylene (δ_H 4.25; 4.38/δ_C 64.7, C-12), four methines (δ_H 6.32/δ_C 164.8, C-2; δ_H 5.97/δ_C 128.1, C-3; δ_H 5.75/δ_C 133.8, C-6, and δ_H 5.22/δ_C 125.9, C-10), two quaternary (δ_C 143.0, C-5; δ_C 132.4, C-9) olefinic carbons, and two methylene groups (δ_H 1.96 and 2.68/δ_C 45.3, C-8; δ_H 1.95 and 2.40/δ_C 40.7, C-11), as well as one quaternary carbon (δ_C 38.0, C-1), one corresponding vinylic methyl (δ_H 1.72/δ_C 20.1, CH$_3$-13) and two methyl singlets (δ_H 1.18/δ_C 24.0, CH$_3$-14; δ_H 1.13/δ_C 29.1, CH$_3$-15). The ^{1}H-^{1}H COSY spectrum enabled three structural fragments to be established: CH=CH-, -CH-CH-CH$_2$-, and -CH-CH$_2$-, which could be connected by observing the correlations in the HMBC experiment (Figure 2). Thus, HMBC correlations from H-3 to C-1, C-4, and C-5, from H-6 to C-12 and C-8, from H-7 to C-5 and C-8, from H-8 to C-7, C-9, C-10, and C-13, from H-11 to C-10, C-9, and C-1, and from CH$_3$-14 and CH$_3$-15 to C-1, C-2, and C-11 established the planar structure of **1** (Figure 2).

The geometry of the $\Delta^{2,3}$ double bond was further determined to be *E* by considering the coupling constant (*J* = 16.3 Hz) displayed in its ^{1}H NMR spectrum. The NOE correlations of CH$_3$-13 to H-2, H-6, and CH$_2$-11, as well as H-10 with H-6 and H-6 with H-2 (Figure 3), suggested that the $\Delta^{5(6)}$ and $\Delta^{9(10)}$ double bonds in **1** were *E* configured.

A modified Mosher ester analysis was obtained, and the negative $\Delta\delta^{SR}$ ($\delta^S - \delta^R$) values of Ha-8, ($\Delta\delta_H$ −0.01), Hb-8, ($\Delta\delta_H$ −0.05) and CH$_3$-13 ($\Delta\delta_H$ −0.01), and positive $\Delta\delta^{SR}$ values of H-6 ($\Delta\delta_H$ +0.04) Ha-12 ($\Delta\delta_H$ +0.01), and Hb-12 ($\Delta\delta_H$ +0.04) (Figure 4) revealed the 7*S* configuration [25]. Thus, the structure of **1** was determined as 4-oxohumula-2*E*,5*E*,9*E*-trien-7*S*,12-diol, as shown in Figure 1, and named leptogorgin A (**1**).

Table 1. ^{1}H (700 MHz) and ^{13}C (175 MHz) NMR spectroscopic data for **1**, **2** and **3** in CDCl$_3$.

Position	1		2		3	
	δ_C	δ_H mult (*J* in Hz)	δ_C	δ_H mult (*J* in Hz)	δ_C	δ_H mult (*J* in Hz)
1	38.0 C	-	38.1 C	-	40.4 * C	-
2	164.8 CH	6.32, d (16.3)	162.8 CH	6.24, d (16.3)	152.7 CH	6.29, d (16.1)
3	128.1 CH	5.97, d (16.3)	128.1 CH	6.07, d (16.3)	128.4 CH	5.76, d (16.1)
4	200.8 C	-	199.4 C	-	204.3 C	-
5	143.0 C	-	145.2 C		48.6 CH	3.38, m
6	133.8 CH	5.75, d (10.6)	129.5 C	5.70, dt (10.6; 1.3)	41.2 CH$_2$	2.43, dd (16.9; 2.9) 2.73, dd (16.9; 9.7)
7	71.7 CH	4.21 td (10.6; 5.4)	72.9 CH	5.28, td (10.6; 5.1)	204.3 C	
8	45.3 CH$_2$	1.96, m 2.68, dd (12.2; 5.4)	42.7 CH$_2$	2.03, m 2.69, dd (12.5; 5.1)	54.1 CH$_2$	3.00, d (12.4) 3.15, d (12.4)
9	132.4 C	-	128.1 C	-	127.8 C	-
10	125.9 CH	5.22, brd (12.5)	127.1 C	5.32, m	129.0 CH	5.37, ddd (10.5; 5.7, 1.2)
11	40.7 CH$_2$	1.95, m 2.40, t (12.5)	40.7	1.97, m 2.39, t (12.6)	40.2 * CH$_2$	2.00, m 2.07, m
12	64.7 CH$_2$	4.25, d (13.3) 4.38, d (13.3)	64.8 CH$_2$	4.26, dd (13.2; 4.6) 4.40, dd (13.2; 6.3)	63.0 CH$_2$	3.78, m 3.89, m
13	20.1 CH$_3$	1.72, s	20.0 CH$_3$	1.73, s	19.0 CH$_3$	1.64, s
14	24.0 CH$_3$	1.18, s	23.9 CH$_3$	1.21, s	28.8 CH$_3$	1.21, s
15	29.1 CH$_3$	1.13, s	29.2 CH$_3$	1.13, s	24.3 CH$_3$	1.09, s
<u>C</u>OCH$_3$			169.7 C	-		
CO<u>C</u>H$_3$			21.2 CH$_3$	1.98, s		

* Signals may be interchangeable.

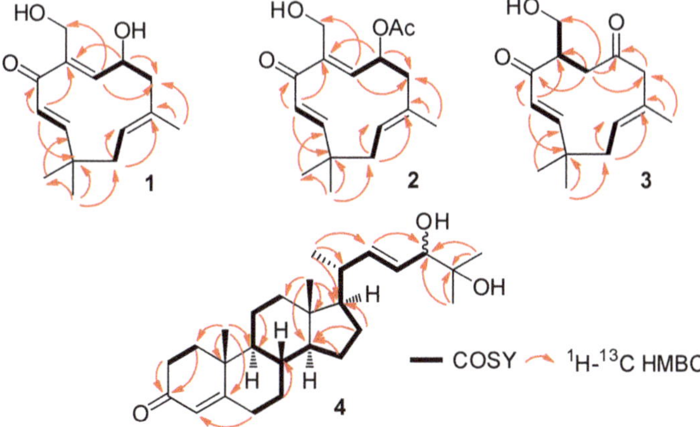

Figure 2. Selected COSY and HMBC correlations for **1–4**.

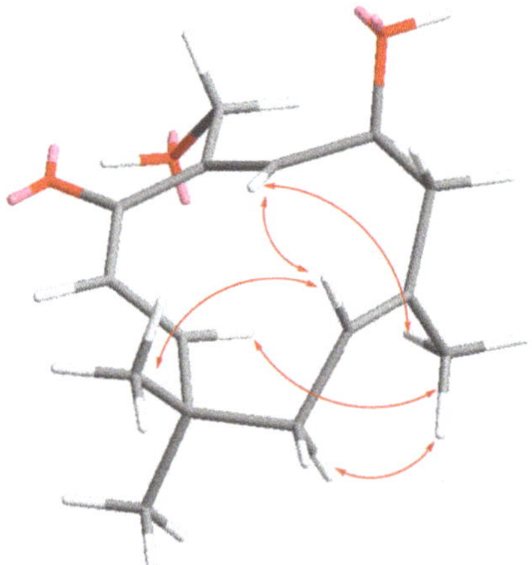

Figure 3. Key NOE correlations for **1**.

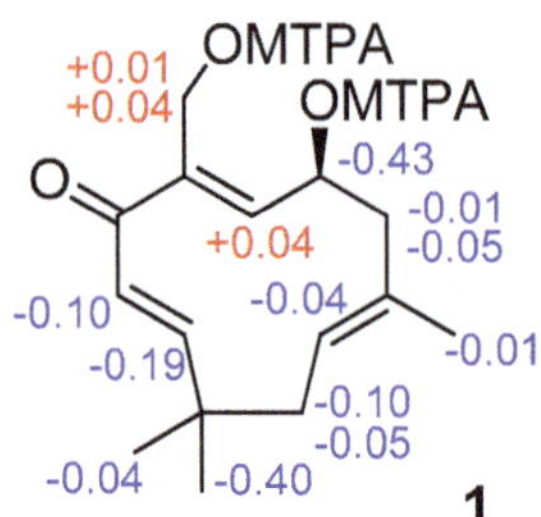

Figure 4. $\Delta\delta$ ($\delta^S - \delta^R$) values (in ppm, CDCl$_3$) for the MTPA esters of **1**.

Compound **2** was obtained as a colourless oil. The HRESIMS of **2** showed an [M + Na]$^+$ ion peak at *m/z* 315.1567 and an [M − H]$^-$ ion peak at *m/z* 291.1602, which indicated a molecular formula of C$_{17}$H$_{24}$O$_4$. The ^{1}H and ^{13}C NMR spectra of **2** (Table 1) were similar to those of **1**, suggesting that this compound possessed the same humulane skeleton. The key differences were in δ_H for H-7 and δ_C for carbon 7 in the spectrum of **2** (δ_H 5.28/δ_C 72.9). The corresponding signals were shifted downfield, compared to those of **1** (δ_H 4.21/δ_C 71.7). This characteristic difference and HRESIMS data were caused by the hydroxy group in **1** being displaced by an acetoxyl group in **2**. The HMBC spectra of **2** demonstrated the expected key correlations. The ECD spectrum of compound **2** was compared with the ECD spectrum of leptogorgin A (**1**), in which the corresponding absolute configuration was established by modification of Mosher's method. Both ECD spectra displayed similar Cotton effects (see Figure S27), allowing us to establish the same 7*S* configuration for compound **2**. From these data, compound **2** was determined to be 4-oxohumula-2*E*,5*E*,9*E*-trien-7*S*-acetate,12-ol, as shown in Figure 1, and named leptogorgin B (**2**).

Compound **3** was isolated as a colourless oil. The HRESIMS of **1** showed an [M + Na]$^+$ ion peak at *m/z* 273.1459 and an [M − H]$^-$ ion peak at *m/z* 249.1496, which indicated a molecular formula of C$_{15}$H$_{22}$O$_3$. The ^{1}H and ^{13}C NMR spectra (Table 1) of **3** were similar to those of **1** and **2**, suggesting that this compound also possessed the same humulane skeleton. Key differences concerned δ_H for protons 6, 7, and 8 and δ_C for carbons 4, 5, 6, 7, and 8 in the spectrum of **3**, which were different compared to those of **1** and **2**. This characteristic difference was caused by an absence of the hydroxy group, as in **1**, or acetyl, as in **2** at position 7, being displaced by a ketogroup in **3**, as well as by the absence of the 5,6 double bound in **3**. The location of the ketogroup was further determined to be at C-7 by COSY, HSQC, and HMBC experiments. Thus, compound **3** was determined to be 4,7-dioxohumula-2*E*,9*E*-dien-12-ol, as shown in Figure 1, and named leptogorgin C (**3**).

Compound **4** was isolated as a colourless powder. The HRESIMS of **4** showed an $[M + Na]^+$ ion peak at *m/z* 437.3026 and an $[M - H]^-$ ion peak at *m/z* 413.3061, which indicated a molecular formula of $C_{27}H_{42}O_3$. The data of 1D- and 2D-NMR spectra of **1** (Table 2) indicated that this compound belonged to steroids. Its spectra contained five methyl groups, including two angular methyl groups in the steroid nucleus (δ_H 0.74/δ_C 12.2, δ_H 1.19/δ_C 17.4) and three methyl groups of the side chain (δ_H 1.04/δ_C 20.3, δ_H 1.15/δ_C 23.8, and δ_H 1.20/δ_C 26.4), eight methylene groups, six methine groups, including one oxygenated methine (δ_H 3.85/δ_C 79.7), two quaternary sp^3 carbons (δ_C 38.6, δ_C 42.5), one quaternary sp^3 oxygenated carbon (δ_C 72.8), one trisubstituted double bond (δ_H 5.72/δ_C 123.8 and 171.4), one disubstituted double bond (δ_H 5.61/δ_C 140.8 and δ_H 5.43/δ_C 126.0), and one conjugated with double bond ketone carbonyl (δ_C 199.5). The geometry of the 22,23 double bond was further determined to be *E* by considering the coupling constant (J = 15.3 Hz) displayed in its ^{1}H NMR spectrum. The HMBC spectra of **4** demonstrated the expected key correlations. From these data, compound **4** was determined to be 3-oxocholesta-4*E*,22*E*-diene-24,25 dienol, as shown in Figure 1, and named leptogorgoid A (**4**).

Table 2. ^{1}H (700 MHz) and ^{13}C (175 MHz) NMR spectroscopic data for **4** in CDCl$_3$.

Position	δ_C	δ_H mult (*J* in Hz)	Position	δ_C	δ_H mult (*J* in Hz)
1	35.7 CH$_2$	1.70, m 2.03, m	16	28.5 CH$_2$	1.29, m 1.70, m
2	34.0 CH$_2$	2.34, m 2.42, m	17	55.6 CH	1.19, m
3	199.5	-	18	12.2 CH$_3$	0.74, s
4	123.8 CH	5.72 s	19	17.4 CH$_3$	1.19, s
5	171.4C	-	20	39.8 CH	2.14, m
6	32.9 CH$_2$	2.27, ddd (14.7; 4.1; 2.4) 2.40, m	21	20.3 CH$_3$	1.04, d (6.6)
7	32.0 CH$_2$	1.02, m 1.84, m	22	140.8 CH	5.61, dd (8.6; 15.3)
8	35.7 CH	1.53, m	23	126.0 CH	5.43, dd (7.3; 15.3)
9	53.8 CH	0.94, m	24	79.7 CH	3.84, d (7.3)
10	38.6 C	-	25	72.8 C	-
11	21.0 CH$_2$	1.44, ddd (13.6; 17.1; 4.2) 1.54, m	26	23.8 CH$_3$	1.15, s
12	39.5 CH$_2$	1.20, m 2.01, m	27	26.4 CH$_3$	1.20, s
13	42.5 C	-			
14	55.8 CH	1.04, m			
15	24.2 CH$_2$	1.11, m			
		1.60, m			

Next, we investigated the effects of the leptogorgins A (**1**) and B (**2**) on the viability of 22Rv1 cells (human drug-resistant prostate cancer cells) as well as on PNT2 cells (human prostate non-cancer cells). MTT assay revealed **1** to exhibit a moderate cytotoxicity to both cell lines (IC$_{50}$ = 31.0 µM and 35.8 µM, respectively), whereas **2** had IC$_{50}$ > 100 µM. Doxorubicine was used as a positive control and exhibited in 22Rv1 and PNT2 cells IC$_{50}$ of 0.084 µM and 1.12 µM, respectively. Interestingly, both compounds were more active in human cancer 22Rv1 cells, in comparison with PNT2 cells (Figure 5). Additionally, we examined the ability of these compounds to inhibit the colony formation of 22Rv1 prostate cancer cells; however, no significant inhibitory activity was observed under the treatment with cytotoxic or non-cytotoxic concentrations of the compounds up to a concentration of 100 µM (data not shown).

The isolated compounds may be considered as prototypes for future anticancer agents capable of selective inhibition of human drug-resistant prostate cancer cells. Note that we could not isolate enough leptogorgins C (**3**) and leptogorgoid A (**4**) to investigate the biological activity of these compounds.

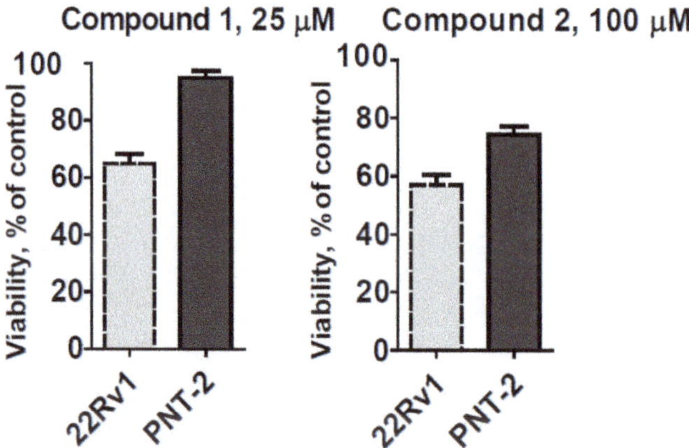

Figure 5. The viability of 22Rv1 and PNT2 cells after 72 h of treatment with the indicated concentrations of the investigated compounds. The viability was evaluated using MTT assay.

3. Materials and Methods

3.1. General Procedures

Optical rotation was measured using a PerkinElmer 343 polarimeter. UV spectra were recorded on a Shimadzu UV-1601 PC spectrophotometer. ECD spectra were recorded with an Applied Photophysics Chirascan plus spectropolarimeter. IR spectroscopic data were measured using an IR spectrometer Equinox 55 (Bruker, Ettlingen, Germany) in $CHCl_3$. The ^1H and ^{13}C NMR spectra were recorded on a Bruker Avance III-700 spectrometer (Bruker, Ettlingen, Germany) at 700 and 175 MHz, respectively, with Me_4Si as an internal standard. ESI mass spectra (including HRESIMS) were obtained on a Bruker maXis Impact II Q-TOF mass spectrometer (Bruker Daltonics, Bremen, Germany) by direct infusion in MeOH. Low-pressure column liquid chromatography was performed using silica gel (Sigma-Aldrich Co., St. Louis, MO, USA) and Sephadex LH-20 (Sigma, Chemical Co., St. Louis, MO, USA) columns. HPLC was performed using a Shimadzu Instrument equipped with the differential refractometer RID-10A, a YMC-Pack ODS-A (250 × 10 mm) column (YM Co., Ltd., Kyoto, Japan), and a silica gel column (SUPELCOSIL^{TM}, 250 × 10 mm, 5 μm) (Sigma-Aldrich Co., USA). TLC was performed on silica gel plates (5–17 μm, Sorbfil, Russia).

3.2. Animal Material

The gorgonian *Leptogorgia* sp. (registration number PIBOC O38-011) was collected by dredging during the 38th scientific cruise of R/V "Academic Oparin", May 2010, South China sea (09°08′2″ N; 109°01′7″ E, depth 134 m), in Vietnamese waters. A voucher specimen of 038-011 sample is stored in the Marine invertebrate collection of the G.B. Elyakov Pacific Institute of Bioorganic Chemistry FEB RAS (Vladivostok, Russia).

3.3. Extraction and Isolation

The EtOH extract of the gorgonian (dry weight 170 g) was concentrated and partitioned between *n*-hexane and aqueous EtOH. The EtOH-soluble material was subjected to column chromatography on a silica gel column using $CHCl_3$-EtOH (stepwise gradient, 1:0 1:1). Fractions eluted with $CHCl_3$:EtOH (20:1) were concentrated and residue (171.3 mg) was subjected to column chromatography on a LH-20 column using CHCl$_3$:EtOH, 2:1 to yield two fractions: F1 (46.6 mg) and F2 (61.4 mg). Preparative HPLC of the fraction F1 (SUPELCOSIL, *n*-hexane:EtOAc, 1:1) gave pure leptogorgin A (**1**, 2.5 mg, 0.002% based on dry weight of gorgonian). Preparative HPLC of the fraction F2 (YMC-Parck ODS-A, EtOH:H$_2$O, 3:2) gave three sub-fractions: F2-1 (2.5 mg), F2-2 (6.4 mg), and F2-3 (8.0 mg). Preparative HPLC of the fraction F2-1 (SUPELCOSIL, *n*-hexane:EtOAc, 2:3) gave pure leptogorgin C (**3**, 1.0 mg, 0.001% based on dry weight of gorgonian). Preparative HPLC of the fraction F2-2 (SUPELCOSIL, *n*-hexane:EtOAc, 1:1) gave pure leptogorgin B (**2**, 0.8 mg, 0.001% based on dry weight of gorgonian). Preparative HPLC of the fraction F2-3 (SUPELCOSIL, *n*-hexane:EtOAc, 1:1) gave pure leptogorgoid A (**4**, 0.6 mg, 0.0006% based on dry weight of gorgonian).

3.4. Compound Characterization Data

Leptogorgin A (**1**): colourless oil; $[\alpha]_D^{22}$ +38.7 (*c* 0.2, CHCl$_3$); UV (EtOH) λ_{max} (log ε) 195 (4.05), 229 (3.75) nm; ECD (*c* 1 × 10^{-3} M, EtOH) λ_{max} ($\Delta\varepsilon$) 194 (7.56), 228 (9.41), 274 (−3.52), 333 (1.30) nm; IR (CHCl$_3$): ν_{max} 3604, 2964, 2928, 2860, 1723, 1641, 1458, 1387, 1365, 1261, 1243, 1104, 1012 cm^{-1}; ^{1}H and ^{13}C NMR data (CDCl$_3$), Table 1; HRESIMS *m/z* 273.1459 [M + Na]$^+$ (calcd for C$_{15}$H$_{22}$O$_3$Na, 273.1461); HRESIMS *m/z* 249.1498 [M − H]$^-$ (calcd for C$_{15}$H$_{21}$O$_3$ 249.1496).

Leptogorgin B (**2**): colourless oil; $[\alpha]_D^{22}$ +16 (*c* 0.1, CHCl$_3$); UV (EtOH) λ_{max} (log ε) 196 (3.23), 229 (3.07) nm; ECD (*c* 3 × 10^{-3} M, EtOH) λ_{max} ($\Delta\varepsilon$) 197 (2.90), 226 (1.41), 254 (−1.06), 336 (0.43) nm; ^{1}H and ^{13}C NMR data (CDCl$_3$) Table 1; HRESIMS *m/z* 315.1571 [M + Na]$^+$ (calcd for C$_{17}$H$_{24}$O$_4$Na, 315.1567); HRESIMS *m/z* 291.1602 [M − H]$^-$ (calcd for C$_{17}$H$_{23}$O$_4$ 291.1602).

Leptogorgin C (**3**): colourless oil; ^{1}H and ^{13}C NMR data (CDCl$_3$) Table 1; HRESIMS *m/z* 273.1463 [M + Na]$^+$ (calcd for C$_{15}$H$_{22}$O$_3$Na, 273.1461); HRESIMS *m/z* 249.1496 [M − H]$^-$ (calcd for C$_{15}$H$_{21}$O$_3$ 249.1496).

Leptogorgoid A (**4**): colourless powder; $[\alpha]_D^{22}$ +33 (*c* 0.05, CHCl$_3$); ^{1}H and ^{13}C NMR data (CDCl$_3$) Table 1. HRESIMS *m/z* 437.3021 [M + Na]$^+$ (calcd for C$_{27}$H$_{42}$O$_3$Na, 437.3026); HRESIMS *m/z* 413.3060 [M − H]$^-$ (calcd for C$_{27}$H$_{41}$O$_3$ 413.3061).

MTPA esterification of **1**. To a part of **1** (0.6 mg) in dry C$_5$H$_5$N (1 µL), *R*-(−)-α-metoxy-α-trifluoromethylphenylacetyl chloride (10 µL) was added. The mixture was stirred on one hour at room temperature.and evaporated in vacuo to give (*S*)-MTPA diester **1a**. By the same procedure, (*R*)-MTPA diester **1b** was prepared.

(*S*)-*MTPA diester* (**1a**): Select ^{1}H NMR data (CDCl$_3$) see Table S1. HRESIMS *m/z* 707.25 [M + Na]$^+$ (calcd for C$_{35}$H$_{38}$F$_6$O$_7$Na, 707.25).

(*R*)-*MTPA diester* (**1b**): Select ^{1}H NMR data (CDCl$_3$) see Table S1. HRESIMS *m/z* 707.25 [M + Na]$^+$ (calcd for C$_{35}$H$_{38}$F$_6$O$_7$Na, 707.25).

3.5. Bioactivity Assay

3.5.1. Reagents

The MTT reagent (Thiazolyl blue tetrazolium bromide) was purchased from Sigma (Taufkirchen, Germany).

3.5.2. Cell Lines and Culture Conditions

The human prostate cancer cells 22Rv1 and human prostate non-cancer cells PNT2 were purchased from ATCC. Cell lines were cultured according to the manufacturer's instructions in 10% FBS/RPMI media (Invitrogen, Carlsbad, CA, USA) and handled as described in [29].

3.5.3. In Vitro MTT-Based Drug Sensitivity Assay

The in vitro cytotoxic activities of the isolated substances were evaluated by MTT assays. The assays were performed as described previously [30]. In brief, cells were seeded in 96-well plates (6 × 10^3 cells/well), incubated overnight, and treated with the tested compounds for 72 h. Next,

10 µL/well of MTT reagent was added and the plates were incubated for 2 h. The media were aspirated and the plates were dried. The formed formazan crystals were dissolved in DMSO and the cell viability was measured using an Infinite F200PRO reader (TECAN, Männedorf, Switzerland). Results were calculated by the GraphPad Prism software v. 7.05 (GraphPad Prism software Inc., La Jolla, CA, USA) and are represented as the IC_{50} of the compounds against the control cells treated with the solvent alone.

3.5.4. Colony Formation Assay

Colony formation assay was performed as described before, with slight modifications [30]. Cells were treated with the drug for 48 h; then, cells were trypsinized and the number of alive cells was counted with the trypan blue exclusion assay as described before [31]. One hundred viable cells were plated into each well of 6-well plates in complete drug-free media (3 mL/well) and were incubated for 14 days. Then, the media were aspirated, surviving colonies were fixed with 100% MeOH, followed by washing with PBS and air-drying at RT. Next, cells were incubated with Giemsa staining solution for 25 min at RT, the staining solution was aspirated, and the wells were rinsed with dH_2O and air-dried. The number of cell colonies was counted with the naked eye.

4. Conclusions

In summary, ^1H NMR-guided chemical investigation led to the isolation of three new humulane-type sesquiterpenoids and one new steroid. The structures of the new compounds were elucidated via analyses of their MS, NMR, and ECD spectroscopic data, as well as using the Mosher's esters analysis. These molecules represent the new humulenes possessing an oxygenation pattern which was significantly different from those found in plants, liverworts, and fungi. Leptogorgin A (**1**) exhibits a moderate cytotoxicity to human prostate cancer 22Rv1 cells.

Supplementary Materials: The following are available online at http://www.mdpi.com/1660-3397/18/6/310/s1. Copies of HRESIMS, 1D- and 2D-NMR spectra of **1–4**.

Author Contributions: I.I.K. isolated the metabolites; T.N.M. elucidated structures; S.A.D. performed the bioactivity assays; A.I.K. performed the NMR spectra; R.S.P. performed the mass spectra; B.B.G. performed species identification of the gorgonian; G.v.A. assisted the results discussion; T.N.M., A.G.G. and V.A.S. wrote the paper, which was revised and approved by all the authors. All authors have read and agreed to the published version of the manuscript.

Funding: Isolation and establishment of chemical structures were partially supported by the RSF grant #20-14-00040 (Russian Science Foundation).

Acknowledgments: The study was carried out on the equipment of the Collective Facilities Center "The Far Eastern Center for Structural Molecular Research (NMR/MS) PIBOC FEB RAS". We would like to thank Ms. Jessica Hauschild (University Medical Center Hamburg-Eppendorf) for technical support of the biological part of this research.

Conflicts of Interest: The authors declare no conflict of interest.

References

1. Carroll, A.R.; Copp, B.R.; Davis, R.A.; Keyzers, R.A.; Prinsep, M.R. Marine Natural Products. *Nat. Prod. Rep.* **2020**, *37*, 175–223. [CrossRef] [PubMed]
2. Dorta, E.; Diaz-Marrero, A.R.; Brito, I.; Cueto, M.; D'Croz, L.; Darias, J. The oxidation profile at C-18 of furanocembranolides may provide a taxonomical marker for several genera of octocorals. *Tetrahedron* **2007**, *63*, 9057–9062. [CrossRef]
3. Ortega, M.J.; Zubia, E.; Sanchez, M.C.; Carballo, J.L. Cembrane diterpenes from the gorgonian *Leptogorgia laxa*. *J. Nat. Prod.* **2008**, *71*, 1637–1639. [CrossRef] [PubMed]
4. Gerhart, D.J.; Coll, J.C. Pukalide, a widely distributed octocoral diterpenoid, includes vomiting in fish. *J. Chem. Ecol.* **1993**, *19*, 2697–2704. [CrossRef] [PubMed]
5. Gutierrez, M.; Capson, T.L.; Guzman, C.M.; Gonzalez, J.; Ortega-Barria, E.; Quinoa, E.; Riguera, R. Leptolide, a new furanocembranolide diterpene from *Leptogorgia alba*. *J. Nat. Prod.* **2005**, *68*, 614–616. [CrossRef]

6.   Diaz-Marrero, A.R.; Porras, G.; Gueto, M.; D'Croz, L.; Lorenzo, M.; San-Martin, A.; Darias, J. Leptogorgolide, a biogenetically interesting 1,4-diketo-cembranoid that reinforcesthe oxidation profile of C-18 as taxonomical marker for octocorals. *Tetrahedron* **2009**, *65*, 6029–6033. [CrossRef]

7.   Gallardo, A.B.; Diaz-Marrero, A.R.; de la Rosa, J.M.; D'Croz, L.; Perdomo, G.; Cozar-Castello, I.; Darias, J.; Gueto, M. Chloro-furanocembranolides from *Leptogorgia* sp. improve pancreatic beta-cell proliferation. *Mar. Drugs* **2018**, *16*, 49. [CrossRef]

8.   Cimino, G.; De Rosa, S.; De Stefano, S.; Scognamiglio, G.; Sodano, G. Cholest-4,14-dien-20ξ-diol-3,16-dione, a novel polyoxygenated marine steroid which easily loses the side chain. *Tetrahedron Lett.* **1981**, *22*, 3013–3016. [CrossRef]

9.   Cimino, G.; De Rosa, S.; De Stefano, S.; Sodano, G. C-18 Hydroxy steroids from the Mediterranean gorgonian *Leptogorgia sarmentosa*. *Experientia* **1984**, *40*, 246–248. [CrossRef]

10. Benvegnu, R.; Cimino, G.; De Rosa, S.; De Stefano, S. Guggulsterol-like steroids from the Mediterranean gorgonian *Leptogorgia sarmentosa*. *Experientia* **1982**, *38*, 1443–1444. [CrossRef]

11. Garrido, L.; Zubia, E.; Ortega, M.J.; Salva, J. Isolation and structure elucidation of new cytotoxic steroids from the gorgonian *Leptogorgia sarmentosa*. *Steroids* **2000**, *65*, 85–88. [CrossRef]

12. Moritz, M.I.G.; Marostica, L.L.; Bianco, E.M.; Almeida, M.T.R.; Carraro, J.L.; Cabrera, G.M.; Palermo, J.A.; Simoes, C.M.O.; Schenkel, E.P. Polyoxygenated steroids from the octocoral *Leptogorgia punicea* and *in vitro* evaluation of their cytotoxic activity. *Mar. Drags* **2014**, *12*, 5864–5880. [CrossRef] [PubMed]

13. Keyzers, R.A.; Gray, C.A.; Schleyer, M.H.; Whibley, C.E.; Hendricks, D.T.; Davies-Coleman, M.T. Malonganenones A–C, novel tetraprenylated alkaloids from the Mozambique gorgonian *Leptogorgia gilchristi*. *Tetrahedron* **2006**, *62*, 2200–2206. [CrossRef]

14. Miralles, J.; Barnathan, G.; Galonnier, R.; Sall, T.; Samb, A.; Gaydou, E.M.; Kornprobst, J.M. New branched-chain fatty acids from the Senegalese gorgonian *Leptogorgia piccola* (white and yellow morphs). *Lipids* **1995**, *30*, 459–466. [CrossRef] [PubMed]

15. Targett, N.M.; Bishop, S.S.; McConnell, O.J.; Yoder, J.A. Antifouling agents against the benthic marine diatom, *Navicula salinicola*. Homarine from the gorgonians *Leptogorgia virgulata* and *L.* setacea and analogs. *J. Chem. Ecol.* **1983**, *9*, 817–829. [CrossRef]

16. Kingsley, R.J.; Corcoran, M.L.; Krider, K.L.; Kriechbaum, K.L. Thyroxine and vitamin D in the gorgonian *Leptogorgia virgulata*. *Comp. Biochem. Physiol.* **2001**, *129*, 897–907. [CrossRef]

17. Otto, A.; Wilde, V. Sesqui- di-, and triterpemoids as chemosystematic markers in extant conifers. A review. *Bot. Rev.* **2001**, *67*, 141–238. [CrossRef]

18. Schifrin, A.; Litzenburger, M.; Ringle, M.; Ly, T.T.B.; Bernhardt, R. New sesqiterpene oxidations with CYP260A1 and CYP264B1 from *Sorangium cellulosum* Soce56. *ChemBioChem* **2015**, *16*, 2624–2632. [CrossRef]

19. Nagashima, F.; Tabuchi, Y.; Ito, T.; Harinantenaia, L.; Asakawa, Y. Terpenoids, Flavonoids, and Acetogenins from some Malagasy plants. *Nat. Prod. Commmun.* **2016**, *11*, 153–157. [CrossRef]

20. Toyota, M.; Omatsu, J.; Braggins, J.; Asakawa, Y. New humulane-type sesquiterpenes from the liverworts *Tylimamthus tenellus* and *Marchantia emarginata* subsp tosana. *Chem. Pharm. Bull.* **2004**, *52*, 481–484. [CrossRef]

21. Wang, J.F.; He, W.J.; Kong, F.D.; Tian, J.P.; Wang, P.; Zhou, X.J.; Liu, Y.H. Ochramecenes A-I, humulane-derived sesquiterpenoids from the Antharctic fungus *Aspergillus ochraceopetaliformis*. *J. Nat. Prod.* **2017**, *80*, 1725–1733. [CrossRef] [PubMed]

22. Chen, H.P.; Liu, J. Secondary metaboilites from higher fungi. *Prog. Chem. Org. Nat. Prod.* **2017**, *106*, 1–201. [CrossRef] [PubMed]

23. Wu, Z.; Liu, D.; Proksch, P.; Guo, P.; Lin, W. Punctaporonins H-M: Caryophyllene-type sesquiterpenoids from the sponge-associated fungus *Hansfordia sinuosae*. *Mar. Drugs* **2014**, *12*, 3904–3916. [CrossRef] [PubMed]

24. Lu, S.Q.; Li, X.W.; Li, S.W.; Cui, Z.; Guo, Y.W.; Hun, G.Y. Sinuhirtins A and B, two uncommon norhumulane-type terpenoids from the South China Sea soft coral *Sinularia hirta*. *Tetrahedron Lett.* **2019**, *60*, 151308. [CrossRef]

25. Chen, S.P.; Su, J.H.; Yeh, H.C.; Ahmed, A.F.; Dai, C.F.; Wu, Y.C.; Sheu, J.H. Novel norhumulene and xeniaphyllane-derived terpenoids from a Formosan Soft Coral *Sinularia gibberosa*. *Chem. Pharm. Bull.* **2009**, *57*, 162–166. [CrossRef]

26. Zhang, R.; Feng, X.; Su, G.; Mu, Z.; Zhang, H.; Zhao, Y.; Jiao, S.; Cao, L.; Chen, S.; Tu, P.; et al. Bioactive sesquiterpenoids from the peeled stems of *Syringa pinnatifolia*. *J. Nat. Prod.* **2018**, *81*, 1711–1720. [CrossRef]

27. Yu, F.; Okomoto, S.; Harada, H.; Yamasaki, K.; Misawa, N.; Utsumi, R. *Zingiber zerumbet* CYP71BA1 catalyses the conversion of α-humulene to 8-hydroxy-α-humulene. *Cell. Mol. Life Sci.* **2011**, *68*, 1033–1040. [CrossRef]

28. Kirana, C.; Mcintosh, G.H.; Record, I.R.; Jones, G.P. Antitumor activity of extract of *Zinger aromaticum* and its bioactive sesquiterpenoid zerumbone. *Nutr. Cancer* **2003**, *45*, 218–225. [CrossRef]

29. Dyshlovoy, S.A.; Menchinskaya, E.S.; Venz, S.; Rast, S.; Amann, K.; Hauschild, J.; Otte, K.; Kalinin, V.I.; Silchenko, A.S.; Avilov, S.A.; et al. The marine triterpene glycoside frondoside A exhibits activity in vitro and in vivo in prostate cancer. *Int. J. Cancer* **2016**, *138*, 2450–2465. [CrossRef]

30. Dyshlovoy, S.A.; Venz, S.; Hauschild, J.; Tabakmakher, K.M.; Otte, K.; Madanchi, R.; Walther, R.; Guzii, A.G.; Makarieva, T.N.; Shubina, L.K.; et al. Antimigrating activity of marine alkaloid monanchocidin A, proteome-based discovery and confirmation. *Proteomics* **2016**, *16*, 1590–1603. [CrossRef]

31. Dyshlovoy, S.A.; Hauschild, J.; Amann, K.; Tabakmakher, K.M.; Venz, S.; Walther, R.; Guzii, A.G.; Makarieva, T.N.; Shubina, L.K.; Fedorov, S.N.; et al. Marine alkaloid Monanchocidin A overcomes drug resistance by induction of autophagy and lysosomal membrane permeabilization. *Oncotarget* **2015**, *6*, 17328–17341. [CrossRef] [PubMed]

 *marine drugs*

Article

αO-Conotoxin GeXIVA Inhibits the Growth of Breast Cancer Cells via Interaction with α9 Nicotine Acetylcholine Receptors

Zhihua Sun [2], Jiaolin Bao [2], Manqi Zhangsun [2], Shuai Dong [2], Dongting Zhangsun [1,2,]* and Sulan Luo [1,2,]*

[1] Medical School, Guangxi University, Nanning 530004, China
[2] Key Laboratory of Tropical Biological Resources of Ministry of Education, Key Laboratory for Marine Drugs of Haikou, School of Life and Pharmaceutical Sciences, Hainan University, Haikou 570228, China; zhihuasun918@163.com (Z.S.); fish1012@hotmail.com (J.B.); zhangsunmanqi@163.com (M.Z.); dongshuai_1024@163.com (S.D.)
* Correspondence: zhangsundt@163.com (D.Z.); luosulan2003@163.com (S.L.)

Received: 4 March 2020; Accepted: 3 April 2020; Published: 7 April 2020

Abstract: The α9-containing nicotinic acetylcholine receptor (nAChR) is increasingly emerging as a new tumor target owing to its high expression specificity in breast cancer. αO-Conotoxin GeXIVA is a potent antagonist of α9α10 nAChR. Nevertheless, the anti-tumor effect of GeXIVA on breast cancer cells remains unclear. Cell Counting Kit-8 assay was used to study the cell viability of breast cancer MDA-MD-157 cells and human normal breast epithelial cells, which were exposed to different doses of GeXIVA. Flow cytometry was adopted to detect the cell cycle arrest and apoptosis of GeXIVA in breast cancer cells. Migration ability was analyzed by wound healing assay. Western blot (WB), quantitative real-time PCR (QRT-PCR) and flow cytometry were used to determine expression of α9-nAChR. Stable MDA-MB-157 breast cancer cell line, with the α9-nAChR subunit knocked out (KO), was established using the CRISPR/Cas9 technique. GeXIVA was able to significantly inhibit the proliferation and promote apoptosis of breast cancer MDA-MB-157 cells. Furthermore, the proliferation of breast cancer MDA-MB-157 cells was inhibited by GeXIVA, which caused cell cycle arrest through downregulating α9-nAChR. GeXIVA could suppress MDA-MB-157 cell migration as well. This demonstrates that GeXIVA induced a downregulation of α9-nAChR expression, and the growth of MDA-MB-157 α9-nAChR KO cell line was inhibited as well, due to α9-nAChR deletion. GeXIVA inhibits the growth of breast cancer cell MDA-MB-157 cells *in vitro* and may occur in a mechanism abolishing α9-nAChR.

Keywords: α9-nicotinic acetylcholine receptors (nAChRs); breast cancer cells; αO-conotoxin GeXIVA; apoptosis; anti-proliferation; targeted therapy

1. Introduction

Cancer is the second leading cause of death worldwide, and it was estimated to account for 9.6 million deaths in 2018 [1]. According to the statistics, breast cancer is the second most common carcinoma in the world after lung cancer, and is also the highest-incidence cancer among women. In a global context, breast cancer remains the leading cause of cancer incidence and mortality, with 2.1 million newly diagnosed cases and 630 thousand deaths in 2018 [1,2]. Recurrence and metastasis are the major cause of these deaths [2]. Additionally, genetic and reproductive risk factors play important roles in susceptibility to breast cancer [3]. Conventional therapeutic strategies usually provide limited specificity, resulting in severe side effects and toxicity to normal organisms. In contrast, targeted cancer therapy could improve the therapeutic potential of anti-tumor agents and reduce adverse side effects [4].

Nicotinic acetylcholine receptors (nAChRs) belong to ligand-gated ion channels, which are composed of transmembrane subunits that share a common evolutionary origin [5]. The different nAChRs subunits can form homo- or hetero-pentamers and enclose a central ion channel. nAChRs are expressed in the cell membrane of all mammalian cells, including cancer cells [6–8]. nAChRs not only mediate normal physiological responses, such as inflammation and pain, but also participate in the regulation of Alzheimer disease, Parkinson's disease, schizophrenia and cancers etc [8–10].

α7-nAChR and α9-nAChR are the major nAChRs in breast cancer cells [11–13]. Increasing evidence suggests that activation of α7-nAChR leads to activation of the ERK/MAPK and JAK2/PI3K signaling cascades in breast cancer cells [7,14]. In addition, it was reported that α9-nAChR was highly correlated with breast cancer [7,11,15], and stimulation of the α9-nAChR led to breast cancer growth [9]. Further research suggests that a low dose of garcinol (1 μM) from the edible fruit *Garcinia indica* could inhibit nicotine-induced breast cancer cell proliferation through the downregulation of α9-nAChR and cyclin D3 expression [7]. Luteolin and quercetin also could inhibit the ability of proliferation by downregulating the expression of α9-nAChRs on the cell surface of human breast cancer cells [15]. Tea polyphenol(-)-epigallocatechin-3-gallate has been found to inhibit nicotine-and estrogen-induced α9-nicotinic acetylcholine receptor upregulation in human breast cancer cells and delay the development of breast cancer cells in vivo [13]. These results implied that α9-containing nAChRs detected in human breast cancer cells could be used as a new therapeutic molecular target for cancer treatment.

As antagonists to nAChRs, α-conotoxins (α-Ctxs) are used to decipher the pharmacological functions of these receptors, and some of them also have therapeutic potential [16,17]. αO-conotoxin GeXIVA is a potent antagonist of α9α10 nAChRs, which was discovered in *Conus generalis*, native to the South China Sea [16]. This peptide is composed of 28 amino acids including four Cys residues that can form three different disulfide bond connection isomers, i.e., GeXIVA[1,2], GeXIVA[1,3], and GeXIVA[1,4]. They were tested for their potency against both rat and human α9α10 nAChRs, as described previously. Among them, GeXIVA[1,2] was the most potent antagonist of α 9 α 10 nAChR (IC 50 = 4.61 nmol at rat α 9 α 10 nAChR) with a high specificity, as described previously [16,18,19]. It displayed potent alleviation of neuropathic pain in rat model in vivo [20]. Recent reports revealed that both GeXIVA, targeting α9α10 nAChRs, and α-conotoxin TxID, targeting α3β4 nAChRs, contributed to the inhibition of cancer cell proliferation [21,22]. Compared with currently widely used macromolecular antibodies, small peptide toxins have a unique advantage to be developed as new drugs to treat various diseases, since small peptides can overcome the limitations of poor tumor penetration and cellular uptake of antibody when introduced in vivo [23,24]. Antagonists of nAChRs may be novel potential drugs for tumor-targeting therapy and diagnosis because of their specific binding activities and other natural properties.

Our previous research found that α9-nAChR was highly expressed in breast cancer MDA-MB-157 cells. In this study, we aimed to investigate the potential of GeXIVA for treatment of α9-nAChR overexpressed breast cancer cell in vitro. Firstly, the stability of GeXIVA in cell culture medium was investigated. Then, the influences of GeXIVA on growth and apoptosis of MDA-MB-157 cells were examined. In addition, we explored the potential carcinogenic effects of the α9-nAChR subunit through a stable α9 nAChR-knockout (KO) cell line of MDA-MB-157 cells.

2. Results

2.1. A Detection of the Stability of GeXIVA in Cell Culture Medium by RP-UPLC

The stability of GeXIVA was tested in cell culture medium (Figure 1). RP-UPLC (Reverse phase high-performance liquid chromatography) was used to determine the extent of the degradation of the peptides, then, a time-changing trend diagram of GeXIVA in culture media with different concentrations of serum was drawn to judge the stability of the GeXIVA in cell cultures. After 24 h incubation in different concentrations of serum (H_2O Control, 0% FBS, 5% FBS, 10% FBS), the remaining amounts of GeXIVA were ~93%, ~73%, ~66% and ~54% respectively. The results showed that the content of

GeXIVA decreased while the serum concentration increased, and as time went on the trend continued. To assure maximum efficacy of drugs, we selected DMEM medium without serum in the process of GeXIVA-treated cells.

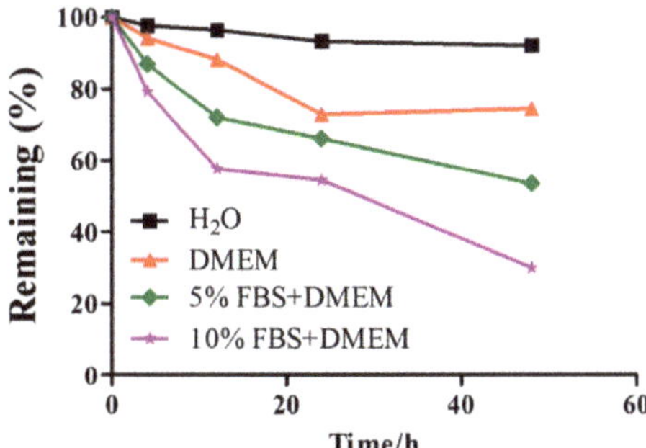

Figure 1. The stability of GeXIVA in cell culture medium. Time course of effects on GeXIVA in cell culture media with different concentration of fetal bovine serum (0%, 5%, 10% FBS), under a 5% CO_2 atmosphere and 37 °C incubation conditions. Each point represents the mean for each time point.

2.2. GeXIVA Affected the Growth of Breast Cancer Cells MDA-MB-157

We compared the effects of GeXIVA on human epithelial cancer cell line MDA-MB-157 and human normal mammary gland epithelial cell line HS578BST (Figure 2). Microscopic examination revealed that cell density and cell morphology were changed significantly after 24 h treatment of GeXIVA at different concentrations, and cancer cells were more sensitive to GeXIVA than normal cells. In the MDA-MB-157 breast cancer cell line, cell proliferation was remarkably decreased after treatment of GeXIVA in a concentration- and time-dependent manner (Figure 2A). In MDA-MB-157 cells, the IC_{50} of 24 h for GeXIVA was approximately 78.31 µM. These results indicated that GeXIVA could effectively inhibit the proliferation of breast cancer cells MDA-MB-157, but had no toxic effects on normal cells HS578BST.

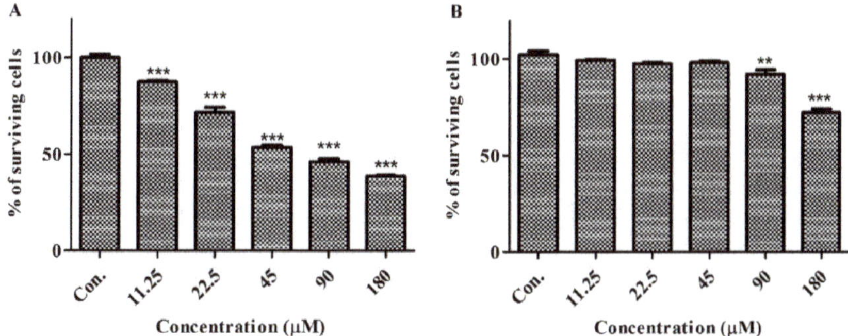

Figure 2. Effects of GeXIVA on the viability of breast cancer cells and normal cells. The cells were treated with various concentrations of GeXIVA for 24 h. Then, the cell viability was determined by CCK-8 assay. Values were expressed as mean ± SD of three independent assays. Statistical analysis was performed with one-way ANOVA and Tukey's test. ** $p < 0.01$, *** $p < 0.001$ indicating a significant difference between the treatments compared to medium control. (**A**) MDA-MB-157; (**B**) Hs578BST.

2.3. GeXIVA Induced Apoptosis in MDA-MB-157 Cells

Apoptosis is a major cause of cancer cell growth inhibition, and previous studies have confirmed that α9-nAChR affects cell proliferation in MDA-MB-157 breast cancer cells. Herein, two-color flow cytometry with Annexin V-FITC and Propidium iodide (PI) labeling showed necrosis to be the predominant mode of cell death in MDA-MB-157 cells treated with various concentrations of GeXIVA (11.25, 22.5, 45 and 90 μM) for 24 h. The significant difference is shown in the representative scatter plots of cells treated by a series of concentrations of GeXIVA (Figure 3A–E). The percentage of early/late apoptosis cells was summarized in Figure 3F. In control group, the proportion of early and late apoptotic cells was 0.73%. After 24 h treatment with 11.25–90 μM GeXIVA, the ratio of early and late apoptotic cells was significantly increased by up to 27.05%. These results showed that GeXIVA inhibits the growth of MDA-MB-157, probably by inducing cell apoptosis.

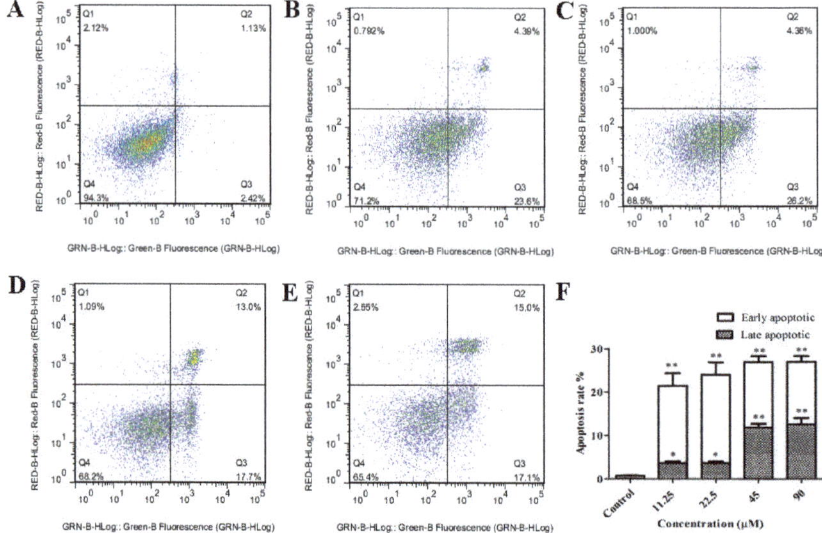

Figure 3. Flow cytometry measurements of apoptosis in MDA-MB-157 cells treated with GeXIVA. Data are presented as dot plots in which the vertical axis represents fluorescence due to PI staining and the horizontal axis represents the fluorescence associated with Annexin V-FITC. The upper left quadrant (Q1) contains necrotic (PI positive) cells, the upper right region (Q2) contains late apoptotic (mixture of PI and Annexin V positive) cells. The lower left region (Q4) contains healthy living (PI and Annexin V negative) cells, and the lower right region (Q3) contains early apoptotic (PI negative and Annexin V positive) cells. Cells were pretreated with 11.25 μM (**B**), 22.5 μM (**C**), 45 μM (**D**), 90 μM (**E**) GeXIVA for 24 h. Then, the cells were washed, harvested, and re-suspended in PBS. The amount of apoptosis cells was measured by flow cytometer. Data were expressed as mean ± SEM of three independent experiments. Significant different was performed by one-way ANOVA. * $p < 0.05$ and ** $p < 0.01$ compared to the control group. **A:** Control. **F:** The inhibition rate was examined by FCM (Flow Cytometry).

2.4. GeXIVA Induced Cell Cycle Arrest in MDA-MB-157 Cells

To elucidate whether GeXIVA treatment induces mitotic inhibition during cell division, we performed cell cycle analysis. Flow cytometry analysis demonstrated that the 24 h incubation of MDA-MB-157 cells with GeXIVA significantly increased the number of cells in the S phase of cell cycle, while the number of cells in the G0/G1 phase was significantly decreased (Figure 4A,B). Regarding G2/M phase, the cells number was significantly decreased when the MDA-MB-157 cells were treated with GeXIVA at the concentration of 90 μM. The results indicated that GeXIVA induced a cell cycle arrest in S phase in MDA-MB-157 cells.

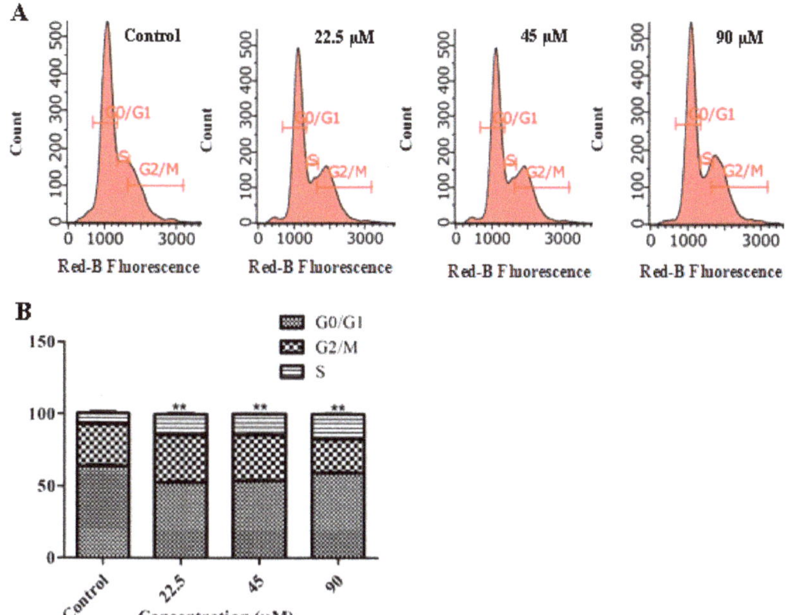

Figure 4. Cell cycle arrest in MDA-MB-157 cells treated with GeXIVA. (**A**) Representative nuclei population distributions of MDA-MB-157 cells before and after incubation with GeXIVA. (**B**) % of cells in each cell cycle phase determined by flow cytometry. The data are presented as % of cells in each cell cycle phase $\pm$ SEM, $n = 3$; ** ($p < 0.01$) indicates a significant difference between the control and GeXIVA-treated groups by one-way ANOVA.

2.5. GeXIVA Decreased the Breast Cancer Cells Abilities of Migration

To test the potential changes in some carcinoma-associated characteristics of the MDA-MB-157 cells after GeXIVA treatment, cell migration ability was measured. The results indicated that cells migrated more slowly to close the scratched wounds after treatment with GeXIVA for 24 h compared with control group (Figure 5). Then, the effects of different concentration on the migration distance of MDA-MB-157 cells were analyzed statistically. This indicated that GeXIVA had inhibition effects on the migration of MDA-MB-157 cells; following the increasing concentrations, cells migration was retarded (Table 1).

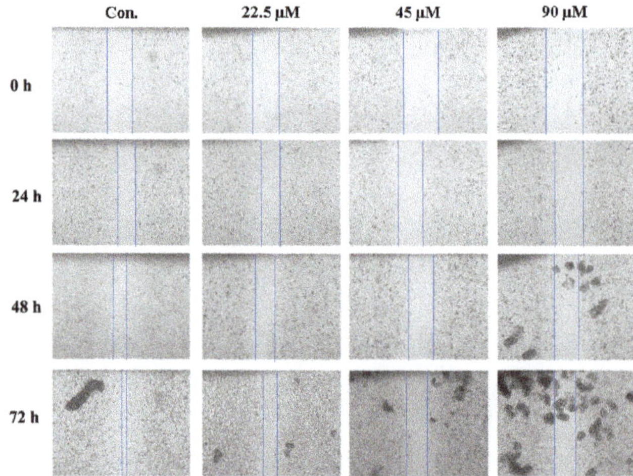

Figure 5. Evaluation of the malignancy of the breast cancer cells after treatment with GeXIVA. Migration assay. MDA-MB-157 cells were pretreated with different concentrations of GeXIVA for 24 h. After treatment, the migration ability of the cells was tested at 24 h, 48 h and 72 h. The blue vertical lines in the images define the open wound areas.

Table 1. The effects of different concentration of on the migration distance of MDA-MB-157 cells in vitro.

Migration Distance	Control		22.5 µM		45 µM		90 µM	
	Mean ± SD	%	Mean ± SD	%	Mean ± SD	%	Mean ± SD	%
0 h	0.52 ± 0.00	–	0.59 ± 0.00	–	0.73 ± 0.00	–	0.74 ± 0.00	–
24 h	0.13 ± 0.01[aAB]	24.52	0.17 ± 0.02[aA]	28.41	0.16 ± 0.03[aAB]	21.56	0.13 ± 0.04[aB]	17.19
48 h	0.24 ± 0.01[bA]	45.81	0.17 ± 0.03[aB]	28.41	0.19 ± 0.02[aA]	25.69	0.21 ± 0.06[aA]	28.05
72 h	0.44 ± 0.02[cA]	84.52	0.27 ± 0.02[bB]	45.45	0.28 ± 0.06[aB]	38.07	0.26 ± 0.02[bB]	34.84

Note: The capital letters A, B and C represent the 95% confidence intervals. One-way ANOVA was performed on the same row (same times) and different columns (different concentrations) in the table. The lowercase letters a, b and c represent the 95% confidence intervals. One-way ANOVA was performed on the same columns (same concentrations) and different columns (different times) in the table. Values with same superscript letters in the same line showed no significant differences ($p > 0.05$), those with different letters showed significant or extreme differences ($p < 0.05$).

2.6. GeXIVA Affected Breast Cancer Cell Line MDA-MB-157 Proliferation through the Inhibition of α9-nAChR-mediated Signals

To test the effect of α9-nAChR while GeXIVA inhibiting breast cancer cells, a stable cell line with *α9-nAChR* gene knock-out was established. The α9-nAChR CRISPR/Cas9 KO plasmid and HDR plasmid were transfected into MDA-MB-157 cells and the transfected cells were selected under the suppression of puromycin. The positive cells were confirmed by detection of the red fluorescent protein (RFP) via fluorescent microscopy (Figure S1). Real-time PCR showed that transfection of MDA-MB-157 cells with α9-nAChR CRISPR/Cas9 KO and HDR plasmid led to more than a 10-fold decrease in α9-nAChR mRNA expression, while the expression of α3 and β4 nAChR subunits mRNA(as control) had no differences (Figure 6A). Flow cytometry showed that knockout of the *α9-nAChR* gene greatly reduced the expression of functional α9-nAChRs on the cell surface (Figure S2 and Figure 6B). We also applied Western blotting analysis to confirm the expression of α9-nAChR protein was significantly silenced in knocked out cell lines (MDA-MB-157 α9-nAChR KO) (Figure 6C).

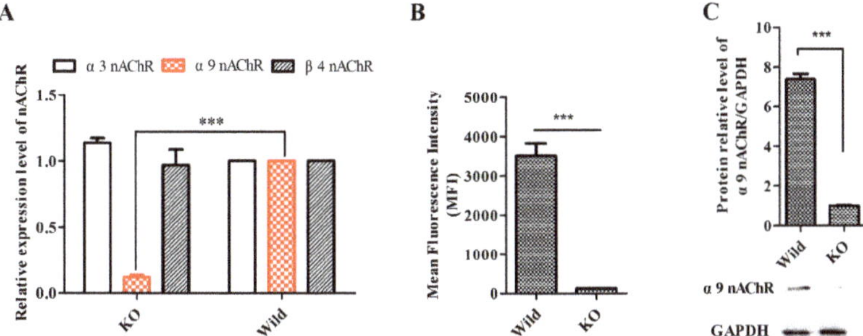

Figure 6. Establishment of a stable α9-nAChR silently expressed MDA-MB-157 breast cancer cell line. **A.** Detection of the mRNA expression of α9-nAChR by qRT-PCR, with the mRNA expression of α3-nAChR and β4-nAChR as control. Data are presented as mean ± SEM, $n = 3$. *** ($p < 0.001$) indicates the significant difference between KO and wild by two-tailed one sample t-test. **B.** Median fluorescence intensities for FITC-labeled antibody against the α9-nAChR for untreated cells (control) and cells with the knocked out α9-nAChR expression by CRISPR/Cas9. Data are presented as mean ± SEM, $n = 3$. *** ($p < 0.001$) indicates the significant difference from control by two-tailed t-test. **C.** Western blot detection of the protein expression of cells after knocked out *CHRNA9* gene treatment. Experiments were repeated three times. *** ($p < 0.001$) indicates the significant difference between KO and wild by one-way ANOVA.

The influence of the α9-nAChR knockout on viability was determined in the breast cancer cell line by commercial Cell Counting Kit-8 (CCK-8) kit. As shown on Figure 7A, MDA-MB-157 WT and MDA-MB-157 α9-nAChR KO cell lines illustrated similar dynamic curves of viability. CCK-8 assay was used to determine the anti-proliferative effect of GeXIVA in both MDA-MB-157 WT and MDA-MB-157 α9-nAChR KO cell lines. GeXIVA displayed cytotoxicity in a concentration-dependent manner in the MDA-MB-157 cell line. However, knocked out *α9-nAChR* gene completely reversed the anti-proliferative effect of GeXIVA (Figure 7B). In addition, GeXIVA produced a survival rate of 1.2% to 18% in MDA-MB-157 wild type cells at high concentrations (90 and 180 μM), while high concentrations of GeXIVA had little effect on the proliferation of MDA-MB-157 α9-nAChR KO cells. Taken together, we can conclude that the anti-proliferative activity of GeXIVA on MDA-MB-157 cells is a result of the peptide interaction with α9-nAChRs.

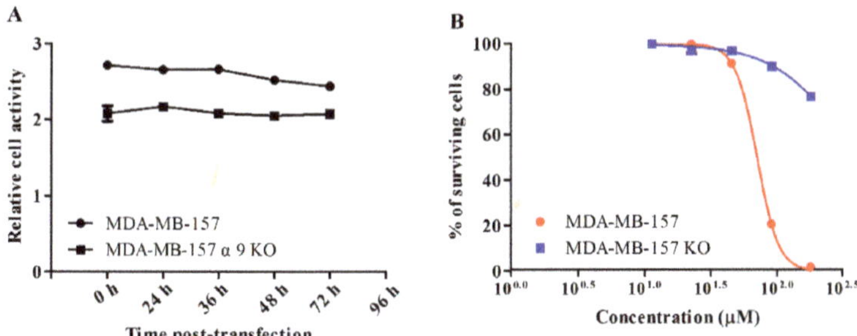

Figure 7. Evaluation of the malignancy of the MDA-MB-157 cells after transfection. (**A**) Cell viability assays. Cells were assayed with Cell Counting Kit-8 kit at various time points after transfection. Each test was repeated three times. Graphical data denote mean ± SD. (**B**) Effects of GeXIVA on cytotoxicity in cells. Breast cancer cells were treated with various concentrations of GeXIVA for 24 h. Then, the anti-proliferation ability of GeXIVA was determined by the CCK-8 assay. Values were expressed as mean ± SD of three independent assays. * $p < 0.001$ for a significant difference in the MDA-MB-157 α9-nAChR KO cell line in comparison to the MDA-MB-157 cell line.

To further test the possible influence of GeXIVA on the expression of α9-nAChR, breast cancer cell line MDA-MB-157 was treated with GeXIVA for 24 h. Cells were harvested and subjected to total RNA extraction. qRT-PCR assays for the α9-nAChR showed significantly reduced transcriptions after cells treated with GeXIVA; however, β4-nAChR was almost unchanged before and after the treatment (Figure 8A). The expression of α9-nAChR was determined by indirect fluorescence staining and flow cytometry analysis, the results revealed a significant decrease in the level of α9-nAChR, after 24 h exposure of cells to GeXIVA at concentrations of 22.5 µM and higher (Figure 8B).

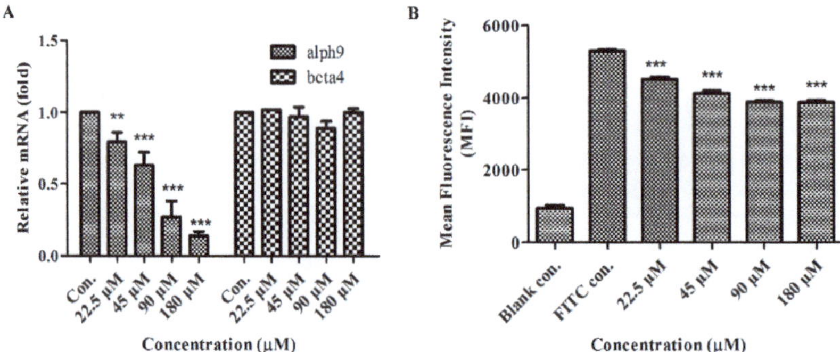

Figure 8. Effects of GeXIVA on the viability of breast cancer cells. **A**, Expression levels of α9 nAChR subunit mRNA were evaluated by real-time PCR analysis. MDA-MB-157 cells were treated with different concentrations of GeXIVA for 24 h. **B**, Flow cytometry analysis of α9-nAChR staining intensity under GeXIVA treatment. Mean staining intensities for α9-nAChR in MDA-MB-157 cells treated with GeXIVA at various concentrations for 24 h compared with the control processed in parallel. Data are expressed as mean ± SD, $n = 3$. ** $p < 0.01$ and *** $p < 0.001$ for significant differences from the control group.

3. Discussion

The α9-nAChR subunit, initially discovered in hair cells of the inner ear, can form a homo-pentamer or assemble into a hetero-pentamer with α10 or α5-nAChR subunits [25,26]. Including α9-nAChR, almost all of nAChRs are expressed not only in neuronal systems, but also in numerous non-neuronal tissues cells such as skin, pancreas and lung, suggesting that nAChRs may have roles in other biological processes in addition to synaptic transmission [8]. It is reported that α9-nAChR is involved in promoting cancer cell proliferation, angiogenesis, cancer metastasis and apoptosis suppression during carcinogenesis in response to tumor microenvironments [7,11,13,27,28]. In a previous study, we determined the expression profile of the α9-nAChR subunit in human breast cancer cell lines and found that α9-nAChR was expressed differently in breast cancer cells. This suggests that α9-nAChR may become an effective target for breast cancer treatment [29].

GeXIVA, a *Conus* peptide, contains 28 amino acids and four Cys residues which have three disulfide bonds [16]. The original study of GeXIVA on α9α10 nAChR expressed in *Xenopus* oocytes revealed that GeXIVA by itself is able to block the α9α10 nAChR current of ACh-evoked currents completely at a 100 nM concentration, and this block was rapidly reversible [16]. GeXIVA is a potent and selective antagonist of α9α10 nAChR subtype [16,30]. In vivo, it also displayed potent alleviation of neuropathic pain in several rat models [18,20]. In vitro, GeXIVA showed an antitumor effect [22]. Therefore, it is worth further investigating GeXIVA as a potential therapeutic agent against breast cancer, etc.

Previous research has indicated that GeXIVA is unstable in serum [30]. In this study, we decided to evaluate the stability of GeXIVA in cell culture medium, and the ability of the GeXIVA to control the growth of MDA-MB-157 cells. The remaining amount of GeXIVA reached ~80% in serum-free DMEM medium after 48 h, so, in the next steps, the cells were co-incubated with GeXIVA in serum-free

DMEM medium to assure maximum efficacy. Prolonged 24 h incubation of MDA-MB-157 cells with GeXIVA resulted in the pronounced concentration-dependent inhibition of the cell growth with IC_{50} of ~78 µM. Moreover, GeXIVA had an obvious inhibitory effect on breast cancer cell line MDA-MB-157 compared with normal breast epithelia cell line HS578BST in vitro. This means that the same doses of GeXIVA could kill more cancer cells than normal cells. In other words, GeXIVA is less toxic to normal cells than cancer cells. Moreover, apoptosis analysis demonstrated that the apoptosis rates in the drug treatment group were significantly higher than in the control group. This result indicated that GeXIVA inhibit the growth of MDA-MB-157, probably by inducing cell apoptosis. We also demonstrated that inhibition of α9-nAChR by GeXIVA could inhibit MDA-MB-157 cancer cell proliferation through the induction of G0/G1 arrest. Metastasis is a major obstacle in clinical cancer therapy [31]. Previous studies have indicated that nicotine can enhance the migratory abilities of cancer cells [31–33]. However, some studies reported the antagonists of nAChRs could reduce the progression of cancer cells [34]. Here, we found that GeXIVA could abolish the abilities of cancer migration.

Previous studies have shown that natural polyphenol compounds can reduce nicotine-mediated carcinogenic effects by inhibiting the expression of α9-nAChR [13,15,35,36]. It is reported that α9-containing nAChRs may play an important role in breast cancer [11,13], but we knew little about the antitumor effect of GeXIVA previously. To determine whether the anti-proliferative effect of GeXIVA is associated with α9-nAChRs, we blocked expression of this receptor by CRISPR/Cas9 knockdown technique. The anti-proliferative effects were profoundly abolished by GeXIVA treatment in α9-nAChR knockdown MDA-MB-157 cells compared with wild-type cells. Thus, we can conclude that the GeXIVA antiproliferative effect on MDA-MB-157 cells is a result of the protein interaction of α9-nAChR with GeXIVA. Therefore, the GeXIVA inhibitory effects might be through downregulating of the protein and gene expression of α9-nAChR. More experiments are needed to gain insight into the underlying molecular mechanisms.

4. Materials and Methods

4.1. Cell Culture

Human breast cancer cell line MDA-MB-157 and human normal mammary gland epithelial cell line HS578BST were purchased from the Kunming Institute of Zoology (Yunnan, China). To make the complete growth medium, all the cells were maintained in Dulbecco's modified eagle medium (DMEM), supplemented with fetal bovine serum (FBS) to a final concentration of 10%, 100 U/mL penicillin and 100 mg/mL streptomycin at 37 °C in an atmosphere of 5% CO_2.

4.2. Cell Culture Medium Stability

Medium stability assay was performed in cell culture media with different concentrations of serum for GeXIVA [1,2]. With setting experimental group (DMEM + GeXIVA, 5% FBS + DMEM + GeXIVA, 10% FBS + DMEM + GeXIVA) and comparison group (GeXIVA was incorporated into the aqueous phase), the sample of each group was placed in an incubator at 37 °C with 5% CO_2 atmosphere and 100% relative humidity for 0, 12, 24 and 48 h. 30 µL triplicate aliquots were taken out at different time points. Each serum aliquot was quenched with 30 µL of 6 M urea and incubated for 10 min at 4 °C. Then, each serum aliquot was treated with 30 µL of 20% trichloroacetic acid (TCA) for another 10 min at 4 °C to precipitate serum proteins. Precipitated serum proteins were then spun down at 14,000 g [19]. Qualitative and quantitative analysis were performed by RP-UPLC, through drawing time-changing trend diagrams of GeXIVA in culture media with different concentrations of serum to judge the stability of the GeXIVA in cell cultures.

4.3. Cell Counting Kit-8 Assay (CCK-8)

To study effects of GeXIVA on cell growth, 5×10^3 cells were seeded in 96-well culture plates. After 24 h, the cells were incubated in a complete medium at 37 °C for adhesion. Thereafter, GeXIVA (dissolved in water) was added to cells at different concentrations (11.25, 22.5, 45, 90 and 180 µM) and grown during 24 h in medium without FBS at 37 °C, 5% CO_2. The medium was removed and the CCK-8 solution was added to cell wells for 1 h at 37 °C. The absorbance at 450 nm (OD 450) each well was obtained using a microplate reader. Do not add any compounds to cells as control. The blank group was supplemented with culture medium and contained no cells.

Cell viability was measured with a commercial Cell Counting Kit-8. Briefly, 2×10^3 cells were seeded onto 96-well plates and the CCK-8 solution was directly added to the cell wells and incubated at 37 °C for 1 h. The absorbance at 450 nm (OD_{450}) for each well was obtained by using a microplate reader with a background control as blank. The cell viability was expressed as the percentage of the untreated control.

4.4. Migration Assay

Wound assay was used to evaluate the migration ability of cells. A total of 1×10^5 cells were seeded in a 12-well plate and grew to confluence followed by scratching the monolayer cells with a 200 µL pipette tip to create a wound. After the medium and floating cells were removed, the cells were rinsed with phosphate-buffered saline three times. The cells were treated with different concentrations of GeXIVA and incubated for 24 h in medium without FBS at 37 °C, 5% CO_2. Plates were washed to remove floating cells and debris, then cell migration images were taken at 0, 24 and 48 hours after drug treatment. Cells without addition of any compounds were used as a control. Each testing group contained at least three independent wells.

4.5. Flow Cytometry

Apoptosis assay was detected with an Annexin V-FITC/PI Apoptosis Detection Kit (KeyGENBioTECH, Nanjing, China) and flow cytometry analysis. A total of 2×10^5 cells were seeded onto 6-well plates. After a 24 h incubation for adhesion, the cells were treated with GeXIVA for 24 h. Cells were collected by trypsinization and dual stained with Annexin V-FITC and PI for 15 min at room temperature in dark. For each sample, data from approximately 1×10^4 cells were recorded in list mode on logarithmic scales. Apoptosis and necrosis were analyzed by quadrant statistics on PI-negative, Annexin V-positive cells and cells positive for both, respectively.

Cells were seeded in 6-well culture plates and incubated with different concentrations of GeXIVA for 24 h. Then the cells were detached from the wells by Trypsin Solution (Without EDTA), washed with PBS containing 3% bovine serum albumin (BSA), and fixed in pre-cooling 70% ethanol for 4 h. After fixation, the cells were washed twice by 3% BAS-PBS, resuspended in 10% RNase-PI staining solution, and analyzed by Guava easyCyte™ flow cytometer.

The expression of α9-nAChR in cells was determined by indirect fluorescence staining and flow cytometry analysis. The treated cells were harvested and washed with PBS twice, fixed with 4% PFA for 20 min and washed twice by centrifugation at 1000 rpm for 5 min each time. Cells were suspended in 0.1% Triton X-100 in 1 × PBS buffer for 10 min and washed twice with PBS by centrifugation at 1300 rpm for 5 min each time. The cells were incubated with 3% BSA in 1 × PBS buffer for 30 min. Primary antibody at an appropriate dilution was added and incubated for 2 hr at room temperature, then washed and incubated with secondary antibody (FITC-conjugated Goat anti-rabbit Ig (G+L)) for 1 hr at room temperature. Finally, the cells were washed twice and re-suspended in 300 µL 1 × PBS and analyzed the expression of α9-nAChRs by flow cytometry. The primary antibody used was as follows: CHRNA9 Rabbit Polychonal Antibody (26025-1-AP, proteintech, Chicago, IL, USA).

4.6. Generation of Stable α9-nAChR-KO Cell Lines

MDA-MB-157 cells were plated in a 6-well culture plate at a density of $1.5–2.5 \times 10^5$ cells in 3 mL antibiotic-free standard growth medium per well. After 24 h, the cells were co-transfected with α9-nAChR CRISPR/Cas9 Knock out (KO) plasmid (Santa Cruz, sc-402753-KO-2, CA, USA) and α9- nAChR homology-directed repair (HDR) plasmid (Santa Cruz, sc-402753-HDR-2, CA, USA). After incubation for 24 h at 37 °C in a CO_2 incubator, the plasmid-containing medium was removed and fresh growth medium containing 10% serum was added in the presence of 6 μg/mL of puromycin. Selected cells, for a minimum of 3–5 days, approximately every 2–3 days, were aspirated and replaced with freshly prepared selective medium. This resulted in the establishment of MDA-MB-157 KO cells in which α9-nAChR gene expression was knocked out.

4.7. RNA Isolation and Quantitative Real-time PCR

Total RNA of cells was isolated using Trizol (Invitrogen, Carlsbad, CA, USA) according to the manufacturer's protocol. Equal amounts of RNA (1 μg) from each sample were reverse-transcribed into first-strand cDNA with a High-Capacity cDNA Reverse Transcription Kit (Applied Biosystems, Carlsbad, CA, USA). The expression of the target gene was evaluated by qRT-PCR on Analytikjena qTOWER^{3}G using specific primers. Primers used were as follows: GAPDH forward: (CAGCCTCAAGATCATCAGCA) and reverse: (TGTGGTCATGAGTCCTTCCA); α9-nAChR forward: (TGGCACGATGCCTATCTCAC) and reverse: (TGATCAGCCCATCATACCGC), α3 nAChR forward: (AACGTGTCTGACCCAGTCATCAT) and reverse: (AGGGGTTCCATTTCAGCTTGTAG); β4-nAChR forward: (TCACAGCTCATCTCCATCAAGCT) and reverse: (CCTGTTTCAGCCAGACATTGGT). The relative expression level of α9, α3 and β4- nAChR mRNAs were determined by the comparative Ct ($2^{-\Delta\Delta Ct}$) method.

4.8. Protein Extraction and Western Blot Analysis

The cultured cells were washed with cold phosphate buffer saline (PBS) three times and harvested using a cell lysis buffer containing protease inhibitors PMSF (Solarbio Life Sciences, Beijing, CHN). Equal amounts of protein from control and treated cell lysates were separated using a 12.5% sodium dodecyl sulfate polyacrylamide gel electrophoresis (SDS-PAGE) gel under reducing conditions and transferred onto polyvinylidene fluoride (PVDF) membranes (Solarbio Life Sciences, Beijing, China), which were subsequently probed with primary antibodies (AChRα9, Santa Cruz, sc-293282, CA, USA). In all Western blots, membranes were additionally probed with an antibody for GAPDH (Santa Cruz, sc-47724, CA, USA) to ensure equal loading of protein between samples. Horseradish peroxidase-conjugated secondary antibodies (m-IgGκ BP-HRP, Santa Cruz, sc-516102, CA, USA) were used with enhanced chemoluminescence reagent (Biosharp, Guangzhou, China) to visualize the protein bands. Images of the films were captured using the Alpha FluorChem E (ProteinSimple, CA, USA).

4.9. Statistical Analysis

Three to five independent repeats were conducted for all experiments. Error bars represent these repeats. Statistical comparisons between groups were performed using ANOVA (Prism GraphPad Software and SPSS 17.0). A Student's *t*-test was used and a *p*-value < 0.05 was considered significant. Analysis was performed with the SPSS software package (version Version 17.0).

5. Conclusions

In this study, the anti-proliferative activity of GeXIVA on breast cancer cells was confirmed *in vitro*, which deserves further study as a potential agent for anti-cancer therapy. Our results demonstrated that α9-nAChR played a key role in breast cancer cells, which was differentially over-expressed in breast cancer cell lines than in normal cells. As an antagonist of α9-nAChR, αO-conotoxin GeXIVA induced cell apoptosis and inhibited the proliferation of breast cancer MDA-MB-157 cells. On the other hand, we found that the proliferation of breast cancer cells was blocked by GeXIVA through S-phase cell cycle arrest. Moreover, GeXIVA induced a downregulation of α9-nAChR expression, and the anti-tumor effect of GeXIVA was abolished by α9-nAChR silencing. These findings provide obvious molecular evidence that GeXIVA is effective drug lead for targeting therapy of breast cancer, which can inhibit the growth of breast cancer cells.

Supplementary Materials: The following are available online at http://www.mdpi.com/1660-3397/18/4/195/s1, Figure S1: The α9-nAChR CRISPR/Cas9 KO plasmid ad HDR plasmid co-transfected into MDA-MB-157 cell line confirmed by detection of the red fluorescent protein (RFP) via fluorescent microscopy. Figure S2. Representative histograms of cell distribution according to intensity of FITC-labeled antibody against the α9-nAChR for untreated cells (left) and cells with the knocked out α9-nAChR expression by CRISPR/Cas9 (right).

Author Contributions: Z.S. and S.L. designed the experiments; Z.S. performed the experiments; Z.S., D.Z., and S.L. analyzed the data; Z.S., S.D., J.B., M.Z., and S.L. wrote the manuscript. All authors have read and agreed to the published version of the manuscript.

Funding: This work was supported in part by National Natural Science Foundation of China (No. 81872794 and No. 41966003) and the Major Science and Technology Project of Hainan Province (No. ZDKJ2016002).

Conflicts of Interest: The authors declare no conflict of interest.

References

1. Bray, F.; Ferlay, J.; Soerjomataram, I.; Siegel, R.L.; Torre, L.A.; Jemal, A. Global cancer statistics 2018: GLOBOCAN estimates of incidence and mortality worldwide for 36 cancers in 185 countries. *CA Cancer J. Clin.* **2018**, *68*, 394–424. [CrossRef]

2. Lee, K.L.; Kuo, Y.C.; Ho, Y.S.; Huang, Y.H. Triple-negative breast cancer: Current understanding and future therapeutic breakthrough targeting cancer stemness. *Cancers* **2019**, *11*, 1334. [CrossRef]

3. Fararjeh, A.F.S.; Tu, S.H.; Chen, L.C.; Cheng, T.C.; Liu, Y.R.; Chang, H.L.; Chang, H.W.; Huang, C.C.; Wang, H.C.R.; Hwang-Verslues, W.W. Long-term exposure to extremely low-dose of nicotine and 4-(methylnitrosamino)-1-(3-pyridyl)-1-butanone (NNK) induce non-malignant breast epithelial cell transformation through activation of the a9-nicotinic acetylcholine receptor-mediated signaling pathw. *Environ. Toxicol.* **2019**, *34*, 73–82. [CrossRef]

4. Shafiee, F.; Aucoin, M.G.; Jahanian-Najafabadi, A. Targeted diphtheria toxin-based therapy: A review article. *Front. Microbiol.* **2019**, *10*, 2340. [CrossRef]

5. Nicolas, L.N.; Pierre-Jean, C.; Jean-Pierre, C. The diversity of subunit composition in nAChRs: Evolutionary origins, physiologic and pharmacologic consequences. *Dev. Neurobiol.* **2010**, *53*, 447–456.

6. Wessler, I.; Kirkpatrick, C.J. Acetylcholine beyond neurons: The non-neuronal cholinergic system in humans. *Br. J. Pharmacol.* **2008**, *154*, 1558–1571. [CrossRef]

7. Ching-Shyang, C.; Chia-Hwa, L.; Chang-Da, H.; Chi-Tang, H.; Min-Hsiung, P.; Ching-Shui, H.; Shih-Hsin, T.; Ying-Jan, W.; Li-Ching, C.; Yu-Jia, C. Nicotine-induced human breast cancer cell proliferation attenuated by garcinol through down-regulation of the nicotinic receptor and cyclin D3 proteins. *Breast Cancer Res. Treat.* **2011**, *125*, 73–87.

8. Dang, N.; Meng, X.; Song, H. Nicotinic acetylcholine receptors and cancer. *Biomed. Rep.* **2016**, *4*, 515–518. [CrossRef]

9. Zhao, Y. The oncogenic functions of nicotinic acetylcholine receptors. *J. Oncol.* **2016**, *2016*, 1–9. [CrossRef]

10. Chen, J.; Cheuk, I.W.Y.; Shin, V.Y.; Kwong, A. Acetylcholine receptors: Key players in cancer development. *Surg. Oncol.* **2019**, *31*, 46–53. [CrossRef]

11. Lee, C.H.; Huang, C.S.; Chen, C.S.; Tu, S.H.; Wang, Y.J.; Chang, Y.J.; Tam, K.W.; Wei, P.L.; Cheng, T.C.; Chu, J.S. Overexpression and activation of the alpha9-nicotinic receptor during tumorigenesis in human breast epithelial cells. *J. Natl. Cancer Inst.* **2010**, *102*, 1322–1335. [CrossRef] [PubMed]

12. Lyukmanova, E.N.; Bychkov, M.L.; Sharonov, G.V.; Efremenko, A.V.; Shulepko, M.A.; Kulbatskii, D.S.; Shenkarev, Z.O.; Feofanov, A.V.; Dolgikh, D.A.; Kirpichnikov, M.P. Human secreted proteins SLURP-1 and SLURP-2 control the growth of epithelial cancer cells via interaction with nicotinic acetylcholine receptors. *Br. J. Pharmacol.* **2018**, *175*, 1973–1986. [CrossRef] [PubMed]

13. Tu, S.-H.; Ku, C.-Y.; Ho, C.-T.; Chen, C.-S.; Huang, C.-S.; Lee, C.-H.; Chen, L.-C.; Pan, M.-H.; Chang, H.-W.; Chang, C.-H.; et al. Tea polyphenol (-)-epigallocatechin-3-gallate inhibits nicotine-and estrogen-induced α9-nicotinic acetylcholine receptor upregulation in human breast cancer cells. *Mol. Nutr. Food Res.* **2011**, *55*, 455–466. [CrossRef]

14. Lin, W.; Hirata, N.; Sekino, Y.; Kanda, Y. Role of α7-nicotinic acetylcholine receptor in normal and cancer stem cells. *Curr. Drug Targets* **2012**, *13*, 656–665.

15. Shih, Y.L.; Liu, H.C.; Chen, C.S.; Hsu, C.H.; Pan, M.H.; Chang, H.W.; Chang, C.H.; Chen, F.C.; Ho, C.T.; Yang, Y.Y.; et al. Combination treatment with luteolin and quercetin enhances antiproliferative effects in

nicotine-treated MDA-MB-231 cells by down-regulating nicotinic acetylcholine receptors. *J. Agric. Food Chem.* **2010**, *58*, 235–241. [CrossRef]

16. Luo, S.; Zhangsun, D.; Harvey, P.J.; Kaas, Q.; Wu, Y.; Zhu, X.; Hu, Y.; Li, X.; Tsetlin, V.I.; Christensen, S.; et al. Cloning, synthesis, and characterization of αO-conotoxin GeXIVA, a potent α9α10 nicotinic acetylcholine receptor antagonist. *Proc. Natl. Acad. Sci. USA* **2015**, *112*, E4026.

17. Harry, K.; Adams, D.J.; Callaghan, B.; Nevin, S.; Alewood, P.F.; Vaughan, C.W.; Mozar, C.A.; Christie, M.J. A novel mechanism of inhibition of high-voltage activated calcium channels by α-conotoxins contributes to relief of nerve injury-induced neuropathic pain. *PAIN* **2011**, *152*, 259–266.

18. Wang, H.; Li, X.; Zhangsun, D.; Yu, G.; Su, R.; Luo, S. The alpha9alpha10 nicotinic acetylcholine receptor antagonist alpha O-conotoxin GeXIVA[1,2] alleviates and reverses chemotherapy-induced neuropathic pain. *Mar. Drugs* **2019**, *17*, E265. [CrossRef]

19. Yu, S.; Wu, Y.; Xu, P.; Wang, S.; Zhangsun, D.; Luo, S. Effects of serum, enzyme, thiol, and forced degradation on the stabilities of alpha O-Conotoxin GeXIVA[1,2] and GeXIVA [1,4]. *Chem. Biol. Drug Des.* **2018**, *91*, 1030–1041. [CrossRef]

20. Li, X.; Hu, Y.; Wu, Y.; Huang, Y.; Yu, S.; Ding, Q.; Zhangsun, D.; Luo, S. Anti-hypersensitive effect of intramuscular administration of αO-conotoxin GeXIVA[1,2] and GeXIVA[1,4] in rats of neuropathic pain. *Prog. Neuropsychopharmacol. Biol. Psychiatry* **2016**, *66*, 112–119. [CrossRef]

21. Qian, J.; Liu, Y.-Q.; Sun, Z.-H.; Zhangsun, D.-T.; Luo, S.-L. Identification of nicotinic acetylcholine receptor subunits in different lung cancer cell lines and the inhibitory effect of alpha-conotoxin TxID on lung cancer cell growth. *Eur. J. Pharmacol.* **2019**, *865*, 172674. [CrossRef]

22. Liu, Y.Q.; Qian, J.; Sun, Z.H.; Zhangsun, D.T.; Luo, S.L. Cervical cancer correlates with the differential expression of nicotinic acetylcholine receptors and reveals therapeutic targets. *Mar. Drugs* **2019**, *17*, E256. [CrossRef] [PubMed]

23. Mei, D.; Zhao, L.; Chen, B.; Zhang, X.; Zhang, Q. α-Conotoxin ImI-modified polymeric micelles as potential nanocarriers for targeted docetaxel delivery to α7-nAChR overexpressed non-small cell lung cancer. *Drug Deliv.* **2018**, *25*, 493–503. [CrossRef] [PubMed]

24. Aina, O.H.; Sroka, T.C.; Chen, M.-L.; Lam, K.S. Therapeutic cancer targeting peptides. *Biopolymers* **2002**, *66*, 184–199. [CrossRef] [PubMed]

25. Elgoyhen, A.B.; Johnson, D.S.; Boulter, J.; Vetter, D.E.; Heinemann, S. Alpha 9: An acetylcholine receptor with novel pharmacological properties expressed in rat cochlear hair cells. *Cell* **1994**, *79*, 705–715. [CrossRef]

26. Hurst, R.; Rollema, H.; Bertrand, D. Nicotinic acetylcholine receptors: From basic science to therapeutics. *Pharmacol. Ther.* **2013**, *137*, 22–54. [CrossRef]

27. Paliwal, A.; Vaissiere, T.; Krais, A.; Cuenin, C.; Cros, M.P.; Zaridze, D.; Moukeria, A.; Boffetta, P.; Hainaut, P.; Brennan, P. Aberrant DNA methylation links cancer susceptibility locus 15q25.1 to apoptotic regulation and lung cancer. *Cancer Res.* **2010**, *70*, 2779–2788. [CrossRef]

28. Wei, P.L.; Chang, Y.J.; Ho, Y.S.; Lee, C.H.; Yang, Y.Y.; An, J.; Lin, S.Y. Tobacco-specific carcinogen enhances colon cancer cell migration through α7-nicotinic acetylcholine receptor. *Ann. Surg.* **2009**, *249*, 978–985. [CrossRef]

29. Sun, Z.H.; Zhangsun, M.Q.; Dong, S.; Liu, Y.Q.; Qian, J.; Zhangsun, D.T.; Luo, S.L. Differential Expression of Nicotine Acetylcholine Receptors Associates with Human Breast Cancer and Mediates Antitumor Activity of αO-Conotoxin GeXIVA. *Mar. Drugs* **2020**, *18*, 61. [CrossRef]

30. Zhangsun, D.; Zhu, X.; Kaas, Q.; Wu, Y.; Craik, D.J.; McIntosh, J.M.; Luo, S. Alpha O-Conotoxin GeXIVA disulfide bond isomers exhibit differential sensitivity for various nicotinic acetylcholine receptors but retain potency and selectivity for the human alpha9alpha10 subtype. *Neuropharmacology* **2017**, *127*, 243–252. [CrossRef]

31. Hung, C.S.; Peng, Y.-J.; Wei, P.-L.; Lee, C.-H.; Su, H.-Y.; Ho, Y.-S.; Lin, S.-Y.; Wu, C.-H.; Chang, Y.-J. The alpha9 nicotinic acetylcholine receptor is the key mediator in nicotine-enhanced cancer metastasis in breast cancer cells. *J. Exp. Clin. Med.* **2011**, *3*, 283–292. [CrossRef]

32. Dasgupta, P.; Rizwani, W.; Pillai, S.; Kinkade, R.; Kovacs, M.; Rastogi, S.; Banerjee, S.; Carless, M.; Kim, E.; Coppola, D. Nicotine induces cell proliferation, invasion and epithelial-mesenchymal transition in a variety of human cancer cell lines. *Int. J. Cancer* **2009**, *124*, 36–45. [CrossRef] [PubMed]

33. Guo, J.; Ibaragi, S.; Zhu, T.; Luo, L.-Y.; Hu, G.-F.; Huppi, P.S.; Chen, C.Y. Nicotine promotes mammary tumor migration via a signaling cascade involving protein kinase c and cdc42. *Cancer Res.* **2008**, *68*, 8473–8481. [CrossRef] [PubMed]

34. Bychkov, M.; Shenkarev, Z.; Shulepko, M.; Shlepova, O.; Kirpichnikov, M.; Lyukmanova, E. Water-soluble variant of human Lynx1 induces cell cycle arrest and apoptosis in lung cancer cells via modulation of alpha7 nicotinic acetylcholine receptors. *PLoS ONE* **2019**, *14*, e0217339. [CrossRef] [PubMed]

35. Chen, C.-S.; Lee, C.-H.; Hsieh, C.-D.; Ho, C.-T.; Pan, M.-H.; Huang, C.-S.; Tu, S.-H.; Wang, Y.-J.; Chen, L.-C.; Chang, Y.-J. Nicotine-induced human breast cancer cell proliferation attenuated by garcinol through down-regulation of the nicotinic receptor and cyclin D3 proteins. *Breast Cancer Res. Treat.* **2010**, *125*, 73–87. [CrossRef] [PubMed]

36. Hsuuw, Y.-D.; Chan, W.-H. Epigallocatechin gallate dose-dependently induces apoptosis or necrosis in human MCF-7 cells. *Ann. N.Y. Acad. Sci.* **2007**, *1095*, 428–440. [CrossRef]

 *marine drugs*

Editorial

Marine Compounds and Cancer: The First Two Decades of XXI Century

Sergey A. Dyshlovoy [1,2,3,4,]* and Friedemann Honecker [2,5,]*

[1] Laboratory of Pharmacology, A.V. Zhirmunski National Scientific Center of Marine Biology, Far Eastern Branch, Russian Academy of Sciences, 690041 Vladivostok, Russia
[2] Department of Oncology, Hematology and Bone Marrow Transplantation with Section Pneumology, Hubertus Wald-Tumorzentrum, University Medical Center Hamburg-Eppendorf, 20251 Hamburg, Germany
[3] Laboratory of Biologically Active Compounds, Department of Bioorganic Chemistry and Biotechnology, School of Natural Sciences, Far Eastern Federal University, 690091 Vladivostok, Russia
[4] Martini-Klinik, Prostate Cancer Center, University Hospital Hamburg-Eppendorf, 20251 Hamburg, Germany
[5] Tumor and Breast Center ZeTuP St. Gallen, 9000 St. Gallen, Switzerland
* Correspondence: dyshlovoy@gmail.com (S.A.D.); friedemann.honecker@zetup.ch (F.H.)

Received: 20 December 2019; Accepted: 24 December 2019; Published: 26 December 2019

In 2019, the scientific and medical community celebrated the 50th anniversary of the introduction of the very first marine-derived drug, Cytarabine, into clinics. Cytarabine (aka Ara-C, Cytosar-U®) was first isolated from a marine sponge and is known to kill cancer cells by blocking DNA polymerase function [1]. In 1969, the FDA approved the drug for the treatment of leukemia. This marine drug, which still belongs to the mainstay of leukemia therapy, and by now has probably saved many thousands of lives, made its way into clinics in less than 20 years after the original prototype molecules, namely spongothymidine and spongouridine, were reported by Bergmann and Feeney in 1951 [2]. Some years later, in 1976, another marine drug, Vidarabine (Ara-A, Vira-A®), was approved for the treatment of *Herpes simplex* virus. However, after that, the clinical development of marine-derived drugs was less successful [1], and for another almost 40 years, no other compounds were approved by drug authorities. Moreover, the rapid development of methods of high-throughput screening, and especially computational approaches to rational drug design, made some scientists believe that the search for new bioactive molecules from natural sources was an activity of the past. Thus, by the end of the last century, a part of the scientific community was rather skeptic on whether new natural products and, in particular, marine natural products, still harbored the potential to make new drugs. However, having passed this transition period, the situation has definitely changed, and at the beginning of the 21st century, marine drugs entered a time of renaissance [3,4]. Nowadays, while many terrestrial animals and plants are already well investigated, marine inhabitants have become the main source of new chemical compounds. Moreover, due to the specific (and often extreme) environmental conditions, marine inhabitants often possess particular biochemistry which results in unique secondary metabolites. These organisms are mainly produced by marine invertebrates like sponges and tunicates (*Ascidia*), and marine fungi and bacteria, the latter often being the true producers of small bioactive molecules of interest.

Currently, the potential for marine natural products as drug candidates has been recognized all over the world, and the field is constantly growing and developing. Additionally, the development of new chemical and physicochemical approaches and tools has led to the isolation and structure elucidation of novel minor marine secondary metabolites, which could not be isolated/detected in the past. The number of structures isolated each year has almost doubled over the past 20 years. This is illustrated by the fact that according to the report of John Faulkner, 869 new structures were isolated from marine organisms during the year 2000 [5]. A decade later, in 2010, the number had risen to 1003 substances per year, as reported by John Blunt et al. [6], and a very recent report by Anthony Carroll et al. described 1490 new molecules isolated in the year 2017 [7]. Moreover, the development of new and improved organic synthesis methods made possible the synthesis of promising active compounds in the amounts required for further preclinical and clinical studies.

Tremendous progress in the clinical development of marine-derived drugs has been achieved over the past 20 years. For example, during this period, six out of nine currently used drugs of marine origin have been approved by their corresponding authorities [8]. Focusing specifically on anti-cancer drugs, two more antineoplastic agents, Plitidepsin [9] and Polatuzumab vedotin [10], have been approved since we reviewed this topic in 2017 [11]. Hence, the full list of the marine-derived drugs used for cancer treatment at the end of 2019 included:

(1) Spongian nucleoside **Cytarabine** (Ara-C, Cytosar-U® (*Pfizer*), see above);

(2) Spongian macrolide **Eribulin mesylate** (E7389, Halaven® (*Eisai Inc.*), first approved in 2010 for the treatment of metastatic breast cancer; mechanism of action—irreversible mitotic blockade);

(3) **Brentuximab vedotin**—an antibody-drug conjugate (ADC) of the CD30-specific monoclonal antibody brentuximab with the antimitotically active monomethyl auristatin E (MMAE), which is a synthetic analog of dolastatin-10, produced by cyanobacteria symbiotic to sea hare *Dolabella auricularia* (SGN-35, Adcetris® (*Seattle Genetics*), which was first approved in 2011 for the treatment of anaplastic large T-cell systemic malignant lymphomas, and Hodgkin's lymphomas; mechanism of action—binding to CD30 (antibody) and inhibition of tubulin polymerization (MMAE));

(4) Ascidian alkaloid **Trabectidine** (ET-743, Yondelis® (*PharmaMar*), first approved in 2015 for the treatment of soft tissue sarcoma and ovarian cancer, mechanism of action—binding to the minor groove of DNA);

(5) The ascidian depsipeptide **Plitidepsin** (dehydrodidemnin B, Aplidin® (*PharmaMar*), first approved in 2018 for the treatment of leukemia, lymphoma, and multiple myeloma, mechanism of action—induction of oxidative stress, binding to eEF1A2). Note: Aplidin® is currently approved by the Australian Regulatory Agency for use in Australia only;

(6) **Polatuzumab vedotin**—another antibody-drug conjugate which consists of a CD76b-targeting antibody and MMAE (DCDS-4501A, Polivy™ (*Genetech, Roche*), and was first approved in 2019 for the treatment of chronic lymphocytic leukemia, B-cell lymphomas, non-Hodgkin lymphomas, mechanism of action—binding to CD76b (antibody) and inhibition of tubulin polymerization (MMAE)).

By the end of 2019, there were an additional 28 drug candidates in different stages of clinical trials. The majority of them (24 substances, 85%) are undergoing clinical trials as anti-cancer drugs. Interestingly, 20 out of 24 molecules (83%) are antibody-drug conjugates (ADC) and therefore consist of a small cytotoxic "warhead" molecule and a specific antibody targeting different cellular membrane proteins.

Currently, five compounds are undergoing testing in phase III clinical trials and are at the final step before approval can be obtained. Apart from the well-known Tetrodotoxin (Halneuron™, *Wex Pharmaceutical Inc.*), which is being developed as a painkiller for chronic, severe to moderate cancer-related pain, the other four candidates are all investigated as potential chemotherapeutic agents. In more detail, these substances are the nectin-4 targeting ADC Enfortumab Vedotin (ASG-22ME, *Seattle Genetics*) which is being tested in several urogenital tumors; the synthetic analog of the fungian diketopiperazine Plinabulin (NPI-2358, *BeyondSpring Pharmaceuticals Inc.*), which is able to inhibit tubulin polymerization and is tested in brain tumors and non-small cell lung cancer (NSCLC); the DNA minor groove binder Lurbinectedin (PM01183, Zepsyre®, *Pharmamar*), a synthetic analog of trabectedin which currently is tested in small cell lung cancer (SCLC), ovarian and breast cancers; and the bacterial proteasome inhibitor β-lactone-γ-lactam Marizomib (NPI-0052, Salinosporamide A, *Triphase*), which has been examined in melanomas, lymphomas, NSCLC, pancreatic cancers, as well as multiple myelomas.

On the other hand, most of the molecules undergoing phase I and phase II trials are monomethyl auristatin E or F (MMAE or MMAF) conjugates with different antibodies [8]. It seems a safe bet that we may expect several MMAE/MMAF based drugs to make it into clinics in the next decade. Excitingly, there are around 2000 marine-derived molecules that show interesting *in vivo* biological activity, and more than 10,000 compounds isolated from marine organisms for which *in vitro* activity has been described [12].

Drug development is a very dynamic field, as new and old drug candidates regularly enter phase I trials for different cancers, go on to the next trial phases, or fail. Therefore, this information should be regularly revised. For those who are interested in the most up-to-date status of different drug candidates, we recommend visiting the Marine Pharmacology web-site (https://www.midwestern.edu/departments/marinepharmacology.xml) ran and regularly updated by former Editor-in-Chief of Marine drugs Prof. Alejandro M. S. Mayer (Midwestern University, IL, USA), and his team. Vital information can also be obtained from www.clinicaltrials.gov, www.accessdata.fda.gov, or directly from the web-pages of pharmaceutical companies involved in marine drug development.

The main obstacle for novel marine-derived molecules on their way to becoming clinically useful drugs (i.e., anti-cancer drugs) is the generation of larger amounts of the active substance. At the beginning of the era of clinical use of marine drugs, naturally harvested or maricultured organisms were used as the main source of an active compound. As an example, the first hundreds of milligrams of halichondrin B and trabectedin were produced from marine invertebrates cultured in marine farms by the US National Cancer Institute [13], and PharmaMar [14], without which preclinical and clinical trials and further development of the meanwhile approved drugs Halaven® and Yondelis® would not have been possible. Nowadays, interesting compounds are largely produced by chemical synthesis or by modifying natural compounds that are mass-produced by microorganisms, either in wild-type or genetically modified cultures. However, macromolecules, such as marine polysaccharides and their derivatives (mainly used as food, as food additives, or in the cosmetic industry), are isolated and purified from natural sources like marine alga. There are some exceptions to this rule, e.g., several small molecules possessing anti-cancer activity isolated directly from the sponge *Aplysina aerophoba* marketed by BromMarin [15].

Marine Drugs (ISSN 1660-3397) is a leading journal on the research, development, and production of biologically and therapeutically active compounds from the sea. To document the dynamic field of marine anti-cancer pharmacology, the Topical Collection "Marine Compounds and Cancer" (http://www.mdpi.com/journal/marinedrugs/special_issues/marine-compounds-cancer) was started three years ago [11]. This Topical Collection covers the whole scope from agents with cancer-preventive activity to novel and previously characterized compounds with anti-cancer activity, both in vitro and in vivo, and the latest status of clinical development from drug trials. Of note, compounds possessing pro-carcinogenic activity or mediating cancer cell survival are also within the scope of the collection. Owing to the importance of trial execution, special attention is given to current shortfalls and possible strategies to overcome obstacles in the area of marine anti-cancer drug development.

These are exciting times for scientists and physicians involved in the investigation and development of marine drugs! We invite the authors to share with us and all our readers your latest and forward-looking discoveries!

Dr. Sergey A. Dyshlovoy and Dr. Friedemann Honecker, Guest Editors, Topical collection "Marine Compounds and Cancer".

References

1. Stonik, V. Marine natural products: A way to new drugs. *Acta Nat.* **2009**, *2*, 15–25. [CrossRef]
2. Bergmann, W.; Feeney, R.J. Contributions to the study of marine products. XXXII. The nucleosides of spongies. I. *J. Org. Chem.* **1951**, *16*, 981–987. [CrossRef]
3. Paterson, I.; Anderson, E.A. The Renaissance of Natural Products as Drug Candidates. *Science* **2005**, *310*, 451. [CrossRef] [PubMed]

4.   Molinski, T.F.; Dalisay, D.S.; Lievens, S.L.; Saludes, J.P. Drug development from marine natural products. *Nat. Rev. Drug Discov.* **2009**, *8*, 69–85. [CrossRef] [PubMed]

5.   Faulkner, D.J. Marine natural products. *Nat. Prod. Rep.* **2002**, *19*, 1–48. [PubMed]

6.   Blunt, J.W.; Copp, B.R.; Keyzers, R.A.; Munro, M.H.; Prinsep, M.R. Marine natural products. *Nat. Prod. Rep.* **2012**, *29*, 144–222. [CrossRef] [PubMed]

7.   Carroll, A.R.; Copp, B.R.; Davis, R.A.; Keyzers, R.A.; Prinsep, M.R. Marine natural products. *Nat. Prod. Rep.* **2019**, *36*, 122–173. [CrossRef] [PubMed]

8.   Mayer, A. Marine Pharmaceutical: the Clinical Pipeline. Available online: https://www.midwestern.edu/departments/marinepharmacology/clinical-pipeline.xml (accessed on 25 December 2019).

9.   *Australian Public Assessment Report*; The Therapeutic Goods Administration, Department of Health, Australian Government: Symonston, Australia for Plitidepsin. Available online: https://www.tga.gov.au/sites/default/files/auspar-plitidepsin-190513.pdf (accessed on 25 December 2019).

10.   U.S. Food and Drug Administrtion. FDA Approves Polatuzumab Vedotin-piiq for Diffuse Large B-Cell Lymphoma. Available online: https://www.fda.gov/drugs/resources-information-approved-drugs/fda-approves-polatuzumab-vedotin-piiq-diffuse-large-b-cell-lymphoma (accessed on 25 December 2019).

11.   Dyshlovoy, S.A.; Honecker, F. Marine Compounds and Cancer: 2017 Updates. *Mar. Drugs* **2018**, *16*, 41. [CrossRef] [PubMed]

12.   Mayer, A. The marine pharmacology and pharmaceuticals pipeline in 2017. In Proceedings of the 10th European conference on Marine Products, Crete, Greece, 3–7 September 2017.

13.   Munro, M.H.; Blunt, J.W.; Dumdei, E.J.; Hickford, S.J.; Lill, R.E.; Li, S.; Battershill, C.N.; Duckworth, A.R. The discovery and development of marine compounds with pharmaceutical potential. *J. Biotechnol.* **1999**, *70*, 15–25. [CrossRef]

14.   Cuevas, C.; Francesch, A. Development of Yondelis (trabectedin, ET-743). A semisynthetic process solves the supply problem. *Nat. Prod. Rep.* **2009**, *26*, 322–337. [CrossRef] [PubMed]

15.   BromMarin. Available online: http://www.brommarin.com/homepage/index.php (accessed on 25 December 2019).

Review

Design and Synthesis of Anti-Cancer Chimera Molecules Based on Marine Natural Products

Min Woo Ha, Bo Reum Song, Hye Jin Chung and Seung-Mann Paek *

College of Pharmacy and Research Institute of Pharmaceutical Sciences, Gyeongsang National University, Jinju Daero 501, Jinju 52828, Gyeongnam, Korea
* Correspondence: million@gnu.ac.kr; Tel.: +82-55-772-2424

Received: 23 July 2019; Accepted: 16 August 2019; Published: 27 August 2019

Abstract: In this paper, the chemical conjugation of marine natural products with other bioactive molecules for developing an advanced anti-cancer agent is described. Structural complexity and the extraordinary biological features of marine natural products have led to tremendous research in isolation, structural elucidation, synthesis, and pharmacological evaluation. In addition, this basic scientific achievement has made it possible to hybridize two or more biologically important skeletons into a single compound. The hybridization strategy has been used to identify further opportunities to overcome certain limitations, such as structural complexity, scarcity problems, poor solubility, severe toxicity, and weak potency of marine natural products for advanced development in drug discovery. Further, well-designed marine chimera molecules can function as a platform for target discovery or degradation. In this review, the design, synthesis, and biological evaluation of recent marine chimera molecules are presented.

Keywords: chimera; chemical conjugation; marine natural products; anticancer agent; hybridization

---

1. Introduction

The struggle for existence in a natural environment occasionally makes a species develop its own unique weapons such as speed, power, or even toxins. As these toxins possess biologically potent activity and unique modes of actions, these natural products have been regarded as a robust platform for further medicinal research [1,2]. Thus far, marine natural products have been particularly highlighted for their extraordinary bioactivity under highly diluted conditions [3–5]. Therefore, it is plausible to utilize marine natural products as a hit or lead compound in drug discovery and its further development [6,7].

Although marine natural products could have the powerful potential for drug discovery, there are also a few obstacles associated with them. First, it is rather difficult to secure a sufficient amount of these products for further study [8]. In most cases, the medicinal study of natural products requires a substantial number of test samples for elucidation of target protein/receptor and following signaling pathway. However, harsh conditions, difficulty in access, and scarcity of the target organism are hindrances to large scalable synthesis of important marine natural products, such as spongistatin 1 (13.8 mg from 400 kg of marine sponge) [9], phorboxazole A (95 mg from 236 g of *Phorbas* sp.) [10], discodermolide (7 mg from 434 g of *Discodermia dissoluta*) [11], bryostatin 1 (0.9~1.8 g from 10,000 gallons of *Bugula neritina*), [12,13] etc. The unmet needs from natural resources occasionally leads to samples being obtained via chemical synthesis, and it usually remains difficult to satisfy substantial supply requirements [14–16] (Figure 1).

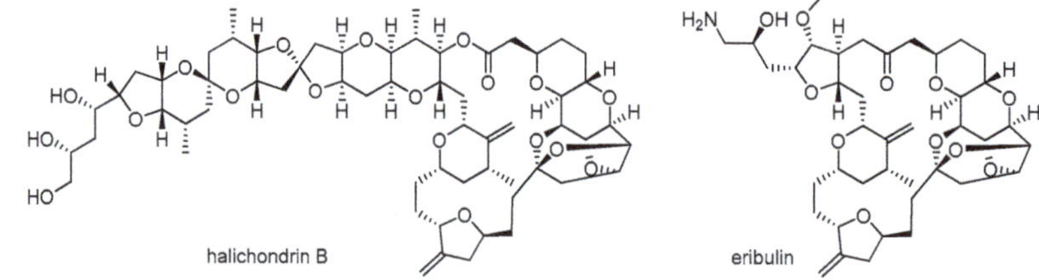

Figure 1. Structure of certain bioactive marine natural products.

Second, structural complexity is another hurdle. The highly complex structure of marine natural products frequently makes it very difficult to modify or synthesize them on a large scale [17,18]. In order to improve the biological activity of these products, both chemical modification and related structure-activity relationship (SAR) study of marine natural products are necessary [19]. However, their highly complex structure hampers efficient modifications and any subsequent systematic research [20]. In order to address this structural complexity, a more simplified analog of marine natural products can be a breakthrough in drug discovery, as evident in the case of halichondrin B [21] and its simplified analog eribulin [22] (Figure 2). Although halichondrin B possesses extraordinary cytotoxicity against B-16 melanoma cells (IC$_{50}$ 0.093 ng/mL) and other tumor cell lines, its structure is too complex to be an advanced drug or drug candidate for related cancer therapy. Consequently, the truncated ketone analog eribulin of halichondrin B was developed through an entirely synthetic approach [23] and approved for treatment of metastatic breast cancer as a mesylate salt.

Figure 2. Halichondrin B [21] and its synthetic analog eribulin [22].

The third hurdle is biological activity and its selectivity. Since marine natural products are derived from marine organisms and not humans, their application to the human body may trigger undesired biological processes [24]. In order to solve this selectivity issue, various analogs are made and tested [25,26]. In addition, systematic study for mechanism of action or ligand-target binding usually follows valuable analogs [27]. Nonetheless, marine natural products continue to serve as a versatile starting point for drug discovery because of their unique structural framework and biological activity. Simultaneously, the strategy for the efficient modification of this complex molecule has also been evolving.

Molecular hybridization can be a good strategy for advanced marine natural products [28–31]. Certain drawbacks of marine natural products such as structural complexity or non-selective biological activities were overcome through a combination with other bioactive molecules. This combination strategy was evolved to develop chimera molecules, pursuing not merely a synergistic sequence but a systematically operating biological sequence. For example, the hybridization of a ligand and ubiquitin recruiting probe enables the degradation of target biomolecules via cellular sequence, such as ubiquitination and following the proteasome pathway. This proteolysis targeting chimera (PROTAC) approach showcases a well-designed chimera strategy from natural products [32–37]. (Figure 3) In addition, biotinylation of active ligand provides affinity column chromatography to enable target protein isolation [38]. In addition to these strategies, an improvement of basic activity can be anticipated. In this regard, advanced chimeric molecules from marine natural products with anticancer activity is reviewed here.

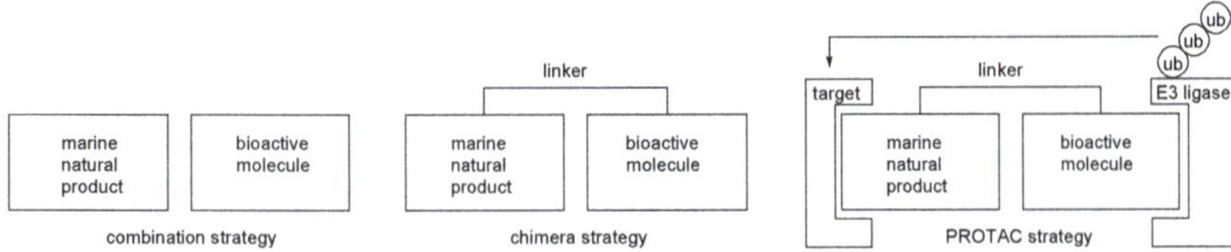

Figure 3. Comparison of the combination and chimera strategies.

2. Results

2.1. Conjugation with Other Active Molecules

The macrosphelide (MS) family [39], comprising 13 natural isomers, are 15- or 16-membered macrolide antitumor agents. They are derived from marine sponge *Periconia byssoides* [40] (MSC **1**, MSE-H, and MSL) and soil fungi *Microsphaeropsis* sp. FO-5050 [41] (MSA-D, MSJ, and MSK) and show cell adhesion inhibitory activity or immunosuppressive activity [42]. In addition, a derivatization of macrosphelide skeleton based on MSA **2** or MSE **3** enables the activation of apoptosis in human lymphoma U937 cells, although their activity is slightly weak [43]. In order to increase this apoptosis-inducing activity in cancer cells, preparation of a chimera compound with another 16-membered anticancer agent, epothilone **4**—which shows not only tubulin-disrupting profiles and anticancer activity, similar to the paclitaxel, but also an apoptosis-inducing property in cancer cells to inhibit tumor cell growth—was pursued [44]. Based on their similar structural features and biological properties, hybridization of MS and epothilone **4** was performed, pursuing an advanced anticancer agent.

Scheme 1 illustrates the synthetic plan of the MS-epothilone chimera. Known intermediate **5** [45] for total synthesis of epothilone was prepared and linked to fragments **6**, **7**, [42] and **8** of MS via iterative esterification and deprotection of secondary alcohol moiety. For an elucidation of the structure-activity relationship, a thiazole side chain of epothilone was introduced in each CH_3 group in the MS skeleton to provide the desired MS-epothilone chimera **9–11**.

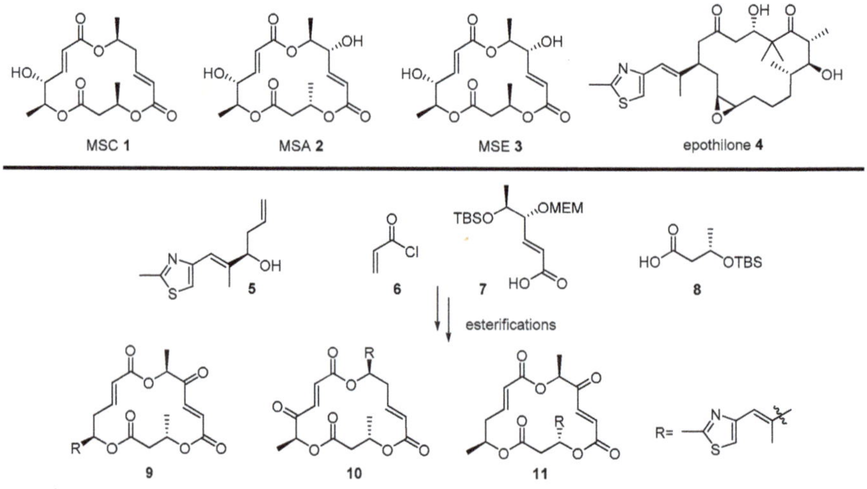

Scheme 1. Hybridization of macrosphelide (MS) skeleton and epothilone side chain [44].

Screening of chimeras **9–11** exhibited an increase in their desired potencies, which is attributed to this hybridization strategy. In particular, chimera 9 showed more potent apoptosis-inducing activity in the U937 cell line as compared to other analogs or parent compounds MSA **2** or MSE **3**. It is noteworthy that other chimeras also showed an increased apoptotic property, while parent MSA **2** or MSE **3** did not have the same activity at the same concentration (1 μM, 12 h incubation). Further, successful hybridization of MS with epothilone **4** presents that the chimeric molecule strategy can be a powerful solution to the limitation of marine natural products themselves.

The complex structure of marine products can hamper efficient design or synthesis of advanced chimera molecules. In such cases, in silico study may offer another solution. Discodermolide **12**, a marine polyketide product from *Discodermia dissolute*, possesses potent antiproliferative activity against various human cancer cell lines [11]. Mechanically, it stabilizes microtubules and finally arrests cells in the mitosis status, just as paclitaxel does [46]. However, unlike paclitaxel **13**, a previous SAR and docking study of discodermolide at the paclitaxel binding site of tubulin revealed that free carbamate of discodermolide could be modified [47]. Based on the following study, conjugated diene in discodermolide resides in an aromatic pocket of tubulin. Moreover, the attachment of a simple aromatic group did not decrease its own anticancer activity. With this early study, the introduction of an aromatic side chain to discodermolide was pursued [47]. Synthetic intermediate **14** was coupled with various amine side chains, such as **15** with photolabile functionality, to finally produce carbamate **16** in good yield (Scheme 2). It was gratifying to note that the biological evaluation of this hybrid **16** showed an improved antiproliferative profile against human cancer cell lines. When it was treated to the lung cancer cell line A549, inhibition of cell growth at IC_{50} 1.21 ± 0.35 nM was demonstrated, while discodermolide **12** showed a value of IC_{50} 9.34 ± 0.56 nM (paclitaxel IC_{50} 3.14 ± 0.09 nM). This pharmacological advance implies that the hybridization strategy based on the docking study could be another option for structurally complex molecules.

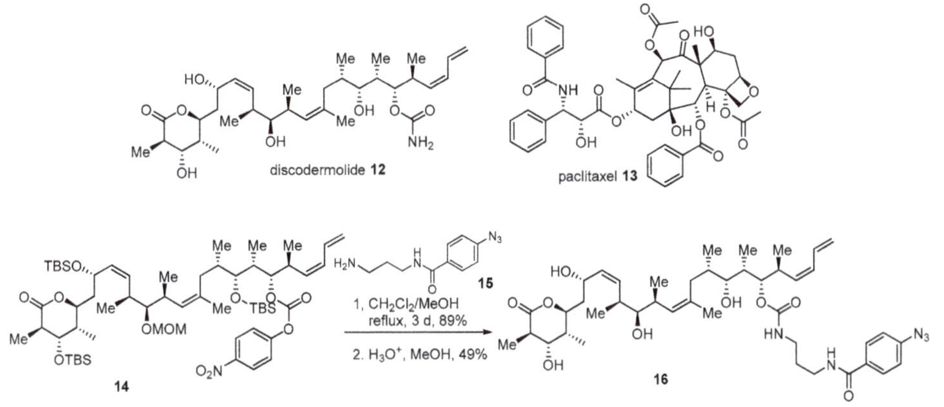

Scheme 2. Hybridization of discodermolide **12** with side chain moiety of paclitaxel **13** [47].

Further, hybridization of discodermolide **12** and dictyostatin **17** was also performed based on their similar structural features and anticancer activities, as depicted in Scheme 3 [48]. Dictyostatin **17** is a 22-membered macrolide natural product from the Indian [49] or Caribbean ocean sponge [50]. Dictyostatin has also garnered substantial attention due to its extraordinary antiproliferative effect (ED_{50} 0.38 nM for P388 leukemia cell) [51]. A comparison of discodermolide and dictyostatin makes it plausible to say that they have similar backbones and substituents, except for the 22-membered lactone of dictyostatin. Based on this observation, hybridization of these two powerful anticancer natural products was performed [48]. During the synthesis and SAR study of discodermoilde **12**, a practical synthetic route to it was developed and applied to chimera preparation. *Para*-methoxybenzyl (PMP)-acetal **18** was coupled with aldehyde **19** using Wittig olefination to construct pivotal Z-alkene of target chimera. After some functional group interconversion and another Wittig olefination with diene **20**, discodermolide-dictyostatin chimera **21** was efficiently prepared. In addition, the biological activity of the chimera was also examined. Although chimera **21** possesses a relatively simple structure compared to discodermolide **12**, it displayed one-third the potency of discodermolide in a displacement test using [^{3}H]-paclitaxel bound to microtubules. This simplified structure with moderate activity might open new possibilities for further development of related natural products.

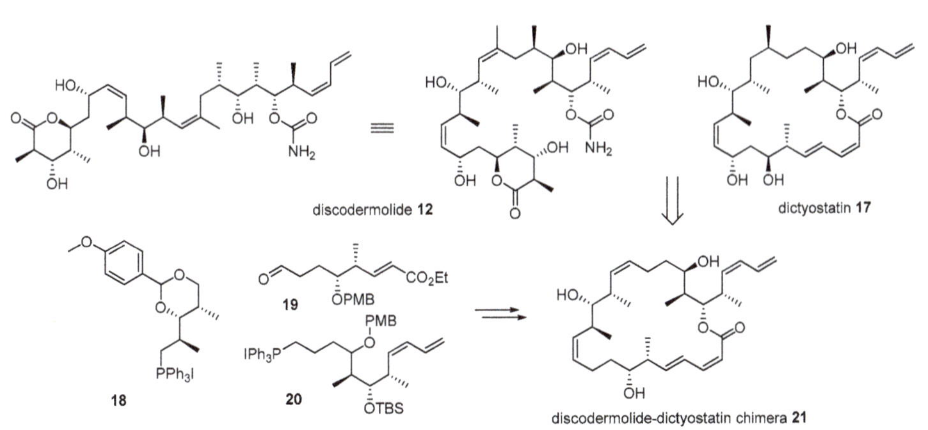

Scheme 3. Structural comparison and hybridization of discodermolide **12** with dictyostatin **17** [48].

Since great cytotoxicity for solid tumors was observed using *Psammocinia* extracts from the waters of Papua New Guinea [52], significant effort has been made to isolate active ingredients produced from moderated cytotoxic marine natural products, such as cyclocinamide A [53], swinholide A [54], and furanosesterpenes [55]. However, these compounds did not explain the extraordinary cytotoxic properties of crude extract. The active ingredient was re-examined only in the early 2000s. Finally, highly anti-proliferative marine natural product psymberin **22** was isolated [52,56], but its absolute configuration and C_4-stereochemistry were not confirmed. An additional research program elucidated its mysterious structure through total synthesis by the De Brabander group [57].

After confirmation of the correct structure, psymberin **22** was compared to classic natural compounds pederin **23** from rove beetles or mycalamide A **24** from marine invertibrates [58]. Although their structure possesses unique aminal-amide and trans-substituted tetrahydropyran skeletons, psymberin **22** has a characteristic dihydroisocoumarin moiety, while pederin/mycalamide A has another tetrahydropyran with the exomethylene group. As pederin/mycalamide A has been well reported as a powerful anticancer natural product based on eukaryotic protein synthesis inhibitors [52,56], a similar skeleton with a different side chain inspired the hybridization of the two natural products. It was anticipated that applying well-known biological properties of pederin to the psymberin-based would lead discovery [59] (Scheme 4).

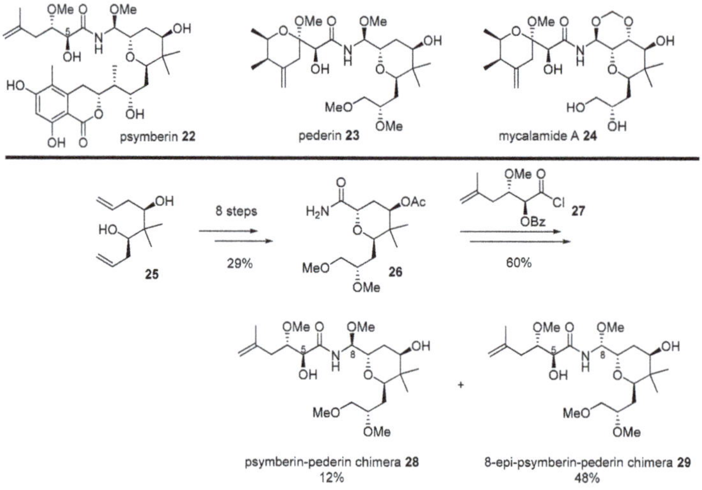

Scheme 4. Hybridization of psymberin **22** with pederin **23** [59].

Bishomoallylic alcohol **25**, a synthetic intermediate for total synthesis of psymberin [57], was transformed into tetrahydropyranyl amide **26** in eight steps, with a 29% yield. It lacks a unique dihydroisocoumarin chain in psymberin, while it possesses a dimethoxy group in pederin. In order to introduce a side chain of psymberin, acyl halide **27** was attached and reduced with NaBH$_4$. Final deprotection using LiOH/MeOH condition afforded separable C_8-diastereomeric mixture of **28** and **29**. These two psymberin-pederin chimeras were used to unveil the SAR of this natural product family. It is interesting to note that the absence of dihydroisocoumarin moiety gave rise to a significant loss of psymberin's own cytotoxic property, while C_8-epimer of the natural product caused a slight loss of this property. When they were treated to colon cancer cell line KM12, psymberin **25** or its C_8-epimer showed a powerful or moderate toxicity (IC_{50} 0.45 nM and 37 nM, respectively). However, its truncated analog **28** or C_8-epi analog **29** showed little or no toxicity to the same cell line (IC_{50} 710 nM and >1000 nM, respectively). These chimeras could be compared to pederin/mycalamide A as well. With a substituted tetrahydropyranyl side chain, which is rather common in natural pederin isomers, mycalamide A showed potent cytotoxicity (IC_{50} 0.95 nM) as well. It was proven that the common substituted side chain, or dihydroisocoumarin unit, plays a pivotal role in its own anticancer activity, employing a direct comparison of mother natural products and daughter chimeras. This case is a good indication of the effectiveness of a chimera strategy for drug development.

Topsentin **30** is a bisindolyl marine alkaloid isolated from several marine sponges in Mediterranean *Topsentia genitrix* or Korean *Spongosorites* sp [60]. As this alkaloid exhibits excellent cytotoxicity to cancer cells and features a rather simple structure when compared to the other active marine natural products [61], numerous researchers have attempted to develop more potent and drug-like analogs [62]. One of the indole side chains in topsentin was changed to an aminothiazole unit, as a thiadiazole in dendrodoine **31**, another marine alkaloid from *Dendroda grossular* [63]. With the first chimera diaminoindolylthiazole (DIT) **32** in hand, additional hybridization with curcumin **33** was performed to yield diaminocinnamoylthiazole (DCT) **34** and its analogs [64] (Scheme 5). Moreover, DITs and DCTs were screened to the cancer cell line. It is interesting that DITs are very effective in the induction of apoptosis in HeLa cells (IC$_{50}$ 1~45 μM), while DCTs play an active role in downregulating TNF-induced NF-κB activation. This research indicates that the hybridization of a simple functional group may also offer an opportunity to improve biological activity.

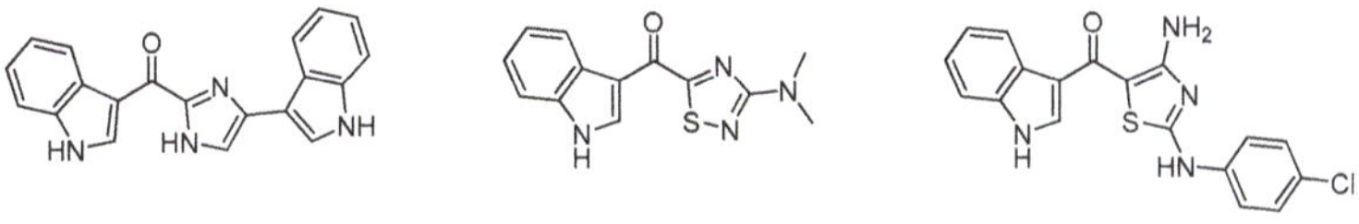

topsentin **30**

dendrodoine **31**

diaminoindolylthiazole 1 **32**
DIT1

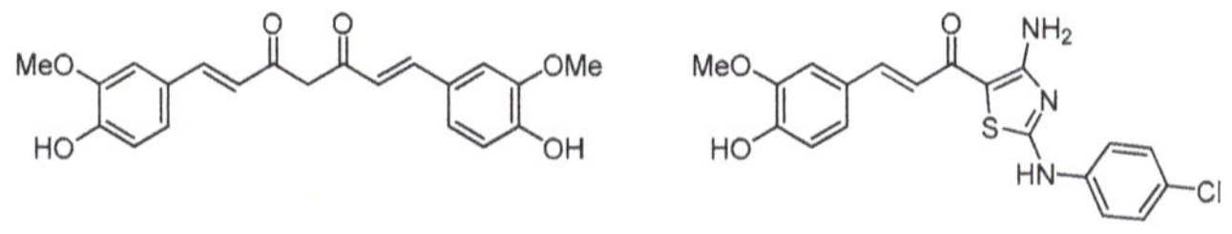

curcumin 1 **33**

diaminocinnamoylthiazole 1 **34**
DCT1

Scheme 5. Hybridization from topsentin **30**, dendrodoine **31**, and curcumin 1 **33** [64].

Neopeltolide **35** is another example of the chimera strategy of marine natural products. Since its isolation from marine sponge *Daedalopelta Sollas* in 2007 [65], urgent synthetic efforts were made to elucidate its correct structure [66]. As this marine macrolide showed various cytotoxic or cytostatic activity for numerous cancer cell lines, substantial research was conducted, including asymmetric synthesis, mechanism study, and analog synthesis. In addition, its antifungal activity was also focused upon. Contrary to its varied anticancer activity, its inhibitory activity to *Candida albicans* was highly potent at the minimum inhibitory concentration (MIC) 0.625 μg/mL [67]. This antifungal inhibition could be used for patients infected with the acquired immune deficiency syndrome (AIDS) or related fungal diseases. Further, the mechanism of action of neopeltolide was also studied to reveal that the agent works as the cytochrome bc$_1$ complex inhibitor and inhibits the adenosine triphosphate (ATP) synthesis in mitochondria [68].

Although systemic research on neopeltolide was undertaken, its application in drug discovery was hampered by its complex structure. Neopeltolide **35** features a cis-substituted tetrahydropyran skeleton with an ansacyclic macrolactone framework. This formidable structure made it difficult to prepare neopeltolide on a large scale or related analogs in a varied manner. An interesting strategy was the chimeric application of neopeltolide with biaryl ether active molecules [69]. An SAR study reported that the oxazolyl carbamate skeleton plays an important role in its activity profile, while a complex macrolide skeleton has little effect on it [70]. Instead of this complex and less important left chain in neopeltolide, simple biaryl moiety was inspired by other similar fungicide metominostrobin **36**, famoxadone **37**, or other biaryl ethers, as illustrated in Scheme 6 [71].

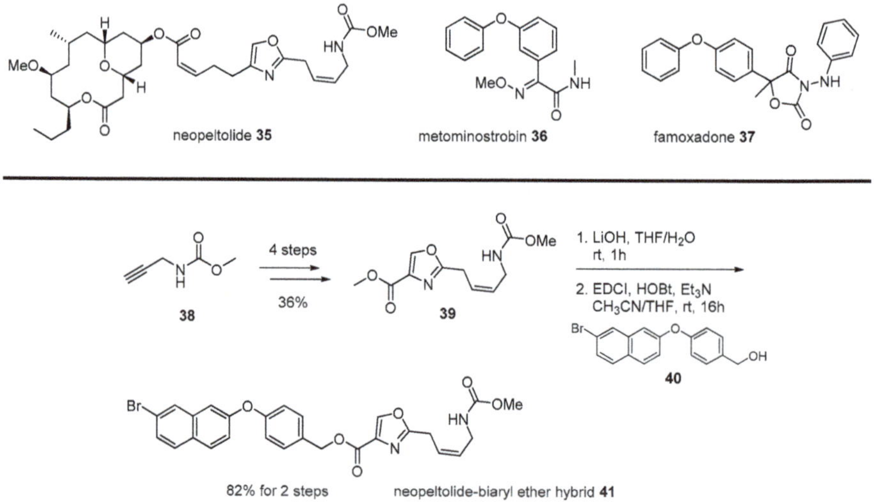

Scheme 6. Hybridization from neopeltolide and biaryl ether hybrid [71].

Methyl propargyl carbamate **38** was converted into oxazolyl ester **39** in four steps and with high yield. Ester **39** was hydrolyzed and esterified with various primary alcohol **40** with biaryl moiety. Among synthesized chimeras, naphthyl benzyl ether **41** showed the most potent inhibitory activity (IC_{50} 12 nM) for porcine succinate cytochrome c reductase (SCR). As this chimera simplified its mother structure, additional development for improved antifungal agent was expected.

Figure 4 presents the hybridization of iejimalide and archazolid skeletons [72]. Iejimalide B **42** and its family compounds were isolated from *Eudistoma cf. rigida* or *Cystodytes* sp., which were collected from an island in Japan [73]. Due to its remarkable anticancer activity and selectivity to NCI 60 cell lines, this polyunsaturated macrolide has been focused upon for its promising role in drug discovery [74]. However, its scarcity in natural resources led to not only total synthesis [75,76] but also structural simplification based on a hybridization strategy [72]. Archazolid A **43**, isolated from terrestrial myxobacteria [77], possesses a very similar structure as that of iejimalide B **42**; however, archazolid A **43** features a relatively simple side chain with thiazole and methyl carbamate moiety. In addition, because archazolid **43** showed potent inhibitory activity to vacuolar-type ATPases as well [77], iejimalide-archazolid chimera **44** was designed and synthesized.

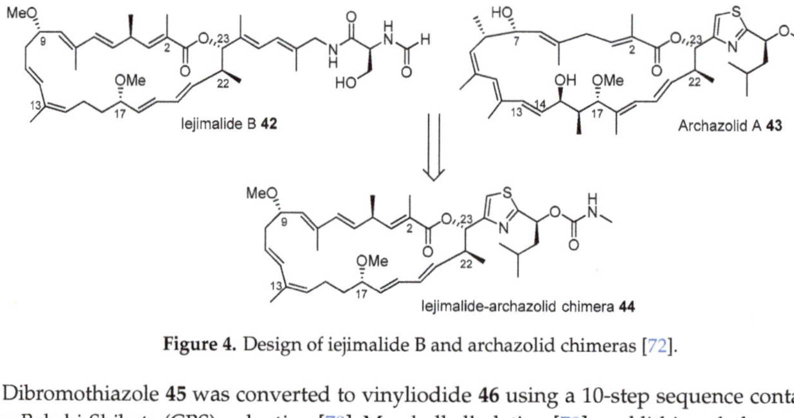

Figure 4. Design of iejimalide B and archazolid chimeras [72].

Dibromothiazole **45** was converted to vinyliodide **46** using a 10-step sequence containing the Corey-Bakshi-Shibata (CBS) reduction [78], Marshall alkylation [79], and lithium-halogen exchange reactions as shown in Scheme 7. It was coupled with known catechol borane [76] **47**, using the Suzuki coupling condition [80] to afford tetraene **48**. Finally, the esterification of tetraene **48** with known carboxylic acid [76] **49** and ring closing metathesis [81] produced the desired chimera **44** in good yield. Biological evaluation of chimera **44** against various cancer cell lines was performed and it was proven that this hybridization yielded reduced toxicity to cancer cell lines, although certain cell lines such as lung adeno (LXFA 629L, IC_{50} 0.32 µM) or colorectal (CSF HT29, IC_{50} 0.49 µM)) were slightly sensitive to chimera **44**.

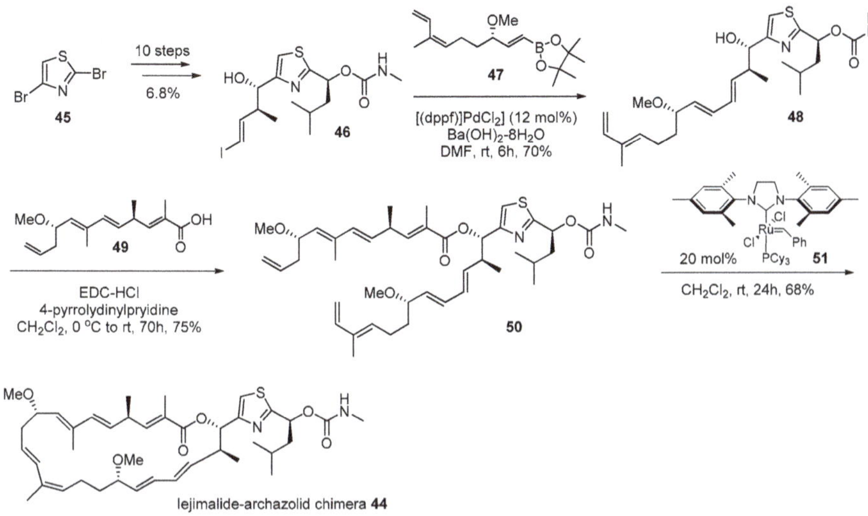

Scheme 7. Synthesis of iejimalide-archazolid chimera **44** [72].

2.2. Conjugation with Other Functional Compounds

For drug discovery and sequential development, marine natural products require more detailed studies, such as on-target protein isolation. The chemical conjugation of marine natural products with other functional compounds provides special opportunities in this regard. Biotin is a good example. Biotin, well known for its strong interaction with streptavidin [82–84], can be utilized to identify the target protein of natural products [85–88]. As this hybridization strategy may lead to the loss of its own binding affinity to natural products, it usually requires preliminary study to identify where to ligate biotin and what to use as a linker between biotin and the natural product. This pre-research requires additional efforts, thereby making it difficult to apply this method to a general natural product. However, in certain cases, well-designed chimeras showcase the usefulness of this hybridization protocol.

Bistramide A **52**, isolated from marine metabolite of *Lissoclinum bistratum* [89–91], is a good example of this hybridization. As the bistramide family has shown potent cytotoxic properties on various cancer cell lines, such as non-small cell broncho-pulmonary carcinoma or HL60 cells as well as unique spiroketal skeleton [92], this marine product has been a captivating target for synthetic chemists. After pioneering research on structural elucidation and synthesis of skeletons or other isomers [93], enantioselective total synthesis of bistramide A was accomplished in 2004 [94]. This synthetic route features iterative cross metathesis of terminal alkenes **53** and **55** with the geometrically strained cyclopropene **54** [95] in the presence of Grubbs second-generation catalyst **55** to construct bisunsaturated ketone **56**. The treatment of H_2 with Pd(OH)₂/C to this ketone induced hydrogenation/hydrogenolysis and spontaneous cyclization to afford the desired spiroketal **57** in 53% yield after the Dess–Martin oxidation sequence. Ketal **57** was converted to bistramide A **52** after employing amidation with other building blocks in six linear steps. This efficient and convergent synthesis made it possible to not only elucidate the C_{37} chiral center [96] but also provide further opportunity for hybridization with other active molecules (Scheme 8).

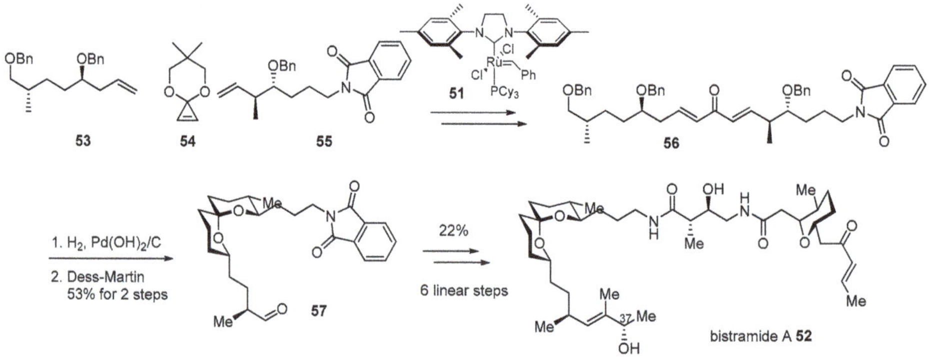

Scheme 8. Total synthesis of bistramide A **52** [94].

Although potent and active, bistramide A **52** has been reported since its first isolation [97]; its mechanism of action was unknown until the preparation of bistramide-biotin chimera, as illustrated in Scheme 8 [98]. Its protein kinase C(PKC) δ inhibitory features caused PKCδ to be a cellular target in HL-60 cells. However, an in vitro study with real-time fluorogenic kinase assay system [99] revealed that bistramide A did not have strong affinity to PKCδ as its inhibitory activity. In order to identify an early-stage target, target protein fishing was planned. Based on the synthetic strategy previously described [94], homologation with carbon linker and final attachment of biotin was accomplished in order to produce desired chimera **58** as well as fragment-biotin chimera **59** for a controlled experiment (Figure 5). The treatment of these two chimeras into whole-cell lysate from A549 cell revealed direct binding of monomeric G-actin (K_d 7nM) as a primary target protein. Using this target fishing study with biotin-chimera enabled the study of a more detailed mechanism of action.

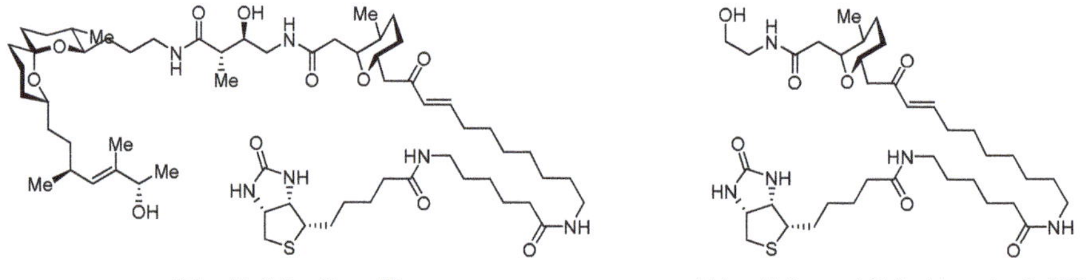

Figure 5. Biotin chimera based on bistramide A [98].

Aplyronine A **60**, isolated from Japanese sea hare *Aplysia kurodai* [100], is another case for the conjugation of marine natural product with biotin to identify the interaction of actin. While bistramide A **52** binds to actin, aplyronine A binds to the actin/tubulin complex, which is a 1:1:1 trimeric complex [101]. The Kigoshi group treated aplyronine A with aqueous HCl to produce the corresponding aldehyde, which was condensed with hydroxylamine **61** to form an oxime bond as an inseparable mixture of aplyronine-biotin-diazine chimera **62** (Scheme 9). When it was treated to HeLa S3 cells, α/β tubulins appeared in SDS-PAGE with some other nonspecific binding proteins. In order to validate this Western blot result, purified actin (from rabbit) or tubulin (from porcine brain) was treated with chimera **62**. This validation study examined that unusual aplyronine/actin/tubulin heterotrimeric complex is responsible for its extraordinary antitumor effect through cell arrest at mitosis in HeLa S3 cells (IC_{50} 0.45 nM) [102].

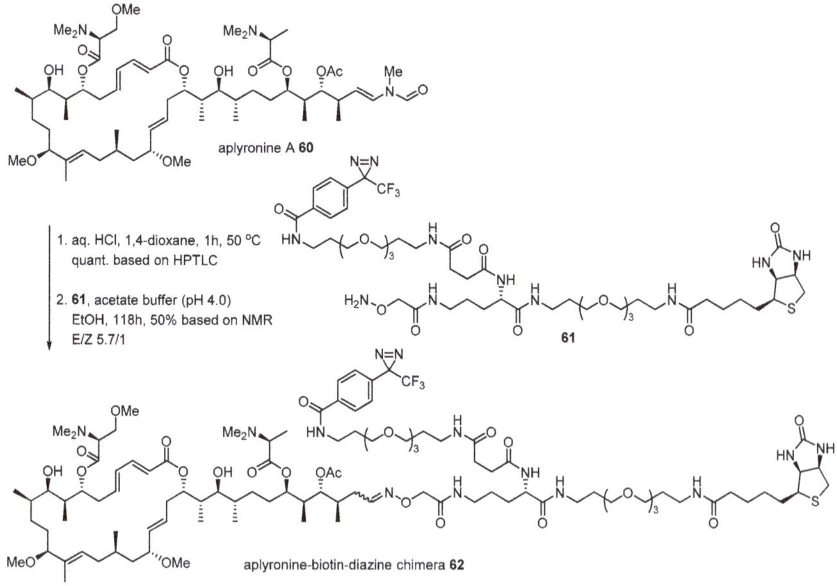

Scheme 9. Preparation of aplyronine-biotin-diazine chimera **62** [101].

Diazonamide A **63** is a marine natural product from *Diazona chinensis*, located on the ceilings of small caves on the Siquijor Islands in the Philippines [103]. Despite its scarcity in natural sources (54 mg from 256 g of *D. chinensis*), its extraordinary antimitotic property in human colon carcinoma and murine melanoma cancer cell lines has garnered substantial interest from chemists and biologists [104–106]. This tremendous study led to structural elucidation, efficient chemical synthesis, and related analog preparation. Its target protein study is interesting. When diazonamide (syndistatin **64**)-biotin chimera **65** was prepared and treated with HeLa extract, two additional bands of an affinity matrix were observed on an SDS/PAGE compared to those in the control experiment. Mass identification indicated that these two bands were ornithine δ-amino transferase, although why the matrix separates is unknown [107]. As ornithine δ-amino transferase is well known as a mitochondrial enzyme in the TCA cycle [108], it was slightly surprising. However, further study using the RNA*i*-mediated knockdown method and additional investigation proved that ornithine δ-amino transferase is essential for mitotic cell division. Thus, target protein and its corresponding pathways of diazonamide were unveiled to lead to the discovery of advanced diazonamide analog DZ2384 **66** [109]. This process shows a 5–50 times higher efficacy without neurotoxicity at an effective dose (Figure 6).

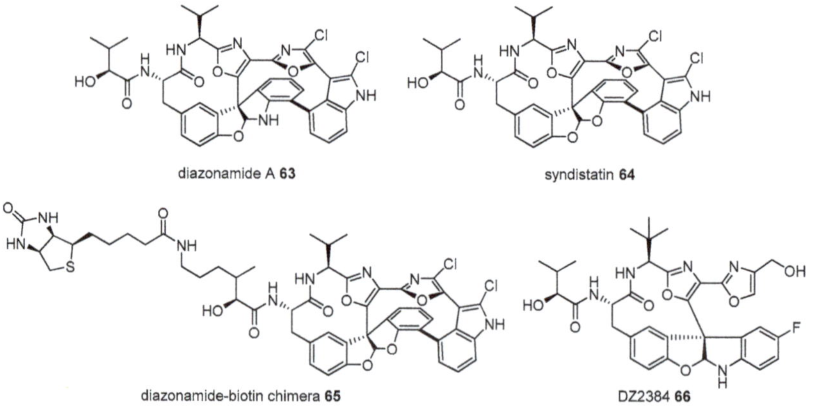

Figure 6. Preparation of biotinylated diazonamide and its advanced analog DZ2384 **66** [107].

Further, macrosphelide (MS) skeleton was also utilized in this chimeric research, as depicted in Scheme 10 [110]. Although it has a structural framework and biological profiles, its mechanism of action is not well studied in the MS family. In order to identify the target protein of MSA **2** and elucidate the following pathway, MSA-biotin hybridization was planned. The protected Weinreb-amide of (*S*)-lactic acid **67** [111] was transformed into allylic alcohol **68** using a three-step protocol—alkylnylatoin, carbonyl reduction, and alkyne reduction—in 42% yield. This key intermediate **68** in hand, allylation, iterative esterification, and deprotection sequence finally afforded allyl-MSA **69** in 7.6% yield and 13 steps. Biotin was linked to allyl-MSA **69** after additional manipulation. For efficient target fishing, a long carbon chain was selected as a linker. Terminal alkene in allyl-MSA was utilized in cross metathesis to introduce a long carbon linker and protect the amine functional group in amino MSA **71**. Finally, acidic deprotection and following amidation with biotin tag **73** produced the desired MSA-biotin chimera **74**. Target fishing studies, employing this chimera, on potent cell-cell adhesion inhibitor are ongoing.

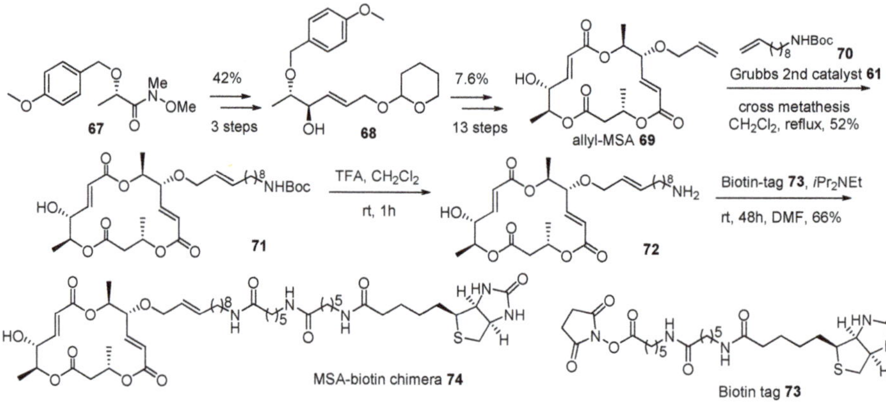

Scheme 10. Preparation of biotinylated marcosphelide chimera **74** [110].

The marine natural product-biotin chimera strategy was also applied to identify the target protein of kahalalide F **75**. Ever since the isolation of kahalalide F from marine mollusk, *Elysia rufescens*, in 1993 [112], this polypeptide marine product has been used in clinical trials because of its powerful anticancer activity and low cellular toxicity [113]. For advanced development, identification of the target protein using biotinylation of kahalalide F was also attempted, as presented in Scheme 11.

Kahalalide F was coupled with biotinylated linker **76** [114], prepared from biotin and tetraethyleneglycol, under a weak basic condition to produce desired chimera **77** in 85% yield. When chimera 77 was treated to T7 cDNA phage for reverse chemical proteomics, human ribosomal protein S25 was found to be responsible for its potent anticancer property in a dose-dependent manner.

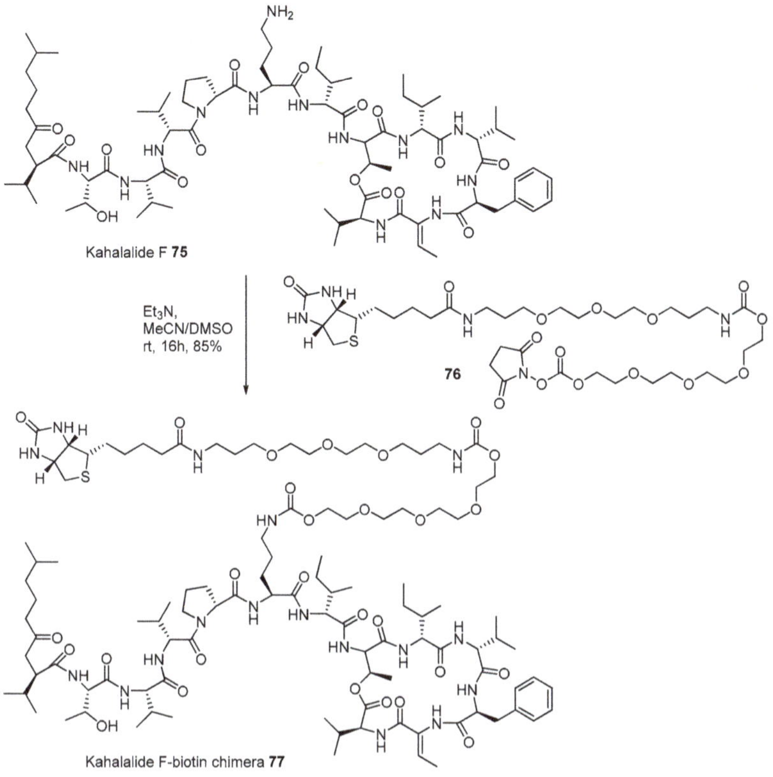

Scheme 11. Preparation of Kahalalide F-biotin chimera **77** [114].

Avrainvillamide is another good example of the biotin chimera strategy [115]. In order to unveil the mysterious target protein of powerful avrainvillamide, isolated from *Aspergillus* sp. CNC358 [116], a biotinylated natural product **79** was designed. Synthesis of chimera **79** is depicted in Scheme 12. Iodoarene **80** [116] was stannylated to yield aromatic stannane **81** via metal/halogen exchange and a substitution reaction. Then, stannane **81** was coupled with vinyl iodide **82** [117] to produce nitroarene **83**, which was converted into biotinylated arene **85** employing cross metathesis with terminal alkene **84** [118]. Reductive cyclization of **85** and HPLC purification afforded the desired chimera **79** (1:1 mixture of diastereomers at the C21 chiral center). Finally, Western blotting and MS/MS sequencing of these chimera-treated T-47 D cells revealed that nucleophosmin was a target protein of avrainvillamide. As nucleophosmin is a multifunctional protein in numerous tumors [119], a more detailed study was subsequently conducted to disclose the mode of action of avrainvillamide and its related natural product stephacidin [120].

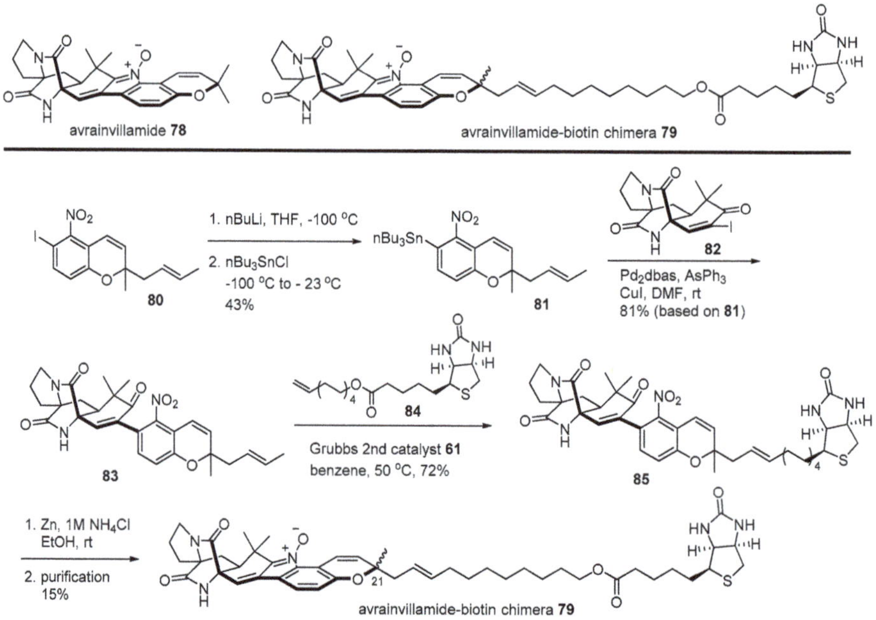

Scheme 12. Preparation of avrainvillamide-biotin chimera **79** [115].

The hybridization strategy was also utilized to identify the target protein of pateamine, which is isolated from the New Zealand marine sponge [121]. As this marine natural product showed not only remarkable cytotoxicity to tumor cell line P338 [121] but also extraordinary immunosuppressive activity [122], target protein identification was pursued, as illustrated in Scheme 13 [123]. Primary iodide **88** [124] was alkylated with truncated pateamine **89** [123], which is more stable than pateamine **86**. Further, corresponding vinyl bromide **90** [124] was coupled with vinyl stannane **91** [122] to produce the desired pateamine-biotin chimera **87** in moderate yield. The pull-down method of this affinity-matrix **87** showed that it bound to and inhibited the association of eIF4A and eIF4B, finally inhibiting the cap-dependent eukaryotic translation sequence [124]. This fundamental study has also opened the way for more detailed research and further applications of pateamine analogs [125].

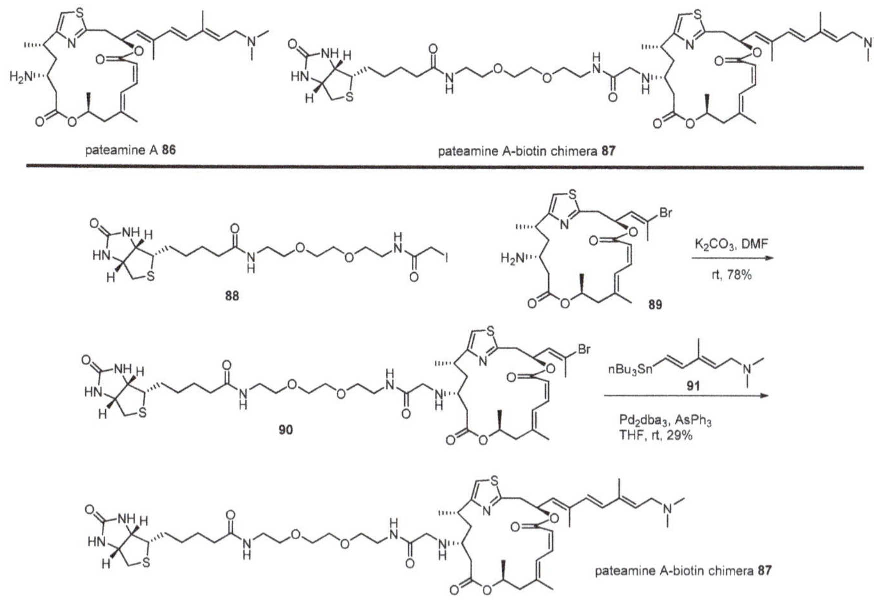

Scheme 13. Preparation of pateamine A-biotin chimera **87** [124].

3. Conclusions

As an endless resource of bioactive molecules, marine natural products stand a reliable chance in the world of drug discovery. However, certain limitations of these products, such as complex structure, toxicity, selectivity, and even potency have driven researchers to identify additional strategies for these natural compounds. Chimeras yielded by the hybridization of two active molecules are one such solution. Changing the active skeleton of the mother framework also added more opportunities, but this process also comes with its own set of unique drawbacks, such as loss of function, scarcity of supply, and complex structure for chemical modification. In order to overcome these limitations, SAR, structural simplification, development of the synthetic route, and the medicinal chemistry approach have been studied thus far. Based on this fundamental progress, currently, hybridization of active natural products is being utilized to improve their own biological properties and elucidate veiled mechanism of actions. The efficient application of this chimera strategy to other active molecules remains a direction for groundbreaking research in the future.

Funding: This research was supported by the Basic Science Research Program through the National Research Foundation of Korea (NRF), funded by the Ministry of Education, Science and Technology (NRF-2016R1C1B2006699).

Acknowledgments: We are grateful to the two reviewers for their constructive input.

Conflicts of Interest: The authors declare no conflict of interest.

References

1. Harvey, A.L. Toxins and drug discovery. *Toxicon* **2014**, *92*, 193–200. [CrossRef] [PubMed]
2. De Souza, J.M.; Goncalves, B.D.; Gomez, M.V.; Vieira, L.B.; Ribeiro, F.M. Animal toxins as therapeutic tools to treat neurodegenerative diseases. *Front. Pharmacol.* **2018**, *9*, 145. [CrossRef] [PubMed]
3. Molinski, T.F.; Dalisay, D.S.; Lievens, S.L.; Saludes, J.P. Drug development from marine natural products. *Nat. Rev. Drug Discov.* **2009**, *8*, 69. [CrossRef] [PubMed]
4. Kiuru, P.; D'Auria, M.V.; Muller, C.D.; Tammela, P.; Vuorela, H.; Yli-Kauhaluoma, J. Exploring marine resources for bioactive compounds. *Planta Med.* **2014**, *80*, 1234–1246. [CrossRef] [PubMed]

5.   Mayer, A.M.; Glaser, K.B.; Cuevas, C.; Jacobs, R.S.; Kem, W.; Little, R.D.; McIntosh, J.M.; Newman, D.J.; Potts, B.C.; Shuster, D.E. The odyssey of marine pharmaceuticals: A current pipeline perspective. *Trends Pharmacol. Sci.* **2010**, *31*, 255–265. [CrossRef] [PubMed]

6.   Li, G.; Lou, H.-X. Strategies to diversify natural products for drug discovery. *Med. Res. Rev.* **2018**, *38*, 1255–1294. [CrossRef] [PubMed]

7.   Gerwick, W.H.; Moore, B.S. Lessons from the past and charting the future of marine natural products drug discovery and chemical biology. *Chem. Biol.* **2012**, *19*, 85–98. [CrossRef] [PubMed]

8.   Montaser, R.; Luesch, H. Marine natural products: A new wave of drugs? *Future Med. Chem.* **2011**, *3*, 1475–1489. [CrossRef] [PubMed]

9.   Pettit, G.R.; Chicacz, Z.A.; Gao, F.; Herald, C.L.; Boyd, M.R.; Schmidt, J.M.; Hooper, J.N. Antineoplastic agents. 257. Isolation and structure of spongistatin 1. *J. Org. Chem.* **1993**, *58*, 1302–1304. [CrossRef]

10.   Smith, A.B.; Razler, T.M.; Meis, R.M.; Pettit, G.R. Synthesis and biological evaluation of phorboxazole congeners leading to the discovery and preparative-scale synthesis of (+)-chlorophorboxazole a possessing picomolar human solid tumor cell growth inhibitory activity. *J. Org. Chem.* **2008**, *73*, 1201–1208. [CrossRef] [PubMed]

11.   Gunasekera, S.P.; Gunasekera, M.; Longley, R.E.; Schulte, G.K. Discodermolide: A new bioactive polyhydroxylated lactone from the marine sponge Discodermia dissoluta. *J. Org. Chem.* **1990**, *55*, 4912–4915. [CrossRef]

12.   Pettit, G.R.; Herald, C.L.; Doubek, D.L.; Herald, D.L.; Arnold, E.; Clardy, J. Isolation and structure of bryostatin 1. *J. Am. Chem. Soc.* **1982**, *104*, 6846–6848. [CrossRef]

13.   Schaufelberger, D.E.; Koleck, M.P.; Beutler, J.A.; Vatakis, A.M.; Alvarado, A.B.; Andrews, P.; Marzo, L.; Muschik, G.; Roach, J.; Ross, J.T. The large-scale isolation of bryostatin 1 from Bugula neritina following current good manufacturing practices. *J. Nat. Prod.* **1991**, *54*, 1265–1270. [CrossRef] [PubMed]

14.   Smith, A.B., III; Tomioka, T.; Risatti, C.A.; Sperry, J.B.; Sfouggatakis, C. Gram-scale synthesis of (+)-spongistatin 1: Development of an improved, scalable synthesis of the F-ring subunit, fragment union, and final elaboration. *Org. Lett.* **2008**, *10*, 4359–4362. [CrossRef] [PubMed]

15.   Smith, A.B.; Beauchamp, T.J.; LaMarche, M.J.; Kaufman, M.D.; Qiu, Y.; Arimoto, H.; Jones, D.R.; Kobayashi, K. Evolution of a gram-scale synthesis of (+)-discodermolide. *J. Am. Chem. Soc.* **2000**, *122*, 8654–8664. [CrossRef]

16.   Mickel, S.J.; Niederer, D.; Daeffler, R.; Osmani, A.; Kuesters, E.; Schmid, E.; Schaer, K.; Gamboni, R.; Chen, W.; Loeser, E. Large-scale synthesis of the anti-cancer marine natural product (+)-Discodermolide. Part 5: Linkage of fragments C1-6 and C7-24 and finale. *Org. Process. Res. Dev.* **2004**, *8*, 122–130. [CrossRef]

17.   Choudhary, A.; Naughton, L.; Montánchez, I.; Dobson, A.; Rai, D. Current status and future prospects of marine natural products (MNPs) as antimicrobials. *Mar. Drugs* **2017**, *15*, 272. [CrossRef] [PubMed]

18.   Lear, M.J.; Hirai, K.; Ogawa, K.; Yamashita, S.; Hirama, M. A convergent total synthesis of the kedarcidin chromophore: 20-years in the making. *J. Antibiot.* **2019**, *72*, 350–363. [CrossRef] [PubMed]

19.   Xiao, Z.; Morris-Natschke, S.L.; Lee, K.-H. Strategies for the optimization of natural leads to anticancer drugs or drug candidates. *Med. Res. Rev.* **2016**, *36*, 32–91. [CrossRef] [PubMed]

20.   Newman, D.J.; Cragg, G.M. Drugs and drug candidates from marine sources: An assessment of the current "state of play". *Planta Med.* **2016**, *82*, 775–789. [CrossRef] [PubMed]

21.   Hirata, Y.; Uemura, D. Halichondrins-antitumor polyether macrolides from a marine sponge. *Pure Appl. Chem.* **1986**, *58*, 701–710. [CrossRef]

22.   McBride, A.; Butler, S.K. Eribulin mesylate: A novel halichondrin B analogue for the treatment of metastatic breast cancer. *Am. J. Health Syst. Pharm.* **2012**, *69*, 745–755. [CrossRef] [PubMed]

23.   Aicher, T.D.; Buszek, K.R.; Fang, F.G.; Forsyth, C.J.; Jung, S.H.; Kishi, Y.; Matelich, M.C.; Scola, P.M.; Spero, D.M.; Yoon, S.K. Total synthesis of halichondrin B and norhalichondrin B. *J. Am. Chem. Soc.* **1992**, *114*, 3162–3164. [CrossRef]

24.   Lindequist, U. Marine-derived pharmaceuticals—Challenges and opportunities. *Biomol. Ther.* **2016**, *24*, 561. [CrossRef] [PubMed]

25.   Bebbington, M.W. Natural product analogues: Towards a blueprint for analogue-focused synthesis. *Chem. Soc. Rev.* **2017**, *46*, 5059–5109. [CrossRef] [PubMed]

26.   Maier, M.E. Design and synthesis of analogues of natural products. *Org. Biomol. Chem.* **2015**, *13*, 5302–5343. [CrossRef]

27.   Ziegler, S.; Pries, V.; Hedberg, C.; Waldmann, H. Target identification for small bioactive molecules: Finding the needle in the haystack. *Angew. Chem. Int. Edit.* **2013**, *52*, 2744–2792. [CrossRef]

28. Tietze, L.F.; Bell, H.P.; Chandrasekhar, S. Natural product hybrids as new leads for drug discovery. *Angew. Chem. Int. Edit.* **2003**, *42*, 3996–4028. [CrossRef]

29. Mehta, G.; Singh, V. Hybrid systems through natural product leads: An approach towards new molecular entities. *Chem. Soc. Rev.* **2002**, *31*, 324–334. [CrossRef]

30. Tsogoeva, S.B. Recent progress in the development of synthetic hybrids of natural or unnatural bioactive compounds for medicinal chemistry. *Mini Rev. Med. Chem.* **2010**, *10*, 773–793. [CrossRef]

31. Fortin, S.; Bérubé, G. Advances in the development of hybrid anticancer drugs. *Expert Opin. Drug Discov.* **2013**, *8*, 1029–1047. [CrossRef]

32. Lai, A.C.; Crews, C.M. Induced protein degradation: An emerging drug discovery paradigm. *Nat. Rev. Drug Discov.* **2017**, *16*, 101. [CrossRef]

33. Salami, J.; Crews, C.M. Waste disposal-An attractive strategy for cancer therapy. *Science* **2017**, *355*, 1163–1167. [CrossRef]

34. Lu, J.; Qian, Y.; Altieri, M.; Dong, H.; Wang, J.; Raina, K.; Hines, J.; Winkler, J.D.; Crew, A.P.; Coleman, K. Hijacking the E3 ubiquitin ligase cereblon to efficiently target BRD4. *Chem. Biol.* **2015**, *22*, 755–763. [CrossRef]

35. Gadd, M.S.; Testa, A.; Lucas, X.; Chan, K.; Chen, W.; Lamont, D.J.; Zengerle, M.; Ciulli, A. Structural basis of PROTAC cooperative recognition for selective protein degradation. *Nat. Chem. Biol.* **2017**, *13*, 514–521. [CrossRef]

36. Bondeson, D.P.; Mares, A.; Smith, I.E.; Ko, E.; Campos, S.; Miah, A.H.; Mulholland, K.E.; Routly, N.; Buckley, D.L.; Gustafson, J.L. Catalytic in vivo protein knockdown by small-molecule PROTACs. *Nat. Chem. Biol.* **2015**, *11*, 611–617. [CrossRef] [PubMed]

37. Schiedel, M.; Herp, D.; Hammelmann, S.; Swyter, S.; Lehotzky, A.; Robaa, D.; Olaáh, J.; Ovaádi, J.; Sippl, W.; Jung, M. Chemically induced degradation of sirtuin 2 (Sirt2) by a proteolysis targeting chimera (PROTAC) based on sirtuin rearranging ligands (SirReals). *J. Med. Chem.* **2017**, *61*, 482–491. [CrossRef]

38. Leslie, B.J.; Hergenrother, P.J. Identification of the cellular targets of bioactive small organic molecules using affinity reagents. *Chem. Soc. Rev.* **2008**, *37*, 1347–1360. [CrossRef]

39. Paek, S. Synthetic advances in macrosphelides: Natural anticancer agents. *Molecules* **2014**, *19*, 15982–16000. [CrossRef]

40. Yamada, T.; Iritani, M.; Doi, M.; Minoura, K.; Ito, T.; Numata, A. Absolute stereostructures of cell-adhesion inhibitors, macrosphelides C, E–G and I, produced by a Periconia species separated from an Aplysia sea hare. *J. Chem. Soc. Perkin Trans. 1* **2001**, 3046–3053. [CrossRef]

41. Hayashi, M.; Kim, Y.; Hiraoka, H.; Natori, M.; Takamatsu, S.; Kawakubo, T.; Masuma, R.; Komiyama, K.; Omura, S. Macrosphelide, a novel inhibitor of cell-cell adhesion molecule. *J. Antibiot.* **1995**, *48*, 1435–1439. [CrossRef]

42. Sunazuka, T.; Hirose, T.; Harigaya, Y.; Takamatsu, S.; Hayashi, M.; Komiyama, K.; Ōmura, S.; Sprengeler, P.A.; Smith, A.B. Relative and absolute stereochemistries and total synthesis of (+)-macrosphelides A and B, potent, orally bioavailable inhibitors of cell—Cell adhesion. *J. Am. Chem. Soc.* **1997**, *119*, 10247–10248. [CrossRef]

43. Paek, S. Development of advanced macrosphelides: Potent anticancer agents. *Molecules* **2015**, *20*, 4430–4449. [CrossRef]

44. Matsuya, Y.; Kawaguchi, T.; Ishihara, K.; Ahmed, K.; Zhao, Q.; Kondo, T.; Nemoto, H. Synthesis of macrosphelides with a thiazole side chain: New antitumor candidates having apoptosis-inducing property. *Org. Lett.* **2006**, *8*, 4609–4612. [CrossRef]

45. Meng, D.; Bertinato, P.; Balog, A.; Su, D.; Kamenecka, T.; Sorensen, E.J.; Danishefsky, S.J. Total syntheses of epothilones A and B. *J. Am. Chem. Soc.* **1997**, *119*, 10073–10092. [CrossRef]

46. Xia, S.; Kenesky, C.S.; Rucker, P.V.; Smith, A.B.; Orr, G.A.; Horwitz, S.B. A photoaffinity analogue of discodermolide specifically labels a peptide in β-tubulin. *Biochemistry* **2006**, *45*, 11762–11775. [CrossRef]

47. Smith, A.B., III; Sugasawa, K.; Atasoylu, O.; Yang, C.H.; Horwitz, S.B. Design and synthesis of (+)-discodermolide–paclitaxel hybrids leading to enhanced biological activity. *J. Med. Chem.* **2011**, *54*, 6319–6327. [CrossRef]

48. Shin, Y.; Choy, N.; Balachandran, R.; Madiraju, C.; Day, B.W.; Curran, D.P. Discodermolide/dictyostatin hybrids: Synthesis and biological evaluation. *Org. Lett.* **2002**, *4*, 4443–4446. [CrossRef]

49. Pettit, G.R.; Cichacz, Z.A.; Gao, F.; Boyd, M.R.; Schmidt, J.M. Isolation and structure of the cancer cell growth inhibitor dictyostatin 1. *J. Chem. Soc. Chem. Commun.* **1994**, 1111–1112. [CrossRef]

50. Isbrucker, R.A.; Cummins, J.; Pomponi, S.A.; Longley, R.E.; Wright, A.E. Tubulin polymerizing activity of dictyostatin-1, a polyketide of marine sponge origin. *Biochem. Pharmacol.* **2003**, *66*, 75–82. [CrossRef]

51. Zanato, C.; Pignataro, L.; Hao, Z.; Gennari, C. A practical synthesis of the C1–C9 fragment of dictyostatin. *Synthesis* **2008**, *2008*, 2158–2162.

52. Cichewicz, R.H.; Valeriote, F.A.; Crews, P. Psymberin, a potent sponge-derived cytotoxin from Psammocinia distantly related to the pederin family. *Org. Lett.* **2004**, *6*, 1951–1954. [CrossRef]

53. Clark, W.D.; Corbett, T.; Valeriote, F.; Crews, P. Cyclocinamide A. An unusual cytotoxic halogenated hexapeptide from the marine sponge Psammocinia. *J. Am. Chem. Soc.* **1997**, *119*, 9285–9286. [CrossRef]

54. Kobayashi, M.; Tanaka, J.I.; Katori, T.; Matsuura, M.; Kitagawa, I. Structure of swinholide A, a potent cytotoxic macrolide from the Okinawan marine sponge Tehonella Swinhoei. *Tetrahedron Lett.* **1989**, *30*, 2963–2966. [CrossRef]

55. Faulkner, D.J. Variabilin, an antibiotic from the sponge, Ircinia variabilis. *Tetrahedron Lett.* **1973**, *39*, 3821–3822. [CrossRef]

56. Pettit, G.R.; Xu, J.; Chapuis, J.; Pettit, R.K.; Tackett, L.P.; Doubek, D.L.; Hooper, J.N.; Schmidt, J.M. Antineoplastic Agents. 520. Isolation and Structure of Irciniastatins A and B from the Indo-Pacific Marine Sponge Ircinia r amosa. *J. Med. Chem.* **2004**, *47*, 1149–1152. [CrossRef]

57. Jiang, X.; García-Fortanet, J.; De Brabander, J.K. Synthesis and complete stereochemical assignment of psymberin/irciniastatin A. *J. Am. Chem. Soc.* **2005**, *127*, 11254–11255. [CrossRef]

58. Perry, N.B.; Blunt, J.W.; Munro, M.H.; Pannell, L.K. Mycalamide A, an antiviral compound from a New Zealand sponge of the genus Mycale. *J. Am. Chem. Soc.* **1988**, *110*, 4850–4851. [CrossRef]

59. Jiang, X.; Williams, N.; De Brabander, J.K. Synthesis of psymberin analogues: Probing a functional correlation with the pederin/mycalamide family of natural products. *Org. Lett.* **2007**, *9*, 227–230. [CrossRef]

60. Bartik, K.; Braekman, J.; Daloze, D.; Stoller, C.; Huysecom, J.; Vandevyver, G.; Ottinger, R. Topsentins, new toxic bis-indole alkaloids from the marine sponge Topsentia genitrix. *Can. J. Chem.* **1987**, *65*, 2118–2121. [CrossRef]

61. Tsujii, S.; Rinehart, K.L.; Gunasekera, S.P.; Kashman, Y.; Cross, S.S.; Lui, M.S.; Pomponi, S.A.; Diaz, M.C. Topsentin, bromotopsentin, and dihydrodeoxybromotopsentin: Antiviral and antitumor bis (indolyl) imidazoles from Caribbean deep-sea sponges of the family Halichondriidae. Structural and synthetic studies. *J. Org. Chem.* **1988**, *53*, 5446–5453. [CrossRef]

62. Gu, X.; Wan, X.; Jiang, B. Syntheses and biological activities of bis (3-indolyl) thiazoles, analogues of marine bis (indole) alkaloid nortopsentins. *Bioorg. Med. Chem. Lett.* **1999**, *9*, 569–572. [CrossRef]

63. Hogan, I.; Sainsbury, M. The synthesis of dendrodoine, 5-[3-(N, N-dimethylamino-1, 2, 4-thiadiazolyl]-3-indolylmethanone, a metabolite of the marine tunicate dendroda grossular. *Tetrahedron* **1984**, *40*, 681–682. [CrossRef]

64. Juneja, M.; Vanam, U.; Paranthaman, S.; Bharathan, A.; Keerthi, V.S.; Reena, J.K.; Rajaram, R.; Rajasekharan, K.N.; Karunagaran, D. 4-amino-2-arylamino-5-indoloyl/cinnamoythiazoles, analogs of topsentin-class of marine alkaloids, induce apoptosis in hela cells. *Eur. J. Med. Chem.* **2013**, *63*, 474–483. [CrossRef]

65. Wright, A.E.; Botelho, J.C.; Guzmán, E.; Harmody, D.; Linley, P.; McCarthy, P.J.; Pitts, T.P.; Pomponi, S.A.; Reed, J.K. Neopeltolide, a macrolide from a lithistid sponge of the family Neopeltidae. *J. Nat. Prod.* **2007**, *70*, 412–416. [CrossRef]

66. Custar, D.W.; Zabawa, T.P.; Scheidt, K.A. Total synthesis and structural revision of the marine macrolide neopeltolide. *J. Am. Chem. Soc.* **2008**, *130*, 804–805. [CrossRef]

67. Custar, D.W.; Zabawa, T.P.; Hines, J.; Crews, C.M.; Scheidt, K.A. Total synthesis and structure—Activity investigation of the marine natural product neopeltolide. *J. Am. Chem. Soc.* **2009**, *131*, 12406–12414. [CrossRef]

68. Ulanovskaya, O.A.; Janjic, J.; Suzuki, M.; Sabharwal, S.S.; Schumacker, P.T.; Kron, S.J.; Kozmin, S.A. Synthesis enables identification of the cellular target of leucascandrolide A and neopeltolide. *Nat. Chem. Biol.* **2008**, *4*, 418. [CrossRef]

69. Zhu, X.L.; Z'hang, R.; Wu, Q.; Song, Y.; Wang, Y.; Yang, J.; Yang, G. Natural Product Neopeltolide as A Cytochrome bc1 Complex Inhibitor: Mechanism of Action and Structural Modification. *J. Agric. Food Chem.* **2019**, *67*, 2774–2781. [CrossRef]

70. Vintonyak, V.V.; Kunze, B.; Sasse, F.; Maier, M.E. Total synthesis and biological activity of neopeltolide and analogues. *Chem. Eur. J.* **2008**, *14*, 11132–11140. [CrossRef]

71. Xiong, L.; Li, H.; Jiang, L.; Ge, J.; Yang, W.; Zhu, X.L.; Yang, G. Structure-based discovery of potential fungicides as succinate ubiquinone oxidoreductase inhibitors. *J. Agric. Food Chem.* **2017**, *65*, 1021–1029. [CrossRef]

72. Moulin, E.; Nevado, C.; Gagnepain, J.; Kelter, G.; Fiebig, H.; Fürstner, A. Synthesis and evaluation of an Iejimalide-archazolid chimera. *Tetrahedron* **2010**, *66*, 6421–6428. [CrossRef]

73. Kobayashi, J.; Cheng, J.; Ohta, T.; Nakamura, H.; Nozoe, S.; Hirata, Y.; Ohizumi, Y.; Sasaki, T. Iejimalides A and B, novel 24-membered macrolides with potent antileukemic activity from the Okinawan tunicate Eudistoma cf. rigida. *J. Org. Chem.* **1988**, *53*, 6147–6150. [CrossRef]

74. Kikuchi, Y.; Ishibashi, M.; Sasaki, T.; Kobayashi, J. Iejimalides C and D, new antineoplastic 24-membered macrolide sulfates from the Okinawan marine tunicate Eudistoma cf. rigida. *Tetrahedron Lett.* **1991**, *32*, 797–798. [CrossRef]

75. Fürstner, A.; Nevado, C.; Tremblay, M.; Chevrier, C.; Teplý, F.; Aïssa, C.; Waser, M. Total synthesis of iejimalide B. *Angew. Chem. Int. Edit.* **2006**, *45*, 5837–5842. [CrossRef]

76. Fürstner, A.; Nevado, C.; Waser, M.; Tremblay, M.; Chevrier, C.; Teplý, F.; Aïssa, C.; Moulin, E.; Müller, O. Total Synthesis of Iejimalide A–D and Assessment of the Remarkable Actin-Depolymerizing Capacity of These Polyene Macrolides. *J. Am. Chem. Soc.* **2007**, *129*, 9150–9161. [CrossRef]

77. Sasse, F.; Steinmetz, H.; Hoefle, G.; Reichenbach, H. Archazolids, New Cytotoxic Macrolactones from Archangium gephyra (Myxobacteria). *J. Antibiot.* **2003**, *56*, 520–525. [CrossRef]

78. Corey, E.J.; Helal, C.J. Reduction of carbonyl compounds with chiral oxazaborolidine catalysts: A new paradigm for enantioselective catalysis and a powerful new synthetic method. *Angew. Chem. Int. Edit.* **1998**, *37*, 1986–2012. [CrossRef]

79. Marshall, J.A. Chiral allylic and allenic metal reagents for organic synthesis. *J. Org. Chem.* **2007**, *72*, 8153–8166. [CrossRef]

80. Miyaura, N.; Suzuki, A. Palladium-catalyzed cross-coupling reactions of organoboron compounds. *Chem. Rev.* **1995**, *95*, 2457–2483. [CrossRef]

81. Ogba, O.; Warner, N.; O'Leary, D.; Grubbs, R. Recent advances in ruthenium-based olefin metathesis. *Chem. Soc. Rev.* **2018**, *47*, 4510–4544. [CrossRef]

82. Laitinen, O.H.; Nordlund, H.R.; Hytönen, V.P.; Kulomaa, M.S. Brave new (strept) avidins in biotechnology. *Trends Biotechnol.* **2007**, *25*, 269–277. [CrossRef]

83. Dundas, C.M.; Demonte, D.; Park, S. Streptavidin–biotin technology: Improvements and innovations in chemical and biological applications. *Appl. Microbiol. Biotechnol.* **2013**, *97*, 9343–9353. [CrossRef]

84. Kim, J.; Cantor, A.B.; Orkin, S.H.; Wang, J. Use of in vivo biotinylation to study protein–protein and protein–DNA interactions in mouse embryonic stem cells. *Nat. Protoc.* **2009**, *4*, 506–517. [CrossRef]

85. Manz, B.; Heubner, A.; Kohler, I.; Grill, H.; Pollow, K. Synthesis of Biotin-Labelled Dexamethasone Derivatives: Novel Hormone-Affinity Probes. *Eur. J. Biochem.* **1983**, *131*, 333–338. [CrossRef]

86. Wright, M.; Sieber, S. Chemical proteomics approaches for identifying the cellular targets of natural products. *Nat. Prod. Rep.* **2016**, *33*, 681–708. [CrossRef]

87. Lomenick, B.; Olsen, R.W.; Huang, J. Identification of direct protein targets of small molecules. *ACS Chem. Biol.* **2010**, *6*, 34–46. [CrossRef]

88. Sato, S.; Murata, A.; Shirakawa, T.; Uesugi, M. Biochemical target isolation for novices: Affinity-based strategies. *Chem. Biol.* **2010**, *17*, 616–623. [CrossRef]

89. Gouiffes, D.; Moreau, S.; Helbecque, N.; Bernier, J.; Henichart, J.; Barbin, Y.; Laurent, D.; Verbist, J. Proton nuclear magnetic study of bistramide A, a new cytotoxic drug isolated from Lissoclinum bistratum Sluiter. *Tetrahedron* **1988**, *44*, 451–459. [CrossRef]

90. Degnan, B.M.; Hawkins, C.J.; Lavin, M.F.; McCaffrey, E.J.; Parry, D.L.; Watters, D.J. Novel cytotoxic compounds from the ascidian Lissoclinum bistratum. *J. Med. Chem.* **1989**, *32*, 1354–1359. [CrossRef]

91. Biard, J.; Roussakis, C.; Kornprobst, J.; Gouiffes-Barbin, D.; Verbist, J.; Cotelle, P.; Foster, M.P.; Ireland, C.M.; Debitus, C. Bistramides A, B, C, D, and K: A new class of bioactive cyclic polyethers from Lissoclinum bistratum. *J. Nat. Prod.* **1994**, *57*, 1336–1345. [CrossRef]

92. Riou, D.; Roussakis, C.; Biard, J.F.; Verbist, J.F. Comparative study of the antitumor activity of bistramides A, D and K against a non-small cell broncho-pulmonary carcinoma. *Anticancer Res.* **1993**, *13*, 2331–2334.

93. Wipf, P.; Uto, Y.; Yoshimura, S. Total synthesis of a stereoisomer of bistramide C and assignment of configuration of the natural product. *Chem. Eur. J.* **2002**, *8*, 1670–1681. [CrossRef]

94. Statsuk, A.V.; Liu, D.; Kozmin, S.A. Synthesis of bistramide A. *J. Am. Chem. Soc.* **2004**, *126*, 9546–9547. [CrossRef]

95. Gagne, M.R.; Grubbs, R.H.; Feldman, J.; Ziller, J.W. Catalytic activity of a well-defined binuclear ruthenium alkylidene complex. *Organometallics* **1992**, *11*, 3933–3935. [CrossRef]

96. Foster, M.P.; Mayne, C.L.; Dunkel, R.; Pugmire, R.J.; Grant, D.M.; Kornprobst, J.M.; Verbist, J.F.; Biard, J.F.; Ireland, C.M. Revised structure of bistramide A (bistratene A): Application of a new program for the automated analysis of 2D INADEQUATE spectra. *J. Am. Chem. Soc.* **1992**, *114*, 1110–1111. [CrossRef]

97. Johnson, W.E.B.; Watters, D.J.; Suniara, R.K.; Brown, G.; Bunce, C.M. Bistratene A induces a microtubule-dependent block in cytokinesis and altered stathmin expression in HL60 cells. *Biochem. Biophys. Res. Commun.* **1999**, *260*, 80–88. [CrossRef]

98. Statsuk, A.V.; Bai, R.; Baryza, J.L.; Verma, V.A.; Hamel, E.; Wender, P.A.; Kozmin, S.A. Actin is the primary cellular receptor of bistramide A. *Nat. Chem. Biol.* **2005**, *1*, 383–388. [CrossRef]

99. Sun, H.; Low, K.E.; Woo, S.; Noble, R.L.; Graham, R.J.; Connaughton, S.S.; Gee, M.A.; Lee, L.G. Real-time protein kinase assay. *Anal. Chem.* **2005**, *77*, 2043–2049. [CrossRef]

100. Yamada, K.; Ojika, M.; Ishigaki, T.; Yoshida, Y.; Ekimoto, H.; Arakawa, M. Aplyronine A, a potent antitumor substance and the congeners aplyronines B and C isolated from the sea hare Aplysia kurodai. *J. Am. Chem. Soc.* **1993**, *115*, 11020–11021. [CrossRef]

101. Kita, M.; Hirayama, Y.; Yoneda, K.; Yamagishi, K.; Chinen, T.; Usui, T.; Sumiya, E.; Uesugi, M.; Kigoshi, H. Inhibition of microtubule assembly by a complex of actin and antitumor macrolide aplyronine A. *J. Am. Chem. Soc.* **2013**, *135*, 18089–18095. [CrossRef]

102. Paterson, I.; Fink, S.J.; Lee, L.Y.; Atkinson, S.J.; Blakey, S.B. Total synthesis of aplyronine C. *Org. Lett.* **2013**, *15*, 3118–3121. [CrossRef]

103. Lindquist, N.; Fenical, W.; Van Duyne, G.D.; Clardy, J. Isolation and structure determination of diazonamides A and B, unusual cytotoxic metabolites from the marine ascidian Diazona chinensis. *J. Am. Chem. Soc.* **1991**, *113*, 2303–2304. [CrossRef]

104. Li, J.; Jeong, S.; Esser, L.; Harran, P.G. Total synthesis of nominal Diazonamides—Part 1: Convergent preparation of the structure proposed for (−)-Diazonamide A. *Angew. Chem. Int. Edit.* **2001**, *40*, 4765–4769. [CrossRef]

105. Li, J.; Burgett, A.W.; Esser, L.; Amezcua, C.; Harran, P.G. Total synthesis of nominal diazonamides—Part 2: On the true structure and origin of natural isolates. *Angew. Chem. Int. Edit.* **2001**, *40*, 4770–4773. [CrossRef]

106. Nicolaou, K.; Bella, M.; Chen, D.Y.; Huang, X.; Ling, T.; Snyder, S.A. Total synthesis of diazonamide A. *Angew. Chem. Int. Edit.* **2002**, *41*, 3495–3499. [CrossRef]

107. Wang, G.; Shang, L.; Burgett, A.W.; Harran, P.G.; Wang, X. Diazonamide toxins reveal an unexpected function for ornithine δ-amino transferase in mitotic cell division. *Proc. Natl. Acad. Sci. USA* **2007**, *104*, 2068–2073. [CrossRef]

108. Seiler, N. Ornithine aminotransferase, a potential target for the treatment of hyperammonemias. *Curr. Drug Targets* **2000**, *1*, 119–154. [CrossRef]

109. Wieczorek, M.; Tcherkezian, J.; Bernier, C.; Prota, A.E.; Chaaban, S.; Rolland, Y.; Godbout, C.; Hancock, M.A.; Arezzo, J.C.; Ocal, O. The synthetic diazonamide DZ-2384 has distinct effects on microtubule curvature and dynamics without neurotoxicity. *Sci. Transl. Med.* **2016**, *8*, 365ra159. [CrossRef]

110. Yun, H.; Sim, J.; An, H.; Lee, J.; Lee, H.S.; Shin, Y.K.; Paek, S.; Suh, Y. Design and synthesis of a macrosphelide A-biotin chimera. *Org. Biomol. Chem.* **2014**, *12*, 7127–7135. [CrossRef]

111. Paek, S.; Seo, S.; Kim, S.; Jung, J.; Lee, Y.; Jung, J.; Suh, Y. Concise syntheses of (+)-macrosphelides A and B. *Org. Lett.* **2005**, *7*, 3159–3162. [CrossRef]

112. Hamann, M.T.; Scheuer, P.J. Kahalalide F: A bioactive depsipeptide from the sacoglossan mollusk Elysia rufescens and the green alga Bryopsis sp. *J. Am. Chem. Soc.* **1993**, *115*, 5825–5826. [CrossRef]

113. Faircloth, G.; Marchante, M.d.C.C. Kahalalide F and ES285: Potent anticancer agents from marine molluscs. In *Molluscs*; Springer: Berlin/Heidelberg, Germany, 2006; pp. 363–379.

114. Piggott, A.M.; Karuso, P. Rapid identification of a protein binding partner for the marine natural product kahalalide F by using reverse chemical proteomics. *ChemBioChem* **2008**, *9*, 524–530. [CrossRef]

115. Wulff, J.E.; Siegrist, R.; Myers, A.G. The natural product avrainvillamide binds to the oncoprotein nucleophosmin. *J. Am. Chem. Soc.* **2007**, *129*, 14444–14451. [CrossRef]

116. Sugie, Y.; Hirai, H.; Inagaki, T.; Ishiguro, M.; Kim, Y.; Kojima, Y.; Sakakibara, T.; Sakemi, S.; Sugiura, A.; Suzuki, Y. A new antibiotic CJ-17, 665 from Aspergillus ochraceus. *J. Antibiot.* **2001**, *54*, 911–916. [CrossRef]
117. Herzon, S.B.; Myers, A.G. Enantioselective synthesis of stephacidin B. *J. Am. Chem. Soc.* **2005**, *127*, 5342–5344. [CrossRef]
118. Wulff, J.E.; Herzon, S.B.; Siegrist, R.; Myers, A.G. Evidence for the rapid conversion of stephacidin B into the electrophilic monomer avrainvillamide in cell culture. *J. Am. Chem. Soc.* **2007**, *129*, 4898–4899. [CrossRef]
119. Chan, W.Y.; Liu, Q.R.; Borjigin, J.; Busch, H.; Rennert, O.M.; Tease, L.A.; Chan, P.K. Characterization of the cDNA encoding human nucleophosmin and studies of its role in normal and abnormal growth. *Biochemistry* **1989**, *28*, 1033–1039. [CrossRef]
120. Myers, A.G.; Herzon, S.B.; Wulff, J.E.; Siegrist, R.; Svenda, J.; Zajac, M.A. Synthesis of Avrainvillamide, Stephacidin B, and Analogues Thereof. U.S. Patent 20110166170A1, 1 July 2011.
121. Northcote, P.T.; Blunt, J.W.; Munro, M.H. Pateamine: A potent cytotoxin from the New Zealand marine sponge, Mycale sp. *Tetrahedron Lett.* **1991**, *32*, 6411–6414. [CrossRef]
122. Romo, D.; Rzasa, R.M.; Shea, H.A.; Park, K.; Langenhan, J.M.; Sun, L.; Akhiezer, A.; Liu, J.O. Total synthesis and immunosuppressive activity of (−)-pateamine A and related compounds: Implementation of a β-lactam-based macrocyclization. *J. Am. Chem. Soc.* **1998**, *120*, 12237–12254. [CrossRef]
123. Low, W.; Dang, Y.; Schneider-Poetsch, T.; Shi, Z.; Choi, N.S.; Merrick, W.C.; Romo, D.; Liu, J.O. Inhibition of eukaryotic translation initiation by the marine natural product pateamine A. *Mol. Cell* **2005**, *20*, 709–722. [CrossRef]
124. Low, W.; Dang, Y.; Schneider-Poetsch, T.; Shi, Z.; Choi, N.S.; Rzasa, R.M.; Shea, H.A.; Li, S.; Park, K.; Ma, G. Isolation and identification of eukaryotic initiation factor 4A as a molecular target for the marine natural product Pateamine A. *Meth. Enzymol.* **2007**, *431*, 303–324.
125. Low, W.; Dang, Y.; Bhat, S.; Romo, D.; Liu, J.O. Substrate-dependent targeting of eukaryotic translation initiation factor 4A by pateamine A: Negation of domain-linker regulation of activity. *Chem. Biol.* **2007**, *14*, 715–727. [CrossRef]

 *marine drugs*

Article

Selective Suppression of Cell Growth and Programmed Cell Death-Ligand 1 Expression in HT1080 Fibrosarcoma Cells by Low Molecular Weight Fucoidan Extract

Kiichiro Teruya [1,2,*], Yoshihiro Kusumoto [2], Hiroshi Eto [3], Noboru Nakamichi [3] and Sanetaka Shirahata [1]

[1]   Faculty of Agriculture, Kyushu University, 744 Motooka, Nishi-ku, Fukuoka 819-0395, Japan
[2]   Graduate School of Bioresource and Bioenvironmental Sciences, Kyushu University, 744 Motooka, Nishi-ku, Fukuoka 819-0395, Japan
[3]   Daiichi Sangyo Co., Ltd., 6-7-2 Nishitenma, Kita-ku, Osaka 530-0047, Japan
*   Correspondence: kteruya@grt.kyushu-u.ac.jp; Tel./Fax: +81-92-802-4728

Received: 28 June 2019; Accepted: 14 July 2019; Published: 19 July 2019

Abstract: Low molecular weight fucoidan extract (LMF), prepared by an abalone glycosidase digestion of a crude fucoidan extracted from *Cladosiphon novae-caledoniae* Kylin, exhibits various biological activities, including anticancer effect. Various cancers express programmed cell death-ligand 1 (PD-L1), which is known to play a significant role in evasion of the host immune surveillance system. PD-L1 is also expressed in many types of normal cells for self-protection. Previous research has revealed that selective inhibition of PD-L1 expressed in cancer cells is critical for successful cancer eradication. In the present study, we analyzed whether LMF could regulate PD-L1 expression in HT1080 fibrosarcoma cells. Our results demonstrated that LMF suppressed PD-L1/PD-L2 expression and the growth of HT1080 cancer cells and had no effect on the growth of normal TIG-1 cells. Thus, LMF differentially regulates PD-L1 expression in normal and cancer cells and could serve as an alternative complementary agent for treatment of cancers with high PD-L1 expression.

Keywords: fucoidan; low molecular weight fucoidan extract; N-Ras; neuroblastoma-rat sarcoma; Cancer; programmed cell death-ligand 1; programmed cell death-ligand 2; human sarcoma cell line (HT1080 cells); human normal diploid fibroblast (TIG-1 cells)

1. Introduction

Various types of cancers evade T-cell responses and host immunity via adaptive negative regulators or co-inhibitory (checkpoint) receptors, such as programmed cell death-1 (PD-1) and their respective ligands, programmed cell death-ligand 1 (PD-L1 or B7-H1), programmed cell death-ligand 2 (PD-L2), cytotoxic T-lymphocyte-associated antigen 4 (CTLA-4), and CD80 (B7.1) or CD86 (B7.2) [1,2]. These regulators react with each other to form an elaborative defense system or host immunity to kill pathogens and cancer cells. In order to acquire cytocidal function, a variety of immune cells, including resting T cells expressing PD-1, must be activated by two stimulating signals from an antigen-presenting cell (APC), as shown in Figure 1 [3–5]. In addition, resting T-cells express T-cell receptor (TCR) and CD28 on their cell surface which interact with the antigen-specific signal presented on the major histocompatibility complex (MHC) and CD80 (B7.1) or CD86 (B7.2) ligand expressed on the surface of an APC, respectively. Interactions of these receptor-ligand pairs are referred to as the early stage of T-cell activation [1,4]. The activated T-cells express CTLA-4, and as the binding affinity of CTLA-4 for CD80/86 is 500- to 2500-fold greater than that for CD28, CTLA-4 (negative regulator) competes with co-stimulatory CD28 signals to bind to CD80/86 [6,7].

Conversely, the interaction of PD-1 expressed on the T-cell surface with its ligand PD-L1 or PD-L2 on APC causes inhibition of T-cell activation, leading to reduced cell proliferation, and induces T-cell cytolysis [6,8]. While a balanced PD-1/PD-L1 interaction is a prerequisite for maintaining the normal T-cell response and assuring normal cell survival, overexpressed PD-L1 on cancer cells in the tumor microenvironment react with the PD-1 expressed on T cells and induce T-cell apoptosis, thereby facilitating tumor survival, progression, and evasion of the host tumor immune surveillance system [4,7,9]. Moreover, higher PD-L1 expression in many tumors corresponds to enhanced immune evasion thereby promoting tumor growth by suppressing the T-cell response [7,10–12]. Therefore, considering the PD-1/PD-L1 axis, the selective inhibition of PD-L1 expressed on cancer cells is expected to maintain the host defense system and consequently contribute towards cancer eradication.

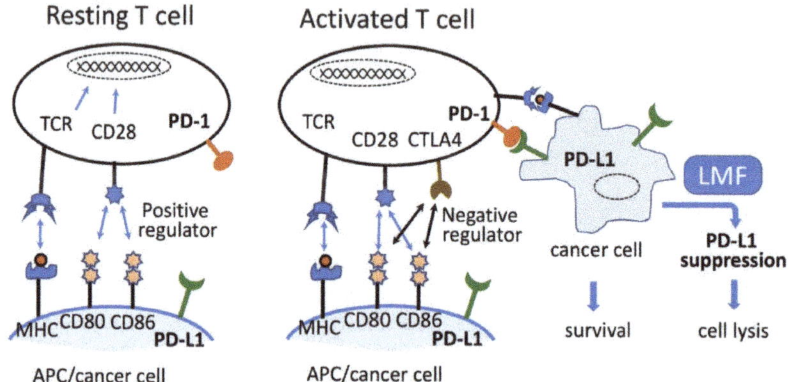

Figure 1. Simplified image of T-cell activation and cancer cell lysis. Resting T-cells are activated by the interaction of TCR:MHC and CD28:CD80 and CD86, leading to the expression of CTLA4. CTLA4 preferentially reacts with CD80 and CD86 causing activated T-cell lysis. Although imbalanced interaction of programmed cell death-1 (PD-1) on activated T-cell and programmed cell death-ligand 1 (PD-L1) on antigen-presenting cell (APC) causes T-cell lysis, such reaction between activated T-cells and cancer cells expressing PD-L1 will lead to the survival of cancer cells and facilitate cancer cell growth. Suppression of PD-L1 by low molecular weight fucoidan extract (LMF) leads to cancer cell lysis. Adapted from [7,13].

PD-L1 expression is upregulated not only by some inflammatory cytokines (e.g., interferon-γ) but also by oncogenic activation of signaling pathways [6,12]. In one of the two main oncogenic signaling pathways, the extracellular epidermal growth factor (EGF) and TGF-α stimulate the EGF receptor (EGFR) which in turn activates wild type rat sarcoma oncogene homolog (Ras) to activate the raf pathway [14]. Moreover, in the second pathway, the constitutively active mutant Ras activates its immediately downstream factors, including phosphatidylinositol 3-kinase (PI3K), Rac1, and Raf which trigger a sequential activation of multiple downstream factors and eventually induce expressions of many genes such as RhoA/B, EGFR, VEGF, and PD-L1 in cancer cells [2,6,15–17].

Other lines of evidence have shown constitutive activation of mutant N-*ras* allele encoded in a human soft tissue sarcoma cell line, HT1080, which regulates Raf and RhoA pathways. Moreover, Raf-activated MEK, an intermediary transducer, is essential in the Ras signaling pathway but not in the PI3K/Akt pathway that includes nuclear factor-kappa B (NF-κB), resulting in the aggressive tumorigenic phenotype in HT1080 cells [18,19]. These data suggest that the Raf regulated pathway is closely associated with carcinogenesis, implying that this pathway regulates PD-L1 expression as one of the terminal steps in HT1080 cells. However, these studies do not address the involvement of Ras-regulated pathways in PD-L1 expression in HT1080 cells. As the PD-1/PD-L1 axis plays a major role in human cancers for immune evasion [17], it might be extremely valuable to cancer patients if prospective agents such as fucoidan devoid of side effects are available to regulate PD-L1 expression exclusively in cancer cells.

Fucoidan can be found mainly in various species of brown algae (brown seaweed) such as wakame (*Undaria pinnatifida*), kombu (*Saccharina japonica*), mozuku (*Cladosiphon okamuranus*), hijiki (*Sargassum fusiforme*), bladderwrack (*Fucus vesiculosus*), and hibamata (*Fucus evanescens*) [20,21]. Fucoidan is fucose containing sulfated polysaccharides exhibiting complicated chemical structures due to various combinations of fucose, uronic acids, galactose, and xylose. [22–24]. The structure of fucoidan from *Cladosiphon okamuranus* belonging to Genus *Cladosiphon* had been determined partly using the nuclear magnetic resonance (NMR) method [25] and we reported that this structure possess similar structural features as the fucoidan from *Cladosiphon novae-caledoniae* Kylin. It has been noted that Quantitative ^{1}H-NMR (qNMR) analysis for the qualitative and quantitative characterization of metabolites may be applied for crude biological extracts such as fucoidan [26]. Just recently, the chemical structure of fucoidan from *Cladosiphon okamuranus* was determined using NMR analyses [27], as shown in Figure 2, and we now also refer to this structure as having a similar structural feature as fucoidan from *Cladosiphon novae-caledoniae* Kylin.

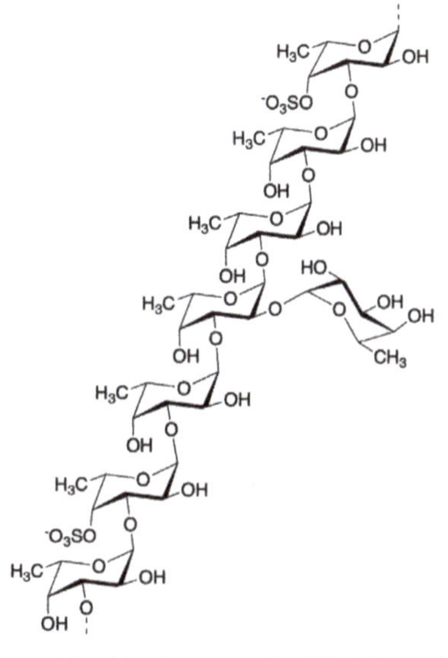

Figure 2. Structure of fucoidan from mozuku (*Cladosiphon okamuranus*) [27].

Fucoidan extracts usually show various sizes, such as 5100 kDa [28] and 1600 kDa [29], depending on the extraction methods and the kinds of seaweeds. Various molecular weight derivatives designated as low-molecular-weight fucoidan (LMWF: up to 40 kDa), intermediate-molecular-weight fucoidan (110–138 kDa), and high-molecular-weight fucoidan (300–330 kDa) [30], and other size classifications up to 10,000 kDa [31], have been prepared for biological and biochemical characterizations. In still other cases, many derivatives having molecular weights ranging between 7.6 kDa and 712 kDa have been assessed for their membrane permeability in vitro, and bioavailability in humans [32–36]. Apart from the above higher molecular weight derivatives, there are a group of fucoidan derivatives possessing much lower molecular weights designated as low molecular weight fucoidan (LMF) also called as oligo-fucoidan having MW of 500–800 Da [37], LMF from L. japonica (LMF-LJ) having an average molecular weight of <667 Da [38], low molecular weight fucoidan (LMWF) having main MW of 760 Da [39], and LMF having an average MW of 800 Da [40]. Finally, low molecular weight fucoidan extract (LMF) having a main MW of <500 Da, prepared by digesting crude fucoidan extracted from mozuku (*Cladosiphon novae-caledoniae* Kylin) with an abalone glycosidase, which was used in this study [41]. These two size groups of fucoidan derivatives have been examined for their health benefits as well as for their therapeutic effects and were shown to exhibit broad biological activities such as anti-tumor, antioxidant, anticoagulant, anti-inflammatory, and immune-modulatory effects in in vitro and in vivo studies [22,23,42–44]. With regards to its anti-tumor activity, fucoidan has been shown to exhibit suppressive effects in lung, breast, liver, colon, prostate, and bladder cancer cells [22,45]. In addition, our previous in vitro studies revealed that LMF (MW < 500 Da) can enhance the anticancer activity of chemotherapeutic agents (such as cisplatin, tamoxifen, and paclitaxel) [46] and it also demonstrated beneficial immunomodulatory effects in a clinical trial [23]. It has been shown that the low molecular weight fucoidan with a MW of 7.6 kDa exhibited higher intestinal absorptivity than the medium molecular weight fucoidan (MW 35 kDa) when assessed in the plasma and urine after oral administration in rats [36]. This result suggests that the lower molecular weight fucoidan derivatives are superior to the ones with higher MW in the intestinal absorptivity which is in line with our long-time notion of using LMF (mainly MW < 500 Da) for clinical application along with collecting supportive basic research data [23,46,47]. In order to further support these conclusions, the low molecular weight fucoidan derivatives mimicking the structural features of the genus *Cladosiphon* have been synthesized and tested for their anti-cancer activities. Results showed that one of the sulfated tetrafucoside synthetic derivatives could reduce MCF-7 and HeLa cell growth while showing no cytotoxic effect on normal WI-38 cells [48]. Moreover, it has been suggested that the most important factor that affects the biological activities of fucoidan is the branching degree of the fucoidan rather than its molecular weight, monosaccharides composition, and sulfate degrees [49]. Thus, it could be envisioned that an efficiently absorbed LMF (main MW < 500 Da) in the small intestine is perhaps also distributed throughout the intercellular environments, including tumor microenvironments, and such localized LMF may suppress cancer cell growth. These data together present the presently used LMF as an advantageous molecule over high molecular weight derivatives in further pursuing studies for basic research and clinical applications.

The molecular mechanism of fucoidan action involves suppression of the growth factor-stimulated pathway involving Ras/Raf cascade. A large amount of data supports the fucoidan inhibition of phosphorylation of intermediary transducers such as ERK1/2, PI3K, Akt, mTOR, c-Jun, c-Fos, EGF receptor, and vascular endothelial growth factor (VEGF) [21,22,39].

Accumulated data suggest that Ras regulates several downstream pathways to upregulate PD-L1 expression, while fucoidan suppresses Ras-regulated downstream signal transducers that coincide with those of the PD-L1 expression pathways. However, none of the preceding studies, as well as the relevant computer literature searches, explored the possibility that fucoidan suppresses PD-L1 expression regulated by the Ras-activated pathways. Therefore, we focused on examining the regulatory effects of LMF for several growth factors, PD-L1/PD-L2 and RhoA/B. We then examined our hypothesis that LMF differentially regulates PD-L1 expression in cancer and normal cells.

2. Results

2.1. Suppression of Cancer Cell Growth by Low Molecular Weight Fucoidan Extract (LMF)

To confirm whether LMF specifically suppresses cancer cell growth, HT1080 fibrosarcoma cells and normal human fetal lung diploid fibroblast TIG-1 cells were treated with varying concentrations (0–500 µg/mL) of LMF for 48 h. WST-1 cell viability assay showed significant dose-dependent suppression of HT1080 cell growth by LMF (** $p < 0.01$, *** $p < 0.001$) (Figure 3A), while there was no statistically significant suppressive effect on TIG-1 cell growth up to 500 µg/mL (Figure 3B). Moreover, the response of HT1080 cells to concentrations lower than 100 µg/mL of LMF showed a similar dose-dependent suppressive effect with both WST-1 and Hoechst 33342 DNA staining assays (Figure 3C). In addition, Hoechst 33342 dye staining method was employed as this dye stains DNA of both live- and apoptotic-cells [50,51] resulting in the measurement of cumulative numbers of both cell types and allows the visualization of the cytocidal effect of LMF when compared with the WST-1 assay, which measures only live or metabolically active cell numbers [52,53]. The results showed a significant dose-dependent reduction in the cumulative cell numbers (Figure 3C). A comparison of IC_{50} of WST-1 as 40.1 µg/mL and IC_{50} of Hoechst staining as 73.7 µg/mL revealed that approximately 12% of HT1080 cells are apoptosis-induced cells by LMF. These results demonstrate that LMF at even lower concentrations suppresses cell growth dose-dependently, and approximately 12% of HT1080 cell apoptosis was observed to be induced by 36.6 µg/mL of LMF up to 100 µg/mL.

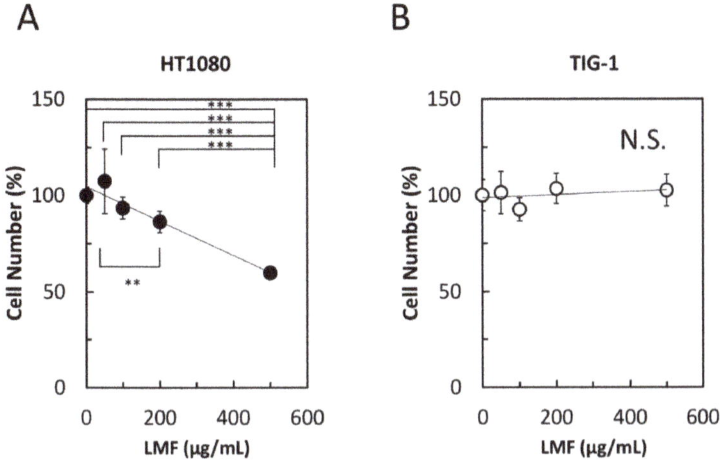

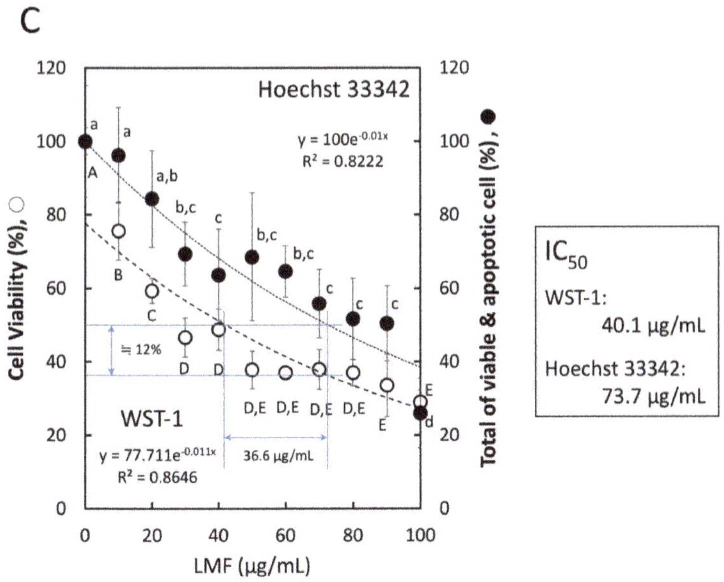

Figure 3. Effects of low molecular weight fucoidan extract (LMF) on cancer cell and normal cell growth. (**A**) HT1080 cells (1×10^4 cells/mL) and (**B**) TIG-1 cells (3×10^4 cells/mL), were plated (100 µL per well) in a 96-well microplate. After 24 h, varying concentrations (0, 50, 100, 200, 500 µg/mL) of LMF were added, the cells were incubated for another 48 h, and live cell numbers in each well were measured by the WST-1 assay. The measured values were used for statistical analyses by Student's *t*-test. (**C**) Effect of lower LMF concentrations on HT1080 cell growth was examined using varying concentrations (0, 20, 40, 60, 80, 100 µg/mL) of LMF. After 48 h of treatment, cellular metabolic activity-based live cell numbers were measured by WST-1 (open circles, ○), and live cell numbers were measured by Hoechst 33342 DNA staining (closed circles, ●). The number of fucoidan untreated HT1080 cells was set to 100% to calculate the relative values of fucoidan-treated HT1080 cells. Statistical significance was determined by ANOVA with Tukey test. Different letters indicate statistically significant differences between each letter ($p < 0.05$). Half-maximal inhibitory concentration (IC$_{50}$) values were calculated using approximation formulae.

2.2. Comparison of PD-L1 Expression Levels in Seven Cell Lines by Immuno-Fluorescent Staining

Prior to examining the effect of LMF on PD-L1 expression, we compared PD-L1 expression levels in seven cell lines, using A549 and PC-9 cells as controls for low and high PD-L1 expression, respectively [12,44,54]. As shown in Figure 4, HT1080 cells expressed 1.06-fold higher PD-L1 compared to PC-9 cells. TIG-1 cells expressed PD-L1 levels comparable to those of the control cells (PC-9). Other cell lines expressed PD-L1 levels similar to or higher than those of the A549 cells. Therefore, we selected the HT1080 cell line for further studies to assess the effects of LMF on PD-L1 expression.

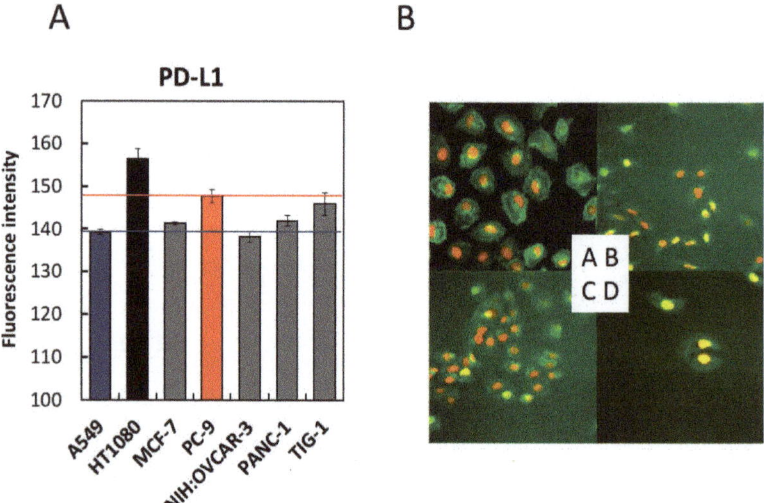

Figure 4. Comparison of programmed cell death-ligand 1 (PD-L1) expression levels in seven cell lines by immunofluorescence staining. Seven cell lines with indicated cell densities were prepared (HT1080 and A549 cells at 2×10^4 cells/mL, MCF-7, PC-9, NIH:OVCAR-3, and PANC-1 cells at 3×10^4 cells/mL, and TIG-1 cells at 4×10^4 cells/mL) and seeded at 100 µL per well of a 96-well black plate followed by 24 h culture. PD-L1 detection was carried out using the anti-PD-L1 antibody, as described in the Methods section. Cells were stained with Hoechst 33342 for 30 min and analyzed using (**A**) IN Cell Analyzer 1000. The fluorescence intensities of each cell line were measured and converted to a numerical form for graphical presentation. For each cell line, six viewing fields per well were analyzed. A549 and PC-9 cells were used as low- and high-level controls for PD-L1 expression, respectively. (**B**) Four representative images acquired in (**A**) were shown. A, HT1080; B, A549; C, PANC-1; D, NIH:OVCAR-3.

2.3. LMF Differentially Regulates Programmed Cell Death-Ligand 1 (PD-L1) and PD-L2 mRNA Expression in HT1080 and TIG-1 Cells

We examined LMF regulation of PD-L1/L2 expression in HT1080 and TIG-1 cell lines. HT1080 cells treated with 1 and 10 µg/mL of LMF for 48 h showed a significant reduction in PD-L1 (all, ** $p < 0.01$) and *PD-L2* (1 µg/mL, * $p < 0.05$; 10 µg/mL, ** $p < 0.01$) mRNA levels (Figure 5A). TIG-1 cells treated with 10 and 100 µg/mL of LMF for 48 h showed significant increases in PD-L1 (10 µg/mL, *** $p < 0.001$; 100 µg/mL, ** $p < 0.01$) and PD-L2 (all: **** $p < 0.0001$) mRNA levels (Figure 5B).

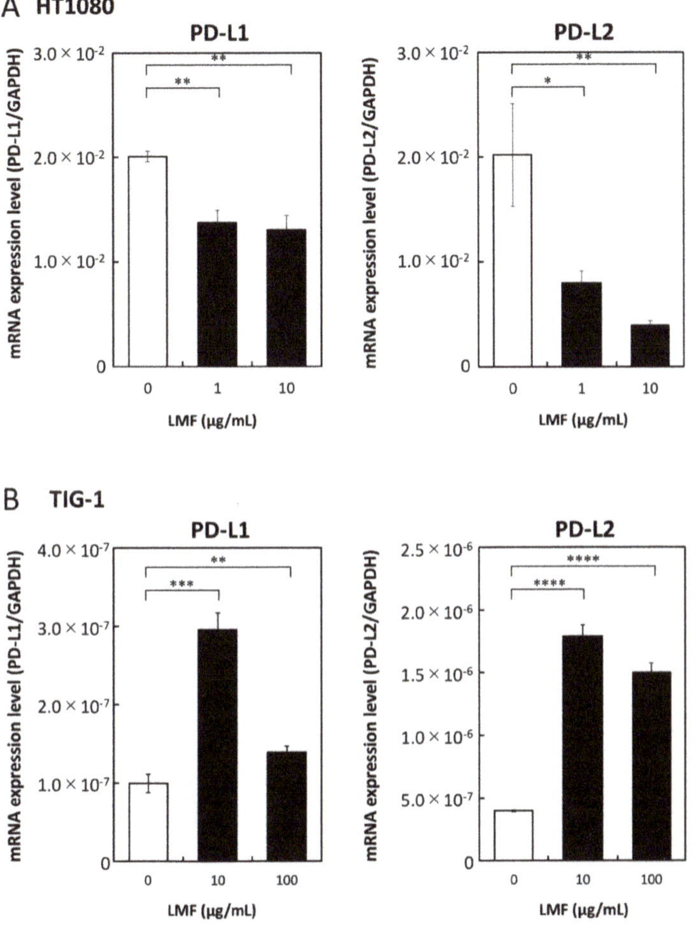

Figure 5. Low molecular weight fucoidan extract (LMF) differentially effects transcription of programmed cell death-ligand 1 (PD-L1) and PD-L2 mRNAs in cancer and normal cells. (**A**) HT1080 and (**B**) TIG-1 cells were treated with varying concentrations (0, 1, 10 µg/mL and 0, 10, 100 µg/mL) of LMF respectively, for 48 h. After treatment, RNAs were isolated and subjected to quantitative reverse transcription polymerase chain reaction (qRT-PCR) analyses. Values obtained from LMF-treated cells were compared with those of untreated cells using Student's *t*-test (* $p < 0.05$; ** $p < 0.01$, *** $p < 0.001$; **** $p < 0.0001$, $n = 3$).

2.4. Suppression of PD-L1 Protein Expression by LMF

To further test the effects of LMF on PD-L1 expression, we measured the PD-L1 protein levels using an anti-PD-L1 antibody to test whether *PD-L1* mRNA suppression reflects similar suppression in protein levels. Western blot analysis of LMF-treated and untreated HT1080 cells was performed as described in the Methods section, and this method has been used by us to see the effects of fucoidan on HT1080 cells as well as other cultured cells [47]. PD-L1 protein band intensities of LMF-treated cells were lower than that of the control cells (Figure 6A). Protein bands in Figure 6A were quantitated for statistical evaluations and results show a significant reduction in PD-L1 protein (*** $p < 0.001$, Figure 6B) in LMF-treated HT1080 cells. Therefore, the PD-L1 protein levels reduced proportionally with a reduction in mRNA and the results suggest that the upstream signal transducers are mostly affected for PD-L1 transcription.

A

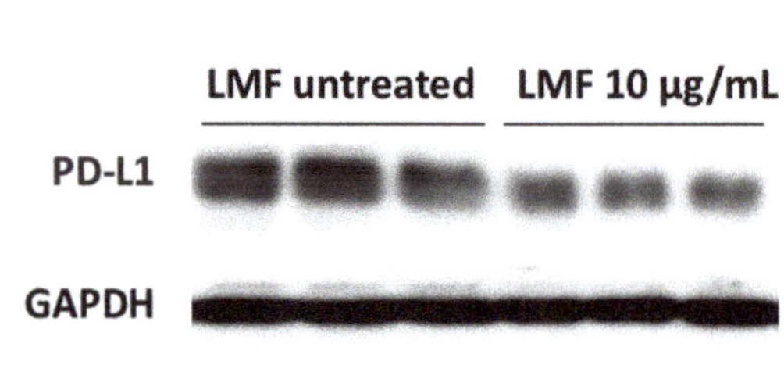

B

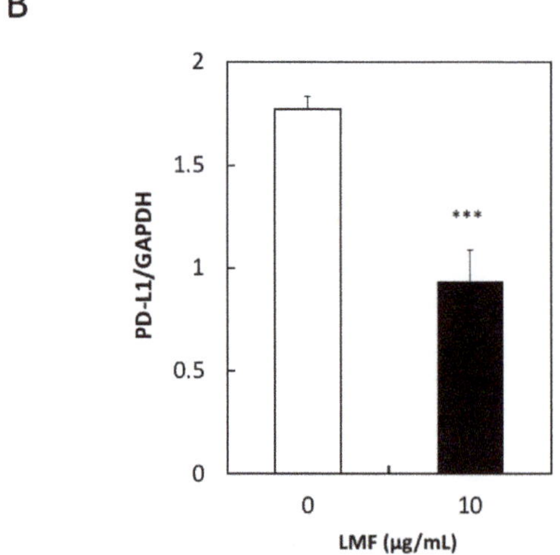

Figure 6. Low concentration of low molecular weight fucoidan extract (LMF) suppresses programmed cell death-ligand 1 (PD-L1) protein expression. (**A**) Western blot detection of PD-L1 protein in total protein lysates from HT1080 cells treated with 10 µg/mL of LMF for 5 days by replacing every 2 days with fresh media containing 10 µg/mL of LMF. Control untreated HT1080 cells treated similarly but without LMF. (**B**) Chemiluminescence protein bands from LMF-treated and untreated samples quantitated and bands from LMF-treated cells compared with that of untreated cells using Student's *t*-test. (*** $p < 0.001$, $n = 3$).

2.5. Suppression of Cell Surface PD-L1 Protein Expression by LMF

Western blot analysis showed reduced PD-L1 protein levels in the HT1080 cell protein lysates proportionally with a reduction in PD-L1 mRNA levels. However, this method detects total PD-L1 proteins including intracellular premature PD-L1. The mature form of PD-L1 protein is anchored to the cell membrane with two extracellular domains [6]. It is necessary to show the level of surface PD-L1 exquisitely, as it is performing the most important and integral role in the immune escape mechanism of cancer cells. For this, we assessed if the effects of LMF are due to changes in the cell surface PD-L1 protein levels using the anti-PD-L1 antibody and analyzed this using flow cytometry, according to the published method. This method has previously been used to measure the PD-L1 expressed on the surface of several cell lines such as A549, H1975, and others using PD-L1 antibody followed by the flow cytometry analysis using a FACScan instrument [12]. Fluorescence histograms showed a small shift towards the left in LMF-treated HT1080 cells (Figure 7A, red line) compared with untreated control cells (Figure 7A, blue line), suggesting reduced levels of membrane-bound PD-L1 protein. To confirm this effect, mean fluorescence intensity (MFI) per cell, derived from histogram data, is presented in Figure 7B. These results show significant reduction in cell surface PD-L1 protein in LMF-treated HT1080 cells compared with the control cells (** $p < 0.01$). Thus, the results show that the PD-L1 mRNA level corresponds with that of the total as well as cell surface PD-L1 protein levels, suggesting an effect on upstream signal transducers regulating PD-L1 transcription.

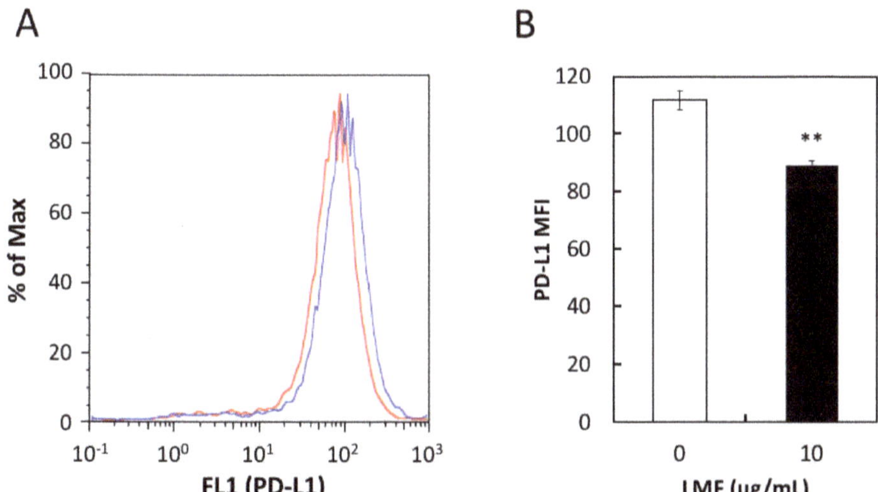

Figure 7. Effect of low amount low molecular weight fucoidan extract (LMF) treatment for 5 days on programmed cell death-ligand 1 (PD-L1) protein expression in HT1080 cells. HT1080 were treated with a low concentration of LMF (10 µg/mL) for 5 days and changes in PD-L1 protein levels were measured by flow cytometer. (**A**) PD-L1 protein expressed on the cell surface of HT1080 cells with (red line) and without (blue line) LMF treatment detected using anti-PD-L1 antibody and histograms were generated. (**B**) Mean fluorescence intensity (MFI) per cell was calculated using histogram data. Student's *t*-test (** $p < 0.01$, $n = 3$) compares MFI of LMF-treated cells with that of untreated cells.

2.6. LMF Does not Interfere with the Binding of PD-1 with PD-L1

The binding of PD-1 and PD-L1 occurs between T-cells and APCs for a normal T-cell response [3]. PD-1 binding with the PD-L1 expressed on cancer cells is an important step to evade the host immune surveillance system. Considering this, if LMF binds to PD-1 or PD-L1, attenuation and eventual inhibition of the normal T-cell response will occur. Therefore, we examined whether LMF inhibits binding of PD-1 with PD-L1. For this, we used PD-1 neutralizing antibody as a positive control which, as expected, showed dose-dependent inhibition of PD-1 and PD-L1 binding (Figure 8A). When LMF replaced the positive control in the same experimental setup, it did not interfere with the PD-1 and PD-L1 binding (Figure 8B). Therefore, the results show that LMF does not interact with PD-1 or PD-L1 and suggest that its suppressive effect is due to regulation of the other cell surface proteins as well as intracellular transducers involved with expression of PD-L1 and other genes.

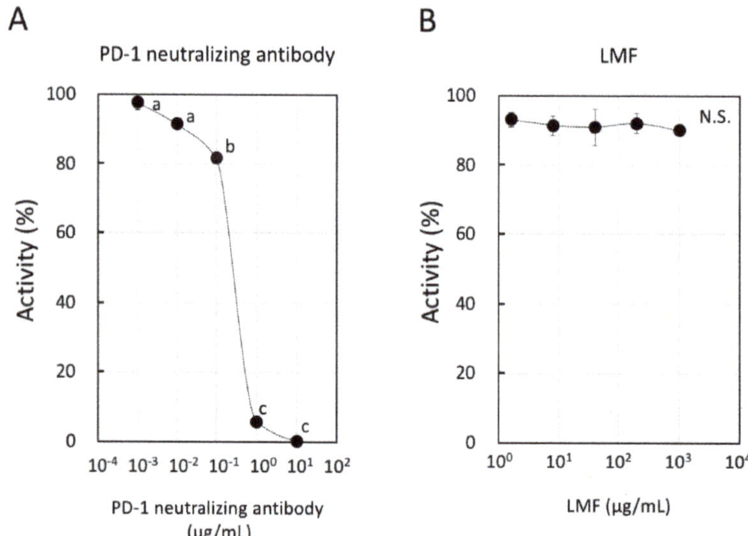

Figure 8. Inhibitory activity of programmed cell death-1 (PD-1): PD-L1 binding by low molecular weight fucoidan extract (LMF). (**A**) We used PD-1 neutralizing antibody as a positive control in dose-dependent inhibition of PD-1 and PD-L1 binding. Statistical significance was by ANOVA with Tukey test. The letters, a, b, and c indicate statistically significant differences between each letter ($p < 0.05$). (**B**) LMF was used to observe a dose-dependent inhibition of PD-1 and PD-L1 binding (N.S.: Not Significant).

2.7. LMF-Suppression of Epidermal Growth Factor Receptor (EGFR) and Vascular Endothelial Growth Factor (VEGF) Expression

Growth factors play a significant role in carcinogenesis and metastasis [2,55]. The effect of LMF on EGFR, VEGF, and cell surface-associated protein expression was tested in HT1080 cells treated with different concentrations of LMF (0, 10, 100 µg/mL) for 24 h and mRNA levels were measured. LMF suppressed both EGFR mRNA (10 µg/mL, * $p < 0.05$; 100 µg/mL, ** $p < 0.01$), and VEGF mRNA (10 µg/mL; ** $p < 0.01$, 100 µg/mL; * $p < 0.05$) in HT1080 cells (Figure 9).

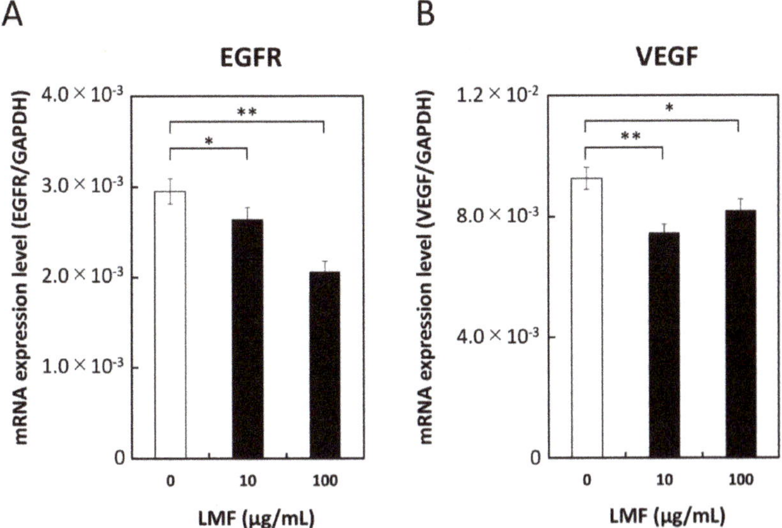

Figure 9. Suppression of epidermal growth factor receptor (EGFR) and vascular endothelial growth factor (VEGF) transcription by low molecular weight fucoidan extract (LMF). HT1080 cells were treated with 0, 10, and 100 μg/mL of LMF for 24 h. Total RNA was isolated and subjected to qRT-PCR analyses. LMF-treated cells were compared with untreated cells, and statistical evaluation was carried out using the Student's *t*-test. (**A**) EGFR, (**B**) VEGF. (* $p < 0.05$, ** $p < 0.01$, $n = 3$).

2.8. Suppression of RhoA and Stimulation of RhoB Expression by LMF

The Ras superfamily contains Rho GTPases which are involved in many cellular processes that regulate various signal transduction cascades that affect gene expression, cell cycle progression, cell migration, and many other cellular events by cycling from GTP-bound active state to GDP-bound inactive state [56]. In addition to their fundamental functions, Rho GTPases such as RhoA and RhoB play a significant role in cancer progression and inflammation [57]. To assess the effect of LMF on RhoA and RhoB expression, HT1080 cells were treated with different concentrations of LMF (0, 10, 100 μg/mL) for 24 h and mRNA levels were measured. LMF suppression of RhoA gene transcription at 10 μg/mL (** $p < 0.001$) and 100 μg/mL (* $p < 0.05$) (Figure 10A) was significant. In contrast, fucoidan caused a significant (all, *** $p < 0.001$) and dose-dependent (10 and 100 μg/mL) upregulation of RhoB gene transcription in HT1080 cells (Figure 10B). All the results obtained are summarized in Table 1.

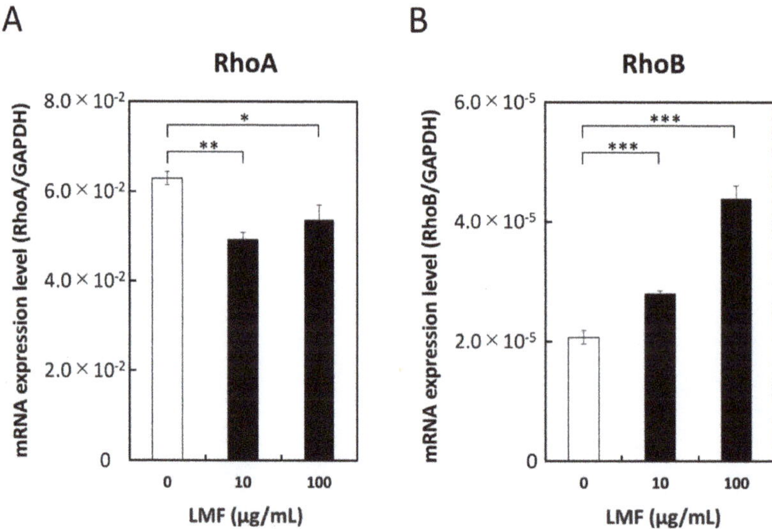

Figure 10. Low molecular weight fucoidan extract (LMF) suppresses RhoA and stimulates RhoB transcription. HT1080 cells were treated with 0, 10, and 100 μg/mL of LMF for 24 h. Total RNAs were isolated and subjected to qRT-PCR analyses. LMF-treated cells were compared with untreated cells and statistically evaluated using the Student's *t*-test. (**A**) RhoA, (**B**) RhoB. (* $p < 0.05$; ** $p < 0.01$; *** $p < 0.001$, $n = 3$).

Table 1. Summary of the results.

Cell Lines	Cell Growth Assay			PD-L1 Levels				mRNA			
				mRNA	Protein						
	Cell Count	WST-1	Hoechst 33342	+LMF *1	Total Surface			EGFR	VEGF	RhoA	RhoB
	+LMF				−LMF	+LMF	+LMF	+LMF			
HT1080	↓	↓	↓	↓	1.06	↓	↓	↓	↓	↓	↑
TIG-1	→	−	−	↑	0.99	−	−	−	−	−	−
A549	−	−	−	−	0.94	−	−	−	−	−	−
PC-9	−	−	−	−	1.00 *2	−	−	−	−	−	−
MCF-7	−	−	−	−	0.96	−	−	−	−	−	−
NIH:OVCAR-3	−	−	−	−	0.94	−	−	−	−	−	−
PANC-1	−	−	−	−	0.96	−	−	−	−	−	−
Figures	3A, B	3C	3C	5A, B	4A	6B	7B	9A	9B	10A	10B

* 1: PD-L2 mRNA was suppressed and increased by LMF in HT-1080 cells and TIG-1 cells, respectively. * 2: PC-9 cell mRNA level without LMF was set to 1.00 and the relative expression levels for other mRNAs were calculated. −: not tested.

3. Discussion

In the present study, we examined the effects of LMF on six gene products including PD-L1, PD-L2, EGFR, VEGF, RhoA, and RhoB, involved in immune evasion, angiogenesis, and malignant transformation, using the human fibrosarcoma HT1080 cell line. We reveal that LMF downregulated five gene products, while upregulating RhoB expression.

One of the important determinants to improve the efficacy of cancer therapy is inhibiting cancer cell evasion from the host tumor immune surveillance system [8]. Various tumors express increased levels of PD-L1 to evade such host defense mechanisms and eventually facilitate tumor growth [2,4,8,9]. Moreover, the tumor growth in terms of Breslow tumor thickness in melanoma specimens correlated with the level of PD-L1 expression, and the survival rate of patients expressing a high-level of PD-L1 was statistically lower than that of low-expressing patients with stage II melanoma [5]. Therefore, we confirmed the effects of LMF on the downregulation of PD-L1 expression in cancer cells. We first examined PD-L1 mRNA levels in six cancer cell lines, HT1080, MCF-7, PC-9, NIH:OVCAR-3, PANC-1, and A549 cells, and in one normal cell line TIG-1. All cell lines express PD-L1 levels similar to or greater than that of the A549 cell line, a low-expressing control [12,58]. The PD-L1 mRNA levels expressed by TIG-1 cells were comparable to those of the high-expressing control cell line PC-9 [12,44,54]. This observation agrees with earlier reports on several human cells, including lung cells and various tissues, which express low- to high-levels of PD-L1 [3,59,60]. These data strongly suggest the importance of selective inhibition of PD-L1 expressed on cancer cells but not on normal cells.

Among the cell lines tested, the HT1080 cell line expressed the highest levels of PD-L1 mRNA and was a suitable model to test the suppressive effects of LMF with emphasis on PD-L1 expression at the mRNA and protein levels. Several inherent characteristics of the HT1080 cell line are known since its establishment in 1974 [61]. HT1080 cells contain both wild-type and mutant N-Ras alleles and are expressed at the protein level. The mutant N-Ras protein induces cell transformation in cells even though it is present in far less amounts compared to that of normal p21 N-Ras in parental HT1080 cells [62]. The mutant N-Ras protein maintains a chronically activated state due to constitutive GTP binding, thereby conferring the constitutive activation of Ras-dependent signaling pathways. Representative Ras-regulated pathways are the Raf, Rac1, RhoA, and PI3K signal cascades, which are known to regulate mitogenesis, motility and invasiveness, actin cytoskeletal architecture, and cell survival, respectively [19,63]. Among the intermediary transducers, activated Raf/MEK is necessary for the aggressive tumorigenic phenotype in HT1080 cells but not the PI3K/Akt pathway [18,19]. However, others reported that the activated mutant N-Ras protein constitutively stimulates the Raf/MEK and the PI3K/Akt pathways to induce PD-L1 gene expression in several cancer cell lines [6,17].

Although, these data suggest that PD-L1 expression could be driven by the Ras regulated pathways in HT1080 cells, direct data proving an association between the Ras regulated pathways and PD-L1 expression in HT1080 cells are not available to date. In the present study, we show significant suppression of mRNA, protein and surface protein levels of PD-L1 in HT1080 cells by LMF. These results show suppression of activated mutant N-Ras regulated pathways for PD-L1 expression and existing data show many transducers in the Ras regulated pathways being suppressed by fucoidan [20–22,39]. LMF down-regulation of PD-L1 expression via Ras pathways in HT1080 cells could be explained by the possible regulatory pathways proposed for PD-L1 gene expression in cancer cells. According to this model, growth factor (GF)-stimulated pathways such as Ras/Raf and Ras/PI3K drive PD-L1 expression [6]. Thus, these pathways for PD-L1 expression coincide mostly with the Ras-regulated pathways proposed in HT1080 cells, as above. Although, the mutant N-Ras protein is constitutively active in regulating the pathways involving Raf, PI3K, and Rac transducers, the normal N-Ras-regulated pathways activated by GFs should be evaluated separately because HT1080 cells express functional wild-type and mutant N-Ras proteins [62].

Normal N-Ras protein is activated by exogenous stimuli such as by the binding of EGF, VEGF, and platelet-derived growth factor (PDGF) with their receptors (EGFR, VEGFR, and PDGFR). These transmembrane receptors mediate GTP-binding with the normal N-Ras protein for activation, enabling regulation of downstream transducers [63]. Cellular activities including proliferation, growth, survival, angiogenesis, metastases, and motility are regulated by the activated EGFR, VEGFR, and PDGFR through pathways such as the mitogen-activated protein kinase (MAPK) pathway: Ras/Raf, the PI3K pathway: PI3K/Akt, and the JAK/STAT pathway: JAK/STAT [64]. Available data shows that EGF-stimulated EGFR upregulates PD-L1 expression through EGFR/PI3K/Akt, EGFR/Ras/Raf, and EGR/PLC-γ signaling pathways [2,12,65,66]. Moreover, tumor cells can evade the immune system or promote their own growth through increased EGFR expression and indoleamine 2,3-dioxygenase 1 or VEGF production [67]. Although ligand-bound-activated EGFR with increased protein expression was present in HT1080 cells [14,68], the supernatant collected from the HT1080 cell culture did not contain secreted EGF but contained moderate concentrations of TGF-α which correlated with pEGFR expression [14]. Thus, an autocrine loop of TGF-α activates wild-type EGFR as well as the downstream factors in HT1080 cells, as both EGF and TGF-α belong to eight members of EGF family ligands, and TGF-α is an EGFR ligand in cancer [69,70]. Similarly, it has been reported that anti-EGF receptor mAbs prevented EGF and TGF-α induced growth stimulation in many human cells [14,71]. Accumulated data suggest that inhibition of EGFR or VEGF could inhibit HT1080 cell growth. In this study, we report significant suppression by LMF of both EGFR transcription and cell growth, suggesting a mechanism to suppress PD-L1 expression through regulation of EGFR expression. A plausible explanation for EGFR suppression by LMF would be that fucoidan binds with EGF causing prevention of EGF-induced phosphorylation of EGFR [72,73]. In support of this, earlier reports showed that fucoidan from *Laminaria guryanovae* exerts a potent inhibitory effect on EGF-induced phosphorylation of EGFR in mouse epidermal JB6 Cl41 cells [72] and fucoidan KW derived from brown algae *Kjellmaniella crassifolia* caused significant reduction in the levels of phosphorylated EGFR and PKCα proteins in A549 cells [43]. Others have shown that fucoidan suppresses growth factor-stimulated pathways involving the Ras/Raf pathway and inhibits the phosphorylation of intermediary transducers such as ERK1/2, PI3K, Akt, mTOR, c-Jun, c-Fos, EGFR, and VEGF [21,22,39]. While future studies are needed to confirm whether LMF binds with TGF-α in reducing EGFR phosphorylation, it is probable to suggest that reduction in the EGFR mRNA levels by LMF decreases the translated and phosphorylated EGFR levels. Contrary to others and ours, an existing report shows that EGFR could not be detected in HT1080 cells and suggested that HT1080 cell oncogenic signaling through Ras/PI3K/Akt is independent of EGFR [74]. In addition, fucoidan exerted a potent inhibitory effect on EGF-induced phosphorylation of EGFR and phosphorylation of ERK and JNK under the control of EGF. Furthermore, EGF-induced c-*fos* and c-*jun* transcriptional activities were inhibited by fucoidan leading to inhibition of AP-1 activity and cell transformation induced by EGF [72].

VEGF is present in a number of cancer cell lines including HT1080 cells and is suggested to play a crucial role in HT1080 cell metastasis based on results obtained using the anti-human VEGF antibody, which reduced spontaneous lung micrometastasis [75]. Moreover, inhibition of tumor-secreted VEGF limits primary tumor growth of sarcoma cell lines by inhibiting host angiogenesis [76]. An inhibitory mechanism of angiogenesis by low molecular weight fucoidan (LMWF), which is distinct from LMF, inhibits hypoxia-induced reactive oxygen species (ROS) formation, HIF-1α expression, VEGF secretion, and downstream VEGFR2/PI3K/Akt/mTOR/p70S6K1/4EBP-1 cascade in human bladder T24 cancer cells, ultimately suppressing HIF-1α/VEGF transcription and angiogenesis [39]. VEGF upregulates both PD-L1 mRNA and protein expression [2,77]. Thus, suppression of both VEGF-enhanced angiogenesis and PD-L1 expression contributes to inhibiting cancer cell evasion. Supporting evidence shows that fucoidan extract inhibits expression and secretion of VEGF which suppresses vascular tubule formation in HeLa cells and suppressing invasion of HT1080 cells [41]. In the present study, because LMF caused significant suppression of both VEGF mRNA and PD-L1 mRNA/protein levels in HT1080 cells, we can extrapolate that cancer cells become more susceptible to the host defense system. The rationale behind the suppression of the PD-L1 level by fucoidan comes from the fact that fucoidan directly binds with VEGF, thereby preventing VEGFR phosphorylation [70,78]. Other growth factors, PDGF and PDGFR, are constitutively expressed in HT1080 cells. PDGF-A is constitutively secreted into the medium, where it binds to and activates PDGFR and downstream PI3K and Akt pathways [19]. Although we did not test the suppressive effect of LMF on PDGF expression, there are studies showing that fucoidan suppresses PDGF expression. Oligo-fucoidan suppresses PDGF-induced cell proliferation and induces G_1/G_0 cell cycle arrest in airway smooth muscle cells [79]. Therefore, LMF is likely to aid in HT1080 cell growth inhibition by suppressing PDGF expression. As mentioned thus far, LMF suppresses VEGF and EGFR expression in HT1080 cells to suppress Ras-regulated pathways for PD-L1/PD-L2 expression.

The Ras superfamily contains a subgroup of Rho GTPases composed of 20 kinds of intracellular signaling molecules including RhoA and RhoB proteins. Constitutively expressed RhoA functions in tumor development. Indeed, RhoA and its downstream pathway is constitutively active in HT1080 cells. The present study shows significant suppression of RhoA mRNA levels by LMF, suggesting that LMF downregulates Ras-controlled oncogenic activities via the RhoA pathway [19]. In contrast, the role of RhoB is not clearly known, it acts as either an oncogene under some conditions or a tumor suppressor gene under others [80,81]. RhoB expression is induced by EGF, PDGF, and many other agents, while several factors including N-Ras and EGFR suppress RhoB promoter activity in NIH3T3 cells and human cancer cell lines derived from lung, pancreatic, and cervical tumors [82]. Fucoidan suppresses PDGF (RhoB inducer) expression indicating downregulation of RhoB expression by fucoidan [79]. Conversely, LMF suppresses EGFR (RhoB suppressor), and fucoidan suppresses phosphorylation of various Ras-regulated transducers. Together these data suggest that LMF/fucoidan could act as an upregulator of RhoB because LMF significantly and dose-dependently upregulated RhoB mRNA levels, supporting the role of RhoB as a tumor suppressor. Thus, LMF performs a significant role in antitumor effect by upregulating RhoB expression in several cancer types [80,81]. Suppressive effects by fucoidan on Ras-regulated pathways are presented in Figure 11.

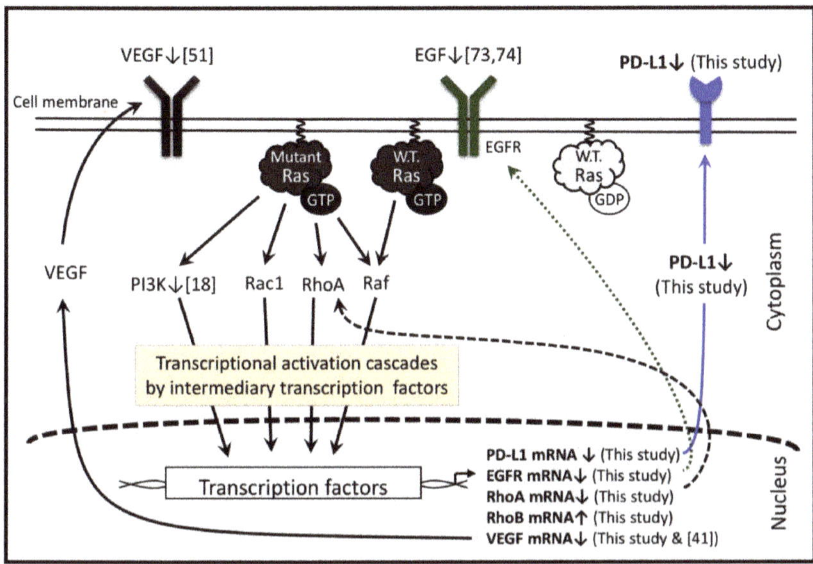

Figure 11. Low molecular weight fucoidan extract (LMF) and several fucoidan species regulate multiple transcription factors in the Ras-controlled pathways in various cells, including HT1080 cells. The numbers in the square bracket indicate the reference numbers, and downward and upward arrows indicate the factors up- and down-regulated by various fucoidan species reported in the references and LMF used in this study. Adapted from [2,6,19].

An important finding that LMF neither inhibited PD-L1 expression nor growth of the TIG-1 cells suggests selective activity towards cancer cells expressing PD-L1. Similar to our results with normal cells, another report shows a lack of effect of fucoidan on normal human mammary epithelial cell growth [83]. These results collectively suggest that LMF/fucoidan treatment against PD-L1 expression is effective because of its selective action on cancer cells. However, the anti-PD-L1 antibody treatment may not be efficacious because not all cancer patients express higher PD-L1 [67]. Similarly, the suppressive effect on PD-L1 expression conferred by LMF/fucoidan may not be efficacious for patients who are not expressing higher PD-L1 [67]. Thus, as both methods rely on the expressed PD-L1 or its expression in cancer cells, these methods have a common limitation in using PD-L1 as a treatment target. On the other hand, in addition to PD-L1 suppression by LMF, extensive data exists demonstrating that fucoidan exhibits various biological activities including the suppressive modulation of the intracellular transducers that regulate proliferation, growth, survival, angiogenesis, metastases, and motility in cancer cells [22,36,39,42,43,84]. Therefore, LMF/fucoidan exhibits anticancer effects due to the broad biological activities independent of its anti-PD-L1 effect.

4. Materials and Methods

4.1. Cell Lines and Cell Culture

Seven established human cell lines and their culture conditions are shown in Table 2.

Table 2. Human cell lines and culture conditions.

Cell Lines	Registry No.	Types of Cells	Sources	Culture Conditions
HT1080	ATCC® CCL-121™	Fibrosarcoma	ATCC[#1]	MEM supplemented with 10 mM HEPES, NEAA, 100,000 Units/L of Penicillin G Potassium, 0.1 g/L of Streptomycin sulfate and 10% fetal bovine serum (FBS) under a 5% CO_2-humidified atmosphere at 37 °C.
A549	ATCC® CCL-185™	Lung adenocarcinoma	ATCC[#1]	
MCF-7	ATCC® HTB-22™	Breast adenocarcinoma	ATCC[#1]	
TIG-1	TKG0276	Normal fetal lung diploid fibroblast	IDAC[#2]	
PC-9	RCB4455	lung adenocarcinoma	RIKEN[#3]	DMEM supplemented with 10 mM HEPES, 100,000 Units/L of Penicillin G Potassium, 0.1 g/L of Streptomycin sulfate and 10% FBS under a 5% CO_2-humidified atmosphere at 37 °C
NIH:OVCAR-3	RCB2135	Ovarian carcinoma	RIKEN[#3]	RPMI 1640 medium supplemented with 10 mM HEPES, 100,000 Units/L of Penicillin G Potassium, 0.1 g/L of Streptomycin sulfate and 10% FBS under a 5% CO_2-humidified atmosphere at 37 °C.
PANC-1	RCB2095	Pancreatic carcinoma	RIKEN[#3]	

MEM (Eagle's Minimum Essential Medium, Nissui Pharmaceutical Co., Ltd., Tokyo, Japan). HEPES (Dojindo Laboratories, Kumamoto, Japan), NEAA (Non-Essential Amino Acids, FUJIFILM Wako Pure Chemical Corporation, Osaka, Japan). Penicillin G Potassium (Meiji Seika Pharma Co., Ltd., Tokyo, Japan). Streptomycin sulfate (Meiji Seika Pharma). FBS: (fetal bovine serum, HyClone, Thermo Fisher Scientific Inc., Waltham, MA, USA). DMEM (Dulbecco's Modified Eagle's Medium, Nissui Pharmaceutical, Tokyo, Japan). RPMI (Nissui Pharmaceutical, Tokyo, Japan). [#1] ATCC: American Type Culture Collection, Manassas, VA, USA; [#2] IDAC: The Cell Resource Center for Biomedical Research, Institute of Development, Aging and Cancer, Tohoku University, Sendai, Japan; [#3] RIKEN: BioResource Research Center, Ibaraki, Japan.

4.2. Fucoidan

The abalone glycosidase-digested LMF, commercially available as "Power fucoidan," was generously donated for this study by the Daiichi Sangyo Corporation (Osaka, Japan). LMF was prepared as previously described [2]. Briefly, high molecular weight fucoidan extract from seaweed of *Cladosiphon novae-caledoniae* Kylin was purified to 85% purity and digested with glycosidases to obtain LMF. LMF consists of a digested small molecular weight fraction (72%, MW < 500 Da) and non-digested fractions (less than 28%, peak MW: 800 kDa). LMF consisted mostly of fucose (73%), xylose (12%), and mannose (7%). The ratio of sulfation was 14.5%.

4.3. WST-1 Assay (Cell Viability Assay)

WST-1 assay is commonly used to measure the number of live cells after treating the cells with culture containing test agents [53]. This assay was used to measure the cytocidal effects of LMF on HT1080 cells. 100 μL each of HT1080 (2×10^4 cells/mL) and TIG-1 (4×10^4 cells/mL) cells were plated in 96-well microplates and cultured for 24 h at 37 °C. The WST-1 measurement was performed according to the standard protocol described by the manufacturer. Briefly, the cells were exposed to various concentrations of LMF for 48 h. Spent medium was removed and cells were washed once with PBS. To each well, 100 μL of culture medium containing 0.5 mM WST-1 (Dojindo) and 0.02 mM 1-methoxy phenazine methosulfate (Dojindo) was added and incubated for 1 h at 37 °C. Cell metabolic activity was measured at 450 nm using a microplate reader (Tecan, Männedorf, Switzerland).

4.4. Hoechst Staining (Cellular DNA Staining for Cell Number Determination)

Hoechst 33342 dye stains both live and apoptotic cell DNA [50,51]. It is possible to estimate the cytocidal effect of LMF by analyzing WST-1 and Hoechst Staining assays. The cells in the same microplates used for WST-1 assay were fixed with 4% formaldehyde solution (Wako) for 30 min at 25 °C. Cells were washed once with PBS and stained with 100 µL PBS per well, containing 2 µg/mL of Hoechst 33342 (Dojindo), for 30 min in the dark at 25 °C, according to the instruction manual. Fluorescence intensity (excitation, 360 nm and emission, 465 nm) was measured by the multimode plate reader, infinite F200 PRO (Tecan).

4.5. Immuno-Fluorescence Staining

Previous studies have used the Immuno-fluorescence staining method to see the effect of fucoidan on T24 cells [39]. The same method was also used to detect PD-L1 in the cultured cells [58]. Therefore, we used this method to detect PD-L1 expressed in seven cell lines. HT1080 and A549 cells at 2×10^4 cells/mL, MCF-7, PC-9, NIH:OVCAR-3, and PANC-1 cells at 3×10^4 cells/mL, and TIG-1 cells at 4×10^4 cells/mL were prepared, and 100 µL of each cell line was seeded per well in a 96-well black plate (Greiner Bio-One International GmbH, Kremsmünster, Oberösterreich, Austria) and cultured for 24 h. Cells were fixed with 4% paraformaldehyde (Wako) for 15 min. After washing with PBS, cells were treated with blocking solution containing 0.1% Tween-20 with 1% BSA in PBS for 1 h. After washing three times with PBS, cells were treated overnight at 4 °C with the anti-PD-L1 Rabbit mAb (E1L3N, Cell Signaling Technology Co., Danvers, MA, USA) diluted 800-fold with blocking solution. Rabbit mAb IgG Isotype Control (DA1E, Cell Signaling Technology) was diluted 800-fold with blocking solution and used as an isotype antibody control. After washing three times with PBS, cells were treated for 30 min with the secondary antibody, anti-Rabbit IgG (H+L), F(ab')2 Fragment (Alexa Fluor® 555 Conjugate) (Cell Signaling Technology), and diluted 1,000-fold with blocking solution. Cells were washed one time with PBS and stained with 2 µg/mL of Hoechst 33342 for 30 min. Fluorescence intensity of each cell line was measured using IN Cell Analyzer 1000 (GE Healthcare UK Ltd., Buckinghamshire, UK).

4.6. Quantitative Reverse Transcription Polymerase Chain Reaction (qRT-PCR)

Polymerase chain reaction (PCR) has been used broadly, to detect a specific gene transcript [85], and to quantitate the target transcript level [86,87]. Quantitative detection and analyses of the expressed gene transcripts have been advanced greatly by incorporating automated instrumentations. Seven gene transcripts were examined by fully utilizing the commercially available instruments and reagents, as described below. Five milliliters of HT1080 cells were seeded at a density of 5×10^4 cells/mL per 60 mm dish (FALCON, Corning Inc., Corning, NY, USA) and cultured for 24 h. Spent medium was replaced with the medium containing various LMF concentrations (0, 1, 10, or 100 µg/mL). After 48 h of treatment, total RNA was isolated using a High Pure RNA Isolation Kit (F. Hoffmann-La Roche Ltd., Basel, Switzerland) following the manufacturer's protocol. cDNA was synthesized using total RNA and ReverTra Ace qPCR RT Kit Master Mix with gDNA Remover (TOYOBO Co., Ltd., Osaka, Japan) and was used for RT-PCR templates. For specific gene transcript amplification, cDNA template, primer pairs (Table 3), and THUNDERBIRDTMSYBR® qPCR Mix (TOYOBO) mixture were placed in a 96-well PCR plate (NIPPON Genetics Co., Ltd., Tokyo, Japan) in triplicate. Primer sequences for amplifying the transcripts of seven genes (EGFR, VEGF, RhoA, RhoB, PD-L1, PD-L2, and GAPDH), shown in Table 1, were purchased from Sigma-Aldrich Co. (St. Louis, MO, USA). Quantitative reverse transcription polymerase chain reaction (qRT-PCR) for seven genes was carried out using Thermal Cycler DiceTM Real Time System TP800 (Takara Bio Inc., Shiga, Japan). The thermal program was as follows: denaturation at 95 °C for 30 s, 45 cycles of denaturation at 95 °C for 5 s, annealing at 60 °C for 10 s, and extension at 72 °C for 20 s. Fluorescence values for each gene were normalized using GAPDH as an internal standard and used for statistical evaluation.

Table 3. Primer sequences used for qRT-PCR analyses.

Target Genes	F: Forward R: Reverse	Primer Sequences	References
EGFR	F R	5′-CGCAAAGTGTGTAACGGAATAG-3′ 5′-CCAGAGGAGGAGTATGTGTGAA-3′	[88]
VEGF	F R	5′-AAGGAGGAGGGCAGAATCAT-3′ 5′-ATCTGCATGGTGATGTTGGA-3′	[89]
RhoA	F R	5′-CCGGCGCGAAGAGGCTGGACT-3′ 5′-GCACATACACCTCTGGGAACT-3′	[90]
RhoB	F R	5′-GGTCCCCTGAGCATGCTTTCTGA-3′ 5′-GCCACACTCCCGCGCCAATCTC-3′	[90]
PD-L1	F R	5′-TATGGTGGTGCCGACTACAA-3′ 5′-TGCTTGTCCAGATGACTTCG-3′	[91]
PD-L2	F R	5′-TGACTTCAAATATGCCTTGTTAGTG-3′ 5′-GAAGAGTTCTTAGTGTGGTTATATG-3′	[91]
GAPDH	F R	5′-ATTGCCCTCAACGACCACTT-3′ 5′-AGGTCCACCACCCTGTTGCT-3′	[88]

4.7. Flow Cytometry (FCM) Analysis

Five milliliters of 3.1×10^3 cells/mL of HT1080 cells were plated on 60 mm dish and incubated for 24 h. The cells were then treated with 5 mL of medium with or without 10 µg/mL of LMF for 5 days. Cells were dislodged from the dish by treating with 0.05% Trypsin-0.02% EDTA solution (Thermo Fisher Scientific) for 3 min at 37 °C. Cells were fixed with 4% paraformaldehyde for 10 min at 37 °C, collected, and treated for 1 h with the drawback Rabbit mAb (E1L3N) or with the corresponding Rabbit mAb IgG isotype control (DA1E) diluted 800-fold with incubation buffer. After washing three times with PBS, cells were treated for 30 min with the secondary antibody, anti-Rabbit IgG (H+L), F(ab′)2 Fragment (Alexa Fluor® 488 Conjugate) (Cell Signaling Technology) diluted 500-fold with incubation buffer. Fluorescence intensity of each cell line was measured using Flow Cytometer EPICS XL System II-JK (Beckman Coulter, Inc., Brea, CA, USA) which integrates flow cytometry technology, thus enabling the study of cellular populations [92]. Data analyses were carried out using Flow Cytometry (FCM) analysis software, FlowJo (Version 4.6.2) (Tree Star, Inc., Ashland, OR, USA) and the results expressed as histograms and MFI.

4.8. Western Blot Analysis

In our previous study, the Western Blot method was used to analyze the expression of Bcl-2 family proteins [47]. We used this method to detect PD-L1 in HT1080 cell lysates treated with LMF. HT1080 cells were cultured for 5 days in the medium supplemented with or without 10 µg/mL of LMF. Cells were washed three times with cold PBS and lysed for 5 min with 400 µL of protein extraction buffer (M-PER Mammalian Protein Extraction Reagent (Thermo Fisher Scientific) supplemented with 1% of Halt Protease Inhibitor Single-Use Cocktail (Thermo Fisher Scientific) and 5 mM EDTA). Lysates were collected in 1.5 mL microtubes and cell lysis was further ensured by repeated sonication for four times at 5 s intervals using the Handy Sonic model UR-20P (TOMY SEIKO CO., LTD., Tokyo, Japan). Protein extracts were centrifuged at $14,000 \times g$ for 15 min at 4 °C. Supernatants were recovered and stored at −80 °C. Prior to use, the protein concentration was determined using the Bradford protein assay. Supernatants were diluted 50-, 100-, 200-, and 400-fold with PBS. A standard solution was prepared using bovine γ globulin (1 mg/mL) by serially diluting with PBS from 90 to 0 µg/mL with 10 µg/mL decrement. To determine the protein concentrations in the dilution series, 100 µL each of the protein extracts and standard solutions were mixed with 100 μL of 2.5-fold diluted Bio-Rad protein assay dye reagent concentrate (Bio-Rad Laboratories, Inc., Hercules, CA, USA) in a 96-well microplate (Nunc, Thermo Fisher Scientific), and absorbance at 595 nm were measured using a microplate reader (TECAN). Protein concentrations in the treated HT1080 cell extracts were determined based on the standard curve derived from bovine γ globulin. Equal amounts of protein from each sample were separated by electrophoresis through SDS-PAGE and transferred to Amersham Hybond P polyvinylidene fluoride 0.45 membranes (GE Healthcare) using The Trans-Blot® SD semi-dry transfer cell apparatus (Bio-Rad) following the manufacturer's protocol. Membranes were blocked for 1 h at room temperature with gentle shaking in Tris-buffered saline containing 5% skim milk powder (Wako). Membrane was cut into two pieces so as to apply different antibodies for detecting closer molecular weight proteins of GAPDH (37 kD) and PD-L1 (40-50 kD) unambiguously. The membrane was incubated overnight at 4 °C with the primary antibody, anti-PD-L1 Rabbit mAb (E1L3N) (Cell Signaling Technology) or anti-GAPDH Rabbit mAb (D16H11) (Cell Signaling Technology), diluted 1000-fold or 4000-fold respectively, in Solution 1 of Can Get Signal® Immunoreaction Enhancer Solution (Toyobo). The membranes were washed twice and incubated for 1 h at room temperature with the Anti-Rabbit IgG, horseradish peroxidase (HRP)-linked secondary antibody (Cell Signaling Technology), diluted 2000-fold with Solution 2 of Can Get Signal® Immunoreaction Enhancer Solution. Chemiluminescence signals of protein bands obtained by ECL prime western blotting detection reagent (GE healthcare), as described by the supplier, were detected with LAS-1000 (FUJIFILM Corporation, Tokyo, Japan), which were quantified and numerated using image J (1.4.6r) software.

4.9. Inhibitory Activity of LMF on PD-1:PD-L1 Binding

Inhibitory activity of LMF on PD-1 and PD-L1 binding was examined using PD-1:PD-L1(Biotinylated) Inhibitor Screening Colorimetric Assay Kit (BPS Bioscience, San Diego, CA, USA) following the manufacturer's protocol. Briefly, PD-1 protein was coated on a 96-well plate and added biotinylated PD-L1 to form complex of PD-1:PD-L1:biotin. The plate was treated with Streptavidin-HRP followed by the addition of a colorimetric HRP substrate to produce color. Thus, this set up will serve as a positive control. To this set up, the addition of PD-1 neutralizing antibody as a test inhibitor will reduce color production serving as a second positive control. When a prospective inhibitor such as LMF is added to this set up, color production is reduced only when LMF binds to PD-1. For ligand control, the well is not coated with PD-1 and for the blank control well, biotinylated PD-L1 is not included. After incubation, these set-ups produce appropriate colors, which can be measured at 450 nm using a microplate reader (TECAN).

4.10. Statistical Analysis

Each experiment was performed at least in triplicate and repeated three times. The results are presented as the mean ± standard deviation (SD) values. The difference between the two groups was analyzed using the two-tailed Student's *t*-test, and differences among three or more groups were analyzed using one-way analysis of variance (ANOVA) with Tukey's multiple comparisons. Statistical analysis was done using Mini StatMate (ATMS Co., Ltd., Tokyo, Japan). A value of $p < 0.05$ represents a significant difference (* $p < 0.05$; ** $p < 0.01$; *** $p < 0.001$).

5. Conclusions

Our findings revealed that LMF exerts its effect by specifically suppressing not only mRNA levels but also cytoplasmic- and surface-protein levels of PD-L1 in HT1080 fibrosarcoma cells. We also provide data for the downregulation of VEGF, EGFR, and RhoA mRNAs, and the upregulation of RhoB mRNA in HT1080 cells. Together, these data suggest that LMF could be used as a complementary agent in the treatment of various types of cancers that express high PD-L1 levels.

Author Contributions: K.T. conceived and designed the experiments; Y.K. and K.T. performed the experiments and analyzed the data; H.E. and N.N. supplied reagents/materials; K.T. and S.S. analyzed the data and wrote the manuscript.

Funding: This research was funded by JSPS KAKENHI, Grant Number 17K07793, and was partially supported by Daiichi Sangyo Corporation (Osaka, Japan).

Conflicts of Interest: This research was partially supported by Daiichi Sangyo Corporation. The funders had no role in the design of the study, in the collection, analyses, or interpretation of data, in the writing of the manuscript, or in the decision to publish the results.

References

1. Topalian, S.L.; Drake, C.G.; Pardoll, D.M. Immune checkpoint blockade: A common denominator approach to cancer therapy. *Cancer Cell* **2015**, *27*, 450–461. [CrossRef] [PubMed]

2. Dong, Y.; Sun, Q.; Zhang, X. PD-1 and its ligands are important immune checkpoints in cancer. *Oncotarget* **2017**, *8*, 2171–2186. [CrossRef] [PubMed]

3. Keir, M.E.; Butte, M.J.; Freeman, G.J.; Sharpe, A.H. PD-1 and its ligands in tolerance and immunity. *Annu. Rev. Immunol.* **2008**, *26*, 677–704. [CrossRef] [PubMed]

4. Alsaab, H.O.; Sau, S.; Alzhrani, R.; Tatiparti, K.; Bhise, K.; Kashaw, S.K.; Iyer, A.K. PD-1 and PD-L1 checkpoint signaling inhibition for cancer immunotherapy: Mechanism, combinations, and clinical outcome. *Front. Pharmacol.* **2017**, *8*, 561. [CrossRef] [PubMed]

5. Hino, R.; Kabashima, K.; Kato, Y.; Yagi, H.; Nakamura, M.; Honjo, T.; Okazaki, T.; Tokura, Y. Tumor cell expression of programmed cell death-1 ligand 1 is a prognostic factor for malignant melanoma. *Cancer* **2010**, *116*, 1757–1766. [CrossRef] [PubMed]

6. Chen, J.; Jiang, C.C.; Jin, L.; Zhang, X.D. Regulation of PD-L1: A novel role of pro-survival signalling in cancer. *Ann. Oncol.* **2016**, *27*, 409–416. [CrossRef]

7. Carosella, E.D.; Ploussard, G.; LeMaoult, J.; Desgrandchamps, F. A systematic review of immunotherapy in urologic cancer: Evolving roles for targeting of CTLA-4, PD-1/PD-L1, and HLA-G. *Eur. Urol.* **2015**, *68*, 267–279. [CrossRef]

8. Mahoney, K.M.; Freeman, G.J.; McDermott, D.F. The next immune-checkpoint inhibitors: PD-1/PD-L1 blockade in melanoma. *Clin. Ther.* **2015**, *37*, 764–782. [CrossRef]

9. Pardoll, D.M. The blockade of immune checkpoints in cancer immunotherapy. *Nat. Rev. Cancer* **2012**, *12*, 252–264. [CrossRef]

10. Soliman, H.; Khalil, F.; Antonia, S. PD-L1 expression is increased in a subset of basal type breast cancer cells. *PLoS ONE* **2014**, *9*, e88557. [CrossRef]

11. Patel, S.P.; Kurzrock, R. PD-L1 expression as a predictive biomarker in cancer immunotherapy. *Mol. Cancer Ther.* **2015**, *14*, 847–856. [CrossRef]

12. Chen, N.; Fang, W.; Zhan, J.; Hong, S.; Tang, Y.; Kang, S.; Zhang, Y.; He, X.; Zhou, T.; Qin, T.; et al. Upregulation of PD-L1 by EGFR activation mediates the immune escape in EGFR-driven NSCLC: Implication for optional immune targeted therapy for NSCLC patients with EGFR mutation. *J. Thorac. Oncol.* **2015**, *10*, 910–923. [CrossRef]

13. Kuol, N.; Stojanovska, L.; Nurgali, K.; Apostolopoulos, V. The mechanisms tumor cells utilize to evade the host's immune system. *Maturitas* **2017**, *105*, 8–15. [CrossRef]

14. Ren, W.; Korchin, B.; Zhu, Q.-S.; Wei, C.; Dicker, A.; Heymach, J.; Lazar, A.; Pollock, R.E.; Lev, D. Epidermal growth factor receptor blockade in combination with conventional chemotherapy inhibits soft tissue sarcoma cell growth In Vitro and In Vivo. *Clin. Cancer Res.* **2008**, *14*, 2785–2795. [CrossRef]

15. Azarova, A.M.; Gautam, G.; George, R.E. Emerging importance of ALK in neuroblastoma. *Semin. Cancer Biol.* **2011**, *21*, 267–275. [CrossRef]

16. Eder, J.P.; Vande Woude, G.F.; Boerner, S.A.; LoRusso, P.M. Novel therapeutic inhibitors of the c-Met signaling pathway in cancer. *Clin. Cancer Res.* **2009**, *15*, 2207–2214. [CrossRef]

17. Coelho, M.A.; de Carné Trécesson, S.; Rana, S.; Zecchin, D.; Moore, C.; Molina-Arcas, M.; East, P.; Spencer-Dene, B.; Nye, E.; Barnouin, K.; et al. Oncogenic RAS signaling promotes tumor immunoresistance by stabilizing PD-L1 mRNA. *Immunity* **2017**, *47*, 1083–1099. [CrossRef]

18. Gupta, S.; Plattner, R.; Der, C.J.; Stanbridge, E.J. Dissection of Ras-dependent signaling pathways controlling aggressive tumor growth of human fibrosarcoma cells: Evidence for a potential novel pathway. *Mol. Cell. Biol.* **2000**, *20*, 9294–9306. [CrossRef]
19. Gupta, S.; Stuffrein, S.; Plattner, R.; Tencati, M.; Gray, C.; Whang, Y.E.; Stanbridge, E.J. Role of phosphoinositide 3-kinase in the aggressive tumor growth of HT1080 human fibrosarcoma cells. *Mol. Cell. Biol.* **2001**, *21*, 5846–5856. [CrossRef]
20. Rui, X.; Pan, H.-F.; Shao, S.-L.; Xu, X.-M. Anti-tumor and anti-angiogenic effects of Fucoidan on prostate cancer: Possible JAKSTAT3 Pathway. *BMC Complement. Altern. Med.* **2017**, *17*, 378. [CrossRef]
21. Senthilkumar, K.; Manivasagana, P.; Venkatesana, J.; Kim, S.-K. Brown seaweed fucoidan: Biological activity and apoptosis, growth signaling mechanism in cancer. *Int. J. Biol. Macromol.* **2013**, *60*, 366–374. [CrossRef]
22. Atashrazm, F.; Lowenthal, R.M.; Woods, G.M.; Holloway, A.F.; Dickinson, J.L. Fucoidan and cancer: A multifunctional molecule with anti-tumor potential. *Mar. Drugs* **2015**, *13*, 2327–2346. [CrossRef]
23. Takahashi, H.; Kawaguchi, M.; Kitamura, K.; Narumiya, S.; Kawamura, M.; Tengan, I.; Nishimoto, S.; Hanamure, Y.; Majima, Y.; Tsubura, S.; et al. An exploratory study on the anti-inflammatory effects of fucoidan in relation to quality of life in advanced cancer patients. *Integr. Cancer Ther.* **2018**, *17*, 282–291. [CrossRef]
24. Jin, W.; Wang, J.; Ren, S.; Song, N.; Zhang, Q. Structural analysis of a heteropolysaccharide from *Saccharina japonica* by electrospray mass spectrometry in tandem with collision-induced dissociation tandem mass spectrometry (ESI-CID-MS/MS). *Mar. Drugs* **2012**, *10*, 2138–2152. [CrossRef]
25. Nagaoka, M.; Shibata, H.; Kimura-Takagi, I.; Hashimoto, S.; Kimura, K.; Makino, T.; Aiyama, R.; Ueyama, S.; Yokokura, T. Structural study of fucoidan from *Cladosiphon okamuranus* TOKIDA. *Glycoconj. J.* **1999**, *16*, 19–26. [CrossRef]
26. Farag, M.A.; Fekry, M.I.; Al-Hammady, M.A.; Khalil, M.N.; El-Seedi, H.R.; Meyer, A.; Porzel, A.; Westphal, H.; Wessjohann, L.A. Cytotoxic effects of *Sarcophyton* sp. soft corals—Is there a correlation to their NMR fingerprints? *Mar. Drugs* **2017**, *15*, 211. [CrossRef]
27. Lim, S.J.; Wan Aida, W.M.; Schiehser, S.; Rosenau, T.; Böhmdorfer, S. Structural elucidation of fucoidan from *Cladosiphon okamuranus* (Okinawa mozuku). *Food Chem.* **2019**, *272*, 222–226. [CrossRef]
28. Yang, C.; Chung, D.; Shin, I.S.; Lee, H.Y.; Kim, J.C.; Lee, Y.J.; You, S.G. Effects of molecular weight and hydrolysis conditions on anticancer activity of fucoidans from sporophyll of *Undaria pinnatifida*. *Int. J. Biol. Macromol.* **2008**, *43*, 433–437. [CrossRef]
29. Rupérez, P.; Ahrazem, O.; Leal, J.A. Potential antioxidant capacity of sulfated polysaccharides from the edible marine brown seaweed *Fucus Vesiculosus*. *J. Agric. Food Chem.* **2002**, *50*, 840–845. [CrossRef]
30. Azuma, K.; Ishihara, T.; Nakamoto, H.; Amaha, T.; Osaki, T.; Tsuka, T.; Imagawa, T.; Minami, S.; Takashima, O.; Ifuku, S.; et al. Effects of oral administration of fucoidan extracted from *Cladosiphon okamuranus* on tumor growth and survival time in a tumor-bearing mouse model. *Mar. Drugs* **2012**, *10*, 2337–2348. [CrossRef]
31. Matsubara, K.; Xue, C.; Zhao, X.; Mori, M.; Sugawara, T.; Hirata, T. Effects of middle molecular weight fucoidans on In Vitro and Ex Vivo angiogenesis of endothelial cells. *Int. J. Mol. Med.* **2005**, *15*, 695–699. [CrossRef]
32. Irhimeh, M.R.; Fitton, J.H.; Lowenthal, R.M.; Kongtawelert, P. A quantitative method to detect fucoidan in human plasma using a novel antibody. *Methods Find. Exp. Clin. Pharmacol.* **2005**, *27*, 705–710. [CrossRef]
33. Kimura, R.; Rokkaku, T.; Takeda, S.; Senba, M.; Mori, N. Cytotoxic Effects of Fucoidan Nanoparticles against Osteosarcoma. *Mar. Drugs* **2013**, *11*, 4267–4278. [CrossRef]
34. Nagamine, T.; Nakazato, K.; Tomioka, S.; Iha, M.; Nakajima, K. Intestinal absorption of fucoidan extracted from the brown seaweed, *Cladosiphon okamuranus*. *Mar. Drugs* **2015**, *13*, 48–64. [CrossRef]
35. Yee, S. In Vitro permeability across Caco-2 cells (colonic) can predict In Vivo (small intestinal) absorption in man-fact or myth. *Pharm. Res.* **1997**, *14*, 763–766. [CrossRef]
36. Zhao, X.; Guo, F.; Hu, J.; Zhang, L.; Xue, C.; Zhang, Z.; Li, B. Antithrombotic activity of oral administered low molecular weight fucoidan from *Laminaria Japonica*. *Thromb. Res.* **2016**, *144*, 46–52. [CrossRef]
37. Chen, L.-M.; Liu, P.-Y.; Chen, Y.-A.; Tseng, H.-Y.; Shen, P.-C.; Hwang, P.-A.; Hsu, H.-L. Oligo-Fucoidan prevents IL-6 and CCL2 production and cooperates with p53 to suppress ATM signaling and tumor progression. *Sci. Rep.* **2017**, *7*, 11864. [CrossRef]
38. Hwang, P.A.; Yan, M.D.; Lin, H.T.V.; Li, K.L.; Lin, Y.C. Toxicological evaluation of low molecular weight fucoidan In Vitro and In Vivo. *Mar. Drugs* **2016**, *14*, 121. [CrossRef]

39. Chen, M.C.; Hsu, W.L.; Hwang, P.A.; Chou, T.C. Low molecular weight fucoidan inhibits tumor angiogenesis through downregulation of HIF-1/VEGF signaling under hypoxia. *Mar. Drugs* **2015**, *13*, 4436–4451. [CrossRef]

40. Lin, H.T.V.; Chen, Y.T.T.; Lu, W.J.; Hwang, P.A. Effects of low-molecular-weight fucoidan and high stability fucoxanthin on glucose homeostasis, lipid metabolism, and liver function in a mouse model of type II diabetes. *Mar. Drugs* **2017**, *15*, 113. [CrossRef]

41. Ye, J.; Li, Y.; Teruya, K.; Katakura, Y.; Ichikawa, A.; Eto, H.; Hosoi, M.; Hosoi, M.; Nishimoto, S.; Shirahata, S. Enzyme-digested fucoidan extracts derived from seaweed *Mozuku of Cladosiphon novae-caledoniae kylin* inhibit invasion and angiogenesis of tumor cells. *Cytotechnology* **2005**, *47*, 117–126. [CrossRef]

42. Boo, H.-J.; Hyun, J.-H.; Kim, S.-C.; Kang, J.-I.; Kim, M.-K.; Kim, S.-Y.; Cho, H.; Yoo, E.-S.; Kang, H.-K. Fucoidan from *Undaria pinnatifida* induces apoptosis in A549 human lung carcinoma cells. *Phytother. Res.* **2011**, *25*, 1082–1086. [CrossRef]

43. Wang, W.; Wu, J.; Zhang, X.; Hao, C.; Zhao, X.; Jiao, G.; Shan, X.; Tai, W.; Yu, G. Inhibition of influenza A virus infection by fucoidan targeting viral neuraminidase and cellular EGFR pathway. *Sci. Rep.* **2017**, *7*, 40760. [CrossRef]

44. Lin, K.; Cheng, J.; Yang, T.; Li, Y.; Zhu, B. EGFR-TKI down-regulates PD-L1 in EGFR mutant NSCLC through inhibiting NF-κB. *Biochem. Biophys. Res. Commun.* **2015**, *463*, 95–101. [CrossRef]

45. Ikeguchi, M.; Yamamoto, M.; Arai, Y.; Maeta, Y.; Ashida, K.; Katano, K.; Miki, Y.; Kimura, T. Fucoidan reduces the toxicities of chemotherapy for patients with unresectable advanced or recurrent colorectal cancer. *Oncol. Lett.* **2011**, *2*, 319–322. [CrossRef]

46. Zhang, Z.; Teruya, K.; Yoshida, T.; Eto, H.; Shirahata, S. Fucoidan extract enhances the anti-cancer activity of chemotherapeutic agents in MDA-MB-231 and MCF-7 breast cancer cells. *Mar. Drugs* **2013**, *11*, 81–98. [CrossRef]

47. Zhang, Z.; Teruya, K.; Eto, H.; Shirahata, S. Fucoidan extract induces apoptosis in MCF-7 cells via a mechanism involving the ROS-Dependent JNK activation and mitochondria-mediated pathways. *PLoS ONE* **2011**, *6*, e27441. [CrossRef]

48. Kasai, A.; Arafuka, S.; Koshiba, N.; Takahashi, D.; Toshima, K. Systematic synthesis of low-molecular weight fucoidan derivatives and their effect on cancer cells. *Org. Biomol. Chem.* **2015**, *13*, 10556–10568. [CrossRef]

49. Oliveira, C.; Ferreira, A.S.; Novoa-Carballal, R.; Nunes, C.; Pashkuleva, I.; Neves, N.M.; Coimbra, M.A.; Reis, R.L.; Martins, A.; Silva, T.H. The key role of sulfation and branching on fucoidan antitumor activity. *Macromol. Biosci.* **2017**, *17*, 1600340. [CrossRef]

50. Chazotte, B. Labeling nuclear DNA with Hoechst 33342. *Cold Spring Harb. Protoc.* **2011**, *2011*, 83–85. [CrossRef]

51. Atale, N.; Gupta, S.; Yadav, U.C.S.; Rani, V. Cell-death assessment by fluorescent and nonfluorescent cytosolic and nuclear staining techniques. *J. Microsc.* **2014**, *255*, 7–19. [CrossRef]

52. Ishiyama, M.; Tominaga, H.; Shiga, M.; Sasamoto, K.; Ohkura, Y.; Ueno, K. A combined assay of cell viability and In Vitro cytotoxicity with a highly water-soluble tetrazolium salt, neutral red and crystal violet. *Biol. Pharm. Bull.* **1996**, *19*, 1518–1520. [CrossRef]

53. Carlson, M.A. Technical note: Assay of cell quantity in the fibroblast-populated collagen matrix with a tetrazolium reagent. *Eur. Cell Mater.* **2006**, *12*, 44–48. [CrossRef]

54. Hong, S.; Chenc, N.; Fang, W.; Zhana, J.; Liu, Q.; Kang, S.; He, X.; Liu, L.; Zhou, T.; Huang, J.; et al. Upregulation of PD-L1 by EML4-ALK fusion protein mediates the immune escape in ALK positive NSCLC: Implication for optional anti-PD-1/PD-L1 immune therapy for ALK-TKIs sensitive and resistant NSCLC patients. *Oncoimmunology* **2016**, *5*, e1094598. [CrossRef]

55. Breier, G.; Blum, S.; Peli, J.; Groot, M.; Wild, C.; Risau, W.; Reichmann, E. Transforming growth factor-β and RAS regulate the VEGF/VEGF-receptor system during tumor angiogenesis. *Int. J. Cancer* **2002**, *97*, 142–148. [CrossRef]

56. Etienne-Manneville, S.; Hall, A. Rho GTPases in cell biology. *Nature* **2002**, *420*, 629–635. [CrossRef]

57. Haga, R.B.; Ridley, A.J. Rho GTPases: Regulation and roles in cancer cell biology. *Small GTPases* **2016**, *7*, 207–221. [CrossRef]

58. Stanciu, L.A.; Bellettato, C.M.; Laza-Stanca, V.; Coyle, A.J.; Papi, A.; Johnston, S.L. Expression of programmed death–1 ligand (PD-L) 1, PD-L2, B7-H3, and inducible costimulator ligand on human respiratory tract epithelial cells and regulation by respiratory syncytial virus and type 1 and 2 cytokines. *J. Infect. Dis.* **2006**, *193*, 404–412. [CrossRef]

59. Eppihimer, M.J.; Gunn, J.; Freeman, G.J.; Greenfield, E.A.; Chernova, T.; Erickson, J.; Leonard, J.P. Expression and regulation of the PD-L1 immunoinhibitory molecule on microvascular endothelial cells. *Microcirculation* **2002**, *9*, 133–145. [CrossRef]

60. Dong, H.; Zhu, G.; Tamada, K.; Chen, L. B7-H1, a third member of the B7 family, co-stimulates T-cell proliferation and interleukin-10 secretion. *Nat. Med.* **1999**, *5*, 1365–1369. [CrossRef]

61. Rasheed, S.; Nelson-Rees, W.A.; Toth, E.M.; Arnstein, P.; Gardner, M.B. Characterization of a newly derived human sarcoma cell line (HT-1080). *Cancer* **1974**, *33*, 1027–1033. [CrossRef]

62. Paterson, H.; Reeves, B.; Brown, R.; Hail, A.; Furth, M.; Bos, J.; Jones, P.; Marshall, C. Activated N-*ras* controls the transformed phenotype of HT1080 human fibrosarcoma cells. *Cell* **1987**, *51*, 803–812. [CrossRef]

63. Takashima, A.; Faller, D.V. Targeting the RAS oncogene. *Expert Opin. Ther. Targets* **2013**, *17*, 507–531. [CrossRef]

64. Zahorowska, B.; Crowe, P.J.; Yang, J.-L. Combined therapies for cancer: A review of EGFR-targeted monotherapy and combination treatment with other drugs. *J. Cancer Res. Clin. Oncol.* **2009**, *135*, 1137–1148. [CrossRef]

65. Zhang, W.; Pang, Q.; Yan, C.; Wang, Q.; Yang, J.; Yu, S.; Liu, X.; Yuan, Z.; Wang, P.; Xiao, Z. Induction of PD-L1 expression by epidermal growth factor receptor–mediated signaling in esophageal squamous cell carcinoma. *OncoTargets Ther.* **2017**, *10*, 763–771. [CrossRef]

66. Soo, R.A.; Lim, S.M.; Syn, N.L.; Teng, R.; Soong, R.; Mok, T.S.K.; Cho, B.C. Immune checkpoint inhibitors in epidermal growth factor receptor mutant non-small cell lung cancer: Current controversies and future directions. *Lung Cancer* **2018**, *115*, 12–20. [CrossRef]

67. Marmarelis, M.E.; Aggarwal, C. Combination immunotherapy in non-small cell lung cancer. *Curr. Oncol. Rep.* **2018**, *20*, 55. [CrossRef]

68. Castoria, G.; Giovannelli, P.; Di Donato1, M.; Hayashi, R.; Arra, C.; Appella, E.; Auricchio, F.; Migliaccio, A. Targeting androgen receptor/Src complex impairs the aggressive phenotype of human fibrosarcoma cells. *PLoS ONE* **2013**, *8*, e76899. [CrossRef]

69. Harris, R.C.; Chung, E.; Coffey, R.J. EGF receptor ligands. *Exp. Cell Res.* **2003**, *284*, 2–13. [CrossRef]

70. Faria, J.A.; de Andrade, C.; Goe, A.M.; Rodrigues, M.A.; Gomes, D.A. Effects of different ligands on epidermal growth factor receptor (EGFR) nuclear translocation. *Biochem. Biophys. Res. Commun.* **2016**, *478*, 39–45. [CrossRef]

71. Mendelsohn, J.; Baselga, J. The EGF receptor family as targets for cancer therapy. *Oncogene* **2000**, *19*, 6550–6565. [CrossRef]

72. Lee, N.Y.; Ermakova, S.P.; Zvyagintseva, T.N.; Kang, K.W.; Dong, Z.; Choi, H.S. Inhibitory effects of fucoidan on activation of epidermal growth factor receptor and cell transformation in JB6 Cl41 cells. *Food Chem. Toxicol.* **2008**, *46*, 1793–1800. [CrossRef]

73. Lee, N.Y.; Ermakova, S.P.; Choi, H.-K.; Kusaykin, M.I.; Shevchenko, N.M.; Zvyagintseva, T.N.; Choi, H.S. Fucoidan from *Laminaria cichorioides* inhibits AP-1 transactivation and cell transformation in the mouse epidermal JB6 cells. *Mol. Carcinog.* **2008**, *47*, 629–637. [CrossRef]

74. Qayum, N.; Muschel, R.J.; Im, J.H.; Balathasan, L.; Koch, C.J.; Patel, S.; McKenna, W.G.; Bernhard, E.J. Tumor vascular changes mediated by inhibition of oncogenic signaling. *Cancer Res.* **2009**, *69*, 6347–6354. [CrossRef]

75. Hanyu, A.; Kojima, K.; Hatake, K.; Nomura, K.; Murayama, H.; Ishikawa, Y.; Miyata, S.; Ushijima, M.; Matsuura, M.; Ogata, E.; et al. Functional In Vivo optical imaging of tumor angiogenesis, growth, and metastasis prevented by administration of anti-human VEGF antibody in xenograft model of human fibrosarcoma HT1080 cells. *Cancer Sci.* **2009**, *100*, 2085–2092. [CrossRef]

76. Kim, K.J.; Li, B.; Winer, J.; Armanini, M.; Gillett, L.; Phillips, H.S.; Ferrara, N. Inhibition of vascular endothelial growth factor-induced angiogenesis suppresses tumour growth In Vivo. *Nature* **1993**, *362*, 841–844. [CrossRef]

77. Koh, Y.W.; Han, J.-H.; Yoon, D.H.; Suh, C.; Huh, J. PD-L1 expression correlates with VEGF and microvessel density in patients with uniformly treated classical Hodgkin lymphoma. *Ann. Hematol.* **2017**, *96*, 1883–1890. [CrossRef]

78. Koyanagi, S.; Tanigawa, N.; Nakagawa, H.; Soeda, S.; Shimeno, H. Oversulfation of fucoidan enhances its anti-angiogenic and antitumor activities. *Biochem. Pharmacol.* **2003**, *65*, 173–179. [CrossRef]

79. Yang, C.-H.; Tsao, C.-F.; Ko, W.-S.; Chiou, Y.-L. The oligo fucoidan inhibits platelet-derived growth factor-stimulated proliferation of airway smooth muscle cells. *Mar. Drugs* **2016**, *14*, 15. [CrossRef]

80. Ridley, A.J. RhoA, RhoB and RhoC have different roles in cancer cell migration. *J. Microsc.* **2013**, *251*, 242–249. [CrossRef]
81. Ju, J.A.; Gilkes, D.M. RhoB: Team oncogene or team tumor suppressor? *Genes* **2018**, *9*, 67. [CrossRef]
82. Jiang, K.; Delarue, F.L.; Sebti, S.M. EGFR, ErbB2 and Ras but not Src suppress RhoB expression while ectopic expression of RhoB antagonizes oncogene-mediated transformation. *Oncogene* **2004**, *23*, 1136–1145. [CrossRef]
83. Yamasaki-Miyamoto, Y.; Yamasaki, M.; Tachibana, H.; Yamada, K. Fucoidan induces apoptosis through activation of caspase-8 on human breast cancer MCF-7 cells. *J. Agric. Food Chem.* **2009**, *57*, 8677–8682. [CrossRef]
84. Zhang, W.; Oda, T.; Yu, Q.; Jin, J.-O. Fucoidan from *Macrocystis pyrifera* has powerful immune-modulatory effects compared to three other fucoidans. *Mar. Drugs* **2015**, *13*, 1084–1104. [CrossRef]
85. Holland, P.M.; Abramson, R.D.; Watson, R.; Gelfand, D.H. Detection of specific polymerase chain reaction product by utilizing the 5′ → 3′ exonuclease activity of *Thermus aquaticus* DNA polymerase. *Proc. Natl. Acad. Sci. USA* **1991**, *88*, 7276–7280. [CrossRef]
86. Du Breuil, R.M.; Patel, J.M.; Mendelow, B.V. Quantitation of β-Actin-specific mRNA transcripts using xeno-competitive PCR. *PCR Methods Appl.* **1993**, *3*, 57–59. [CrossRef]
87. Karge, W.H.; Schaefer, E.J.; Ordovas, J.M. Quantification of mRNA by Polymerase Chain Reaction (PCR) using an internal standard and a nonradioactive detection method. In *Lipoprotein Protocols*; Ordovas, J.M., Ed.; Humana Press: Totowa, NJ, USA, 1998; Volume 110. [CrossRef]
88. Huang, B.; Sun, L.; Cao, J.; Zhang, Y.; Wu, Q.; Zhang, J.; Ge, Y.; Fu, L.; Wang, Z. Downregulation of the GnT-V gene inhibits metastasis and invasion of BGC823 gastric cancer cells. *Oncol. Rep.* **2013**, *29*, 2392–2400. [CrossRef]
89. Hong, K.-J.; Hsu, M.-C.; Hou, M.-F.; Hung, W.-C. The tumor suppressor RECK interferes with HER-2/Neu dimerization and attenuates its oncogenic signaling. *FEBS Lett.* **2011**, *585*, 591–595. [CrossRef]
90. Lartey, J.; Smith, M.; Pawade, J.; Strachan, B.; Mellor, H.; López Bernal, A. Up-regulation of myometrial RHO effector proteins (PKN1 and DIAPH1) and CPI-17 (PPP1R14A) phosphorylation in human pregnancy is associated with increased GTP-RHOA in spontaneous preterm labor. *Biol. Reprod.* **2007**, *76*, 971–982. [CrossRef]
91. Sadun, R.E.; Sachsman, S.M.; Chen, X.; Christenson, K.W.; Morris, W.Z.; Hu, P.; Epstein, A.L. Immune signatures of murine and human cancers reveal unique mechanisms of tumor escape and new targets for cancer immunotherapy. *Clin. Cancer Res.* **2007**, *13*, 4016–4025. [CrossRef]
92. Picot, J.; Guerin, C.L.; Le Van Kim, C.; Boulanger, C.M. Flow cytometry: Retrospective, fundamentals and recent instrumentation. *Cytotechnology* **2012**, *64*, 109–130. [CrossRef]

 *marine drugs*

Article

Actinomycin V Inhibits Migration and Invasion via Suppressing Snail/Slug-Mediated Epithelial-Mesenchymal Transition Progression in Human Breast Cancer MDA-MB-231 Cells In Vitro

Shiqi Lin [1], Caiyun Zhang [1], Fangyuan Liu [1], Jiahui Ma [1], Fujuan Jia [1], Zhuo Han [1], Weidong Xie [1] and Xia Li [1,2,3,*]

[1] School of Ocean, Shandong University, Weihai 264209, China; lsqsd@outlook.com (S.L.); caiyun617@outlook.com (C.Z.); fangyuan617@outlook.com (F.L.); sdumjh@hotmail.com (J.M.); jfj1996@outlook.com (F.J.); hanzhuo1013@gmail.com (Z.H.); wdxie@sdu.edu.cn (W.X.)
[2] School of Pharmaceutical Sciences, Shandong University, Jinan 250012, China
[3] The Key Laboratory of Chemistry for Natural Product of Guizhou Province, Chinese Academy of Science, Guiyang 550002, China
* Correspondence: xiali@sdu.edu.cn; Tel.: +86-531-88382612

Received: 23 April 2019; Accepted: 22 May 2019; Published: 24 May 2019

Abstract: Actinomycin V, an analog of actinomycin D produced by the marine-derived actinomycete *Streptomyces* sp., possessing a 4-ketoproline instead of a 4-proline in actinomycin D. In this study, the involvement of snail/slug-mediated epithelial-mesenchymal transition (EMT) in the anti-migration and -invasion actions of actinomycin V was investigated in human breast cancer MDA-MB-231 cells in vitro. Cell proliferation effect was evaluated by 3-(4,5-Dimethylthiazol)-2,5-diphenyltetrazolium bromide (MTT) assay. Wound-healing and Transwell assay were performed to investigate the anti-migration and -invasion effects of actinomycin V. Western blotting was used to detect the expression levels of E-cadherin, N-cadherin, vimentin, snail, slug, zinc finger E-box binding homeobox 1 (ZEB1), and twist proteins and the mRNA levels were detected by rt-PCR. Actinomycin V showed stronger cytotoxic activity than that of actinomycin D. Actinomycin V up-regulated both of the protein and mRNA expression levels of E-cadherin and down-regulated that of N-cadherin and vimentin in the same cells. In this connection, actinomycin V decreased the snail and slug protein expression, and consequently inhibited cells EMT procession. Our results suggest that actinomycin V inhibits EMT-mediated migration and invasion via decreasing snail and slug expression, which exhibits therapeutic potential for the treatment of breast cancer and further toxicity investigation in vivo is needed.

Keywords: breast cancer; actinomycin; EMT; migration; invasion

---

1. Introduction

Breast cancer has the highest diagnosis and death rates among women followed by lung and colorectal cancer and has generally increased in recent years [1]. Since many therapies including surgery, radiotherapy, chemotherapy, and human epidermal growth factor receptor 2 (HER2) targeted therapy have been used in clinical trials, the overall survival rate of patients still remained poor [2]. Metastasis is the majority reason for breast cancer death. Epithelial-mesenchymal transition (EMT) is a key process in cancer invasion and metastasis [3]. EMT is a dynamic and complex multistep process that cells attach to and degrade the surrounding extracellular matrix (ECM) to escape from the primary tumor site, invade the stromal tissues, and establish the distant secondary tumor [4].

Therefore, attention has been focused on blocking the migration and invasion processes of breast cancer cells to inhibit metastasis [5–7].

Many effective anti-cancer compounds that were originally isolated from marine invertebrates, such as discodermolide, halichondrin B, bryostatin1, and phorboxazole A, were actually produced by marine microorganisms [8–10]. Marine microorganisms were considered to be the promising sources for discovering new effective anti-cancer drugs and many efforts have been made in recent years. Actinomycins, produced by the marine-derived actinomycete *Streptomyces* sp., is a class of cyclic chromopeptide consisting of two cyclic pentapeptide lactones attached to a central phenoxazinone chromophore via amide bonds [11–13]. Typically, actinomycin D has been well used in treatment of a variety of cancers such as Wilms tumor, rhabdomycosarcoma, and Ewing's sarcoma through binding and intercalating double-stranded DNA and inhibiting RNA synthesis [14]. A report has shown that actinomycin D inhibited the migration of MDA-MB-231 cells but the underlying mechanism was not mention [15]. Moreover, the toxicity development, particularly in hepatotoxicity of actinomycin D, is associated with use in an anti-tumor process, which severely limits its effectiveness in clinic. Actinomycin V possesses 4-ketoproline as a substitute for 4-proline of actinomycin D. Studies have reported that actinomycin V's inhibitory effects on the F5-5 friend leukemia cells, MCF-7, A549, and K562 cells were superior to actinomycin D and the effect on human breast cancer cells MCF-7 was the most obvious [16,17]. However, the anti- migration and invasion actions of actinomycin V and the underlying mechanism remains unclear. In this present study, we examined that actinomycin V treatment inhibited the proliferation of human breast cancer cells and suppressed the EMT process by down-regulating the snail and slug protein expression. These results may provide a new theoretical basis for the treatment of breast cancer with actinomycin V and the useful information for developing novel marine-derived anti-cancer drugs.

2. Results

2.1. Actinomycin V Inhibits the Proliferation of Human Breast Cancer Cells

To compare the effects of three actinomycins (actinomycin D, actinomycin V, and actinomycin X_{ob} as shown in Figure 1) on breast cancer cells, we initially performed 3-(4,5-Dimethylthiazol)-2,5-diphenyltetrazolium bromide (MTT) analysis in several subtypes of human breast cancer (MDA-MB-231, BT474, and MCF-7) and noncancer (HMLE and MCF-10A) cell lines. We found that actinomycin V had the excellent activity on breast cancer cells among three actinomycins. Actinomycin V displayed the strongest sensitivity on MDA-MB-231 cells and with lower cytotoxicity in noncancer breast cells as shown in the Table 1. The IC_{50} of actinomycin V for 48 h treatment to MDA-MB-231, MCF-7, BT474, HMLE, and MCF-10A were 0.83 ± 0.32 nmol/L, 1.92 ± 0.25 nmol/L, 4.16 ± 0.25 nmol/L, 3.49 ± 0.31 nmol/L, and 4.07 ± 0.26 nmol/L, respectively, and actinomycin D treatment for 48 h to those cell lines were 15.15 ± 0.52 nmol/L, 8.23 ± 0.50 nmol/L, 37.00 ± 3.15 nmol/L, 30.22 ± 0.50 nmol/L, and 34.01 ± 0.25 nmol/L, respectively. Further experiments showed that actinomycin V dose- and time-dependently inhibited the proliferation of human breast cancer cells (Figure 2).

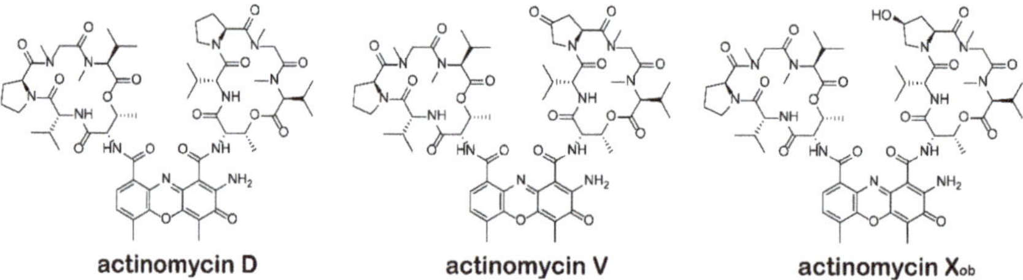

Figure 1. Chemical structure of actinomycin D, actinomycin V, and actinomycin X_{ob}.

Table 1. Growth inhibitory activities of three actinomycins and Adriamycin in different cell lines.

Compounds IC$_{50}$ (nmol/L)	MCF-7	MDA-MB-231	BT474	MCF-10A	HMLE
Actinomycin V	1.92 ± 0.25	0.83 ± 0.32	4.16 ± 0.25	4.07 ± 0.26	3.49 ± 0.31
Actinomycin D	8.23 ± 0.50	15.15 ± 0.52	37.00 ± 3.15	34.01 ± 0.25	30.22 ± 0.50
Actinomycin X$_{ob}$	149.40 ± 4.03	127.33 ± 4.49	369.90 ± 0.14	248.57 ± 14.69	105.83 ± 8.44
Adriamycin	885.38 ± 13.50	942.60 ± 22.50	584.70 ± 50.00	1489.13 ± 25.50	1627 ± 15.50

[1] The effects of three actinomycins and Adriamycin on various human breast cancer cells (MCF-7, MDA-MB-231, and BT474) and normal human breast epithelial cell lines (HMLE and MCF-10A) were examined by 3-(4,5-Dimethylthiazol)-2,5-diphenyltetrazolium bromide (MTT) assay. The cells were treated with various concentrations of compounds for 48 h and the IC$_{50}$ values were then calculated. Results were obtained from three independent experiments. Data were expressed as the mean $\pm$ SD (Standard Deviation) of triplicate experiments.

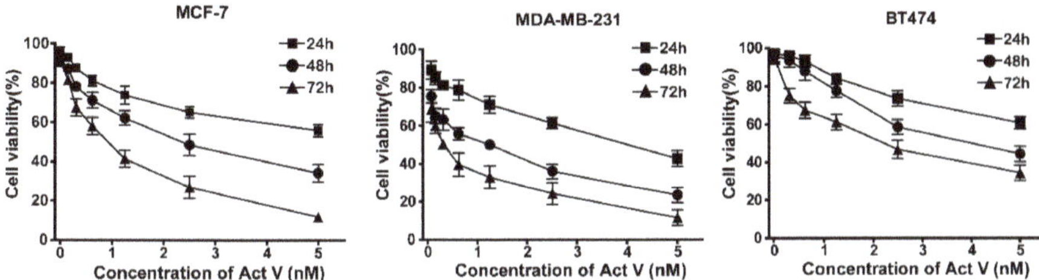

Figure 2. Actinomycin V does- and time- dependently inhibited the proliferation of various cell lines. Cells were treated with various concentrations of actinomycin V (0 to 5 nmol/L) for 24 h to 72 h. Cell viability was denoted as a percentage of the untreated control (actinomycin V 0 nmol/L) at the concurrent time point. Results were obtained from three independent experiments.

2.2. Actinomycin V's Effects on Morphological Changes in Breast Cancer Cells

We observed the effects of actinomycin V on morphological changes in breast cancer cells (MCF-7, MDA-MB-231, and BT474) (Figure 3). After 1 nmol/L actinomycin V 24 h treatment cells performed in different levels of morphological changes. Especially the metastatic MDA-MB-231 cells, it displayed in a scattered distribution in the culture and a spindle or star-like structure before actinomycin V treated but after 24 h treatment cells changed to a round cell shape and tend to an aggregation distribution, indicating cells appeared to have improved intercellular adhesion and reduced cells migration and invasion.

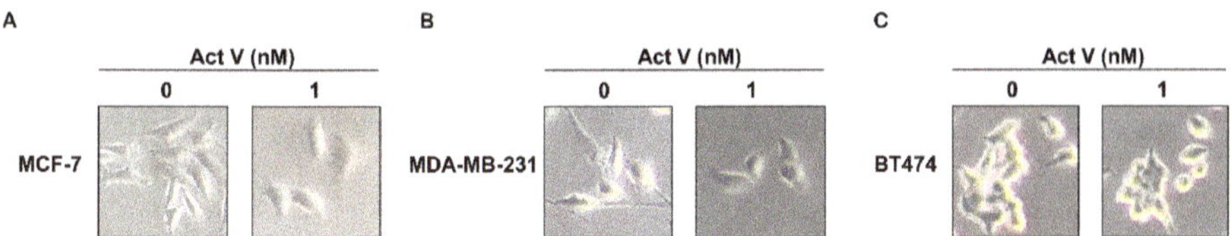

Figure 3. Changes in morphology of MCF-7, MDA-MB-231, and BT474 cells upon actinomycin V treatment. (**A–C**) Cells were treated with 1 nmol/L actinomycin V for 24 h and phase contrast images of the cultured cells were taken under 200× objective lenses.

2.3. Actinomycin V Inhibits the Migration and Invasion of Human Breast Cancer Cells

Considering that migration and invasion are the major drivers of breast cancer metastasis, wound-healing assay and Transwell assay were used to further confirm the actinomycin V effects on anti-migration and anti-invasion in human breast cancer cells. In the wound-healing assay (Figure 4), the untreated MDA-MB-231 cells wound-healing rate was 52% in 24 h. However, actinomycin V treatment apparently retarded the wound healing, the healing rates were only about 23% and 10%, respectively, in the presence of 1–2 nmol/L actinomycin V. In the Transwell assay (Figure 5), actinomycin V dose-dependent treatment significantly suppressed the invasion of MDA-MB-231 cells. Similarly, actinomycin V treatment inhibited the migration and invasion of MCF-7 and BT474 cells but the inhibitory effects were not as significant when compared with its anti-migration and anti-invasion effects on the metastatic MDA-MB-231 cells. These results indicated that actinomycin V inhibited the migration and invasion of human breast cancer cells.

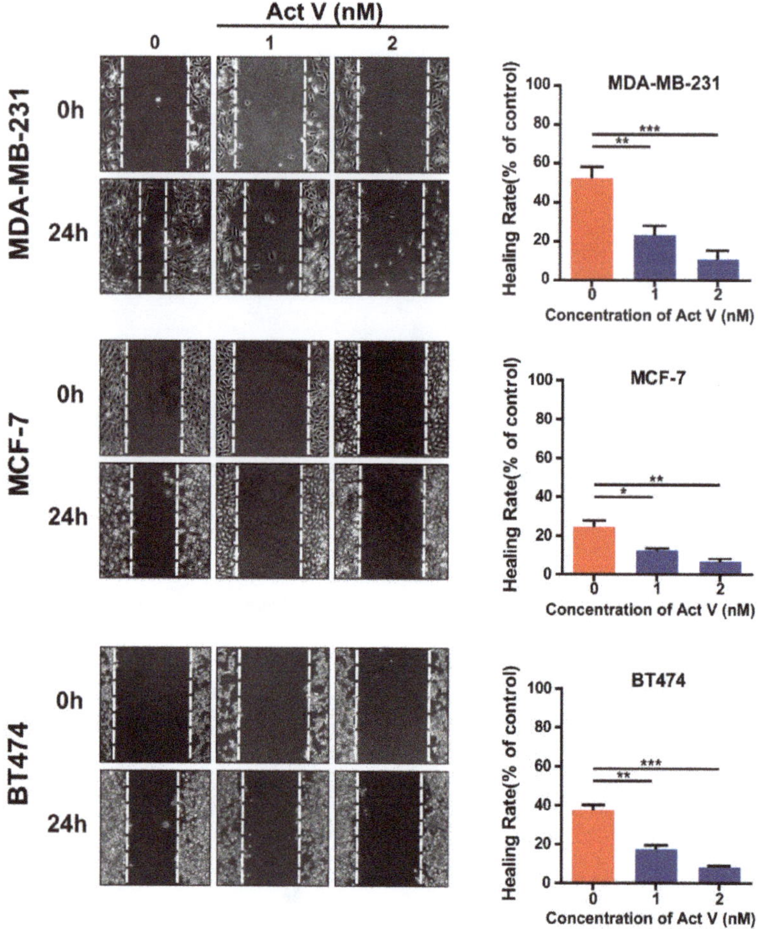

Figure 4. Effect of actinomycin V on the motility of MDA-MB-231, MCF-7, and BT474 cells. The wound-healing rate of each cell line was analyzed (magnification: ×200). Results were obtained from three independent experiments. * $p < 0.05$, ** $p < 0.01$, *** $p < 0.001$.

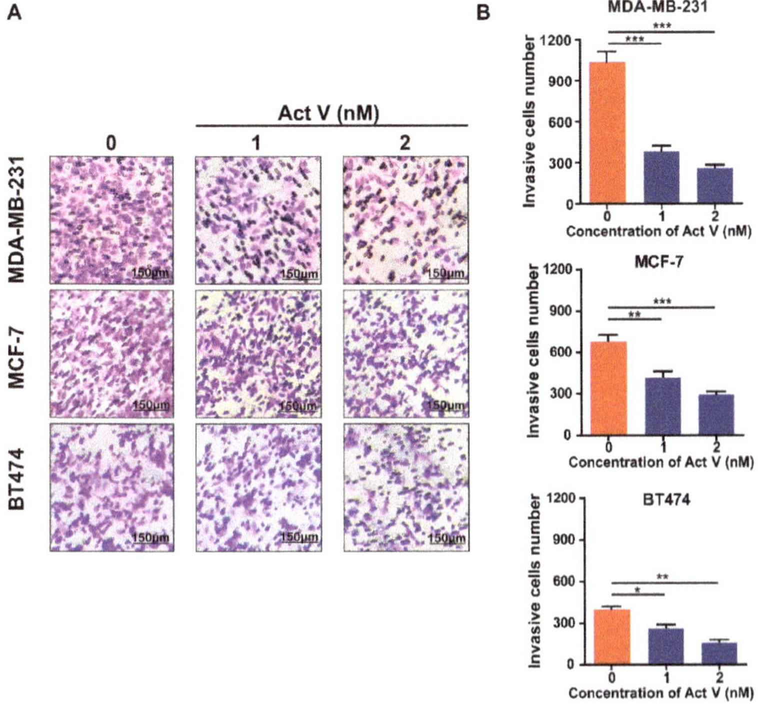

Figure 5. Effects of actinomycin V on the invasion of MDA-MB-231, MCF-7, and BT474 cells. (**A**,**B**) Cells were treated with 0–2 nmol/L actinomycin V for 24 h, and the invasion in these two cell lines was analyzed (magnification: ×200). Results were obtained from three independent experiments. * $p < 0.05$, ** $p < 0.01$, *** $p < 0.001$.

2.4. Effects of Actinomycin V on the Expression of EMT-Associated Proteins and mRNA in MDA-MB-231 Cells

To investigate the actinomycin V effects on Epithelial-mesenchymal transition inhibition in human breast cancer cells, we preformed the Western blot analysis to evaluate the expression levels of the epithelial marker E-cadherin, and mesenchymal markers N-cadherin and Vimentin in MDA-MB-231 cells (Figures 6 and 7). We found that actinomycin V treatment increased the expression levels of E-cadherin while decreased that of N-cadherin and Vimentin. Immunofluorescence staining also showed that compared with untreated cells, actinomycin V treatment significantly decreased intracellular N-cadherin and vimentin content while increased the E-cadherin content in cancer cells.

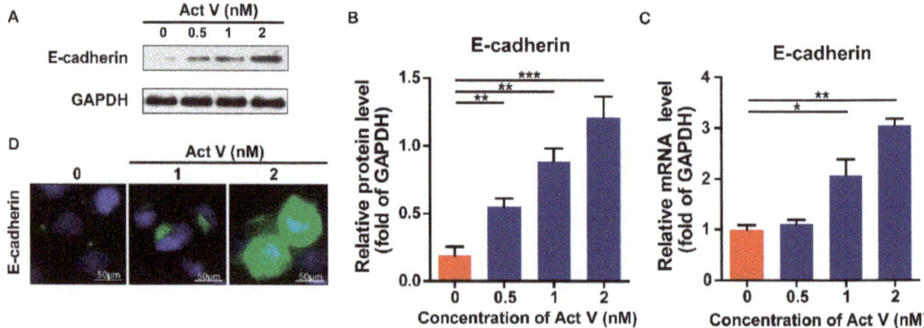

Figure 6. Effect of actinomycin V on the expression of E-cadherin. (**A**,**B**) MDA-MB-231 cells were treated with 0–2 nmol/L actinomycin V for 24 h then the protein expression of E-cadherin was measured by Western blot. (**C**) Relative expression of E-cadherin mRNA in the MDA-MB-231 cells was analyzed by real-time PCR. RNA levels are represented as fold increase relative to the level of the control (normalized to glyceraldehyde-3-phosphate dehydrogenase (GAPDH) mRNA level). Results were obtained from three independent experiments. * $p < 0.05$, ** $p < 0.01$, *** $p < 0.001$. (**D**) cells were treated with 0–2 nmol/L actinomycin V for 6 h and analyzed by E-cadherin fluorescent signals. Cells were stained with anti- E-cadherin (green) and 4′,6-diamidino-2-phenylindole (DAPI, blue). Magnification ×200. E-cadherin levels were increased in the cells, consistent with the Western blot results.

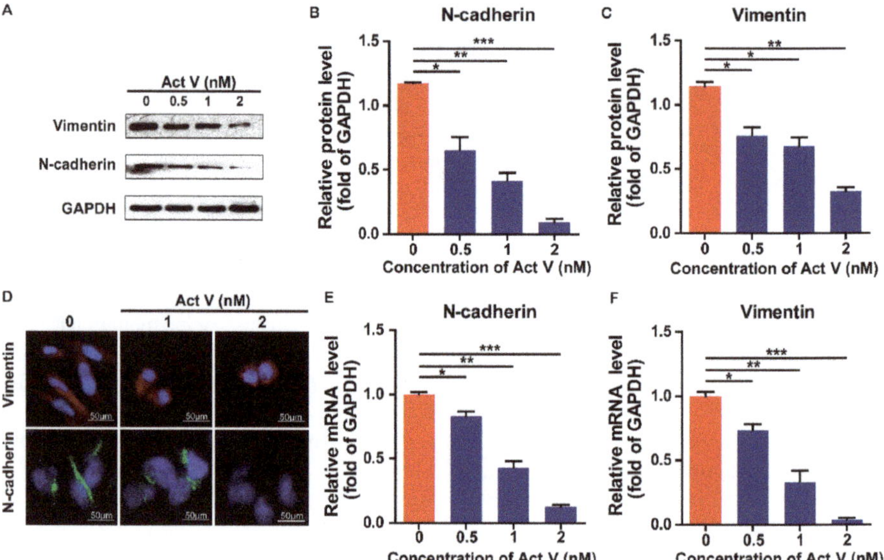

Figure 7. Effects of actinomycin V on the N-cadherin and vimentin expression. (**A–C**) MDA-MB-231 cells were treated with 0–2 nmol/L actinomycin V for 24 h then the protein expression of vimentin and N-cadherin was measured by Western blot. (**D**) cells were treated with 0–2 nmol/L actinomycin V for 6 h and analyzed by vimentin and N-cadherin fluorescent signals. Cells were stained with anti-vimentin (red), N-cadherin (green), and DAPI (blue). Magnification: ×200. Vimentin and N-cadherin levels were decreased in the cells, consistent with the Western blot results. Results were obtained from three independent experiments. * $p < 0.05$, ** $p < 0.01$, *** $p < 0.001$. (**E**,**F**) Relative expression of vimentin and N-cadherin mRNA in the MDA-MB-231 cells were analyzed by real-time PCR. RNA levels are represented as fold increase relative to the level of the control (normalized to GAPDH mRNA level).

Real-time PCR (rt-PCR) showed that the mRNA level of E-cadherin was upregulated while the N-cadherin and Vimentin were downregulated significantly in response to actinomycin V treatment. These results confirmed that actinomycin V inhibited the epithelial-mesenchymal transition process in MDA-MB-231 cells.

2.5. Actinomycin V Suppresses the EMT Process by Reducing the Expression of SNAIL and SLUG

EMT is a complex process which required a complicated network of transcription factors such as zinc finger and basic helix loop helix to participate in. Snail family transcriptional repressor 1 (also known as SNAIL) and Snail family transcriptional repressor 2 (SLUG) are proved to be the E-cadherin repressors and act as inducers of the invasion process when they are overexpressed in epithelial cell lines [18,19]. As shown in Figure 8, we found that actinomycin V treatment down-regulated the expression levels of snail and slug protein while the expression levels of zinc finger E-box binding homeobox 1 (ZEB1) and twist remained unchanged in MDA-MB-231 cells. Immunofluorescence staining also showed the same reduction effects after 1–2 nmol/L actinomycin V treatment. Therefore, we suggest the actinomycin V inhibitory effect on EMT process may through down-regulate the expression levels of the snail and slug protein.

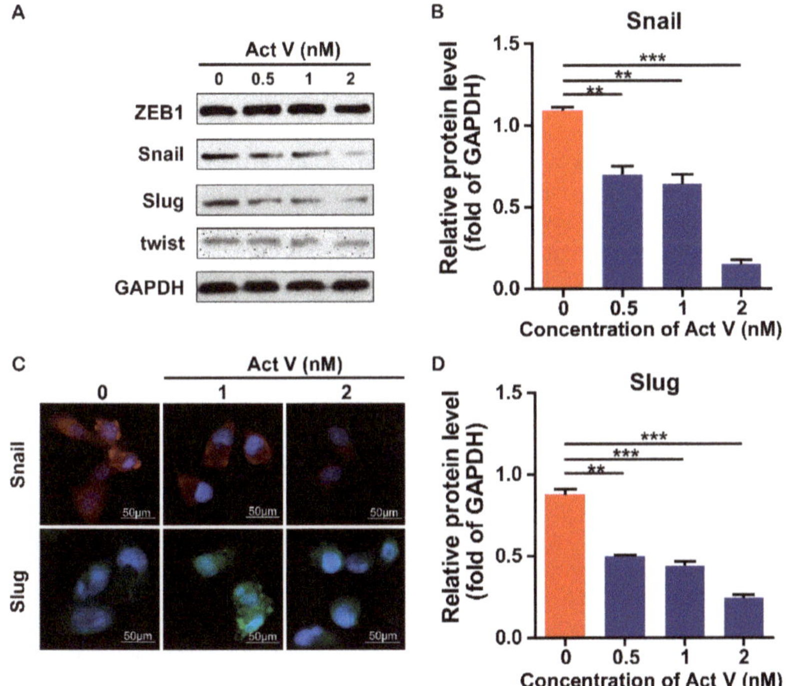

Figure 8. Effects of actinomycin V on the expression levels of snail and slug. (**A,B,D**) MDA-MB-231 cells were treated with 0 to 2 nmol/L actinomycin V for 24 h then the protein expression of snail and slug was measured by Western blot. Results were obtained from three independent experiments. * $p < 0.05$, ** $p < 0.01$, *** $p < 0.001$. (**C**) cells were treated with 0 to 2 nmol/L actinomycin V for 6 h and analyzed by snail and slug fluorescent signals. Cells were stained with anti-snail (red), slug (green) and DAPI (blue). Magnification ×200. Snail and slug levels were decreased in the cells, consistent with the Western blot results.

3. Discussion

The majority reason of the high incidence and death rates of breast cancer among women is metastasis, which is initiated by the processes of adhesion, invasion and migration [20]. First, adhesion, it plays a vital role in maintaining cellular structure and the disruption of cellular adhesion may led to cell invasion and migration [6,7]. Cellular adhesion is mediated by a variety of membrane molecules, for example, cadherins, the Ca^{2+}-dependent transmembrane proteins, is one of the major members of adhesion related membrane molecules [21]. More specially, reports have been confirmed that E-cadherin was required to initiate cell-cell adhesion and the procession of epithelial-mesenchymal transition (EMT) is closely related to its deletion [22,23]. In this paper, we found that the actinomycin V was effective in suppressing migration and invasion of breast cancer cells as revealed by morphological changes, wound-healing and Transwell assays. Moreover, actinomycin V treatment upregulated the protein and mRNA levels of E-cadherin, indicating that actinomycin V enhanced cell-cell adhesion in MDA-MB-231 cells. Next, we measured the N-cadherin and Vimentin, the other two important maker molecules of EMT, by Western blotting and immunofluorescent staining. E-cadherin serves as an invasion suppressor, N-cadherin, which is frequently upregulated in cancer cells, functions as an invasion promoter. The expression of N-cadherin induces epithelial cells transform the morphological into a fibroblastic phenotype and the biological function also changes to make epithelial cells more motile and invasive in the EMT process [24]. Vimentin is one of the major members of intermediate filament family and frequently overexpresses in a variety of cancer cells such as prostate cancer, breast cancer and lung cancer. The overexpression of vimentin in cancer cells promotes cellular proliferation, invasion and migration [25]. Our results showed that actinomycin V treatment significantly down-regulated both of the protein and mRNA expression levels of N-cadherin and vimentin, which may inhibit EMT in cells by reducing cells invasion and migration. These findings suggest that actinomycin V inhibits MDA-MB-231 cells migration and invasion via suppressing the EMT process.

A complicated network of transcription factors (TFs) are involved in EMT process. Factors from Snail, ZEB and helix-loop-helix families are acted as the inducers of epithelial-mesenchymal transition (EMT) and potent repressors of E-cadherin expression [26,27]. The activation of these factors initiates the EMT process. Then we examined the effects of actinomycin V on these family members including ZEB1, twist, slug, and snail. We found that the actinomycin V treatment down-regulated the expression levels of snail and slug protein while the expression levels of ZEB1 and twist remained unchanged. Immunofluorescence staining also showed us that the expression of snail and slug were reduced in MDA-MB-231 cells after actinomycin V treatment. It has been reported that snail and slug active during EMT and down-regulate the expression of adhesion genes like E-cadherin. Snail and slug also involved in the expression of the invasion promoter such as vimentin, N-cadherin, and fibronectin [27–30]. Therefore, the reduction of snail and slug may lead to the up-regulation of E-cadherin expression while the down-regulation of vimentin and N-cadherin and suppress the EMT process, which is consistent with the results showed in our studies. As a result, we conclude that the actinomycin V suppresses EMT process via down-regulating the snail family members snail and slug and subsequently inhibits the proliferation, migration, and invasion of breast cancer cells. On the other hand, it is noticed that the selectivity index of actinomycin V between cancer (such as MDA-MB-231) and noncancer cell lines (HMLE/MCF-10A) still remains low (around 4–5) as shown in Table 1. Toxicity investigation should be especially warranted in vitro/vivo for actinomycin V drug development.

4. Materials and Methods

4.1. Chemicals and Reagents

Actinomycin V (>98%), actinomycin X_{ob} (>98%) and actinomycin D (>98%) were provided by Dr. Xie (Shandong University, Weihai, China) isolated from *Streptomyces* sp. N1510.2 and identified by ESI-MS, ^{1}H and ^{13}C NMR data [31]. Compounds were dissolved in dimethylsulfoxide (DMSO) as 10 μmol/L stock solution and diluted according to experimental requirement. MTT

(3-(4,5-Dimethylthiazol)-2,5-diphenyltetrazolium bromide) was purchased from Sigma-Aldrich Corp. (St. Louis, MO, USA). Antibodies against E-cadherin, N-cadherin, Vimentin, snail and ZEB1 were purchased from Cell Signaling Technology (CST, Inc., Beverly, MA, USA). Slug, twist, GAPDH antibodies and goat/donkey anti- mouse/rabbit IgG-Alexa Fluor 488/647 secondary antibodies were purchased from Abcam, Inc. (Cambridge, MA, USA). The primers used in this study were bought from Sangon Biotech Co Ltd. (Shanghai, China). All chemicals used in this study were commercial reagent grade products.

4.2. Cell Lines and Cell Culture

Human breast cancer cell lines MCF-7, MDA-MB-231, BT474 and normal human breast epithelial cell lines HMLE and MCF-10A were purchased from the Shanghai Institute for Biological Sciences (SIBS), Chinese Academy of Sciences (Shanghai, China). MCF-7, MDA-MB-231 and BT474 were cultured in RPMI 1640 Medium (Hyclone Laboratories, Inc., Logan, UT, USA) and HMLE cells were cultured in DMEM/F-12 1:1 (Hyclone Laboratories, Inc.) containing 10% fetal bovine serum supplemented with 100 units/mL of penicillin and 100 µg/mL of streptomycin. MCF-10A cells were cultured in DMEM/F-12 1:1 containing insulin, hydrocortisone, EGF and 10% horse serum supplemented with 100 units/mL of penicillin and 100 µg/mL of streptomycin. All cell lines were cultured in a humidified incubator at 37 °C with 5% CO_2.

4.3. MTT Assay

Cell proliferation inhibitory effect of actinomycin V was evaluated by MTT assay. Cells were seed in 96-well plates at a density of 5×10^3 cells per well. After 24 h incubating, cells were treated with actinomycin V, actinomycin X_{ob} and actinomycin D at various concentrations while adriamycin was used as the comparison. After 24–72 h of continuous culture, 15 µL MTT (5 g/L) was added to each well and incubated for another 4 h at 37 °C. The medium with MTT was removed and 150 µL per well of DMSO was added to dissolve the formazan. The absorbance of each well was measured by a microplate reader at 570 nm. IC_{50} values (concentration resulting in 50% inhibition of cell growth) for each cell line were calculated by plotting the untreated cells, which were considered to be 100% cell survival. Each text was performed in triplicate independently.

4.4. Wound-Healing Assay

Cells in logarithmic growth phase were seeded into 6-well plates at a density of 3×10^4 cells per well. When the cells had adhered to 90%, they were scratched by micropipette tips vertically in the middle of each well. Each plate was washed with 3 mL PBS three times to remove the suspension cells. The cells in plate were starved for 12 h to eliminate the interference of proliferation in advance and then treated with 0 to 2 nmol/L actinomycin V for 24 h. The scratch width changes of each group were observed under a microscope, and the wound healing rate (%) was calculated as (1 − (scratch width of the actinomycin V group/scratch width of the control group)) × 100% to evaluate the migration ability of cells. Each experiment was performed in triplicate independently.

4.5. Transwell Assay

We used transwell assays to measure the invasive capacity of cells. We took 50 µL Matrigel diluted with 1:8 of serum-free culture medium and added onto the upper chamber membrane of the insert. After solidifying into gel, cells were suspended in 200 µL serum-free culture medium at a destiny of 1×10^5 cells/mL and then added into the upper chamber of the insert, while the lower chamber were filled with 600 µL culture medium containing 20% FBS. Cells were treated with varying concentration of actinomycin V ranging from 0 to 2 nmol/L and incubated with 5% CO_2 at 37 °C for 24 h. The medium in chamber was removed and the cells in the upper chamber were wiped by cotton swabs. Cells in lower chamber were fixed in formaldehyde for 20 min, then stained with 0.5% crystal violet for 15 min. After three times washing, five fields were taken randomly under a Nikon TE2000 microscope

(Nikon Instruments Inc., Tokyo, Japan) and the number of transmembrane cells were counted. Each experiment was performed in triplicate independently.

4.6. Immunofluorescence Staining

Cells were seeded onto 12-mm round, glass over slips in 24-well plates and treated with 0 to 2 nmol/L actinomycin V for 6 h. Cells were washed and fixed with 4% Paraformaldehyde for 15 min, washed three times with cold PBS, and permeabilized in 0.1% Triton X-100 for 20 min at room temperature. To prevent non-specific antibody binding, the cells were washed and incubated with 3% goat serum for 30 min. Then cells were incubated with the E-cadherin, N-cadherin, vimentin, snail and slug antibody at 4 °C overnight, washed and incubated with donkey anti-rabbit IgG-Alexa Fluor 647 antibody or goat anti-mouse IgG-Alexa Fluor 488 antibody for 1 h. After washing three times, cells were counterstained with 4 µg/mL 4′,6-diamidino-2-phenylindole (DAPI) for 10 min at room temperature. After washing three times with PBS-TX, cover slips containing the cells were then mounted using mounting medium (PBS:glycerol = 1:1 (v/v)). Fluorescence images were captured by using a fluorescence microscope.

4.7. Western Blot Analysis

Cells were treated with indicated concentrations of actinomycin V for 24 h, then the protein from whole cell lysates was analyzed by Western blot assay. Samples were denatured by boiling with 1 × Laemmli buffer, subjected to 10% SDS-polyacrylamide gel electrophoresis and transferred to nitrocellulose membranes. The membranes were washed with distilled water and then blocked with 5% non-fat milk in TBS-T buffer (10 mmol/L Tris-HCl, 150 mmol/L NaCl and 0.05% Tween-20 (v/v), pH 7.8) for at least 1 h at room temperature. After a short wash in TBS-T buffer, the membranes were incubated in a solution containing monoclonal antibodies specific for GAPDH, Vimentin, E-cadherin, N-cadherin, snail, slug, twist and ZEB1 for 2 h at room temperature or overnight at 4 °C. Then, the membranes were washed with TBS-T buffer for 6 min three times and once in TBS buffer. Membranes were incubated in secondary HRP-conjugated goat anti-mouse IgG or donkey anti-rabbit IgG for 1 h at room temperature. Membranes were washed with TBS-T buffer three times and once in TBS buffer. Proteins on the membranes were visualized using the enhanced chemiluminescence detection system (ECL®, Amersham Biosciences). Proteins expression density values were quantified by gray analysis using the Image J 2.0 software (National Institutes of Health, Bethesda, MD, USA).

4.8. RNA Extraction and Relative Quantification by Real Time PCR

Cells were seeded in 6-well plates and treated with 0 to 2 nmol/L actinomycin V for 24 h. Total RNA was extracted using RNeasy kit under the manufacturer's instructions and the purity of RNA was detected by OD260/280 of RNA sample (>1.8). Then we converted the RNA into cDNA using the ReverTra Aoe® qPCR RT Kit (QIAGEN, Hilden, Germany). Primers were designed by primer premier 5 for the E-cadherin gene (forward primer: 5′-CGAGAGCTACACGTTCACGG-3′ and reverse primer: 5′-GGGTGTCGAGGGAAAAATAGG-3′), N-cadherin gene (forward primer: 5′-CCTTTCAAACACAGCCACGG-3′ and reverse primer: 5′-TGTTTGGGTCGGTCTGGATG-3′), and vimentin gene (forward primer: 5′-GACGCCATCAACACCGAGTT-3′ and reverse primer: 5′-CTTTGTCGTTGGTTAGCTGGT-3′). GAPDH gene (forward primer: 5′-CATCAAGAAGGTGGTGAAGCAGG-3′ and reverse primer: 5′-TCAAAGGTGGAGGAGTGGGTGTCGC-3′), was used as a control for the amount of RNA. Real-time PCR assay was used to detect the expression levels of the Vimentin, N-cadherin and E-cadherin genes and the amplification was performed in triplicate. For each reaction, we prepared 1×SYBR Green Realtime PCR Master Mix (TOYOBO, Osaka, Japan), 1 µL forward primer and reverse primer and 1 µg of template cDNA in a final volume of 20 µL. The condition of the amplification were 45 cycles of sequential denaturation (95 °C, 2 min), annealing (60 °C, 15 s), and extension (72 °C,

20 s). The ΔΔCT value was used to analyze the rt-qPCR data. All samples were measured in triplicate experiments independently.

4.9. Statistical Analysis

Data are presented as Mean ± Standard deviation from triplicate experiments. All data were analyzed using One-way Analysis of Variance (ANOVA) for multiple comparisons by SPSS 16.0 (SPSS Inc., Chicago, IL, USA). Significant differences are indicated as follows: * $p < 0.05$; ** $p < 0.01$; *** $p < 0.001$.

Author Contributions: Conceptualization, X.L.; Formal analysis, S.L.; Funding acquisition, X.L.; Investigation, S.L., C.Z., F.L., J.M., F.J., and Z.H.; Resources, W.X. and X.L.; Writing—original draft, S.L.; Writing—review & editing, S.L. and X.L.

Funding: This work was supported in part by grants from Shandong Provincial Natural Science Foundation (No. ZR2019MH001), the Program for Changjiang Scholars and Innovative Research Team in University (PCSIRT, No IRT_17R68), the Fundamental Research Funds for the Central Universities (No.2019ZRJC004), and National Natural Science Foundation of China (No. 81273532).

Acknowledgments: The authors thank Weidong Xie (Shandong university, weihai) for providing actinomycin V, actinomycin X_{ob} and actinomycin D.

Conflicts of Interest: The authors declare no conflict of interest.

References

1. Bray, F.; Ferlay, J.; Soerjomataram, I.; Siegel, R.L.; Torre, L.A.; Jemal, A. Global cancer statistics 2018: GLOBOCAN estimates of incidence and mortality worldwide for 36 cancers in 185 countries. *CA Cancer J. Clin.* **2018**, *68*, 394–424. [CrossRef] [PubMed]
2. Miller, K.D.; Goding Sauer, A.; Ortiz, A.P.; Fedewa, S.A.; Pinheiro, P.S.; Tortolero-Luna, G.; Martinez-Tyson, D.; Jemal, A.; Siegel, R.L. Cancer Statistics for Hispanics/Latinos, 2018. *CA Cancer J. Clin.* **2018**, *68*, 425–445. [CrossRef] [PubMed]
3. Mani, S.A.; Guo, W.; Liao, M.-J.; Eaton, E.N.; Ayyanan, A.; Zhou, A.Y.; Brooks, M.; Reinhard, F.; Zhang, C.C.; Shipitsin, M.; et al. The epithelial-mesenchymal transition generates cells with properties of stem cells. *Cell* **2008**, *133*, 704–715. [CrossRef] [PubMed]
4. Park, S.Y.; Gonen, M.; Kim, H.J.; Michor, F.; Polyak, K. Cellular and genetic diversity in the progression of in situ human breast carcinomas to an invasive phenotype. *J. Clin. Investig.* **2010**, *120*, 636–644. [CrossRef] [PubMed]
5. Fehm, T.; Muller, V.; Alix-Panabieres, C.; Pantel, K. Micrometastatic spread in breast cancer: Detection, molecular characterization and clinical relevance. *Breast Cancer Res.* **2008**, *10*, S1. [CrossRef] [PubMed]
6. Luo, M.; Guan, J.L. Focal adhesion kinase: A prominent determinant in breast cancer initiation, progression and metastasis. *Cancer Lett.* **2010**, *289*, 127–139. [CrossRef]
7. Xu, Y.J.; Bismar, T.A.; Su, J.; Xu, B.; Kristiansen, G.; Varga, Z.; Teng, L.H.; Ingber, D.E.; Mammoto, A.; Kumar, R.; Alaoui-Jamali, M.A. Filamin A regulates focal adhesion disassembly and suppresses breast cancer cell migration and invasion. *J. Exp. Med.* **2010**, *207*, 2421–2437. [CrossRef] [PubMed]
8. Zhang, X.; Ye, X.; Chai, W.; Lian, X.-Y.; Zhang, Z. New Metabolites and Bioactive Actinomycins from Marine-Derived *Streptomyces* sp. ZZ338. *Mar. Drugs* **2016**, *14*, 181. [CrossRef] [PubMed]
9. Piel, J. Metabolites from symbiotic bacteria. *Nat. Prod. Rep.* **2004**, *21*, 519–538. [CrossRef] [PubMed]
10. Taylor, M.W.; Radax, R.; Steger, D.; Wagner, M. Sponge-associated microorganisms: Evolution, ecology, and biotechnological potential. *Microbiol. Mol. Biol. Rev.* **2007**, *71*, 295–347. [CrossRef]
11. Brockmann, H. actinomycines. *Angew. Chem. Int. Edit.* **1960**, *72*, 447–939.
12. Muller, W.; Crothers, D.M. Studies of binding of actinomycin and related compounds to DNA. *J. Mol. Biol.* **1968**, *35*, 251–290. [CrossRef]
13. Cai, W.L.; Wang, X.C.; Elshahawi, S.I.; Ponomareva, L.V.; Liu, X.D.; McErlean, M.R.; Cui, Z.; Arlinghaus, A.L.; Thorson, J.S.; Van Lanen, S.G. Antibacterial and Cytotoxic Actinomycins Y-6–Y-9 and Zp from *Streptomyces* sp. Strain Go–GS12. *J. Nat. Prod.* **2016**, *79*, 2731–2739. [CrossRef] [PubMed]

14. Farber, S.; Dangio, G.; Evans, A.; Mitus, A. Clinical significance.3. clinical studies of actinomycin–D with special reference to wilms tumor in children. *Ann. N. Y. Acad. Sci.* **1960**, *89*, 421–425. [CrossRef]

15. Sliva, D.; Rizzo, M.T.; English, D. Phosphatidylinositol 3-kinase and NF-kappa B regulate motility of invasive MDA-AM-231 human breast cancer cells by the secretion of urokinase-type plasminogen activator. *J. Biol. Chem.* **2002**, *277*, 3150–3157. [CrossRef] [PubMed]

16. Morioka, H.; Takezawa, M.; Shibai, H. Actinomycin-V as a potent differentiation inducer of F5-5 friend-leukemia cells. *Agric. Biol. Chem.* **1985**, *49*, 2835–2842.

17. Wang, D.; Wang, C.; Gui, P.; Liu, H.; Khalaf, S.M.H.; Elsayed, E.A.; Wadaan, M.A.M.; Hozzein, W.N.; Zhu, W. Identification, Bioactivity, and Productivity of Actinomycins from the Marine-Derived *Streptomyces heliomycini*. *Front. Microbiol.* **2017**, *8*, 1147–1158. [CrossRef]

18. Blanco, M.J.; Moreno-Bueno, G.; Sarrio, D.; Locascio, A.; Cano, A.; Palacios, J.; Nieto, M.A. Correlation of Snail expression with histological grade and lymph node status in breast carcinomas. *Oncogene* **2002**, *21*, 3241–3246. [CrossRef] [PubMed]

19. Wong, A.S.T.; Gumbiner, B.M. Adhesion-independent mechanism for suppression of tumor cell invasion by E-cadherin. *J. Cell Biol.* **2003**, *161*, 1191–1203. [CrossRef]

20. Geho, D.H.; Bandle, R.W.; Clair, T.; Liotta, L.A. Physiological mechanisms of tumor-cell invasion and migration. *Physiology* **2005**, *20*, 194–200. [CrossRef] [PubMed]

21. Ito, H.; Kobayashi, E.; Takamatsu, Y.; Li, S.H.; Hatano, T.; Sakagami, H.; Kusama, K.; Satoh, K.; Sugita, D.; Shimura, S.; Itoh, Y.; Yoshida, T. Polyphenols from *Eriobotrya japonica* and their cytotoxicity against human oral tumor cell lines. *Chem. Pharm. Bull.* **2000**, *48*, 687–693. [CrossRef] [PubMed]

22. Pinho, S.S.; Oliveira, P.; Cabral, J.; Carvalho, S.; Huntsman, D.; Garter, F.; Seruca, R.; Reis, C.A.; Oliveira, C. Loss and Recovery of Mgat3 and GnT-III Mediated E-cadherin N-glycosylation Is a Mechanism Involved in Epithelial-Mesenchymal-Epithelial Transitions. *PLoS ONE* **2012**, *7*, e33191. [CrossRef] [PubMed]

23. Wang, S.; Yan, Y.; Cheng, Z.; Hu, Y.; Liu, T. Sotetsuflavone suppresses invasion and metastasis in non–small-cell lung cancer A549 cells by reversing EMT via the TNF-alpha/NF-kappaB and PI3K/AKT signaling pathway. *Cell Death Discov.* **2018**, *4*, 26–36. [CrossRef] [PubMed]

24. Derycke, L.D.M.; Bracke, M.E. N-cadherin in the spotlight of cell-cell adhesion, differentiation, embryogenesis, invasion and signalling. *Int. J. Dev. Biol.* **2004**, *48*, 463–476. [CrossRef]

25. Satelli, A.; Li, S.L. Vimentin in cancer and its potential as a molecular target for cancer therapy. *Cell. Mol. Life Sci.* **2011**, *68*, 3033–3046. [CrossRef] [PubMed]

26. Peinado, H.; Olmeda, D.; Cano, A. Snail, ZEB and bHLH factors in tumour progression: An alliance against the epithelial phenotype? *Nat. Rev. Cancer* **2007**, *7*, 415–428. [CrossRef]

27. Alidadiani, N.; Ghaderi, S.; Dilaver, N.; Bakhshamin, S.; Bayat, M. Epithelial mesenchymal transition Transcription Factor (TF): The structure, function and microRNA feedback loop. *Gene* **2018**, *674*, 115–120. [CrossRef]

28. Wei, Z.; Shan, Z.; Shaikh, Z.A. Epithelial-mesenchymal transition in breast epithelial cells treated with cadmium and the role of Snail. *Toxicol. Appl. Pharm.* **2018**, *344*, 46–55. [CrossRef] [PubMed]

29. Cano, A.; Perez-Moreno, M.A.; Rodrigo, I.; Locascio, A.; Blanco, M.J.; del Barrio, M.G.; Portillo, F.; Nieto, M.A. The transcription factor Snail controls epithelial–mesenchymal transitions by repressing E-cadherin expression. *Nat. Cell Biol.* **2000**, *2*, 76–83. [CrossRef]

30. Batlle, E.; Sancho, E.; Franci, C.; Dominguez, D.; Monfar, M.; Baulida, J.; de Herreros, A.G. The transcription factor Snail is a repressor of E-cadherin gene expression in epithelial tumour cells. *Nat. Cell Biol.* **2000**, *2*, 84–89. [CrossRef]

31. Lu, D.D.; Ren, J.W.; Du, Q.Q.; Song, Y.J.; Lin, S.Q.; Li, X.; Xie, W.D. p-Terhenyls and actinomycins from a *Streptomyces* sp. associated with the Iarva of mud dauber wasp. *Nat. Prod. Res.* **2019**, in press.

Article

Divergolides T–W with Apoptosis-Inducing Activity from the Mangrove-Derived Actinomycete *Streptomyces* sp. KFD18

Li-Man Zhou [1,2,†], Fan-Dong Kong [2,†], Qing-Yi Xie [2], Qing-Yun Ma [2], Zhong Hu [3], You-Xing Zhao [2,*] and Du-Qiang Luo [1,*]

[1] College of Life Science, Key Laboratory of Medicinal Chemistry and Molecular Diagnosis of Ministry of Education, Hebei University, Baoding 071002, China; zhouliman88@126.com

[2] Hainan Key Laboratory for Research and Development of Natural Product from Li Folk Medicine, Institute of Tropical Bioscience and Biotechnology, Chinese Academy of Tropical Agricultural Sciences, Haikou 571101, China; kongfandong@itbb.org.cn (F.-D.K.); xieqingyi@itbb.org.cn (Q.-Y.X.); maqingyun@itbb.org.cn (Q.-Y.M.)

[3] Guangdong Provincial Key Laboratory of Marine Biotechnology, Department of Biology, Shantou University, Shantou 515063, China; hzh@stu.edu.cn

* Correspondence: zhaoyouxing@itbb.org.cn (Y.-X.Z.); duqiangluo@163.com (D.-Q.L.); Tel.: +86-139-5169-2350 (Y.-X.Z.)

† These authors contributed equally to this paper.

Received: 19 March 2019; Accepted: 8 April 2019; Published: 11 April 2019

Abstract: Four new ansamycins, named divergolides T–W (**1–4**), along with two known analogs were isolated from the fermentation broth of the mangrove-derived actinomycete *Streptomyces* sp. KFD18. The structures of the compounds, including the absolute configurations of their stereogenic carbons, were determined by spectroscopic data and single-crystal X-ray diffraction analysis. Compounds **1–4** showed cytotoxic activity against the human gastric cancer cell line SGC-7901, the human leukemic cell line K562, the HeLa cell line, and the human lung carcinoma cell line A549, with **1** being the most active while compounds **5** and **6** were inactive against all the tested cell lines. Compounds **1** and **3** showed very potent and specific cytotoxic activities (IC$_{50}$ 2.8 and 4.7 μM, respectively) against the SGC-7901 cells. Further, the apoptosis-inducing effect of **1** and **3** against SGC-7901 cells was demonstrated by two kinds of staining methods for the first time.

Keywords: mangrove-derived actinomycete; ansamycins; divergolides; apoptosis-inducing activity

---

1. Introduction

Ansamycins are a class of bioactive macrolides that have been isolated from actinomycetes [1–4]. The most representatives of them are geldanamycin with HSP90 inhibitory activity [1], rifamycin with antibacterial activity [2], and maytansinoid with anticancer activity [3]. Divergolides represent a family of ansamycins with a 19-membered naphthalenic ansamacrolactam skeleton, which was first discovered from *Streptomyces* sp. HKI0576 and reported in 2011. Until now, a total of 19 members (divergolides A–S) of this family has been reported [5–7]. Many divergolides have shown cytotoxic and antibacterial activities [5–8].

As part our ongoing search for new bioactive secondary metabolites from marine microorganisms [9–12], *Streptomyces* sp. KFD18 attracted our attention for its ability to produce a series of metabolites with UV absorption bands around 275 and 305 nm, detected by HPLC analysis. Subsequent chemical investigations on the EtOAc extract from the fermentation broth of this strain led to the isolation and identification of four new ansamycins, named divergolides T–W (**1–4**), as well as two known analogues 6,7-*epi*-24,25-dihydro-divergolide U (**5**) [8] and divergolide E (**6**) [7] (Figure 1). Herein, the structures and bioactivities of these compounds are reported.

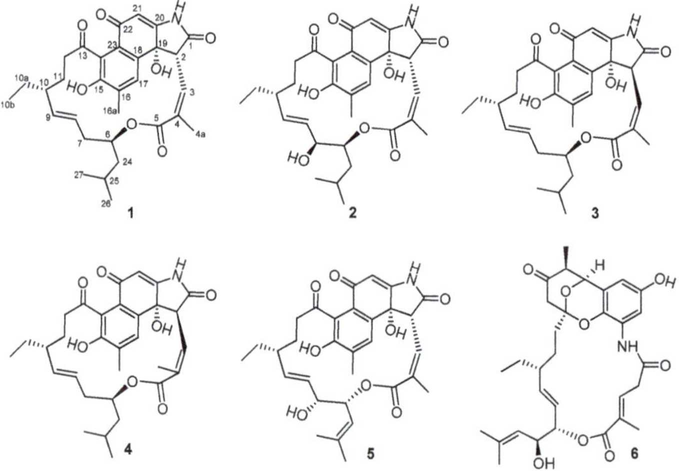

Figure 1. Structures of compounds **1–6**.

2. Results and Discussion

Compound **1** was obtained as a yellow crystal, and was found to have the molecular formula $C_{31}H_{37}NO_7$ from the HRESIMS *m/z* 536.2641 [M + H]⁺. The UV spectrum showed characteristic absorption bands around 221 and 240 nm. The IR absorptions at 3414 and 1663 cm⁻¹ revealed the presence of a hydroxy and carbonyl group, respectively. The ¹H and ¹³C NMR spectra (Supplementary materials, Figures S2-1 and S2-2) along with the HSQC spectra (Supplementary materials, Figure S2-4) revealed the presence of five methyls, five sp³ methylenes, nine methines (including five sp² and one oxygenated sp³), twelve non-protonated carbons (including two ketone carbonyls, two ester or amide carbonyls, seven aromatic or olefinic carbons, and one hydroxylated carbon). Comparison of the above data with those of the known analogue **5** [8] suggested that their planar structures were quite similar, except that the hydroxy at C-7 was absent, and the Δ^{24} double bond of **5** was hydrogenated in **1**. In the ¹H-¹H COSY spectrum (Figure 2) of **1**, correlations of H-26/H-25/H-27 and H-25/H-24/H-6/H-7 were observed, which further confirmed the above deduction. The remaining substructure of **1** was found to be identical to that of **5** by analysis of the 2D NMR data.

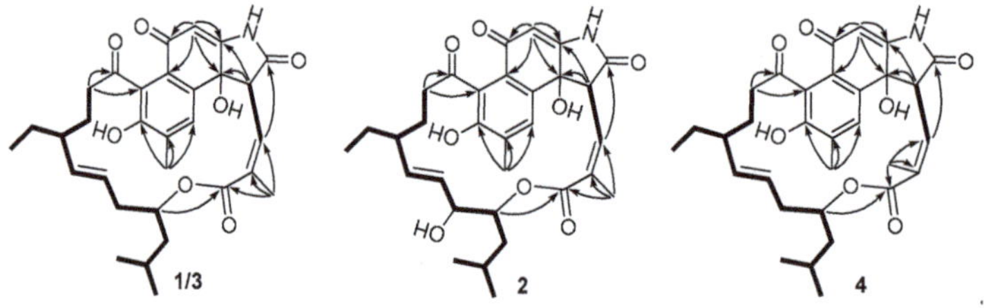

Figure 2. Key COSY (—) and HMBC (→) correlations of **1–4**.

The large *J* value (15.6 Hz) of H-8/H-9 (Table 1) suggested the *E* configuration of the Δ^8 double bond, while the relative downfield shift ($\delta_{C/H}$ 21.4/2.17) of the allylic methyl C-4a [13] and ROESY cross-peak (Figure 3) between H-4a and H-3 (δ_H 6.60) suggested the *Z* configuration of the Δ^3 double bond. Additionally, in the ROESY spectrum (Figure 3), correlations of H-10/H-8/H-24/H-2 and

H-9/H-10a led to the assignment of the full relative configuration of compound **1**, as shown in Figure 3. To support the above assignment and determine the absolute configuration of **1**, a single-crystal X-ray diffraction pattern was obtained using the anomalous scattering of Cu Kα radiation (Figure 4), allowing an explicit assignment of the absolute structure as 2*R*, 6*S*, 10*R*, and 19*R* based on the Flack parameter of −0.05(8).

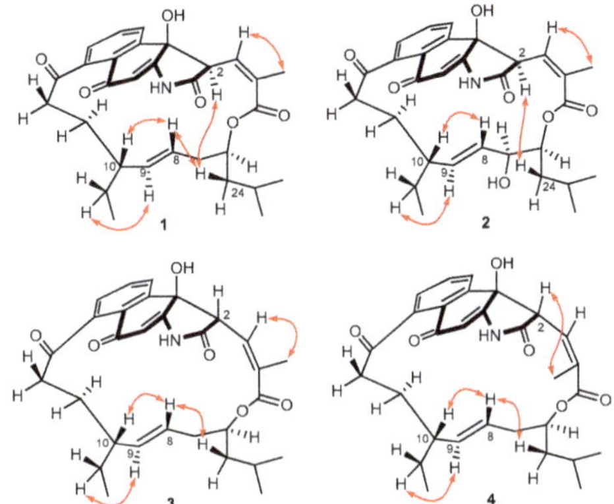

Figure 3. Key ROESY correlations of **1–4**.

Table 1. ^{13}C NMR data for **1–4** in CD$_3$OD.

Position	1 δ_C	2 δ_C	3 δ_C	4 δ_C
1	177.2, C	177.1, C	177.2, C	176.8, C
2	55.2, CH	55.3, CH	55.8, CH	55.3, CH
3	131.7, CH	132.9, CH	126.7, CH	132.5 CH
4	136.8, C	136.1, C	138.4, C	135.0, C
4a	22.0, CH$_3$	22.1, CH$_3$	21.4, CH$_3$	13.5, CH$_3$
5	168.0, C	167.7, C	169.5, C	167.9, C
6	74.6, CH	76.7, CH	74.5, CH	74.0, CH
7	36.1, CH$_2$	70.5, CH	36.6, CH$_2$	36.3, CH$_2$
8	125.1, CH	128.2, CH	126.3, CH	125.3, CH
9	139.6, CH	136.1, CH	138.8, CH	138.6, CH
10	46.0, CH	45.9, CH	44.2, CH	43.2, CH
10a	26.9, CH$_2$	27.0, CH$_2$	29.3, CH$_2$	25.6, CH$_2$
10b	13.2, CH$_3$	13.2, CH$_3$	12.7, CH$_3$	11.1, CH$_3$
11	31.7, CH$_2$	31.5, CH$_2$	34.8, CH$_2$	31.6, CH$_2$
12	40.5, CH$_2$	40.5, CH$_2$	42.5, CH$_2$	42.0, CH$_2$
13	212.4, C	212.4, C	212.1, C	212.1, C
14	130.1, C	130.2, C	127.8, C	130.6, C
15	153.5, C	153.5, C	153.4, C	152.9, C
16	132.6, C	132.9, C	133.1, C	133.4, C
16a	17.0, CH$_3$	17.0, CH$_3$	17.0, CH$_3$	16.9, CH$_3$
17	130.8, CH	130.8, CH	132.8, CH	131.9, CH
18	134.3, C	136.1, C	134.3, C	135.0, C
19	73.7, C	73.8, C	73.6, C	75.2, C
20	164.8, C	164.7, C	164.6, C	164.6, C
21	103.9, CH	104.0, CH	103.4, CH	104.3, CH
22	185.4, C	185.4, C	185.7, C	185.7, C
23	129.8, C	129.9, C	130.4, C	130.6, C
24	42.6, CH$_2$	38.4, CH$_2$	41.5, CH$_2$	41.9, CH$_2$
25	25.5, CH	25.7, CH	25.3, CH	25.6, CH
26	22.7, CH$_3$	22.2, CH$_3$	22.1, CH$_3$	22.4, CH$_3$
27	23.0, CH$_3$	23.8, CH$_3$	23.7, CH$_3$	23.6, CH$_3$

Compound **2** was determined to have a molecular formula of $C_{31}H_{37}NO_8$ based on HRESIMS data, with one oxygen atom more than that of **1**. The UV spectrum of **2** was nearly identical to that of **1**, suggesting that **2** was a homologue of **1**. Their NMR data (Tables 1 and 2) were also quite similar, except for the replacement of CH_2-7 signals in **1** by signals for a hydroxylated sp^3 methine ($\delta_{C/H}$ 70.5/3.90) in **2**. In the COSY spectrum (Supplementary materials, Figure S3-6), correlations of this hydroxylated sp^3 methine with H-8 (δ_H 4.06) and H-6 (δ_H 4.99) were observed, further confirming that CH_2-7 in **1** was oxidized to a hydroxylated methine in **2**. The similar J values (Table 1) and ROESY data (Figure 3) between **1** and **2** suggested that both compounds had the same configuration at the stereogenic centers C-2, C-6, C-10, and C-19 and double bonds Δ^3 and Δ^8. The syn orientation between H-6 and H-7 was deduced from their small vicinal coupling constant (J = 2.6 Hz) [12].

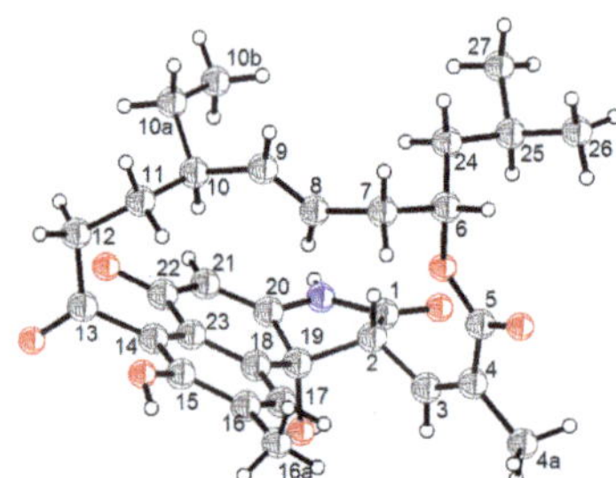

Figure 4. ORTEP diagram of **1**.

Table 2. ^1H NMR data for **1–4** in CD_3OD.

Position	1 δ_H (J in Hz)	2 δ_H (J in Hz)	3 δ_H (J in Hz)	4 δ_H (J in Hz)
2	4.74, d (10.9)	4.84, d (10.6)	4.09, d (10.9)	4.06, d (8.4)
3	6.60, dq (10.9, 1.6)	6.67, dq (10.6, 1.6)	6.36, dq (10.8, 1.6)	5.89, dq (8.4, 1.6)
4a	2.20, d (1.6)	2.21, d (1.6)	2.17, d (1.6)	2.08, d (1.0)
6	5.05, m	4.99, m	5.04, m	4.87, m
7	1.96, m	3.90, ddd (2.74, 2.6, 2.6)	2.15, m	2.25, m
	2.15, m		2.15, m	2.14, m
8	3.93, ddd (15.3, 10.2, 3.6)	4.06, dd (15.6, 2.8)	3.78, ddd (15.6, 6.0, 6.0)	4.77, ddd (15.6, 9.1, 4.9)
9	5.01, dd (15.3, 9.3)	5.24, dd (15.6, 9.3)	4.87, dd (15.6, 9.4)	5.24, dd (15.6, 7.7)
10	1.32, overlap	1.37, overlap	1.46, overlap	1.78, m
10a	0.89, m	0.92, m	1.02, m	1.43, overlap
	0.89, m	1.49, overlap	1.34, overlap	1.18, overlap
10b	0.66, t (7.4)	0.65, t (7.4)	0.73, t (7.4)	0.77, t (7.5)
11	1.35, m	1.37, overlap	1.68, m	1.29, overlap
	1.46, overlap	1.49, overlap	1.25, m	1.57, m
12	2.61, m	2.62, m	2.64, ddd (14.0, 11.3, 2.8)	2.46, m
	2.90, m	2.99, m	2.46, ddd (14.0, 7.4, 2.9)	2.77, m
16a	2.22, s	2.21, s	2.30, s	2.31, s
17	7.41, s	7.38, s	7.57, s	7.28, s
21	5.82, s	5.82, s	5.80, s	5.85, s
24	1.13, m	1.14, m	1.15, m	1.20, overlap
	1.32, overlap	1.31, overlap	1.32, overlap	1.30, overlap
25	1.46, overlap	1.49, overlap	1.47, overlap	1.45, overlap
26	0.81 d (6.6)	0.83 d (6.6)	0.88 d (6.5)	0.88 d (6.6)
27	0.81, d (6.6)	0.79, d (6.6)	0.82, d (6.5)	0.83, d (6.6)

Compounds **3** and **4** had the same molecular formula of $C_{31}H_{37}NO_7$ as that of **1**. The ^1H and ^{13}C NMR data (Supplementary materials, Figures S4-1, S4-2, S5-1, and S5-2) of **3** and **4** were also quite similar to those of **1**. Detailed analysis of the ^1H-^1H COSY and HMBC data (Supplementary materials, Figures S4-5, S4-6, S5-5, and S5-6) of **3** and **4** revealed the same H/H and H/C correlational relationship as those of **1**, indicating that **3** and **4** shared the same planar structure with **1**. However, unlike the ROESY data of **1** and **2**, the absence of correlations (Supplementary materials, Figures S4-7 and S5-7) between H-2 and H-24 (δ_H 1.15 and 1.20, respectively) in **3** and **4** revealed the H-2 protons had opposite orientations as compared to those of **1** and **2**. The syn orientation of H-2 and OH-19 in **3** and **4** was deduced by comparison of the NMR data with those of hygrocins D and F [13]. The above assignment was further supported by the phenomenon that H-2 signals (δ_H 6.36 and 5.89, respectively) of **3** and **4** resonated upfield [13] compared to those (δ_H 6.60 and 6.67, respectively) of **1** and **2**. Further, in the ROESY spectra (Figure 3), correlations of H-4a/H-3 of **3** while H-4a/H-2 of **4** were observed, revealing the *Z* and *E* configuration of Δ^3 double bond in **3** and **4**, respectively.

Compounds **1–6** were tested for their cytotoxic activity against the human gastric cancer cell line SGC-7901, the human leukemic cell line K562, the HeLa cell line, and the human lung carcinoma cell line A549. The results (Table 3) showed that compounds **1–4** exhibited cytotoxic activity against SGC-7901 (IC_{50} = 2.8, 9.8, 4.7, and 20.9 µM, respectively), K562 (IC_{50} = 6.6, 9.0, 7.6, and 16.3 µM, respectively), HeLa (IC_{50} = 9.6, >50, 14.1, and 29.5 µM, respectively), and A549 (IC_{50} = 14.9, 24.7, 20.9, and 33.2 µM, respectively) cell lines, with **1** being the most active while compounds **5** and **6** were inactive against all the tested cell lines. The above data showed that hydroxylation at C-7 or inversion of the configuration at C-2 or Δ^3 double bond in compound **1** could significantly reduce cytotoxic activity.

Table 3. Cytotoxic activities of compounds **1–6**.

Compound	IC_{50} (µM)			
	SGC-7901	**K562**	**Hela**	**A549**
1	2.8	6.6	9.6	14.9
2	9.8	9.0	>50	24.7
3	4.7	7.6	14.1	20.9
4	20.9	16.3	29.5	33.2
5	>50	>50	>50	>50
6	>50	>50	>50	>50
Imatinib	86.8	0.2	18.8	45.6
Adriamycin	6.9	10.7	11.4	5.5

In order to determine whether the compounds could induce apoptosis, we used two kinds of staining methods. Double staining with acridine orange-ethidium bromide (AOEB) allows for differentiation of live, apoptotic, and necrotic cells [14]; live cells have green, regular-sized nuclei. Green or yellow-green nuclear condensation or fragmentation identifies early apoptotic cells, and orange or red staining identifies late apoptotic or necrotic cells. DAPI staining can reveal the typical apoptotic feature: a condensed nucleus and apoptotic body formation [15]. After SGC-7901 cells were cultured with compounds **1** and **3** at double the IC_{50} concentration for 48 h. AOEB staining showed us that the cells were dyed yellow-green or orange. DAPI staining showed that many cells had typical apoptotic features (Figure 5). All staining results indicated that compounds **1** and **3** had apoptosis-inducing activity against SGC-7901. The apoptosis-inducing activity of divergolides has been reported for the first time.

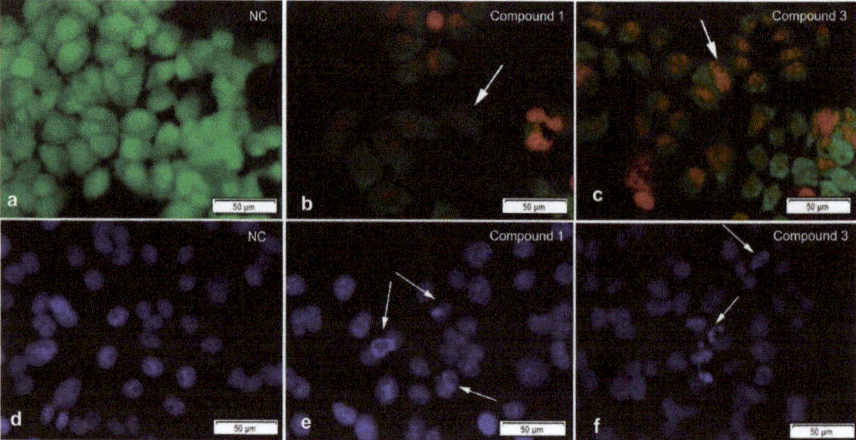

Figure 5. The staining results of compounds **1** and **3** on SGC-7901. Acridine orange-ethidium bromide (AOEB) staining on SGC-7901 cells at 48 h after compound addition (**a–c**). DAPI staining on SGC-7901 cells at 48 h after compound addition (**d–f**). The concentrations of compounds **1** and **3** were 5.6 µM and 9.4 µM, respectively. NC: Negative control, DMSO of the same volume.

3. Experimental Section

3.1. General Experimental Procedure

Optical rotations were measured with a JASCO P-1020 digital polarimeter. The IR spectra were obtained with a Nicolet Nexus 470 spectrophotometer as KBr discs. The UV spectra were obtained with a Beckman DU 640 spectrophotometer. The NMR spectra were recorded on a Bruker AV-500 spectrometer, with a CD_3OD solvent peak signal as the chemical shift reference. All compounds isolated underwent NMR analysis using about 500 µL CD_3OD solvent. HREIMS data were acquired on a Micromass Autospec-Ultima-TOF, API QSTAR Pulsar 1, or Waters Autospec Premier spectrometer. Semi-preparative HPLC separation used octadecyl silane (ODS) columns (YMC-pack ODS-A, 10 × 250 mm, 5 µm, 4 mL/min) for separation. Thin-layer chromatography (TLC) and column chromatography (CC) were carried out on precoated silica gel GF_{254} (10–40 µm, Qingdao Marine Chemical Inc., Qingdao, China) and silica gel (200–300 mesh, Qingdao Marine Chemical Inc., Qingdao, China), respectively.

3.2. Strain and Fermentation

The strain *Streptomyces* sp. KFD18 was isolated from Mangrove sediment, collected from Danzhou, Hainan province, in China, which was identified based on the 16S rRNA gene sequences (GenBank accession No. MK478900, Supporting Information) of the single colonies. A reference culture of *Streptomyces* sp. KFD18 was deposited in our laboratory and was maintained at −80 °C. *Streptomyces* sp. KFD18 was cultured in seawater medium containing 1% starch, 0.1% peptone, and 0.2% $CaCO_3$ on a rotary shaker (180 rpm) at 28 °C for 4 d to afford a seed culture. Fermentation (30 L) was performed using the same medium on a rotary shaker (180 rpm) at 28 °C for 10 d.

3.3. Extraction and Isolation

The fermented cultures were extracted with three-fold volumes of EtOAc, then the EtOAc solutions were combined and evaporated under reduced pressure to produce a dark brown, solid, crude extract (2.9 g). The extract was fractionated by a silica gel VLC column using different solvents of increasing polarity, from MeOH/H_2O (1:4) to MeOH/H_2O (1:0), to yield seven fractions (Frs. 1−7). Fr. 5 (87 mg) was subjected to semipreparative HPLC (YMC-pack ODS-A, 5 µm; 10 × 250 mm;

50% MeCN/H$_2$O; containing 0.1% TFA; 4 mL/min) to afford compounds **1** (t_R 19.4 min; 14.2 mg) and **4** (t_R 23.4 min; 4.3 mg). Fr. 6 (264 mg) was subjected to semipreparative HPLC (YMC-pack ODS-A, 5 μm; 10 × 250 mm; 70% MeCN/H$_2$O; containing 0.1% TFA; 4 mL/min) to afford compound **3** (t_R 13.1 min; 6.4 mg). Fr. 4 (124 mg) was purified by semipreparative HPLC (YMC-pack ODS-A, 5 μm; 10 × 250 mm; 40% MeCN/H$_2$O; containing 0.1% TFA; 4 mL/min) to afford compound **2** (t_R 9.6 min; 3.1 mg) and compound **5** (t_R 11.3 min; 7.7 mg). Fr. 3 (214 mg) was purified by Sephadex LH-20 chromatography and eluted with MeOH to afford compound **6** (13.8 mg)

Divergolide T (**1**): Colorless crystal; $[\alpha]_D^{25}$ −190 (*c* 0.1, MeOH); UV (MeOH) λ_{max} (log ε): 305.0 (3.70), 275.0 (3.68) nm; IR (KBr) ν_{max} (cm^{-1}): 3414, 2957, 2855, 1663, 1573, 1194, and 1144. ^{1}H and ^{13}C NMR data, Tables 1 and 2; HRESIMS *m/z* 536.2641 [M + H]$^+$ (calculated for C$_{31}$H$_{38}$O$_7$N, 536.2643).

Divergolide U (**2**): White powder; $[\alpha]_D^{25}$ +60 (*c* 0.1, MeOH); UV (MeOH) λ_{max} (log ε): 305.0 (3.72), 275.0 (3.69) nm; IR (KBr) ν_{max} (cm^{-1}): 3444, 2925, 2855, 1677, 1442, 1199, and 1141. ^{1}H and ^{13}C NMR data, Tables 1 and 2; HRESIMS *m/z* 550.2438 [M − H]$^+$ (calculated for C$_{31}$H$_{36}$O$_8$N, 550.2446).

Divergolide V (**3**): White powders; $[\alpha]_D^{25}$ +118 (*c* 0.1, MeOH); UV (MeOH) λ_{max} (log ε): 305.0 (3.75), 275.0 (3.70) nm; IR (KBr) ν_{max} (cm^{-1}): 3413, 2926, 1649, 1583, 1334, 1243, 1146, and 1058. ^{1}H and ^{13}C NMR data, Tables 1 and 2; HRESIMS *m/z* 536.2640 [M + H]$^+$ (calculated for C$_{31}$H$_{38}$O$_7$N, 536.2643).

Divergolide W (**4**): White powders; $[\alpha]_D^{25}$ +72 (*c* 0.1, MeOH); UV (MeOH) λ_{max} (log ε): 305.0 (3.68), 275.0 (3.66) nm; IR (KBr) ν_{max} (cm^{-1}): 3442, 2926, 2961, 1673, 1577, 1199, and 1138. ^{1}H and ^{13}C NMR data, Tables 1 and 2; HRESIMS *m/z* 534.2490 [M − H]$^-$ (calculated for C$_{31}$H$_{36}$O$_7$N, 534.2497).

X-ray Crystal Data for 1: Colorless crystals of **1** were obtained in the mixed solvent of MeOH. Crystal data of **1** were obtained on a Bruker D8 QUEST diffractometer (Bruker) with graphite monochromated Cu Kα radiation (λ = 1.54178 Å). Crystallographic data for **1** were deposited in the Cambridge Crystallographic Data Center as supplementary publication number CCDC 1893418. These data can be obtained free of charge from The Cambridge Crystallographic Data Centre via www.ccdc.cam.ac.uk/data_request/cif.

Crystal data for **1**. Monoclinic, C$_{31}$H$_{37}$NO$_7$; space group P 1 21 1 with a = 12.5723(5) Å, b = 14.6723(6) Å, c = 17.4900(8) Å, V = 3226.3(2) Å^3, Z = 1, D_{calcd} = 1.109 g/cm^3, μ = 0.691 mm^{-1}, and $F(000)$ = 1077. T = 296.15 K. $R1$ = 0.0526 (I > 2σ(I)), $wR2$ = 0.1480 (all data), S = 1.021. Absolute structure parameter: −0.05(8). The structures were solved using ShelXS. The structural solutions were found by direct methods and refined using the ShelXL package by least squares minimization. The final structures were examined using the Addsym subroutine of PLATON to assure that no additional symmetry could be applied to the models. All non-hydrogen atoms were refined with anisotropic thermal factors.

3.4. Bioassays for Cytotoxic and Apoptosis-Inducing Activity

The cytotoxic activities of compounds **1–6** were tested in vitro by using the MTT method optimized by Chuan et al. [16]. Imatinib and adriamycin were used as the positive controls, and a medium with 4‰ DMSO was used as the negative control in the bioassay test. For AOEB staining, SGC-7901 cells were cultured in 96-well cell culture plates. After 48 h incubation, the culture medium was removed and washed with PBS three times. AO and EB were added to a final concentration of 2 μg/mL each. For DAPI staining, cells were fixed with 4% paraformaldehyde solution for 10 min, incubated with 0.1% TritonX-100 on ice for 30 min, and then washed with PBS three times. DAPI was added to a final concentration of 1 μg/mL each. The pictures were taken using a fluorescence microscope.

4. Conclusions

In conclusion, four new ansamycins (**1–4**) and two known analogs (**5** and **6**) were isolated from the fermentation broth of mangrove-derived actinomycete *Streptomyces* sp. KFD18. Compounds **1–4** exhibited cytotoxic activity against SGC-7901(IC_{50} = 2.8, 9.8, 4.7, and 20.9 µM, respectively), K562 (IC_{50} = 6.6, 9.0, 7.6, and 16.3 µM, respectively), HeLa (IC_{50} = 9.6, >50, 14.1, and 29.5 µM, respectively), and A549 (IC_{50} = 14.9, 24.7, 20.9, and 33.2 µM, respectively) cell lines, with **1** being the most active while compounds **5** and **6** were inactive against all the tested cell lines. The two most active compounds, **1** and **3**, could induce apoptosis of SGC-7901 cells.

Supplementary Materials: The following are available online at http://www.mdpi.com/1660-3397/17/4/219/s1, Figures S1–S5-9: HRESIMS, IR and 2D NMR spectra of the new compounds **1**–**4**, and the 16S rRNA gene sequence of *Streptomyces* sp. KFD18 are supplied.

Author Contributions: L.-M.Z. contributed to the fermentation, compound purification, and the bioassay. F.-D.K. was responsible for structural elucidation and preparation of the paper. Q.-Y.X. contributed to Actinomycete strain isolation. Q.-Y.M. identified the strain. Y.-X.Z. and D.-Q.L. designed the work and revised the paper.

Acknowledgments: This work was supported by Natural Science Foundation of Hainan Province (2019CXTD411), the Natural Science Foundation of China (81741157, 31672070), Financial Fund of the Ministry of Agriculture and Rural Affairs, P. R. of China (NFZX2018), Foundation of Guangdong Provincial Key Laboratory of Marine Biotechnology (No. GPKLMB201704), Central Public-interest Scientific Institution Basal Research Fund for Chinese Academy of Tropical Agricultural Sciences (17CXTD-15, 1630052016008), and the National Key Research and Development Program of China (2017YFD0201400 and 2017YFD0201401).

Conflicts of Interest: The authors declare no conflict of interest.

References

1. Fukuyo, Y.; Hunt, C.R.; Horikoshi, N. Geldanamycin and its anti-cancer activities. *Cancer Lett.* **2010**, *290*, 24–35. [CrossRef] [PubMed]

2. Floss, H.G.; Yu, T.W. Rifamycin mode of action, resistance, and biosynthesis. *Chem. Rev.* **2005**, *105*, 621–632. [CrossRef] [PubMed]

3. Cassady, J.M.; Chan, K.K.; Floss, H.G.; Leistner, E. Recent developments in the maytansinoid antitumor agents. *Chem. Pharm. Bull.* **2004**, *52*, 1–26. [CrossRef] [PubMed]

4. Higashide, E.; Asai, M.; Ootsu, K.; Tanida, S.; Kozai, Y.; Hasegawa, T.; Kishi, T.; Sugino, Y.; Yoneda, M. Ansamitocin, a group of novel maytansinoid antibiotics with antitumour properties from Nocardia. *Nature* **1977**, *270*, 721–722. [CrossRef] [PubMed]

5. Ding, L.; Maier, A.; Fiebig, H.H.; Görls, H.; Lin, W.H.; Peschel, G.; Hertweck, C. Divergolides A–D from a mangrove endophyte reveal an unparalleled plasticity in ansa-macrolide biosynthesis. *Angew. Chem.* **2011**, *123*, 1668–1672. [CrossRef]

6. Ding, L.; Franke, J.; Hertweck, C. Divergolide congeners illuminate alternative reaction channels for ansamycin diversification. *Org. Biomol. Chem.* **2015**, *13*, 1618–1623. [CrossRef] [PubMed]

7. Xu, Z.; Baunach, M.; Ding, L.; Peng, H.; Franke, J.; Hertweck, C. Biosynthetic code for divergolide assembly in a bacterial mangrove endophyte. *ChemBioChem* **2014**, *15*, 1274–1279. [CrossRef] [PubMed]

8. Zhao, G.; Li, S.; Guo, Z.; Sun, M.; Lu, C. Overexpression of div 8 increases the production and diversity of divergolides in *Streptomyces* sp. W112. *RSC Adv.* **2015**, *5*, 98209–98214. [CrossRef]

9. Kong, F.D.; Ma, Q.Y.; Huang, S.Z.; Wang, P.; Wang, J.F.; Zhou, L.M.; Yuan, J.Z.; Dai, H.F.; Zhao, Y.X. Chrodrimanins K-N and related meroterpenoids from the fungus *Penicillium* sp. SCS-KFD09 isolated from a marine worm, *Sipunculus nudus*. *J. Nat. Prod.* **2017**, *80*, 1039–1047. [CrossRef] [PubMed]

10. Kong, F.D.; Zhang, R.S.; Ma, Q.Y.; Xie, Q.Y.; Wang, P.; Chen, P.W.; Zhou, L.M.; Dai, H.F.; Luo, D.Q.; Zhao, Y.X. Chrodrimanins O–S from the fungus *Penicillium* sp. SCS-KFD09 isolated from a marine worm, *Sipunculusnudus*. *Fitoterapia* **2017**, *122*, 1–6. [CrossRef] [PubMed]

11. Kong, F.D.; Huang, X.L.; Ma, Q.Y.; Xie, Q.Y.; Wang, P.; Chen, P.W.; Zhou, L.M.; Yuan, J.Z.; Dai, H.F.; Luo, D.Q. Helvolic acid derivatives with antibacterial activities against *Streptococcus agalactiae* from the marine-derived fungus *Aspergillus fumigatus* HNMF0047. *J. Nat. Prod.* **2018**, *81*, 1869–1876. [CrossRef] [PubMed]

12. An, C.L.; Kong, F.D.; Ma, Q.Y.; Xie, Q.Y.; Yuan, J.Z.; Zhou, L.M.; Dai, H.F.; Yu, Z.F.; Zhao, Y.X. Chemical Constituents of the Marine-Derived Fungus *Aspergillus* sp. SCS-KFD66. *Mar. Drugs* **2018**, *16*, 468. [CrossRef] [PubMed]

13. Lu, C.; Li, Y.; Deng, J.; Li, S.; Shen, Y.; Wang, H.; Shen, Y. Hygrocins C–G, cytotoxic naphthoquinone ansamycins from gdmAI-disrupted *Streptomyces* sp. LZ35. *J. Nat. Prod.* **2013**, *76*, 2175–2179. [CrossRef] [PubMed]
14. Braun, J.S.; Novak, R.; Murray, P.J.; Eischen, C.M.; Susin, S.A.; Kroemer, G.; Halle, A.; Weber, J.R.; Tuomanen, E.I.; Cleveland, J.L. Apoptosis-inducing factor mediates microglial and neuronal apoptosis caused by pneumococcus. *J. Infect. Dis.* **2001**, *184*, 1300–1309. [CrossRef] [PubMed]
15. Yuan, Z.F.; Tang, Y.M.; Xu, X.J.; Li, S.S.; Zhang, J.Y. 10-Hydroxycamptothecin induces apoptosis in human neuroblastoma SMS-KCNR cells through p53, cytochrome c and caspase 3 pathways. *Neoplasma* **2016**, *63*, 72–79. [CrossRef] [PubMed]
16. Chen, C.; Liang, F.; Chen, B.; Sun, Z.Y.; Xue, T.D.; Yang, R.L.; Luo, D.Q. Identification of demethylincisterol A3 as a selective inhibitor of protein tyrosine phosphatase Shp2. *Eur. J. Pharmacol.* **2017**, *795*, 124–133. [CrossRef] [PubMed]

 *marine drugs*

Article

Discovery of Natural Dimeric Naphthopyrones as Potential Cytotoxic Agents through ROS-Mediated Apoptotic Pathway

Kuo Xu [1,†], Chuanlong Guo [1,2,†], Jie Meng [1,3], Haiying Tian [4], Shuju Guo [1,*] and Dayong Shi [1,5,*]

[1] Chinese Academy of Sciences Key Laboratory of Experimental Marine Biology, Institute of Oceanology, Chinese Academy of Sciences, Qingdao 266071, China; xukuoworld@126.com (K.X.); gcl_cpu@126.com (C.G.); mengjie@qibebt.ac.cn (J.M.)

[2] Department of Pharmacy, College of Chemical Engineering, Qingdao University of Science and Technology, Qingdao 266042, China

[3] College of Resources and Environment, Qingdao Agricultural University, Qingdao 266109, China

[4] Technology Center, China Tobacco Henan Industrial Co., Ltd., Zhengzhou, 450000, China; 13623810925@126.com

[5] State Key Laboratory of Microbial Technology, School of Life Science, Shandong University, No. 72 Binhai Road, Qingdao 266237, China

*   Correspondence: guoshuju@qdio.ac.cn (S.G.); shidayong@qdio.ac.cn (D.S.); Tel.: +86-532-82898741 (S.G.); +86-532-82898719 (D.S.)

†   These authors contributed equally to this work.

Received: 11 March 2019; Accepted: 28 March 2019; Published: 2 April 2019

Abstract: A study on the secondary metabolites of *Aspergillus sp.* XNM-4, which was derived from marine algae *Leathesia nana* (Chordariaceae), led to the identification of one previously undescribed (**1**) and seventeen known compounds (**2–18**). Their planar structures were established by extensive spectroscopic analyses, while the stereochemical assignments were defined by electronic circular dichroism (ECD) calculations. The biological activities of the compounds were assessed on five human cancer cell lines (PANC-1, A549, MDA-MB-231, Caco-2, and SK-OV-3), and one human normal cell line (HL-7702) using an MTT [3-(4,5-dimethyl-2-thiazolyl)-2,5-diphenyl tetrazolium bromide] assay. Among them, the dimeric naphthopyrones **7**, **10** and **12** exhibited potent cytotoxicity. Further mechanism studies showed that **12** induced apoptosis, arrested the cell cycle at the G0/G1 phase in the PANC-1 cells, caused morphological changes and generated ROS; and it induces PANC-1 cells apoptosis via ROS-mediated PI3K/Akt signaling pathway.

Keywords: *Aspergillus*; naphthopyrones; cytotoxicity; endophytic fungus; *Leathesia nana*

1. Introduction

Marine-derived endophytic fungi have drawn considerable attention for their surprising potential in drug discovery [1–5]. These endophytic fungi can be distributed in every possible marine host, such as plants, invertebrates and vertebrates [6]. In the interactional process of symbiosis and evolution, the host provides suitable living conditions to the endophytes, while the endophytes contribute bioactive secondary metabolites that provide protection and, ultimately, survival value to their hosts [7,8]. As one of the most prevalent sources of microorganisms, marine algae offer an abundant amount of endophytic fungi for chemical studies. Hundreds of natural products have been identified from the algal-derived fungi [9].

In our ongoing efforts to discover the bioactive secondary metabolites of endophytic fungi from the marine brown algae *Leathesia nana* (Chordariaceae), eighteen compounds were isolated from an *Aspergillus sp.* XNM-4 strain (Figure 1). The planar structures of the metabolites were established by HRESIMS, UV, IR, one- and two-dimensional (1D and 2D) NMR spectroscopic data, while the stereochemistry of compounds **1** and **12** were assigned by a comparison of the calculated and experimental electronic circular dichroism (ECD). All compounds were assessed for inhibitory effects on five human cancer cell lines (PANC-1, A549, MDA-MB-231, Caco-2, and SK-OV-3) and one human normal cell line (HL-7702). Notably, as the most promising candidate, the cytotoxic mechanism of compound **12** in PANC-1 cells was studied preliminarily. These experimental results may be beneficial for the development of naturally occurring dimeric naphthopyrones as anti-tumor agents.

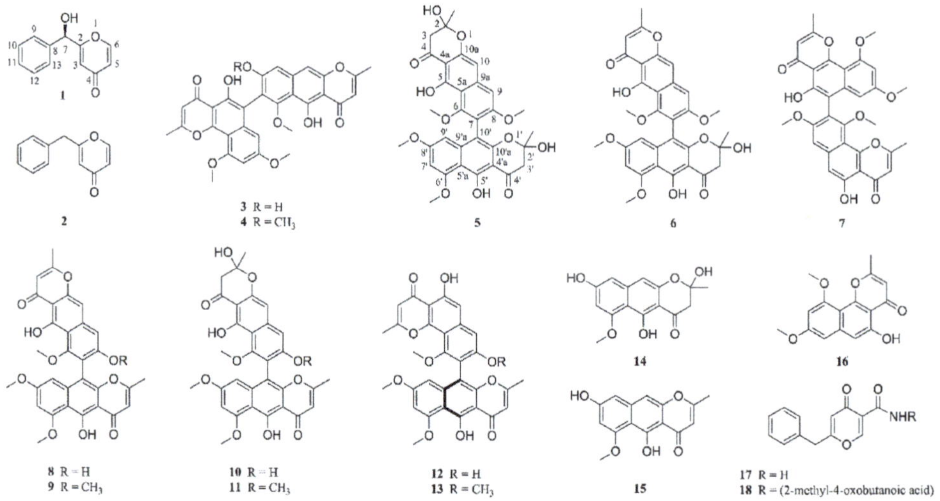

Figure 1. Compounds **1–18** isolated from *Aspergillus sp.* XNM-4.

2. Results and Discussion

2.1. Structural Elucidation

Compound **1** was isolated as a white amorphous powder. Its molecular formula ($C_{12}H_{10}O_3$) was determined on the HRESIMS (m/z 203.0707 [M + H]$^+$, calcd 203.0708, Figure S3) in association with ^{13}C NMR data. The IR spectrum (Figure S4) exhibited absorption bands for carbonyl (1653 cm^{-1}) and olefinic (1384, 1602 cm^{-1}) groups. The ^{1}H NMR data of **1** (Table 1, Figure S5) revealed an ABX spin-system assignable to three pyrone protons [δ_H 8.05 (^{1}H, d, J = 6.0 Hz), 6.21 (^{1}H, dd, J = 2.4, 6.0 Hz), and 6.42 (^{1}H, d, J = 2.4 Hz)]; however, their coupling constants were different from those of the benzene ring. The ^{1}H NMR spectrum also showed five pyrone protons [δ_H 7.41 (2H, overlap), 7.37 (2H, overlap), and 7.32 (^{1}H, m)] and an oxygenated methyne δ_H 5.48 (^{1}H, s). In combination with the five resonance peaks at δ_H 7.32–7.41, the pyrone resonances in the ^{13}C NMR spectrum [δ_C 140.6 (1C), 128.4 (2C), 128.0 (1C), and 126.8 (2C), Figure S6] supported the existence of a mono-substituted benzene ring [10]. The remaining five resonance peaks [δ_C 177.9 (1C), 170.2 (1C), 156.2 (1C), 116.2 (1C), and 112.3 (1C)] indicated a skeleton of 4H-pyran-4-one [11], except for the oxygenated carbon resonance at δ_C 71.2 (1C). In the 2D NMR experiment (Figures S7 and S8), the hydrogen resonance at δ_H 6.21 (H-5) has a homonuclear correlation with the resonance at δ_H 8.05 (H-6), and the hydrogen resonance at δ_H 5.48 (H-7) has long-range heteronuclear correlations with the carbon resonances at δ_C 112.3 (C-3) and 126.8 (C-9, 13), which confirmed that a benzyl group was substituted at C-2. Thus, the structure of compound **1** was identified as (hydroxy(phenyl)methyl)-4H-pyran-4-one.

Table 1. ^{1}H, ^{13}C NMR and HMBC spectroscopic data of **1** (600 MHz, ppm in DMSO-d_6).

Position	δ_H (*J* in Hz)	δ_C (m)	Key HMBC (H→C)
2		170.2	
3	6.42, d (*J* = 2.4)	112.3	C-5, 7
4		177.9	
5	6.21, dd (*J* = 2.4, 6.0)	116.2	C-3
6	8.05, d (*J* = 6.0)	156.2	C-4
7	5.48, s	71.2	C-3, 9, 13
8		140.6	
9	7.37, overlap	126.8	C-7, 11
10	7.41, overlap	128.4	C-8
11	7.32, m	128.0	C-9, 13
12	7.41, overlap	128.4	C-8
13	7.37, overlap	126.8	C-7, 11

The configurational assignment of C-7 was defined by ECD calculations using a MMFF94 force field and time-dependent density functional theory (TDDFT) at the B3LYP/6-311+G(d, p) level. The overall calculated ECD curve of (7*R*)-**1** were produced by Boltzmann weighting of their lowest energy conformers, matching well with the corresponding experimental ECD data (Figure 2, the procedure was detailed in Supplementary Materials, S34–S36). Thus, the structure of compound **1** was finally established as (7*R*)-(hydroxy(phenyl)methyl)-4*H*-pyran-4-one.

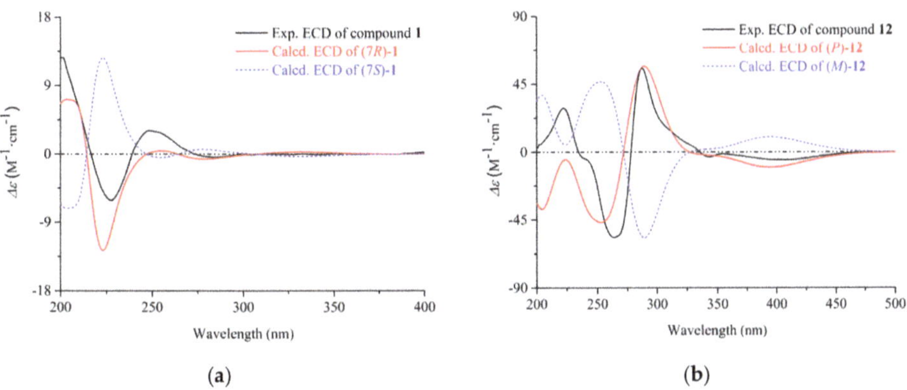

Figure 2. The experimental and calculated ECD of compounds **1** (**a**) and **12** (**b**).

By comparing the spectroscopic data (HRESIMS, ^{1}H and ^{13}C NMR, Figures S9–S60) with those reported in the literature, the remaining nineteen known compounds (**2–21**) were identified as 2-benzyl-4*H*-pyran-4-one (**2**) [11], asperpyrone D (**3**) [12], asperpyrone C (**4**) [13], aurosperone B (**5**) [14], fonsecinone B (**6**) [14], asperpyrone B (**7**) [15], dianhydro-aurasperone C (**8**) [12], isoaurasperone A (**9**) [15], aurasperone F (**10**) [16], fonsecinone D (**11**) [14], asperpyrone A (**12**) [12], fonsecinone A (**13**) [15], fonsecin (**14**) [14], TMC 256 A1 (**15**) [17], flavasperone (**16**) [14], carbonarone A (**17**) [18], pestalamide A (**18**) [19]. In addition, the *p* configuration of **12** was defined by ECD calculation at the B3LYP/6-311+G(d, p) level (Figure 2, Supplementary Materials, S34–36).

2.2. Cytotoxic Activities of Compounds **1–18**

Natural naphthopyrones have been previously reported for their anticancer potential [13,15,17]. Therefore, the present study evaluated the inhibitory effects of the isolated compounds on five human cancer cell lines (PANC-1, A549, MDA-MB-231, Caco-2, and SK-OV-3), and one human normal cell line (HL-7702) at a concentration of 50 μM. As a result, the dimeric naphthopyrones **7**, **10**, and especially **12**, exhibited potent cytotoxicity on PANC-1, A549, MDA-MB-231, Caco-2, SK-OV-3 and HL-7702 cells (Figure 3). The IC_{50} values of compound **12** on the different cells were further measured, and it possessed the greatest inhibitory effects against PANC-1, with an IC_{50} value of 8.25±2.20 μM (Figure 4A).

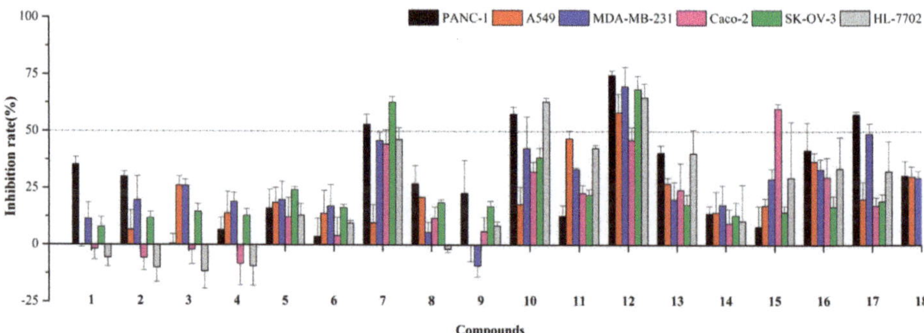

Figure 3. Cytotoxic activities of compounds **1–18** at the concentration of 50 μM.

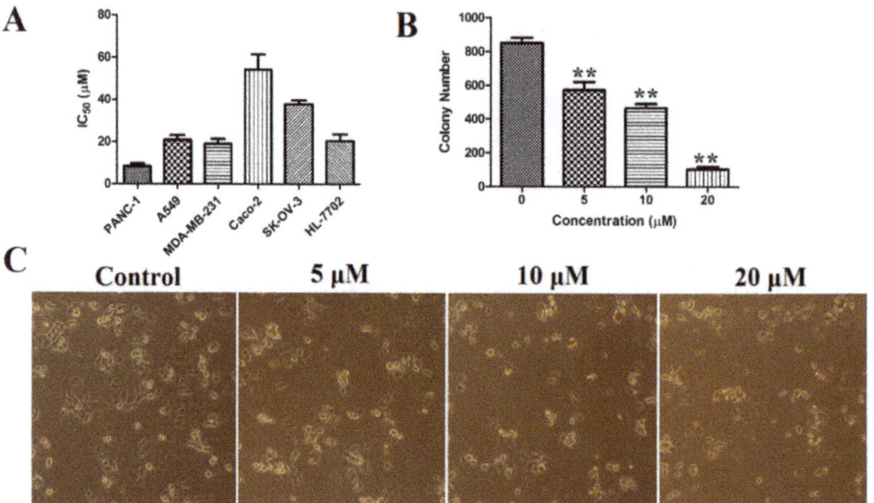

Figure 4. (**A**) The IC_{50} values of compound **12** on PANC-1, A549, MDA-MB-231, Caco-2, SK-OV-3, and HL-7702 cells; (**B**) colony formation in PANC-1 cells was determined by staining with crystal violet; (**C**) cell morphology was observed using inverted microscope. ** $p < 0.01$ vs. control group.

2.3. Pharmacological Mechanism of Compound 12 on PANC-1 Cells

2.3.1. Morphological Changes

It is well known that cytotoxic agents often cause changes in cell morphology, such as irregular cell morphology, increased cell debris, and reduced cell numbers. As shown in Figure 4, after treatment with compound **12**, the PANC-1 cells showed morphological changes such as cell shrinkage, deformation and a reduced number of viable cells.

2.3.2. Colony Formation

A 10-day colony formation experiment was performed to explore the long-term impact of compound **12** on the PANC-1 cells growth. PANC-1 cells were seeded in 6-well plates (1000 cells/well) and were treated with various concentrations of compound **12** (0, 5, 10, 20 μM) for 10 days to allow colony formation. As shown in Figure 4B, 852 ± 43 colonies were present in the control group, whereas the colony numbers decreased to 574 ± 65, 421 ± 30 and 105 ± 21 after treatment with compound **12** (5, 10, and 20 μM, respectively). These results showed that compound **12** could inhibit the colony formation of PANC-1 cells.

2.3.3. Cell Apoptosis

To explore whether the abovementioned reduction in cell viability was caused by the induction of apoptosis, PANC-1 cells were treated with compound **12** (5, 10, and 20 μM) for 72 h. The cells were then stained with fluorescein isothiocyanate (Annexin-V FITC) and propidium iodide (PI) and were analyzed by flow cytometry. The results indicated that compound **12** could induce cell apoptosis in a concentration-dependent manner. As shown in Figure 5A,B, 11.07 ± 2.43% of the apoptotic cells were present in the control, whereas the apoptotic population increased to 19.93 ± 65, 26.43 ± 3.81 and 40.43 ± 3.27 after treatment with **12** (5, 10, and 20 μM, respectively).

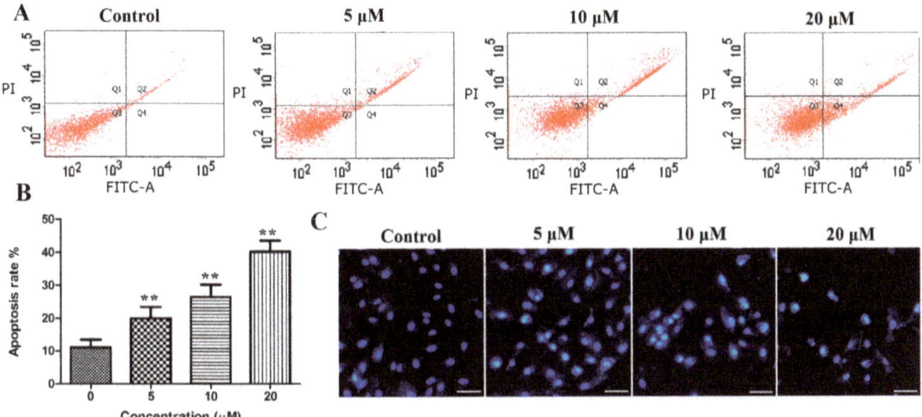

Figure 5. (**A,B**) PANC-1 cells were stained with Annexin-V FITC and propidium iodide (PI), and then analyzed using flow cytometry, both early and late apoptotic cells were analyzed; (**C**) PANC-1 cells were stained with Hoechst 33258 and photographed using a fluorescence microscopy (bar = 50 μm). ** $p < 0.01$ vs. control group.

Apoptosis often causes morphological changes, which can be observed by Hoechst 33258 staining the apoptotic cells. Thus, the PANC-1 cells were treated with compound **12** (5, 10, and 20 μM) for 72 h, stained with Hoechst 33258 and analyzed by fluorescence microscopy; significant morphological changes were observed. As shown in Figure 5C, nuclear pyknosis and chromosome condensation were observed in PANC-1 treated with compound **12**, and no apoptosis was found in the control group.

2.3.4. Cell Cycle

To explore the influence of this compound on the cell cycle distribution, PANC-1 cells were treated with compound **12** (5, 10, and 20 μM) for 72 h. Next, the cell cycle distribution was analyzed by flow cytometry after staining with PI. As shown in Figure 6A,B, the G0/G1 phase was increased in a concentration-dependent manner in the PANC-1 cells. Compared with the control group, the population in the G1 phase increased from 45.97% to 70.94% at a concentration of 20 μM. Moreover, the sub-G1 group significantly increased after the cells were cultured with compound **12** (Figure 6C). These results indicated that compound **12** could induce apoptosis and arrested the cell cycle at the G0/G1 phase in PANC-1 cells in a concentration-dependent manner.

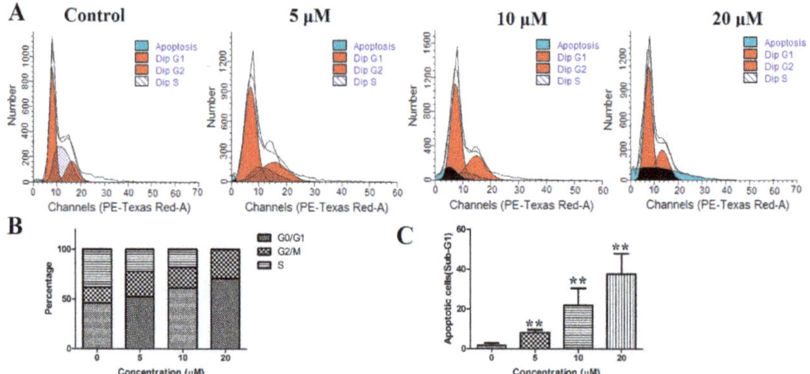

Figure 6. (**A**) PANC-1 cells were stained with propidium iodide (PI) and analyzed using flow cytometry; (**B**) proportion of PANC-1 cells in each phase; (**C**) the sub-G1 was increased after treatment of compound **12**. ** $p < 0.01$ vs. control group.

2.3.5. ROS Generation

ROS (reactive oxygen species) plays an important role in cell proliferation or apoptosis [20,21], and it can induce cell death in a variety of ways. When intracellular ROS accumulates in cells, it causes the mitochondrial membrane potential damage and eventually leads to apoptosis [22,23]. To explore whether this compound triggers ROS generation, PANC-1 cells were stained with a fluorescent probe, 2′,7′-dichlorodihydrofluorescein in diacetate (DCFH-DA), which can detect intracellular ROS. The result showed that a rapid production of ROS could be detected in the PANC-1 cells after the treatment of compound **12**. As shown in Figure 7A,B, compared with that of the control, the ROS content in the experimental group increased to 120.09%, 336.99% and 449.09%. The ROS-mediated effects may be modulated by antioxidants such as *N*-acetylcysteine (NAC). Next, PANC-1 cells were treated with 10 μM compound **12** combined with/without 5 mM NAC (Beyotime, Nanjing, China), a ROS scavenger, for 72 h, cells were harvested and analyzed after staining with DCFH-DA. The results showed that compound **12**-induced ROS generation was blocked by NAC in PANC-1 cells (Figure 7C). These data indicated that compound **12** could induce ROS generation, and this might be a mechanism of apoptosis.

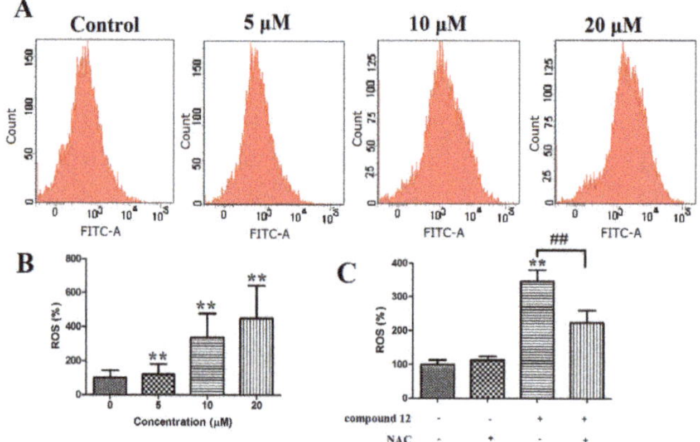

Figure 7. (**A**,**B**) PANC-1 cells were stained with 2′,7′-dichlorodihydrofluorescein in diacetate (DCFH-DA) and then analyzed using flow cytometry; (**C**) PANC-1 cells were treated with 10 μM compound **12** alone or in combination with NAC (5 mM) for 72 h. ** $p < 0.01$ vs. control group. ## $p < 0.01$ vs. compound **12**(+)/NAC(−) group.

2.3.6. Mechanism Study of Compound **12**

Apoptosis serves a key role in the regulation of cells. It mainly comprised two apoptotic pathways: The death receptor-mediated apoptosis pathway and the mitochondria-mediated apoptosis pathway [24]. In the mitochondria-mediated apoptosis pathway, proteins from the Bcl-2 family, such as Bax and Bcl-2, are the main components that regulate mitochondrial permeability [25]. In this study, it was demonstrated that compound **12** treatment could increase the ratio of Bax/Bcl-2 as well as activate Caspase-3 and PARP (Figure 8A).

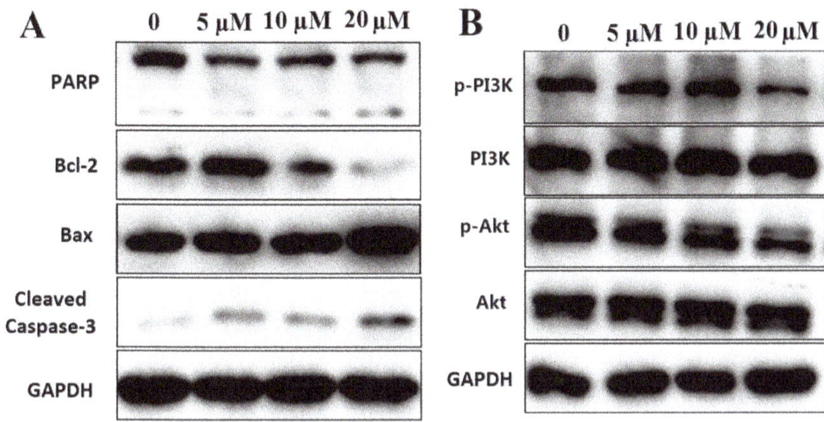

Figure 8. (**A**) Western blot analysis of apoptosis-related proteins, including PARP, Bcl-2, Bax and Cleaved-Caspase-3; (**B**) western blot analysis of PI3K/Akt pathway proteins, including PI3K and Akt. GAPDH was used to normalize protein content (n =3).

PI3K/Akt signaling pathway plays an important role in the process of apoptosis especially in the ROS-mediated apoptotic pathway [26,27]. In this study, the phosphorylation of PI3K and Akt were decreased after treatment with compound **12** (Figure 8B). Based on the above studies, our results indicated that compound **12**-induced PANC-1 apoptosis may be through ROS-mediated PI3K/Akt signaling pathway.

3. Materials and Methods

3.1. General Experimental Procedures

The HRESIMS analyses were performed on a Waters Xevo G2-XS QTof mass spectrometer (Waters Corp., Milford Massachusetts, America). NMR spectra were recorded on an Ascend 600 MHz instrument (Bruker-Biospin, Billerica, MA, America). The analytical experiments were performed on a Shimadzu LC-20AT HPLC system (Shimadzu Corp., Kyoto, Japan) equipped with a Shimadzu InertSustain C18 column (4.6 I.D. × 250 mm, 5 μm, S/N 6LR98081). A Hanbon NP700 semipreparative HPLC (Hanbon Sci. & Tech., Jiangsu, China) equipped with a Shimadzu InertSustain C18 column (10 I.D. × 250 mm, 5 μm, S/N 7ER43006) was used for purifying compounds. Biological assays were monitored on a BioTek ELx808 microplate spectrophotometer (BioTek Instruments, Inc., Winooski, America), a BD FACSCalibur flow cytometry (BD Biosciences, Franklin Lakes, America) and an Olympus BX-51 Fluorescence Microscopy (Olympus Corporation, Tokyo, Japan).

3.2. Fungal Material and Fermentation

The fungus strain *Aspergillus* sp. XNM-4 was isolated from *Leathesia nana*, which was collected in April 2017 in Weihai, Shandong Province, China (Latitude: 37°31′57.58″N; Longitude: 122°02′52.85″E).

The strain was identified according to 18S rDNA gene sequence analysis by the Beijing Genomics Institute (Shenzhen, China).

The fungus *Aspergillus* sp. XNM-4 was fermented on Malt Extract Medium (130.0 g/L malt extract, 0.1 g/L chloramphenicol, 15.0 g/L agar, pH 5.6 ± 0.2) under static conditions at 25 °C for 10 days. A total of 300 culture dishes (90 * 15 mm) were used in the experiment.

3.3. Extraction and Isolation

The agar blocks with mycelium were collected in a 2 L beaker, and ultrasonically extracted with 1.5 L of ethyl acetate (three times and for 30 min each). The crude extract (10.3 g) was chromatographed on a silica gel column (4*40 cm, 200–300 mesh) and successively eluted with petroleum ether (0.5 L), petroleum-EtOAc (4:1, 2.5 L), petroleum-EtOAc (1:2, 2 L), EtOAc (1 L), and EtOAc-MeOH (1:2, 2 L). The eluents were concentrated by reduced pressure at 40 °C, and then merged in nine fractions under HPLC analysis, including Fractions A (985.5 mg), B (364.6 mg), C (41.7 mg), D (90.4 mg), E (220.2 mg), F (344.6 mg), G (384.7 mg), H (8.3 mg), and I (1611.4 mg). These subfractions were further purified by semipreparative HPLC using a continuous gradient of MeOH-H_2O (60–100%, 20 min, 3 mL/min). The obtained eluents were extracted by ethyl acetate (v/v, 1:2) twice. After being dried by anhydrous Na_2SO_4, the organic phase was concentrated under a reduced pressure at 40 °C and then freeze-dried to yield compounds **1**–**18**. As a result, compounds **2** (3.1 mg,), **15** (5.2 mg), and **16** (4.4 mg) were from Fr. D, compounds **3** (3.0 mg), **4** (3.8 mg), **5** (9.2 mg), **6** (7.0 mg), **7** (5.3 mg), **9** (13.4 mg), **11** (4.2 mg), **13** (8.9 mg), and **14** (11.6 mg) were from Fr. E, compounds **8** (7.4 mg), **10** (8.2 mg), **12** (8.8 mg), **17** (29.2 mg), **18** (5.4 mg) were from Fr. F, compound **1** (5.0 mg) was from Fr. G. The purities of all isolated compounds was determined to be >95% under two solvent conditions by analytical HPLC recorded on a Shimadzu LC-20A system. Solvent conditions A: CH_3OH/H_2O with 0.1% trifluoroacetic acid 60–100% (20 min); Solvent conditions B: CH_3CN/H_2O with 0.1% trifluoroacetic acid 30–100% (20 min); UV detection, 254 nm; flow rate, 1.0 mL/min; temperature, 40 °C; injection volume, 30 μL. The analytical HPLC spectra were listed on page S37–S54 in the Supplementary Materials.

(7R)-(hydroxy(phenyl)methyl)-4H-pyran-4-one (**1**): white amorphous powder; $[\alpha]_D^{20}$ +78.1° (*c* 0.10, MeOH); UV (MeOH) λ_{max} (log ε) 248 (4.14) nm; ECD (MeOH) λ_{max} ($\Delta\varepsilon$) 201 (+12.65), 228 (−6.13), 250 (+3.01) nm; IR (KBr) ν_{max} 3447, 1653, 1602, 1384 cm^{-1}; ^{1}H; for ^{13}C NMR data see Table 1; HRESIMS (*m/z*): 203.0707 [M + H]$^+$ (calcd for $C_{12}H_{11}O_3$, 203.0708).

3.4. Biological Activity Test

3.4.1. Cell Culture

PANC-1, A549, MDA-MB-231, Caco-2, SK-OV-3 and HL-7702 were supplied by Cell Bank, Chinese Academy of Sciences (Shanghai, China). These cells were separately maintained in DMEM medium, F-12K medium, L15 medium, MEM medium, McCoy's 5A (Modified) medium, and RPMI-1640 medium. All media were supplemented with 10% FBS, 100 U/mL penicillin and 100 μg/mL streptomycin. Cells were cultured at 37 °C in a humidified CO_2 (5%).

3.4.2. Determination of Cell Viability

Cell viability was evaluated by a 3-(4,5-dimethyl-2-thiazolyl)-2,5-diphenyl tetrazolium bromide (MTT) assay [28]. For the preliminary anti-tumor activity screening, the cells were plated in 96-well plates (3×10^3 cells/well for A549 and PANC-1, A549, MDA-MB-231 and Caco-2, 5×10^3 cells/well for SK-OV-3 and GL-7702) and incubated with the tested compounds at a concentration of 50 μM for 72 h. For detection of the IC$_{50}$, cells were treated with varying concentrations of **12** (0, 6.25, 12.5, 25, 50 μM) for 72 h. After incubation, MTT (5 mg/mL) was added and incubated at 37 °C for 4 h. The formazan was dissolved by DMSO and measured using a microplate reader at 490 nm.

3.4.3. Colony Forming Assay

The PANC-1 cells were seeded in 6-well plates (1000 cells/well) and treated with varying concentrations of **12** (0, 5, 10, 20 μM). These cells were further incubated for 10 days to allow colony formation; then, the cells were fixed with 4% paraformaldehyde for 10 min. After three washes, the cells were finally stained with crystal violet for 10 min. Cells > 50 were scored as colonies [29].

3.4.4. Analysis of Apoptosis

The PANC-1 cells were seeded in 6-well plates (2×10^5/well). After 24 h of incubation, the cells were treated with compound **12** (0, 5, 10, 20 μM) for 72 h. After being harvested and washed with PBS, PANC-1 cells were stained with Annexin V/PI for 15 min. Finally, the cells were detected and analyzed by flow cytometry.

3.4.5. Hoechst 33258 Staining

The PANC-1 cells were seeded in 6-well plates (2×10^5/well). After 24 h of incubation, the cells were treated with compound **12** (0, 5, 10, 20 μM) for 72 h. Next, the cells were stained with Hoechst dye 33258 for five min at room temperature and assessed by a fluorescence microscopy.

3.4.6. Analysis of Cell Cycle

The PANC-1 cells were seeded in 6-well plates (2×10^5/well). After 24 h of incubation, the cells were treated with varying concentrations of compound **12** (0, 5, 10, 20 μM) for 72 h. After fixed in cold 75% ethanol at -20 °C overnight, the cells were washed twice with PBS and stained with a PI solution containing 20 μg/mL of RNaseA and 50 μg/mL of PI for 30 min. Finally, the cells were detected and analyzed by flow cytometry.

3.4.7. Measurement of Intracellular ROS

The PANC-1 cells were seeded in 6-well plates (2×10^5/well). After 24 h of incubation, the cells were treated with varying concentrations of compound **12** (0, 5, 10, 20 μM) for 72 h. After being stained with 10 μM of DCFH-DA at 37 °C for 30 min, the cells were washed with media and were detected and analyzed by flow cytometry.

3.4.8. Western Blot Analysis

PANC-1 cells were harvested and seeded in 6-cell plates and allowed to settle overnight. Cells were treated with compound **12** (0, 5, 10 and 20 μM) for 72 h. Proteins were harvested and separated by SDS-PAGE and transferred onto PVDF membranes. Membranes were blocked in a blocking solution (containing 5% non-fat milk) and subsequently probed with primary antibodies at 4 °C overnight. After 15 min washes in TBST, the membranes were incubated with a secondary antibody for 1 h at room temperature. Antibodies against Bcl-2, Bax, PARP, Cleaved-Caspase-3, phosphorylation-Akt, Akt and GAPDH were purchased from Cell Signaling Technology (Beverly, MA, USA). Phosphorylation-PI3K and PI3K were purchased from Abcam (Cambridge, UK). The anti-mouse IgG and anti-rabbit secondary antibodies raised from goat were obtained from Abcam (Cambridge, UK). The bands were detected using an enhanced chemiluminescence system BeyoECL Plus (Beyotime, Nanjing, China).

4. Conclusions

In the present study, eighteen metabolites, including one new pyrone derivative (**1**), were identified from the culture of an algae-derived endophytic fungus *Aspergillus* sp. XNM-4. Among them, compounds **1**, **17**, and **18** were first reported from the genus *Aspergillus*. The pharmacological experiments showed that dimeric naphthopyrones **7**, **10**, and especially **12**, possessed potent cytotoxicity on five human cancer cell lines (PANC-1, A549, MDA-MB-231, Caco-2, and SK-OV-3), and one human normal cell line (HL-7702). Further studies indicated that compound **12** induced apoptosis, arrested the cell cycle at the G0/G1 phase in PANC-1 cells, caused morphological changes and generated ROS. Mechanism studies found that compound **12** induced PANC-1 apoptosis was via ROS-mediated PI3K/Akt signaling pathway. These experimental results may be beneficial for the development of naturally occurring dimeric naphthopyrones as anti-tumor agents.

Supplementary Materials: The following are available online at http://www.mdpi.com/1660-3397/17/4/207/s1, including the HPLC, UV, IR, HRESIMS, 1D NMR, HSQC, and HMBC spectra for compound **1**, the HPLC, HRESIMS and 1D NMR spectra for compounds **2–18**, and the ECD calculation for compounds **1** and **12**.

Author Contributions: This paper was written by K.X. and C.G., and they contributed equally to this work. S.G. isolated the fungus *Aspergillus sp.* XNM-4 from *Leathesia nana*, while J.M. and H.T. identified and cultivated it. K.X. purified the secondary metabolites and defined their structures. C.G. tested their cytotoxic effects in vitro. D.S. approved the final version.

Funding: This research was supported by the National Natural Science Foundation of China (81803375), National Postdoctoral Program for Innovative Talents (BX201700247), China Postdoctoral Science Foundation (2018M630804), Natural Science Foundation of Shandong Province (ZR2016DQ07), and Shandong Provincial Natural Science Foundation for Distinguished Young Scholars (JQ201722).

Conflicts of Interest: The authors declare no conflict of interest.

References

1. Bugni, T.S.; Ireland, C.M. Marine-derived fungi: A chemically and biologically diverse group of microorganisms. *Nat. Prod. Rep.* **2004**, *21*, 143–163. [CrossRef] [PubMed]

2. Rateb, M.E.; Ebel, R. Secondary metabolites of fungi from marine habitats. *Nat. Prod. Rep.* **2011**, *28*, 290–344. [CrossRef]

3. Kjer, J.; Debbab, A.; Aly, A.H.; Proksch, P. Methods for isolation of marine-derived endophytic fungi and their bioactive secondary products. *Nat. Protoc.* **2010**, *5*, 479–490. [CrossRef]

4. Uzma, F.; Mohan, C.D.; Hashem, A.; Konappa, N.M.; Rangappa, S.; Kamath, P.V.; Singh, B.P.; Mudili, V.; Gupta, V.K.; Siddaiah, C.N.; et al. Endophytic fungi-alternative sources of cytotoxic compounds: A review. *Front. Pharmacol.* **2018**, *9*, 309–345. [CrossRef]

5. Corinaldesi, C.; Barone, G.; Marcellini, F.; Dell'Anno, A.; Danovaro, R. Marine microbial-derived molecules and their potential use in cosmeceutical and cosmetic products. *Mar. Drugs* **2017**, *15*, 118. [CrossRef]

6. Debbab, A.; Aly, A.H.; Proksch, P. Endophytes and associated marine derived fungi-ecological and chemical perspectives. *Fungal Diversity* **2012**, *57*, 45–83. [CrossRef]

7. Schulz, B.; Boyle, C.; Draeger, S.; Römmert, A.K.; Krohn, K. Endophytic fungi: A source of novel biologically active secondary metabolites. *Mycol. Res.* **2002**, *06*, 996–1004. [CrossRef]

8. Strobel, G.; Daisy, B.; Castillo, U.; Harper, J. Natural products from endo-phytic microorganisms. *J. Nat. Prod.* **2004**, *67*, 257–268. [CrossRef]

9. Zhang, P.; Li, X.M.; Wang, B.G. Secondary metabolites from the marine algal-derived endophytic fungi: chemical diversity and biological activity. *Planta Med.* **2016**, *82*, 832–842. [CrossRef] [PubMed]

10. Chen, X.Y.; Han, J.X.; Liu, Y.S.; Hajiakber, A.; Yuan, T. Chemical constituents from traditional Uighur herbal medicine *Elaeagnus angustifolia* flowers. *China J. Chin. Mater. Med.* **2018**, *43*, 1749–1753.

11. Huang, Z.J.; Shao, C.L.; Chen, Y.G.; She, Z.G.; Lin, Y.C. Pyrones in metabolites of marine mangrove endophytic fungus(No. ZZF79) from the South China Sea. *Acta Sci. Nat. Univ. Sunyatseni.* **2007**, *46*, 113–115.

12. Zhan, J.; Gunaherath, G.M.; Wijeratne, E.M.; Gunatilaka, A.A. Asperpyrone D and other metabolites of the plant-associated fungal strain *Aspergillus tubingensis*. *Phytochemistry* **2007**, *68*, 368–372. [CrossRef]

13. Akiyama, K.; Teraguchi, S.; Hamasaki, Y.; Mori, M.; Tatsumi, K.; Ohnishi, K.; Hayashi, H. New dimeric naphthopyrones from *Aspergillus niger*. *J. Nat. Prod.* **2003**, *66*, 136–139. [CrossRef]

14. Priestap, H.A. New naphthopyrones from *Aspergillus fonsecaeus*. *Tetrahedron* **1984**, *40*, 3617–3624. [CrossRef]

15. Li, D.H.; Han, T.; Guan, L.P.; Bai, J.; Zhao, N.; Li, Z.L.; Wu, X.; Hua, H.M. New naphthopyrones from marine derived fungus *Aspergillus niger* 2HLM-8 and their in vitro antiproliferative activity. *Nat. Prod. Res.* **2016**, *30*, 1116–1122. [CrossRef] [PubMed]

16. Bouras, N.; Mathieu, F.; Coppel, Y.; Lebrihi, A. Aurasperone F-a new member of the naphthogamma -pyrone class isolated from a cultured microfungus, *Aspergillus niger* C-433. *Nat. Prod. Res.* **2005**, *19*, 653–659. [CrossRef] [PubMed]

17. Huang, H.B.; Xiao, Z.E.; Feng, X.J.; Huang, C.H.; Zhu, X.; Ju, J.H.; Li, M.F.; Lin, Y.C.; Liu, L.; She, Z.G. Cytotoxic naphtho-γ-pyrones from the mangrove endophytic fungus *Aspergillus tubingensis* (GX1-5E). *Helv. Chim. Acta* **2011**, *94*, 1732–1740. [CrossRef]

18. Zhang, Y.; Zhu, T.; Fang, Y.; Liu, H.; Gu, Q.; Zhu, W. Carbonarones A and B, new bioactive γ-Pyrone and α-Pyridone derivatives from the marine-derived fungus *Aspergillus carbonarius*. *J. Antibiot.* **2007**, *60*, 153–157. [CrossRef] [PubMed]

19. Ding, G.; Jiang, L.; Guo, L.; Chen, X.; Zhang, H.; Che, Y. Pestalazines and Pestalamides, bioactive metabolites from the plant pathogenic fungus *Pestalotiopsis theae*. *J. Nat. Prod.* **2008**, *71*, 1861–1865. [CrossRef]

20. Guo, Y.X.; Lin, Z.M.; Wang, M.J.; Dong, Y.W.; Niu, H.M.; Young, C.Y.; Lou, H.X.; Yuan, H.Q. Jungermannenone A and B induce ROS- and cell cycle-dependent apoptosis in prostate cancer cells in vitro. *Acta Pharmacol. Sin.* **2016**, *37*, 814–824. [CrossRef]

21. Li, Z.; Qin, B.; Qi, X.; Mao, J.; Wu, D. Isoalantolactone induces apoptosis in human breast cancer cells via ROS-mediated mitochondrial pathway and downregulation of SIRT1. *Arch. Pharmacal Res.* **2016**, *39*, 1441–1453. [CrossRef]

22. Zhang, J.X.; Wang, X.L.; Vikash, V.; Ye, Q.; Wu, D.D.; Liu, Y.L.; Dong, W.G. ROS and ROS-mediated cellular signaling. *Oxid. Med. Cell. Longevity* **2016**, *2016*, 4350965. [CrossRef]

23. Jiang, Y.; Wang, X.Q.; Hu, D.D. Furanodienone induces G0/G1 arrest and causes apoptosis via the ROS/MAPKs-mediated caspase-dependent pathway in human colorectal cancer cells: A study in vitro and in vivo. *Cell Death Dis.* **2017**, *8*, 2815–2828. [CrossRef]

24. Hanahan, D.; Weinberg, R.A. Hallmarks of cancer: the next generation. *Cell* **2011**, *144*, 646–674. [CrossRef]

25. Gibson, C.J.; Davids, M.S. BCL-2 antagonism to target the intrinsic mitochondrial pathway of apoptosis. *Clin. Cancer Res.* **2015**, *21*, 5021–5029. [CrossRef]

26. Ghosh, S.; Sarkar, A.; Bhattacharyya, S.; Sil, P.C. Silymarin Protects mouse liver and kidney from thioacetamide induced toxicity by scavenging reactive oxygen species and activating PI3K-Akt pathway. *Front. Pharmacol.* **2016**, *7*, 481–495. [CrossRef]

27. Zhao, Y.; Wang, X.P.; Sun, Y.; Zhou, Y.X.; Yin, Y.H.; Ding, Y.X.; Li, Z.Y.; Guo, Q.L.; Lu, N. LYG-202 exerts antitumor effect on PI3K/Akt signaling pathway in human breast cancer cells. *Apoptosis* **2015**, *20*, 1253–1269. [CrossRef]

28. Guo, C.L.; Wang, L.J.; Zhao, Y.; Liu, H.; Li, X.Q.; Jiang, B.; Luo, J.; Guo, S.J.; Wu, N.; Shi, D.Y. A novel bromophenol derivative BOS-102 induces cell cycle arrest and apoptosis in human A549 lung cancer cells via ROS-mediated PI3K/Akt and the MAPK signaling pathway. *Mar. Drugs* **2018**, *16*, 43. [CrossRef]

29. Wang, L.J.; Guo, C.L.; Li, X.Q.; Wang, S.Y.; Jiang, B.; Zhao, Y.; Luo, J.; Xu, K.; Liu, H.; Guo, S.J.; et al. Discovery of novel bromophenol hybrids as potential anticancer agents through the ROS-mediated apoptotic pathway: Design, synthesis and biological evaluation. *Mar. Drugs* **2017**, *15*, 343. [CrossRef]

Article

λ-Carrageenan Oligosaccharides of Distinct Anti-Heparanase and Anticoagulant Activities Inhibit MDA-MB-231 Breast Cancer Cell Migration

Hugo Groult, Rémi Cousin, Caroline Chot-Plassot, Maheva Maura, Nicolas Bridiau, Jean-Marie Piot, Thierry Maugard and Ingrid Fruitier-Arnaudin *

Equipe BCBS (Biotechnologies et Chimie des Bioressources pour la Santé), Université de La Rochelle, UMR CNRS 7266 LIENSs, 17000 La Rochelle, France; hugo.groult@univ-lr.fr (H.G.); remi.cousin1@univ-lr.fr (R.C.); caroline.chotplassot@etudiant.univ-lr.fr (C.C.-P.); maheva.maura1@univ-lr.fr (M.M.); nicolas.bridiau@univ-lr.fr (N.B.); jean-marie.piot@univ-lr.fr (J.-M.P.); thierry.maugard@univ-lr.fr (T.M.)
* Correspondence: ingrid.fruitier@univ-lr.fr; Tel.: +33-(0)5-4645-8562

Received: 8 January 2019; Accepted: 22 February 2019; Published: 27 February 2019

Abstract: In tumor development, the degradation of heparan sulfate (HS) by heparanase (HPSE) is associated with cell-surface and extracellular matrix remodeling as well as the release of HS-bound signaling molecules, allowing cancer cell migration, invasion and angiogenesis. Because of their structural similarity with HS, sulfated polysaccharides are considered a promising source of molecules to control these activities. In this study, we used a depolymerisation method for producing λ-carrageenan oligosaccharides (λ-CO), with progressive desulfation over time. These were then used to investigate the influence of polymeric chain length and degree of sulfation (DS) on their anti-HPSE activity. The effects of these two features on λ-CO anticoagulant properties were also investigated to eliminate a potential limitation on the use of a candidate λ-CO as a chemotherapeutic agent. HPSE inhibition was mainly related to the DS of λ-CO, however this correlation was not complete. On the other hand, both chain length and DS modulated λ-CO activity for factor Xa and thrombin IIa inhibition, two enzymes that are involved in the coagulation cascade, and different mechanisms of inhibition were observed. A λ-carrageenan oligosaccharide of 5.9 KDa was identified as a suitable anticancer candidate because it displayed one of the lowest anticoagulant properties among the λ-CO produced, while showing a remarkable inhibitory effect on MDA-MB-231 breast cancer cell migration.

Keywords: λ-carrageenan; heparanase; anticoagulant; depolymerisation; cell migration

---

1. Introduction

In order to identify an alternative to conventional cancer chemotherapy, which is based on the inhibition of mitosis in malignant cells, medical research has become interested in the specific biological attributes of the tumor microenvironment [1,2]. One of the distinctive features of this microenvironment is the overexpression of numerous hydrolytic enzymes, including an endo-β-D-glucuronidase called Heparanase (HPSE) [3]. HPSE is the only enzyme able to hydrolyse the heparan sulfate (HS) chains of proteoglycans at specific glycosidic bonds that are components of the extracellular matrix and cell-surfaces [4]. This activity governs matrix 1 integrity and basement membrane degradation, allowing cancer cell migration and invasion. It concurrently releases sequestered HS-binding growth factors, cytokines or enzymes, leading to inflammatory and angiogenic signalling activation [5,6]. The identification of drugs targeting HPSE that could be used as promising anticancer therapeutics has therefore been the subject of many studies [7].

Sulfated polysaccharides are widely appreciated as a potent class of inhibitors because they are structurally related to HS, a natural substrate of HPSE [8]. However, their high molecular weight (MW), complex structure and broad range of bioactivities, which could lead to unforeseen events, have limited their clinical development for this precise application. For instance, heparin (UFH), a close analogue of HS, is considered a gold standard for HPSE inhibition, however its well-known anticoagulant activity limits its use in oncology due to the risk of internal bleeding [9,10]. Consequently, efforts have been made to screen oligosaccharide derivatives suitable for in vivo use and with a potentiated anti-HPSE specificity. The synthesis of oligosaccharides derivatives is essentially based on depolymerisation processes as well as chemical modifications to vary the degree of sulfation (DS) and acetylation or to open carbohydrate rings using the glycol-split method [9,11]. Four compounds are currently being tested in clinical trials: two are derived from heparin (Roneparstat® and Necuparanib®), one is a heterogeneous mixture of sulfomannan oligosaccharides (Muparfostat®) and the last is a synthetic tetrasaccharide conjugated to a steroid moiety (PG 545) [12]. However, their mechanism of action at the molecular level is not yet clearly understood [13]. Although the recent report showing for the first time the crystal structure of HPSE definitely constitutes a major breakthrough in the field that will lead to molecular modelling simulations [14], the structural heterogeneity of the candidates together with the challenging chemical carbohydrate synthesis remain obstacles to a detailed understanding of the specific units that mediate enzyme inhibition [15]. For instance, the role of glycol-split sugars used in several candidates is still under debate [16,17], while recent works have suggested a complex unusual mode of inhibition that varies with ligand concentration [18]. In addition, the bioactivities of oligosaccharide derivatives are clearly not restricted to the anti-HPSE activity and their anticancer effects are due to their interactions with other macromolecules, most notably growth factors [19,20]. Finally, in the case of heparin-based inhibitors, the difficulty of producing native heparin at a high yield constitutes a further complication, and the fact that they are of animal origin means that production has an environmental cost [21,22]. Thus, it is still necessary to screen natural polysaccharide derivatives and determine different production conditions in order to improve our understanding of the relationship between the carbohydrate architecture and the anti-HPSE function [23].

Carrageenans are a family of high MW sulfated galactans that are extracted from red seaweed in an environmentally-friendly way and are already extensively used in the food industry for their rheological properties [24]. They are characterised by long homogeneous linear chains of repeated disaccharide units consisting of a 1,3-linked β-D-galactopyranose (G unit) alternating with a 1,4-linked α-D-galactopyranose (D unit), differently sulfated depending on the species [25]. The G2S-D2S,6S disaccharide unit, bearing three sulfate groups, forms λ-carrageenans, which are known as the most sulfated plant-based polysaccharides with an ester sulfate content of about 35% in weight (Figure 1). Like other polysaccharides, carrageenans have many pharmacological properties, including their anticoagulant, antiviral, antioxidant or anticancer activities; these are summarized in the excellent review by Pangestuti et al. [26]. Again, this diversity limits their use to specific pharmacological applications in the clinic due to potentially serious adverse effects [27]. As a matter of fact, there is currently some debate over a pro-inflammatory and possible toxicity of the carrageenans used as food additives [28]. Fortunately, methods for modulating their bioactivities have been applied, with the main one involving various depolymerisation strategies [29]. In particular, three groups have reported anticancer activity of low MW λ-carrageenan derivatives [30–32]. These effects are mainly explained by a stimulation of the immune response [33–35], a direct cytotoxic effect [36,37] or an interaction with growth factors or associated receptors [31,38,39]. Initially hypothesized in some studies [40,41], an anti-HPSE activity has also been investigated recently [31,42,43]. However, most studies assessing the structure/activity of λ-carrageenan oligosaccharides have focused on the effect of polymer MW, however the exact role of each sulfate group in the anticancer effect, especially in HPSE inhibition, remains to be clarified. Moreover, the influence of the depolymerisation process on other bioactivities of λ-carrageenans that can limit their use as anticancer agents has barely been studied in parallel [41]. For instance, the anticoagulant properties of carrageenans, especially of the λ type, have been widely reviewed and could be one of these limitations [44].

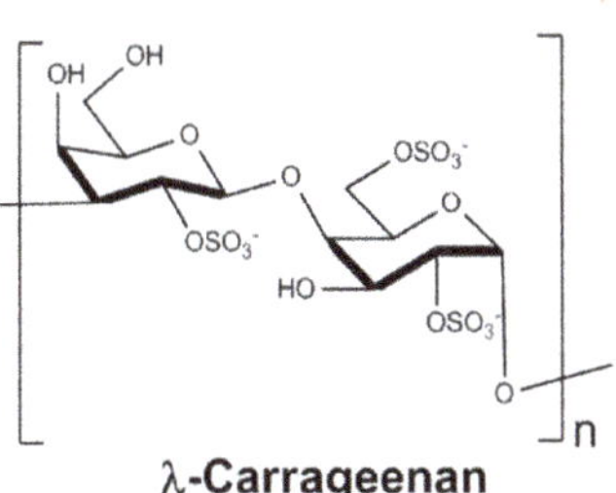

Figure 1. λ-carrageenan structure.

In this work, we developed a scaling-up method for the depolymerisation of λ-carrageenan under two temperature conditions, with a partial desulfation according to the reaction time. The anti-HPSE and anticoagulant properties were assessed to determine the influence of λ-carrageenan polymeric chain length and the role of the sulfate groups. We identified a potential anticancer candidate that was assessed in vitro for the inhibition of MDA-MB-231 breast cancer cell migration.

2. Results and Discussion

2.1. Depolymerisation of λ-Carrageenan

λ-carrageenans were depolymerised using a radical hydrolysis method with H_2O_2 at 40 and 60 °C. The H_2O_2/λ-carrageenan (w/w) ratio was set at 1.5, a ratio previously described as being suitable for obtaining low MW carrageenans [43]. The Mn of λ-CO produced over time was measured by SEC (size exclusion chromatography)-HPLC analysis (Table 1 and Appendix A, Figure A1). A calibration curve of pullulans was selected because, to the best of our knowledge, no carrageenan oligosaccharide standards are commercially available (Appendix A, Figure A2). Native λ-carrageenans are characterized by a high Mn of about 2586 kDa, corresponding to a DP of 4463, based on a reference weight of 579 Da per disaccharide unit. Depolymerisation was effective in both cases: there was an initial sudden fall in Mn within the first hours followed by a more progressive decrease, as shown by the logarithm scale used on the y-axis of Figure 2A. As expected, the kinetics of chain length reduction was faster at the highest temperature. Indeed, the Mn was reduced to 4.3 kDa after 48 h of reaction at 60 °C, which corresponds to a mean of 7 disaccharides units, while 240 h at 40 °C were needed to reach a similar level of depolymerisation (5.5 KDa). These results confirm previous work by our group showing that radical hydrolysis is an advantageous method for λ-carrageenan depolymerisation (Poupard, Groult, et al., 2017) and complements other techniques proposed such as acidic hydrolysis or the use of microwave protocols.

The effect of depolymerisation on the DS of λ-CO was then assessed (Table 1). Surprisingly, most sulfates were removed when the Mn was less than 20 kDa once the initial rapid kinetics of depolymerisation had slowed down. Thus, for an Mn ≥ 20 kDa, the effects of λ-CO Mn on various biological activities can be investigated separately from the influence of the sulfate groups. For an Mn below 20 kDa, a significant desulfation was observed. Indeed, native λ-carrageenans are characterized by a DS of ~30%, which is reduced to 8.7% for species with the lowest Mn, corresponding to a ~70% loss. This corresponds to a mean of 0.87 sulfate groups per disaccharide unit compared to the three sulfate groups bound to the native polymer. Overall, the DS was related to the Mn of the derivatives whatever the temperature condition although, in some cases, there was slightly less sulfate removal for derivatives prepared at 40 °C compared to those prepared at 60 °C (Figure 2B). It was concluded that the preservation of the DS was difficult to control by changes in the temperature condition with this depolymerisation method. Nevertheless, the slight differences will be used in subsequent structure/activity studies. This is of interest because the high DS of λ-carrageenans is often suggested to account for their better anticancer activity compared to the other types of carrageenans, however this aspect remains to be thoroughly investigated.

Table 1. Physicochemical properties of λ-CO. Molecular weight (Mn and Mw), Degree of polymerisation (DP) and degree of sulfation (DS).

Condition	Time (h)	Mn (KDa)	Mw (KDa)	DP	I	DS% (*w/w*)
	0	2585.9	2722.2	4463	1.1	~30
	2	931.3	1120.2	1607	1.2	29.5 ± 0.5
	4	752.5	987.2	1299	1.3	29.7 ± 0.5
	8	261.4	425.8	451	1.6	30.9 ± 0.7
Depolymerisation at 60 °C	12	31.7	42.7	55	1.4	28.0 ± 0.6
	16	17.8	23.8	31	1.3	22.3 ± 0.7
	24	10.0	12.7	17	1.3	20.3 ± 0.8
	36	5.9	8.4	10	1.4	17.8 ± 0.9
	48	4.3	6.8	7	1.6	15.9 ± 0.8
	72	2.8	3.7	5	1.4	8.7 ± 0.6
	0	2585.9	2722.2	4463	1.1	~30
	24	130.3	197.2	225	1.5	27.6 ± 0.5
	48	74.5	110.4	129	1.5	27.6 ± 0.7
	72	23.9	36.2	41	1.5	29.2 ± 0.8
Depolymerisation at 40 °C	96	13.0	19.2	22	1.5	23.8 ± 0.6
	120	11.7	17.4	20	1.5	24.6 ± 0.8
	144	9.5	14.3	16	1.5	23.7 ± 0.8
	168	7.6	11.2	13	1.5	22.5 ± 0.8
	240	5.5	8.0	10	1.4	19.0 ± 1.4

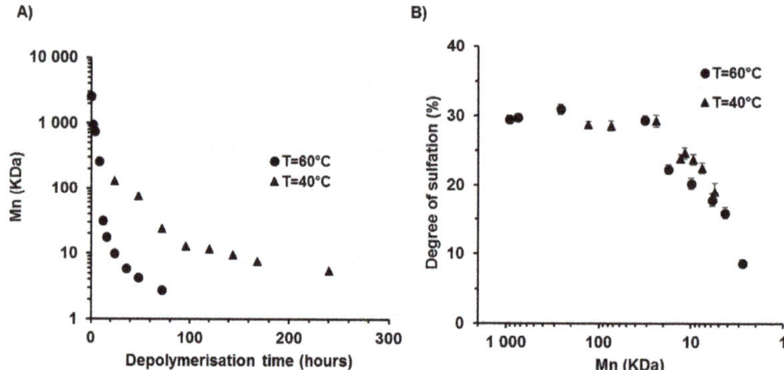

Figure 2. (**A**) Effect of the H_2O_2-assisted radical depolymerisation method on λ-carrageenan number average molecular weight (Mn) at 60 °C and 40 °C. (**B**) Degree of sulfation of λ-CO produced at 40 °C and 60 °C as a function of the Mn.

Thus, most previously published structure-activity relationship studies have focused on the influence of the length of oligosaccharide derivatives, while the outcome of the sulfation pattern has barely been addressed [30–32]. Navarro et al. previously reported that the 2-O sulfate position of the G unit is more resistant to hydrolysis [45]. Thus, we may assume that the desulfation observed may be distinguished by specific sulfate substitutions.

2.2. Effects of the Degree of Depolymerisation and Sulfation on the Anti-Heparanase Activity of λ-CO

We then studied the impact of polymer chain length and sulfate level on the anti-HPSE activity of λ-carrageenan derivatives. HPSE cleaves the β-glycosidic bonds of HS chains between a glucuronic acid (GlcA) and a glucosamine (GlcN) through a general acid catalysis mechanism which involves two glutamate residues. Two patches of basic amino acids at either side of the catalytic site, known as the heparin binding domains HBD-1 and -2, coordinate the interaction with HS sites of a specific sulfation pattern [4]. The inhibition of a recombinant HPSE by each oligosaccharide from a labelled-HS hydrolysis was monitored using a FRET (fluorescence resonance energy transfer)-based assay at a fixed concentration of 1.25×10^{-3} mg·mL^{-1} (Figure 3A). The anti-HPSE activity was maintained initially before it was to be reduced at ~20 kDa, which corresponds to the value at which desulfation starts. This indicated that the inhibitory activity was not correlated with the Mn of the derivatives, at least for those greater than 20 kDa. In fact, a rare lengthy structural pattern cannot be excluded as an explanation for the reduced activity of λ-CO below this Mn, although it is more likely to be the result of the decrease in the DS. Indeed, when the anti-HPSE activity was plotted against the DS, there was a clear correlation (Figure 3B). Interestingly, this activity seemed to be more drastically impaired when the DS was less than 20%, which corresponds to a mean loss of one sulfate group per disaccharide unit. In case of discriminated desulfation, this could indicate that the first preferentially removed sulfate substitution is not essential for enzyme inhibition. In line with this result, previous work on heparin has shown that the concurrent presence of sulfate groups at the O-2 position of IdoA and at the O-6 position of GlcN was not mandatory for an effective inhibition of HPSE [11].

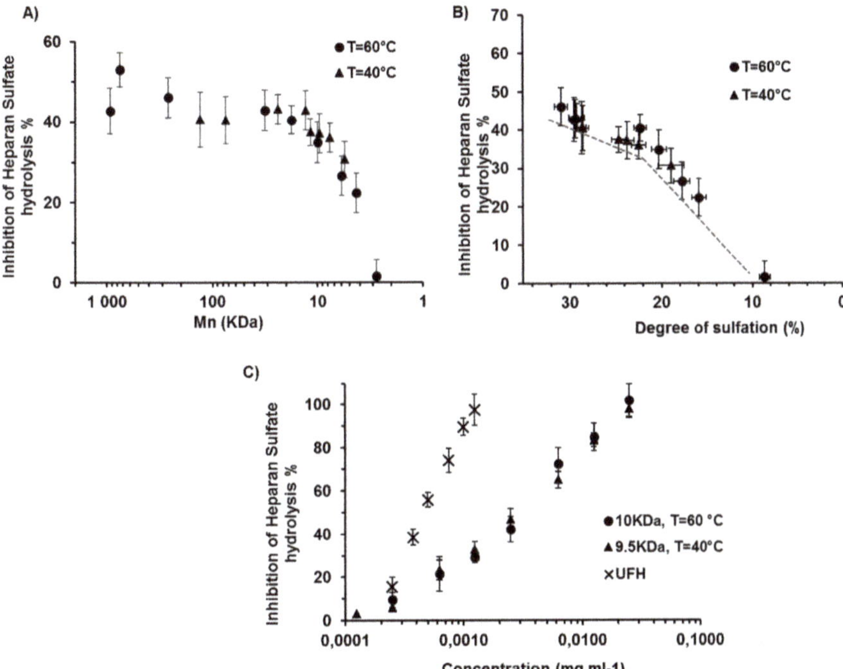

Figure 3. (**A**) Anti-HPSE (heparanase) activity of λ-CO produced at 40 °C and 60 °C as a function of their Mn for the inhibition of heparan sulfate hydrolysis by heparanase at 1.25×10^{-3} mg·mL^{-1}. (**B**) Correlation between the DS of λ-CO produced at 40 °C and 60 °C and their inhibitory activity against heparan sulfate hydrolysis by heparanase. (**C**) IC$_{50}$ of 10 kDa (produced at 60 °C, after 24 h) and 9.5 kDa λ-CO (produced at 40 °C, after 144 h) compared to native UFH (heparin); (HPSE = 100 ng·mL^{-1} and HS = 0.5 µg·mL^{-1}).

We then compared the half maximal inhibitory concentrations (IC_{50}) of two of our derivatives of about 10 kDa (24 h at 60 °C and 144 h at 40 °C) to that of native heparin, a gold standard for heparanase inhibition (Figure 3C). λ-CO had a moderate ability to inhibit HPSE with an IC_{50} about six times higher than that of native heparin, 3.1 mg·L^{-1} and 3.0 mg·L^{-1} versus 0.47 mg·L^{-1}, respectively. Although they were less potent HPSE inhibitors than heparin, they may represent a promising alternative given heparin's disadvantages, which include a costly and low-yield production, strong possibility of contamination and the fact that it is of animal origin, with the environmental issue that entails.

2.3. Effects of the Degree of Depolymerisation and Sulfation on the Anticoagulant Activity of λ-CO

A λ-carrageenans have been shown to have a higher anticoagulant activity compared to other members of the carrageenan family, although this activity is much weaker than that of heparin, a reference in this field [24,46]. Their anticoagulant activity has been attributed mainly to the inhibition of thrombin IIa and Factor Xa mediated by anti-thrombin III (AT-III) and/or heparin cofactor II [44]. The interactions between carrageenans and these plasma cofactors are complex and the functions of various parameters have been extensively discussed, including saccharide composition, MW, charge density, DS and sulfate position [47–49]. The anticoagulant properties of the carrageenan derivatives prepared in this study were thus studied through their ability to inhibit Factors Xa and IIa via AT-III activation. These key factors intervene at the end of the coagulation cascade to activate fibrin formation, which will then polymerize to form blood clots. To discuss the results, a mechanism similar to that of heparin was hypthesized. This consists of an AT-III conformational activation through binding with a specific pentasaccharide sequence present in the heparin chain, and the resulting complex inhibits Factors Xa and IIa. Regarding Factor IIa, an additional steric hindrance effect caused by the other sugars of a sufficiently long heparin chain is involved [50]. The anti-Xa activity of each oligosaccharide was measured at 0.025 mg·mL^{-1} by following the initial velocity for converting a chromogenic substrate compared to a control (vehicle). An increase in anti-Xa activity associated with the initial reduction in chain length was first observed for the derivatives (Figure 4A). This could be due to a higher steric freedom gained from the start of the depolymerisation of the very long native polymer that allows a better accessibility to the probable binding sequence of carrageenan to AT-III. Then, from 20 kDa, a decrease in anti-Xa activity was observed that was clearly related to the desulfation of λ-CO (Appendix A, Figure A3). For example, the 17.8 kDa and 7.5 kDa derivatives with an equivalent DS of ~22.5% had the same anti-Xa activity of about 70%. This result suggests that sugars with key sulfate substitutions are included in the potential binding sequence of λ-carrageenans to AT-III. Moreover, it appeared that in this case, the three sulfate positions had to be present because the correlation showed that anti-Xa activity was linearly impaired from the start of desulfation (Appendix A, Figure A3). In the literature, the role of the sulfate positions about the anticoagulant properties of λ-carrageenan is still under debate. Two previous studies have shown that sulfation at the C2 position of the D unit were beneficial to the anticoagulant activity [51,52].

Regarding factor IIa, overall the activities were lower than those of factor Xa and monitoring was performed at 0.125 mg·mL^{-1}. As shown in Figure 4B, anti-IIa activity was first maintained before it rapidly decreased from 50 kDa, well before the start of desulfation. This result could be explained by the fact that depolymerisation decreases the additional steric effect needed for factor IIa inhibition, which is due to the long polymeric chains. Similarly, the thorough study by Melo et al. stressed that a MW higher than 45 kDa is required for the interaction between galactan oligosaccharides and factor IIa during the time of binding to AT-III [49]. Thus, although desulfation contributed to a decrease in anti-IIa activity by modifying the sequence by which λ-carrageenans bind to AT-III, the role of the polymeric chain length/Mn appeared to be more significant (Appendix A, Figure A3). Finally and somewhat surprisingly, anti-IIa activity was recovered for the smallest oligosaccharide that was produced at 60 °C (2.77 kDa; 31.5% inhibition) and was almost completely desulfated. λ-CO did not inhibit factor Xa or factor IIa in the absence of AT III (negative control) except for derivatives of less than 6 kDa. This suggests that an oligosaccharide chemically defined by very low sulfate substitutions on the galactopyranose units was obtained and that it had a direct anti-thrombin activity. Other studies have shown that other oligosaccharides were likely to interact directly with the proteins of the coagulation cascades without potentiation of AT-III [53,54]. Taken together, the results showed that, although anti-IIa activity was rapidly abolished for λ-CO of less than 50 kDa, the Mn range of 20–500 kDa induced anti-Xa properties and should not be used for the development of a λ-carrageenan-based anticancer candidate to avoid any adverse effects like internal bleeding.

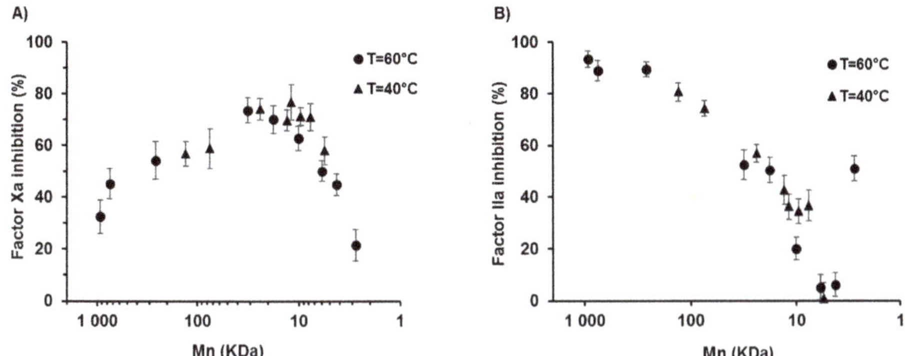

Figure 4. (**A**) Anti-Factor Xa activity of λ-CO produced at 40 °C and 60 °C as a function of their Mn, at 0.025 mg·mL^{-1} and (**B**) Anti-Factor IIa activity of λ-CO produced at 40 °C and 60 °C as a function of their Mn, at 0.125 mg·mL^{-1} (AT III = 0.625 µg·µL^{-1} and factor Xa or IIa = 11.25 nK at·mL^{-1}).

2.4. Effect of λ-CO Candidates on the Migration of MDA-MB-231 Breast Cancer Cells In Vitro

Since HPSE activity has been shown to be involved in the metastatic potential of cancer cells, the effect of a λ-CO derivative (5.9 KDa produced at 60 °C) on HPSE-associated migration were compared to native heparin using a transwell assay [55,56]. The highly motile MDA-231 breast cancer cells with high-level expression of HPSE were selected for this experiment [42,57]. As shown in Figure 5A, large number of control cells (treated with vehicle) passed through the pores towards the lower chambers in response to a 10% FBS (Fetal Bovine Serum) solution that was used as a chemoattractant. Treatment with native heparin led to a moderate 7.6% inhibition rate in accordance with previously published results [58–60]. Treatment with the λ-CO candidate at 100 µg·mL^{-1} significantly reduced the migration of MDA-MB-231 cells. The inhibition rate (32.8%) was higher than with heparin, while heparin had the best IC50 against HPSE activity (Figure 5B). The experiment was also repeated on the more quiescent MCF-7 breast cancer cell line (Appendix A, Figure A4) [61]. In this case, heparin slightly promoted the MCF-7 cells migration. Though, here, the λ-CO candidate again displayed an inhibitory activity of 12% on MCF-7 cell migration, however this was non-significant. As anticipated, this revealed that in addition to HPSE, the λ-CO candidate probably interacts with other molecules involved in the motility of cancer cells. It is known that migration is regulated by a complex interplay between varied glycosaminoglycans (e.g., syndecan-4 at the cell surface), protein expressions and degradative enzymes, in which heparin may have a different impact [62].

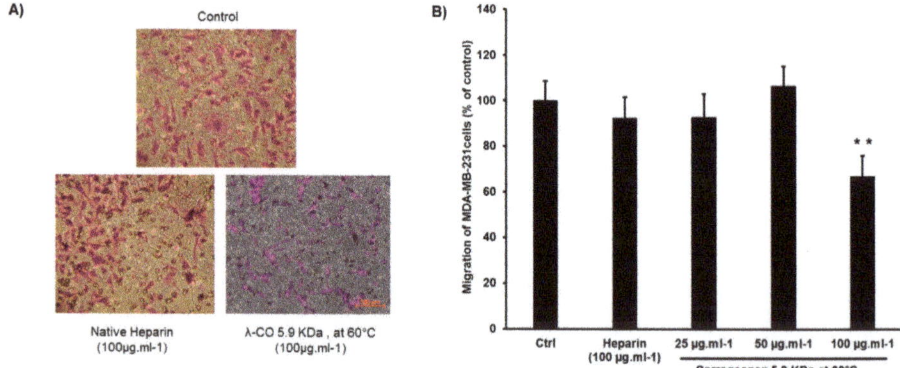

Figure 5. (**A**) Images showing the effects and (**B**) the inhibition activity at three doses 25, 50 and 100 µg·mL^{-1} of the λ-CO candidate 5.9 KDa prepared at 60 °C, compared with native heparin (at 100 µg·mL^{-1}) on the migration ability of human breast MDA-MB-231 cells. The migration ability was assessed in a transwell assay after 24 h as described in Materials and Methods. The data are representative as the mean (±SEM of the errors mean) from three independent experiments, with at least four replicates. ** $p < 0.01$, test Anova-Two way Bonneferonni with mean ± SD of the three independent experiments.

3. Materials and Methods

All reagents, unless otherwise specified, were purchased from Sigma Aldrich (Saint Louis, MO, USA). Native λ-carrageenans were purchased from FMC Biopolymer (Villefranche-Sur-Saône, France).

3.1. Depolymerisation of λ-Carrageenan for the Production of Oligosaccharides (λ-CO)

Native λ-carrageenans were dissolved in 200 mL of Milli-Q water at a concentration of 5 mg·mL^{-1}. The solution was rapidly heated to 40–50 °C to completely dissolve the polysaccharide and then purged under argon. Then, 30% hydrogen peroxide (H_2O_2) (5 mL) was added and the reaction mixture was immediately sealed and placed in an incubator at 40 °C or 60 °C under 200 rpm stirring. Aliquots were taken at different time points and were dry frozen prior to analysis.

3.2. Structural and Quantitative Analysis of λ-CO by Size Exclusion Chromatography (SEC)

Structural and quantitative analysis of λ-CO by size exclusion chromatography (SEC) were performed using a LC/MS-ES system from Agilent (Santa Clara, CA, USA) (1100 LC/MSD Trap VL mass spectrometer) with two columns, TSK-GEL G5000PW and TSK-GEL G4000PW (30 cm × 7.5 mm), mounted in series. The columns were maintained at 30 °C and the products were eluted with 0.1 M Sodium nitrate ($NaNO_3$) at a flow rate of 0.5 mL·min^{-1}. The products were detected and quantified by differential refractometry using HP Chemstation software (Agilent, Santa Clara, CA, USA). Pullulans of different molecular weights ranging from 1.3 to 805 kDa purchased from Polymer Standards Service GmbH (Mainz, Germany) were used as calibrants for the standard curve and to determine the size of the carrageenan derivatives. The number-average molecular weights (Mn), weight-average molecular weights (Mw) and polydispersity index (PI) were calculated according to a previously published method [63] using the following equations:

$$M_n = \frac{(\sum N_i \times M_i)}{\sum N_i} \tag{1}$$

$$M_w = \frac{(\sum N_i \times M_i^2)}{(\sum N_i \times M_i)} \tag{2}$$

$$PI = \frac{M_w}{M_n} \tag{3}$$

with N_i representing the number of moles of polymer species and M_i the MW of the polymer species. The degree of polymerization (DP) was calculated as follows:

$$DP = \frac{M_n}{M_0} \tag{4}$$

with M_0 representing the G2S-D2S,6S disaccharide unit MW set at 579 Da.

3.3. Quantification of the Sulfation Degree of λ-CO

The DS was monitored using (7-aminophenothiazin-3-ylidene)-dimethylazanium chloride (Azure A), which binds the sulfated groups on the sugar backbone to form a colored complex [64]. In a 96-well plate, 20 µL of three dilutions (0.03, 0.04 and 0.05 mg·mL^{-1}) of λ-CO samples were added to 200 µL of a 10 mg·L^{-1} aqueous Azure A solution. Absorbance was measured at 640 nm after 10 min of incubation. DSs were calculated from a calibration curve constructed using absorbance values obtained from a serial dilution (0–0.03 mg·mL^{-1}) of a dextran sulfate standard with a known sulfur content of 17%.

3.4. Anti-Heparanase Activity of λ-CO

The inhibition of HPSE activity was assessed using the heparanase assay toolbox (Cisbio Assay, Codolet, France) and heparanase purchased from R&D systems (HPSE-1 human recombinant heparanase). Briefly, upon excitation at 337 nm, a HS substrate labeled with both biotin and Eu3+ cryptate can produce a fluorescent emission at 665 nm by fluorescence resonance energy transfer (FRET) to streptavidin-XL665 (SA-XL665), which is added during the detection step. During hydrolysis, HPSE cleaves the substrate, resulting in a loss of possible energy transfer and thus, a reduction in SA-XL665 emissions. The enzymatic reaction was performed in white 96-well half-area plates (Corning® #3693) and was monitored using a spectrofluorometer (BMG Labtech FLUOstar Omega, Champigny-sur-Marne, France) with a high time resolved fluorescence (HTRF) module. First, 15 µL of λ-CO or heparin solutions in Milli-Q water were added into the wells followed by 15 µL of heparanase solution (HPSE-1, 400 ng·mL^{-1} in Tris-HCl at pH 7.5, 0.15 M NaCl and 0.1% CHAPS). After a 10-min pre-incubation at 37 °C, an enzyme reaction was initiated by adding 30 µL of a Biotin-HS-Eu(K) solution (1.0 ng·µL^{-1} in 0.2 M sodium acetate buffer, pH 4.5) and the plate was incubated at 37 °C for 15 min. At the end of the reaction, the detection step consisted of adding 30 µL of streptavidin-XL665 solution (SA-XL665, 10 ng·µL^{-1} in NaPO$_4$ 0.1 M buffer, pH 7.4, 0.8 M KF, 0.1% BSA, 1 mg·mL^{-1} heparin). The fluorescence was measured after 5 min at λ_{em1} = 620 nm and λ_{em2} = 665 nm, after 60 µs of excitation at λ_{ex} = 337 nm. The Delta F (%) was calculated using the following equation according to the manufacturer's instructions:

$$\text{Delta F } (\%) = \frac{(F665/F620)\ sample - (F665/F620)\ blank}{(F665/F620)\ blank} \times 100 \tag{5}$$

with F665 and F620 representing the fluorescence signals measured at 665 nm and 620 nm, respectively. The percentage of inhibition was calculated based on the Delta F(%) of the maximum heparanase activity measured in the absence of inhibitor. The HPSE activity of each λ-CO was measured at a final concentration of 1.25×10^{-3} mg·mL^{-1}. For the IC$_{50}$ calculations, a curve-fitting tool from SigmaPlot software (Systat Software Inc, San Jose, CA, USA) was applied using a sigmoidal, logistic three-parameter equation.

3.5. Anticoagulant Activity of λ-CO

For anti-Xa and anti-IIa activity assays, 25 µL of λ-CO solution in Milli-Q water were incubated with anti-thrombin III (25 µL, 0.625 µg·µL^{-1}) at 37 °C in 96-well plates for 2 min. Then, factor Xa or factor IIa was added at a final concentration of 11.25 nKat·mL^{-1} (25 µL). After 2 min of incubation, 3.25 nM (25 µL) of factor Xa chromogenic substrate (CBS 31.39; CH2SO2-D205 Leu-Gly-Arg-pNA, AcOH) for the anti-Xa activity assay or 1.4 nM (25 µL) of factor IIa chromogenic substrate (CBS 61.50; EtM-SPro-Arg-pNA, AcOH) for the anti-IIa activity assay were added. Absorbance of the reaction mixture was read for 3 min at 405 nm every 8 s with an absorbance reader (FLUOstar Omega BMG Labtech, Champigny-sur-Marne, France). The initial velocity was determined as the slope of the linear segment of the kinetics curve and the % of inhibition was calculated based on the initial velocity of an inhibitor-free blank (Milli-Q water). Anti-Xa and anti-IIa activities of each λ-CO were measured at a final concentration of 0.025 mg·mL^{-1} and 0.125 mg·mL^{-1}, respectively. Controls used to assess a direct inhibition of Factors Xa or IIa were performed with the same protocol, however the anti-thrombin III solution was replaced by a vehicle solution (Milli-Q water). Native heparin at the same concentrations has been used as a positive control for 100% inhibition to validate the assay.

3.6. MDA-MB-231 Cell Migration Assay

A total of 1×10^5 MDA-MB-231 or MCF-7 breast tumor cells were seeded in serum-free DMEM medium in the upper chambers of a transwell plate (8-µm pore size; Stardest). After 24 h, the medium was replaced by fresh serum-free DMEM medium treated with the polysaccharides candidates at concentration between 25 and 100 µg·mL^{-1} or the vehicle (Milli-Q water); meanwhile, a complete DMEM medium with 10% FBS was added to the lower chambers. After a 24 h culture at 37 °C, the cells that had migrated to the lower chambers were fixed with cold ethanol and stained with 0.1% crystal violet. Then, the non-migrating cells were removed from the upper chambers by wiping the membrane with cotton swabs. The remaining cells were photographed and eluted with 10% acetic acid solution. The absorbance of the resulting dilution was measured at 600 nm with an absorbance reader (FLUOstar Omega BMG Labtech, Champigny-sur-Marne, France). The percentage of inhibition was calculated by the following equation:

$$\text{Inhibition (\%)} = \left(1 - \frac{A_{sample}}{A_{blank}}\right) \times 100 \tag{6}$$

with A_{sample} representing the absorbance of 10% acetic acid solutions obtained from the cells treated with the polysaccharides candidates and A_{blank} representing the absorbance obtained from the cells treated with the vehicle.

4. Conclusions

We developed a radical hydrolysis method for an easy large-scale production of structurally varied low MW λ-carrageenan derivatives. According to the depolymerisation temperature and time, λ-CO with distinct Mn and DS can be obtained. This is relevant for structure/activity relationship studies since previous studies have only focused on the Mn of the derivatives without really discussing the role of the sulfate substitutions of λ-carrageenans. HPSE inhibition was assessed and it was mainly correlated with the DS of λ-CO without a clear correlation with the Mn. It also appeared that one of the three sulfates of the λ-carrageenan disaccharide unit may not be essential for HPSE inhibition. Further experiments are needed to assess not only the effect of the overall DS of species, however also the specific role of each sulfate substitution. The development of λ-carrageenan-based anticancer candidates could be limited by unexpected anticoagulant properties, which were therefore assessed. Unlike HPSE inhibition, interactions between λ-CO and coagulation factors Xa and IIa were correlated with both the Mn and the DS. Moreover, while most species inhibited factors Xa and IIa through their interaction with AT-III, the smallest desulfated λ-CO was identified as a direct thrombin inhibitor. It would be interesting to screen other bioactivities of λ-carrageenans that could be related to adverse effects in oncology, including their pro-inflammatory properties.

We identified a suitable anticancer drug candidate with an anti-heparanase activity, the 5.9 kDa λ-CO produced after 36 h at 60 °C. At 60 °C, its production was more rapid than at 40 °C and its size is small enough to be appropriated for in vivo applications. Although its DS is slightly below the threshold DS for effective HPSE inhibition, at the same time, its partial desulfation (17%) and size guarantee not to fall within specifications for optimum λ-CO anticoagulant properties. Moreover, in a preliminary anticancer biological assessment, it showed very promising activity against MDA-MB-231 cell migration. To complete this finding, the study should be extended to other targets involved in tumor development and that are likely to interact with λ-CO, in particular the HS-sequestered molecules, such as growth factors.

Author Contributions: Conceptualization, H.G., T.M. and I.F.-A.; methodology, H.G., C.C.-P. and M.M.; validation, N.B., J.-M.P., T.M. and I.F.-A.; formal analysis, H.G. and R.C.; investigation, H.G., R.C., C.C.-P. and M.M.; data curation, H.G. and R.C.; writing—original draft preparation, H.G.; writing—review and editing, J.-M.P., T.M. and I.F.-A.; visualization, N.B.; supervision, T.M. and I.F.-A.; funding acquisition, J.-M.P., T.M. and I.F.-A.

Funding: This work was supported by the Ligue Contre le Cancer (Comité CD17 Charente Maritime and CSIRGO) and the Region Nouvelle Aquitaine ("Nanovect" Project). The authors would like to acknowledge the financial support.

Acknowledgments: The authors acknowledge Beatrice Colin for her help in the statistical analysis.

Conflicts of Interest: The authors declare no conflict of interest.

Appendix A

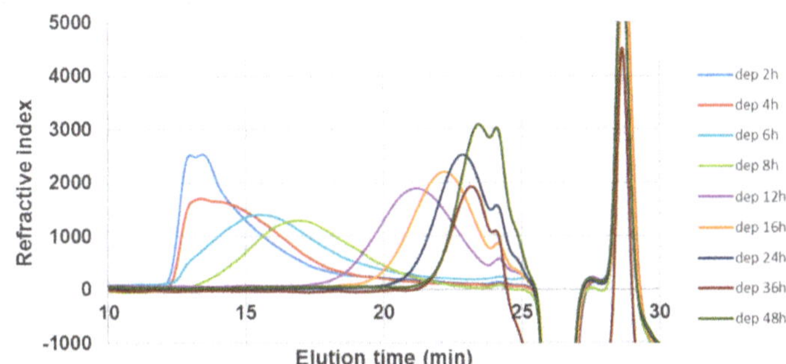

Figure A1. SEC-HPLC analysis with refractive detector of λ-CO produced along time. The SEC (size exclusion chromatography) separation was performed on a TSK-GEL G5000PW column in series with a TSK-GEL G4000PW at a flow rate of 0.5 mL/min using 0.1 M sodium nitrate (NaNO3) as the eluent.

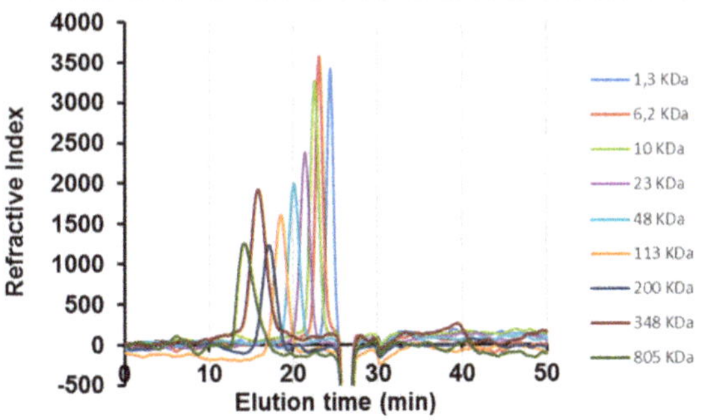

Figure A2. SEC-HPLC analysis with a refractive detector of pullulan standards of different molecular weights to construct the calibration curve. The SEC separation was performed on a TSK-GEL G5000PW column in series with a TSK-GEL G4000PW at a flow rate of 0.5 mL/min using 0.1 M sodium nitrate (NaNO3) as the eluent.

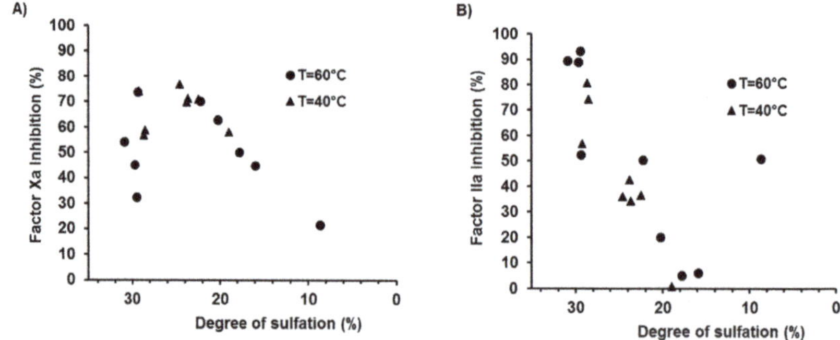

Figure A3. (**A**) Correlation between the DS of λ-CO produced at 40 °C and 60 °C and their inhibitory activity against Factor Xa and (**B**) Correlation between the DS of λ-CO produced at 40 °C and 60 °C and their inhibitory activity against Factor IIa.

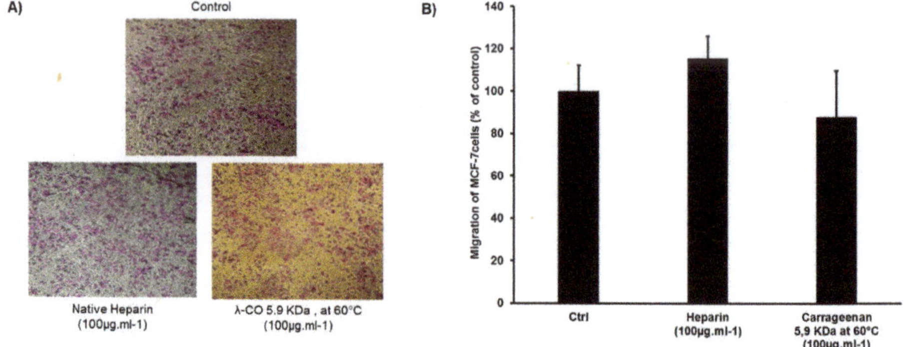

Figure A4. (**A**) Images showing the effects and (**B**) the inhibition activity of the λ-CO candidate 5.9 KDa prepared at 60 °C compared with native heparin (at 100 μg·mL^{-1}) on the migration ability of human breast MCF-7 cells. The migration ability was assessed in a transwell assay after 24 h, as described in Materials and Methods. For Heparin, the data are representative as the mean (±SEM of the errors mean) from three independent experiments, with at least four replicates. For λ-CO, the data are representative as the mean ± SD of one experiment, with at least four replicates.

References

1. Belli, C.; Trapani, D.; Viale, G.; D'Amico, P.; Duso, B.A.; Della Vigna, P.; Orsi, F.; Curigliano, G. Targeting the microenvironment in solid tumors. *Cancer Treat. Rev.* **2018**, *65*, 22–32. [CrossRef] [PubMed]

2. Chen, F.; Zhuang, X.; Lin, L.; Yu, P.; Wang, Y.; Shi, Y.; Hu, G.; Sun, Y. New horizons in tumor microenvironment biology: Challenges and opportunities. *BMC Med.* **2015**, *13*, 45. [CrossRef] [PubMed]

3. Balkwill, F.R.; Capasso, M.; Hagemann, T. The tumor microenvironment at a glance. *J. Cell Sci.* **2012**, *125*, 5591–5596. [CrossRef] [PubMed]

4. Rivara, S.; Milazzo, F.M.; Giannini, G. Heparanase: A rainbow pharmacological target associated to multiple pathologies including rare diseases. *Future Med. Chem.* **2016**, *8*, 647–680. [CrossRef] [PubMed]

5. Vlodavsky, I.; Friedmann, Y. Molecular properties and involvement of heparanase in cancer metastasis and angiogenesis. *J. Clin. Investig.* **2001**, *108*, 341–347. [CrossRef] [PubMed]

6. Nadir, Y.; Brenner, B. Heparanase multiple effects in cancer. *Thromb. Res.* **2014**, *133*, S90–S94. [CrossRef]

7. Masola, V.; Secchi, M.F.; Gambaro, G.; Onisto, M. Heparanase as a target in cancer therapy. *Curr. Cancer Drug Targets* **2014**, *14*, 286–293. [CrossRef] [PubMed]

8. Jia, L.; Ma, S. Recent advances in the discovery of heparanase inhibitors as anti-cancer agents. *Eur. J. Med. Chem.* **2016**, *121*, 209–220. [CrossRef] [PubMed]

9. Vlodavsky, I.; Ilan, N.; Naggi, A.; Casu, B. Heparanase: Structure, biological functions, and inhibition by heparin-derived mimetics of heparan sulfate. *Curr. Pharm. Des.* **2007**, *13*, 2057–2073. [CrossRef] [PubMed]

10. Warkentin, T.E.; Levine, M.N.; Hirsh, J.; Horsewood, P.; Roberts, R.S.; Gent, M.; Kelton, J.G. Heparin-Induced Thrombocytopenia in Patients Treated with Low-Molecular-Weight Heparin or Unfractionated Heparin. *N. Engl. J. Med.* **1995**, *332*, 1330–1336. [CrossRef] [PubMed]

11. Naggi, A.; Casu, B.; Perez, M.; Torri, G.; Cassinelli, G.; Penco, S.; Pisano, C.; Giannini, G.; Ishai-Michaeli, R.; Vlodavsky, I. Modulation of the Heparanase-inhibiting Activity of Heparin through Selective Desulfation, Graded *N*-Acetylation, and Glycol Splitting. *J. Biol. Chem.* **2005**, *280*, 12103–12113. [CrossRef] [PubMed]

12. Arvatz, G.; Weissmann, M.; Ilan, N.; Vlodavsky, I. Heparanase and cancer progression: New directions, new promises. *Hum. Vaccines Immunother.* **2016**, *12*, 2253–2256. [CrossRef] [PubMed]

13. Loka, R.S.; Yu, F.; Sletten, E.T.; Nguyen, H.M. Design, synthesis, and evaluation of heparan sulfate mimicking glycopolymers for inhibiting heparanase activity. *Chem. Commun.* **2017**, *53*, 9163–9166. [CrossRef] [PubMed]

14. Wu, L.; Viola, C.M.; Brzozowski, A.M.; Davies, G.J. Structural characterization of human heparanase reveals insights into substrate recognition. *Nat. Struct. Mol. Biol.* **2015**, *22*, 1016–1022. [CrossRef] [PubMed]

15. Roy, S.; El Hadri, A.; Richard, S.; Denis, F.; Holte, K.; Duffner, J.; Yu, F.; Galcheva-Gargova, Z.; Capila, I.; Schultes, B.; et al. Synthesis and Biological Evaluation of a Unique Heparin Mimetic Hexasaccharide for Structure–Activity Relationship Studies. *J. Med. Chem.* **2014**, *57*, 4511–4520. [CrossRef] [PubMed]

16. Alekseeva, A.; Casu, B.; Cassinelli, G.; Guerrini, M.; Torri, G.; Naggi, A. Structural features of glycol-split low-molecular-weight heparins and their heparin lyase generated fragments. *Anal. Bioanal. Chem.* **2014**, *406*, 249–265. [CrossRef] [PubMed]

17. Ni, M.; Elli, S.; Naggi, A.; Guerrini, M.; Torri, G.; Petitou, M. Investigating Glycol-Split-Heparin-Derived Inhibitors of Heparanase: A Study of Synthetic Trisaccharides. *Molecules* **2016**, *21*, 1602. [CrossRef] [PubMed]

18. Pala, D.; Rivara, S.; Mor, M.; Milazzo, F.M.; Roscilli, G.; Pavoni, E.; Giannini, G. Kinetic analysis and molecular modeling of the inhibition mechanism of roneparstat (SST0001) on human heparanase. *Glycobiology* **2016**, *26*, 640–654. [CrossRef] [PubMed]

19. Zhou, H.; Roy, S.; Cochran, E.; Zouaoui, R.; Chu, C.L.; Duffner, J.; Zhao, G.; Smith, S.; Galcheva-Gargova, Z.; Karlgren, J.; et al. M402, a Novel Heparan Sulfate Mimetic, Targets Multiple Pathways Implicated in Tumor Progression and Metastasis. *PLoS ONE* **2011**, *6*, e21106. [CrossRef] [PubMed]

20. Liang, X.-J.; Yuan, L.; Hu, J.; Yu, H.-H.; Li, T.; Lin, S.-F.; Tang, S.-B. Phosphomannopentaose sulfate (PI-88) suppresses angiogenesis by downregulating heparanase and vascular endothelial growth factor in an oxygen-induced retinal neovascularization animal model. *Mol. Vis.* **2012**, *18*, 1649–1657. [PubMed]

21. Guerrini, M.; Shriver, Z.; Bisio, A.; Naggi, A.; Casu, B.; Sasisekharan, R.; Torri, G. The tainted heparin story: An update. *Thromb. Haemost.* **2009**, *102*, 907–911. [CrossRef] [PubMed]

22. Szajek, A.Y.; Chess, E.; Johansen, K.; Gratzl, G.; Gray, E.; Keire, D.; Linhardt, R.J.; Liu, J.; Morris, T.; Mulloy, B.; et al. The US regulatory and pharmacopeia response to the global heparin contamination crisis. *Nat. Biotechnol.* **2016**, *34*, 625–630. [CrossRef] [PubMed]

23. de Jesus Raposo, M.; de Morais, A.; de Morais, R. Marine Polysaccharides from Algae with Potential Biomedical Applications. *Mar. Drugs* **2015**, *13*, 2967–3028. [CrossRef] [PubMed]

24. Necas, J.; Bartosikova, L. Carrageenan: A review. *Veterinární Med.* **2013**, *58*, 187–205. [CrossRef]

25. Campo, V.L.; Kawano, D.F.; da Silva, D.B.; Carvalho, I. Carrageenans: Biological properties, chemical modifications and structural analysis—A review. *Carbohydr. Polym.* **2009**, *77*, 167–180. [CrossRef]

26. Pangestuti, R.; Kim, S.-K. Biological Activities of Carrageenan. In *Advances in Food and Nutrition Research*; Elsevier: Amsterdam, The Netherlands, 2014; Volume 72, pp. 113–124. ISBN 978-0-12-800269-8.

27. Zia, K.M.; Tabasum, S.; Nasif, M.; Sultan, N.; Aslam, N.; Noreen, A.; Zuber, M. A review on synthesis, properties and applications of natural polymer based carrageenan blends and composites. *Int. J. Biol. Macromol.* **2017**, *96*, 282–301. [CrossRef] [PubMed]

28. Tobacman, J.K. Review of harmful gastrointestinal effects of carrageenan in animal experiments. *Environ. Health Perspect.* **2001**, *109*, 983–994. [CrossRef] [PubMed]

29. Sun, Y.; Yang, B.; Wu, Y.; Liu, Y.; Gu, X.; Zhang, H.; Wang, C.; Cao, H.; Huang, L.; Wang, Z. Structural characterization and antioxidant activities of κ-carrageenan oligosaccharides degraded by different methods. *Food Chem.* **2015**, *178*, 311–318. [CrossRef] [PubMed]

30. Chen, H.; Yan, X.; Lin, J.; Wang, F.; Xu, W. Depolymerized products of lambda-carrageenan as a potent angiogenesis inhibitor. *J. Agric. Food Chem.* **2007**, *55*, 6910–6917. [CrossRef] [PubMed]

31. Niu, T.-T.; Zhang, D.-S.; Chen, H.-M.; Yan, X.-J. Modulation of the binding of basic fibroblast growth factor and heparanase activity by purified λ-carrageenan oligosaccharides. *Carbohydr. Polym.* **2015**, *125*, 76–84. [CrossRef] [PubMed]

32. Zhou, G. In vivo antitumor and immunomodulation activities of different molecular weight lambda-carrageenans from *Chondrus ocellatus*. *Pharmacol. Res.* **2004**, *50*, 47–53. [CrossRef] [PubMed]

33. Luo, M.; Shao, B.; Nie, W.; Wei, X.-W.; Li, Y.-L.; Wang, B.-L.; He, Z.-Y.; Liang, X.; Ye, T.-H.; Wei, Y.-Q. Antitumor and Adjuvant Activity of λ-carrageenan by Stimulating Immune Response in Cancer Immunotherapy. *Sci. Rep.* **2015**, *5*, 11062. [CrossRef] [PubMed]

34. Zhou, G.; Sheng, W.; Yao, W.; Wang, C. Effect of low molecular λ-carrageenan from Chondrus ocellatus on antitumor H-22 activity of 5-Fu. *Pharmacol. Res.* **2006**, *53*, 129–134. [CrossRef] [PubMed]

35. Zhou, G.; Xin, H.; Sheng, W.; Sun, Y.; Li, Z.; Xu, Z. In vivo growth-inhibition of S180 tumor by mixture of 5-Fu and low molecular λ-carrageenan from. *Pharmacol. Res.* **2005**, *51*, 153–157. [CrossRef] [PubMed]

36. Chen, H.-M.; Yan, X.-J.; Mai, T.-Y.; Wang, F.; Xu, W.-F. Lambda-carrageenan oligosaccharides elicit reactive oxygen species production resulting in mitochondrial-dependent apoptosis in human umbilical vein endothelial cells. *Int. J. Mol. Med.* **2009**, *24*, 801–806. [PubMed]

37. Jazzara, M.; Ghannam, A.; Soukkarieh, C.; Murad, H. Anti-Proliferative Activity of λ-Carrageenan Through the Induction of Apoptosis in Human Breast Cancer Cells. *Iran. J. Cancer Prev.* **2016**, *9*, e3836. [CrossRef] [PubMed]

38. Hoffman, R. Carrageenans inhibit growth-factor binding. *Biochem. J.* **1993**, *289 Pt 2*, 331–334. [CrossRef]

39. Hoffman, R.; Burns, W.W.; Paper, D.H. Selective inhibition of cell proliferation and DNA synthesis by the polysulphated carbohydrate l-carrageenan. *Cancer Chemother. Pharm.* **1995**, *36*, 325–334. [CrossRef]

40. Kosir, M.A.; Wang, W.; Zukowski, K.L.; Tromp, G.; Barber, J. Degradation of basement membrane by prostate tumor heparanase. *J. Surg. Res.* **1999**, *81*, 42–47. [CrossRef] [PubMed]

41. Parish, C.R.; Coombe, D.R.; Jakobsen, K.B.; Bennett, F.A.; Underwood, P.A. Evidence that sulphated polysaccharides inhibit tumour metastasis by blocking tumour-cell-derived heparanases. *Int. J. Cancer* **1987**, *40*, 511–518. [CrossRef] [PubMed]

42. Poupard, N.; Badarou, P.; Fasani, F.; Groult, H.; Bridiau, N.; Sannier, F.; Bordenave-Juchereau, S.; Kieda, C.; Piot, J.-M.; Grillon, C.; et al. Assessment of Heparanase-Mediated Angiogenesis Using Microvascular Endothelial Cells: Identification of λ-Carrageenan Derivative as a Potent Anti Angiogenic Agent. *Mar. Drugs* **2017**, *15*, 134. [CrossRef] [PubMed]

43. Poupard, N.; Groult, H.; Bodin, J.; Bridiau, N.; Bordenave-Juchereau, S.; Sannier, F.; Piot, J.-M.; Fruitier-Arnaudin, I.; Maugard, T. Production of heparin and λ-carrageenan anti-heparanase derivatives using a combination of physicochemical depolymerization and glycol splitting. *Carbohydr. Polym.* **2017**, *166*, 156–165. [CrossRef] [PubMed]

44. *Marine Glycobiology: Principles and Applications*; Kim, S.-K. (Ed.) CRC Press: Boca Raton, FL, USA; Taylor & Francis Group: London, UK; New York, NY, USA, 2017; ISBN 978-1-4987-0961-3.

45. Navarro, D.A.; Flores, M.L.; Stortz, C.A. Microwave-assisted desulfation of sulfated polysaccharides. *Carbohydr. Polym.* **2007**, *69*, 742–747. [CrossRef]

46. Sokolova, E.V.; Byankina, A.O.; Kalitnik, A.A.; Kim, Y.H.; Bogdanovich, L.N.; Solov'eva, T.F.; Yermak, I.M. Influence of red algal sulfated polysaccharides on blood coagulation and platelets activation in vitro: Influence of Red Algal Sulfated Polysaccharides. *J. Biomed. Mater. Res. Part A* **2014**, *102*, 1431–1438. [CrossRef] [PubMed]

47. Ciancia, M.; Quintana, I.; Cerezo, A.S. Overview of Anticoagulant Activity of Sulfated Polysaccharides from Seaweeds in Relation to their Structures, Focusing on those of Green Seaweeds. *Curr. Med. Chem.* **2010**, *17*, 2503–2529. [CrossRef] [PubMed]

48. *Hb25_Springer Handbook of Marine Biotechnology*; Kim, S.-K. (Ed.) Springer: Berlin/Heidelberg, Germany, 2015; ISBN 978-3-642-53970-1.

49. Melo, F.R.; Pereira, M.S.; Foguel, D.; Mourão, P.A.S. Antithrombin-mediated Anticoagulant Activity of Sulfated Polysaccharides: Different mechanisms for heparin and sulfated galactans. *J. Biol. Chem.* **2004**, *279*, 20824–20835. [CrossRef] [PubMed]

50. Fu, L.; Suflita, M.; Linhardt, R.J. Bioengineered heparins and heparan sulfates. *Adv. Drug Deliv. Rev.* **2016**, *97*, 237–249. [CrossRef] [PubMed]

51. de Araújo, C.A.; Noseda, M.D.; Cipriani, T.R.; Gonçalves, A.G.; Duarte, M.E.R.; Ducatti, D.R.B. Selective sulfation of carrageenans and the influence of sulfate regiochemistry on anticoagulant properties. *Carbohydr. Polym.* **2013**, *91*, 483–491. [CrossRef] [PubMed]

52. Liang, W.; Mao, X.; Peng, X.; Tang, S. Effects of sulfate group in red seaweed polysaccharides on anticoagulant activity and cytotoxicity. *Carbohydr. Polym.* **2014**, *101*, 776–785. [CrossRef] [PubMed]

53. Fernández, P.V.; Quintana, I.; Cerezo, A.S.; Caramelo, J.J.; Pol-Fachin, L.; Verli, H.; Estevez, J.M.; Ciancia, M. Anticoagulant activity of a unique sulfated pyranosic (1->3)-β-L-arabinan through direct interaction with thrombin. *J. Biol. Chem.* **2013**, *288*, 223–233. [CrossRef] [PubMed]

54. Matsubara, K.; Matsuura, Y.; Bacic, A.; Liao, M.; Hori, K.; Miyazawa, K. Anticoagulant properties of a sulfated galactan preparation from a marine green alga, *Codium cylindricum. Int. J. Biol. Macromol.* **2001**, *28*, 395–399. [CrossRef]

55. Dai, X.; Yan, J.; Fu, X.; Pan, Q.; Sun, D.; Xu, Y.; Wang, J.; Nie, L.; Tong, L.; Shen, A.; et al. Aspirin Inhibits Cancer Metastasis and Angiogenesis via Targeting Heparanase. *Clin. Cancer Res.* **2017**, *23*, 6267–6278. [CrossRef] [PubMed]

56. Ma, X.M.; Shen, Z.H.; Liu, Z.Y.; Wang, F.; Hai, L.; Gao, L.T.; Wang, H.S. Heparanase promotes human gastric cancer cells migration and invasion by increasing Src and p38 phosphorylation expression. *Int. J. Clin. Exp. Pathol.* **2014**, *7*, 5609–5621. [PubMed]

57. Li, Y.; Liu, H.; Huang, Y.Y.; Pu, L.J.; Zhang, X.D.; Jiang, C.C.; Jiang, Z.W. Suppression of endoplasmic reticulum stress-induced invasion and migration of breast cancer cells through the downregulation of heparanase. *Int. J. Mol. Med.* **2013**, *31*, 1234–1242. [CrossRef] [PubMed]

58. Mellor, P.; Harvey, J.R.; Murphy, K.J.; Pye, D.; O'Boyle, G.; Lennard, T.W.J.; Kirby, J.A.; Ali, S. Modulatory effects of heparin and short-length oligosaccharides of heparin on the metastasis and growth of LMD MDA-MB 231 breast cancer cells in vivo. *Br. J. Cancer* **2007**, *97*, 761–768. [CrossRef] [PubMed]

59. Ettelaie, C.; Fountain, D.; Collier, M.E.W.; Beeby, E.; Xiao, Y.P.; Maraveyas, A. Low molecular weight heparin suppresses tissue factor-mediated cancer cell invasion and migration in vitro. *Exp. Ther. Med.* **2011**, *2*, 363–367. [CrossRef] [PubMed]

60. Ponert, J.; Gockel, L.; Henze, S.; Schlesinger, M. Unfractionated and Low Molecular Weight Heparin Reduce Platelet Induced Epithelial-Mesenchymal Transition in Pancreatic and Prostate Cancer Cells. *Molecules* **2018**, *23*, 2690. [CrossRef] [PubMed]

61. Nurcombe, V.; Smart, C.E.; Chipperfield, H.; Cool, S.M.; Boilly, B.; Hondermarck, H. The Proliferative and Migratory Activities of Breast Cancer Cells Can Be Differentially Regulated by Heparan Sulfates. *J. Biol. Chem.* **2000**, *275*, 30009–30018. [CrossRef] [PubMed]

62. Viola, M.; Brüggemann, K.; Karousou, E.; Caon, I.; Caravà, E.; Vigetti, D.; Greve, B.; Stock, C.; De Luca, G.; Passi, A.; et al. MDA-MB-231 breast cancer cell viability, motility and matrix adhesion are regulated by a complex interplay of heparan sulfate, chondroitin−/dermatan sulfate and hyaluronan biosynthesis. *Glycoconj. J.* **2017**, *34*, 411–420. [CrossRef] [PubMed]

63. Mulloy, B.; Hogwood, J. Chromatographic molecular weight measurements for heparin, its fragments and fractions, and other glycosaminoglycans. *Methods Mol. Biol.* **2015**, *1229*, 105–118. [PubMed]

64. Gao, G.; Jiao, Q.; Ding, Y.; Chen, L. Study on quantitative assay of chondroitin sulfate with a spectrophotometric method of azure A. *Guang Pu Xue Yu Guang Pu Fen Xi* **2003**, *23*, 600–602. [PubMed]

Article

Tetracenomycin X Exerts Antitumour Activity in Lung Cancer Cells through the Downregulation of Cyclin D1

Xinran Qiao †, Maoluo Gan †, Chen Wang, Bin Liu, Yue Shang, Yi Li and Shuzhen Chen *

Institute of Medicinal Biotechnology, Chinese Academy of Medical Science & Peking Union Medical College, 1# Tiantan Xili, Chong Wen District, Beijing 100050, China; qxinran_yss@126.com (X.Q.); ganml@imb.pumc.edu.cn (M.G.); wangc1014@163.com (C.W.); bin0629bin@163.com (B.L.); shyue5775@163.com (Y.S.); liyi0108@163.com (Y.L.)

* Correspondence: bjcsz@imb.pumc.edu.cn

† These authors contributed to equally to this works.

Received: 2 December 2018; Accepted: 15 January 2019; Published: 18 January 2019

Abstract: Tetracenomycin X (Tcm X) has been reported to have antitumour activity in various cancers, but there have not been any studies on its activity with respect to lung cancer to date. Therefore, this study aims to investigate the anti-lung cancer activity of Tcm X. In this study, we found that tetracenomycin X showed antitumour activity in vivo and selectively inhibited the proliferation of lung cancer cells without influencing lung fibroblasts. In addition, apoptosis and autophagy did not contribute to the antitumour activity. Tetracenomycin X exerts antitumour activity through cell cycle arrest induced by the downregulation of cyclin D1. To explore the specific mechanism, we found that tetracenomycin X directly induced cyclin D1 proteasomal degradation and indirectly downregulated cyclin D1 via the activation of p38 and c-JUN proteins. All these findings were explored for the first time, which indicated that tetracenomycin X may be a powerful antimitotic class of anticancer drug candidates for the treatment of lung cancer in the future.

Keywords: tetracenomycin X; cell cycle arrest; cyclin D1; proteasomal degradation; p38; c-JUN

---

1. Introduction

Lung cancer is one of the fastest growing malignancies with respect to morbidity and mortality worldwide. It is estimated that lung cancer caused 1.59 million deaths across the world in 2012, and the number of lung cancer deaths is forecasted to increase to 3 million in 2035 [1,2]. Traditional radiotherapy and chemotherapy do not have a successful impact on many lung cancer patients. Recent studies have found that there was only a 16% survival rate in lung cancer patients over the last five years [3,4]. Therefore, new drugs are urgently needed to treat lung cancer.

Tetracenomycinsare, a discrete group of aromatic polyketide antibiotics produced by the *Streptomyces* and *Nocardia* species, shows antibacterial activities mainly against Gram-positive bacteria and moderate antitumour activities [5,6]. The representative members of this group of antibiotics consist of tetracenomycins C and X and elloramycins A–F. As part of our screening program for new antibiotics from marine-derived microorganisms, tetracenomycin X with a high yield (31.8 mg/L), together with the novel *seco*-tetracenomycin analogues, saccharothrixones A–I, from the marine-derived rare actinomycete *Saccharothrix* sp. 10-10 were isolated [7–9]. In the previous work, tetracenomycin X was found to show significant in vitro cytotoxic activities in leukaemia and liver and breast cancer cell lines [7,9]. Although tetracenomycins showed cytotoxic activities in many kinds of cancer cells, there have been few studies that have investigated their in vivo activity. To the best of our knowledge, the only member reported to have in vivo antitumour activity was tetracenomycin C, which displayed antitumour effects against leukaemia cells (P388) in mice [10]. However, their in vivo activities against lung cancer and antitumour mechanisms have not been investigated thus far. This has encouraged us to further explore the anti-lung cancer and antitumour mechanisms of tetracenomycins.

In the current study, we examine the antitumour activity of tetracenomycin X in lung cancer cells and further explored its anticancer mechanisms.

2. Results

2.1. Tetracenomycin X Exerts Antitumour Activity in H460 Xenografts in BALB/c Nude Mice

Because there have been few studies investigating the antitumour activity of tetracenomycin X in vivo, we first detected its antitumour activity in nude mice. As shown in Figure 1A, there were no deaths or significant weight changes in the two study groups, which suggested that the dose of tetracenomycin X was tolerated. Compared to the control group, tetracenomycin X significantly inhibited the growth of H460 xenografts (Figure 1B). The antitumour rate of the tetracenomycin X group was 42%.

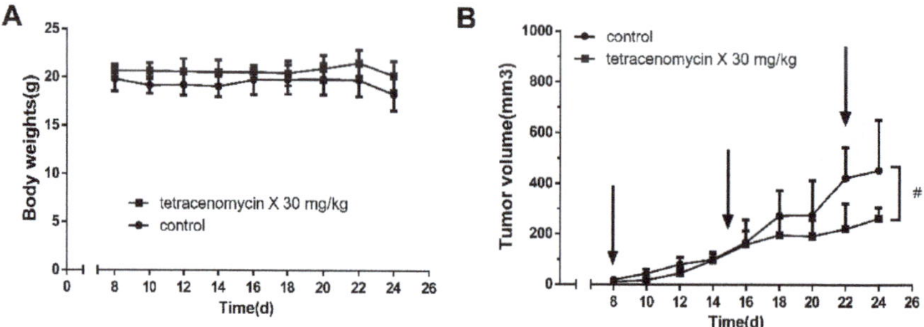

Figure 1. Tetracenomycin X exerts antitumour activity in H460 xenografts in BALB/c nude mice. (**A**) The body weight of H460 xenograft-bearing nude mice (n = 6). (**B**) The volume of H460 xenografts in nude mice (n = 6). $^{\#}$ $p < 0.05$ compared with the control group.

2.2. Tetracenomycin X Selectively Inhibits Human Lung Cancer Cell Proliferation

Tetracenomycin X, an aromatic polyketide antibiotic, was identified from the marine-derived actinomycete *Saccharothrix* sp. 10-10 by Professor Maoluo Gan at our institute [9]. Its structure is shown in Figure 2A. As tetracenomycin X is similar to Adriamycin in structure, Adriamycin was selected as a positive drug with which to compare the antitumour activity of tetracenomycin X. The anti-proliferative activity of tetraccenomycin X and Adriamycin were tested against five lung cancer cells (H157, H1975, HCC827, H460 and A549) and one lung fibroblasts (MRC-5). As shown in Figure 2B, tetracenomycin X hardly inhibited the proliferation of MRC-5 compared with the other lung cancer cells, whereas it significantly inhibited the growth of the A549 cells and H460 cells in a dose-dependent manner among the five lung cancer cells. The half-inhibitory concentration (IC_{50}) upon 24-h treatment was 6.41 ± 0.87 μmol/L and 5.42 ± 1.17 μmol/L in the A549 and H460 cells, respectively. Adriamycin, on the other hand, showed good anti-proliferation activity not only in the four types of lung cancer cells, but also in the normal lung MRC-5 fibroblasts. The IC_{50} upon 24-h treatment of Adriamycin was 7.58 ± 2.21 μmol/L and 0.60 ± 0.26 μmol/L in the A549 and H460 cells, respectively. It was obvious that tetracenomycin X selectively targeted lung cancer cells without inducing cytotoxicity in normal cells, but this advantage was not available to Adriamycin. Furthermore, we found that there was no change in the A549 and H460 cell morphology, while the cell density decreased clearly as the concentration of tetracenomycin X increased (Figure 2C).

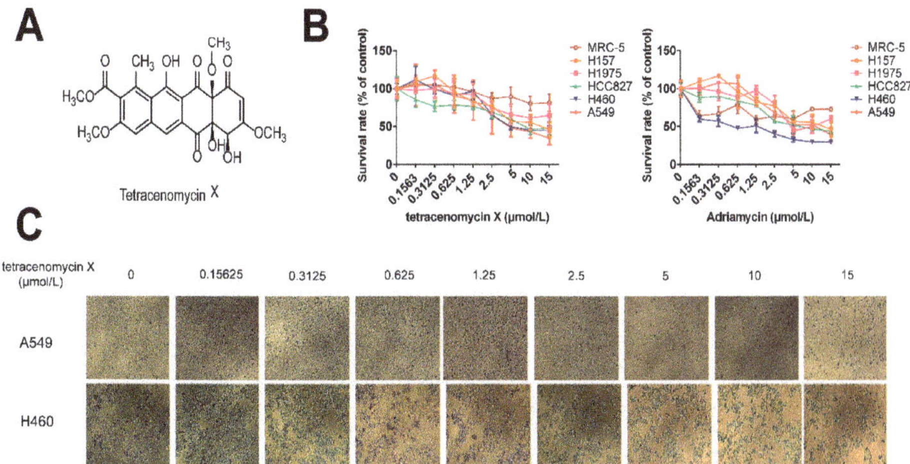

Figure 2. Tetracenomycin X selectively inhibits the cell proliferation of lung cancer cells. (**A**) The structure of tetracenomycin X. (**B**) The proliferative activity of the five lung cancer cells and lung fibroblasts after being treated with tetracenomycin X and Adriamycin (0.1563, 0.3125, 0.625, 1.25, 2.5, 5, 10 and 15 µmol/L) for 24 h was assessed by sulforhodamine B (SRB) assay. The survival rates were calculated as a ratio compared with the control group (untreated cells). The values represent the mean ± SD of three independent samples. Each experiment was repeated three times under each condition. (**C**) The cell morphology of the A549 and H460 cells under a 4 *0.1 microscope after treatment with tetracenomycin X for 24 h.

2.3. The Antitumour Activity of Tetracenomycin X Is Independent of Apoptosis and Autophagy

Annexin V-FITC/PI (Annexin V-fluoresceine isothiocyanate/Propidium Iodide) staining and flow cytometry analysis were performed to detect apoptosis. As shown in Figure 3A, the apoptosis rate in the A549 cells was (4.13 ± 0.94)% in the control group and (5.43 ± 0.21)% in the tetracenomycin X (10 µmol/L) group. Similarly, the apoptosis rate in the H460 cells was (10.28 ± 2.05)% in the control group and (15.39 ± 3.34)% in the tetracenomycin X (10 µmol/L) group. The apoptosis rate was too low in the tetracenomycin X group in two cells. Furthermore, western blotting was used to examine the changes in the protein levels involved in the apoptosis (PARP (poly ADP-ribose polymerase), p53, DR4/5 (death receptor 4/5), Bcl-2 (B-cell lymphoma-2) and Bax (Bcl-2 Associated X Protein) and autophagy (p62 and LC-3B) pathways. As shown in Figure 3B, tetracenomycin X at a concentration of 2.5 or 5 µmol/L did not induce the cleavage of PARP in the A549 and H460 cells and increased the expression of p53 and DR4 in the A549 cells, which were decreased in the H460 cells. In addition, the expression of the DR5, Bcl-2 and Bax proteins also showed different changes in these two cells under the tetracenomycin X treatment. Interestingly, tetracenomycin X reduced the expression of the LC-3B protein but had no effect on the p62 protein (Figure 3C). Tetracenomycin X induced different changes or no change in the expression of the same protein in different cells, indicating that this protein was not the key effector protein in the antitumour activity of tetracenomycin X.

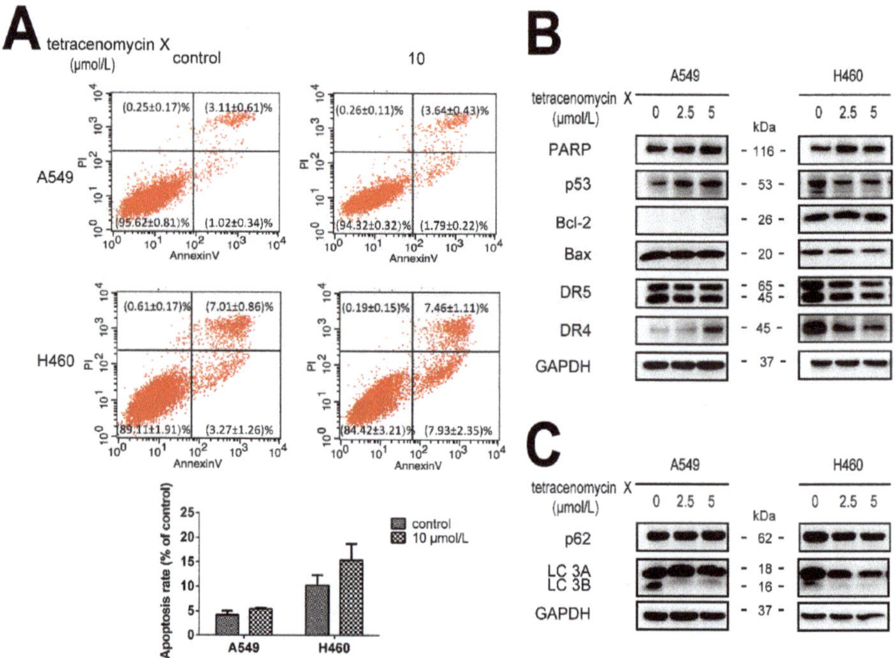

Figure 3. The antitumour activity of tetracenomycin X is independent of apoptosis and autophagy. (**A,C**) The A549 and H460 cells were treated with tetracenomycin X (2.5 and 5 μmol/L) for 8 h. The expression levels of the proteins were determined using western blotting. (**B**) The A549 and H460 cells were treated with tetracenomycin X (10 μmol/L) for 24 h. Apoptosis was detected by annexin V-FITC/PI staining and flow cytometry analysis.

2.4. Tetracenomycin X Induces Cell Cycle Arrest in the G0/G1 Phase and Decreases the Expression Levels of Cell Cycle-Related Proteins in Lung Cancer Cells

To elucidate the antitumour mechanism of tetracenomycin X in lung cancer cells, flow cytometry analysis was used to observe the cell cycle distribution in different groups. We found that the percentage of the G0/G1 phase in the control and tetracenomycin X (5 μmol/L) groups in the A549 cells was (55.51 ± 5.71)% and (65.94 ± 2.24)%, respectively. Additionally, the percentage of the G0/G1 phase in the H460 cells was (52.25 ± 7.38)% in the control group and (64.48 ± 7.66)% in the tetracenomycin X (10 μmol/L) group (Figure 4A). We next performed western blotting to assess the expression levels of the cell cycle-related proteins. As shown in Figure 4B, doses of tetracenomycin X at 2.5 and 5 μmol/L abated the levels of cyclin D1 and CDK4 (cyclin-dependent kinase 4) in the five lung cancer cells. Furthermore, tetracenomycin X (5 μmol/L) decreased the expression of cyclin D1 and CDK4 in 4–16 h in the A549 and H460 cells.

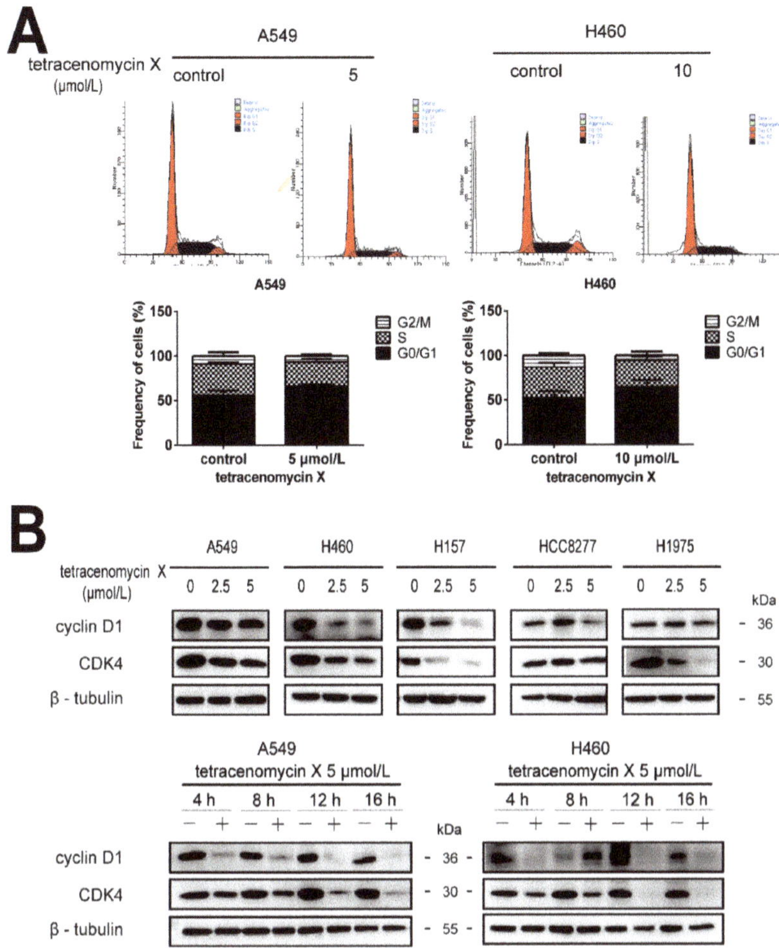

Figure 4. Tetracenomycin X-induced cell cycle arrest in the G0/G1 phase and decreased the expression levels of cell cycle-related proteins in the lung cancer cells. (**A**) the A549 and H460 cells were treated with tetracenomycin X (5 or 10 µmol/L) for 16 h. The cell cycle was detected by flow cytometry analysis. (**B**) The A549 and H460 cells were treated with tetracenomycin X for the indicated times or treated with various concentrations of tetracenomycin X for 8 h. The expression levels of proteins were determined using western blotting.

2.5. Tetracenomycin X Induces the Proteasomal Degradation of Cyclin D1

To verify whether the downregulation of cyclin D1 by tetracenomycin X is mediated by suppression at the mRNA level or by proteasomal degradation, we first used real-time PCR to assess the mRNA level of cyclin D1 in the A549 cells and H460 cells. However, the cyclin D1 mRNA relative expression was increased under doses of the tetracenomycin X treatment (Figure 5A). Then, MG132 and cycloheximide (CHX) were used to investigate the proteasomal degradation of cyclin D1. MG132 is a proteasome inhibitor, and CHX is usually used to inhibit protein synthesis by inhibiting peptidyl transferase activity in eukaryotic cells [11]. As shown in Figure 5B, we found that MG132 at the concentration of 10 µmol/L attenuated the tetracenomycin X-mediated downregulation of cyclin D1, and the combination of tetracenomycin X and CHX enhanced the degradation of cyclin D1 compared with the CHX-added DMSO group. All of these indicated that tetracenomycin X induced the degradation of cyclin D1 through the proteasomal degradation pathway rather than through the suppression of the mRNA level.

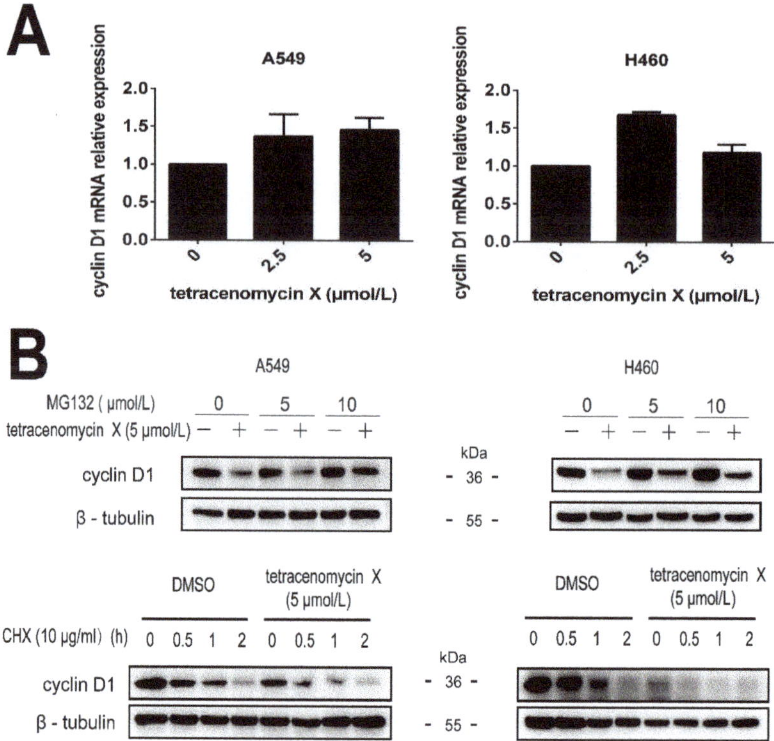

Figure 5. Tetracenomycin X induces the proteasomal degradation of cyclin D1. (**A**) RT-PCR analysis of the gene expression of cyclin D1 in the A549 cells and H460 cells. (**B**) The A549 cells and H460 cells were pre-treated with MG132 at the indicated concentration for 2 h and then co-treated with tetracenomycin X (5 μmol/L) for 12 h. The A549 cells and H460 cells were pre-treated with DMSO or tetracenomycin X (5 μmol/L) and then co-treated with 10 μg/mL of cycloheximide (CHX) for the indicated times.

2.6. Tetracenomycin X Decreases the Expression of Cyclin D1 by the Activation of p38 and c-JUN

Studies have showed that the changes in the expression of cyclin D1 are associated with Erk1/2, p38 and Akt [12–14]. To further elucidate the antitumour mechanism of tetracenomycin X, we tested its impact on the MAPK and Akt signalling pathways. As shown in Figure 6A, tetracenomycin X activated the phosphorylation of Akt in the A549 cells but inactivated it in the H460 cells. We found that doses of tetracenomycin X elevated the phosphorylation levels of the p38 and c-JUN proteins in the five lung cancer cells, and tetracenomycin X (5 μmol/L) also active the p38 and c-JUN proteins in 4–16 h in the A549 and H460 cells. Accordingly, we used SP600125 (a JNK inhibitor) and SB203580 (a p38 MAPK inhibitor) in combination with tetracenomycin X to treat the A549 and H460 cells. The western blotting results showed that both the inhibitors reversed the tetracenomycin X-induced degradation of cyclin D1, while they did not increase the expression of CDK4 (Figure 6B).

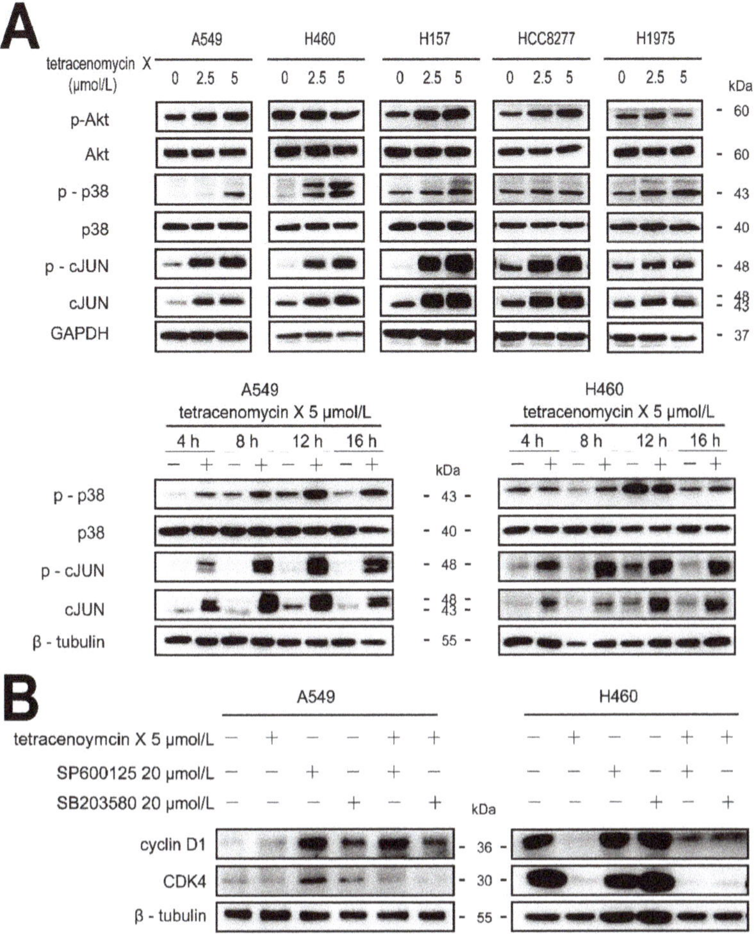

Figure 6. Tetracenomycin X decreases the expression of cyclin D1 by the activation of p38 and c-JUN. (**A**) The A549 and H460 cells were treated with tetracenomycin X for the indicated times or treated with various concentrations of tetracenomycin X for 8 h. The expression levels of the proteins were determined using western blotting. (**B**) The viability of the A5459 and H460 cells was determined after a 4-h pre-treatment with SP600125 or SB203580 and an 8-h treatment with tetracenomycin X. The expression levels of the proteins were determined using western blotting.

3. Discussion

Tetracenomycins, a family of tetracyclic haphthacenequinones, have large anti-Gram-positive effects and significant antitumour activity. Tetracenomycin X, as a member of this family, was only reported to have antitumour activity in leukaemia cells, liver cancer cells and breast cancer cells [6–9,15,16]. This study is the first to show that tetracenomycin X possesses antitumour activity in H460 xenografts in BALB/c nude mice, so we studied the antitumour mechanism in depth. Experiments in vitro showed that tetracenomycin X selectively attenuates the proliferation of lung cancer cells without inhibiting the lung fibroblasts, while this inhibitory effect was significant in the A549 cells and H460 cells. At the same time, there was no obvious difference in the effect of Adriamycin on the anti-proliferation activity between the lung cancer cells and normal lung fibroblasts. Furthermore, few studies have described the antitumour mechanism of tetracenomycin X, so we needed to examine the cell death pathways one by one. First, we tested the changes in apoptosis and autophagy. The results showed that neither apoptosis nor autophagy was involved in the antitumour mechanism. Then, we performed flow cytometry analysis to observe cell cycle distribution. We finally found that tetracenomycin X could increase cell cycle arrest at the G0/G1 phase, which suggested that the induction of cell cycle arrest was the major antitumour mechanism in the lung cancer cells.

The cell cycle, which includes four phases (G1, S, G2 and M), is the basic process to guarantee cell proliferation. The four phases transition in sequence under exogenous and endogenous regulations [14]. Exogenous regulation is mainly caused by cytokines, and endogenous regulation is induced by the cyclin–CDK network. Cyclins and cyclin-dependent kinases (CDKs) are sensitive to oncogenic stimulation, and their excessive activation can make cancer cells grow strongly [17]. Cyclin D1 together with CDK4 plays an important role in promoting the cell cycle from the G0/G1 phase to the S phase. Accordingly, changes in the complex of cyclin D1 and CDK4 may be responsible for the tetracenomycin X-induced cell cycle arrest. As expected, decreases in cyclin D1 and CDK4 were found in the lung cancer cells after the tetracenomycin X treatment.

Recent studies have shown that cyclin D1 expression can be regulated through transcription and proteasomal degradation, and the proteasomal degradation of cyclin D1 has been thought to be an important antitumour pathway [18]. The current study shows that tetracenomycin X increases the mRNA level of cyclin D1, while MG132 attenuates tetracenomycin X-induced cyclin D1 downregulation and CHX enhances the degradation of cyclin D1 after tetracenomycin X treatment. These findings suggest that cyclin D1 proteasomal degradation may be one of the important molecular targets for the antitumour activity of tetracenomycin X.

It has been reported that cyclin expression is regulated by upstream kinases, such as p38, Akt and JNK [13,19]. Our results show that tetracenomycin X only activates the phosphorylation levels of p38 and c-JUN proteins in lung cancer cells, which leads to the downregulation of cyclin D1. It appears that p38 and c-JUN activation by tetracenomycin X, at least in part, contributes to tetracenomycin X-mediated cyclin D1 downregulation.

In conclusion, tetracenomycin X selectively exerts high anti-proliferation activity in lung cancer cells without inhibiting normal cells. Cyclin D1 proteasomal degradation-mediated cell cycle arrest at the G0/G1 phase may be the important antitumour mechanism of tetracenomycin X in lung cancer cells. In addition, p38 and c-JUN activation by tetracenomycin X partly results in cyclin D1 downregulation (Figure 7). Thus, tetracenomycin X may have the potential to become an antimitotic class of drugs against lung cancer.

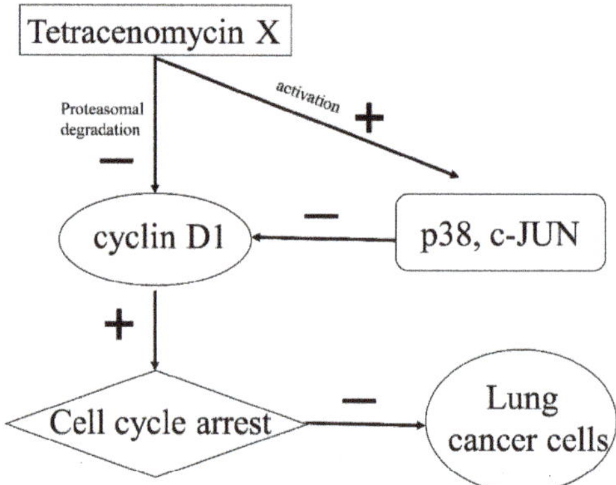

Figure 7. Schematic presentation of the antitumour mechanism. In lung cancer cells, tetracenomycin X directly induces cyclin D1 proteasomal degradation and indirectly downregulates cyclin D1 via the activation of p38 and c-JUN. The decrease in cyclin D1 results in cell cycle arrest at the G0/G1 phase and eventually inhibits the proliferation of lung cancer cells.

4. Materials and Methods

4.1. Cell Culture

The human lung cancer A549, H460, H157, HCC827 and H1975 cell lines were kept in our laboratory and maintained in RPMI-1640 medium (ThermoFisher Scientific, Waltham, MA, USA) supplemented with 10% foetal bovine serum (Gibco, Carlsbad, CA, USA) and penicillin (100 U/mL)/streptomycin (100 µg/mL) (National China Pharmaceutical Inc., Beijing, China) at 37 °C in a 5% CO_2 incubator. The human embryonic lung MRC-5 fibroblasts were purchased from National Institutes for Food and Drug Control (Beijing, China) and cultured in MEM-EBSS (Minimum Essential Medium-Earle's Balanced Salts Solution) medium supplemented with 10% bovine serum and penicillin (100 U/mL)/streptomycin (100 µg/mL).

4.2. Reagents and Antibodies

Tetracenomycin X (Tcm X) was isolated from the marine-derived actinomycete *Saccharothrix* sp. 10-10 by Professor Maoluo Gan at our institute. The molecular weight of Tcm X was 486 Da, and it was dissolved in dimethyl sulfoxide (DMSO). The anti-p53, anti-PARP, anti-Bcl-2, anti-Bax, anti-p38 MAPK, anti-phospho-p38 MAPK, anti-cJUN, anti-phospho-cJUN, anti-cyclin D1 and anti-CDK4 antibodies were purchased from Cell Signaling Technology (Danvers, MA, USA). The anti-LC3B antibody was obtained from Sigma (St. Louis, MO, USA). The anti-Human CD261 Azide Free (DR4) antibody was obtained from Diaclone (Besancon, France), and the DR5 antibody was purchased from ProSci (San Diego, CA, USA). β-Actin (6G3), β-tubulin and GAPDH (glyceraldehyde-3-phosphate dehydrogenase) (1C4) mouse monoclonal antibodies were obtained from AmeriBiopharma (Wulmington, DE, USA). For the second antibodies, peroxidase-conjugated affiniPure goat anti-mouse IgG (H + L) and anti-rabbit IgG (H + L) were obtained from ZSGB-BIO (Beijing, China). MG132, cycloheximide (CHX), SB203580 and SP600125 were purchased from Selleck.cn (Houston, TX, USA). Sulforhodamine B (SRB) was obtained from Sigma (St. Louis, MO, USA). Trichloroacetic acid (TCA) was obtained from Solarbio (Beijing, China).

4.3. Cell Survival Assay

A sulforhodamine B (SRB) method was used to detect cell survival. The tested cell lines were seeded in 96-well plates at 4×10^3 cells/well in 100 µl of culture medium for 24 h. Different concentrations of tetracenomycin X were then added to the 96-well plates and incubated for 24 h. The SRB method was described previously [20]. The cell survival was calculated as the ratio of the experimental groups to control group.

4.4. Western Blot Analysis

The cells were seeded in 6-well plates and exposed to drugs for specific times. The preparation of cell samples and western blot analyses were described previously [20].

4.5. Annexin V-FITC/PI Staining

The tested cells were plated on 6-well plates overnight and treated with drugs for 24 h. Then, cell apoptosis was examined by an Annexin V-FITC/PI apoptosis detection kit (Bio Friend, Beijing, China), according to the manufacturer's instructions. The Annexin V-FITC/PI staining was detected by flow cytometry analysis.

A schematic plot was applied to analyse the results, in which the lower left quadrant represented live cells, the lower right and upper right quadrants represented early and late apoptotic cells, respectively, and the upper left quadrant represented necrotic cells. The cell apoptosis rate was the sum of the percentages of the early and late apoptotic cells.

4.6. Cell Cycle Analysis

The A549 cells and H460 cells were plated on 6-well plates and grown overnight. Then, the cells were treated with tetracenomycin X for 16 h. The cell cycle was determined using a PI (including RNase A) cell cycle detection kit (Bio Friend, Beijing, China) according to the manufacturer's instructions and then analysed by a flow cytometer COULTER EPICS XL (Beckman Coulter, Inc., Brea, CA, USA).

4.7. Reverse Transcriptase-Polymerase Chain Reaction

The A549 cells were plated on 6-well plates and exposed to tetracenomycin X for 12 h. The total RNA was extracted by a RNA fast 200 Kit (FASTAGEN, Shanghai Fastagen Biotechnology, Shanghai, China), and 1 µg of RNA was used with a Prime Script^{TM} RT reagent kit with a gDNA Eraser (TaKaRa, Tokyo, Japan) according to the manufacturer's instructions for cDNA synthesis. RT-PCR was carried out using a qPCR Mix Kit (TOYOBO, Osaka, Japan) with human primers for cyclin D1 and β-actin as follows: cyclin D1: forward 5′-aactacctggaccgcttcct-3′ and reverse 5′-ccacttgagcttgttcacca-3′, β-actin: forward 5′-agcgagcatcccccaaagtt-3′ and reverse 5′-gggcacgaaggctcatcatt-3′.

4.8. Lung Cancer Xenograft

The animal experiments were performed in the Experimental Animal Center of our institute and in accordance with all the relevant guidelines and regulations. First, the H460 cells (5×10^6 cells in 200 µL of PBS) were injected into the right armpits of 6-week-old female BALB/c nude mice (Anikeeper, Beijing, China). Then, 21 days after inoculation, tumours were removed and cut into 2 mm × 2 mm × 2 mm prisms and transplanted into the right armpits of mice. On the 8th day, the mice were randomly divided into two groups: one was a control group and the other was a group given tetracenomycin X (30 mg/kg/day, dissolved in 10% alcohol) via intraperitoneal injection for 3 consecutive days in a week. The tumour volumes and body weights were monitored once every 2 days. The tumour volume was calculated by the following formula: $V = ab^2/2$, where a represents the length and b represents the width. The animal experiment lasted for a total of 24 days.

4.9. Statistical Analysis

All the experiments were repeated at least 2 times. The data are expressed as the mean ± SD. The statistical analysis was performed via Student's *t* test for the independent samples.

Author Contributions: S.C. designed the experiments and revised the manuscript. X.Q. performed all the experiments. X.Q. and M.G. wrote the manuscript. M.G. and B.L. provided tetracenomycin X. C.W. participated in cell survival assay and the animal experiments. Y.S. and Y.L. participated in the animal experiments. All the authors read and approved the final manuscript.

Funding: The manuscript was supported by grants from the National Natural Science Foundation of China (81621064, 81702934), CAMS Innovation Fund for Medical Sciences (CIFMS, 2016-I2M-02-002), and the "Significant New Drug Development" Major Science and Technology Development Projects of China (No. 2018ZX09711001-007-002).

References

1. Torre, L.A.; Bray, F.; Siegel, R.L.; Ferlay, J.; Lortet-Tieulent, J.; Jemal, A. Global cancer statistics, 2012. *CA A Cancer J. Clin.* **2015**, *65*, 87–108. [CrossRef] [PubMed]

2. McIntyre, A.; Ganti, A.K. Lung cancer-A global perspective. *J. Surg. Oncol.* **2017**, *115*, 550–554. [CrossRef] [PubMed]

3. Sharma, D.; Newman, T.G.; Aronow, W.S. Lung cancer screening: History, current perspectives, and future directions. *Arch. Med. Sci. AMS* **2015**, *11*, 1033–1043. [PubMed]

4. Jiang, J.; Huang, J.; Wang, X.R.; Quan, Y.H. MicroRNA-202 induces cell cycle arrest and apoptosis in lung cancer cells through targeting cyclin D1. *Eur. Rev. Med. Pharm. Sci.* **2016**, *20*, 2278–2284.

5. Lazar, G.; Zähner, H.; Breiding, S.; Damberg, M.; Zeeck, A. 3-Demethoxy-3-ethoxy-tetracenomycin C. *J. Antibiot.* **1981**, *34*, 1067–1068. [CrossRef] [PubMed]

6. Egert, E.; Noltemeyer, M.; Siebers, J.; Rohr, J.; Zeeck, A. The structure of tetracenomycin C. *J. Antibiot.* **1992**, *45*, 1190–1192. [CrossRef] [PubMed]

7. Liu, B.; Li, J.; Chen, M.; Hao, X.; Cao, F.; Tan, Y.; Ping, Y.; Wang, Y.; Xiao, C.; Gan, M. Seco-Tetracenomycins from the Marine-Derived Actinomycete *Saccharothrix* sp. 10-10. *Mar. Drugs* **2018**, *16*, 245. [CrossRef] [PubMed]

8. Gan, M.; Liu, B.; Tan, Y.; Wang, Q.; Zhou, H.; He, H.; Ping, Y.; Yang, Z.; Wang, Y.; Xiao, C. Saccharothrixones A-D, Tetracenomycin-Type Polyketides from the Marine-Derived Actinomycete *Saccharothrix* sp. 10-10. *J. Nat. Prod.* **2015**, *78*, 2260–2265. [CrossRef] [PubMed]

9. Liu, B.; Tan, Y.; Gan, M.L.; Zhou, H.X.; Wang, Y.G.; Ping, Y.H.; Li, B.; Yang, Z.Y.; Xiao, C.L. [Identification of tetracenomycin X from a marine-derived *Saccharothrix* sp. guided by genes sequence analysis]. *Yao Xue Xue Bao* **2014**, *49*, 230–236. [PubMed]

10. Weber, W.; Zähner, H.; Damberg, M.; Siebers, J.; Schroder, K.; Zeeck, A. Tetracenomycins—New antibiotics from Streptomyces glaucescens. In Proceedings of the International Symposium on Actinomycete Biology, Cologne, Germany, 3–7 September 1979.

11. Wang, Y.; Gao, W.; Svitkin, Y.V.; Chen, A.P.; Cheng, Y.C. DCB-3503, a Tylophorine Analog, Inhibits Protein Synthesis through a Novel Mechanism. *PLoS ONE* **2010**, *5*, e11607. [CrossRef] [PubMed]

12. Park, S.B.; Park, G.H.; Song, H.M.; Son, H.J.; Um, Y.; Kim, H.S.; Jeong, J.B. Anticancer activity of calyx of Diospyros kaki Thunb. through downregulation of cyclin D1 via inducing proteasomal degradation and transcriptional inhibition in human colorectal cancer cells. *BMC Complement. Altern. Med.* **2017**, *17*, 445. [CrossRef] [PubMed]

13. Kim, M.K.; Park, G.H.; Eo, H.J.; Song, H.M.; Lee, J.W.; Kwon, M.J.; Koo, J.S.; Jeong, J.B. Tanshinone I induces cyclin D1 proteasomal degradation in an ERK1/2 dependent way in human colorectal cancer cells. *Fitoterapia* **2015**, *101*, 162–168. [CrossRef] [PubMed]

14. Li, T.; Zhao, X.; Mo, Z.; Huang, W.; Yan, H.; Ling, Z.; Ye, Y. Formononetin promotes cell cycle arrest via downregulation of Akt/Cyclin D1/CDK4 in human prostate cancer cells. *Cell. Physiol. Biochem. Int. J. Exp. Cell. Physiol. Biochem. Pharmacol.* **2014**, *34*, 1351–1358. [CrossRef] [PubMed]

15. Anderson, M.G.; Khoo, C.L.; Rickards, R.W. Oxidation processes in the biosynthesis of the tetracenomycin and elloramycin antibiotics. *J. Antibiot.* **1989**, *42*, 640–643. [CrossRef] [PubMed]

16. Hutchinson, C.R. Biosynthetic Studies of Daunorubicin and Tetracenomycin C. *Chem. Rev.* **1997**, *97*, 2525–2536. [CrossRef] [PubMed]

17. Vermeulen, K.; Bockstaele, D.R.V.; Berneman, Z.N. The cell cycle: A review of regulation, deregulation and therapeutic targets in cancer. *Cell Prolif.* **2010**, *36*, 131–149. [CrossRef]

18. Mukhopadhyay, A.; Banerjee, S.; Stafford, L.J.; Xia, C.; Liu, M.; Aggarwal, B.B. Curcumin-induced suppression of cell proliferation correlates with down-regulation of cyclin D1 expression and CDK4-mediated retinoblastoma protein phosphorylation. *Oncogene* **2002**, *21*, 8852–8861. [CrossRef] [PubMed]

19. Thoms, H.C.; Dunlop, M.G.; Stark, L.A. p38-mediated inactivation of cyclin D1/cyclin-dependent kinase 4 stimulates nucleolar translocation of RelA and apoptosis in colorectal cancer cells. *Cancer Res.* **2007**, *67*, 1660–1669. [CrossRef] [PubMed]

20. Qiao, X.; Wang, X.; Shang, Y.; Li, Y.; Chen, S.Z. Azithromycin enhances anticancer activity of TRAIL by inhibiting autophagy and up-regulating the protein levels of DR4/5 in colon cancer cells in vitro and in vivo. *Cancer Commun.* **2018**, *38*, 43. [CrossRef] [PubMed]

 *marine drugs*

Article

Fragment-Based Structural Optimization of a Natural Product Itampolin A as a p38α Inhibitor for Lung Cancer

Jing-wei Liang, Ming-yang Wang, Shan Wang, Xin-yang Li and Fan-hao Meng *

School of Pharmacy, China Medical University, Liaoning 110122, China; liangjw89@163.com (J.-w.L.);
wmy940623@163.com (M.-y.W.); 18341653600@139.com (S.W.); 13125424162@163.com (X.-y.L.)
* Correspondence: fhmeng@cmu.edu.cn; Tel.: +86-133-8688-7639

Received: 1 December 2018; Accepted: 7 January 2019; Published: 12 January 2019

Abstract: Marine animals and plants provide abundant secondary metabolites with antitumor activity. Itampolin A is a brominated natural tyrosine secondary metabolite that is isolated from the sponge *Iotrochota purpurea*. Recently, we have achieved the first total synthesis of this brominated tyrosine secondary metabolite, which was found to be a potent p38α inhibitor exhibiting anticancer effects. A fragment-based drug design (FBDD) was carried out to optimize itampolin A. Forty-five brominated tyrosine derivatives were synthesized with interesting biological activities. Then, a QSAR study was carried out to explore the structural determinants responsible for the activity of brominated tyrosine skeleton p38α inhibitors. The lead compound was optimized by a FBDD method, then three series of brominated tyrosine derivatives were synthesized and evaluated for their inhibitory activities against p38α and tumor cells. Compound **6o** (IC_{50} = 0.66 μM) exhibited significant antitumor activity against non-small cell lung A549 cells (A549). This also demonstrated the feasibility of the FBDD method of structural optimization.

Keywords: itampolin A; FBDD; antitumor; p38α; novel inhibitor

---

1. Introduction

Sponges are the most abundant source of bioactive compounds in marine invertebrates [1]. About half of the current FDA-approved marine drugs come from sponges or derivatives based on sponge metabolites [2–4]. Most of the natural products isolated from sponges belong to bromotyrosine derivatives and exhibit extensive biological activities [5–8]. Itampolin A was the first brominated tyrosine alkaloid isolated from the sponge *Iotrochota purpurea* [9]. In our previous study, the total synthesis of itampolin A was achieved for the first time. In an evaluation of the pharmacological properties, the alkaloid showed potent concentration-dependent antitumor activity against non-small cell lung cancer. In addition, its enantiomer (−)-itampolin A exhibited significant inhibitory activity against non-small cell lung A549 cells. The underlying anticancer mechanism was associated with its binding to DFG-out conformation of p38α and then decreasing the phospho-p38 expression in a time-dependent manner [10].

In recent years, the design of more selective kinase inhibitors, targeting inactive conformations, has been explored. Relative to type I inhibitors that compete directly with ATP, type II inhibitors bind to the inactive DFG-out conformation of kinase induced by the conformational transition of DFG-loop. This opened up a second hydrophobic subcavity formed by the catalytic amino acid triad Asp168, Phe169, and Glu71. A number of studies have demonstrated the advantages of targeting the DFG-out binding mode of kinases in general and p38 MAP kinase (p38 MAPK) in particular such as low toxicity [11].

Fragment-based drug design (FBDD) is now widely used in academia and industry to obtain small molecule inhibitors for a given target. Moreover, it is established for many fields of research including antimicrobials and oncology [12–14]. Many molecules derived from fragment-based approaches are already in clinical trials and two—Vemurafenib and Venetoclax—are on the market [13]. Unlike other computer aided drug design (CADD) methods, the FBDD theory maintains that the active pockets of the drug target are made up of multiple subcavities, and the fragments are units that combine with these subcavities. Finding these fragments and linking them together often leads to higher active compounds [15].

For the purpose of improving the activity of itampolin A, and increasing the structural diversity of type II inhibitors, we here reported the optimization of (−)-itampolin A as novel p38α inhibitors by using the FBDD method. This strategy involved interrogation of structural information that was available for different in-house chemotypes [16]. The challenge included three aspects. The first one was deconstruction of known p38α inhibitors to identify highly efficient interactions in the binding site. The second one was screening out suitable units that fit the second hydrophobic subcavity. The last one was exploring the effects on the activity of some atom or fragment substitutions of the brominated tyrosine skeleton.

2. Results

2.1. Fragment-Based Drug Design

The conformation of inactive p38α bound with type II inhibitors was screened out from the PDB website as 3HV3, 3IW5, 3L8S, 3IW7, 3IW8, 4FA2, 2KV2, and 2PUU. The conformation of itampolin A overlapped with BIRB-796 was obtained in a previous work. The above conformations were superposed together after alignment (Figure 1a). An FBDD-based BREED technique was adopted as a novel fragment-based drug design method, which was based on sets of aligned 3D ligand structures binding to the same target or target family. The implementation comprises two steps. Firstly, a superposition of ligands (Figure 1a). Secondly, a ligand fragmentation based on interatomic distance and bonding angle. This was followed by a scoring scheme assigning individual scores to each fragment, and the incremental construction of novel ligands based on a greedy search algorithm guided by the calculated fragment scores (Figure 1b). These small molecules were then screened by pharmacophore models and the lipinsiki rule of five. The BREED results generated by MOE software are described in the Supplementary Materials, Table S1.

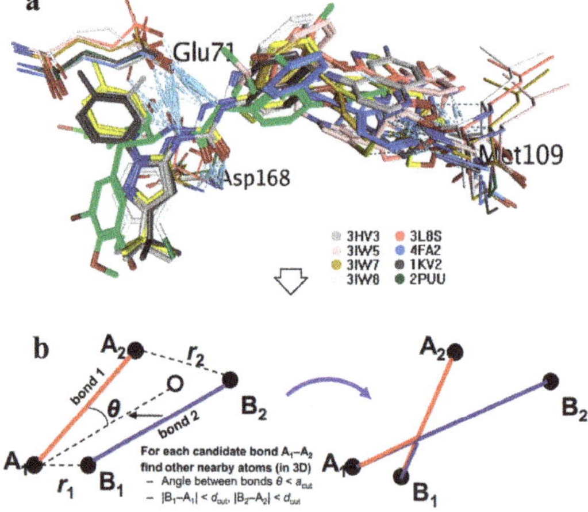

Figure 1. Workflow of the BREED method; (**a**) The lead compound (green) overlapped with type II inhibitors; (**b**) ligand fragmentation based on interatomic distance and bonding angle.

2.2. Synthesis

The synthesis of brominated tyrosine derivatives followed the synthetic route outlined in Scheme 1. The chemical synthesis method to access the parent compounds (+)-itampolin A and (−)-itampolin A were reported previously, as well as 2p, 2q and 3a–3o [10]. The other important intermediates 1a-1o for synthesizing the brominated tyrosine derivatives were substituted benzoyl azide. Thearomatic hydrazines were used as raw material to obtain aryl azides by diazotization. The substituted benzoyl azides produced corresponding substituted isocyanatobenzenes through Curtius rearrangement in DCE at 80 °C.

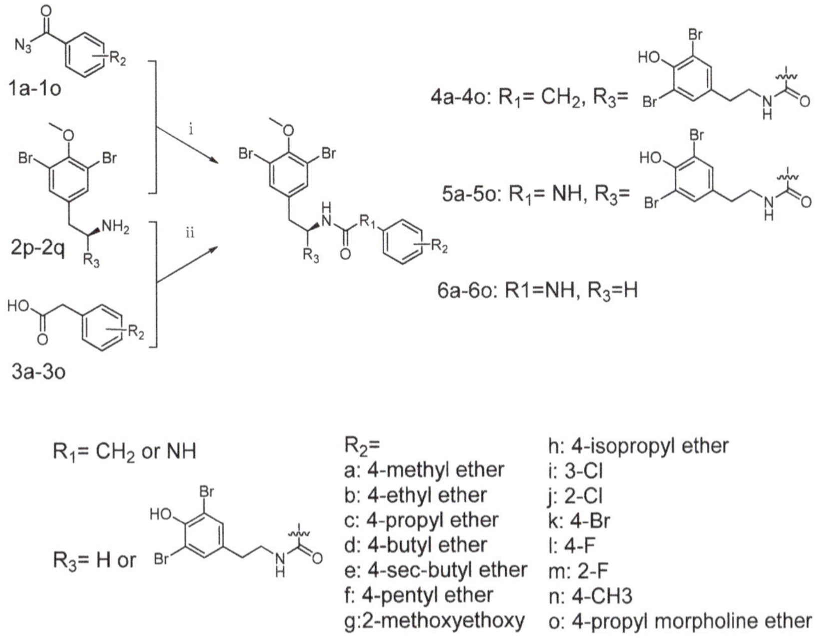

Scheme 1. Synthesis of novel brominated tyrosine derivatives. Conditions: (i) 1.1 eq. 1a–1o, DCE, 80 °C, 39–51%; (ii) 1.05 eq. 2p, 2q, EDCI, HOBt, Et3N, DMF, rt., 52–57%.

2.3. Inhibitory Activities of (−)-Itampolin A Skeleton Brominated Tyrosine Derivatives

The capacity of the synthesized brominated tyrosine derivatives for inhibiting p38α activity was evaluated. The evaluation results are summarized in Table 1.

Table 1. p38α MAP kinase inhibitory activity of brominated tyrosine derivatives

No.	IC$_{50}$ (nM)	No.	IC$_{50}$ (nM)	No.	IC$_{50}$ (nM)
4a	–	5a	112.3 ± 12.0	6a	262.7 ± 3.3
4b	492.5 ± 46.2	5b	346.4 ± 18.4	6b	221.4 ± 4.2
4c	–	5c	167.2 ± 23.9	6c	701.4 ± 7.4
4d	805.6 ± 139.4	5d	690.6 ± 38.4	6d	–
4e	–	5e	–	6e	920.3 ± 6.1
4f	–	5f	–	6f	–
4g	214.8 ± 64.0	5g	223.4 ± 10.3	6g	11.7 ± 3.0
4h	–	5h	–	6h	329.4 ± 44.2
4i	91.0 ± 11.1	5i	51.3 ± 4.5	6i	17.5 ± 2.4
4j	–	5j	447.7 ± 64.1	6j	71.2 ± 4.5
4k	832.1 ± 12.1	5k	90.3 ± 2.7	6k	169.0 ± 2.6

Table 1. *Cont.*

No.	IC$_{50}$ (nM)	No.	IC$_{50}$ (nM)	No.	IC$_{50}$ (nM)
4l	617.7 ± 87.7	5l	110.4 ± 18.8	6l	21.5 ± 4.6
4m	78.6 ± 59.2	5m	71.4 ± 19.6	6m	13.6 ± 3.0
4n	–	5n	278.6 ± 28.9	6n	299.6 ± 11.7
4o	137.5 ± 13.6	5o	134.2 ± 26.3	6o	7.9 ± 1.7
				BIRB-796	11.3 ± 0.2

In the assay of inhibiting p38α activity, the compound **6o** exhibited the best activity. The compoundwas selected for further cell proliferation inhibition experiments by MTT assay. Because the lead compound (−)-itampolin A showed an inhibitory effect on A549, this cell line was still selected to evaluate **6o** activity. Compound **6o** inhibited A549 cell proliferation in a concentration-dependent manner with IC$_{50}$ = 0.66 μM. The experimental results are shown in Figure 2.

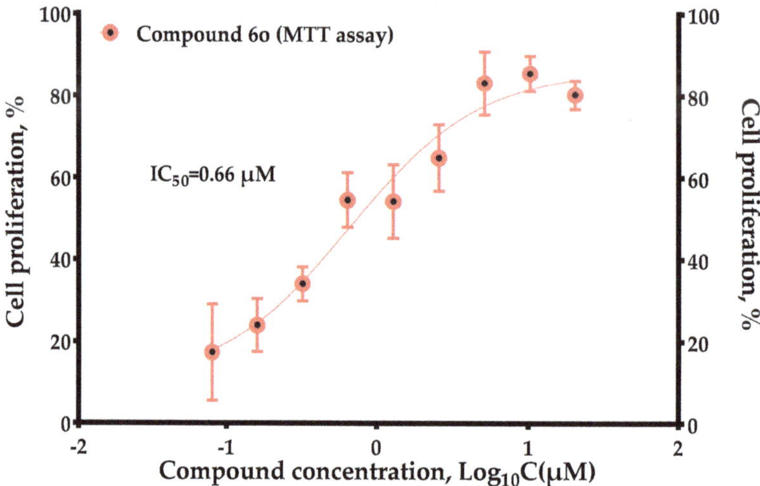

Figure 2. Dependence of concentration versus the activity of compound **6o**.

2.4. 3D-QSAR Study

In order to clarify the structure–activity relationship between the bromine tyrosine derivatives and antitumor activity, a 3D-QSAR study on the derivatives was carried out using the CoMFA and CoMSIA method in Sybyl-X2.0.

The statistical results of the Topomer CoMFA model were as follows: q^2 value of 0.700; r^2 value of 0.954, with five optimum components. With the optimal number of components being 4 in the CoMSIA model, q^2, r^2, and SEE were found to be 0.615, 0.897, and 0.124 respectively. The statistical results proved that the QSAR model of Topomer CoMFA and CoMSIA has precise predictability. The experimental and predicted activities of both the training set and test set are shown in Figure 3. The Topomer CoMFA and CoMSIA models gave the correlation coefficient (r^2) value of 0.9271 and 0.909, respectively, which demonstrated the internal robustness and external high prediction of the QSAR models.

A 3D-QSAR contour map was utilized to exhibit the Topomer CoMFA model and CoMSIA model properties in 3D-space, as well as to obtain information in ligand-receptor conformation. After the visualization of favorable and unfavorable regions of fields in a 3D-QSAR contour map, several fields (steric fields, electrostatic fields, hydrophobic fields, hydrogen bond donor atom fields, and hydrogen bond acceptor atom fields) were used to realize the relationship between the biological activities and structures. Steric and electrostatic contour maps of the Topomer CoMFA QSAR model are shown in Figure 4a,b respectively. HBA and hydrophobic contour maps of the CoMSIA QSAR model are shown in Figure 4c,d respectively. Compound **60** has the most complex chemical structure for the visual clarity of analyzing the QSAR, hence it was chosen as the reference structure for the generation of the Topomer CoMFA and CoMSIA contour map.

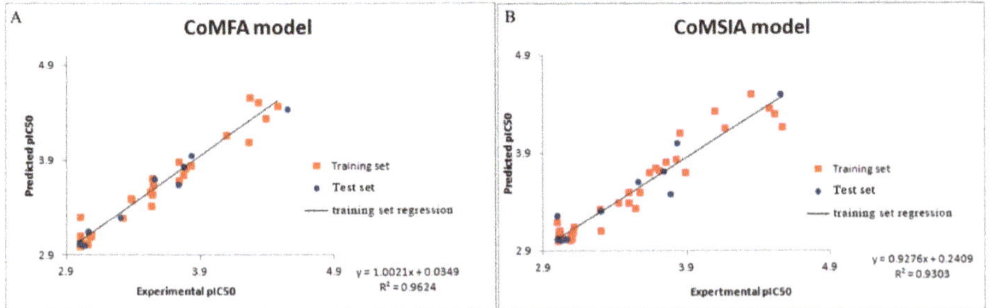

Figure 3. Experimental versus predicted activity of the training and test set based on the Topomer CoMFA model (**A**) and CoMSIA model (**B**).

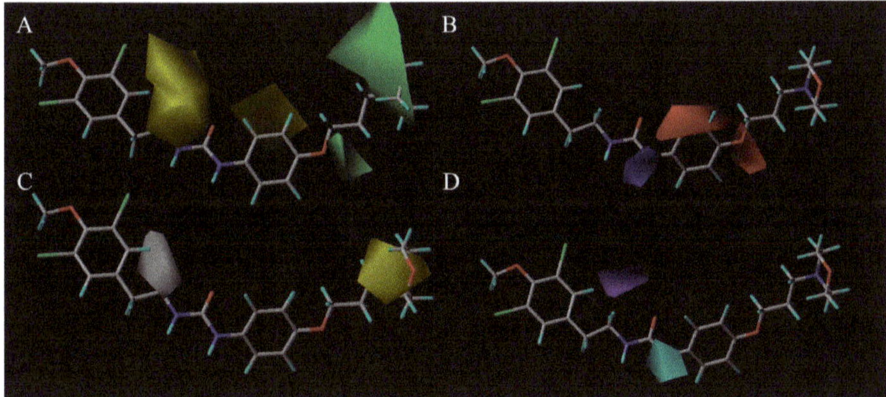

Figure 4. (**A**) Steric contour maps of Topomer CoMFA around the brominated tyrosine skeleton. Green contours refer to sterically favourable regions and yellow contours refer to sterically unfavorable regions. (**B**) Electrostatic contour maps of Topomer CoMFA around the brominated tyrosine skeleton. Red contours refer to regions where electropositive groups were favorable and blue contours refer to regions where electropositive groups were unfavourable. (**C**) Hydrophobic field contour maps for CoMSIA. Yellow contours represent regions where hydrophobic substituents were favorable, white contours represent regions where hydrophilic substituents were unfavourable. (**D**) H-bond donor field contour maps for CoMSIA. Cyan contours indicate regions where hydrogen bond donor substituents increase activity, and purple contours indicate the unfavorable regions for hydrogen bond donor groups.

3. Discussion

Based on the aligned 3D ligand structures binding to p38α, a novel skeleton was generated based on the structure of (−)-itampolin A (Figure 5). The carbon atom of 14-position was changed to a nitrogen atom, making the amide unit (12-position, 13-position) convert to a disubstituted urea group. As a result, the new skeleton was capable of binding Glu71 residue. It stabilized the conformation of the molecule in the active site. This binding was crucial, because the interaction between the urea unit and Glu71 ensured that bromoaromatic groups enter the second hydrophobic pocket unique to DFG-out conformation. The atoms between 23-position and 26-position were replaced with a morpholine unit (there were also other units such as fluorine-substituted aromatic ring that are listed in Table S1). The urea group probably came from the inhibitors which contained the urea unit. The morpholine unit came from BIRB-796. The fluorine-substituted aromatic came from the inhibitor of 3IW5 or 4FA2. Most inhibitors supported the change to the urea unit. However, each inhibitor supported different changes in the atoms between 23-position and 26-position (Figure 5 shows a more active one with the morpholine unit). Therefore, it was suggested that it is necessary to explore more units to replace the atoms between 23-position and 26-position. The results of molecular docking showed that most of the brominate tyramine fragment was exposed outside the active pocket. Moreover, excessive hydrophobic fragments might hinder the binding of the molecule to Glu71 (Figure 5a). After the discovery of defects in the skeleton, a filter was added to iterate the process of FBDD. This filter included the Lipinski's rule of five and a structure-based pharmacophore, worked out by MOE 2015. The screening results showed that cutting off the fragment can further improve the activity of the compounds with a novel skeleton.

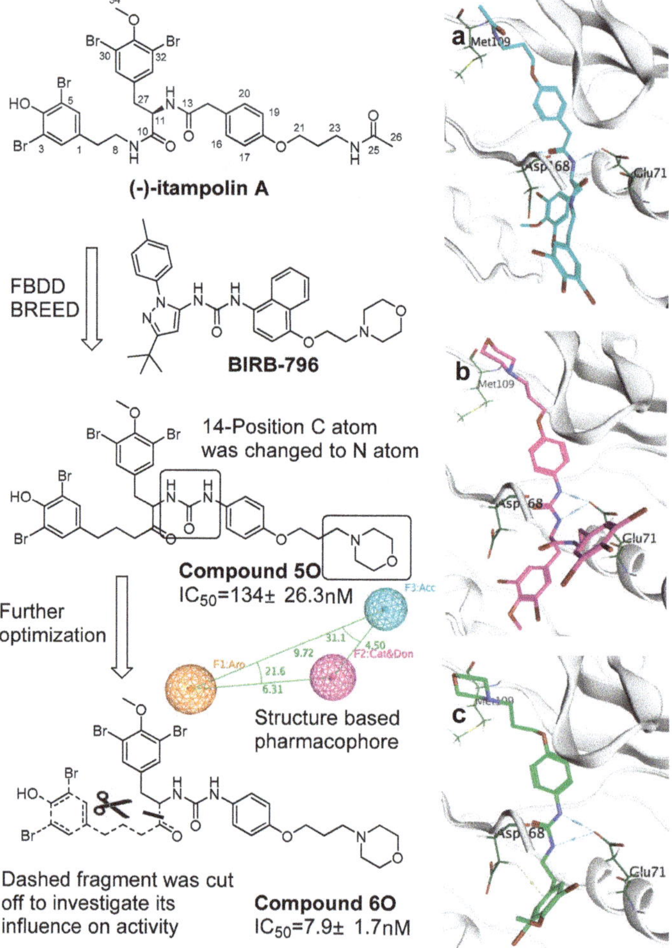

Figure 5. Structural optimization of (−)-itampolin A; (**a**) the conformation of (−)-itampolin A docked to p38α; (**b**) the conformation of compound **5o** docked to p38α; (**c**) the conformation of compound **6o** docked to p38α.

In the case of CoMFA analysis, the contour maps around **60** (Figure 4a,b) were generated from the CoMFA model. The steric contour map of **60** is shown in Figure 4a. Two green regions near the *N*-substituents of the morpholine proved that this length of hydrocarbon chain preferred sterically favorable functional groups. In contrast, a yellow region over the 11-C showed that a large group such as segment II leads to a decrease in activity. Consistent with this result, after cutting off the brominate tyramine fragment, the activities of compounds 3a–3o were found to be improved. This may be attributable to the excessive volume of the fragment. It was exposed to the outside of the active site pocket, when (−)-itampolin A was docked into p38α, as shown in Figure 5a. The electrostatic contour map of **60** is shown in Figure 4b. The red region over the benzene ring revealed that substituents on the benzene ring preferred electronegative groups, such as halogen atoms. The blue region around the 14-N showed that electropositive groups possessed good anticancer activity. Consistent with the QSAR analysis result, the molecular docking study showed that the urea group forms one more hydrogen bond to Glu71 than the amide group (Figure 5b). These results explained the fact that almost all of the compounds in the second series exhibited more potent inhibitory effects than those in the first series after the 14-position carbon atom was changed to a nitrogen atom.

The CoMSIA contour hydrophobic map is shown in Figure 4c. The yellow contour map in the morpholine indicated the region where addition of the hydrophobic groups would increase the inhibitory activity while the white contour map over the 11-C proved that segment II in this position will decrease the inhibitory activity. The contour map of the CoMSIA hydrogen bond donor (HBD) is shown in Figure 4d; the purple region above 11-C represents the HBD atom substitution which was adverse to the activity. This might be attributed to the hydroxyl of segment II. The cyan region around the 14-N atom indicates that the addition of the HBD atom was conducive to the activities. The CoMSIA HBD contour map can be validated by the fact that almost all of the compounds in the second series exhibited more potent inhibitory effects than those in the first series after the 14-position carbon atom was changed to a nitrogen atom.

Finally, we combined the ligand-based 3D-QSAR analysis with the structure-based molecular docking study, to identify the necessary moiety of the brominated tyrosine derivatives (Figure 6).

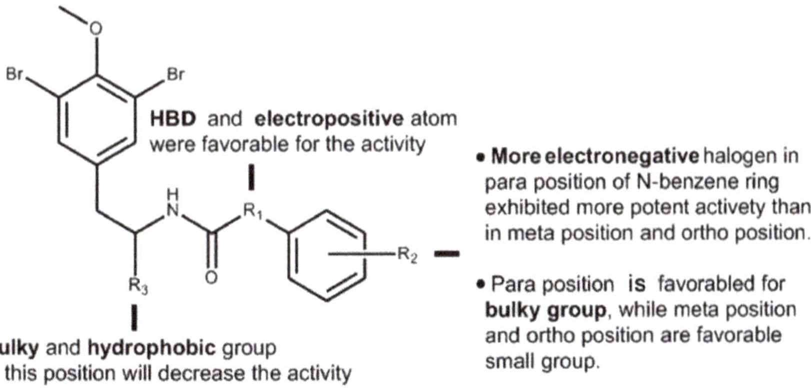

Figure 6. Important structural requirements of the novel brominated tyrosine skeleton by means of the ligand-based 3D-QSAR and structure-based molecular docking study.

In conclusion, (−)-itampolin A, as a lead compound, was optimized by the FBDD method based on the interaction between type II inhibitors and p38α. The 14-position carbon atom was changed to a nitrogen atom, making the amide group change to a diaryl urea unit. The brominate tyramine fragment and the substituent groups on the benzene ring were also modified. Thus, 45 brominated tyrosine derivatives of three series were designed and synthesized. Not all the compounds were reported in previous studies, and their structures were confirmed by ^{1}H-NMR and ^{13}C-NMR.

Through the analysis of 3D-QSAR, it was found that the conversion of the 14-position carbon atom to a nitrogen atom contributed to improving the activity of the derivatives. Cleavage of the brominate tyramine fragment helped to improve the activities of **6a–6o**. The groups on the benzene ring preferred electron-withdrawing groups, followed by meta-substituted halogen atoms. The alkane chains on the aromatic ring were not beneficial to increasing inhibitory activity. Moreover, we discovered that a novel derivative, **6o** (IC_{50} = 0.66 µM), exhibited significant antitumor activity against A549.

4. Experimentation

4.1. Fragment-Based Drug Design

Fragment-based drug design was performed using the BREED module in Molecular Operating Environment package (MOE 2015.1001.). Eight molecules (ligands in 3HV3, 3IW5, 3L8S, 3IW7, 3IW8, 4FA2, 2KV2, 2PUU) after alignment were set as novel bond producers. (−)-Itampolin A was set as the parent compound. During the generation, a filter included the Lipinski's rule of five and a structure-based pharmacophore was toggled ON. The novel compounds were generated based on candidate bonds.

4.2. Molecular Docking

Molecular docking was performed using MOE. The ligands were docked into p38α after energy minimization. Their conformations were generated with the bond rotation method. In order to validate the docking protocol, a self-docking of BIRB-796 into the binding pocket was firstly performed. The triangle matcher was selected as the ligand placement method, the London dG was selected as the scoring function, the rigid receptor was used as the type of post-placement refinement, the GBVI/WSA dG was used as the refinement scoring function. The one with the best docking conformation was selected from 30 poses according to E-strain. The RMSD value between the docking conformation and actual conformation was 0.9695, as shown in Figure S1.

4.3. The p38α MAP Kinase Activity Based on the Rate of Phosphorylation of ATF-2

Kinase reaction buffer composition: 50 mM HEPES (pH 7.5), 10 mM $MgCl_2$, 0.01% BRIJ-35 and 1 mM EGTA. Titration MAPK14/p38α at 90 µM ATP: Prepare MAPK14/p38α in kinase buffer at a concentration of 500 ng/mL. Perform two-fold serial dilution using kinase buffer from 500 ng/mL, 16 dose points. Add 5 µL of the serially diluted MAPK14/p38α into the 384-well plate in triplicate. Prepare 1 mL of 0.8 µM substrate GFP-ATF2 (19–96) and 180 µMATP in kinase reaction buffer. The reaction was carried out by adding 5 µL of the substrate GFP-ATF2 (19–96) and ATP solution into each well of the assay plate. Seal the assay plate and incubate for 1 h at room temperature (RT). Add 10 µL of the antibody solution prepared in TR-FRET dilution buffer (20 mM EDTA and 4 nM Tb-antipATF2 (pThr71), and mix gently. The final EDTA and Tb-antipATF2 (pThr71) concentrations were 10 mM and 2 nM, respectively. Seal the assay plate and incubate for 30 min at RT. Read the TR-FRET signal on the Envision 2104 plate reader.

Determination of IC_{50} values of inhibitors: add 2 µL/well of inhibitor in 0.5% DMSO at five-fold the final assay concentration to the 384-well assay plate. For the first cycle inhibitor screening, the final concentrations of inhibitors were 1117, 279.25, 69.81, 17.45, and 4.36 nM (three-fold dilution, five dose points, two replicates for each dose). Adjust the inhibitor concentration based on the results of the first cycle. An amount of 4 µL MAPK14/p38α was added to each well of the 384-well assay plate. Incubate for 15 min at RT. Finally, 4 µL substrate GFP-ATF2 (19–96) and ATP in kinase reaction buffer were added to start the reaction. The final concentrations of MAPK14/p38α, Substrate, and ATP were 1 ng/mL, 0.4 µM, and 90 µM, respectively. Incubate for 1 h at RT. Add 10 µL/well of antibody solution. The final concentrations of EDTA and antibody were 10 mM and 2 nM respectively. Incubate for 30 min at RT. Read the TR-FRET signal on the Envision 2104 plate reader.

4.4. MTT Assay

The A549 cell lines were cultured using DMEM and RPMI, respectively, in standard humidified incubation conditions at 37 °C in 5% CO_2. Then, the cancer cells (1000/well) were placed on 96-well plates in triplicate. After 24 h incubation, cells were treated with a series of concentrations of the brominated tyrosine derivatives (1, 10, 20, 80 μmol/L). The positive control contained the same concentrations of BIRB-796, and the negative control received the same volume of DMSO. After 48h incubation, cells were stained with 5% MTT, and the optical density was recorded at an absorbance wavelength of 570 nm.

4.5. 3D-QSAR Study

The SKETCH function of Sybyl-X2.0 was utilized for drawing the structure and charges were calculated by the Gasteiger–Huckel method, and tripos force field was utilized for energy minimization of these molecules. These 45 synthetic molecules were divided into the training set and test set in the ratio of 80:20. The training set was used to build the 3D-QSAR model, and the test set was used to test the predictions of the model [17,18]. In the case of building the Topomer CoMFA QSAR model, a carbon sp3 probe was applied for calculating steric and electrostatic parameters. The database alignment was used to build the CoMSIA QSAR model [19].

4.6. Chemsitry

^{1}H-NMR and ^{13}C-NMR spectra were recorded on Varian NMR spectrometers operating at 600 MHz for ^{1}H, and 150 MHz for ^{13}C. All chemical shifts were measured in DMSO-*d6* as solvent. All chemicals were purchased from Sinoreagent Chemcal Reagent (Beijing, China) and were used as received, unless stated otherwise. Analytical TLC was performed on Haiyang (Qingdao Haiyang Chemcal Co., Ltd. (Qingdao, China)) silica gel 60 F254 plates and visualized by UV and potassium permanganate staining. Flash column chromatography was performed on Haiyang (Qingdao Haiyang Chemcal Co., Ltd.) gel 60 (40–63 mm). HPLC was performed on Agilent 1260 Infinity II. Chromatographic separation was performed on a C18 column (3.0 × 50 mm, 1.7 μm) and the column temperature was maintained at 25 °C. The mobile phase consisted of methanol (50%) and water (50%).

The synthesis methods of the derivatives are described in the Supplementary Materials.

Supplementary Materials: The following are available online at http://www.mdpi.com/1660-3397/17/1/53/s1. Figure S1: Comparison of the BIRB-796 molecular docking conformation (green) with the actual conformation (blue).

Author Contributions: J.-w.L. conceived and designed the experiments, M.-y.W. and S.W. performed the experiments; S.W. and X.-y.L. analyzed the data; J.-w.L. and F.-h.M. wrote and reviewed the final manuscript.

Funding: This work was supported by the National Natural Science Foundation of China (grant no. 81573687, 81274182).

Conflicts of Interest: The authors declare no conflict of interest.

References

1. Mayer, A.M.S.; Rodríguez, A.D.; Taglialatela-Scafati, O.; Fusetani, N. Marine pharmacology in 2009–2011: Marine compounds with antibacterial, antidiabetic, antifungal, anti-inflammatory, antiprotozoal, antituberculosis, and antiviral activities; affecting the immune and nervous systems, and other miscellaneous mechanisms of action. *Mar. Drugs* **2013**, *11*, 2510–2573. [PubMed]

2. Mayer, A.M.S.; Glaser, K.B.; Cuevas, C.; Jacobs, R.S.; Kem, W.; Little, R.D.; McIntosh, J.M.; Newman, D.J.; Potts, B.C.; Shuster, D.E. The odyssey of marine pharmaceuticals: A current pipeline perspective. *Trends Pharmacol. Sci.* **2010**, *31*, 255–265. [CrossRef] [PubMed]

3. Sagar, S.; Kaur, M.; Minneman, K.P. Antiviral lead compounds from marine sponges. *Mar. Drugs* **2010**, *8*, 2619–2638. [CrossRef] [PubMed]

4.   Yasuhara-Bell, J.; Lu, Y. Marine compounds and their antiviral activities. *Antiviral Res.* **2010**, *86*, 231–240. [CrossRef] [PubMed]

5.   He, W.F.; Liang, L.F.; Cai, Y.S.; Gao, L.X.; Li, Y.F.; Li, J.; Liu, H.L.; Guo, Y.W. Brominated polyunsaturated lipids with protein tyrosine phosphatase-1B inhibitory activity from Chinese marine sponge *Xestospongia testudinaria*. *J. Asian Nat. Prod. Res.* **2015**. [CrossRef] [PubMed]

6.   Kunze, K.; Niemann, H.; Ueberlein, S.; Schulze, R.; Ehrlich, H.; Brunner, E.; Proksch, P.; Van Pée, K.H. Brominated skeletal components of the marine demosponges, *Aplysina cavernicola* and *Ianthella basta*: Analytical and biochemical investigations. *Mar. Drugs* **2013**, *11*, 1271–1287. [CrossRef] [PubMed]

7.   Mani, L.; Jullian, V.; Mourkazel, B.; Valentin, A.; Dubois, J.; Cresteil, T.; Folcher, E.; Hooper, J.N.A.; Erpenbeck, D.; Aalbersberg, W.; et al. New antiplasmodial bromotyrosine derivatives from *suberea ianthelliformis lendenfeld*, 1888. *Chem. Biodivers.* **2012**. [CrossRef] [PubMed]

8.   Tohme, R.; Darwiche, N.; Gali-Muhtasib, H. A journey under the sea: The quest for marine anti-cancer alkaloids. *Molecules* **2011**, *16*, 9665–9696. [CrossRef] [PubMed]

9.   Sorek, H.; Rudi, A.; Aknin, M.; Gaydou, E.; Kashman, Y. Itampolins A and B, new brominated tyrosine derivatives from the sponge *Iotrochota purpurea*. *Tetrahedron Lett.* **2006**. [CrossRef]

10.   Liang, J.; Li, X.; He, X.; Sun, Q.; Zhang, T.; Meng, F. (+)- and (−)-itampolin A: First total synthesis, anticancer effect through inhibition of phospho p38 expression. *Curr. Org. Synth.* **2017**. [CrossRef]

11.   Badrinarayan, P.; Sastry, G.N. Journal of Molecular Graphics and Modelling Virtual screening filters for the design of type II p38 MAP kinase inhibitors: A fragment based library generation approach. *J. Mol. Graph. Model.* **2012**. [CrossRef] [PubMed]

12.   Mashalidis, E.H.; Śledå, P.; Lang, S.; Abell, C. A three-stage biophysical screening cascade for fragment-based drug discovery. *Nat. Protoc.* **2013**. [CrossRef] [PubMed]

13.   Erlanson, D.A.; Fesik, S.W.; Hubbard, R.E.; Jahnke, W.; Jhoti, H. Twenty years on: The impact of fragments on drug discovery. *Nat. Rev. Drug Discov.* **2016**, *15*, 605. [CrossRef] [PubMed]

14.   Murray, C.W.; Rees, D.C. The rise of fragment-based drug discovery. *Nat. Chem.* **2009**. [CrossRef] [PubMed]

15.   Scott, D.E.; Coyne, A.G.; Hudson, S.A.; Abell, C. Fragment-based approaches in drug discovery and chemical biology. *Biochemistry* **2012**. [CrossRef] [PubMed]

16.   Gibson, T.S.; Johnson, B.; Fanjul, A.; Halkowycz, P.; Dougan, D.R.; Cole, D.; Swann, S. Structure-based drug design of novel ASK1 inhibitors using an integrated lead optimization strategy. *Bioorganic. Med. Chem. Lett.* **2017**. [CrossRef] [PubMed]

17.   Lim, H.Y.; Heo, J.; Choi, H.J.; Lin, C.Y.; Yoon, J.H.; Hsu, C.; Rau, K.M.; Poon, R.T.P.; Yeo, W.; Park, J.W.; et al. A phase II study of the efficacy and safety of the combination therapy of the MEK inhibitor refametinib (BAY 86-9766) plus sorafenib for Asian patients with unresectable hepatocellular carcinoma. *Clin. Cancer Res.* **2014**. [CrossRef]

18.   Klebe, G.; Abraham, U.; Mietzner, T. Molecular Similarity Indices in a Comparative Analysis (CoMSIA) of Drug Molecules To Correlate and Predict Their Biological Activity. *J. Med. Chem.* **1994**. [CrossRef]

19.   Cichero, E.; Bruno, O.; Fossa, P. Docking-based CoMFA and CoMSIA analyses of tetrahydro-β-carboline derivatives as type-5 phosphodiesterase inhibitors. *J. Enzyme Inhib. Med. Chem.* **2012**. [CrossRef] [PubMed]

 *marine drugs*

Article

Nile Tilapia Derived TP4 Shows Broad Cytotoxicity toward to Non-Small-Cell Lung Cancer Cells

Chen-Hung Ting [1] and Jyh-Yih Chen [1,2,*]

[1] Marine Research Station, Institute of Cellular and Organismic Biology, Academia Sinica, Ilan 262, Taiwan; koichiting@gmail.com

[2] The iEGG and Animal Biotechnology Center, National Chung Hsing University, Taichung 402, Taiwan

* Correspondence: zoocjy@gate.sinica.edu.tw; Tel.: +886-920802111; Fax: +886-39871035

Received: 16 November 2018; Accepted: 12 December 2018; Published: 13 December 2018

Abstract: Non-small cell lung cancer (NSCLC) is among the leading causes of human mortality due to a lack of effective treatments. Conventional chemotherapies affect healthy cells and cause multidrug resistance, while tumors may eventually develop resistance to less-toxic targeted therapies. Thus, the need to develop novel therapies for NSCLC is urgent. Here, we show that Nile tilapia-derived Tilapia piscidin (TP) 4 is cytotoxic to a panel of NSCLC cells with different genetic profiles. We observed that TP4 triggers NSCLC cell death through the necrosis and combining TP4 with potent Epidermal growth factor receptor (EGFR)- tyrosine kinase inhibitors (TKI)s, Erlotinib, and Gefitinib, improved drug responses in EGFR-mutated NSCLC cells, but not in EGFR-wild-type NSCLC cells. This work provides novel insights into potential NSCLC treatments, which may utilize antimicrobial peptide TP4 as monotherapy or in combination with EGFR-TKIs.

Keywords: Antimicrobial peptide (AMP); Tilapia piscidin 4 (TP4); non-small cell lung cancer (NSCLC)

1. Introduction

Lung cancer is the leading cause of cancer mortality worldwide [1]. Most lung cancer patients die within one year of diagnosis, and the five-year-survival rate is around 18.6% [2]. Among these patients, over 85% are diagnosed with non-small cell lung cancer (NSCLC), while 15% have small cell lung cancer (SCLC). About 50% of NSCLCs are phenotypically characterized as adenocarcinomas (ADCs), of which gene expression profiles are consistent with a distal lung cell origin. Meanwhile, about 40% of NSCLCs are squamous cell carcinomas (SCCs), which are thought to arise from proximal tracheal basal cells in the lung [3,4]. Chemotherapies are routinely used for NSCLC treatment; however, chemotherapeutic drugs can cause serious side-effects and multidrug resistance. Advanced transcriptome analyses unveiled several gene mutations that are associated with NSCLC, including Epidermal growth factor receptor (EGFR) [5–8]. Constitutively activating EGFR mutations have been observed in 10% and 35% of NSCLC patients in the U.S. and East Asia, respectively [9,10]. Due to the tumorigenic nature of these mutations, EGFR-directed tyrosine kinase inhibitors (TKIs) have become valuable tools for the treatment of NSCLC [6–8,11–14]. The first and second-generation EGFR-TKIs (e.g., Erlotinib, Gefitinib, and Afatinib) were developed to target proteins derived from exon 19 deletions or a point mutation in exon 21 (L858R) [9]. These early EGFR-TKIs have produced remarkable results in the clinic; however, the median progression-free survival (PFS) is still less than 12 months due to the occurrence of drug-resistant cancer cells [15,16]. One critical secondary mutation that confers drug resistance is a gatekeeper point mutation in exon 20 (T790M) of EGFR, which is observed in about 50%–60% of all patients [11,17–19]. The third-generation EGFR-TKIs (e.g. Osimertinib) were developed to selectively target mutants (particularly the T790M mutation) and show low affinity to wild-type EGFR, thereby significantly reducing toxicity [13,20]. Unfortunately, other novel mutations may still occur in TKI-treated tumors, leading to drug resistance [21,22]. In order to overcome this resistance, combinations of EGFR-TKIs with chemotherapy or other targeted agents are recommended [20,23,24]. Thus, alternative strategies and novel drugs for treatment of NSCLC are urgently needed.

Antimicrobial peptides (AMPs) are evolutionarily conserved peptides that function to combat microbial infections [25] and have been suggested as potential anti-cancer agents [26–28]. The defensive utility of cationic AMPs is derived from a structural amphipathic property, which enables electrostatic interactions with anionic molecules on the plasma membrane of microbes or cancer cells [26–28]. Unlike chemotherapeutic drugs, which damage healthy cells, AMPs selectively target cancer cells with lower toxicity to non-cancerous cell types [26–28]. We previously found that the Nile tilapia (*Oreochromis niloticus*)-derived cationic AMP, tilapia piscidin (TP4) [29], binds to the negatively charged membrane of breast cancer cells and subsequently triggers cancer cell death [27]. NSCLC cells are also good candidate targets for TP4 because extensive glycosylation and low cholesterol levels enhance the negative charge of the membrane [30,31].

In this study, we aimed to investigate the therapeutic potential of TP4 in NSCLC. We show that TP4 is highly cytotoxic to multiple NSCLC cell lines with different EGFR status. Moreover, NSCLC cells with wild-type EGFR were equally susceptible to TP4 alone and in combination with EGFR-TKIs, but combined treatment of TP4 with EGFR-TKIs showed enhanced cytotoxicity over TKIs or TP4 alone in EGFR-mutated NSCLC cells. Furthermore, TP4 was found to induce necrotic death in NSCLC cells. Combining TP4 with EGFR-TKIs enhanced necrosis in NSCLC cells with EGFR mutations. These findings support the notion that TP4 is a promising candidate drug for treatment of various NSCLC types.

2. Results

2.1. Cationic TP4 is Toxic to NSCLC Cells

Cancerous cells with negatively charged membranes might be attacked by AMPs [26–28]. To investigate whether NSCLC cells are potential targets for TP4, cytotoxicity was evaluated in a panel of NSCLC cell lines. Control cell lines, BEAS-2B and MRC-5, treated with a range of TP4 concentrations (0.838 to 6.710 μM) showed limited cytotoxicity, with 50% inhibitory concentrations (IC_{50}) of over 15.48 and 26.50 μM, respectively (Figure 1A,B, Table 1). Notably, low TP4 concentrations, from 0.830 to 6.710 μM, enhanced cell proliferation in normal BEAS-2B cells, as evidenced by the gradual increase in relative adenosine triphoshapte (ATP) level of TP4-treated cells compared to mock-treated controls at 12 and 24 h post-TP4 treatment (Figure 1B). The IC_{50} values among NSCLC cells at different time points after TP4 treatment were 1.922–27.62 μM in A549 cells, 3.769–14.17 μM in NCI-H661 cells, 1.241–5.472 μM in NCI-H1975 cells, and 10.61–18.52 μM in HCC827 cells (Figure 1C–F, and Table 2).

Table 1. IC_{50} values for TP4 in normal cells at various treatment times. Results are presented as mean $\pm$ SD.

Treatment	IC_{50} (μM)	95% Confidence Interval
BEAS-2B		
3 h	33.100 $\pm$ 1.032	31.11 to 35.22
6 h	32.040 $\pm$ 1.048	29.21 to 35.15
12 h	29.760 $\pm$ 1.074	25.88 to 34.22
24 h	26.500 $\pm$ 1.086	22.53 to 31.17
MRC-5		
3 h	46.440 $\pm$ 1.049	42.32 to 50.96
6 h	28.000 $\pm$ 1.068	24.65 to 31.80
12 h	18.220 $\pm$ 1.079	15.67 to 21.17
24 h	15.480 $\pm$ 1.063	13.72 to 17.46

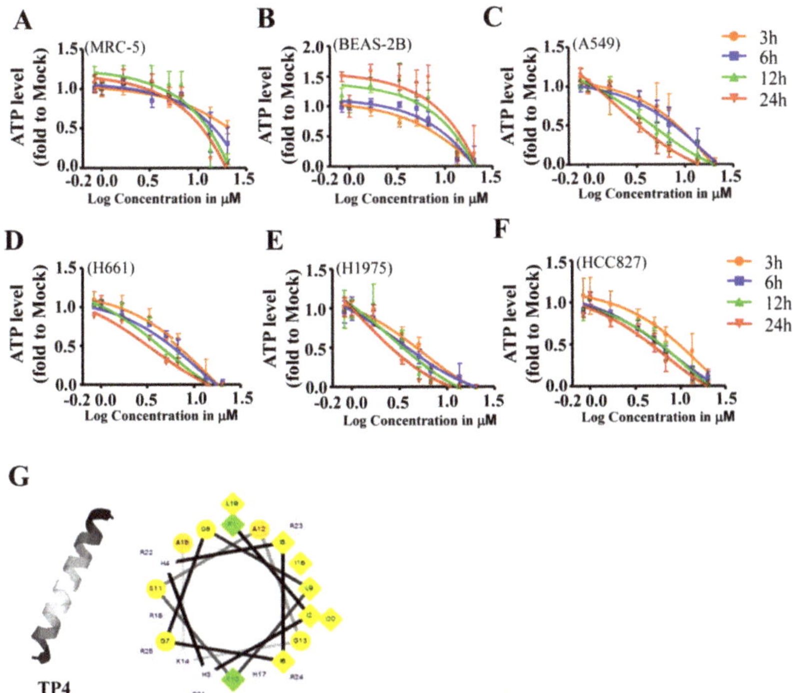

Figure 1. Broad cytotoxicity of Tilapia piscidin (TP) 4 toward non-small cell lung cancer (NSCLC) cells. (**A–F**) Cell viability in BEAS-2B (**A**), MRC-5 (**B**), and NSCLC cells (**C–F**) was determined by an adenosine triphosphate (ATP) assay, following treatment with varying doses of TP4 (0.83–20.12 µM) at the indicated time-points (3–24 h). The x-axis shows the logarithm of the TP4 concentration. Multiple wells were analyzed for each assay and compiled with other independent assays. Quantitative results represent the mean ± SD (statistical analyses are shown in Supplementary Tables S1 and S2). (**G**) The α-helical (left) and three-dimensional structures (right) of TP4. Relative position of each amino acid residues along the helix are numbered and connected with lines. Diamonds are presented as the hydrophobic residues; circles are hydrophilic residues. Triangles and pentagons are presented as residues with negatively charged and positively charged, respectively. Hydrophobicity is color-coded. The most hydrophobic residue is shown in green and the most hydrophilic residue is shown in red. Amino acid with zero hydrophobicity is shown in yellow.

2.2. Combining TP4 with Potent EGFR Tyrosine Kinase Inhibitors (TKIs) Enhances Toxicity

We next evaluated whether combining TP4 with EGFR-TKIs improves cytotoxicity in NSCLC cells. Cell lines with different EGFR status (EGFR-wild-type A549, EGFR-mutated NCI-H1975, and HCC827) were used to test the efficacy of combined treatments. In A549 cells, a combination of 10 µM Erlotinib (Erlo) or Gefitinib (Gef) with a range of TP4 concentrations (3.35–6.71 µM) showed enhanced cytotoxicity (Figure 2A). In NCI-H1975 cells, combinations of 10 µM of TKIs with TP4 (3.35–6.71 µM) showed enhanced cytotoxicity compared to TKI or TP4 treatments alone (0.11 ± 0.04, 0.06 ± 0.01, and 0.02 ± 0.01 fold in the Erlo + TP4 group; 0.14 ± 0.04, 0.06 ± 0.03, and 0.03 ± 0.01 fold in the Gef + TP4 group; and 0.32 ± 0.03, 0.07 ± 0.01, and 0.05 ± 0.02 fold in the TP4 alone group) (Figure 2B). In the HCC827 cells, a combination of 1 µM of TKIs with TP4 (3.35–6.71 µM) showed enhanced cytotoxicity compared to TKIs or TP4 treatment alone (0.32 ± 0.03, 0.15 ± 0.02 and 0.03 ± 0.01 fold in the Erlo + TP4 group; 0.26 ± 0.02, 0.14 ± 0.02 and 0.04 ± 0.01 fold in the Gef + TP4 group; and 0.65 ± 0.07, 0.47 ± 0.08 and 0.24 ± 0.07 fold in the TP4 alone group) (Figure 2C). These results showed that combination treatment markedly improved cytotoxicity in EGFR-mutated cells. However, combination treatment did not show better efficacy than TP4 alone in NSCLC cells with wild-type EGFR.

Table 2. IC_{50} values for TP4 in NSCLC cells at various treatment times. Results are presented as mean $\pm$ SD.

Treatment	IC_{50} (μM)	95% Confidence Interval
A549		
3 h	27.620 $\pm$ 1.402	14.25 to 53.55
6 h	17.080 $\pm$ 1.236	11.27 to 25.89
12 h	4.089 $\pm$ 1.130	3.216 to 5.200
24 h	1.922 $\pm$ 1.112	1.560 to 2.367
NCI-H661		
3 h	14.170 $\pm$ 1.169	10.44 to 19.24
6 h	11.890 $\pm$ 1.102	9.824 to 14.38
12 h	5.276 $\pm$ 1.111	4.289 to 6.489
24 h	3.769 $\pm$ 1.113	3.054 to 4.652
NCI-H1975		
3 h	5.472 $\pm$ 1.112	4.445 to 6.737
6 h	3.262 $\pm$ 1.143	2.513 to 4.236
12 h	2.755 $\pm$ 1.202	1.920 to 3.954
24 h	1.241 $\pm$ 1.174	0.9065 to 1.698
HCC827		
3 h	18.520 $\pm$ 1.304	11.00 to 31.16
6 h	18.370 $\pm$ 1.153	13.89 to 24.29
12 h	16.070 $\pm$ 1.154	12.12 to 21.29
24 h	10.610 $\pm$ 1.146	8.118 to 13.86

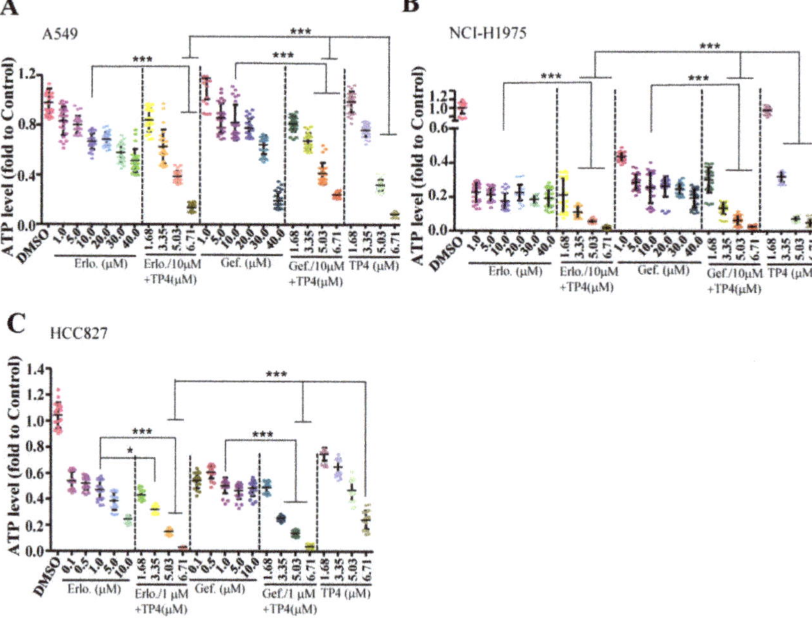

Figure 2. Combining TP4 and EGFR-TKIs enhances cytotoxicity. (**A–C**) Cell viability of A549 (**A**), H1975 cells (**B**), and HCC827 (**C**) were determined by the ATP assay 24 h after treatment with varying doses of EGFR-TKIs, TP4, or combinations thereof. At least six wells were analyzed for each condition in a single repeat (n = 3). Quantitative results represent the mean $\pm$ SD (One-way ANOVA: * $p < 0.05$; *** $p < 0.001$ versus control, ns: Not significant).

2.3. TP4 Induces Necrotic Death in NSCLC Cells

We next examined the cell death pathway triggered by TP4 in NSCLC cells. Treatment of TP4 for six and 24 h induced lactate dehydrogenase (LDH) release from NSCLC cells (Figure 3A,B), suggesting the occurrence of necrotic death. To evaluate whether apoptotic death may also be induced at early time-points after TP4 treatment (6.71 µM), we assayed caspase three activation and Lamin cleavage at 1.5 and three hours post drug treatment. The results showed no obvious changes in the levels of cleaved Lamin A/C, Lamin B1 or caspase three upon TP4 treatment (Supplementary Figure S1A,B). Moreover, treatment of cells with Necrox-2 (10 µM, Necrosis inhibitor) but not Z-VAD-FMK (50 µM, pan-caspase inhibitor) blocked TP4-induced cell death (Figure 3C). Together, these results indicate that TP4 robustly induces necrotic cell death in NSCLC cells. Furthermore, we asked whether combined TP4/TKI treatments also induce necrosis in NSCLC cells. The results showed that no significant difference in LDH production was observed in A549 cells after combined treatment (10 µM TKIs + 6.71 µM TP4); while a significant increase of LDH level was measured in H1975 or HCC827 cells with combined treatment (10 µM or 1 µM TKIs + 6.71 µM TP4) (Figure 4A–C). These results are consistent with the findings showing improved cellular toxicity of combination treatments in EGFR-mutated cells but not in EGFR-wild-type cells.

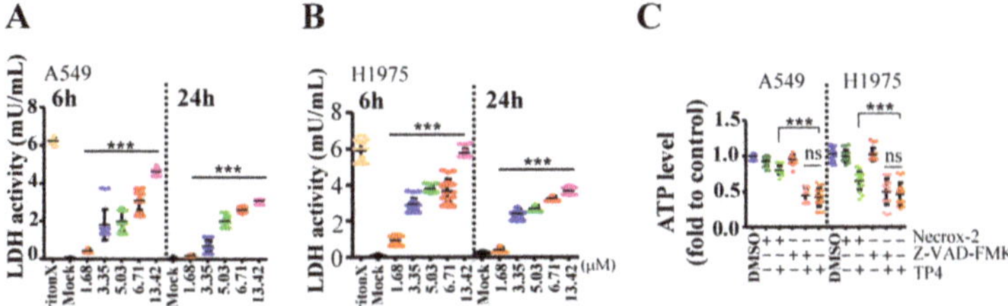

Figure 3. TP4 triggers NSCLC death by necrosis. (**A**,**B**) lactate dehydrogenase (LDH) release in A549 (**A**) and NCI-H1975 (**B**) cultures was determined 6 h or 24 h after treatment with varying doses of TP4 (1.68−13.42 µM). t-Octylphenoxypolyethoxyethanol (Triton-X) was used as a positive control. Each independent replicate was measured at least in triplicate (n = 3). Quantitative results represent the mean ± SD (One-way *ANOVA*:* $p < 0.05$; *** $p < 0.001$ versus control, ns: not significant). (**C**) Cell viability of A549 and H1975 cells were determined by the ATP assay 24 h after treatment with Dimethyl sulfoxide (DMSO), Necrox-2 (10 µM, Necrosis inhibitor), Z-VAD-FMK (50 µM, pan-caspase inhibitor), TP4 (6.71 µM), or combinations thereof. At least six wells were analyzed for each condition in a single repeat (n = 3). Quantitative results represent the mean ± SD (two-tailed Student's t-test: *** $p < 0.001$ versus control, ns: Not significant).

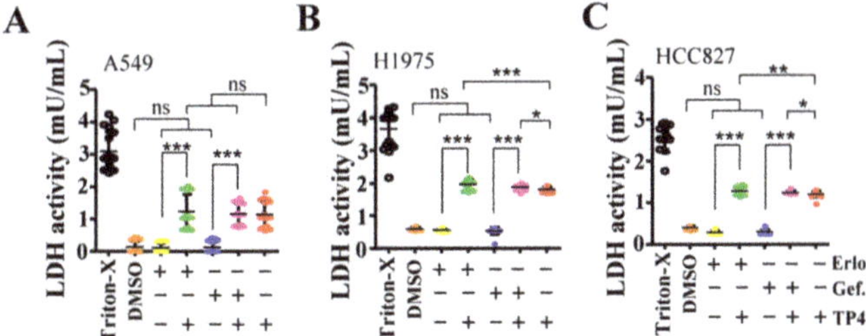

Figure 4. Combining TP4 with EGFR-TKIs enhances necrosis in EGFR-mutated NSCLC cells. (**A–C**) LDH release in A549 (**A**), NCI-H1975 (**B**), and HCC827 (**C**) cultures was determined 24 h after treatment with Triton-X, DMSO, EGFR-TKIs, TP4, or combinations thereof. Triton-X was used as a positive control. Each independent replicate was measured at least in triplicate (n = 3). Quantitative results represent the mean ± SD (two-tailed Student's t-test: * $p < 0.05$; *** $p < 0.001$ versus control, ns: Not significant).

3. Discussion

In this work, we show that the antimicrobial peptide, TP4, shows excellent cytotoxicity toward NSCLC cells with different EGFR status, and combining TP4 with potent EGFR-TKIs enhanced cytotoxicity in EGFR-mutated cells. The ratio of surviving EGFR-mutated H1975 cells and HCC827 cells was decreased from 17.6–25.6% (10 μM TKIs) to 1.7%–13.6% (10 μM TKIs + 3.35–6.71 μM of TP4) and 47.1–50.7% (1 μM TKIs) to 3%–25.5% (1 μM TKIs + 3.35–6.71 μM of TP4) (Figure 1E,F), suggesting that these combinations may be considered as a potential therapeutic strategy for EGFR-mutated NSCLC. Similar responses were not observed in EGFR-wild-type A549 cells, where TP4 alone was sufficient to cause maximal cell death (Figure 1C). Furthermore, enhanced necrosis was observed in EGFR-mutated NSCLC cells after combination treatment (Figure 4B,C). While TKIs are known to induce apoptosis in cultured NSCLC cells, it has been reported that combined SU11274 (c-Met inhibitor) with Erlotinib resulted in tumor necrosis [22]. Dual effects induced by AMP in cancer cells have been reported [32]. High concentrations of AMPs may directly lyse membranes, while low concentrations of AMPs can induce controlled cell death (i.e., apoptosis, necroptosis, or others). Here, we found that TP4 mainly induced necrosis in NSCLC cells and not apoptosis (Figure 3A,B, and Supplementary Figure S1), since TP4-induced death was blocked by Necrorex-2 but not Z-VAD-FMK (Figure 3C). This result is similar to our earlier findings showing that TP4 induces necrotic death in triple-negative-breast-cancer (TNBC) cells [27]. Interestingly, combining EGFR-TKIs with TP4 enhanced cytotoxicity to EGFR-activated NSCLC cells (Figure 1E,F), suggesting that blockage of EGFR signaling contributes to TP4-mediated cytotoxicity. However, it remains unclear whether TP4 can modulate EGFR signaling in NSCLC cells. The neutrophil antimicrobial peptide LL37/hCAP-18 was shown to induce DNA breaks in A549 cells [33], with low concentrations (1 μg/mL) inducing EGFR transactivation to promote keratinocyte migration [34]. Thus, our study suggests a link between AMP action and EGFR activity.

The mechanism of TP4-mediated cytotoxicity may differ depending on the cancer cell type, transcriptome, or regulation of specific genes. For example, TP4 was found to induce FBJ murine osteosarcoma viral oncogene homolog B (FOSB) activation in all tested TNBC cells. Interestingly, in MDA-MB-231 and MCF7 cells, full-length FOSB is induced by TP4, while the alternatively spliced form (truncated) of FOSB (FOSΔB) is not; conversely, TP4 mainly induces FOSΔB in MB453 cells [27]. Moreover, either FOSB or FOSΔB overexpression triggered cell death in all tested breast cancer cells, and *FOSB* knockdown attenuated TP4-mediated cytotoxicity [27]. TP4 induction of *FOSB* (or *FOSΔB*) requires Ca^{2+} signaling and mitochondrial dysfunction to trigger necrotic death [27], suggesting that Ca^{2+}-dependent FOSB signaling is involved in TP4-induced cytotoxicity. In addition, TP4 was shown to induce apoptosis in an osteosarcoma cell-line, MG63, through activation of extrinsic Fas/FasL- and intrinsic mitochondria-mediated pathways [35]. Pretreating these cells with Z-IETD-FMK (caspase-8 inhibitor) or Z-LEHD-FMK (caspase-9 inhibitor) significantly attenuated caspase-3 activation and prevented apoptosis [35]. These findings indicated that TP4 stimulates distinctive cytotoxic pathways among different cancer types. In NSCLC, we observed TP4-induced necrotic but not apoptotic death (Figure 3, Figure 4, and Figure S1), which is similar to what was previously found in TNBC cells. Whether FOSB signaling and Ca^{2+} homeostasis are also involved in TP4-mediated cytotoxicity in NSCLC cells remains to be further addressed.

The therapeutic efficacy of some AMPs, such as Xenopus skin-derived hymenochirin-1B and magainin, has been studied in lung cancer [36,37]. Hymenochirin-1B shows potent cytotoxicity in A549 cells with a lethal concentration $(LC)_{50}$ of 2.5 μM. Meanwhile, magainin analogs (magainin A and G) were shown to exhibit antitumor activity in SCLC cell lines but were less-toxic to normal human fibroblast cell lines [38]. The average IC_{50} of MAG A and G in the SCLC cell lines were 8.64 and 8.82 μM, respectively, while the average IC_{50} of MAG A and G against normal human fibroblast cell lines were 21.1 and 29.2 μM, respectively. In addition, two sequence-modified magainin analogues (MSI-136 and MSI-238), which were designed to enhance the amphiphilic structure, showed more toxicity to A549 cells compared to magainin 2 [39]. The hymenochirin-1B and magainin analogues harbor seven positively charged amino acid residues, while TP4 contains ten positive charged residues. The positively charged residues of cationic AMPs electrostatically interact with anionic molecules on the cell membrane [40]. In addition, the number of positively charged residues has been correlated with amphipathicity and antimicrobial activity [40]. The in vitro cytotoxicity (IC_{50}) of all $_L$-form MSI-136 and all $_D$-form MSI-238 to the A549 cells are 6 µg/mL (2.44 µM) and 10 µg/mL (4.06 µM), respectively. The all $_L$-form of TP4 has an IC_{50} value about 1.922 µM (Table 2), indicating that TP4 is a more potent peptide drug than the magainin analogues and hymenochirin-1B. Intratumoral injection of TP4 in TNBC cell-xenotransplanted mice showed no adverse side-effects [27], however, strong hemolytic activity has been described for the molecule [29]. $_D$-form amino acid-modified MSI-238 has been used to enhance proteolytic stability and to extend the survival of P388 tumor-bearing mice in vivo [40], providing a feasible approach for the development of modified TP4 isoforms for NSCLC therapy.

Overall, this study suggests that the tilapia-derived AMP, TP4, as a highly potent peptide drug that may be useful for NSCLC treatment. TP4 monotherapies can be used in NSCLC with wild-type EGFR, and combination therapies of TP4 with EGFR-TKIs may be applicable in EGFR-mutated NSCLC.

4. Material and Methods

4.1. Reagents and Peptide Sequence Analysis

TP4 (FIHHIIGGLFSAGKAIHRLIRRRRR) were synthesized and purified by GL Biochem Ltd. (Shanghai, China) as previously described [27]. Necrosis inhibitor, Necrox-2, and pan-Caspase inhibitor, Z-VAD(OMe)-FMK, were purchased from Santa Cruz Biotechnology Inc. (Dallas, TX, USA) *N*-(3-Chloro-4-fluoro-phenyl)-7-methoxy-6-(3-morpholin-4-ylpropoxy)quinazolin-4-amine (Gefitinib) and *N*-(3-ethynylphenyl)-6,7-bis(2-methoxyethoxy)-4-quinazolinamine hydrochloride (Erlotinib) were purchased from Sigma-Aldrich (St.Louis, MO, USA). Primary antibodies were purchased from the Cell Signaling (Danvers, MA, USA) (Lamin A/C, clone 4C11; Lamin B1, clone D4Q4Z; Caspase 3; cleaved Caspase 3) and EMD Millipore (GAPDH, clone 6C5) (Temecula, CA, USA). The Structural model and the helical projection of TP4 were generated, as previously described [41].

4.2. Cell Culture and Cell Viability Assay

Cell-lines (A549 (BCRC 60074), NCI-H661 (BCRC 60125), NCI-H209 (BCRC 60123), NCI-H1975 (ATCC CRL-5908), BEAS-2B (ATCC CRL-9609), and MRC-5 (BCRC 60023)) were purchased from the Bioresource Collection and Research Center (BCRC) (Taipei, Taiwan) and the American Type Culture Collection (ATCC) (Manassas, VA, USA) and cultured as suggested by BCRC and ATCC. For the cell viability assay, 4–5 × 10^3 cells were seeded into the wells of a 96-well plate and cultured overnight. During the drug treatment assay, inhibitors were added 30 minutes prior to TP4, and cell viability was determined at indicated time-points. LDH release was quantified with a Cytotoxicity Detection Kit^{PLUS} (LDH) (Roche Applied Science, Basel, Switzerland), as previously described [27].

4.3. Western Blotting

For Western blots, sample preparations were performed, as previously described [28], and were electrophoresed and transferred onto polyvinylidene fluoride (PVDF) membrane. The membranes were incubated in blocking buffer (0.1 M PBS, 5% non-fat milk, 0.2% Tween-20) for 1 hour at room temperature and then incubated in the same solution with primary or secondary antibodies (GE Healthcare Life Science, Buckinghamshire, UK). Signals were detected by enhanced chemiluminescence (Immobilon Western Chemiluminescent HRP substrate, Merck Millipore, Billerica, MA, USA) on an imaging system (UVP, BioSpectrumTM 500 (Upland, CA, USA)).

4.4. Statistical Analysis

Cells were plated at least in quadruplicate for the multi-well based assay. Data were collected from repeated experiments (n ≥ 3) and were analyzed by Prism5 software (GraphPad Inc., (La Jolla,

CA, USA)). Statistical significance was determined by two-tailed Student's t-test or one-way analysis of variance (ANOVA) with Bonferroni post hoc test. Differences were considered statistically significant at $p < 0.05$.

Supplementary Materials: Supplementary Figure S1, Table S1, and Table S2 are published online alongside the manuscript. Figure S1: TP4 does not induce caspase 3 activation and Lamin cleavage. (A, B) Total lysates from A549 (A) and NCI-H1975 (B) without (M) or with TP4 treatment (T) for 1.5 and 3 h were analyzed by Western blot using antibodies against GAPDH, caspase 3, and cleaved-caspase 3. Red dotted lines demarcate samples collected from three independent assays. Table S1: TP4 toxicity in normal BEAS-2B and MRC-5 cells was evaluated by the ATP assay. Statistical test results related to Figure 1A and B are shown. Eight wells were analyzed for each experiment (n = 24 per dose). Results represent the mean $\pm$ SD from independent experiments. Statistical comparisons between mock and each dose of TP4 were performed using one-way ANOVA analysis with Bonferroni post hoc test: ns, not significant; * $p < 0.05$; ** $p < 0.01$; *** $p < 0.001$. Table S2: TP4 toxicity in lung cancer cells was evaluated by the ATP assay. Statistical test results related to Figure 1C$-$F are shown. Multiple wells were analyzed for each experiment ($n = 24$ per dose in A549, H661, H1975 cells and $n \geq 17$ per dose in HCC827 cells, respectively). Results represent the mean $\pm$ SD from independent experiments. Statistical comparisons between mock and each dose of TP4 were performed using one-way ANOVA analysis with Bonferroni post hoc test: ns, not significant; * $p < 0.05$; ** $p < 0.01$; *** $p < 0.001$.

Author Contributions: C.H.T. designed the study, performed experiments, and wrote the manuscript; J.Y.C. supervised the study and finalized the manuscript.

Funding: The funding sources played no part in study design, data collection and analysis, decision to publish, or preparation of the manuscript. This work was financially supported by the iEGG and Animal Biotechnology Center from The Feature Areas Research Center Program within the framework of the Higher Education Sprout Project by the Ministry of Education (MOE-107-S-0023-A) in Taiwan.

Acknowledgments: We would like to thank Marcus J. Calkins for language editing. This work was supported by a PI quota to Jyh-Yih Chen from the Marine Research Station, Institute of Cellular and Organismic Biology.

Conflicts of Interest: The authors declare no conflict of interest.

References

1. McGuire, S. World Cancer Report 2014. Geneva, Switzerland: World Health Organization, International Agency for Research on Cancer, WHO Press, 2015. *Adv. Nutr.* **2016**, *7*, 418–419. [CrossRef] [PubMed]

2. Noone, A.M.; Howlader, N.; Krapcho, M.; Miller, D.; Brest, A.; Yu, M.; Ruhl, J.; Tatalovich, Z.; Mariotto, A.; Lewis, D.R.; et al. SEER Cancer Statistics Review, 1975-2015, National Cancer Institute. Bethesda, MD. Available online: https://seer.cancer.gov/csr/1975_2015/ (accessed on 10 September 2018).

3. Langer, C.J.; Besse, B.; Gualberto, A.; Brambilla, E.; Soria, J.C. The evolving role of histology in the management of advanced non-small-cell lung cancer. *J. Clin. Oncol.* **2010**, *28*, 5311–5320. [CrossRef]

4. Davidson, M.R.; Gazdar, A.F.; Clarke, B.E. The pivotal role of pathology in the management of lung cancer. *J. Thorac. Dis.* **2013**, *5* (Suppl. 5), S463–S478.

5. Davies, H.; Bignell, G.R.; Cox, C.; Stephens, P.; Edkins, S.; Clegg, S.; Teague, J.; Woffendin, H.; Garnett, M.J.; Bottomley, W.; et al. Mutations of the BRAF gene in human cancer. *Nature* **2002**, *417*, 949–954. [CrossRef] [PubMed]

6. Lynch, T.J.; Bell, D.W.; Sordella, R.; Gurubhagavatula, S.; Okimoto, R.A.; Brannigan, B.W.; Harris, P.L.; Haserlat, S M.; Supko, J.G.; Haluska, F.G.; et al. Activating mutations in the epidermal growth factor receptor underlying responsiveness of non-small-cell lung cancer to gefitinib. *N. Engl. J. Med.* **2004**, *350*, 2129–2139. [CrossRef] [PubMed]

7. Paez, J.G.; Janne, P.A.; Lee, J.C.; Tracy, S.; Greulich, H.; Gabriel, S.; Herman, P.; Kaye, F.J.; Lindeman, N.; Boggon, T.J.; et al. EGFR mutations in lung cancer: correlation with clinical response to gefitinib therapy. *Science* **2004**, *304*, 1497–1500. [CrossRef] [PubMed]

8. Pao, W.; Miller, V.; Zakowski, M.; Doherty, J.; Politi, K.; Sarkaria, I.; Singh, B.; Heelan, R.; Rusch, V.; Fulton, L.; et al. EGF receptor gene mutations are common in lung cancers from "never smokers" and are associated with sensitivity of tumors to gefitinib and erlotinib. *Proc. Natl. Acad. Sci. USA* **2004**, *101*, 13306–13311. [CrossRef]

9. Shigematsu, H.; Lin, L.; Takahashi, T.; Nomura, M.; Suzuki, M.; Wistuba, I.I.; Fong, K.M.; Lee, H.; Toyooka, S.; Shimizu, N.; et al. Clinical and biological features associated with epidermal growth factor receptor gene mutations in lung cancers. *J. Natl. Cancer Inst.* **2005**, *97*, 339–346. [CrossRef]

10. Shigematsu, H.; Takahashi, T.; Nomura, M.; Majmudar, K.; Suzuki, M.; Lee, H.; Wistuba, I.I.; Fong, K.M.; Toyooka, S.; Shimizu, N.; et al. Somatic mutations of the HER2 kinase domain in lung adenocarcinomas. *Cancer Res.* **2005**, *65*, 1642–1646. [CrossRef]

11. Pao, W.; Miller, V.A.; Politi, K.A.; Riely, G.J.; Somwar, R.; Zakowski, M.F.; Kris, M.G.; Varmus, H. Acquired resistance of lung adenocarcinomas to gefitinib or erlotinib is associated with a second mutation in the EGFR kinase domain. *PLoS Med.* **2005**, *2*, e73. [CrossRef]

12. Pao, W.; Wang, T.Y.; Riely, G.J.; Miller, V.A.; Pan, Q.; Ladanyi, M.; Zakowski, M.F.; Heelan, R.T.; Kris, M.G.; Varmus, H.E. KRAS mutations and primary resistance of lung adenocarcinomas to gefitinib or erlotinib. *PLoS Med.* **2005**, *2*, e17. [CrossRef]

13. Janne, P.A.; Yang, J.C.; Kim, D.W.; Planchard, D.; Ohe, Y.; Ramalingam, S.S.; Ahn, M.J.; Kim, S.W.; Su, W.C.; Horn, L.; Haggstrom, D.; Felip, E.; Kim, J.H.; Frewer, P.; Cantarini, M.; Brown, K.H.; Dickinson, P.A.; Ghiorghiu, S.; Ranson, M. AZD9291 in EGFR inhibitor-resistant non-small-cell lung cancer. *N. Engl. J. Med.* **2015**, *372*, 1689–1699. [CrossRef]

14. Sequist, L.V.; Soria, J.C.; Goldman, J.W.; Wakelee, H.A.; Gadgeel, S.M.; Varga, A.; Papadimitrakopoulou, V.; Solomon, B.J.; Oxnard, G.R.; Dziadziuszko, R.; et al. Rociletinib in EGFR-mutated non-small-cell lung cancer. *N. Engl. J. Med.* **2015**, *372*, 1700–1709. [CrossRef]

15. Jackman, D.; Pao, W.; Riely, G.J.; Engelman, J.A.; Kris, M.G.; Janne, P.A.; Lynch, T.; Johnson, B.E.; Miller, V.A. Clinical definition of acquired resistance to epidermal growth factor receptor tyrosine kinase inhibitors in non-small-cell lung cancer. *J. Clin. Oncol.* **2010**, *28*, 357–360. [CrossRef]

16. Romanidou, O.; Landi, L.; Cappuzzo, F.; Califano, R. Overcoming resistance to first/second generation epidermal growth factor receptor tyrosine kinase inhibitors and ALK inhibitors in oncogene-addicted advanced non-small cell lung cancer. *Ther. Adv. Med. Oncol.* **2016**, *8*, 176–187. [CrossRef]

17. Kobayashi, S.; Boggon, T.J.; Dayaram, T.; Janne, P.A.; Kocher, O.; Meyerson, M.; Johnson, B.E.; Eck, M.J.; Tenen, D.G.; Halmos, B. EGFR mutation and resistance of non-small-cell lung cancer to gefitinib. *N. Engl. J. Med.* **2005**, *352*, 786–792. [CrossRef]

18. Shih, J.Y.; Gow, C.H.; Yang, P.C. EGFR mutation conferring primary resistance to gefitinib in non-small-cell lung cancer. *N. Engl. J. Med.* **2005**, *353*, 207–208. [CrossRef]

19. Balak, M.N.; Gong, Y.; Riely, G.J.; Somwar, R.; Li, A.R.; Zakowski, M.F.; Chiang, A.; Yang, G.; Ouerfelli, O.; Kris, M.G.; Ladanyi, M.; Miller, V.A.; Pao, W. Novel D761Y and common secondary T790M mutations in epidermal growth factor receptor-mutant lung adenocarcinomas with acquired resistance to kinase inhibitors. *Clin. Cancer Res.* **2006**, *12*, 6494–6501. [CrossRef]

20. Mok, T.S.; Wu, Y.L.; Ahn, M.J.; Garassino, M.C.; Kim, H.R.; Ramalingam, S.S.; Shepherd, F.A.; He, Y.; Akamatsu, H.; Theelen, W.S.; et al. Osimertinib or Platinum-Pemetrexed in EGFR T790M-Positive Lung Cancer. *N. Engl. J. Med.* **2017**, *376*, 629–640. [CrossRef]

21. Thress, K.S.; Paweletz, C.P.; Felip, E.; Cho, B.C.; Stetson, D.; Dougherty, B.; Lai, Z.; Markovets, A.; Vivancos, A.; Kuang, Y.; et al. Acquired EGFR C797S mutation mediates resistance to AZD9291 in non-small cell lung cancer harboring EGFR T790M. *Nat. Med.* **2015**, *21*, 560–562. [CrossRef]

22. Tang, Z.; Du, R.; Jiang, S.; Wu, C.; Barkauskas, D.S.; Richey, J.; Molter, J.; Lam, M.; Flask, C.; Gerson, S.; et al. Dual MET-EGFR combinatorial inhibition against T790M-EGFR-mediated erlotinib-resistant lung cancer. *Br. J. Cancer* **2008**, *99*, 911–922. [CrossRef]

23. Janjigian, Y.Y.; Smit, E.F.; Groen, H.J.; Horn, L.; Gettinger, S.; Camidge, D.R.; Riely, G.J.; Wang, B.; Fu, Y.; Chand, V.K.; et al. Dual inhibition of EGFR with afatinib and cetuximab in kinase inhibitor-resistant EGFR-mutant lung cancer with and without T790M mutations. *Cancer Discov.* **2014**, *4*, 1036–1045. [CrossRef]

24. Soria, J.C.; Wu, Y.L.; Nakagawa, K.; Kim, S.W.; Yang, J.J.; Ahn, M.J.; Wang, J.; Yang, J.C.; Lu, Y.; Atagi, S.; et al. Gefitinib plus chemotherapy versus placebo plus chemotherapy in EGFR-mutation-positive non-small-cell lung cancer after progression on first-line gefitinib (IMPRESS): A phase 3 randomised trial. *Lancet Oncol.* **2015**, *16*, 990–998. [CrossRef]

25. Zasloff, M. Antimicrobial peptides of multicellular organisms. *Nature* **2002**, *415*, 389–395. [CrossRef]

26. Hilchie, A.L.; Doucette, C.D.; Pinto, D.M.; Patrzykat, A.; Douglas, S.; Hoskin, D.W. Pleurocidin-family cationic antimicrobial peptides are cytolytic for breast carcinoma cells and prevent growth of tumor xenografts. *Breast Cancer Res.* **2011**, *13*, R102. [CrossRef]

27. Ting, C.H.; Chen, Y.C.; Wu, C.J.; Chen, J.Y. Targeting FOSB with a cationic antimicrobial peptide, TP4, for treatment of triple-negative breast cancer. *Oncotarget* **2016**, *7*, 40329–40347. [CrossRef]

28. Ting, C.H.; Huang, H.N.; Huang, T.C.; Wu, C.J.; Chen, J.Y. The mechanisms by which pardaxin, a,natural cationic antimicrobial peptide, targets the endoplasmic reticulum and induces c-FOS. *Biomaterials* **2014**, *35*, 3627–3640. [CrossRef]

29. Peng, K.C.; Lee, S.H.; Hour, A.L.; Pan, C.Y.; Lee, L.H.; Chen, J.Y. Five Different Piscidins from Nile Tilapia, Oreochromis niloticus: Analysis of Their Expressions and Biological Functions. *PLoS ONE* **2012**, *7*, e50263. [CrossRef]

30. Ayyub, A.; Saleem, M.; Fatima, I.; Tariq, A.; Hashmi, N.; Musharraf, S.G. Glycosylated Alpha-1-acid glycoprotein 1 as a potential lung cancer serum biomarker. *Int. J. Biochem. Cell Biol.* **2016**, *70*, 68–75. [CrossRef]

31. Kucharska-Newton, A.M.; Rosamond, W.D.; Schroeder, J.C.; McNeill, A.M.; Coresh, J.; Folsom, A.R. HDL-cholesterol and the incidence of lung cancer in the Atherosclerosis Risk in Communities (ARIC) study. *Lung Cancer* **2008**, *61*, 292–300. [CrossRef]

32. Paredes-Gamero, E.J.; Martins, M.N.; Cappabianco, F.A.; Ide, J.S.; Miranda, A. Characterization of dual effects induced by antimicrobial peptides: Regulated cell death or membrane disruption. *Biochim. Biophys. Acta* **2012**, *1820*, 1062–1072. [CrossRef]

33. Aarbiou, J.; Tjabringa, G.S.; Verhoosel, R.M.; Ninaber, D.K.; White, S.R.; Peltenburg, L.T.; Rabe, K.F.; Hiemstra, P.S. Mechanisms of cell death induced by the neutrophil antimicrobial peptides alpha-defensins and LL-37. *Inflamm. Res.* **2006**, *55*, 119–127. [CrossRef]

34. Tokumaru, S.; Sayama, K.; Shirakata, Y.; Komatsuzawa, H.; Ouhara, K.; Hanakawa, Y.; Yahata, Y.; Dai, X.; Tohyama, M.; Nagai, H.; et al. Induction of keratinocyte migration via transactivation of the epidermal growth factor receptor by the antimicrobial peptide LL-37. *J. Immunol.* **2005**, *175*, 4662–4668. [CrossRef]

35. Kuo, H.M.; Tseng, C.C.; Chen, N.F.; Tai, M.H.; Hung, H.C.; Feng, C.W.; Cheng, S.Y.; Huang, S.Y.; Jean, Y.H.; Wen, Z.H. MSP-4, an Antimicrobial Peptide, Induces Apoptosis via Activation of Extrinsic Fas/FasL- and Intrinsic Mitochondria-Mediated Pathways in One Osteosarcoma Cell Line. *Mar. Drugs* **2018**, *16*, 8. [CrossRef]

36. Attoub, S.; Arafat, H.; Mechkarska, M.; Conlon, J.M. Anti-tumor activities of the host-defense peptide hymenochirin-1B. *Regul. Pept.* **2013**, *187*, 51–56. [CrossRef]

37. Mechkarska, M.; Attoub, S.; Sulaiman, S.; Pantic, J.; Lukic, M.L.; Conlon, J.M. Anti-cancer, immunoregulatory, and antimicrobial activities of the frog skin host-defense peptides pseudhymenochirin-1Pb and pseudhymenochirin-2Pa. *Regul. Pept.* **2014**, *194-195*, 69–76. [CrossRef]

38. Liu, S.; Yang, H.; Wan, L.; Cai, H.W.; Li, S.F.; Li, Y.P.; Cheng, J.Q.; Lu, X.F. Enhancement of cytotoxicity of antimicrobial peptide magainin II in tumor cells by bombesin-targeted delivery. *Acta Pharmacol. Sin.* **2011**, *32*, 79–88. [CrossRef]

39. Baker, M.A.; Maloy, W.L.; Zasloff, M.; Jacob, L.S. Anticancer Efficacy of Magainin2 and Analog Peptides. *Cancer Res.* **1993**, *53*, 3052–3057.

40. Jiang, Z.Q.; Vasil, A.I.; Hale, J.D.; Hancock, R.E.W.; Vasil, M.L.; Hodges, R.S. Effects of net charge and the number of positively charged residues on the biological activity of amphipathic alpha-helical cationic antimicrobial peptides. *Biopolymers* **2008**, *90*, 369–383. [CrossRef]

41. Ting, C.H.; Liu, Y.C.; Lyu, P.C.; Chen, J.Y. Nile Tilapia Derived Antimicrobial Peptide TP4 Exerts Antineoplastic Activity Through Microtubule Disruption. *Mar. Drugs* **2018**, *16*, 462. [CrossRef]

Article

The Anti-Angiogenic Activity of a Cystatin F Homologue from the Buccal Glands of *Lampetra morii*

Mingru Zhu †, Bowen Li †, Jihong Wang and Rong Xiao *

School of Life Sciences, Liaoning Normal University, Dalian 116081, China; zhumingru_72@163.com (M.Z.); libw@mail.dlut.edu.cn (B.L.); y.y.200@163.com (J.W.)
* Correspondence: liulangmao1980@126.com; Tel.: +1-364-494-6728
† These authors contributed equally.

Received: 12 November 2018; Accepted: 27 November 2018; Published: 29 November 2018

Abstract: Cystatins are a family of cysteine protease inhibitors which are associated with a variety of physiological and pathological processes in vivo. In the present study, the cDNA sequence of a cystatin F homologue called Lm-cystatin F was cloned from the buccal glands of *Lampetra morii*. Although Lm-cystatin F shares a lower homology with cystatin superfamily members, it is also composed of a signal peptide and three highly conserved motifs, including the G in the N-terminal, QXVXG, as well as the PW in the C-terminal of the sequence. After sequence optimization and recombination, the recombinant protein was expressed as a soluble protein in *Escherichia coli* with a molecular weight of 19.85 kDa. Through affinity chromatography and mass spectrometry analysis, the purified protein was identified as a recombinant Lm-cystatin F (rLm-cystatin F). Additionally, rLm-cystatin F could inhibit the activity of papain. Based on MTT assay, rLm-cystatin F inhibited the proliferation of human umbilical vein endothelial cells (HUVECs) dose dependently with an IC_{50} of 5 µM. In vitro studies show that rLm-cystatin F suppressed the adhesion, migration, invasion, and tube formation of HUVECs, suggesting that rLm-cystatin F possesses anti-angiogenic activity, which provides information on the feeding mechanisms of *Lampetra morii* and insights into the application of rLm-cystatin F as a potential drug in the future.

Keywords: *Lampetra morii*; buccal gland; cystatin F; anti-angiogenesis; cystatin superfamily

1. Introduction

Cystatins are a group of inhibitors, which could suppress the activity of C1 cysteine proteases and play important roles in a variety of physiological process to protect our tissues from inappropriate proteolysis [1,2]. Regarded as inhibitors, cystatins would also cooperate with cathepsins to regulate a series of events, including the maturation of dendritic cells, antigen processing and presentation, phagocytosis, as well as the expression of cytokines [1,3–6]. At any moment, cystatins are expressed or secreted to control the activity of cathepsin B, L, C, S, H, or other C1 cysteine proteases strictly. Once the dynamic equilibrium between the levels of cystatins and their substrates is destroyed, people might suffer diseases, such as malignant tumor, neurodegenerative diseases, cardiovascular disease, and chronic kidney disease [1,7–10].

In bloodsucking animals, such as ticks, cystatins are usually expressed in their salivary glands or the midgut, which suggests that cystatins might participate in the feeding process of bloodsuckers [11–15]; while in snakes, the cystatins usually exist in their venom glands, which have been reported to suppress the growth, invasion, and metastasis of B16F10 cells and MHCC97H cells, as well as to inhibit tumor angiogenesis [16–19]. At present, more cystatins have been identified from venomous insects, snakes, fishes, or mollusks through gene cloning, and transcriptomic and proteomic approaches [20–24]. Regrettably, their biological functions still need further studies.

In our previous study, the protein components of the buccal gland secretion from the fasting and feeding *Lampetra morii* (*L. morii*), which also suck the blood of fishes to survive, were compared through proteomic assays [25]. Among the diverse proteins emerged, a cystatin F homologue (also called as Lm-cystatin F), was identified in the buccal gland secretion of *L. morii* that had been fed on the blood of a catfish for 60 min, which suggests that Lm-cystatin F is closely related to the parasitic mechanisms of the *L. morii* (Figure S1).

To date, cystatin F from the other vertebrates and invertebrates was extensively studied [6,26]. However, little is known about the cystatin F from the buccal glands of *L. morii*, which are one of the most primitive vertebrates still alive. In the present study, a cystatin F homologue from the buccal glands of *L. morii* was cloned, recombined, and expressed. Additionally, its effects on the activity of papain and the endothelial cells (human umbilical vein endothelial cells, HUVECs) were also investigated.

2. Results

2.1. A Cystatin F Homologue was Identified from the Buccal Glands of L. morii

As shown in Figure 1, the open reading frame (ORF) sequence of Lm-cystatin F is 459 bp, which encodes 152 amino acids. The predicted molecular weight and theoretical isoelectric point of Lm-cystatin F are 17.1 kDa and 10.31, respectively. Noticeably, the sequence of Lm-cystatin F contains eight rare codons, including four codons for arginines (AGG, AGA, CGA), three for prolines (CCC), and one for leucine (CTA). Based on the analysis on the website (http://www.cbs.dtu.dk/services/SignalP/), the signal peptide sequence of Lm-cystatin F is MSRVASLSLLLCGLCYFCCEA, which indicated that Lm-cystatin F might be secreted extracellularly (Figure 1, green). Similar to the cystatin F from the other species, Lm-cystatin F also contains three highly conserved motifs, which could interact with the cysteine proteases, including the G in the N-terminal, QXVXG, as well as the PW in the C-terminal of the sequence (Figures 1 and 2). Furthermore, the amino acid sequence of Lm-cystatin F possesses eight cysteines, and four cysteines located at the signal peptide region (Figure 1). The nucleotide sequence of *Lm-cystatin F* has been submitted to the GenBank database (accession number: MG902948). Based on the three-dimensional structure of human cystatin F reported in the previous study, the mimetic structure of Lm-cystatin F was performed and it contains three α helixes, 5 β sheets, two loops (L1 and L2), and two disulfide bonds (Figure 2b) [27].

```
  1 ATG TCC CGT GTG GCA TCG TTG TCC CTA CTG CTG TGT GGC CTC TGC TAC TTC TGC 54
  1  M   S   R   V   A   S   L   S   L   L   L   C   G   L   C   Y   F   C  18
 55 TGC GAG GCG CTG CCC GAA ACT CGC TGG AGG CCC ATG CTG GGT GCC CCG ACC GCC 108
 19  C   E   A   L   P   E   T   R   W   R   P   M   L   G   A   P   T   A  36
109 ATC CAG ACC ACG GAC CCG GGC CTC CAC ACG GCG GCC GCG GAG GCC ACT CGC CTC 162
 37  I   Q   T   T   D   P   G   L   H   T   A   A   A   E   A   T   R   L  54
163 TTC AAC TCG GGC CTC AAC AGC CGC ACG GTC TAC AGA CTC GAC AGG GTG ACG AAG 216
 55  F   N   S   G   L   N   S   R   T   V   Y   R   L   D   R   V   T   K  72
217 GCC ACA CGC CAG ATT GTG AGC GGT CTC AAG TAC ATC TTT GAG GCA GAC CTG AAG 270
 73  A   T   R   Q   I   V   S   G   L   K   Y   I   F   E   A   D   L   K  90
271 AGC ACG GAA TGC TTG AAA ATG GAG GAA CGA GTG GCT GAG GAC TGC AAC TTT CAC 324
 91  S   T   E   C   L   K   M   E   E   R   V   A   E   D   C   N   F   H 108
325 GAC GAT GGT GTA GCC ACT GTT TTC AAG TGC CGC TTT GAG GTG TGG ACA ATC CCC 378
109  D   D   G   V   A   T   V   F   K   C   R   F   E   V   W   T   I   P 126
379 TGG CGC AAG CAG ACC AAG GTG CTC AGT CAA AGC TGC GAG CAA AAC AAC CCA ATC 432
127  W   R   K   Q   T   K   V   L   S   Q   S   C   E   Q   N   N   P   I 144
433 AAA CCT TCT ACC ATG CCA AAC GCC TAA 459
145  K   P   S   T   M   P   N   A   *  152
```

Figure 1. The ORF sequence of Lm-cystatin F and its deduced amino acid sequences. The upper lines show the ORF sequence of Lm-cystatin F, and the lower lines show its deduced amino acid sequence. The sequences are numbered from methionine, and terminated with stop codon. The signal peptide is shown in green; while the three conserved motifs are shown in purple, respectively. Except the cysteines in the signal peptide, the other four cysteines are indicated with orange.

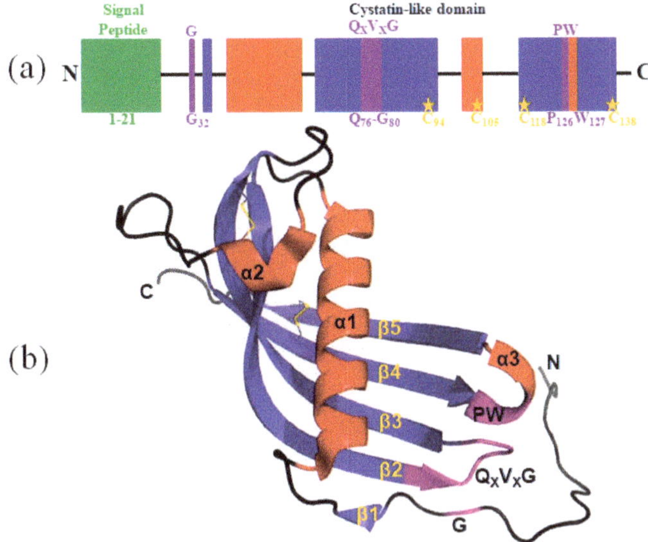

Figure 2. A schematic diagram of Lm-cystatin F and its predicted three-dimensional structure. (a) The diagrammatic structure of Lm-cystatin F. The signal peptide and the three conserved motifs are shown in green and purple, respectively. Except the cysteines in the signal peptide, the other four cysteines are labeled with orange. (b) The spatial structure of Lm-cystatin F was simulated with the three-dimensional structure of human cystatin F reported in the previous study [27]. The three α helixes and five β sheets are shown with red and blue, respectively. The two disulfide bonds are shown with orange.

2.2. Sequence Alignment and Phylogenetic Tree

As shown in Figure 3, multiple sequence alignment showed that the three motifs of Lm-cystatin F are highly conserved. In addition to the three conserved motifs, the homology between Lm-cystatin F and cystatin F from the other species is not very high. As shown in Table 1, Lm-cystatin F shares 26–38% homology with the cystatin F from nematodas, fishes, amphibians, reptiles, aves, and mammals. Phylogenetic tree showed that the cystatin F from the 20 species is mainly clustered into two groups (Figure 4). One is from the invertebrates, while the other is mainly from the vertebrates (Figure 4). Furthermore, cystatin F in the vertebrate cluster is classified into two groups (Figure 4). One group is from fishes, amphibians, reptiles, aves, and mammals, and the other group is from agnathans (Figure 4). Phylogenetic analysis showed Lm-cystatin F was clustered as the out group of the cystatin F from fishes, amphibians, reptiles, aves, and mammals.

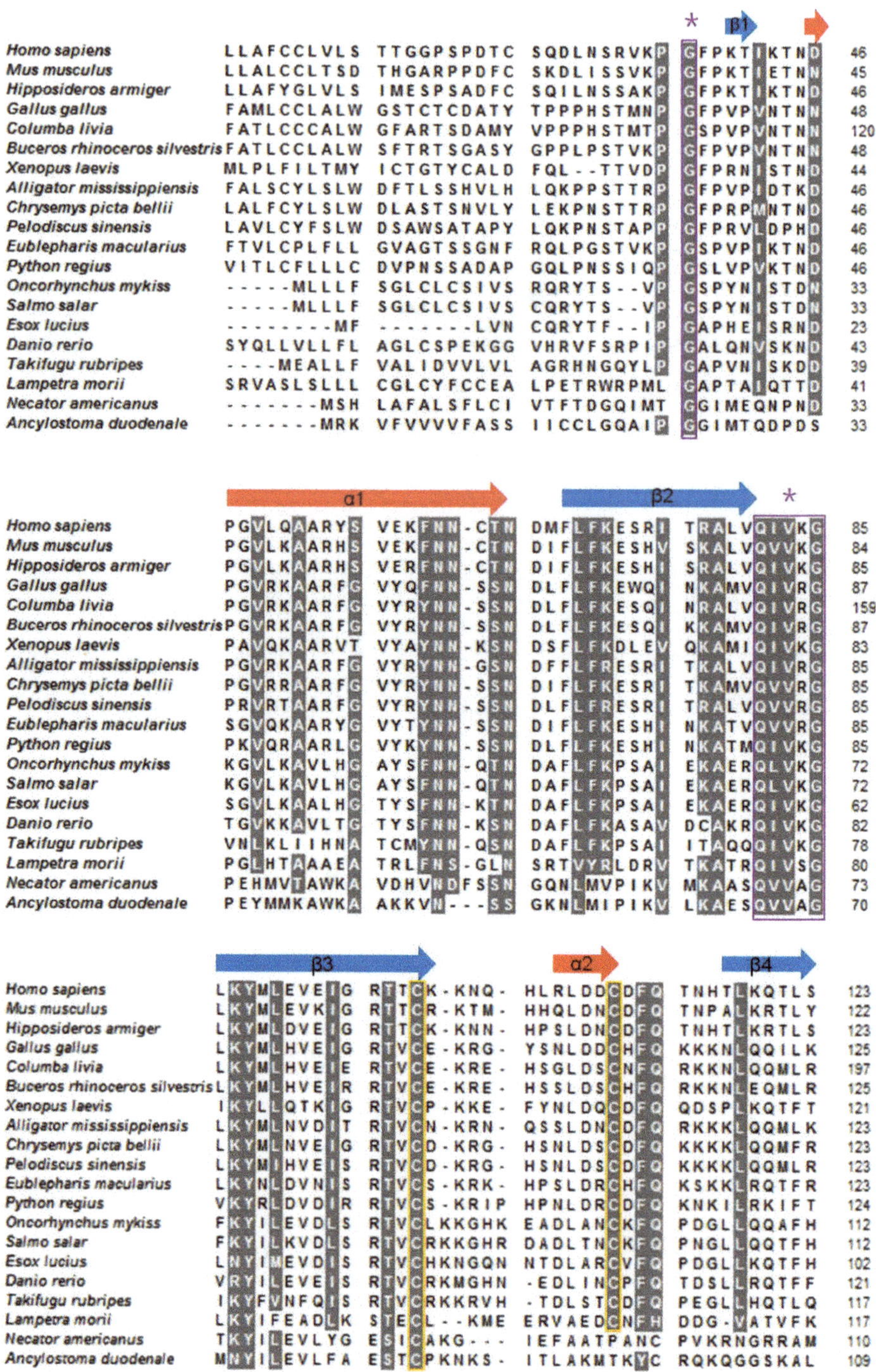

Figure 3. *Cont.*

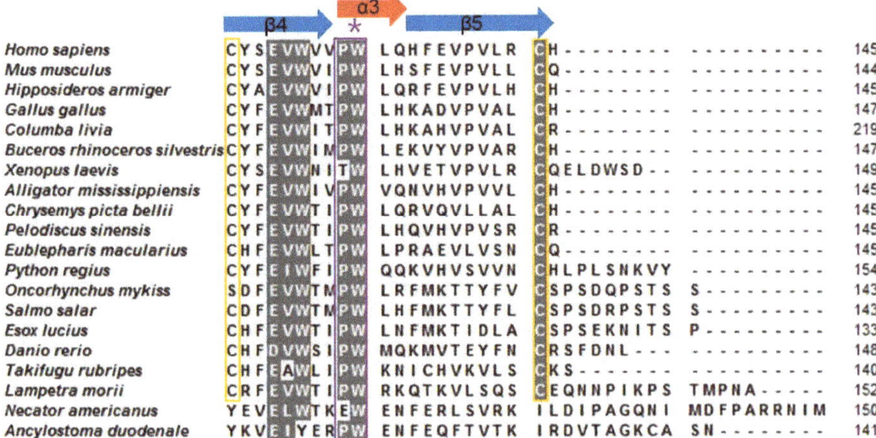

Figure 3. Sequence alignment of 20 cystatin F from the species mentioned previously. Except Lm-cystatin F, the sequences of 19 cystatin F were obtained from the EXPASY database and their accession numbers are listed as followed: *Homo sapiens*, CAD52872.1; *Mus musculus*, NP_034107.2; *Hipposideros armiger*, XP_019488972; *Gallus gallus*, NP_001186323; *Columba livia*, PKK20598.1; *Buceros rhinoceros silvestris*, XP_010137948.1; *Xenopus laevis*, NP_001091281; *Alligator mississippiensis*, KYO18358.1; *Chrysemys picta bellii*, XP_005293586; *Pelodiscus sinensis*, XP_006136954; *Eublepharis macularius*, JAC94872; *Python regius*, JAC94922; *Oncorhynchus mykiss*, XP_021420522.1; *Salmo salar*, NP_001134364.1; *Esox lucius*, NP_001297968.1; *Danio rerio*, NP_001082882; *Takifugu rubripes*, XP_011601516; *Necator americanus*, ETN77353.1; *Ancylostoma duodenale*, KIH58790.1. Additionally, the nucleotide sequence of Lm-cystatin F from *L. morii* was submitted to Genbank (accession number: MG902948). Dashes (-) indicate gaps inserted into the alignment. Asterisks (*) indicate the identical residues. The highly conserved motifs and cysteines are covered with purple and yellow frames, respectively.

Table 1. The sequence identity of Lm-cystatin F with the other cystatin F.

Category	Amino Acids	Species	Accession Number	Identity (%)
Mammals	145	*Homo sapiens*	CAD52872.1	33
Mammals	144	*Mus musculus*	NP_034107.2	32
Mammals	145	*Hipposideros armiger*	XP_019488972	31
Aves	147	*Gallus gallus*	NP_001186323	32
Aves	147	*Columba livia*	PKK20598.1	31
Aves	147	*Buceros rhinoceros silvestris*	XP_010137948.1	30
Amphibians	149	*Xenopus laevis*	NP_001091281	32
Amphibians	145	*Alligator mississippiensis*	KYO18358.1	38
Amphibians	145	*Chrysemys picta bellii*	XP_005293586	35
Amphibians	145	*Pelodiscus sinensis*	XP_006136954	31
Reptiles	145	*Eublepharis macularius*	JAC94872	32
Reptiles	154	*Python regius*	JAC94922	31
Fishes	143	*Oncorhynchus mykiss*	XP_021420522.1	35
Fishes	143	*Salmo salar*	NP_001134364.1	34
Fishes	133	*Esox lucius*	NP_001297968.1	38
Fishes	128	*Danio rerio*	NP_001082882	36
Fishes	140	*Takifugu rubripes*	XP_011601516	32
Agnathans	152	*Lampetra morii*	MG902948	100
Nematodas	136	*Necator americanus*	ETN77353.1	27
Nematodas	141	*Ancylostoma duodenale*	KIH58790.1	26

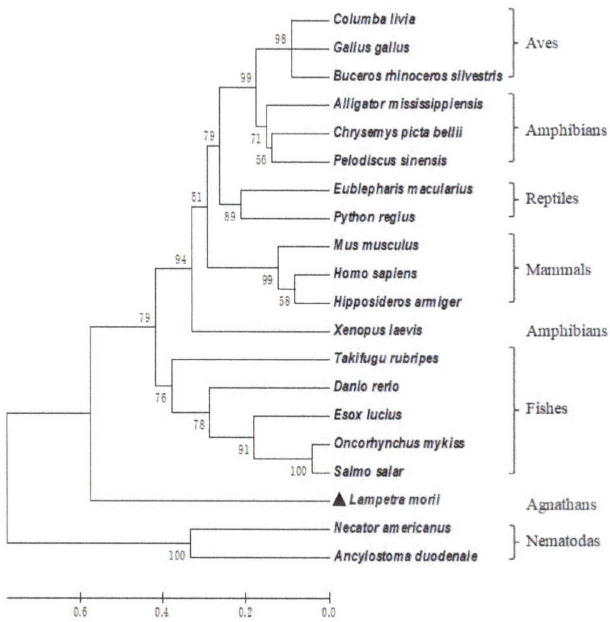

Figure 4. Phylogenetic tree of 20 cystatin F, including Lm-cystatin F. A phylogenetic tree was constructed according to the amino acid sequences of cystatin F listed in Figure 3. The number at each node indicates the percentage of bootstrapping after 1000 replications. The scale bar indicates the average number of amino acid substitutions per site.

2.3. Lm-cystatin F was Expressed as a His-Tag Fusion Protein

In order to further reveal the functions of cystatin F in *L. morii*, Lm-cystatin F was recombinant and expressed in the present study. After cloning the optimized sequence into a pCold I vector, Lm-cystatin F was expressed with the induction of 0.5 mg/mL L-Arabinos and 0.1 mM isopropyl-β-d-thiogalactoside (IPTG). As shown in Figure 5, the recombinant protein was detected in both the supernatant and precipitate of the chaperone competent cells. Additionally, the recombinant protein was further identified as a recombinant Lm-cystatin F (rLm-cystatin F) through the analysis from matrix-assisted laser desorption/ ionization time of flight (MALDI-TOF/TOF) mass spectrometry (Figure S2). Due to the His-tag, the rLm-cystatin F was purified through an affinity column and migrated as a single band on 12% sodium dodecyl sulfate-polyacrylamide gel electrophoresis (SDS-PAGE) (Figure 5). Bicinchoninic Acid (BCA) detection showed the concentration of the rLm-cystatin F was 1 mg/mL. Additionally, the residual endotoxin (lipopolysaccharide, LPS) in the 3 and 7 μM rLm-cystatin F was about 0.2 EU/mL (0.08 ng) and 0.42 EU/mL (0.17 ng), respectively. Furthermore, we also obtained a recombinant protein in *Escherichia coli* (*E. coli*) named rLj-26, which is a mutant of rLj-RGD3, and it also migrated as a single protein band on 16.5% Tricine SDS-PAGE [28]. The residual LPS in 7 μM rLj-26 was about 0.35 EU/mL. This protein was used as a control protein to exclude the effects of the His-tag and residual LPS in rLm-cystatin F.

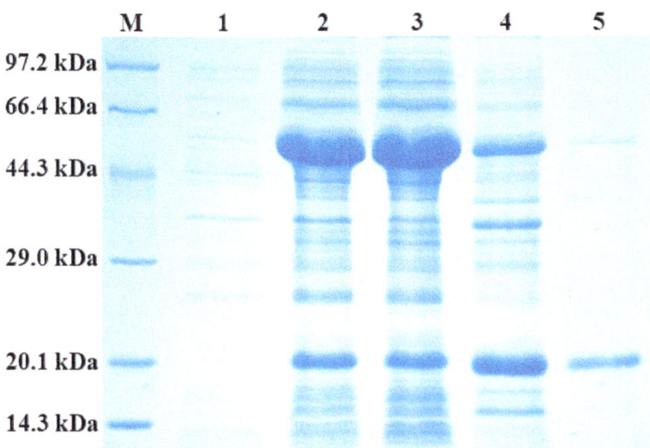

Figure 5. The expressed rLm-cystatin F was detected by 12% SDS-PAGE. After recombination, the rLm-cystatin F was expressed in the presence of pTf16. The expressed rLm-cystatin F was purified through an affinity chromatography column and detected by 12% SDS-PAGE. M, low molecular weight protein standard; 1, the chaperone competent cells were not induced with L-Arabinose (0.5 mg/mL) and IPTG (0.1 mM); 2, the chaperone competent cells were induced with L-Arabinose (0.5 mg/mL) and IPTG (0.1 mM) at 15 °C for 24 h; 3, after ultrasonication and centrifugation, the supernatant of the induced cells; 4, after ultrasonication and centrifugation, the precipitate of the induced cells; 5, the purified rLm-cystatin F.

2.4. rLm-cystatin F Blocked the Activity of Papain and the Proliferation of HUVECs

In order to detect whether rLm-cystatin F has biological functions, we firstly analyzed the inhibitory effect of rLm-cystatin F on the activity of papain with casein as a substrate. As shown in Figure 6, the degradation of casein catalyzed by papain was inhibited in a dose-dependent manner as the concentration of rLm-cystatin F increased. As the concentration of rLm-cystatin F reached 160 µg/mL, the papain activity was almost not detected. As previously reported, recombinant snake venom cystatin (sv-cystatin) from Taiwan cobra (*Naja naja atra*) showed anti-angiogenic activity [18,29]. Whether Lm-cystatin F is also capable of disturbing the angiogenic process has not been reported yet. In the present study, various concentrations (0, 1.9, 3.8, 5.7, 7.5, 9.4, and 11.3 µM) of rLm-cystatin F were incubated with HUVECs at 37 °C for 24 h. According to our 3-(4,5-dimethylthiazol-2-yl)-2, 5-diphenyltetrazolium bromide (MTT) assay, rLm-cystatin F showed inhibitory effects on the proliferation of HUVECs dose-dependently (Figure 7). The half inhibitory concentration (IC_{50}) of rLm-cystatin F on HUVECs' proliferation was 5 µM (Figure 7). Compared with the phosphate buffered saline (PBS, negative control) group, less than 10% HUEVCs were alive after treating with 11.3 µM rLm-cystatin F (Figure 7). Furthermore, 0.08 and 0.17 ng LPS, which are, respectively, equal to the content of the residual endotoxin in the 3 and 7 µM rLm-cystatin F, did not inhibit the proliferation rate of HUVECs at the same conditions (Figure S3).

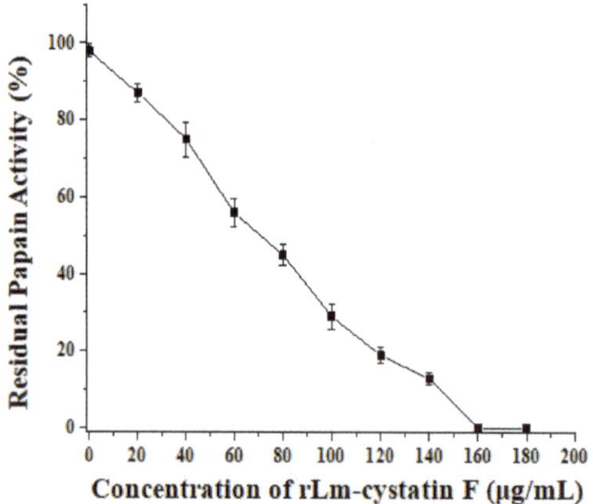

Figure 6. rLm-cystatin F inhibited the activity of papain. Different concentrations of rLm-cystatin F (0, 20, 40, 60, 80, 100, 120, 140, 160, and 180 μg/mL, final concentration) were respectively added into the casein solution in the presence of papain. Residual papain activity was calculated based on the formula mentioned in the methods.

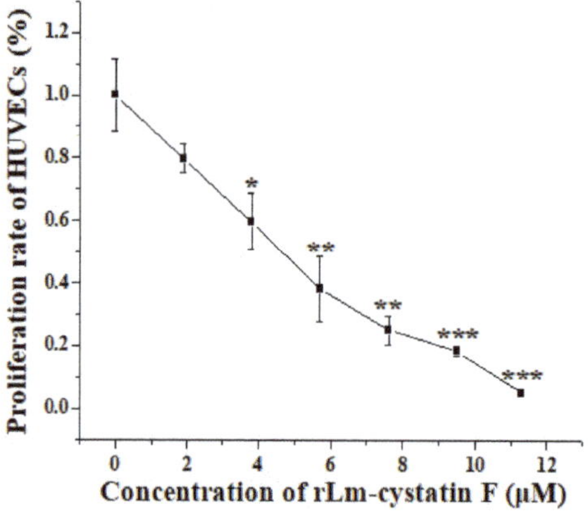

Figure 7. MTT assay showed the inhibitory effects of rLm-cystatin F on the HUVEC's proliferation. PBS was used as a negative control. The same volume of rLm-cystatin F (0, 1.9, 3.8, 5.7, 7.5, 9.4, and 11.3 μM, final concentration) was added into the HUVECs in the 96-well plates at 37 °C for 24 h. Relative to the PBS group, * $p < 0.05$; ** $p < 0.01$, and *** $p < 0.001$.

2.5. rLm-cystatin F Suppressed the Adhesive, Migrated, and Invasive Processes of HUVECs

Besides properties that caused inhibitory effects on the proliferation of endothelial cells, other anti-angiogenic factors were also reported to affect the adhesive, migrated, and invasive abilities of the endothelial cells [30]. Therefore, we used three extracellular matrix proteins, MTT, and Transwells, the classic assays to detect whether rLm-cystatin F is also able to affect the adhesive, migrated and invasive activities of HUVECs. In the presence of rLm-cystatin F, the HUVECs that were adhered to the three extracellular matrix proteins were thwarted significantly (Figure 8). When fibronectin was used as an adhesive molecule, the inhibitory rates of 3 and 7 μM rLm-cystatin F on HUVECs adhesion were $25 \pm 7\%$ ($p < 0.05$) and $38 \pm 9.8\%$ ($p < 0.01$), respectively. Similarly, 3 and 7 μM rLm-cystatin F suppressed HUVECs adhered to laminin by $27 \pm 5.6\%$ ($p < 0.01$) and $45 \pm 7.6\%$ ($p < 0.01$), respectively, and thwarted HUVECs adhered to collagen IV by $12 \pm 4.3\%$ ($p < 0.05$) and $31 \pm 3.5\%$ ($p < 0.01$), respectively. Furthermore, 0.1 and 0.2 ng LPS did not affect the adhesion rate of HUVECs when collagen IV, fibronectin, and laminin were used as adhesive molecules (Figure S3). Even though the lower chambers of the Transwells were full of RPMI 1640 medium with 15% fetal bovine serum (FBS) in order to stimulate the cell migration or invasion, rLm-cystatin F reduced the number of the migrated and invasive HUVECs on the polycarbonate filter of the Transwells (Figure 9). The inhibitory migration rates were $39.3 \pm 1.9\%$ ($p < 0.05$) and $75.6 \pm 3.6\%$ ($p < 0.01$) in the 3 and 7 µM rLm-cystatin F treating groups, respectively (Figure 9a); while the inhibitory invasion rates were $51.4 \pm 3.2\%$ ($p < 0.05$) and $78.2 \pm 2.8\%$ ($p < 0.01$) in the 3 and 7 µM rLm-cystatin F treating groups, respectively (Figure 9b). Furthermore, 0.1 and 0.2 ng LPS did not inhibit the migration and invasion of HUVECs (Figure S4). The above results suggested that rLm-cystatin F decreased the abilities of HUVECs on their adhesion, migration, and invasion in a dose-dependent manner.

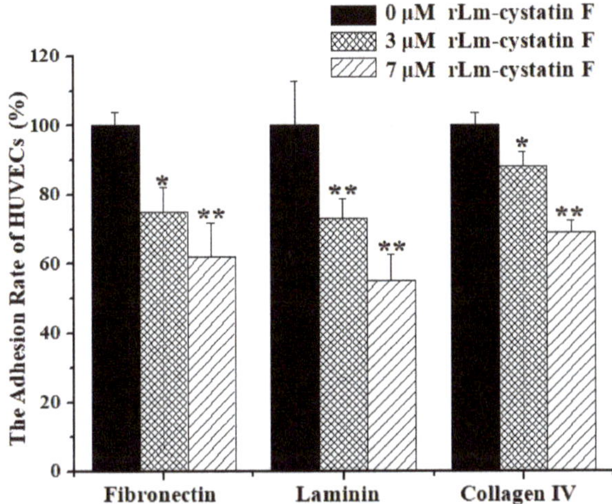

Figure 8. MTT assay showed that rLm-cystatin F thwarted HUVECs adhered to fibronectin, laminin, and collagen IV. PBS (0 µM rLm-cystatin F) was used as a negative control. Relative to the PBS group, $* p < 0.05$ and $** p < 0.01$.

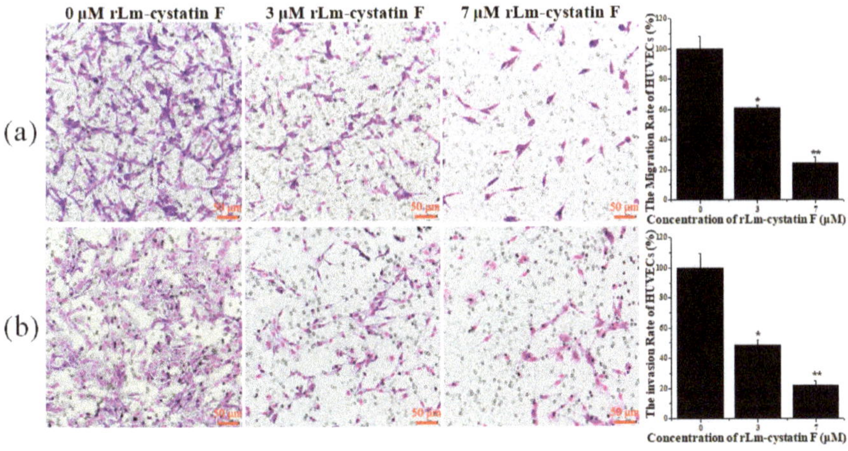

Figure 9. Transwell assays showed the inhibitory effects of rLm-cystatin F on the HUVEC's migration (**a**) and invasion (**b**). PBS was used as a negative control. Relative to the PBS group, $* p < 0.05$ and $** p < 0.01$.

2.6. rLm-cystatin F Reduced the Abilities of Tube Formation from HUVECs

In the present study, the anti-angiogenic activity of rLm-cystatin F was analyzed in the classic tube formation assay in vitro. As shown in Figure 10, the HUVECs were able to form tube-like structures on the Matrigel obviously. However, after treating with rLm-cystatin F, the abilities of HUVECs to form the tube-like structures decreased in a dose-dependent manner (Figure 10). The inhibitory rates on the surface of the formed tubes were $23.1 \pm 6\%$ ($p < 0.01$) and $58.2 \pm 6.23\%$ ($p < 0.001$) in the 3 and 7 µM rLm-cystatin F treating groups, respectively (Figure 10). Furthermore, 0.1 and 0.2 ng LPS did not affect the tube formation of HUVECs in vitro (Figure S5). Similarly, the control protein, rLj-26, which was also expressed in *E. coli* and contained a His-tag as well as the residual LPS, did not inhibit the tube formation of HUVECs in vitro (Figure S6).

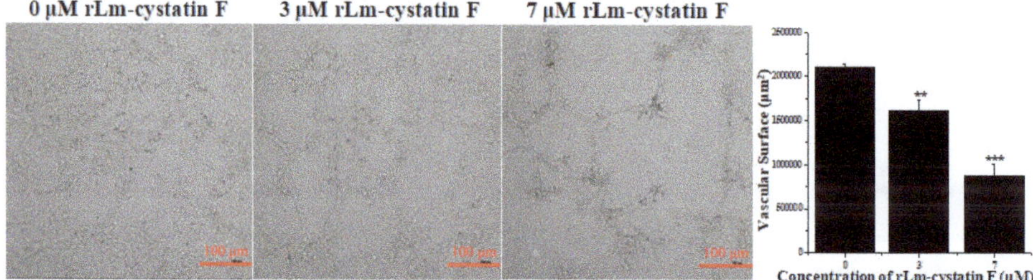

Figure 10. The inhibitory effects of rLm-cystatin F on the tube formation from HUVECs in vitro. PBS was used as a negative control. Relative to the PBS group, ** $p < 0.01$ and *** $p < 0.001$.

3. Discussion

In the present study, a cystatin F homologue was cloned from the buccal gland secretion of *L. morii* for the first time. Based on the characterization, cystatins are usually classified into three groups, including stefins (family 1), cystatins (family 2), and kininogens (family 3) [1]. As the amino acid sequence of Lm-cystatin F contains a signal peptide and two disulfide bonds at the carboxyl terminal, which are the classic characterizations of family 2, Lm-cystatin F should be classified into the family 2 in the cystatin superfamily [1]. Although sequence alignments displayed that Lm-cystatin F has relatively lower sequence identity with the cystatin F from the other species, Lm-cystatin F still possesses the highly-conserved motifs, which indicated that Lm-cystatin F might also exert the typical functions of the cystatin superfamily. Furthermore, phylogenetic analysis showed Lm-cystatin F was clustered as the out group of the cystatin F from the other vertebrates. This is also in accordance with the evolutional pattern, as agnathans (*L. morii*) are one of the most primitive vertebrates. Actually, lots of proteins identified from agnathans are clustered as a single group, which is located at the bottom of the phylogenetic tree [31].

To date, previous studies used affinity chromatography, reverse-phase chromatography, and ion-exchange chromatography to purify the native cystatin from the venom of the snakes to reveal its biological functions [29,32]. The shortcomings of this method are that a lot of venom from the snakes is required to be collected and the relatively lower yield of the proteins, as well as the multiple steps. At present, the cystatins are usually expressed in the *Pichia pastoris* (*P. pastoris*) or insect cells probably due to the relatively complex structures, which might affect their expression and folding [13,16]. Actually, cystatins were usually expressed as inclusion bodies in the *E. coli* [33,34]. Although some studies used pEGX-4T-1, pSmart-I, or pET32a vector to obtain the soluble cystatins in the *E. coli*, the recombinant cystatins usually contain a tag with a relatively larger molecular weight that needs to be removed or set in an empty plasmid control to eliminate its effects on the functions of the recombinant cystatins [35–37]. At first, we subcloned the Lm-cystatin F into a pCold I vector, however, the expressed rLm-cystatin F was hardly detected in the *E. coli* (our unpublished data). As Lm-cystatin F contains eight rare codons in its cDNA sequence, we speculated that the rare codons might lead to the difficult expression of Lm-cystatin F in *E. coli*. Thus, we optimized the sequence of Lm-cystatin F to replace the rare codons without changing its amino acid sequence. With the help of a plasmid named pTf16, which would help the recombinant protein form correct folding, rLm-cystatin F was obtained as a soluble protein with only a His-tag. Importantly, the yield of rLm-cystatin F was also improved and it showed inhibitory effects on papain activity and angiogenesis. This method helped us acquire the rLm-cystatin F in a relatively simple step, which provided novel information for the expression of the other members in the cystatin superfamily in the *E. coli* system. Actually, the expression of recombinant proteins in *E. coli* usually encounters aggregation or degradation due to their inability to form correct folding [38]. At present, the synergetic expression of trigger factor (TF) with the target proteins would help solve this problem [38]. In the future, we would also try to use virus or eukaryotic systems to express Lm-cystatin F and compare its activity with rLm-cystatin F expressed in *E. coli* to check whether rLm-cystatin F formed the same folding as that expressed in virus or eukaryotic systems.

In the present study, the His-tag in rLm-cystatin F was not removed. According to the previous studies, a recombinant protein named rLj-GRIM19, which was identified in the buccal glands of *Lampetra japonica*, was expressed in *E. coli* with a His-tag (unpublished data). Additionally, rLj-GRIM19 was proved not to be able to inhibit novel blood vessel generation in chorioallantoic membrane (CAM) models, suggesting that rLj-GRIM19 did not possess anti-angiogenic activity [39]. In our present study, a control protein (rLj-26), which was also expressed in *E. coli* and contained a His-tag, was also proved not to suppress tube formation of HUVECs in vitro. This further indicated that the His-tag would not affect the functions of rLm-cystatin F. Furthermore, the residual imidazole in the purified rLm-cystatin F was removed through ultrafiltration as the molecular weight of imidazole is only 68 Da, which is far smaller than that of rLm-cystatin F (19.85 kDa). During the expression in *E. coli*, LPS might be accompanied with the production of recombinant proteins. Additionally, it needs to be removed in extremely special conditions. In the present study, we did not remove the residual LPS in the purified rLm-cystatin F, as LPS was reported to promote angiogenesis, which was contrary to the functions of rLm-cystatin F [40,41]. Furthermore, the residual LPS in rLm-cystatin F did not inhibit the proliferation, adhesion, migration, invasion, and tube formation of HUVECs. In addition, the residual LPS was also not removed from the purified rLj-26, and rLj-26 did not possess anti-angiogenic activity. This further confirmed that the residual LPS did not affect the functions of rLm-cystatin F. In the future, we would remove residual LPS to meet the criterion of genetic engineering drugs.

According to the previous studies, cystatins participate in a variety of physiological activities and are closely related to the initiation of several diseases in abnormal conditions [1–10]. Besides, lots of studies also focused on the cystatins in the venom glands of snakes, as well as the cystatins in the salivary glands and the midgut of bloodsuckers. In snakes, cystatins have been extensively identified from a variety of snake species, including the African puff adder (*Bitis arietans*), Japanese Habu (*Trimeresurus flavoviridis*), Taiwan cobra (*Naja naja atra*), pit viper (*Bothrops jararaca*), Brown Treesnake (*Boiga irregularis*), elapid, etc. [21,29,32,42–44]. Additionally, some of the cystatins have been reported to inhibit the activity of cysteine proteases in the papain family [29]. In the present study, rLm-cystatin F was found to inhibit the activity of papain, which is a classic cysteine protease (EC 3.4.22.2), suggesting that rLm-cystatin F possessed the basic functions of cystatins [45]. At the similar conditions, rLm-cystatin F showed a better inhibitory effect on the activity of papain when compared with rEsCystatin, which was a recombinant protein identified from the Chinese mitten crab (*Eriocheir sinensis*) [34]. In 2011, Xie and colleagues showed that sv-cystatin from *Naja naja atra* was able to suppress the growth, invasion, and metastasis of B16F10 cells and MHCC97H cells by reducing the activity of cathepsin B, matrix metalloproteinase-2 (MMP-2), and matrix metalloproteinase-9 (MMP-9), and inhibition of epithelial-mesenchymal transition [16,17]. Two years later, the same authors put forward that sv-cystatin was also able to suppress tumor angiogenesis by reducing the level of vascular endothelial growth factor (VEGF)-A165, basic fibroblast growth factor (bFGF), and fms-related tyrosine kinase 1 (Flt-1) [18]. The authors concluded that family 2 cystatins, such as sv-cystatin, might target papain-type cysteine proteases and MMPs to suppress angiogenesis [18]. This is coincident with our observations, as cystatin F from the buccal glands of *L. morii* was also able to repress angiogenesis based on our in vitro studies. Although Lm-cystatin F shares 29% identity with sv-cystatin, it also suppressed crucial steps of angiogenesis, including the proliferation, adhesion, migration, invasion, and tube formation of endothelial cells (HUVECs), probably due to the highly-conserved motifs in its amino acid sequence. In addition, previous studies have reported that papain and some cysteine proteases could promote angiogenesis [46,47]. This suggests that rLm-cystatin F might inhibit angiogenesis by interaction with papain or other cysteine proteases probably due to its inhibitory effects on papain activity. As a member of family 2 cystatins, rLm-cystatin F might also affect the activity of MMPs to block angiogenesis. During the bloodsucking period, the anti-angiogenic activity of Lm-cystatin F might help *L. morii* inhibit the wound healing process of the host fishes. However, further studies are still required to clarify the detailed mechanisms of Lm-cystatin F on anti-angiogenesis. In addition, previous studies have shown that angiogenesis is closely related to tumor progression as novel blood vessels would provide nutrition for the tumor cells, thus, rLm-cystatin F might also affect the activity of tumors [30]. However, further studies of the effects of rLm-cystatin F on tumor cells are still required in the near future.

4. Materials and Methods

4.1. Cloning of a Cystatin F Homologue (Lm-cystatin F) from the Buccal Glands of L. morii

The handling of live animals was approved by the Animal Welfare and Research Ethics Committee of the Institute of Dalian Medical University (Permit Number: SYXK2004—0029). In the present study, the *L. morii* were captured in December of 2015 in Yalu River in Liaoning province of China. After collecting the buccal glands of *L. morii* in an RNase-free tube, total RNA was immediately extracted through a TakaRa MiniBEST Universal RNA extraction Kit (TaKaRa, Dalian, China). According to the instructions of the manufacturer, the total RNA was used to synthesize cDNA templates by a PrimeScript™ RT-PCR Kit (TaKaRa, Dalian, China). Bases on the ORF sequence (Accession number: ENSPMAP00000007215) in the Ensembl database (www.ensembl.org), the primers for Lm-cystatin F, were designed and listed as follows: 5′-ATGTCCCGTGTGGCATCGTTGTC-3′; 5′-TTAGGCGTTTGGCATGGTAGAAGGT-3′. After PCR amplifications, Lm-cystatin F was subcloned into a pMD® 19-T Vector and sequenced by a PRISMTM 3730XL DNA Analyzer (ABI, Carlsbad, CA, USA).

4.2. Sequence Analysis, Alignment, and Phylogenetic Tree Construction

The amino acid sequence and bioinformatic analysis of Lm-cystatin F were performed on the website listed in Table S1. Except Lm-cystatin F, additional 19 cystatin F sequences were obtained from the nematodas, fishes, amphibians, reptiles, aves, and mammals on ExPASy (http://www.expasy.ch/tools/blast). The multiple sequence alignments of cystatin F were performed by ClustalX 1.83 software. A neighbor-joining tree was constructed by MEGA 4.0 software based on the pair-wise deletion of gaps/missing data and a p-distance matrix of an amino acid model with 1000 bootstrapped replicates.

4.3. Expression, Purification, and Identification of rLm-cystatin F

Firstly, the sequence of Lm-cystatin F was synthesized by substituting the rare codons in the original sequence of Lm-cystatin F (TaKaRa, Dalian, China). Then, the primers with *EcoR* I and *Pst* I restriction sites were designed based on the optimized sequence of Lm-cystatin F and listed as follows: 5′-GGAATTCCTGCCGGAAACCCGTTG-3′; 5′-AACTGCAGTTAAGCGTTCGGCATGG TAG-3′. Subsequently, the optimized sequence of Lm-cystatin F was subcloned into a pCold I vector and transformed into chaperone competent cells with a plasmid named pTf16 (Chaperone Competent Cell BL21 Series Kit, TaRaKa, Dalian, China) at the same time. After induction with 0.5 mg/mL L-Arabinose and 0.1 mM IPTG at 15 °C for 24 h, the cells were collected through centrifugation at 4 °C and washed with 10 mM Tris-HCl buffer containing 25 mM NaCl and 10 mM imidazole. The rLm-cystatin F was purified through a HisTrap affinity column (GE, Boston, MA, USA) equilibrated with the above Tris-HCl buffer and eluted with the 40–400 mM imidazole in a gradient concentration. Next, 30 mL rLm-cystatin F was firstly concentrated to 2 mL rLm-cystatin F through Amicon® Ultra-15 10K Centrifugal Filter Devices (Millipore, Billerica, MA, USA). Then, 10 mL PBS was added into the above filter and further ultrafiltrated to 2 mL. This step was repeated for three times. After ultrafiltration, the concentration and the purity of rLm-cystatin F were, respectively, detected by a BCA Protein Assay kit (Thermo SCIENTIFIC, Waltham, MA, USA) and 12% SDS-PAGE. The protein band of rLm-cystatin F on 12% SDS-PAGE was digested in-gel by trypsin (25 mM, Promega, Madison, WI, USA) and analyzed by MALDI-TOF/TOF mass spectrometry (Bruker, Billerica, MA, USA). Furthermore, a control protein, rLj-26, was obtained based on the methods reported in the previous study [28]. The residual endotoxin in rLm-cystatin F and rLj-26 was, respectively, detected according to the instructions of the manufacturer (ToxinSensor™ Chromogenic LAL Endotoxin Assay Kit, Nanjing, China).

4.4. Enzyme Activity Assay

Based on the previous studies, the inhibitory effect of rLm-cystatin F on the activity of cysteine proteases was analyzed by using casein as a substrate [34]. Briefly, casein (5 mg/mL, final concentration) was firstly dissolved in 50 mM Tris-HCl buffer (pH 7.6) in the presence of 2 mM cysteine-HCl and 0.1 mM EDTA, and then incubated at 37 °C for 10 min. Subsequently, papain (0.1 mg/mL, final concentration) and rLm-cystatin F with various concentrations (0, 20, 40, 60, 80, 100, 120, 140, 160, and 180 µg/mL, final concentration) were, respectively, added into the above reactions and further incubated at 37 °C for 10 min. After mixing with coomassie brilliant blue G250 for 20 min at room temperature, the absorbance at 595 nm was detected. The residual papain activity was calculated based on the following formula:

$$[1 - (OD595 - OD595')/OD595'] \times 100\% \qquad (1)$$

OD595 indicated the absorbance of the reactions which contained casein, papain, and rLm-cystatin F; while OD595' indicated the absorbance of the reaction, which contained only casein and papain. Triplicate experiments were performed independently.

4.5. HUVECs' Culture and MTT Assay

HUVECs were obtained from Dr. Jihong Wang in Liaoning Normal University. HUVECs were cultured in the medium named RPMI 1640 (GIBCO, Grand Island, NY, USA) in the presence of 10% FBS (GIBCO, Grand Island, NY, USA) in a CO₂ incubator (Thermo, Waltham, MA, USA). After trypsin digestion, the HUVECs were put into 96-well plates and cultured in the incubator for 24 h. rLm-cystatin F was diluted to different concentrations from 0–11.3 µM with the above medium without FBS. The same volumes of PBS and rLm-cystatin F were incubated with the HUVECs in the incubator for 24 h. Then, 5 mg/mL MTT solution was put into the above HUVECs for 4 h. After removing the medium without drawing the formazan, 100 µL dimethyl sulfoxide (DMSO) was added into the HUVECs and incubated at 37 °C for 10 min. The absorbance at 492 nm of the HUVECs in the presence of PBS and rLm-cystatin F was recorded by a microplate reader (Thermo SCIENTIFIC, Waltham, MA, USA). The proliferative rate of HUVECs was calculated according to our previous studies [30]. At the same conditions, the effects of LPS on the proliferation rate of HUVECs were also detected by MTT assay.

4.6. Adhesion, Migration, and Invasion Assays

After digestion, the HUVECs were collected and resuspended with the FBS-free 1640 medium. PBS, 3, and 7 µM rLm-cystatin F were then added into the HUVECs. During the adhesive assays, fibronectin, laminin, as well as collagen IV, were dissolved in PBS buffer with final concentration of 0.1 mg/mL and then were, respectively, used to coat the 96-well plates at low temperatures. After removal of the extracellular matrix proteins, the pretreated HUVECs were put into the above 96-well plates and incubated in the CO_2 incubator for 2 h (collagen IV) or 3 h (fibronectin and laminin). Subsequently, the HUVECs without adhesion were washed with PBS buffer and the adhesive HUVECs were measured with MTT assay. During the migration assays, the pretreated HUVECs were put into the upper chamber of the Transwells. Meanwhile, the 1640 medium with 15% FBS was added into the lower chamber of the Transwells. Twenty hours later, the polycarbonate filters were fixed with the fixative. After removal from the Transwells, the filters were put onto the slides and stained with the Wright-Giemsa solution according to the previous studies [48]. Four views were randomly selected and captured with an inverted fluorescence microscope (Nikon, Tokyo, Japan). The migrated HUVECs were counted with the NIS-Elements D software and analyzed according to the method reported by Qi Jiang and colleagues [30]. In addition, the first step in the invasion assays was to add 4 mg/mL Matrigel (BD Bioscience, New York, NY, USA) in the upper chamber of the Transwells. After incubation at 37 °C for 30 min, the upper and lower chambers of the Transwells were, respectively, added with the PBS or rLm-cystatin F (3 and 7 µM) treated HUVECs and 1640 medium with 15% FBS. The subsequent procedures were similar to the migration assays. The time for invasion assays was 36 h. At the same conditions, the effects of LPS on the adhesion, migration, and invasion rate of HUVECs were also detected by MTT and Transwell assays.

4.7. Anti-Angiogenic Activity Assay

Firstly, both the 96-well plates and tips were put at 4 °C overnight. Subsequently, the Matrigel (4 mg/mL, final concentration) was diluted with the 1640 medium in the absence of FBS and then added into the 96-well plates. After incubation at 37 °C for 40 min, the PBS, 3, and 7 µM rLm-cystatin F were, respectively, mixed with the HUVECs and then added into the above 96-well plates. After 10 h, the inverted fluorescence microscope was used to observe the tube formation from the HUVECs. Four views were randomly selected and captured. The total area of the tubes was analyzed by the NIS-Elements D software. At the same conditions, the effects of LPS and rLj-26 on the tube formation ability of HUVECs were also observed with PBS as a control according to our previous studies [30].

4.8. Statistical Analysis

The experiments were performed three times, each in triplicate. Student's *t*-test was used to analyze the differences between the PBS and rLm-cystatin F treating groups. A statistical significance was shown as followed: * $p < 0.05$; ** $p < 0.01$, and *** $p < 0.001$.

5. Conclusions

This is the first report to show the characterization of a cystatin F homologue (Lm-cystatin F) identified from the buccal glands of *L. morri*. After cloning and recombination, Lm-cystatin F was successfully expressed as a soluble protein in the *E. coli* system with a His-tag. Although Lm-cystatin F shares low sequence identity with the members from the cystatin superfamily, rLm-cystatin F still inhibited papain activity and showed anti-angiogenic activity, which suggested that Lm-cystatin F might be an important protein to suppress the wound healing process of host fishes and might be used as a potential anti-angiogenic drug in the future.

Supplementary Materials: The following are available online at http://www.mdpi.com/1660-3397/16/12/477/s1, Figure S1: The identification of a cystatin F homologue from the buccal glands of feeding *L. morii*. Figure S2: The expressed rLm-cystatin F was identified by mass spectrometry. Figure S3: LPS did not inhibit the proliferation and adhesion rate of HUVECs. Relative to the PBS group, * $p < 0.05$. Figure S4: LPS did not inhibit the migration and invasion rate of HUVECs. Relative to the PBS group, * $p < 0.05$. Figure S5: LPS did not affect the tube formation of HUVECs in vitro. Figure S6: rLj-26 did not inhibit tube formation of HUVECs in vitro. Table S1: The websites and softwares were used to analyze the amino acid sequence and bioinformatic information of Lm-cystatin F.

Author Contributions: Conceptualization, R.X.; Methodology, M.Z. and B.L.; Software, B.L.; Validation, M.Z., B.L., J.W. and R.X.; Formal Analysis, R.X.; Investigation, M.Z. and B.L.; Resources, R.X.; Data Curation, M.Z. and B.L.; Writing-Original Draft Preparation, R.X.; Writing-Review & Editing, R.X.; Visualization, M.Z. and B.L.; Supervision, R.X.; Project Administration, R.X.; Funding Acquisition, R.X.

Funding: This research was funded by Liaoning Provincial Natural Science Foundation of China [20180550829] and Dalian high level talent innovation support plan [2015R067].

Acknowledgments: We thank Jihong Wang for her kind providing HUVECs.

Conflicts of Interest: The authors declare no conflict of interest. The funders had no role in the design of the study; in the collection, analyses, or interpretation of data; in the writing of the manuscript, and in the decision to publish the results.

References

1. Ochieng, J.; Chaudhuri, G. Cystatin superfamily. *J. Health Care Poor Underserved* **2010**, *21*, 51–70. [CrossRef] [PubMed]
2. Kim, J.T.; Lee, S.J.; Kang, M.A.; Park, J.E.; Kim, B.Y.; Yoon, D.Y.; Yang, Y.; Lee, C.H.; Yeom, Y.I.; Choe, Y.K.; et al. Cystatin SN neutralizes the inhibitory effect of cystatin C on cathepsin B activity. *Cell Death Dis.* **2013**, *4*. [CrossRef] [PubMed]
3. Zavasnik-Bergant, T.; Repnik, U.; Schweiger, A.; Romih, R.; Jeras, M.; Turk, V.; Kos, J. Differentiation- and maturation-dependent content, localization, and secretion of cystatin C in human dendritic cells. *J. Leukoc Biol.* **2005**, *78*, 122–134. [CrossRef] [PubMed]
4. Zavasnik-Bergant, T. Cystatin protease inhibitors and immune functions. *Front. Biosci.* **2008**, *13*, 4625–4637. [CrossRef] [PubMed]
5. Magister, S.; Kos, J. Cystatins in immune system. *J. Cancer* **2013**, *4*, 45–56. [CrossRef] [PubMed]
6. Ao, J.; Li, Q.; Yang, Z.; Mu, Y. A cystatin F homologue from large yellow croaker (*Larimichthys crocea*) inhibits activity of multiple cysteine proteinases and Ii chain processing *in vitro*. *Fish Shellfish Immunol.* **2016**, *48*, 62–70. [CrossRef] [PubMed]
7. Mathews, P.M.; Levy, E. Cystatin C in aging and in Alzheimer's disease. *Ageing Res. Rev.* **2016**, *32*, 38–50. [CrossRef] [PubMed]
8. Xu, Y.; Ding, Y.; Li, X.; Wu, X. Cystatin C is a disease-associated protein subject to multiple regulation. *Immunol. Cell Biol.* **2015**, *93*, 442–451. [CrossRef] [PubMed]
9. Oh, S.S.; Park, S.; Lee, K.W.; Madhi, H.; Park, S.G.; Lee, H.G.; Cho, Y.Y.; Yoo, J.; Dong Kim, K. Extracellular cystatin SN and cathepsin B prevent cellular senescence by inhibiting abnormal glycogen accumulation. *Cell Death Dis.* **2017**, *8*. [CrossRef] [PubMed]
10. Gevorgyan, M.M.; Voronina, N.P.; Goncharova, N.V.; Kozaruk, T.V.; Russkikh, G.S.; Bogdanova, L.A.; Korolenko, T.A. Cystatin C as a marker of progressing cardiovascular events during coronary heart disease. *Bull. Exp. Biol. Med.* **2017**, *162*, 421–424. [CrossRef] [PubMed]
11. Horka, H.; Staudt, V.; Klein, M.; Taube, C.; Reuter, S.; Dehzad, N.; Andersen, J.F.; Kopecky, J.; Schild, H.; Kotsyfakis, M.; et al. The tick salivary protein sialostatin L inhibits the Th9-derived production of the asthma-promoting cytokine IL-9 and is effective in the prevention of experimental asthma. *J. Immunol.* **2012**, *188*, 2669–2676. [CrossRef] [PubMed]
12. Schwarz, A.; Valdés, J.J.; Kotsyfakis, M. The role of cystatins in tick physiology and blood feeding. *Ticks Tick Borne Dis.* **2012**, *3*, 117–127. [CrossRef] [PubMed]
13. Zavašnik-Bergant, T.; Vidmar, R.; Sekirnik, A.; Fonović, M.; Salát, J.; Grunclová, L.; Kopáček, P.; Turk, B. Salivary tick cystatin OmC2 targets lysosomal cathepsins S and C in human dendritic cells. *Front. Cell Infect. Microbiol.* **2017**, *7*, 288. [CrossRef] [PubMed]
14. Cardoso, T.H.S.; Lu, S.; Gonzalez, B.R.G.; Torquato, R.J.S.; Tanaka, A.S. Characterization of a novel cystatin type 2 from *Rhipicephalus microplus* midgut. *Biochimie* **2017**, *140*, 117–121. [CrossRef] [PubMed]
15. Chmelař, J.; Kotál, J.; Langhansová, H.; Kotsyfakis, M. Protease inhibitors in tick saliva: The role of serpins and cystatins in tick-host-pathogen interaction. *Front. Cell Infect. Microbiol.* **2017**, *7*, 216. [CrossRef] [PubMed]
16. Xie, Q.; Tang, N.; Wan, R.; Qi, Y.; Lin, X.; Lin, J. Recombinant snake venom cystatin inhibits the growth, invasion and metastasis of B16F10 cells and MHCC97H cells *in vitro* and *in vivo*. *Toxicon* **2011**, *57*, 704–711. [CrossRef] [PubMed]

17. Tang, N.; Xie, Q.; Wang, X.; Li, X.; Chen, Y.; Lin, X.; Lin, J. Inhibition of invasion and metastasis of MHCC97H cells by expression of snake venom cystatin through reduction of proteinases activity and epithelial-mesenchymal transition. *Arch. Pharm. Res.* **2011**, *34*, 781–789. [CrossRef] [PubMed]

18. Xie, Q.; Tang, N.; Wan, R.; Qi, Y.; Lin, X.; Lin, J. Recombinant snake venom cystatin inhibits tumor angiogenesis *in vitro* and *in vivo* associated with downregulation of VEGF-A165, Flt-1 and bFGF. *Anticancer Agents Med. Chem.* **2013**, *13*, 663–671. [CrossRef] [PubMed]

19. Xie, Q.; Tang, N.; Lin, Y.; Wang, X.; Lin, X.; Lin, J. Recombinant adenovirus snake venom cystatin inhibits the growth, invasion, and metastasis of B16F10 cells *in vitro* and *in vivo*. *Melanoma Res.* **2013**, *23*, 444–451. [CrossRef] [PubMed]

20. Walker, A.A.; Madio, B.; Jin, J.; Undheim, E.A.; Fry, B.G.; King, G.F. Melt with this kiss: Paralyzing and liquefying venom of the assassin bug *Pristhesancus plagipennis* (Hemiptera: Reduviidae). *Mol. Cell. Proteomics.* **2017**, *16*, 552–566. [CrossRef] [PubMed]

21. Pla, D.; Petras, D.; Saviola, A.J.; Modahl, C.M.; Sanz, L.; Pérez, A.; Juárez, E.; Frietze, S.; Dorrestein, P.C.; Mackessy, S.P.; et al. Transcriptomics-guided bottom-up and top-down venomics of neonate and adult specimens of the arboreal rear-fanged Brown Treesnake, *Boiga irregularis*, from Guam. *J. Proteomics* **2017**, *174*, 71–84. [CrossRef] [PubMed]

22. Li, S.; Yang, Z.; Ao, J.; Chen, X. Molecular and functional characterization of a novel stefin analogue in large yellow croaker (*Pseudosciaena crocea*). *Dev. Comp. Immunol.* **2009**, *33*, 1268–1277. [CrossRef] [PubMed]

23. Premachandra, H.K.; Wan, Q.; Elvitigala, D.A.; De Zoysa, M.; Choi, C.Y.; Whang, I.; Lee, J. Genomic characterization and expression profiles upon bacterial infection of a novel cystatin B homologue from disk abalone (*Haliotis discus discus*). *Dev. Comp. Immunol.* **2012**, *38*, 495–504. [CrossRef] [PubMed]

24. Xiao, P.P.; Hu, Y.H.; Sun, L. *Scophthalmus maximus* cystatin B enhances head kidney macrophage-mediated bacterial killing. *Dev. Comp. Immunol.* **2010**, *34*, 1237–1241. [CrossRef] [PubMed]

25. Li, B.; Gou, M.; Han, J.; Yuan, X.; Li, Y.; Li, T.; Jiang, Q.; Xiao, R.; Li, Q. Proteomic analysis of buccal gland secretion from fasting and feeding lampreys (*Lampetra morii*). *Proteome Sci.* **2018**, *16*, 9. [CrossRef] [PubMed]

26. Perišić Nanut, M.; Sabotič, J.; Švajger, U.; Jewett, A.; Kos, J. Cystatin F affects natural killer cell cytotoxicity. *Front. Immunol.* **2017**, *8*, 1459. [CrossRef] [PubMed]

27. Schüttelkopf, A.W.; Hamilton, G.; Watts, C.; van Aalten, D.M. Structural basis of reduction-dependent activation of human cystatin F. *J. Biol. Chem.* **2006**, *281*, 16570–16575. [CrossRef] [PubMed]

28. Wu, C.; Lu, L.; Zheng, Y.; Liu, X.; Xiao, R.; Wang, J.; Li, Q. Novel anticandidal activity of a recombinant *Lampetra japonica* RGD3 protein. *J. Microbiol. Biotechnol.* **2014**, *24*, 905–913. [CrossRef] [PubMed]

29. Brillard-Bourdet, M.; Nguyên, V.; Ferrer-di Martino, M.; Gauthier, F.; Moreau, T. Purification and characterization of a new cystatin inhibitor from Taiwan cobra (*Naja naja atra*) venom. *Biochem. J.* **1998**, *331*, 239–244. [CrossRef] [PubMed]

30. Jiang, Q.; Liu, Y.; Duan, D.; Gou, M.; Wang, H.; Wang, J.; Li, Q.; Xiao, R. Anti-angiogenic activities of CRBGP from buccal glands of lampreys (*Lampetra japonica*). *Biochimie* **2016**, *123*, 7–19. [CrossRef] [PubMed]

31. Xiao, R.; Zhang, Z.; Wang, H.; Han, Y.; Gou, M.; Li, B.; Duan, D.; Wang, J.; Liu, X.; Li, Q. Identification and characterization of a cathepsin D homologue from lampreys (*Lampetra japonica*). *Dev. Comp. Immunol.* **2015**, *49*, 149–156. [CrossRef] [PubMed]

32. Evans, H.J.; Barrett, A.J. A cystatin-like cysteine proteinase inhibitor from venom of the African puff adder (*Bitis arietans*). *Biochem. J.* **1987**, *246*, 795–797. [CrossRef] [PubMed]

33. Kotsyfakis, M.; Sá-Nunes, A.; Francischetti, I.M.; Mather, T.N.; Andersen, J.F.; Ribeiro, J.M. Antiinflammatory and immunosuppressive activity of sialostatin L, a salivary cystatin from the tick *Ixodes scapularis*. *J. Biol. Chem.* **2006**, *281*, 26298–26307. [CrossRef] [PubMed]

34. Li, F.; Gai, X.; Wang, L.; Song, L.; Zhang, H.; Qiu, L.; Wang, M.; Siva, V.S. Identification and characterization of a Cystatin gene from Chinese mitten crab *Eriocheir sinensis*. *Fish Shellfish Immunol.* **2010**, *29*, 521–529. [CrossRef] [PubMed]

35. Wang, Y.; Yu, X.; Cao, J.; Zhou, Y.; Gong, H.; Zhang, H.; Li, X.; Zhou, J. Characterization of a secreted cystatin from the tick *Rhipicephalus haemaphysaloides*. *Exp. Appl. Acarol.* **2015**, *67*, 289–298. [CrossRef] [PubMed]

36. Yu, Y.; Zhang, G.; Li, Z.; Cheng, Y.; Gao, C.; Zeng, L.; Chen, J.; Yan, L.; Sun, X.; Guo, L.; et al. Molecular cloning, recombinant expression and antifungal activity of BnCPI, a Cystatin in Ramie (*Boehmeria nivea* L.). *Genes (Basel)* **2017**, *8*, 265. [CrossRef] [PubMed]

37. Wang, Y.; Wu, L.; Liu, X.; Wang, S.; Ehsan, M.; Yan, R.; Song, X.; Xu, L.; Li, X. Characterization of a secreted cystatin of the parasitic nematode *Haemonchus contortus* and its immune-modulatory effect on goat monocytes. *Parasit. Vectors.* **2017**, *10*, 425. [CrossRef] [PubMed]

38. Nishihara, K.; Kanemori, M.; Yanagi, H.; Yura, T. Overexpression of trigger factor prevents aggregation of recombinant proteins in *Escherichia coli*. *Appl. Environ. Microbiol.* **2000**, *66*, 884–889. [CrossRef] [PubMed]

39. Wang, J.; Zhang, Y.; Lu, L.; Liu, X.; Li, Q. Anti-angiogenic activities of Lj-RGD3 toxin protein from *Lampetra japonica* and its mutation protein Lj-112 without RGD motifs. *Sheng Wu Gong Cheng Xue Bao* **2011**, *27*, 1428–1437. [PubMed]

40. Li, Y.; Zhu, H.; Wei, X.; Li, H.; Yu, Z.; Zhang, H.; Liu, W. LPS induces HUVEC angiogenesis *in vitro* through miR-146a-mediated TGF-β1 inhibition. *Am. J. Transl. Res.* **2017**, *9*, 591–600. [PubMed]

41. Shin, M.R.; Kang, S.K.; Kim, Y.S.; Lee, S.Y.; Hong, S.C.; Kim, E.C. TNF-α and LPS activate angiogenesis via VEGF and SIRT1 signalling in human dental pulp cells. *Int. Endod. J.* **2015**, *48*, 705–716. [CrossRef] [PubMed]

42. Yamakawa, Y.; Omori-Satoh, T. Primary structure of the antihemorrhagic factor in serum of the Japanese Habu: a snake venom metalloproteinase inhibitor with a double-headed cystatin domain. *J. Biochem.* **1992**, *112*, 583–589. [CrossRef] [PubMed]

43. Valente, R.H.; Dragulev, B.; Perales, J.; Fox, J.W.; Domont, G.B. BJ46a, a snake venom metalloproteinase inhibitor. Isolation, characterization, cloning and insights into its mechanism of action. *Eur. J. Biochem.* **2001**, *268*, 3042–3052. [CrossRef] [PubMed]

44. Richards, R.; St Pierre, L.; Trabi, M.; Johnson, L.A.; de Jersey, J.; Masci, P.P.; Lavin, M.F. Cloning and characterisation of novel cystatins from elapid snake venom glands. *Biochimie* **2011**, *93*, 659–668. [CrossRef] [PubMed]

45. Otto, H.H.; Schirmeister, T. Cysteine proteases and their inhibitors. *Chem. Rev.* **1997**, *97*, 133–172. [CrossRef] [PubMed]

46. Figueiredo Azevedo, F.; Santanna, L.P.; Bóbbo, V.C.; Libert, E.A.; Araújo, E.P.; Abdalla Saad, M.; Lima, M.H.M. Evaluating the effect of 3% papain gel application in cutaneous wound healing in mice. *Wounds* **2017**, *29*, 96–101. [PubMed]

47. Premzl, A.; Turk, V.; Kos, J. Intracellular proteolytic activity of cathepsin B is associated with capillary-like tube formation by endothelial cells in vitro. *J. Cell. Biochem.* **2006**, *97*, 1230–1240. [CrossRef] [PubMed]

48. Jiang, Q.; Li, Q.; Han, J.; Gou, M.; Zheng, Y.; Li, B.; Xiao, R.; Wang, J. rLj-RGD3 induces apoptosis via the mitochondrial-dependent pathway and inhibits adhesion, migration and invasion of human HeyA8 cells via FAK pathway. *Int. J. Biol. Macromol.* **2017**, *96*, 652–668. [CrossRef] [PubMed]

 *marine drugs*

Article

Anticancer Activity of Fascaplysin against Lung Cancer Cell and Small Cell Lung Cancer Circulating Tumor Cell Lines

Barbara Rath, Maximilian Hochmair, Adelina Plangger and Gerhard Hamilton *

Department of Surgery, Medical University of Vienna, A-1090 Vienna, Austria; barbara.rath@gmx.eu (B.R.);
maximilian.hochmair@wienkav.at (M.H.); a01331326@unet.univie.ac.at (A.P.)
* Correspondence: gerhard.hamilton@meduniwien.ac.at; Tel./Fax: +43-40400-41700

Received: 15 September 2018; Accepted: 10 October 2018; Published: 14 October 2018

Abstract: Lung cancer is a leading cause of tumor-associated mortality. Fascaplysin, a bis-indole of a marine sponge, exhibit broad anticancer activity as specific CDK4 inhibitor among several other mechanisms, and is investigated as a drug to overcome chemoresistance after the failure of targeted agents or immunotherapy. The cytotoxic activity of fascaplysin was studied using lung cancer cell lines, primary Non-Small Cell Lung Cancer (NSCLC) and Small Cell Lung Cancer (SCLC) cells, as well as SCLC circulating tumor cell lines (CTCs). This compound exhibited high activity against SCLC cell lines (mean IC_{50} 0.89 µM), as well as SCLC CTCs as single cells and in the form of tumorospheres (mean IC_{50} 0.57 µM). NSCLC lines showed a mean IC_{50} of 1.15 µM for fascaplysin. Analysis of signal transduction mediators point to an ATM-triggered signaling cascade provoked by drug-induced DNA damage. Fascaplysin reveals at least an additive cytotoxic effect with cisplatin, which is the mainstay of lung cancer chemotherapy. In conclusion, fascaplysin shows high activity against lung cancer cell lines and spheroids of SCLC CTCs which are linked to the dismal prognosis of this tumor type. Derivatives of fascaplysin may constitute valuable new agents for the treatment of lung cancer.

Keywords: fascaplysin; lung cancer; circulating tumor cells; signal transduction; cytotoxicity; cisplatin

---

1. Introduction

Among malignant diseases, lung cancer is the leading cause of mortality [1]. NSCLC constitutes the most common subtype with approximately 85% of cases and a 5-year survival rate ranging from 50–17%, depending on the stage of the disease [2]. SCLC accounts for the rest of the cases; it is associated with smoking and has a poor prognosis upon dissemination [3]. NSCLC tumors feature a similar poor prognosis, except for those variants amenable to specific therapies directed to mutated epidermal growth factor receptor (EGFR), anaplastic lymphoma kinase (ALK), and other kinases [4,5]. Targeted therapies in the form of tyrosine kinase inhibitors (TKIs) and immunotherapy directed to checkpoint proteins have successfully changed the treatment of NSCLC; however, patients lacking markers for precision medicine or eventually progressing after specific regimens are nevertheless referred to classical chemotherapy consisting of platinum-drug-based combinations [6]. Cisplatin/carboplatin combined with either etoposide, docetaxel, or pemetrexed have limited clinical activity, and new agents may lead to increased responses and survival. The dismal prognosis of SCLC seems to be linked to the formation of large spheroidal aggregates, termed tumorospheres, which are difficult to eliminate due to poor drug perfusion and to the existence of quiescent and hypoxic tumor cells in the interior layers of the 3D-structures [7]. A host of diverse drugs have failed to provide clinical improvements for SCLC in recent decades [3].

The marine drug fascaplysin (12,13-Dihydro-13-oxopyrido[1,2-a:3,4-b′] diindol-5-ium chloride) is a red bis-indole alkaloid of the *Fascaplysinopsis Bergquist* sp. sponge which was isolated by Roll et al. in 1988 [8]. The structure of fascaplysin is shown in Figure 1. Novel derivatives comprise 3-bromo-fascaplysin, 4-chloro-fascaplysin, and 7-phenyl-fascaplysin, among others. Fascaplysin possesses antibacterial, antifungal, and antiviral properties as well as antiangiogenic and antiproliferative activity against a range of cancer cell lines [9–11]. Cyclin-dependent kinase 4 (CDK4) was reported as the main target of fascaplysin (IC$_{50}$ of 0.35 µM), and accordingly, drug-treated cancer cell lines arrested preferentially in the G0/1 cell cycle phase [12–14]. Minor activity of fascaplysin was observed against other CDKs with IC$_{50}$ of >100 µM for CDK1, >50 µM for CDK2 as well as 20 µM for CDK5 [14]. In addition, fascaplysin was demonstrated to exhibit DNA-intercalating capability with an affinity similar to those of other typical DNA intercalators [15]. Non-planar derivatives of fascaplysin have been developed in order to possibly reduce non-CD4-mediated cytotoxic effects [16].

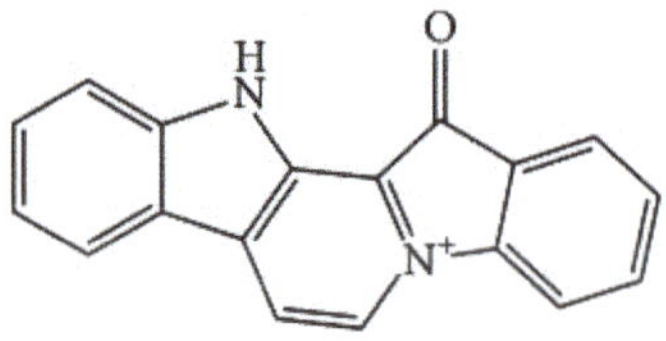

Figure 1. Structure of fascaplysin.

Cytotoxicity tests showed broad activity of fascaplysin towards a panel of 36 cancer cell lines (IC$_{50}$ values 0.6–4 µM) [9]. Anticancer activities of fascaplysin in cell lines in vitro resulted in reduced expression of CDK4, cyclin D1 and downregulation of CDK4-specific Ser795 retinoblastoma (Rb) phosphorylation in HeLa cells [17]. Fascaplysin-induced apoptosis was characterized by the activation of effector caspases, relocalization of cytochrome c into cytosol, and reduced expression of Bcl-2. Cytotoxicity of fascaplysin in chemosensitive promyelocytic HL-60 cancer cells activated both pro-apoptotic events like PARP-1 cleavage/caspase activation and triggered autophagy, as shown by the increased expression of LC3-II, ATG7 and beclin [17]. In experimental animal models, fascaplysin suppressed tumor growth in a murine sarcoma S180 through apoptosis as well as antiangiogenesis, and HCT-116 colon tumors showed reduced size in the absence of drug toxicity [18]. Angiogenesis was blocked by fascaplysin by the inhibition of vascular endothelial growth factor (VEGF) and apoptosis of endothelial cells [19].

SCLC responds to first-line chemotherapy with platinum-based drugs/etoposide, but relapses early with topotecan remaining as the single approved therapeutic agent [3]. We have previously assessed cytotoxic activity of fascaplysin against SCLC cell lines, not covered by the NCI60 cell line panel, a tumor entity that accounts for a significant fraction of lung cancer deaths [20]. Fascaplysin was found to show high cytotoxicity against SCLC cells and to induce cell cycle arrest in G1/0 at lower and S-phase at higher concentrations, respectively. The compound generated reactive oxygen species (ROS) and induced apoptotic cell death in the chemoresistant NCI-H417 SCLC cell line. Furthermore, fascaplysin revealed marked synergism with camptothecines [21,22]. Fascaplysin IC$_{50}$ values measured in SCLC cell lines were found to be similar to the two chemoresistant NSCLC cell lines H1299 and A549 and the chemosensitive H23 cell line, respectively.

In the present work, the investigation of the cytotoxic effects of fascaplysin is extended to include single cell suspensions and spheroids of SCLC circulating tumor cells (CTCs) and several NSCLC cell lines. Our lab has established a panel of 6 CTC SCLC cell lines derived from the blood samples of distinct patients with extended disease SCLC [7]. Furthermore, the effects of fascaplysin on the main pathways of cellular signal transduction and stress response were assessed employing phosphoprotein Western blot arrays and the NCI-H526 SCLC and the A549 NSCLC cell line, respectively.

2. Results

2.1. Fascaplysin Cytotoxicity against SCLC, NSCLC and Non-lung Cancer Cell Lines

The chemosensitivity of a range of cancer cell line to fascaplysin was measured in MTT cytotoxicity assays. Figure 2 shows the IC_{50} values of breast cancer and ovarian cell lines (range: 0.48–1.21 μM), SCLC cell lines (range: 0.2–1.48 μM) and NSCLC cell lines (range: 0.63–2.04 μM). Whereas SCLC and breast/ovarian cancer cell lines exhibited similar mean IC_{50} values (0.96 ± 0.5 versus 0.89 ± 0.45 μM), NSCLC cell lines proved to be less sensitive (1.15 ± 0.59 μM). SCLC26A and S457 are primary SCLC cell lines derived from pleural effusions of patients before and after therapy failure, respectively. The nonmalignant HEK293 cell line showed an IC_{50} value of 1.6 ± 0.42 μM. BH295 and IVIC-A are primary NSCLC cell lines derived from pleural effusions of patients with ALK and EGFR TKI resistance. Numrical values of the IC50 data are presented in Supplementary Table S1.

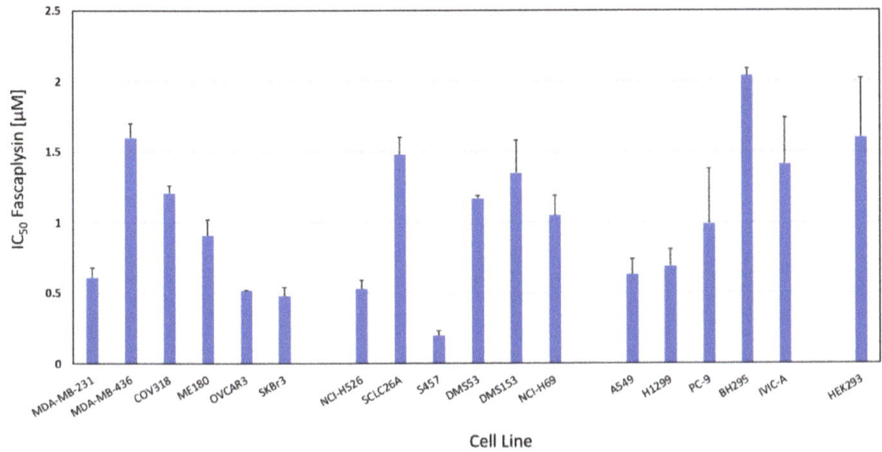

Figure 2. Fascaplysin chemosensitivity of a panel of SCLC, NSCLC and a panel of non-lung cancer lines. IC_{50} values are presented as mean values ± SD. Non-lung cancer cells used for comparison are breast and ovarian cancer cell lines and nonmalignant HEK293 cells are shown as normal tissue control.

2.2. Fascaplysin Cytotoxicity against SCLC CTC Single Cells and Tumorospheres

The SCLC CTCs form spontaneously large spheroids which are markedly chemoresistant to cisplatin and other drugs used for the treatment of SCLC patients in comparison to CTCs in form of single cell suspensions. The chemosensitivity of such single cell suspensions and tumorospheres against fascaplysin were compared in MTT tests (Figure 3). With exception of BHGc26 and 27 CTC lines, fascaplysin IC_{50} values of the other lines were equal or below 0.5 μM. A comparison of the ratios of IC_{50} values of single cell suspensions and tumorospheres for cisplatin and fascaplysin demonstrates that for fascaplysin, the differences in chemosensitivities between these 2D- and 3D-cultures are much less than for the platinum drug (Table 1) indicating superior anticancer activity for spheroids.

Fascaplysin versus cisplatin showed a 1.5 fold increased cytotoxic activity for tumorospheres for BHGc10 and BHGc27, 2.5 fold for BHGc7 and UHGc5, and 6.7 fold for BHGc16 and 26, respectively (Table 1). The mean cytotoxicity ratios between fascaplysin and cisplatin are significantly different for all SCLC CTC cell lines.

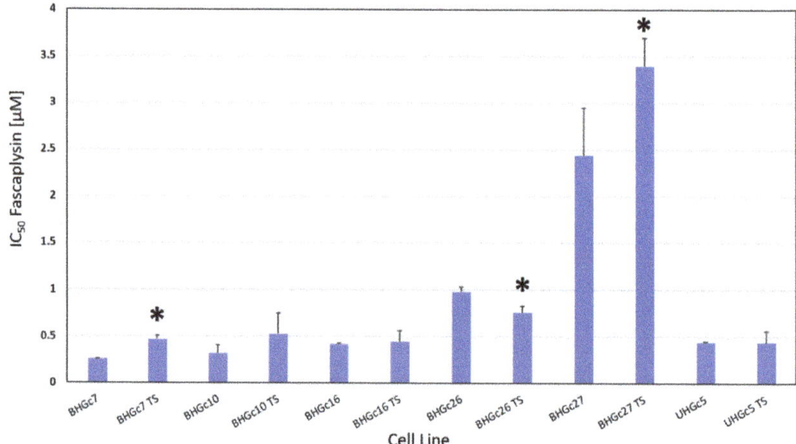

Figure 3. Chemosensitivity of SCLC CTC lines for fascaplysin. The CTC lines were tested in form of single cell suspensions and as tumorospheres. IC_{50} values are presented as mean values $\pm$ SD and significant differences between 2D- and 3D-cultures are indicated by an asterisk.

Table 1. Mean values of IC_{50} ratios for SCLC CTC tumorospheres versus single cell suspensions for fascaplysin and cisplatin, respectively (mean values $\pm$ SD). All ratios for fascaplysin, cisplatin, and the CTC lines are significantly different.

CTC Cell Line	Fascaplysin		Cisplatin	
	Mean Ratio (TS/SC)	SD	Mean Ratio (TS/SC)	SD
BHGc7	1.83	0.1	4.31	0.2
BHGc10	1.63	0.2	2.32	0.4
BHGc16	1.06	0.1	7.22	0.3
BHGc26	0.77	0.1	5.20	0.3
BHGc27	1.39	0.5	2.17	0.2
UHGc5	0.99	0.1	4.8	1.0

2.3. Alterations of Selected Phosphoproteins of NCI-H526 and A549 in Response to Fascaplysin

Figure 4 shows the first part of the phosphoproteins assayed with the ARY003 human proteome profiler kit for fascaplysin-treated NCI-H526 and A549, respectively. In contrast to the cytotoxicity assays, incubation time for phosphoprotein analysis was reduced to 72 h to prevent cell death. In NCI-H526 SCLC cells fascaplysin induced significantly increased phosphorylation of src kinases (Hck, Fyn, Yes and Fgr), CHK-2 and FAK, whereas phosphorylation of mTOR, CREB and p38α was significantly decreased compared to untreated controls. In contrast, A549 NSCLC cells revealed increased phosphorylation of CHK-2 in combination with CREB, HSP27, and STAT5b, with decreased phosphorylation of src kinases (except Fgr) and FAK.

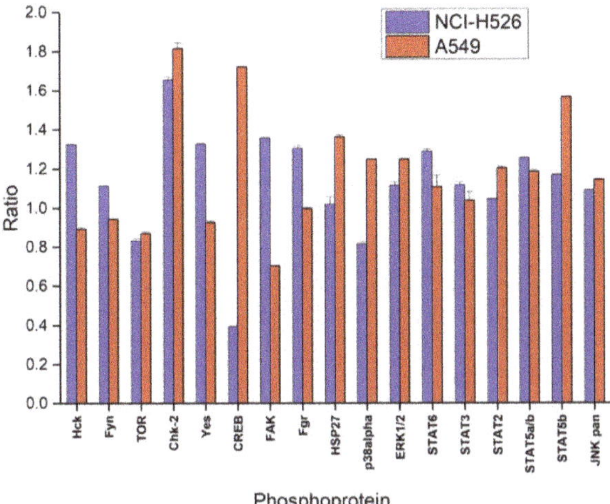

Figure 4. Relative phosphorylation (ratio of treatment:control) of selected components (part A of the array) of the signal transduction system (mean ± SD) of NCI-H526 and A549 cells treated with 0.5 µM fascaplysin for 72 h (NCI-H526: significantly different to controls, except for HSP27 and STAT2; A549 significantly different to controls, except Fgr and STAT3).

Analysis of the second part of phosphoproteins of the ARY003 kit yielded decreased phosphorylation of Akt, p53(S46/S392) and increased phosphorylation of STAT4, eNOS, c-Jun, and p27(T157) in the case of fascaplysin-pretreated NCI-H526, and numerous increases of phosphorylation in A549 cells, with the exception of decreases in p70 S6 kinase, STAT4, and p53(S392) (Figure 5). Phosphoproteins of the ARY003 blots which exhibited no significant changes for NCI-H526 or A549 cells in response to treatment with fascaplysin were not included in these figures.

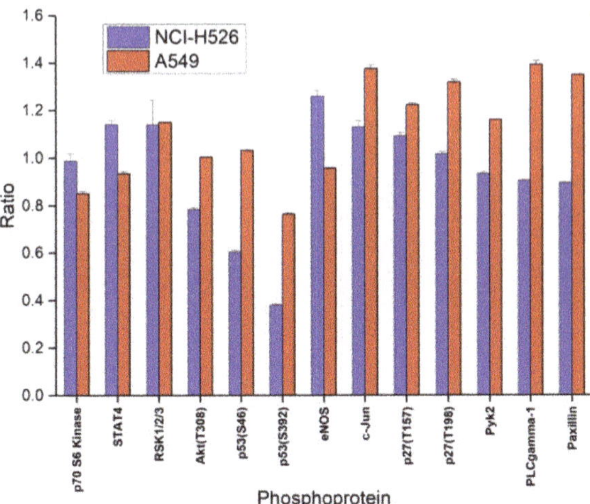

Figure 5. Relative phosphorylation (ratio treatment: control) of selected components (part B of the array) of the signal transduction system (mean ± SD) of NCI-H526 and A549 cells treated with 0.5 µM fascaplysin for 72 h (NCI-H526: significantly different to controls, except forp70 S6 kinase and p27; A549 significantly different to controls, except Akt, p53/S46, and eNOS).

2.4. Signaling Pathways Affected by Fascaplysin in NCI-H526 and A549 Lung Cancer Cells

The signal transduction mediators related to fascaplysin-induced alterations in NCI-H526 and A549 cells are depicted schematically in Figure 6. The schemes start with fascaplysin-induced DNA damage (left) and receptors/src kinases (right), respectively. Chk2 is activated by upstream DNA damage-sensing ATM and modulate functions of CREB, p53, CDC25, and stress kinases (left). Src kinases are activated by a number of connected membrane receptors (X) or oncogenic mutation, and regulate the activities of Stat5, FAK, and the Akt–mTOR axis (right).

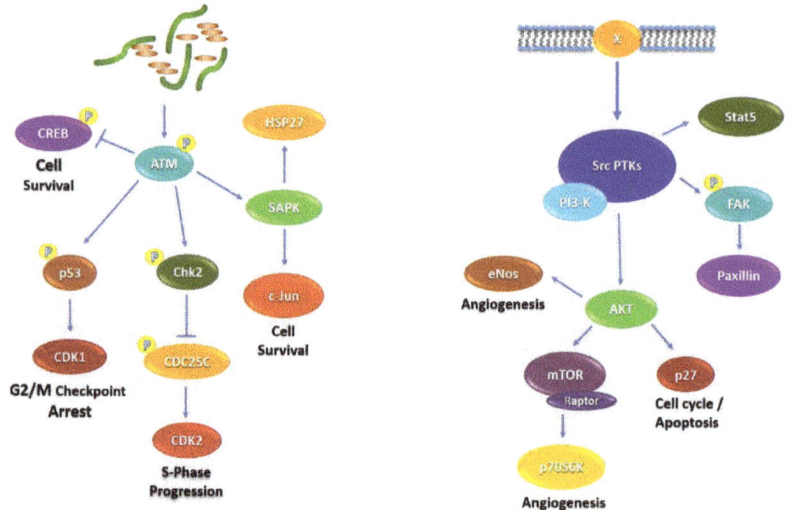

Figure 6. Schematic presentation of the signaling pathways involved in fascaplysin treatment of NCI-H526 and A549 lung cancer cells.

2.5. Combination of Cisplatin/Etoposide with Fascaplysin in Cytotoxicity Assays for NCI-H526 and A549 Cell Lines

Combination indices (CI) were calculated using the Chou-Talaly method, indicating synergy at values <1 [23]. For NCI-H526, CIs < 0.49 were found for cisplatin concentrations 0.625–5 µg/mL and 0.125–1 µM fascaplysin (fixed ratio of 2.5:1), similar to A549 with CI < 0.62 for cisplatin concentrations 1.25–10 µg/mL and 0.25–2 µM fascaplysin (fixed ratio of 5:1). For the NSCLC lines PC-9 and A549, the synergistic effects of fascaplysin with cisplatin or etoposide are shown in Figure 7.

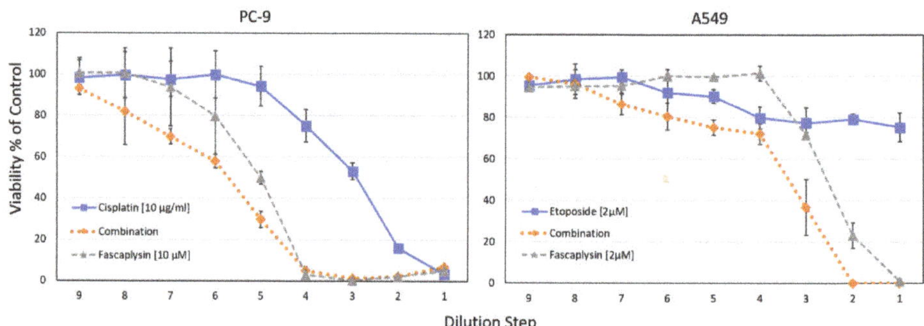

Figure 7. Cytotoxicity tests investigating fascaplysin combinations with chemotherapeutics.

For the NSCLC cell lines PC-9 and A549, combinations of fascaplysin and cisplatin or etoposide were tested in cytotoxicity tests. For both tests, IC_{50} values for the combinations using the concentrations as indicated revealed synergistic interactions. The CI values ranged from 0.26–0.76 for PC-9 fascaplysin-cisplatin and from 0.1–0.94 for A549 fascaplysin-etoposide, respectively. Data are shown as mean $\pm$ SD, the initial concentrations were titrated in 9 two-fold dilution steps.

3. Discussion

Lung cancer is the leading cause of cancer-related mortality in both men and women worldwide [1]. Targeted therapy is applicable to a minor fraction of NSCLC patients [4]. Patients with advanced lung cancer exhibit low survival rates and novel modes of chemotherapy need to be developed [2]. Deregulated proliferation of tumor cells is accomplished by alterations of the cell cycle and checkpoint controls amenable to inhibition by targeting of cell cycle and checkpoint kinases (CDKs) [24]. In particular, CDK4/6 inhibitors seem to present suitable targets in a majority of patients with advanced cancer [25,26]. Besides CDK4/6 inhibitors palbociclib and LY2835219, which have shown high activity in breast cancer, a host similar drugs are under development, and fascaplysin and derivatives share the same target [27]. Proteins in this cell proliferative pathway include p16, an endogenous suppressor of CDK4/6, cyclin D1, the regulatory subunit of CDK4/6, and retinoblastoma (Rb) protein, a tumor suppressor [28]. Both CDK4 and CDK6 encode cyclin-dependent kinases which complexed with cyclines of the D-type phosphorylate the Rb protein. Rb in turn triggers the expression of gene products for G1-S phase cell cycle progression. Rb inactivation is a common event in lung cancer, and is more frequent in SCLC than in NSCLC [29]. In SCLCs, Rb alterations can be found in a high percentage of cases, i.e., from 88% to 100% of the biopsy samples [30]. Therefore, in the present study we compared the effects of fascaplysin in the A549 Rb-wildtype NSCLC cell line to the Rb-mutated NCI-H526 SCLC cell line. Although, both cell lines have a similar chemosensitivity to fascaplysin, analysis of the intracellular signal transduction by Western blotting of selected phosphoproteins revealed marked differences in response to this drug.

DNA damage response is triggered when sensor proteins ATM (ataxia telangiectasia mutated) and ATR (also called ataxia telangiectasia and Rad3-related protein) detect structural distortions or breaks [31]. After DNA damage, CHK2 is phosphorylated by ATM on the priming site T68, and in turn, phosphorylates more than 24 proteins to induce apoptosis, DNA repair, or tolerance of the damage [32]. In wildtype cells, CHK2 phosphorylates Rb which enhances the formation of the transcriptionally-inactive pRb/E2F-1 complex causing G1/S arrest and suppression of apoptosis. Pronounced activation of CHK-2 in NCI-H526 and A549 cells indicates direct damage of DNA by fascaplysin and activation of the corresponding cellular responses in both cell lines. The cyclic AMP response element-binding protein (CREB) initiates transcriptional responses associated with cell survival to a wide variety of stimuli following its phosphorylation on Ser-133. Whereas fascaplysin treatment resulted in decreased phosphorylation of CREB in NCI-H526 cells, this transcription factor is hyperphosphorylated in A549 cells, possibly indicating anti- and pro-survival signaling, respectively [33,34]. Furthermore, cisplatin-induced activation of FAK has been linked to increased chemoresistance in ovarian cancer cells and FAK inhibitors induce tumor cell apoptosis [35]. Activated FAK forms a complex with Src family kinases and seems to provide a prosurvival signal in NCI-H526 cells, in contrast to fascaplysin-treated A549 cells [36]. In addition, overexpression of Src in cancer accelerates metastasis and is responsible for chemoresistance via multiple downstream signaling pathways, concerning Akt, MAPKs, STAT3, cytokines, etc. [37]. Therefore, activation of a number of Src kinases in NCI-H526 cells (Hck, Fyn, Yes and Fgr) may counteract fascaplysin toxicity and retard cell death; possibly contributing to the observed slower rate of loss of viability in the presence of increasing doses of fascaplysin in these cells. The stress kinases p38 and JNK are generally activated by inflammatory cytokines and different stressors, including DNA-damaging compounds [38]. p38 MAPK signaling results in the phosphorylation of CREB at Ser133, which seem to occur in A549 cells, contrary to NCI-H526 which shoes decreased p38 activity and phosphorylation of CREB [39]. Clearly, fascaplysin is an inhibitor of CDC25, and this pathway is expected to be inhibited in NCI-H526 cells [40].

In conclusion, fascaplysin shows marked anticancer activity in NSCLC and SCLC cells independently of the function of the CDK4 pathway, thus pointing to direct effects on DNA and the transcription of various proteins. The mechanisms of the antitumor effect of fascaplysin demonstrated on several carcinoma models indicate that fascaplysin is close to some drug groups such as intercalating agents, inhibitors of serine-threonine, and tyrosine kinases. Additionally, fascaplysin increases phosphorylation of AKT/PKB and adenosine monophosphate-activated protein kinase (AMPK), which feature anti-apoptotic or pro-survival functions in cancer [41]. In detail, fascaplysin abolishes the phosphorylation of mTOR, 4EBP1, and p70S6K1, which trigger the cap-dependent translation machinery and affect the expression of oncoproteins, such as survivin, c-myc, cyclin D1, VEGF, and HIF-1α. Similarly, 7-chloro-fascaplysin inhibited cell survival through interference with the PI3K/Akt/mTOR pathway, which in turn modulates HIF-1α, eNOS and MMP-2/9 in a breast cancer cell line [42]. The cytotoxicity of 4-chlorofascaplysin (4-CF) was reversed by co-treatment with the VEGF and Akt inhibitors or in response to neutralizing VEGF antibodies. Fascaplysin has stronger anti-cancer effects than other CDK4 inhibitors on lung cancer cells that are wild-type or null for Rb, indicating that unknown target molecules might be involved in the antitumor activity of fascaplysin [43]. In good accordance with the results of Oh et al. and Sharma et al., our results show alterations of phosphoproteins altering the Akt-mTor pathway which are triggered mainly by upstream stress and src kinases.

Relapsed SCLC is resistant to a wide range of drugs and clinical trials have not led to improvements in survival rates over recent decades. Chemoresistance of SCLC seems to be related to the formation of large spheroids, termed tumorospheres, which limit drug access and contain quiescent and hypoxic tumor cells which are less sensitive to chemotherapeutics. Such 3D-structures were demonstrated to show increased resistance to cisplatin, etoposide, topotecan, and epirubicin when compared to the same SCLC CTC cells in form of single cell suspensions. In particular, fascaplysin is cytotoxic against SCLC CTC tumorospheres which exhibit high chemoresistance against a range of commonly-administered chemotherapeutics. Fascaplysin-induced cell death of outer SCLC CTC cell layers seems to trigger the elimination of the whole spheroid. Especially in SCLC cells, the induction of ROS by fascaplysin is expected to exert increased damage due to the small volume of the cytoplasmic fraction [7]. It should be noted that spheroids are similarly observed in pleural effusions of NSCLC patients. Although the parent drug fascaplysin seems too toxic for clinical application, derivatives such as 3-bromofascaplysin and 7-phenylfascaplysin were demonstrated to possess higher cytotoxic efficiency and different profiles [44–46]. Furthermore, the alkaloid derivative 4-CF exhibits five times higher cytotoxic IC$_{50}$ value in normal cells, as well as no apparent toxicities in murine xenograft models at therapeutic doses [42].

4. Cell Culture and Methods

4.1. Chemicals

Unless otherwise noted, all chemicals were obtained from Sigma-Aldrich (St. Louis, MO, USA). Dulbecco's phosphate buffered saline (PBS) was purchased from Gibco/Invitrogen (Carlsbad, CA, USA). Compounds were prepared as stock solutions of 2 mg/mL in either DMSO or in 0.9% NaCl solution (cisplatin), sterilized by filtration in case of cisplatin, and aliquots stored at $-20\,^{\circ}$C.

4.2. Cell Culture

The A549 NSCLC A549 (Rb/p53 wild-type) and NCI-H526 SCLC A549 (Rb protein not expressed/p53 wild-type) cell lines were obtained from the American Type Culture Collection (Rockville, MD, USA), as well as the other cell lines except primary lines and all SCLC CTCs established in our lab [7]. Cells were grown in RPMI-1640 bicarbonate medium (Seromed, Berlin, Germany), supplemented with 10% FBS (Seromed), 4 mM glutamine, and antibiotics (final concentrations: 50 U/mL of penicillin, 50 µg/mL of streptomycin, and 100 µg/mL neomycin; Sigma-Aldrich, St. Louis, MO, USA), and subcultivated twice a week. A549 is p53 wildtype and DNA profiling by short tandem repeat analysis of the NCI-H526 cells proved their identity to the American Type Culture Collection specifications, and the yeast p53 functional assay revealed expression of fully active p53 (functional assay of separated alleles in yeast FASAY; data not shown).

4.3. Phosphokinase Array

Relative protein phosphorylation levels of 38 selected proteins were obtained by analysis of 46 specific phosphorylation sites using the Proteome Profiler Human Phospho-Kinase Array Kit ARY003 (R&D Systems, Minneapolis, MN, USA) in duplicate tests according to the manufacturer's instructions. Briefly, cells were rinsed with PBS, 1×10^7 cells/mL lysis buffer were solubilized under permanent shaking at 4 °C for 30 min, and aliquots of the lysates were stored frozen at −80 °C. After blocking, membranes with spotted catcher antibodies were incubated with diluted cell lysates at 4 °C overnight. Thereafter, cocktails of biotinylated detection antibodies were added at room temperature for 2 h. Phosphorylated proteins were revealed using streptavidin-HRP/chemiluminescence substrate (SuperSignal West Pico, Thermo Fisher Scientific, Rockford, IL, USA) and detection with a Molecular Imager VersaDoc MP imaging system (Bio-Rad, Hercules, CA, USA). Images were quantified using the ImageQuant TL v2005 software (Amersham Biosciences, Buckinghamshire, UK) and Microsoft Excel software (Microsoft, Redmond, WA, USA). The different Western blot membranes were normalized using the 3 calibration spots included. Signaling pathways affected by fascaplysin in NCI-H526 and A549 lung cancer cells were produced using Power Point software (Microsoft, Redmond, WA, USA).

4.4. Cytotoxicity Assay

Aliquots of 1×10^4 cells in 200 µL medium were treated for four days with twofold dilutions of fascaplysin or cisplatin, respectively in 96-well microtiter plates in quadruplicate (Greiner, Kremsmuenster, Austria). For SCLC CTC tumorospheres, an equivalent number of cells in form of spheroids were tested as described [7]. The plates were incubated under tissue culture conditions, and cell viability was measured using a modified MTT (3-(4,5-dimethylthiazol-2-yl)-2,5-diphenyltetrazolium bromide) assay (EZ4U, Biomedica, Vienna, Austria). Optical density was measured using a microplate reader at 450 nm with an empty well as reference. Values obtained from control wells containing cells and media alone were set to 100% proliferation. For the assessment of the interaction of fascaplysin with cisplatin, tests were performed comprising the individual drugs alone and in combination, followed by analysis using the Chou-Talalay method with help of the Compusyn software (ComboSyn, Inc. Paramus, NJ, USA).

4.5. Statistics

Statistical analysis was performed using Student's *t* test for normally distributed samples (* $p < 0.05$ was regarded as statistically significant). Values are shown as mean ± SD.

5. Conclusions

CDKs are a group of serine/threonine kinases which are critical in the regulation of the cell cycle. A major role of CDK-4 is the phosphorylation of Rb, which is inhibited by fascaplysin and a range of other compounds. Mutations in Rb, along with those of cyclin D and p16(INK4a), has been seen frequently during tumorigenesis of cancers. Investigation of a part of the kinome of NCI-H526 SCLC and A549 NSCLC cell lines reveals different responses to treatment with fascaplysin, most likely to be connected to the Rb phenotype. In NCI-H526 cells, fascaplysin sensitivity is determined by the absence of the CDK4–Rb pathway and DNA damage in combination with putative CDC25 inhibition, whereas in A549, inhibition of CDK4 seems to be the major effect with distinct and small effects on phosphoproteins. Fascaplysin exhibits marked anticancer activity against permanent and primary

SCLC and NSCLC cells, with cytotoxic effects against SCLC CTC tumorospheres that are far superior to those of other therapeutics. Therefore, fascaplysin and derivatives with a better clinical profile may constitute valuable agents for lung cancer therapy.

Supplementary Materials: The following are available online at http://www.mdpi.com/1660-3397/16/10/383/s1, Table S1: Fascaplysin activity against a panel of cell lines.

Author Contributions: B.R. and A.P. were involved in experimental work and preparation of the manuscript, M.H. in design of the experiments and G.H. in all aspects of this project.

Funding: This research received no external funding.

Conflicts of Interest: The authors declare no conflict of interest.

References

1. Bray, F.; Ferlay, J.; Soerjomataram, I.; Siegel, R.L.; Torre, L.A.; Jemal, A. Global Cancer Statistics 2018: GLOBOCAN Estimates of Incidence and Mortality Worldwide for 36 Cancers in 185 Countries. *CA Cancer J. Clin.* **2018**, in press. [CrossRef] [PubMed]

2. Herbst, R.S.; Morgensztern, D.; Boshoff, C. The biology and management of non-small cell lung cancer. *Nature* **2018**, *553*, 446–454. [PubMed]

3. Kalemkerian, G.P. Small Cell Lung Cancer. *Semin. Respir. Crit. Care Med.* **2016**, *37*, 783–796. [CrossRef] [PubMed]

4. Parums, D.V. Current status of targeted therapy in non-small cell lung cancer. *Drugs Today (Barc)* **2014**, *50*, 503–525. [CrossRef] [PubMed]

5. Nguyen, K.S.; Neal, J.W.; Wakelee, H. Review of the current targeted therapies for non-small-cell lung cancer. *World J. Clin. Oncol.* **2014**, *5*, 576–587. [CrossRef] [PubMed]

6. Fennell, D.A.; Summers, Y.; Cadranel, J.; Benepal, T.; Christoph, D.C.; Lal, R.; Das, M.; Maxwell, F.; Visseren-Grul, C.; Ferry, D. Cisplatin in the modern era: The backbone of first-line chemotherapy for non-small cell lung cancer. *Cancer Treat. Rev.* **2016**, *44*, 42–50. [CrossRef] [PubMed]

7. Klameth, L.; Rath, B.; Hochmaier, M.; Moser, D.; Redl, M.; Mungenast, F.; Gelles, K.; Ulsperger, E.; Zeillinger, R.; Hamilton, G. Small cell lung cancer: model of circulating tumor cell tumorospheres in chemoresistance. *Sci. Rep.* **2017**, *7*, 5337. [CrossRef] [PubMed]

8. Roll, D.M.; Ireland, C.M.; Lu, H.S.M.; Clardy, J. Fascaplysin, an unusual antimicrobial pigment from the marine sponge Fascaplysinopsis sp. *J. Org. Chem.* **1988**, *53*, 3276–3278. [CrossRef]

9. Segraves, N.L.; Robinson, S.J.; Garcia, D.; Said, S.A.; Fu, X.; Schmitz, F.J.; Pietraszkiewicz, H.; Valeriote, F.A.; Crews, P. Comparison of fascaplysin and related alkaloids: A study of structures, cytotoxicities, and sources. *J. Nat. Prod.* **2004**, *67*, 783–792. [CrossRef] [PubMed]

10. Segraves, N.L.; Lopez, S.; Johnson, T.A.; Said, S.A.; Fu, X.; Schmitz, F.J.; Pietraszkiewicz, H.; Valeriote, F.A.; Crews, P. Structures and cytotoxicities of fascaplysin and related alkaloids from two marine phyla—Fascaplysinopsis sponges and Didemnum tunicates. *Tetrahedron Lett.* **2003**, *44*, 3471–3475. [CrossRef]

11. Bharate, S.B.; Manda, S.; Mupparapu, N.; Battini, N.; Vishwakarma, R.A. Chemistry and biology of fascaplysin, a potent marine-derived CDK-4 inhibitor. *Mini Rev. Med. Chem.* **2012**, *12*, 650–664. [CrossRef] [PubMed]

12. Soni, R.; Muller, L.; Furet, P.; Schoepfer, J.; Stephan, C.; Zumstein-Mecker, S.; Fretz, H.; Chaudhuri, B. Inhibition of cyclindependentkinase 4 (Cdk4) by fascaplysin, a marine natural product. *Biochem. Biophys. Res. Commun.* **2000**, *275*, 877–884. [CrossRef] [PubMed]

13. Soni, R.; O'Reilly, T.; Furet, P.; Muller, L.; Stephan, C.; Zumstein-Mecker, S.; Fretz, H.; Fabbro, D.; Chaudhuri, B. Selective in vivo and in vitro effects of a small molecule inhibitor of cyclindependent kinase-4. *J. Natl. Cancer Inst.* **2001**, *93*, 436–446. [CrossRef] [PubMed]

14. Shafiq, M.I.; Steinbrecher, T.; Schmid, R. Fascaplysin as a specific inhibitor for CDK4: Insights from molecular modelling. *PLoS ONE* **2012**, *7*, e42612. [CrossRef] [PubMed]

15. Hormann, A.; Chaudhari, B.; Fretz, H. DNA binding properties of marine sponge pigment fascaplysin. *Bioorg. Med. Chem.* **2001**, *9*, 917–921. [CrossRef]

16. Mahale, S.; Bharate, S.B.; Manda, S.; Joshi, P.; Bharate, S.S.; Jenkins, P.R.; Vishwakarma, R.A.; Chaudhuri, B. Biphenyl-4-carboxylic Acid [2-(1H-Indol-3-yl)-ethyl]-methylamide (CA224), a nonplanar analogue of Fascaplysin, inhibits Cdk4 and tubulin polymerization: evaluation of in vitro and in vivo anticancer activity. *J. Med. Chem.* **2014**, *57*, 9658–9672. [CrossRef] [PubMed]

17. Kumar, S.; Kumar Guru, S.; Pathania, A.S.; Manda, S.; Kumar, A.; Bharate, S.B.; Vishwakarma, R.A.; Malik, F.; Bhushan, S. Fascaplysin induces caspase mediated crosstalk between apoptosis and autophagy through the inhibition of PI3K/AKT/mTOR signaling cascade in human leukemia HL-60 cells. *J. Cell Biochem.* **2015**, in press. [CrossRef] [PubMed]

18. Yan, X.; Chen, H.; Lu, X.; Wang, F.; Xu, W.; Jin, H.; Zhu, P. Fascaplysin exert anti-tumor effects through apoptotic and anti-angiogenesis pathways in sarcoma mice model. *Eur. J. Pharm. Sci.* **2011**, *43*, 251–259. [CrossRef] [PubMed]

19. Zheng, Y.L.; Lu, X.L.; Lin, J.; Chen, H.M.; Yan, X.J.; Wang, F.; Xu, W.F. Direct effects of fascaplysin on human umbilical vein endothelial cells attributing the anti-angiogenesis activity. *Biomed. Pharmacother.* **2010**, *64*, 527–533. [CrossRef] [PubMed]

20. Hamilton, G. Cytotoxic effects of fascaplysin against small cell lung cancer cell lines. *Mar Drugs* **2014**, *12*, 1377–1389. [CrossRef] [PubMed]

21. Hamilton, G.; Olszewski, U.; Klameth, L.; Ulsperger, E.; Geissler, K. Synergistic anticancer activity of topotecan—Cyclin-dependent kinase inhibitor combinations against drug-resistant small cell lung cancer (SCLC) cell lines. *J. Cancer Ther.* **2013**, *4*, 47–53. [CrossRef]

22. Hamilton, G.; Klameth, L.; Rath, B.; Thalhammer, T. Synergism of cyclin-dependent kinase inhibitors with camptothecin derivatives in small cell lung cancer cell lines. *Molecules* **2014**, *19*, 2077–2088. [CrossRef] [PubMed]

23. Chou, T.C. Drug combination studies and their synergy quantification using the Chou-Talalay method. *Cancer Res.* **2010**, *70*, 440–446. [CrossRef] [PubMed]

24. Aarts, M.; Linardopoulos, S.; Turner, N.C. Tumor selective targeting of cell cycle kinases for cancer treatment. *Curr. Opin. Pharmacol.* **2013**, *13*, 529–535. [CrossRef] [PubMed]

25. Whittaker, S.R.; Mallinger, A.; Workman, P.; Clarke, P.A. Inhibitors of cyclin-dependent kinases as cancer therapeutics. *Pharmacol. Ther.* **2017**, *173*, 83–105. [CrossRef] [PubMed]

26. Graf, F.; Mosch, B.; Koehler, L.; Bergmann, R.; Wuest, F.; Pietzsch, J. Cyclin-Dependent Kinase 4/6 (Cdk4/6) Inhibitors: Perspectives in Cancer Therapy and Imaging. *Mini Rev. Med. Chem.* **2015**, *15*. in press. [CrossRef]

27. Jaganathan, H.; Overstreet, K.; Reed, E. Improving breast cancer therapy with CDK4/6 inhibitors. *Clin. Breast Cancer* **2014**, *14*, 379–380. [CrossRef] [PubMed]

28. Beasley, M.B.; Lantuejoul, S.; Abbondanzo, S.; Chu, W.S.; Hasleton, P.S.; Travis, W.D.; Brambilla, E. The P16/cyclin D1/Rb pathway in neuroendocrine tumors of the lung. *Hum. Pathol.* **2003**, *34*, 136–142. [CrossRef] [PubMed]

29. Shapiro, G.I.; Edwards, C.D.; Kobzik, L.; Godleski, J.; Richards, W.; Sugarbaker, D.J.; Rollins, B.J. Reciprocal Rb inactivation and p16INK4 expression in primary lung cancers and cell lines. *Cancer Res.* **1995**, *55*, 505–509. [PubMed]

30. Coe, B.P.; Lockwood, W.W.; Girard, L.; Chari, R.; Macaulay, C.; Lam, S.; Gazdar, A.F.; Minna, J.D.; Lam, W.L. Differential disruption of cell cycle pathways in small cell and non-small cell lung cancer. *Br. J. Cancer* **2006**, *94*, 1927–1935. [CrossRef] [PubMed]

31. Ciccia, A.; Elledge, S.J. The DNA damage response: making it safe to play with knives. *Mol. Cell* **2010**, *40*, 179–204. [CrossRef] [PubMed]

32. Zannini, L.; Delia, D.; Buscemi, G. CHK2 kinase in the DNA damage response and beyond. *J. Mol. Cell Biol.* **2014**, *6*, 442–457. [CrossRef] [PubMed]

33. Trinh, A.T.; Kim, S.H.; Chang, H.Y.; Mastrocola, A.S.; Tibbetts, R.S. Cyclin-dependent kinase 1-dependent phosphorylation of cAMP response element-binding protein decreases chromatin occupancy. *J. Biol. Chem.* **2013**, *288*, 23765–23775. [CrossRef] [PubMed]

34. Rolli, M.; Kotlyarov, A.; Sakamoto, K.M.; Gaestel, M.; Neininger, A. Stress-induced stimulation of early growth response gene-1 by p38/stress-activated protein kinase 2 is mediated by a cAMP-responsive promoter element in a MAPKAP kinase 2-independent manner. *J. Biol. Chem.* **1999**, *274*, 19559–19564. [CrossRef] [PubMed]

35. Villedieu, M.; Deslandes, E.; Duval, M.; Héron, J.F.; Gauduchon, P.; Poulain, L. Acquisition of chemoresistance following discontinuous exposures to cisplatin is associated in ovarian carcinoma cells with progressive alteration of FAK, ERK and p38 activation in response to treatment. *Gynecol. Oncol.* **2006**, *101*, 507–519. [CrossRef] [PubMed]

36. Warmuth, M.; Damoiseaux, R.; Liu, Y.; Fabbro, D.; Gray, N. SRC family kinases: potential targets for the treatment of human cancer and leukemia. *Curr. Pharm. Des.* **2003**, *9*, 2043–2059. [CrossRef] [PubMed]

37. Lieu, C.; Kopetz, S. The SRC family of protein tyrosine kinases: a new and promising target for colorectal cancer therapy. *Clin. Colorectal Cancer* **2010**, *9*, 89–94. [CrossRef] [PubMed]

38. Silvers, A.L.; Bachelor, M.A.; Bowden, G.T. The role of JNK and p38 MAPK activities in UVA-induced signaling pathways leading to AP-1 activation and c-Fos expression. *Neoplasia* **2003**, *5*, 319–329. [CrossRef]

39. Gao, J.; Wagnon, J.L.; Protacio, R.M.; Glazko, G.V.; Beggs, M.; Raj, V.; Davidson, M.K.; Wahls, W.P. A stress-activated, p38 mitogen-activated protein kinase-ATF/CREB pathway regulates posttranscriptional, sequence-dependent decay of target RNAs. *Mol. Cell Biol.* **2013**, *33*, 3026–3035. [CrossRef] [PubMed]

40. Lazo, J.S.; Wipf, P. Is Cdc25 a druggable target? *Anticancer Agents Med. Chem.* **2008**, *8*, 837–842. [CrossRef] [PubMed]

41. Oh, T.I.; Lee, Y.M.; Nam, T.J.; Ko, Y.S.; Mah, S.; Kim, J.; Kim, Y.; Reddy, R.H.; Kim, Y.J.; Hong, S.; et al. Fascaplysin Exerts Anti-Cancer Effects through the Downregulation of Survivin and HIF-1α and Inhibition of VEGFR2 and TRKA. *Int. J. Mol. Sci.* **2017**, *18*, E2074. [CrossRef] [PubMed]

42. Sharma, S.; Guru, S.K.; Manda, S.; Kumar, A.; Mintoo, M.J.; Prasad, V.D.; Sharma, P.R.; Mondhe, D.M.; Bharate, S.B.; Bhushan, S. A marine sponge alkaloid derivative 4-chloro fascaplysin inhibits tumor growth and VEGF mediated angiogenesis by disrupting PI3K/Akt/mTOR signaling cascade. *Chem. Biol. Interact.* **2017**, *275*, 47–60. [CrossRef] [PubMed]

43. Oh, T.I.; Lee, J.H.; Kim, S.; Nam, T.J.; Kim, Y.S.; Kim, B.M.; Yim, W.J.; Lim, J.H. Fascaplysin Sensitizes Anti-Cancer Effects of Drugs Targeting AKT and AMPK. *Molecules* **2017**, *23*, E42. [CrossRef] [PubMed]

44. Calcabrini, C.; Catanzaro, E.; Bishayee, A.; Turrini, E.; Fimognari, C. Marine Sponge Natural Products with Anticancer Potential: An Updated Review. *Mar. Drugs* **2017**, *15*, E310. [CrossRef] [PubMed]

45. Mahale, S.; Bharate, S.B.; Manda, S.; Joshi, P.; Jenkins, P.R.; Vishwakarma, R.A.; Chaudhuri, B. Antitumour potential of BPT: A dual inhibitor of cdk4 and tubulin polymerization. *Cell Death Dis.* **2015**, *6*, e1743. [CrossRef] [PubMed]

46. Lyakhova, I.A.; Bryukhovetsky, I.S.; Kudryavtsev, I.V.; Khotimchenko, Y.S.; Zhidkov, M.E.; Kantemirov, A.V. Antitumor Activity of Fascaplysin Derivatives on Glioblastoma Model In Vitro. *Bull. Exp. Biol. Med.* **2018**, *164*, 666–672. [CrossRef] [PubMed]

Article

Manzamine A Exerts Anticancer Activity against Human Colorectal Cancer Cells

Li-Chun Lin [1,†], Tzu-Ting Kuo [2,†], Hsin-Yi Chang [3], Wen-Shan Liu [2], Shih-Min Hsia [4,5,6] and Tsui-Chin Huang [1,2,7,*]

[1]   Ph.D. Program for Cancer Molecular Biology and Drug Discovery, College of Medical Science and Technology, Taipei Medical University and Academia Sinica, Taipei 11031, Taiwan; lisa81318@gmail.com
[2]   Graduate Institute of Cancer Molecular Biology and Drug Discovery, College of Medical Science and Technology, Taipei Medical University, Taipei 11031, Taiwan; taros71526@hotmail.com (T.-T.K.); cindy20028@yahoo.com.tw (W.-S.L.)
[3]   Graduate School of Pharmaceutical Sciences, Kyoto University, Kyoto 6068501, Japan; hsinyi.chang@pharm.kyoto-u.ac.jp
[4]   School of Nutrition and Health Sciences, College of Nutrition, Taipei Medical University, Taipei 11031, Taiwan; bryanhsia@tmu.edu.tw
[5]   Graduate Institute of Metabolism and Obesity Sciences, College of Nutrition, Taipei Medical University, Taipei 11031, Taiwan
[6]   School of Food Safety, College of Nutrition, Taipei Medical University, Taipei 11031, Taiwan
[7]   TMU Research Center of Cancer Translational Medicine, Taipei Medical University, Taipei 11031, Taiwan
*   Correspondence: tsuichin@tmu.edu.tw; Tel.: +886-2-2736-1661 (ext. 7675)
†   These authors contributed equally to this work.

Received: 11 June 2018; Accepted: 27 July 2018; Published: 29 July 2018

Abstract: Marine sponges are known to produce numerous bioactive secondary metabolites as defense strategies to avoid predation. Manzamine A is a sponge-derived β-carboline-fused pentacyclic alkaloid with various bioactivities, including recently reported anticancer activity on pancreatic cancer. However, its cytotoxicity and mode of action against other tumors remain unclear. In this study, we exhibit that manzamine A reduced cell proliferation in several colorectal cancer (CRC) cell lines. To further investigate the manzamine A triggered molecular regulation, we analyzed the gene expression with microarray and revealed that pathways including cell cycle, DNA repair, mRNA metabolism, and apoptosis were dysregulated. We verified that manzamine A induced cell cycle arrest at G_0/G_1 phase via inhibition of cyclin-dependent kinases by p53/p21/p27 and triggered a caspase-dependent apoptotic cell death through mitochondrial membrane potential depletion. Additionally, we performed bioinformatics analysis and demonstrated that manzamine A abolished epithelial–mesenchymal transition process. Several mesenchymal transcriptional factors, such as Snail, Slug, and Twist were suppressed and epithelial marker E-cadherin was induced simultaneously in HCT116 cells by manzamine A, leading to the epithelial-like phenotype and suppression of migration. These findings suggest that manzamine A may serve as a starting point for the development of an anticancer drug for the treatment of metastatic CRC.

Keywords: manzamine A; cell cycle; apoptosis; epithelial–mesenchymal transition; colorectal cancer

---

1. Introduction

Colorectal cancer (CRC) is the third-most common cancer in both genders and the fourth leading cause of cancer related mortality, responsible for 9.7% of cancer-related deaths worldwide [1–3]. Distant metastasis caused by disease recurrence and development of drug resistance is the main cause of death in CRC patients [4,5]. In 2017, an estimated 135,430 new cases will be diagnosed with colorectal cancer and about 50,260 people will die from this disease in the US [6].

Accumulated evidences of genetic mutations and epigenetic instability on oncogene activation and tumor suppressor inactivation have been reported to be central molecular and pathophysiological events to initiate CRC [7,8]. The most frequent aberrations found in CRC patients are mutations in adenomatous polyposis coli (APC), catenin-β1 (CTNNB1), family with sequence similarity 123B (FAM123B; also known as AMER1), kirsten rat sarcoma viral oncogene homolog (KRAS), B-Raf proto-oncogene, serine/threonine kinase (BRAF), erb-b2 receptor tyrosine kinase 2 E (RBB2), SMAD family member 4 (SMAD4), phosphatidylinositol-4,5-bisphosphate 3-kinase catalytic subunit-α (PIK3CA), transforming growth factor-β receptor 2 (TGFBR2), AT-rich interactive domain 1A (ARID1A), and tumor Protein P53 (TP53). These mutations promote tumorigenesis by perturbing the key signaling pathways, such as WNT–β-catenin, epidermal growth factor (EGF)–mitogen-activated protein kinase (MAPK), phosphatidylinositol 3-kinase (PI3K)–AKT, and TGFβ signaling pathways, or affecting genes that regulate fundamental processes, such as DNA repair, cell cycle progression and proliferation [2,9].

Several therapeutic strategies are developed to overcome CRC. However, server side effects due to drug toxicity towards normal tissues and development of drug resistance are occurred in CRC patients receiving conventional chemotherapy. Therefore, development of novel anticancer therapeutic alternatives for CRC is still urgently required. Manzamine A (Manz A), comprising a pentacyclic core coupled to a β-carboline alkaloid, is isolated from sponges of the genera *Haliclona* sp., *Xestospongia* sp. and *Pellina* sp. [10–12]. Manz A was first isolated from *Haliclona* sp. in Okinawa sea and reported to display anticancer activity against leukemia cells [13]. It possesses a diverse range of potent bioactivities including insecticidal, antibacterial, and antileishmaniasis effects, and anti-malarial, anti-inflammatory, and antiviral activities [10,14–19]. A previous study demonstrated that Manz A decreased single cell formation, abrogated cell migration, and sensitized AsPC-1 pancreatic adenocarcinoma cells towards TRAIL induced apoptosis [20]. A recent work also reported that Manz A targeted vacuolar ATPases and inhibited autophagy in pancreatic cancer cells, suggesting a promising strategy for the treatment of cancer [21]. Nevertheless, the effects of Manz A on CRC and their mechanisms remain unclear. In the present study, we attempted to investigate the anti-tumor properties of Manz A in HCT116 human colorectal carcinoma cells. We showed that Manz A significantly inhibited the proliferation of several CRC cell lines. By combining enrichment analysis and network analysis on microarray data, we found that Manz A reduced the expression of genes involved in several fundamental pathways and activated the apoptotic gene expression. The effects of MA on cell cycle progression, apoptosis, epithelial–mesenchymal transition (EMT) process, and cell migration were further validated.

2. Results

2.1. Manz A Inhibits Cell Proliferation in Human Colorectal Carcinoma Cells

To evaluate the effects of Manz A on the proliferation of human colorectal carcinoma cells, we performed the MTS assay on HCT116, HT-29, and DLD-1 cells in a dose-dependent manner at concentrations of 0, 0.5, 1, 2.5, 5, and 10 μM for 24 h. We found that Manz A significantly decreased the cell viability of all colorectal carcinoma cells and showed a higher efficacy on HCT116 compared with HT-29 and DLD-1 (Figure 1A). The IC50 values were 4.5 ± 1.7 μM in HCT116 cells and more than 10 μM in DLD-1 and HT-29. To determine the long-term inhibitory effect of Manz A on the proliferation of HCT116 cells, we performed a colony formation assay. The 24-h treated cells were seeded in 6-well plates and cultured in drug-free medium for one week. We showed that Manz A significantly reduced the number and size of colonies without continued exposure to the drug (Figure 1B), suggesting that Manz A caused an irreversible cell proliferation inhibition.

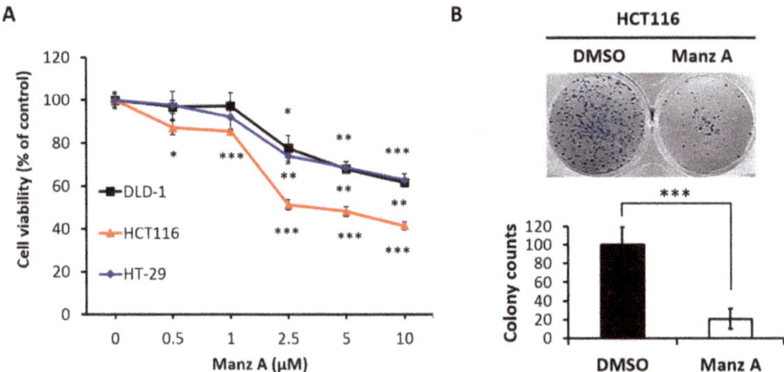

Figure 1. Manz A reduced cell proliferation in colorectal cancer cells. (**A**) HCT116, DLD-1, and HT-29 cells were treated with Manz A at various concentrations of 0, 0.5, 1, 2.5, and 5 μM for 24 h. Cell viability (%) was measured using MTS cell proliferation assay and data was expressed as percentage of absorbance from Manz A treated cells compared to DMSO treated ones; (**B**) Colony formation assay was performed to determine the long-term effects of Manz A on the growth of HCT116 cells. Cells were pre-treated with 0.1% DMSO or 5 μM Manz A for 24 h and left for 7 days to grow. Colonies were then stained with Giemsa. The data were expressed as the mean ± SD of three independent experiments. * $p < 0.05$, ** $p < 0.01$, *** $p < 0.001$.

2.2. Manz A Reduces Gene Expressions Involved in Several Fundamental Pathways

To comprehensively elucidate the regulation of Manz A on HCT116 cells, we profiled the expression of ~40,000 genes by microarray analysis. A total of 1574 genes were significantly regulated ($p < 0.05$) and 639 of them were more than 2-fold changes (FC) with 308 upregulated and 311 downregulated transcripts. To find out the connection between gene expression and biological function, we used two enrichment analyses, over-representation analysis (ORA), and functional class sorting (FCS), against pre-defined sets of gene lists. In ORA, we revealed that Manz A caused the downregulation in several pathways including cell cycle, DNA repair, mRNA metabolism, mitochondrial electron transport chain (ETC), and transcription (Figure S1). For FCS, we applied gene set enrichment analysis (GSEA) and integrated the influences of Manz A in biological process and activity of transcriptional regulation. With the consistency with ORA result, most downregulated gene ontology terms of biological process (GOBP) were belonged to cell cycle, DNA replication, mRNA metabolism, and DNA repair, while one upregulated GOBP term related to apoptosis was enriched (Figure 2). In transcriptional activity analysis, we found that E2F Transcription Factor 1 (E2F) was the most downregulated transcription factor, while cAMP response element binding protein (CREB), activating transcription factor (ATF) and activator protein 1 (AP1) were upregulated in transcriptional activity (Figure S2). Together, we suggest that Manz A inhibition of cell proliferation might be due to the diminishment in cell cycle progression and the initiation of apoptotic cell death.

2.3. Manz A Induces Cell Cycle Arrest at G_0/G_1 Phase

According to the findings from microarray analysis, we designed a series of experiments to validate the mode of cell death induced by Manz A. First, we performed flow cytometry analysis to measure the changes in cell population distribution whinin cell cycle. We revealed that Manz A increased the cell population at G_0/G_1 phase by 1.46 fold and decreased that at S phase by 1.65 fold simultaneously (Figure 3A,B), indicating that Manz A markedly induced cell cycle arrest at G_0/G_1 phase.

Aberrant cell division and uncontrolled proliferation are often associated with deregulation of cell cycle kinases in cancer cells. Cell cycle progression is regulated by several cyclin-dependent kinases (CDKs) that act in complex with cyclins. Cyclin Ds–CDK4/6 complexes drive cell cycle progression from G_0 or

G_1 phase into S phase, in which DNA replication occurs. The complex sequester CDK inhibitors (CKIs), $p21^{CIP1}$ and $p27^{KIP1}$, bind to and prevent activation of cyclin E–CDK2 kinase, driving cell cycle progression from G_0 or G_1 phase into S phase. While DNA damage occurs, p21 is activated by p53 to arrest cell cycle progression [22]. To determine the effect of Manz A on protein expression of cell cycle regulators, we conducted western blot analysis and found that Manz A obviously decreased the protein expression of CDK2, CDK4 and cyclin D_1 and increased levels of CKIs including $p21^{CIP1}$, $p27^{KIP1}$, and p53 (Figure 3C). We imply that cell cycle arrest was involved in Manz-A-inhibited cell proliferation.

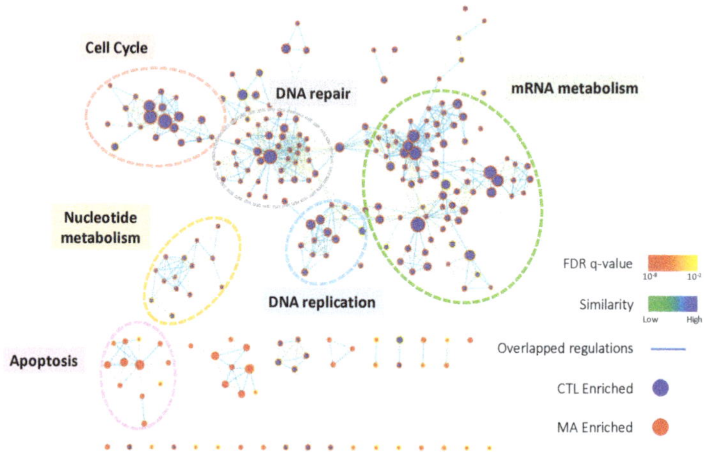

Figure 2. The enrichment map of gene ontology biological processes (GOBPs) enriched in GSEA result. GSEA was performed on microarray data to enrich terms in GOBP. Enrichment results were visualized with Cytoscape 3.0 using Enrichment Map plugin. Each node indicates an enriched term and edges represent overlapped genes between terms. The color of node boarder and edge were shown according to enrichment FDR and similarity, respectively. Blue and red nodes refer to enriched functions in DMSO vehicle control (CTL) and Manz A treatment. The size of node and edge width represents the number of genes enriched in each term and overlapped between terms, respectively.

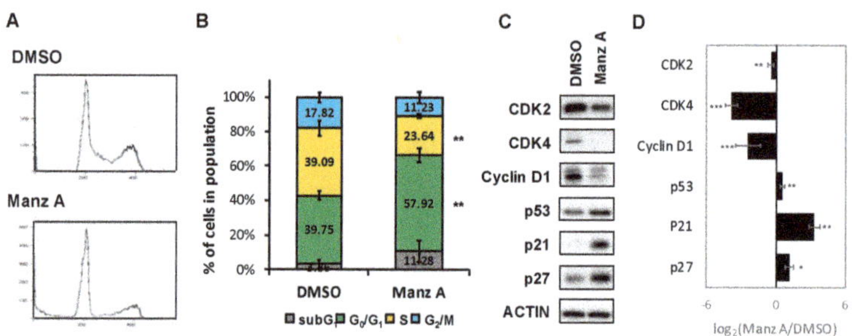

Figure 3. Manz A induced G_0/G_1 phase arrest. (**A**) HCT116 cells were treated with 5 µM Manz A for 24 h and then subjected to DNA content analysis by flow cytometry. The cell cycle distribution was quantified with model fitting in FlowJo; (**B**) Cell cycle distributions from three independent experiments is shown. * $p < 0.05$, ** $p < 0.01$, *** $p < 0.001$; (**C**) Representative western blot analyses of the cell cycle regulator levels in response to Manz A treatment in HCT116 cells. ACTIN was used as an internal control; (**D**) Relative expression level of each protein was densitometrically estimated and normalized to that of ACTIN under the same treatment. Data is presented in $\log_2$ ratio of the protein level in Manz A treated cells to that in DMSO treated ones from three independent results.

2.4. Manz A Induces Caspase-Dependent Apoptotic Cell Death

We next examined whether Manz A induced apoptosis in HCT116 cells. FITC-conjugated annexin V and PI double stained assay by flow cytometry was conducted. We found that Manz A significantly increased the early apoptotic cell population (annexin V positive cells) from 2% to 16% (Figure 4A). Mitochondrial membrane potential (MMP) is an important indicator for mitochondrial integrity and widely used to study apoptosis. Here, we used JC-1 staining and flow cytometry to detect the MMP. Tetraethylbenzimi-dazolylcarbocyanine iodide (JC-1), a lipophilic cationic dye, enters and accumulates in the electronegative interior of mitochondria where the probe aggregated and its fluorescent property changes. The dye forms complexes known as aggregates that yield a red fluorescence in healthy cells. During the early stage of programmed cell death, mitochondrial membrane potential decreases due to the opening of the mitochondrial permeability transition pores. This mitochondrial disruption results in the release of cytochrome c into the cytosol, which in turn triggers the downstream apoptotic cascades. Due to the loss of MMP, JC-1 remains as monomer which exhibits green fluorescence. In the present results, we showed that Manz A significantly reduced the fluorescence ratio of red to green, indicating that depolarization of MMP occurred during apoptosis (Figure 4B).

We then examined whether caspase cascade was triggered in the Manz A induced apoptosis. We applied a DNA fluorescent dye containing caspase 3/7 cleaved sequence DEVD to specify the activation of caspase-3/7 in apoptotic cells. We demonstrated that Manz A treatment caused a significant increase in fluorescent positive cells, showing that caspase-3/7 was activated by Manz A (Figure 4C). Collectively, our findings suggest that Manz A induce apoptotic cells death through mitochondria- and caspase-dependent pathways in HCT116 cells.

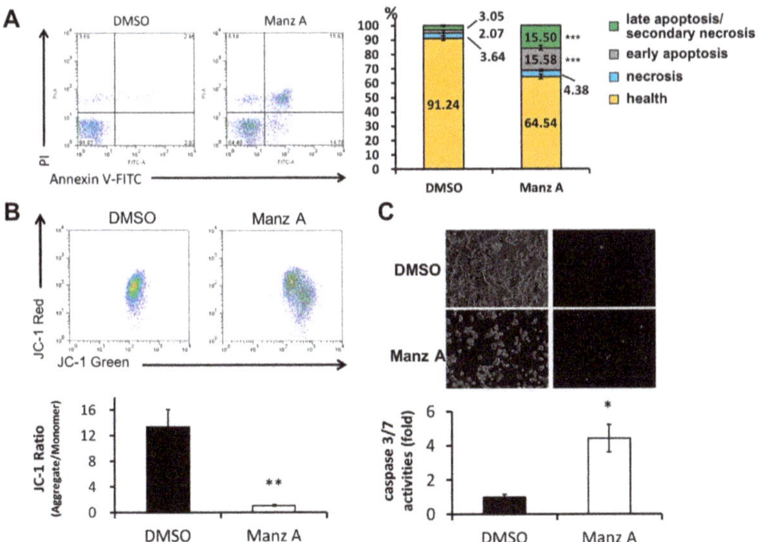

Figure 4. Manz A induced apoptosis in HCT116 cells. Cells were treated with 5 µM Manz A for 24 h. (**A**) Cells were harvested and stained with Annexin V-FITC and propidium iodide (PI). The fluorescent signal was measured by flow cytometry. A representative result is shown in the left panel and statistics analysis is at the right panel; (**B**) Cells were subjected to mitochondrial membrane potential detection by JC-1 and flow cytometry; (**C**) Cells were subjected to caspase 3/7 activity assay by DNA fluorescent dye containing caspase 3/7 cleavage site. Values are expressed as the mean ± SD from three independent experiments. * $p < 0.05$, ** $p < 0.01$, *** $p < 0.001$.

2.5. Epithelial–Mesenchymal Transition (EMT) Is Inactivated in Manz A Treated HCT116 Cells

The migratory capacity of cancer cells is necessary for metastasis and the acquisition of metastatic capability is associated with EMT in cancer cells. During the process of EMT, epithelial cells lose cell-to-cell contacts and gain expression of mesenchymal factors, enabling migration and invasion into surrounding stroma to facilitate metastasis. Because epithelial-like changes in cell morphology was observed during Manz A treatment, we performed an enrichment analysis on EMT regulators to evaluate whether Manz A also altered the EMT at gene expression levels. We divided the EMT-related genes according to their function into two portions, the activating and inactivating genes. By applying GSEA, we revealed that genes involved in EMT inactivation was significantly enriched in Manz A treated condition (Figure 5A, FDR = 0.02) and genes for activating EMT was unchanged (FDR = 0.84). We subsequently compared the mean expression of genes involved in EMT regulation and found that genes inactivating EMT were significantly higher than those activating EMT (Figure 5B), implying that Manz A inhibited the EMT process through gene expression regulation and might cause the cells to lose their migratory ability. Additionally, we compared the expression of EMT regulating genes in clinical subjects. The genes inactivating EMT were enriched in healthy controls (Figure 5C) and in lower expression than genes activating EMT in CRC patients (Figure 5D), suggesting that CRC patients who exhibiting lower expression in EMT inactivating genes are potent responders to Manz A treatment due to the effect that Manz A activates these genes.

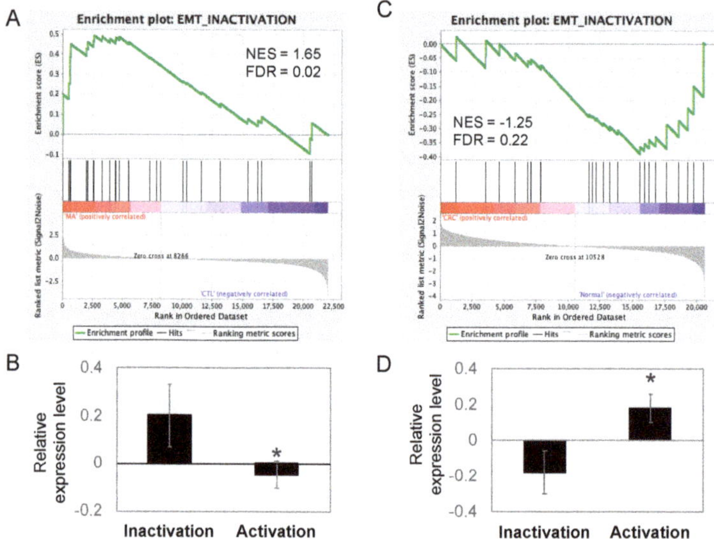

Figure 5. EMT was inactivated in Manz A treated HCT116 cells and normal colorectal tissues. Genes involved in EMT inactivation and activation were collected as two gene sets for GSEA to test the gene expression pattern between (**A**) MA-treated and DMSO-treated HCT116 cells or (**C**) the pattern between CRC and healthy control clinical tissues. Relative gene expression was ranked in descending order and colored from red to blue in response to MA (**A**) or occurrence of disease (**C**). Green line indicates the profile of running ES score and black lines represent the positions of gene set members on the rank ordered list according to their relative levels in Manz A (MA) compared to DMSO vehicle control (CTL) or CRC tumors (CRC) compared to biopsies from healthy controls (Normal). Relative expression levels of genes involved in EMT inactivation and activation were compared in (**B**) MA-treated HCT116 and (**D**) clinical tissues. Data is shown in mean expression of Z-transformed expression level. * $p < 0.05$.

2.6. Manz A Suppresses EMT Markers and Migration of Colorectal Carcinoma Cells

Loss of epithelial cadherin (E-cadherin) is considered to be a critical molecular feature of EMT. E-cadherin acts as a tumor suppressor which inhibits invasion and metastasis. Several signaling pathways including TGFβ, PI3K/AKT, and WNT signaling promote EMT by inhibiting glycogen synthase kinase-3β (GSK3β) to stabilize β-catenin, which translocates to the nucleus to regulate the transcription factors and promote a gene expression profile that favors EMT [23]. Therefore, we assessed the effects of Manz A on cell migration and EMT makers in HCT116 cells. We found that Manz A significantly reduced the migratory ability (Figure 6A). Subsequently, we revealed that Manz A increased E-cadherin expression and inhibited the nuclear translocation of β-catenin (Figure 6B). Moreover, Manz A induced the protein expression of several epithelial markers including E-cadherin and Zona occludens-1 (ZO-1) and reduced that of β-catenin, tight junction protein claudin-1, and EMT transcriptional regulators Snail, Slug, and Twist1/2 (Figure 6C). Simultaneously, CDH1 (encoding E-cadherin), SNAI1, TWIST1, and CTNNB1 (encoding β-catenin) were consistently regulated at the mRNA levels (Figure 6D). Collectively, these results demonstrate that Manz A reverses EMT and drives an epithelial-like phenotype, thereby suppressing the migration of colorectal carcinoma cells.

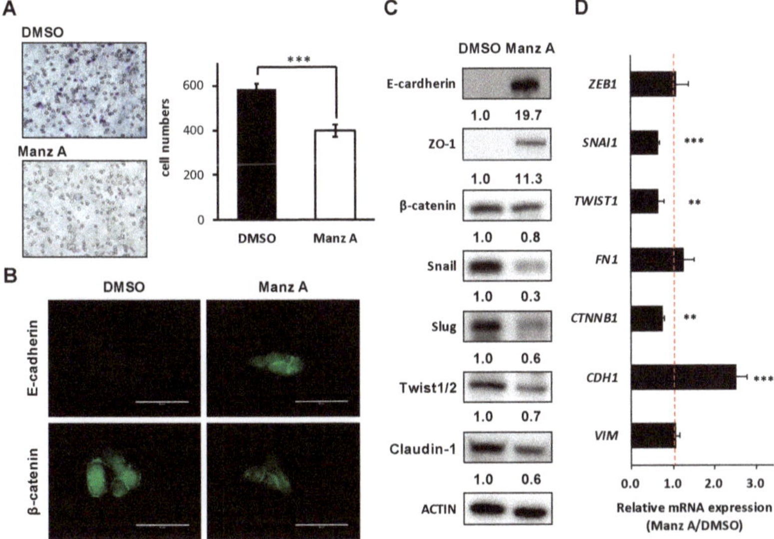

Figure 6. Manz A reversed epithelial-mesenchymal transition (EMT) and decreased cell mobility in HCT116 cells. (**A**) Transwell assay of HCT116 cells. Cells were pre-treated with 5 μM Manz A or DMSO for 24 h before seeded onto 24-well transwells. Migrated cell were stained and counted after 8 h; (**B**) Immunofluorescence of stained E-cadherin and β-catenin in HCT116 cells were monitored in the presence or absence of Manz A. Bars indicate 40 μm (**C**) Western blot analyses of the EMT regulator levels in response to Manz A treatment in HCT116 cells. ACTIN was used as an internal control. Relative expression level of each protein was densitometrically estimated and normalized to that of ACTIN under the same treatment; (**D**) Relative mRNA expression of EMT-related genes in HCT116 cells in response to Manz A treatment. The mRNA level of GAPDH was used as an internal control. ** $p < 0.01$, *** $p < 0.001$.

3. Discussion

Manz A was reported to exhibit anticancer activities in pancreatic cancer cells [13,20,21], yet the role of Manz A in CRC remains poorly understood. Consistent with prior studies, we demonstrated that Manz A significantly inhibited the proliferation of colorectal carcinoma cells (Figure 1). To comprehensively understand the molecular mechanism induced by Manz A, we obtained gene expression profile using microarray analysis and implemented a series of bioinformatics analyses. Using functional enrichment analysis on the transcriptome data, we revealed that Manz A caused downregulation in several fundamental pathways including cell cycle, DNA repair, and mRNA metabolism, and triggered apoptosis to inhibit the cell survival of CRC cells (Figure 2). Manz A caused cell cycle arrest and apoptosis was subsequently confirmed (Figures 3 and 4). Based on the results from GSEA, we found that genes involved in EMT inactivation was significantly enriched in Manz A treated condition, explaining the epithelial-like phenotype we observed in cells treated with Manz A (Figure 5). We further validated the results by assessing the expression of EMT markers at both transcriptional and translational levels and showed the evidence of EMT reversal at molecular level which lead to diminishment of cell migration (Figure 6).

In mammalian cells, G_1–S transcription depends on the transcription factor E2F family and their heterodimerization partner proteins during cell cycle process. Dysregulation of E2F function is frequently observed in cancer. In addition to E2F proteins, phosphorylation of their pocket proteins such as RB by cyclin D–CDK4/6 induces the transcription of G_1–S target genes, such as the gene encoding cyclin E. Then cyclin E–CDK2 phosphorylates their pocket proteins, thus generating a positive feed ward loop to ensure the progression of cell cycle [24,25]. In the present study, we performed transcriptional activity analysis and found that E2F was downregulated by Manz A treatment (Figure S2). We further verified that Manz A induced cell cycle arrest at G_0/G_1 phase and downregulated the expression of CDK2, CDK4 and cyclin D1 through p53/p21 and p27 pathway (Figure 3).

Here, we also found that CREB, ATF and AP-1 were enriched in transcriptional activity analysis under Manz A treatment (Figure S2). AP-1 transcription factor can exert both oncogenic and tumor suppressive effects by regulating genes involved in cell proliferation, apoptosis, angiogenesis, and tumor invasion. Forming a dimeric complex, AP-1 interacts with JUN, FOS, ATF, or MAF (musculoaponeurotic fibrosarcoma) protein families. JUNB and JUND can repress crucial regulators of cell-cycle progression including gene encoding cyclin D1 [26]. Accumulating evidence indicates that increased AP-1 activity leads to apoptosis. c-Jun and c-Fos regulate the gene encoding Fas ligand, which triggers apoptosis [27], suggesting that Manz A not only activate mitochondria-mediated apoptosis (Figure 4) but also trigger an extrinsic apoptotic pathway through TRAIL activation as previously reported [20]. Inactivation of JunB in myeloid cells leads to reduced apoptosis with increased expression of the anti-apoptotic Bcl2 and Bcl-xl genes [28]. The activating transcription factor/cyclic AMP response element binding (ATF/CREB) family is involved in various cellular processes including cell stress responses, cell survival, and cell growth in cancer. Similar to AP-1, the ATF/CREB family also play roles as both a tumor suppressor and oncogene. ATF3, an adaptive-response gene, exhibits tumor suppressor function in the development of human colorectal cancer [29]. ATF3 can enhance the activation of p53 [30] and down-regulate cyclin D1 [31]. ATF3/CREB activation induces apoptosis in HCT116 human colorectal cancer cells [32,33]. Collectively, our results indicate that Manz A might induce apoptosis by upregulating transcriptional activities of AP-1 and ATF (Figure 2, Figure 4 and Figure S2). Nevertheless, the precise role of Manz-A-upregulated ATF and CREB in human colorectal cancer needs to be investigated further in the future.

In CRC patients, metastases are the main cause of cancer-related mortality. Approximately half of CRC patients will develop liver or lung metastasis during disease progression [34–36]. By comparing the expression pattern of EMT regulators in CRC patients and healthy controls, we revealed that inactivation of EMT in healthy control subjects is significantly enriched and might be critical to prevent the disease occurrence (Figure 5C,D). EMT has been considered to be a fundamental event in cancer metastasis. We showed that Manz A markedly induced E-cadherin expression with surface-surrounding localization, decreased EMT transcriptional regulators such as Snail, Slug, and Twist1/2, and prevented the nuclear translocation of β-catenin. Furthermore, a decreased expression of Claudin-1 was observed. The claudins are a family of integral membrane proteins forming the tight junction to maintain the barrier function that exists between epithelial cells. Claudin-1 is one of the most dysregulated claudins in human cancers and functions as a cancer promoter and tumor suppressor depending on cancer type [37]. Claudin-1 has been reported to overexpress in colon cancer, in particular in metastatic cells with mislocalization from the cell membrane to the cell nucleus and cytoplasm, and regulate cellular transformation and metastasis in xenograft tumor model through its effects on E-cadherin and β-catenin/Tcf signaling [38]. As a result, we hypothesized that Manz A might disrupted the EMT upstream signaling through preventing nuclear localization of β-catenin to diminish the activity of EMT transcription factors and reestablish the tight junction and adherent junction between epithelial cells. Based on the effects on activation of genes inactivating EMT, we assumed that CRC patients with tumors appearing in poor histological differentiation are potentially benefit from Manz A treatment. Tumors with dedifferentiated phenotypes are known to have the acquired capability for metastasis through the induction of epithelial–mesenchymal transition (EMT) [39]. Our results suggest that Manz A could be expected to have a potential role in a personalized selectivity in CRC treatment, for patients experienced recurrence and distal metastasis.

4. Materials and Methods

4.1. Reagents and Chemicals

Manzamine A (Manz A), with a purity of more than 98%, was obtained from Enzo Life Sciences (Farmingdale, NY, USA). The stock solution of Manz A was prepared at a concentration of 10 mM in dimethyl sulfoxide (DMSO, Sigma-Aldrich, St. Louis, MO, USA) and stored at $-20\,°C$. Manz A was diluted in culture medium to obtain the desired concentration. DNase-free RNase A, propidium iodide (PI), and Triton X-100 were purchased from Sigma-Aldrich. Annexin V: FITC Apoptosis Detection Kit I was purchased from BD Biosciences (San Diego, CA, USA). 0.05% trypsin-EDTA (1X) were purchased from Caisson Labs (Smithfield, UT, USA). Trypan blue, M-PER™ Mammalian Protein Extraction Reagent, Halt™ Protease and Phosphatase Inhibitor Cocktail, and Electrochemiluminescence (ECL) HRP substrate were purchased from Thermo Fisher Scientific (Boston, MA, USA). CellTiter 96® AQueous One Solution Cell Proliferation (MTS) assay was purchased from Promega (Madison, WI, USA). Bicinchoninic acid (BCA) assay kit was purchased from T-Pro Biotechnology (New Taipei County, Taiwan). Antibodies against E-cadherin, β-catenin, Snail, Slug, Claudin-1, and ZO-1 were purchased from Cell Signaling Technology (Beverley, MA, USA). Antibodies against Cyclin D1, p21$^{\text{cip1}}$, p27$^{\text{kip1}}$, and Twist1/2 were purchased from GeneTex (Irvine, CA, USA). Antibodies against p53 and ACTIN were purchased from EMD Millipore (Billerica, MA, USA). Antibodies against CDK2 and CDK4, and the goat anti-rabbit/mouse antibody IgG were purchased from Abcam (Cambridge, UK). TRI Reagent was purchased from Invitrogen (Carlsbad, CA, USA). TURBO DNA-free™ Kit was purchased from Ambion (Austin, TX, USA). Low Input Quick-Amp Labeling kit and CyDye Cy3 were purchased from Agilent Technologies (Santa Clara, CA, USA).

4.2. Cell Culture

Human colorectal carcinoma cell lines HCT116, HT-29, and DLD-1 were obtained from the American Type Culture Collection (ATCC, Manassas, VA, USA). All cell lines were cultured in RPMI 1640 supplemented with 10% fetal bovine serum in a humidified atmosphere containing 95% air and 5% CO_2 at 37 °C.

4.3. Cell Proliferation Analysis and Colony Formation Assay

The effects of Manz A on cell proliferation were assessed by using the MTS assay (CellTiter 96 Aqueous One Solution cell proliferation assay; Promega, Madison, WI, USA). Briefly, HCT116, DLD-1, and HT-29 cells were seeded onto 96-well plates at a density of 7000 cells/well with 100 µL of complete medium/well for 24 h prior to Manz A treatment. MTS reagent was added to each well at the indicated incubation times and incubated at 37 °C for 90 min in the dark. The spectrophotometric absorbance of colored formazan dye generated by viable cells was measured on an Epoch Microplate Spectrophotometer (BioTek, Winooski, VT, USA) at 490 nm. The relative cell viability is defined as the ratio of the absorbance at 490 nm of Manz A treated cells to that of DMSO treated ones. All assays were performed in at least three independent experiments.

To evaluate the long-term inhibitory effects of Manz A on cell proliferation, a colony formation assay was performed as described previously [40] with modifications. HCT116 cells were seeded onto 6-well plates at a density of 2.5×10^5 cells/well. After attachment, cells were treated with 5 µM Manz A for 24 h. The cells were then detached, reseeded onto 6-well plates at a density of 1000 cells/well, and cultured with drug-free medium. Cells were cultured for 7 days to allow the colonies to form. Culture medium was changed every two days to ensure the supplement of nutrients. Formed colonies were fixed in methanol, stained with 1:10 Giemsa stain (Sigma-Aldrich), photographed under a microscope and counted.

4.4. Microarray Analysis

Total RNA was isolated using the TRI Reagent according to the manufacturer's instructions and treated with the TURBO DNA-free™ Kit to remove genomic DNA contamination. 0.2 µg of total RNA was amplified by a Low Input Quick-Amp Labeling kit and labeled with Cy3 during the in vitro transcription process. 0.6 µg of Cy3-labled cRNA was fragmented to an average size of about 50–100 nucleotides by incubation with fragmentation buffer at 60 °C for 30 mins. Correspondingly fragmented labeled cRNA is then pooled and hybridized to Agilent SurePrint G3 Human GE 8 × 60 K Microarray (Agilent Technologies) at 65 °C for 17 h. After washing and drying by nitrogen gun blowing, microarrays are scanned with an Agilent microarray scanner (Agilent Technologies) at 535 nm for Cy3. Scanned images are analyzed by an image analysis and normalization software Feature extraction 10.5.1.1 software (Agilent Technologies) to quantify signal and background intensity for each feature.

4.5. Functional Enrichment Analysis and Data Visualization

To interpret the results from the microarray analysis, we applied functional enrichment analysis in two ways, over-representation analysis (ORA) and functional class sorting (FCS). For ORA, significantly regulated genes ($p < 0.05$) with more than 2-fold changes were functionally enriched using DAVID (https://david.ncifcrf.gov) [41]. Data from Reactome pathway database [42] was used for interpreting the involved pathways. In parallel, we utilized gene set enrichment analysis (GSEA, http://software. broadinstitute.org/gsea/index.jsp) [43] for FCS according to online documentation. In brief, raw signal data from microarray was used as the input expression file. The genes were ranked according to the expression difference between MA-treated and DMSO control cells using the signal to noise parameter. 615 gene lists from C3 (TFT: transcription factor targets) in MSigDB was assign as the gene set. A non-parametric running sum statistic termed the enrichment score (ES) was measured to obtain the association between gene set and expression ranking. Permutation testing was applied to assess the statistical significance of the maximum ES, which is calculated as the fraction of the 1000 random permutations on the gene set. The unadjusted nominal p value estimates the statistical significance of a gene set and the false discovery rate (FDR) statistic adjusting for gene set size and multiple hypothesis testing were reported. A FDR of 0.25 was selected as a threshold as recommended. To visualize the functional enrichment results, a Cytoscape [44] plugin Enrichment Map [45] was applied using default setting of p-value cutoff = 0.005, FDR cutoff = 0.1, and overlap coefficient = 0.5. For tissue expression data (GSE9348 [46]), the genes were ranked according to the expression difference between CRC patients and healthy controls and the enrichment analysis in EMT genes were analyzed by GSEA with the same parameters for Manz A dataset.

4.6. Flow Cytometry Analysis

HCT116 cells were seeded onto 6-well plates at a density of 2.5×10^5/well for 24 h and treated with 5 µM Manz A for 24 h. Cells were then washed with PBS and harvested with trypsin-EDTA solution. For the cell cycle analysis, cells were fixed with 70% ethanol at -20 °C overnight, washed with ice-cold PBS, incubated with 0.1 mg/mL DNAse-free RNAse A, and stained with 50 mg/mL PI. DNA content of PI-stained single-cell suspensions was acquired with a FACSCalibur flow cytometer equipped with CellQuest software (BD Biosciences, San Jose, CA, USA). For the apoptosis assay, Annexin V: FITC Apoptosis Detection Kit I was used. The cells were suspended with 100 µL of binding buffer (10 mM HEPES/NaOH, 140 mM NaCl, 2.5 mM CaCl2, pH 7.4) and stained with 2 µL of FITC-conjugated Annexin V and 2 µL of PI (50 µg/mL) for 15 mins at room temperature in the dark and analyzed with a FACSCalibur flow cytometer equipped with CellQuest software (BD Biosciences).

For flow cytometric detection of mitochondrial membrane potential (MMP, $\Delta\psi$m), cells were washed with PBS once and incubated with 2 µM JC-1 in 5% CO_2 incubator at 37 °C for 30 min after being treated with 5 µM Manz A for 24 h. Then the stained cells were harvested with trypsin-EDTA solution into plastic tubes fitted with the flow cytometer and centrifuged at 1200 rpm for 5 min to remove the dye and resuspended with PBS buffer. The MMP of Manz A treated cells was quantified by FACSCalibur flow cytometer equipped with CellQuest software (BD Biosciences) using 488 nm excitation. Cells with red JC-1 aggregates and apoptotic cells with collapsed mitochondria containing green JC-1 monomers were detected in FL2 and FL1 channels, respectively.

4.7. Caspases 3/7 Activities Assay

Detection of caspases 3 and 7 activation was performed by using the CellEvent™ Caspase 3/7 Green Detection Reagent which is a four-amino acid peptide (DEVD) conjugated to a nucleic acid-binding dye. Upon the activation of caspase-3/7 in apoptotic cells, the DEVD peptide is cleaved and the dye is released from the peptide to bind DNA and generate a bright green fluorescence in nucleus. 10^5 HCT116 cells were seeded onto 12-well plates for 24 h. After attachment, cells were treated with DMSO control or 5 µM Manz A for 24 h and then washed with PBS once. CellEvent™ Caspase 3/7 Green Detection Reagent was added to the wells at the final concentration of 5 µM. After incubation for 30 min at 37 °C in the dark, cells were then imaged by a EVOS® FL Auto Cell Imaging System (Thermo Fisher Scientific) with a FITC filter.

4.8. Transwell Migration Assay

Migratory capacity of colorectal carcinoma cells was assayed using modified Boyden chamber assay with transwell consisting of a polycarbonate membrane at the bottom with 8 µm pore (Corning Inc., Acton, MA, USA). HCT116 cells were seeded overnight and treated with DMSO control or 5 µM Manz A for 24 h. Cells were then washed and harvested. The cells suspension was diluted to an optimal density and loaded into the transwell placed in wells of 24-well plates without touching the membrane or introducing air bubbles. The lower chamber contained Manz A-free RPMI medium supplied by 10% FBS as a chemoattractant. Following incubation for 8 h, cells in the upper chamber were scraped off and the membranes were fixed in methanol for 20 min. Migrated cells remaining on the bottom surface of the membrane were counted after stained with 0.1% crystal violet (Sigma-Aldrich) for 10 min.

4.9. Western Blot Analysis

HCT116 cells treated with DMSO or 5 µM Manz A for 24 h were lysed for 30 min in ice-cold lysis buffer containing protease and phosphatase inhibitor cocktail. The protein concentration was determined by Pierce bicinchoninic acid (BCA) protein assay kit (Thermo Scientific). Thirty µg protein was subjected to SDS-PAGE, resolved on a 7.5–15% polyacrylamide gel, and transferred onto a polyvinylidene difluoride (PVDF) membrane. The membrane was then blocked in 5% bovine serum albumin (BSA) for 1 h and incubated with the appropriate primary antibody overnight at 4 °C. After three washes with Tris-buffered saline containing 0.05% Tween-20 (TBST), the membrane was incubated with secondary anti-rabbit or anti-mouse IgG antibodies (1:15,000) for 1 h at room temperature. The immunoreaction was visualized using the ECL HRP substrate and detected using a Luminescent Image Analyzer Amersham Imager 600 (GE Healthcare Life Sciences, MA, USA). The band intensity was quantified using ImageJ software.

4.10. RNA Isolation, Reverse Transcription and Quantitative Real-Time PCR (qPCR) Analysis

Total RNA from DMSO- or Manz-A-treated HCT116 cells was extracted using TRIzol™ Reagent (Invitrogen) followed by purification with Direct-zol RNA MiniPrep (Thermo Scientific). cDNA was reverse transcribed from 2 µg total RNA using RevertAid RT Reverse Transcription Kit (Thermo Scientific). mRNA level of genes of interests was measured with gene-specific primers and Power SYBR Green Master Mix (Thermo Fisher Scientific). In brief, 1 ng of cDNA was added to the mix containing appropriate primer sets (400 nM) and SYBR green in a 10 µL reaction volume. All samples were analyzed in triplicate. Real-time PCR analyses were performed with a Applied Biosystems StepOnePlus™ Real-Time PCR System (Thermo Fisher Scientific). Amplification of all genes was performed under the following cycling conditions: denaturation at 95 °C for 10 mins followed by 40 cycles for 15 s at 95 °C and 30 s at 60 °C. Synthesis of DNA product of the expected size was confirmed by melting curve analysis and DNA electrophoresis. Relative quantification was done by $\Delta\Delta Ct$ method as fold change normalized to internal control GAPDH and vehicle control. Primers used for qPCR analysis are listed in Table 1.

Table 1. Sequences of qPCR primers.

Gene	Forward (5′ to 3′)	Reverse (5′ to 3′)
VIM	AGTCCACTGAGTACCGGAGAC	CATTTCACGCATCTGGCGTTC
CDH1	ATTTTTCCCTCGACACCCGAT	TCCCAGGCGTAGACCAAGA
CTNNB1	GTCTGAGGACAAGCCACAAGA	TCCCTGGGCACCAATATCAAG
FN1	GGCCAGTCCTACAACCAGTAT	TCGGGAATCTTCTCTGTCAGC
TWIST1	GCTGAGCAAGATTCAGACCCT	TCCATCCTCCAGACCGAGAA
SNAI1	AAGGGACTGTGAGTAATGGCTG	TAGTTCTGGGAGACACATCGGT
ZEB1	CAGCTTGATACCTGTGAATGGG	TATCTGTGGTCGTGTGGGACT

4.11. Immunocytochemistry

HCT116 cells were cultured on sterile glass coverslips and treated with 5 µM of Manz A for 24 h. Cells were fixed with 4% paraformaldehyde at room temperature for 15 min followed by permeablized with 0.025% Triton X-100/PBS for 10 min. Cells were then blocked in blocking solution consisting of 5% (*w/v*) BSA in TBST at room temperature for 1 h and incubated with primary antibodies overnight at 4 °C. After being washed three times with PBS containing 0.1% Tween-20 (PBST), cells were further incubated with secondary antibodies (Alexa Fluor® 488 dye, LifeTechnologies, Gaithersburg, MD, USA) for 1 h at room temperature in the dark. After three times of PBST washes, cells were mounted with ProLong® Gold Antifade Mountant with 4′,6-diamidino-2-phenylindole (DAPI). Fluorescent images were taken using an EVOS® microscope (Thermo Fisher Scientific).

4.12. Statistical Analysis

The data was shown as the means ± standard deviations (SD). A two-tailed Student's *t*-test was used to determine the significance of differences between DMSO control- and Manz-A-treated groups. Statistical significance was reached when the *p*-value less than 0.05. Experiments were repeated independently at least three times ($n = 3$).

5. Conclusions

In summary, we demonstrated that Manz A exhibits an antiproliferative effect on human colorectal carcinoma cells and displays broad effects on gene expression to downregulate fundamental maintenances of cell survival and induce apoptotic cell death and EMT inactivation. We validated that Manz A causes cell cycle arrest at G_0/G_1 phase through the activation of p53, p21, and p27 CKIs and triggers caspase-dependent apoptosis via mitochondrial membrane potential depletion. Specifically, we found that the EMT and migratory ability are inhibited, suggesting Manz A can serve as a potential anticancer drug for CRC patients baring tumors undergoing EMT process and developing distal metastasis.

Supplementary Materials: The following are available online at http://www.mdpi.com/1660-3397/16/8/252/s1, Figure S1: over-representation analysis of differentially expressed genes obtained from microarray analysis, Figure S2: Enrichment map of functional enrichment in transcription factor targets from GSEA result.

Author Contributions: T.-C.H. conceived and designed the experiments; L.-C.L., T.-T.K., and W.-S.L. performed the experiments; L.-C.L., T.-T.K., H.-Y.C., W.-S.L., and S.-M.H. analyzed the data; L.-C.L., T.-T.K., H.-Y.C., and T.-C.H. wrote the paper.

Funding: This study was supported by the Ministry of Science and Technology, Taiwan (MOST105-2320-B-038-004-; MOST106-2320-B-038-064-MY3), Council of Agriculture, Taiwan, (106AS-16.4.1-ST-a4) the health and welfare surcharge of tobacco products (MOHW106-TDU-B-212-144001; MOHW107-TDU-B-212-114020), the Shuang Ho Hospital-Taipei Medical University grant (105TMU-SHH-14) and the University System of Taipei Joint Research Program (USTP-NTOU-TMU-106-03). This work was also financially supported by the "TMU Research Center of Cancer Translational Medicine" from The Featured Areas Research Center Program within the framework of the Higher Education Sprout Project by the Ministry of Education (MOE) in Taiwan.

Acknowledgments: The authors acknowledge TMU Core Facility for technical support on flow cytometry analysis.

Conflicts of Interest: The authors declare no conflict of interest.

References

1. Williams, T.G.; Cubiella, J.; Griffin, S.J.; Walter, F.M.; Usher-Smith, J.A. Risk prediction models for colorectal cancer in people with symptoms: A systematic review. *BMC Gastroenterol.* **2016**, *16*, 63. [CrossRef] [PubMed]
2. Kuipers, E.J.; Grady, W.M.; Lieberman, D.; Seufferlein, T.; Sung, J.J.; Boelens, P.G.; van de Velde, C.J.; Watanabe, T. Colorectal cancer. *Nat. Rev. Dis. Primers* **2015**, *1*, 15065. [CrossRef] [PubMed]
3. Marmol, I.; Sanchez-de-Diego, C.; Pradilla Dieste, A.; Cerrada, E.; Rodriguez Yoldi, M.J. Colorectal Carcinoma: A General Overview and Future Perspectives in Colorectal Cancer. *Int. J. Mol. Sci.* **2017**, *18*, 197. [CrossRef] [PubMed]
4. Center, M.M.; Jemal, A.; Ward, E. International trends in colorectal cancer incidence rates. *Cancer Epidemiol. Biomark. Prev.* **2009**, *18*, 1688–1694. [CrossRef] [PubMed]
5. Welch, J.P.; Donaldson, G.A. The clinical correlation of an autopsy study of recurrent colorectal cancer. *Ann. Surg.* **1979**, *189*, 496–502. [PubMed]
6. Siegel, R.L.; Miller, K.D.; Jemal, A. Cancer statistics, 2016. *CA Cancer J. Clin.* **2016**, *66*, 7–30. [CrossRef] [PubMed]
7. Colussi, D.; Brandi, G.; Bazzoli, F.; Ricciardiello, L. Molecular pathways involved in colorectal cancer: Implications for disease behavior and prevention. *Int. J. Mol. Sci.* **2013**, *14*, 16365–16385. [CrossRef] [PubMed]
8. Grady, W.M.; Carethers, J.M. Genomic, and epigenetic instability in colorectal cancer pathogenesis. *Gastroenterology* **2008**, *135*, 1079–1099. [CrossRef] [PubMed]
9. Grady, W.M.; Pritchard, C.C. Molecular alterations and biomarkers in colorectal cancer. *Toxicol. Pathol.* **2014**, *42*, 124–139. [CrossRef] [PubMed]
10. Edrada, R.A.; Proksch, P.; Wray, V.; Witte, L.; Muller, W.E.; Van Soest, R.W. Four new bioactive manzamine-type alkaloids from the Philippine marine sponge *Xestospongia ashmorica*. *J. Nat. Prod.* **1996**, *59*, 1056–1060. [CrossRef] [PubMed]
11. Ichiba, T.; Corgiat, J.M.; Scheuer, P.J.; Kelly-Borges, M. 8-Hydroxymanzamine A, a beta-carboline alkaloid from a sponge, *Pachypellina* sp. *J. Nat. Prod.* **1994**, *57*, 168–170. [CrossRef] [PubMed]

12. Watanabe, D.; Tsuda, M.; Kobayashi, J. Three new manzamine congeners from *Amphimedon* sponge. *J. Nat. Prod.* **1998**, *61*, 689–692. [CrossRef] [PubMed]

13. Sakai, R.; Higa, T.; Jefford, C.W.; Bernardinelli, G. Manzamine A, a novel antitumor alkaloid from a sponge. *J. Am. Chem. Soc.* **1986**, *108*, 6404–6405. [CrossRef]

14. Rao, K.V.; Donia, M.S.; Peng, J.; Garcia-Palomero, E.; Alonso, D.; Martinez, A.; Medina, M.; Franzblau, S.G.; Tekwani, B.L.; Khan, S.I.; et al. Manzamine B and E and Ircinal A Related Alkaloids from an Indonesian *Acanthostrongylophora* Sponge and Their Activity against Infectious, Tropical Parasitic, and Alzheimer's Diseases. *J. Nat. Prod.* **2006**, *69*, 1034–1040. [CrossRef] [PubMed]

15. Rao, K.V.; Santarsiero, B.D.; Mesecar, A.D.; Schinazi, R.F.; Tekwani, B.L.; Hamann, M.T. New Manzamine Alkaloids with Activity against Infectious and Tropical Parasitic Diseases from an Indonesian Sponge. *J. Nat. Pzrod.* **2003**, *66*, 823–828. [CrossRef] [PubMed]

16. Ang, K.K.H.; Holmes, M.J.; Higa, T.; Hamann, M.T.; Kara, U.A.K. In Vivo Antimalarial Activity of the Beta-Carboline Alkaloid Manzamine A. *Antimicrob. Agents Chemother.* **2000**, *44*, 1645–1649. [CrossRef] [PubMed]

17. Rao, K.V.; Kasanah, N.; Wahyuono, S.; Tekwani, B.L.; Schinazi, R.F.; Hamann, M.T. Three new manzamine alkaloids from a common Indonesian sponge and their activity against infectious and tropical parasitic diseases. *J. Nat. Prod.* **2004**, *67*, 1314–1318. [CrossRef] [PubMed]

18. Yousaf, M.; Hammond, N.L.; Peng, J.; Wahyuono, S.; McIntosh, K.A.; Charman, W.N.; Mayer, A.M.S.; Hamann, M.T. New Manzamine Alkaloids from an Indo-Pacific Sponge. Pharmacokinetics, Oral Availability, and the Significant Activity of Several Manzamines against HIV-I, AIDS Opportunistic Infections, and Inflammatory Diseases. *J. Med. Chem.* **2004**, *47*, 3512–3517. [CrossRef] [PubMed]

19. Peng, J.; Hu, J.F.; Kazi, A.B.; Li, Z.; Avery, M.; Peraud, O.; Hill, R.T.; Franzblau, S.G.; Zhang, F.; Schinazi, R.F.; et al. Manadomanzamines A and B: A novel alkaloid ring system with potent activity against mycobacteria and HIV-1. *J. Am. Chem. Soc.* **2003**, *125*, 13382–13386. [CrossRef] [PubMed]

20. Guzman, E.A.; Johnson, J.D.; Linley, P.A.; Gunasekera, S.E.; Wright, A.E. A novel activity from an old compound: Manzamine A reduces the metastatic potential of AsPC-1 pancreatic cancer cells and sensitizes them to TRAIL-induced apoptosis. *Investig. New Drugs* **2011**, *29*, 777–785. [CrossRef] [PubMed]

21. Kallifatidis, G.; Hoepfner, D.; Jaeg, T.; Guzman, E.A.; Wright, A.E. The marine natural product manzamine A targets vacuolar ATPases and inhibits autophagy in pancreatic cancer cells. *Mar. Drugs* **2013**, *11*, 3500–3516. [CrossRef] [PubMed]

22. Otto, T.; Sicinski, P. Cell cycle proteins as promising targets in cancer therapy. *Nat. Rev. Cancer* **2017**, *17*, 93–115. [CrossRef] [PubMed]

23. Lamouille, S.; Xu, J.; Derynck, R. Molecular mechanisms of epithelial-mesenchymal transition. *Nat. Rev. Mol. Cell Biol.* **2014**, *15*, 178–196. [CrossRef] [PubMed]

24. Bertoli, C.; Skotheim, J.M.; de Bruin, R.A. Control of cell cycle transcription during G1 and S phases. *Nat. Rev. Mol. Cell Biol.* **2013**, *14*, 518–528. [CrossRef] [PubMed]

25. Trimarchi, J.M.; Lees, J.A. Sibling rivalry in the E2F family. *Nat. Rev. Mol. Cell Biol.* **2002**, *3*, 11–20. [CrossRef] [PubMed]

26. Eferl, R.; Wagner, E.F. AP-1: A double-edged sword in tumorigenesis. *Nat. Rev. Cancer* **2003**, *3*, 859–868. [CrossRef] [PubMed]

27. Kasibhatla, S.; Brunner, T.; Genestier, L.; Echeverri, F.; Mahboubi, A.; Green, D.R. DNA damaging agents induce expression of Fas ligand and subsequent apoptosis in T lymphocytes via the activation of NF-kappa B and AP-1. *Mol. Cell* **1998**, *1*, 543–551. [CrossRef]

28. Passegue, E.; Jochum, W.; Schorpp-Kistner, M.; Mohle-Steinlein, U.; Wagner, E.F. Chronic myeloid leukemia with increased granulocyte progenitors in mice lacking junB expression in the myeloid lineage. *Cell* **2001**, *104*, 21–32. [CrossRef]

29. Bottone, F.G., Jr.; Moon, Y.; Kim, J.S.; Alston-Mills, B.; Ishibashi, M.; Eling, T.E. The anti-invasive activity of cyclooxygenase inhibitors is regulated by the transcription factor ATF3 (activating transcription factor 3). *Mol. Cancer Ther.* **2005**, *4*, 693–703. [CrossRef] [PubMed]

30. Yan, C.; Lu, D.; Hai, T.; Boyd, D.D. Activating transcription factor 3, a stress sensor, activates p53 by blocking its ubiquitination. *EMBO J.* **2005**, *24*, 2425–2435. [CrossRef] [PubMed]

31. Lu, D.; Wolfgang, C.D.; Hai, T. Activating transcription factor 3, a stress-inducible gene, suppresses Ras-stimulated tumorigenesis. *J. Biol. Chem.* **2006**, *281*, 10473–10481. [CrossRef] [PubMed]

32. Eo, H.J.; Kwon, T.H.; Park, G.H.; Song, H.M.; Lee, S.J.; Park, N.H.; Jeong, J.B. In Vitro Anticancer Activity of Phlorofucofuroeckol A via Upregulation of Activating Transcription Factor 3 against Human Colorectal Cancer Cells. *Mar. Drugs* **2016**, *14*, 69. [CrossRef] [PubMed]

33. Park, G.H.; Song, H.M.; Jeong, J.B. Kahweol from Coffee Induces Apoptosis by Upregulating Activating Transcription Factor 3 in Human Colorectal Cancer Cells. *Biomol. Ther. (Seoul)* **2017**, *25*, 337–343. [CrossRef] [PubMed]

34. Van Cutsem, E.; Oliveira, J. Advanced colorectal cancer: ESMO clinical recommendations for diagnosis, treatment and follow-up. *Ann. Oncol.* **2009**, *20* (Suppl. S4), 61–63. [CrossRef] [PubMed]

35. Biasco, G.; Derenzini, E.; Grazi, G.; Ercolani, G.; Ravaioli, M.; Pantaleo, M.A.; Brandi, G. Treatment of hepatic metastases from colorectal cancer: Many doubts, some certainties. *Cancer Treat. Rev.* **2006**, *32*, 214–228. [CrossRef] [PubMed]

36. Vatandoust, S.; Price, T.J.; Karapetis, C.S. Colorectal cancer: Metastases to a single organ. *World J. Gastroenterol.* **2015**, *21*, 11767–11776. [CrossRef] [PubMed]

37. Kwon, M.J. Emerging roles of claudins in human cancer. *Int. J. Mol. Sci.* **2013**, *14*, 18148–18180. [CrossRef] [PubMed]

38. Dhawan, P.; Singh, A.B.; Deane, N.G.; No, Y.; Shiou, S.R.; Schmidt, C.; Neff, J.; Washington, M.K.; Beauchamp, R.D. Claudin-1 regulates cellular transformation and metastatic behavior in colon cancer. *J. Clin. Investig.* **2005**, *115*, 1765–1776. [CrossRef] [PubMed]

39. Karagiannis, G.S.; Poutahidis, T.; Erdman, S.E.; Kirsch, R.; Riddell, R.H.; Diamandis, E.P. Cancer associated fibroblasts drive the progression of metastasis through both paracrine and mechanical pressure on cancer tissue. *Mol. Cancer Res.* **2012**, *10*, 1403–1418. [CrossRef] [PubMed]

40. Chang, H.Y.; Ye, S.P.; Pan, S.L.; Kuo, T.T.; Liu, B.C.; Chen, Y.L.; Huang, T.C. Overexpression of miR-194 reverses HMGA2-driven signatures in colorectal cancer. *Theranostics* **2017**, *7*, 3889–3900. [CrossRef] [PubMed]

41. Huang, D.W.; Sherman, B.T.; Lempicki, R.A. Systematic and integrative analysis of large gene lists using DAVID bioinformatics resources. *Nat. Protoc.* **2009**, *4*, 44–57. [CrossRef] [PubMed]

42. Joshi-Tope, G.; Gillespie, M.; Vastrik, I.; D'Eustachio, P.; Schmidt, E.; de Bono, B.; Jassal, B.; Gopinath, G.R.; Wu, G.R.; Matthews, L.; et al. Reactome: A knowledgebase of biological pathways. *Nucleic Acids Res.* **2005**, *33*, D428–D432. [CrossRef] [PubMed]

43. Subramanian, A.; Tamayo, P.; Mootha, V.K.; Mukherjee, S.; Ebert, B.L.; Gillette, M.A.; Paulovich, A.; Pomeroy, S.L.; Golub, T.R.; Lander, E.S.; et al. Gene set enrichment analysis: A knowledge-based approach for interpreting genome-wide expression profiles. *Proc. Natl. Acad. Sci. USA* **2005**, *102*, 15545–15550. [CrossRef] [PubMed]

44. Shannon, P.; Markiel, A.; Ozier, O.; Baliga, N.S.; Wang, J.T.; Ramage, D.; Amin, N.; Schwikowski, B.; Ideker, T. Cytoscape: A software environment for integrated models of biomolecular interaction networks. *Genome Res.* **2003**, *13*, 2498–2504. [CrossRef] [PubMed]

45. Isserlin, R.; Merico, D.; Voisin, V.; Bader, G.D. Enrichment Map—A Cytoscape app to visualize and explore OMICs pathway enrichment results. *F1000Research* **2014**, *3*, 141. [CrossRef] [PubMed]

46. Hong, Y.; Downey, T.; Eu, K.W.; Koh, P.K.; Cheah, P.Y. A 'metastasis-prone' signature for early-stage mismatch-repair proficient sporadic colorectal cancer patients and its implications for possible therapeutics. *Clin. Exp. Metastasis* **2010**, *27*, 83–90. [CrossRef] [PubMed]

 *marine drugs*

Review

Review of Chromatographic Bioanalytical Assays for the Quantitative Determination of Marine-Derived Drugs for Cancer Treatment

Lotte van Andel [1,2,*], Hilde Rosing [1], Jan HM Schellens [2,3,4] and Jos H Beijnen [1,2,4]

1   Department of Pharmacy & Pharmacology, Antoni van Leeuwenhoek, The Netherlands Cancer Institute and MC Slotervaart, 1066 CX Amsterdam, The Netherlands; h.rosing@nki.nl (H.R.); j.beijnen@nki.nl (J.H.B.)
2   Division of Pharmacology, Antoni van Leeuwenhoek, The Netherlands Cancer Institute, 1066 CX Amsterdam, The Netherlands; j.schellens@nki.nl
3   Department of Clinical Pharmacology, Division of Medical Oncology, The Netherlands Cancer Institute, 1066 CX Amsterdam, The Netherlands
4   Department of Pharmaceutical Sciences, Faculty of Science, Division of Pharmacoepidemiology and Clinical Pharmacology, Utrecht University, 3584 CG Utrecht, The Netherlands
*   Correspondence: l.v.andel@nki.nl; Tel.: +31-20-512-9086

Received: 15 June 2018; Accepted: 18 July 2018; Published: 23 July 2018

Abstract: The discovery of marine-derived compounds for the treatment of cancer has seen a vast increase over the last few decades. Bioanalytical assays are pivotal for the quantification of drug levels in various matrices to construct pharmacokinetic profiles and to link drug concentrations to clinical outcomes. This review outlines the different analytical methods that have been described for marine-derived drugs in cancer treatment hitherto. It focuses on the major parts of the bioanalytical technology, including sample type, sample pre-treatment, separation, detection, and quantification.

Keywords: marine-derived drugs; cancer; bioanalysis; chromatography

---

1. Introduction

For years, researchers have roamed the seas and oceans in search of organisms possessing chemicals that could exhibit therapeutic effects. These include a wide variety of creatures, such as tunicates, mollusks, sponges, bacteria, seaweeds, chordates, mangroves, sea hares, and sharks [1–3]. Therapeutic indications include cancer, Alzheimer's disease, thrombosis, hypertension, psoriasis, asthma, and pain [4–7]. Ruiz-Torres et al. (2017) have outlined the increase in manuscripts published about marine-derived compounds, and more specifically, the substantial increase in these compounds studied for cancer treatment over the last several years, with about 125 publications in 2016 [8]. This nicely demonstrates the increasing information and insights researchers have gathered with regards to marine-derived anticancer drugs.

Sea-based organisms have obtained the ability to produce chemicals, of which some might be cytotoxic, because they have learned to live under relatively bitter and hostile conditions in the marine environment. Biological evolution has therefore enabled them to acquire and to select these chemicals to survive [1] and to compensate for their lack of physical protection [9]. The extreme conditions involve high salt, high pressure, lack of oxygen and light, and extremely high or low temperatures [7,10]. Furthermore, these organisms suffer fierce survival competition, and the chemicals, or secondary metabolites, help marine organisms to survive, reproduce, and resist predators [7,10]. The therapeutic agents derived from these chemicals can be unaltered natural products, chemically altered derivatives, or their structure might have been inspired by natural marine products [11].

Unfortunately, the supply of organisms is not endless, and ecological constraints are often limiting research in this field. The challenges with regard to the limited availability of the compounds in nature are demonstrated by the narrative of trabectedin. Despite early proof that trabectedin seemed to be effective in cancer treatment, development delays were encountered due to its low natural abundance, and efforts had to be made to produce sufficient amounts on a large scale [9]. In fact, this limitation of the scaleup process has oftentimes been the reason for the discontinuation of clinical development [12]. Drug development is in itself a long journey but even more so if having to deal with ecological considerations when accessing the marine source of the compounds.

Many marine-derived compounds have complex molecular structures and therefore different challenges arise. Their structural elucidation and chemical synthesis might be difficult, and the complex molecules may include many potential bioactive moieties, which make it difficult to identify a single mechanism of action [13,14].

Despite these challenges, the discovery of new cancer treatments originating from marine-derived animals or plants is very much in the spotlight today. From the numerous discoveries of the new chemical entities done so far, some compounds have progressed into clinical trials. Clinical trials investigating safety and toxicity also require the characterization of the pharmacokinetic profile of these potential therapeutics. In order to accurately describe the pharmacokinetic profiles in biological fluids, such as plasma, whole blood, serum, and urine, (validated) bioanalytical methods are indispensable.

The first marine-derived compound approved by the United States (US) Food and Drug Administration (FDA) was cytarabine (Cytosar U®) in 1969, followed by trabectedin (Yondelis®), eribulin mesylate (Halaven®), brentuximab vedotin (Adcetris®), and midostaurin (Rydapt®) in the years 2007, 2010, 2011, 2013, and 2017, respectively.

Most marine-derived compounds possess a few defining characteristics. First of all, often very low doses are given to patients, and they are administered almost exclusively via intravenous infusion (i.v.). The compounds are typically large molecules (400–1700 Da), especially the depsipeptides, and they are often hydrophobic. These attributes make them fascinating compounds in terms of pharmacokinetics but also from a bioanalytical point of view. For instance, the hydrophobic drugs tend to distribute massively into peripheral tissues, which is demonstrated by high volumes of distribution. This means that plasma levels can be rather low at all times and that bioanalytical methods to quantify the drug in the plasma thus need to be highly sensitive.

Focus of This Review

The demands for sensitive and specific analytical technologies for the determination and quantification of cytotoxic drugs in biological matrices are high. These methods provide the means to construct pharmacokinetic profiles, to determine drug exposure during treatment, and to evaluate if target concentrations are reached. Moreover, they can be used to quantify drugs and their metabolites in excreta to investigate excretion profiles and metabolism. This information is pivotal in drug development and clinical application but also required by regulatory agencies, such as the FDA and the European Medicines Agency (EMA). This review outlines the available methods for analyses of marine-derived anticancer agents in biological matrices. Furthermore, information concerning the sensitivity capabilities of each method is provided. This review provides a comprehensive list of analyses of marine-derived compounds in various biological matrices published over the last few decades, which could potentially also serve as a guide for newly discovered or synthesized drugs from marine origin.

2. Marine-Derived Anticancer Drugs

A huge number of marine-derived compounds are described in over 700 pages in the book titled "Handbook of Anticancer Drugs from Marine Origin" (2015), which far exceeds the practicality of this review. Instead, successful anticancer compounds, those that have progressed into clinical development, described in three relatively recent reviews by Mudit and El Sayed (2016), Newman and Cragg (2016), and Palanisamy et al. (2017) were used as basis of this review and cover the relevant anticancer compounds of current interest [2,15,16].

Bioanalysis refers to the analysis of compounds, particularly drugs, in biological matrices and encompasses therefore a large range of applications. Manuscripts were included in this review if they described some type of chromatography for bioanalysis, because this is the most widely applied and available method for drug analysis. Also, compounds that have progressed into clinical trials but failed were included in our literature search, because knowledge about the analysis of these agents might be valuable, especially as it is not uncommon to chemically synthesize derivatives of the potentially failed drugs. Manuscripts published before 1990 were excluded from the review. Publications were also excluded if the method described was not applied to biological samples.

Cytarabine, although being the first identified anticancer drug of marine origin, is excluded from the review. Cytarabine is a nucleoside analogue that is structurally very different but also much simpler than the other marine-derived agents. The bioanalysis of this group of compounds has been extensively reviewed elsewhere [17]. Moreover, the antibody-drug conjugates (ADCs) with payloads (monomethyl auristatin E [MMAE]) derived from the marine-derived cytotoxic agent auristatin were excluded as well, because bioanalysis of this class of drugs is a subject of its own. Drugs belonging to this group are ASG-15ME, ASG-67E, brentuximab vedotin, depatuxizumab mafodotin, denintuzumab mafodotin, enfortumab vedotin, glembatumumab vedotin, GSK2857916, indusatumab vedotin, ladiratuzumab vedotin, lifastuzumab vedotin, pinatuzumab vedotin, polatuzumab vedotin, tisotumab vedotin, and vandortuzumab vedotin.

Within the PubMed database, keyword searches for 'drug name AND bioanalysis', 'drug name AND quantification', 'drug name AND assay', and 'drug name AND determination', where 'drug name' included all synonyms, both generic and brand name, were applied. The careful analysis of titles and abstracts led to a selection of manuscripts describing chromatographic bioanalytical methods to determine marine-derived anticancer drug concentrations in biological matrices.

3. Bioanalysis of Marine-Derived Anticancer Drugs

3.1. Drugs

In total, 35 marine-derived anticancer drugs are included in the three aforementioned reviews [2,15,16]. Table 1 lists the marine-derived drugs approved for cancer treatment or still under investigation in clinical trials. It includes 20 compounds, of which 5 are approved, 6 are still being investigated in phase I clinical trials at escalating doses, 6 are currently in phase II clinical trials, and 3 have entered phase III clinical trials. Some drugs mentioned in the reviews by Mudit and El Sayed [2], Newman and Cragg [15], and Palanisamy et al. [16] have been discontinued from clinical trials or no active clinical trials can be found in the ClinicalTrials.gov database. These compounds can be found in Table 2, and the corresponding bioanalytical methods were still included in the present review whenever available. Seven of these drugs are natural products (didemnin B, marizomib, plitidepsin, PM060184, pipecolidepsin A, stellatolide A, and trabectedin); all others are derivatives or analogues thereof.

Table 1. Overview of marine-derived compounds approved or under investigation in clinical trials.

Name	Synonym (s)	Natural Product or Derivative	Origin	Administration	Dose	Indication	Status
AGS-67E		Derivative	Cyanobacterium *Caldora penicillata* [15]	intravenous infusion (i.v.)	Escalating doses	Lymphoid malignancies [18]	Phase I
Brentuximab vedotin (Adcetris®)	SGN-35/cAC10-vcMMAE	Derivative	Cyanobacterium *Caldora penicillata* [15]	i.v.	1.8 mg/kg [19]	Hodgkin lymphoma, Systemic anaplastic large cell lymphoma [19]	Approved
Cytarabine (Cytosar-U®/DepoCyte®)	Ara-C/Cytosine arabinoside/1-ß-D-Arabinofuranosylcytosine	Derivative	Sponge *Cryptotethya crypta* [12]	i.v./intrathecal	75–200 mg/m^2 [20]	Acute myelogenous leukemia, Chronic myelogenous leukemia, Acute lymphoblastic leukemia non-Hodgkin's lymphoma [20]	Approved
Denintuzumab mafodotin	SGN-CD19A/SGN-19A	Derivative	Cyanobacterium *Caldora penicillata* [15]	i.v.	3 mg/kg	Diffuse large B-cell lymphoma [21]	Phase II
Depatuxizumab mafodotin	ABT-414	Derivative	Cyanobacterium *Caldora penicillata* [15]	i.v.	1.25 mg/kg	Glioblastoma/Pediatric brain tumors [22]	Phase II
Enfortumab vedotin	ASG-22ME/ASG-22CE	Derivative	Cyanobacterium *Caldora penicillata* [15]	i.v.		Urothelial cancer [23]	Phase II
Eribulin mesylate (Halaven®)	E7389	Derivative	Sponge *Halichodria okadai* [12]	i.v.	1.23 mg/m^2 [24]	Breast cancer, Liposarcoma [24]	Approved
Glembatumumab vedotin	CDX-011/CR011-vcMMAE	Derivative	Cyanobacterium *Caldora penicillata* [15]	i.v.	1.9, 2.2 mg/kg	Breast cancer, Metastatic melanoma [25]	Phase II
GSK2857916	J6M0-mcMMAF	Derivative	Cyanobacterium *Caldora penicillata* [15]	i.v.	Escalating doses	Multiple myeloma [26]	Phase I
Ladiratuzumab vedotin	SGN-LIV1A	Derivative	Cyanobacterium *Caldora penicillata* [15]	i.v.	Escalating doses	Human epidermal growth factor receptor 2 (HER2)-negative breast cancer, triple negative breast cancer	Phase I
Lurbinectedin (Zepsyre®)	PM01183	Trabectedin analogue	Tunicate *Ecteinascidia turbinata* [12]	i.v.	3.2 mg/mL	Platinum-resistant ovarian cancer [27]	Phase III
Marizomib	Salinosporamide A/NPI-0052	Natural product	Marine actinomycete *Salinispora tropica* [12]	i.v.	0.5 mg/m^2	Multiple myeloma, glioblastoma [28]	Phase II
Midostaurin (Rydapt®)	PKC412/CGP41251/N-benzoylstaurosporine	Staurosporine analogue	Bacterium *Streptomyces staurosporeus* [29]	oral	50 mg twice daily [30]	FLT3+ Acute myeloid leukemia [30]	Approved
Pinatuzumab vedotin	DCDT-2980S/RG7593	Derivative	Cyanobacterium *Caldora penicillata* [15]	i.v.	1.8, 2.4 mg/kg	Follicular lymphoma, Diffuse large B-cell lymphoma	Phase I
Plinabulin	NPI2358	Derivative	Marine fungus *Aspergillus* sp. [12]	i.v.	30 mg/m^2	Non-small cell lung cancer [31]	Phase III
Plitidepsin (Aplidin®)	Dehydrodidemnin B	Natural product	Tunicate *Aplidium albicans* [12]	i.v.	5 mg/m^2	Multiple Myeloma, Lymphoma [27]	Phase III
PM060184	PM0184/Plocabulin	Natural product	Sponge *Lithoplocamia lithistoides* [12]	i.v.	Starting dose 4 mg/m^2	Breast cancer, Solid tumors [27]	Phase I

Table 1. *Cont.*

Name	Synonym (s)	Natural Product or Derivative	Origin	Administration	Dose	Indication	Status
Polatuzumab vedotin	DCDS-4501A	Derivative	Cyanobacterium *Caldora penicillata* [15]	i.v.	1.8, 2.4 mg/kg	Non-Hodgkin's lymphoma, B-cell lymphoma [32]	Phase II
Tisotumab vedotin	HuMax-TF-ADC/HuMab-TF-011-vcMMAE/TF-011-vcMMAE	Derivative	Cyanobacterium *Caldora penicillata* [15]	i.v.	Escalating doses	Solid tumors [33]	Phase I
Trabectedin (Yondelis®)	ET-743	Natural product	Tunicate *Ecteinascidia turbinate* [12]	i.v.	1.1/1.5 mg/m^2 [34]	Soft tissue sarcoma [34]	Approved

Table 2. Marine-derived anticancer drugs for which trials were discontinued or for which no active trials (NAT) were found in the ClinicalTrials.gov database.

Name	Synonym (s)	Natural Product or Derivative	Origin	Discontinued/Inactive	Reason for Discontinuation
ASG-15ME	AGS15E	Derivative	Cyanobacterium *Caldora penicillata* [15]	Discontinued	Unspecified [35]
Becatecarin	XL-119/NSC 655649/BMY 27557/BMS 181176	Rebeccamycin analogue	Marine actinomycete *Saccharothrix aerocolonigenes* [36]	Discontinued	Not superior to existing therapies [37]
CEP-2563	KT-8391	Staurosporine derivative		NAT	
Didemnin B		Natural product	Tunicate *Trididemnin cyanophorum* [38]	Discontinued	Toxicity [39]
Edotecarin	J-107088/PF-804950/PHA-782615/ED-749	Derivative		NAT	
Elisidepsin (Irvalec®)	PM02734	Structural analogue	Mollusk *Elysia rufescens* [40]	Discontinued	Strategic [41]
Enzastaurin	LY317615	Staurosporine derivative	Bacterium *Streptomyces staurosporeus*	Discontinued	Lack of efficacy [42]
Indusatumab vedotin	MLN-0264/TAK-0264	Derivative	Cyanobacterium *Caldora penicillata* [15]	Discontinued	Lack of efficacy [43]
Lestaurtinib	CEP-701	Staurosporine derivative	Bacterium *Streptomyces staurosporeus*	Discontinued	Strategic [41]
Lifastuzumab vedotin	DNIB0600A/Anti-NaPi2B ADC/RG7599	Derivative	Cyanobacterium *Caldora penicillata* [15]	Discontinued	Lack of efficacy [44]
Pipecolidepsin A		Natural product	Sponge *Homophymia lamellose* [13]	NAT	
Stellatolide A		Natural product	Sponge *Ecionemia acervus* [45]	NAT	
UCN-01	7-hydroxystaurosporine	Staurosporine analogue	Bacterium *Streptomyces staurosporeus* [29]	NAT	
Vandortuzumab vedotin	DSTP-3086 S/RG-7450/thio-antiSTEAP1-MC-vc-PAB-MMAE	Derivative	Cyanobacterium *Caldora penicillata* [15]	NAT	
Zalypsis	PM00104/PM-10450	Derivative	Sponge *Netropsia* sp. [15]	NAT	

Of those that are still under investigation or approved for cancer treatment, all but one are administered via i.v. Midostaurin, approved for acute myeloid leukemia in May 2017, is an oral drug given at a dose of 50 mg twice daily, the highest dosage compared to the other selected compounds in Table 1. This could be related to limited oral bioavailability, but unfortunately this remains unknown [30]. Plinabulin is given at 30 mg/m^2, and all other drugs are administered at dosages between 0.5–5 mg/m^2. Most compounds distribute widely to peripheral tissues, with volumes of distribution at steady state (V_{ss}) ranging from approximately 18–5000 L [19,24,30,34,46–58]. UCN-01 and CEP-2536 are the exceptions, with lower V_{ss} ranging from 4 to 14.3 L and 0.35 to 44 L, respectively [59–61].

Table 3 includes all manuscripts covering bioanalytical methods developed to quantify marine-derived anticancer therapeutics in biological matrices. In total, 23 manuscripts describing bioanalysis were included. Compounds that were included in the above three reviews but for which no bioanalytical methods were found include becatecarin, CEP-2563, didemnin B, edotecarin, enzastaurin, lestaurtinib, marizomib, pipecolidepsin A, stellatolide A, plinabulin, and PM060184.

Table 3. Overview of published bioanalytical methods to quantify marine-derived anticancer drugs in biological matrices.

Compound	Matrix	Sample Pre-Treatment	Stationary Phase	Mobile Phase	Detection	Internal Standard	Linear Range	LOD	Metabolites	Ref.
Elisidepsin	Plasma (Dog)	LLE (ethyl acetate)	YMC Pro C18, S-5 (50 × 2.0 mm, 120 Å)	A: 5 mM ammonium acetate, 0.1% FA in H_2O B: 0.1% FA in MeOH	MS/MS	$^2H_8$-PM02734 (SIL)	0.05–100 ng/mL		No	[62]
Eribulin mesylate	Plasma Urine (Human)	LLE (ethyl acetate/MeOH/EtOH)	Polaris® C18 (30 × 2.0 mm, 3 μm)	A: 0.1% FA in H_2O-ACN (87:13, *v/v*) B: 0.1% FA in THF-ACN (30:70, *v/v*)	MS/MS	ER-076349 (Structural analogue)	0.2–100 ng/mL		No	[63]
Eribulin mesylate	Plasma Whole blood Urine Feces (Human)	P/WB/U: LLE (ethyl acetate/MeOH/EtOH) F: dilution (ACN)	Polaris® C18-A (30 × 2.0 mm, 3 μm)	A: 0.1% FA in H_2O B: 0.1% FA in ACN	MS/MS	ER-076349 (Structural analogue)	0.2–100 ng/mL (P); 0.5–100 ng/mL (WB/U); 100–25,000 ng/mL (F)		No	[64]
Lurbinectedin	Plasma (Cynomolgus monkey Dog Mice Mini-pig Rat)	SLE (TBME)	ACE C18 PFP (30 × 2.1 mm, 3 μm)	A: 0.1% FA in H_2O B: 0.1% FA in ACN	MS/MS	PM040038 (SIL)	0.1–100 ng/mL	0.025 ng/mL	No	[65]
Midostaurin	Plasma (Human)	LLE (diisopropyl ether)	μBondapak RP-18 (300 × 3.9 mm, 10 μm)	ACN-0.001 M ammonium acetate in H_2O, pH 4.0 (45:55, *v/v*)	FLD 286/386 nm	CGP 41 126 (Structural analogue)	1–1000 ng/mL	0.5 ng/mL	CGP 50 723; CGP 50 750; CGP 52 421	[66]
Midostaurin	Plasma (Human)	LLE (diisopropyl ether)	RP LiChrospher C18 end-capped (125 × 4.0 mm, 5 μm)	A: ACN B: 445 μL TEA in 1 L phosphate buffer, pH 3.6	FLD 286/386 nm	CGP 41 126 (Structural analogue)	0.2–1000 ng/mL	0.1 ng/mL	CGP 50 673; CGP 50 723; CGP 50 750; CGP 52 421	[67]
Midostaurin	Plasma (Human)	LLE (diethyl ether)	Prodigy ODS-2 (150 × 3.2 mm, 5 μm)	0.4 mL Titriplex III-solution in MeOH-ACN-0.05 M ammonium acetate in water (40:26:34, *v/v/v*)	FLD 286/386 nm	N-phenyl-1-naphthylamine	10–10,000 ng/mL	10 ng/mL	CGP 52421e1; CGP 52421e2; CGP 62221; CGP 62221e1; CGP 62221e2;	[68]
Midostaurin	Plasma (Human)	PP (MeOH)	SunFire bonded and end-capped C18 silica (150 × 2.1 mm, 3.5 μm)	A: 10 mM ammonium formate, 0.1% FA in H_2O B: 0.1% FA in ACN	MS/MS	Midostaurin-d5 (SIL)	75–2500 ng/mL		No	[69]
Plitidepsin	Plasma Urine (Rat)	PP (0.1% FA in ACN) + LLE (chloroform)	RP C18 Hypersil-5 ODS (100 × 3.0 mm; 5 μm, 120 Å)	A: 0.5% FA in H_2O B: 0.5% FA in ACN	MS/MS	Didemnin B (Structural analogue)	5–100 ng/mL (P); 1.25–125 ng/mL (U)	1 ng/mL (P); 0.5 ng/mL (U)	No	[70]

Table 3. *Cont.*

Compound	Matrix	Sample Pre-Treatment	Stationary Phase	Mobile Phase	Detection	Internal Standard	Linear Range	LOD	Metabolites	Ref.
Plitidepsin	Plasma (Mice)	Derivatization + SPE	Symmetry C18 (100 × 4.6 mm, 3.5 μm)	ACN–water–TFA (47:52.9:0.1, $v/v/v$)	FLD 410/560 nm	None	2–100 ng/mL		No	[71]
Plitidepsin	Plasma (Human)	LLE (TBME)	Zorbax Bonus-RP (50 × 2.0 mm, 5 μm)	0.1% FA in ACN-5 mM ammonium acetate, 0.1% FA in H_2O (80:20, v/v)	MS/MS	Didemnin B (Structural analogue)	0.05–50 ng/mL		No	[72]
Plitidepsin	Plasma Whole Blood Urine (Human)	PP (0.1% FA in ACN) + LLE (chloroform)	Hypersil-5 ODS (100 × 3.0 mm, 5 μm, 120 Å)	A: 0.5% FA in ACN B: 0.5% FA in H_2O	MS/MS	Didemnin B (Structural analogue)	1–250 ng/mL	0.25 ng/ml	No	[73]
Plitidepsin	Whole Blood (Human)	Derivatization + SPE	Symmetry C18 (100 × 4.6 mm, 3.5 μm)	ACN-0.1% TFA in H_2O (50:50, v/v)	FLD 410/560 nm	None	2–100 ng/mL		No	[74]
Plitidepsin	Plasma Whole Blood Urine (Human)	LLE (TBME)	SunFire C18 (50 × 2.1 mm, 5 μm)	A: 5 mM ammonium acetate, 0.1% FA in H_2O B: 0.1% FA in ACN	MS/MS	(PM130461) $^{13}C_5$-^{15}N-plitidepsin (SIL)	0.1–100 ng/mL		No	[75]
Trabectedin	Plasma (Human)	SPE	Zorbax SB-C18 column (75 × 4.6 mm, 3.5 μm)	ACN–25 mM phosphate buffer, pH 5.0 (70:30, v/v)	UV 210 nm	POB	1–50 ng/mL		No	[76]
Trabectedin	Plasma (Human)	SPE	Zorbax Rx-C18 (150 × 2.1 mm, 5 μm)	MeOH-5 mM ammonium acetate, 0.4% FA in H_2O (75:25, v/v)	MS/MS	ET-729 (Structural analogue)	0.01–2.5 ng/mL		No	[77]
Trabectedin	Plasma (Human)	PP (MeOH)	Zorbax Rx-C18 (150 × 2.1 mm, 5 μm)	MeOH–H_2O (85:15, v/v)	MS/MS	$^2H_3$-ET-743 (SIL)	0.05–2.5 ng/mL		No	[78]
Trabectedin	Plasma (Human)	PP (HCl in MeOH)	Accucore XL C18 (50 × 2.1 mm, 4 μm)	A: 10 mM ammonium acetate in H_2O, pH 6.8 B: MeOH	MS/MS	$^2H_3$-ET-743 (SIL)	0.025–1.0 ng/ml		No	[79]
Trabectedin	Liver cells Tumor cells (Mice)	Lysis	Accucore XL C18 (50 × 2.1 mm, 4 μm)	A: 10 mM ammonium acetate in H_2O, pH 6.8B: MeOH	MS/MS	$^2H_3$-ET-743 (SIL)	0.1–3 ng/mL (TC); 0.25–6 ng/mL (LC)		No	[80]
UCN-01	Plasma Urine (Human)	PP (ACN)	AM-312 ODS (150 × 6.0 mm S-5mm, 120 Å)	ACN–0.1% TEA in 0.05 M phosphate buffer, pH 7.3 (50:50, v/v).	FLD 310/410 nm	Staurosporine (Structural analogue)	0.2–100 ng/mL (P); 1–400 ng/mL (U)		No	[81]
UCN-01	Plasma Saliva (Human)	PP (ACN)	Nova-Pak Phenyl (150 × 3.9 mm, 4 μm)	A: 0.05 M ammonium acetate in H_2O, pH 4.15 B: ACN	UV (P) 295 nm FLD (Sal) 290/400 nm	Umbelliferone	200–20,000 ng/mL (P); 4–200 ng/mL (S)		No	[82]

Table 3. *Cont.*

Compound	Matrix	Sample Pre-Treatment	Stationary Phase	Mobile Phase	Detection	Internal Standard	Linear Range	LOD	Metabolites	Ref.
UCN-01	Plasma (Human)	PP (ACN)	Nova-Pak Phenyl (150 × 3.9 mm, 4 µm, 60 Å)	ACN–0.5 M ammonium acetate, 0.2% TEA in H_2O (45:55, v/v)	FLD 310/410 nm	Staurosporine (Structural analogue)	200–30,000 ng/mL	0.1 µg/mL	No	[83]
Zalypsis	Plasma (Dog Human Mice Rat)	LLE (TBME)	Zorbax SB-C18 (50 × 2.1 mm, 5 µm, 80 Å)	A: 5 mM ammonium acetate, 0.1% FA in H_2O B: 0.1% FA in MeOH	MS/MS	$^{13}C_2,^{2}H_3$-PM00104 (SIL)	0.01–5 ng/mL		No	[84]

ACN: acetonitrile; EtOH: ethanol; F: Feces; FA: formic acid; FLD: fluorescence detection with excitation and emission wavelengths; HCl: hydrochloric acid; H_2O: water; LC: liver cells; LLE: liquid-liquid extraction; LOD: limit of detection; MeOH: methanol; MS/MS: tandem mass spectrometry; P: plasma; PFP: pentafluorophenyl; POB: propyl-p-hydroxybenzoate; PP: protein precipitation; RP: reversed phase; S: saliva; SIL: stable isotopically labeled; SLE: supported liquid extraction; SPE: solid phase extraction; TBME: *tert*-butyl methyl ether; TC: tumor cells; TEA: triethylamine; TFA: trifluoroacetic acid; THF: tetrahydrofuran; U: urine; UV: ultraviolet detection with detector wavelength; and WB: whole blood.

The compounds included in the present review can be grouped together according to their chemical structures. Eribulin belongs to the macrolides [15] (Figure 1), marizomib is a salinosporamide [85] (Figure 2), plinabulin is a piperazine derivative [86] (Figure 3), and PM060184 is a polyketide [15] (Figure 4). The isoquinoline alkaloids include zalypsis, trabectedin, and its structural analogue lurbinectedin [15] (Figure 5). The largest groups are the depsipeptides (didemnin B, elisidepsin, pipecolidepsin A, plitidepsin, and stellatolide A) [15] (Figure 6) and the indolocarbazoles (CEP-2563, enzastaurin, lestaurtinib, midostaurin and UCN-01), all derivatives of staurosporine [87,88], (Figure 7) and becatecarin and edotecarin, derivatives of rebeccamycin [89] (Figure 8).

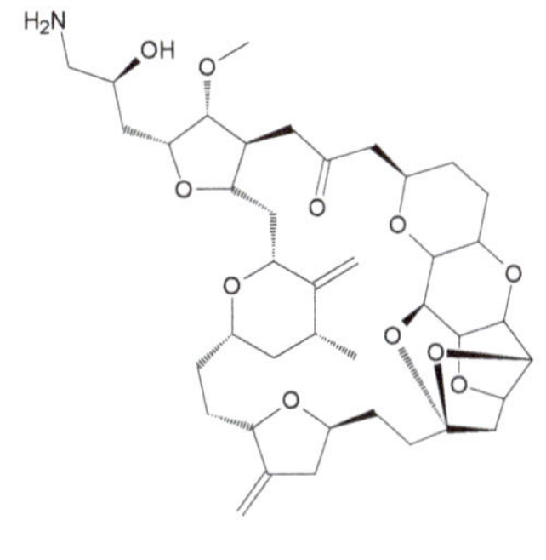

Figure 1. Chemical structure of the macrolide eribulin [64].

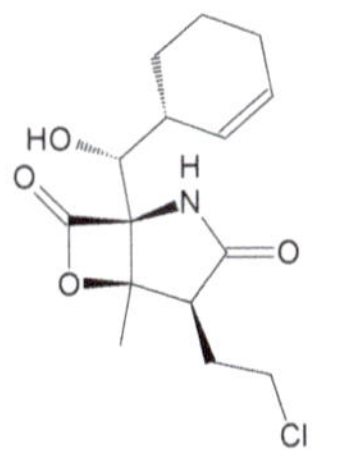

Figure 2. Chemical structure of marizomib, a salinosporamide [39].

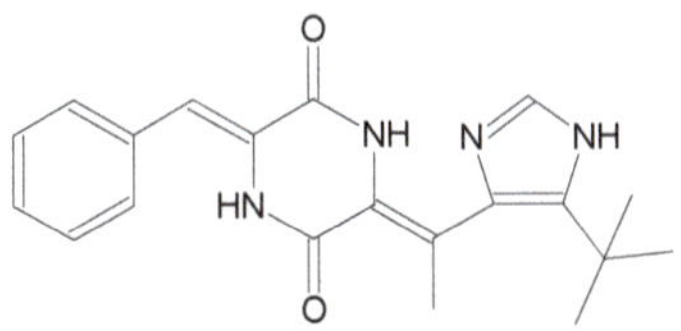

Figure 3. Chemical structure of the piperazine derivative plinabulin [86].

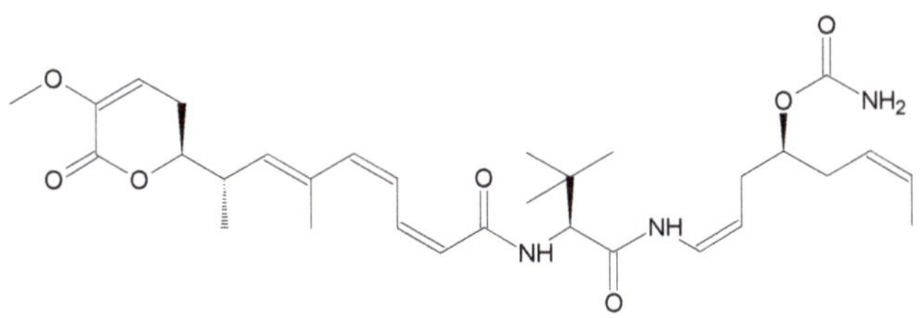

Figure 4. Chemical structure of PM060184, a polyketide [39].

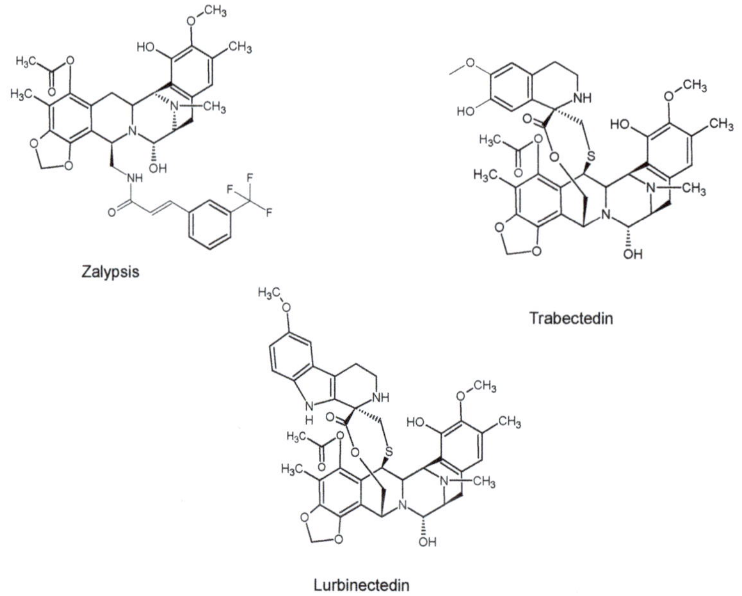

Figure 5. Chemical structures of the isoquinolines zalypsis [39], trabectedin [90], and lurbinectedin [65].

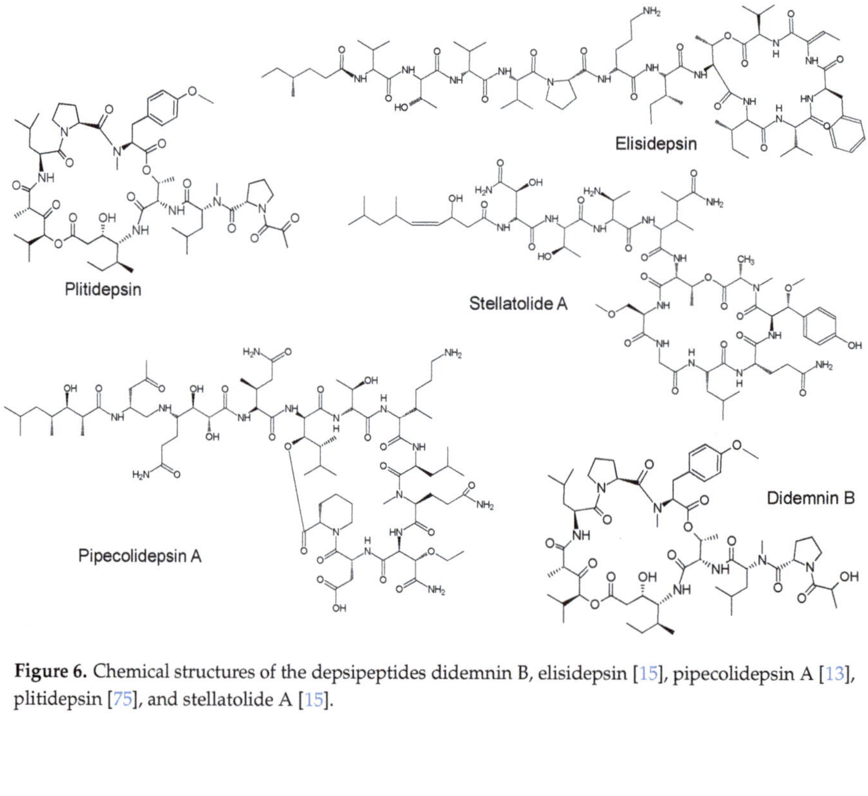

Figure 6. Chemical structures of the depsipeptides didemnin B, elisidepsin [15], pipecolidepsin A [13], plitidepsin [75], and stellatolide A [15].

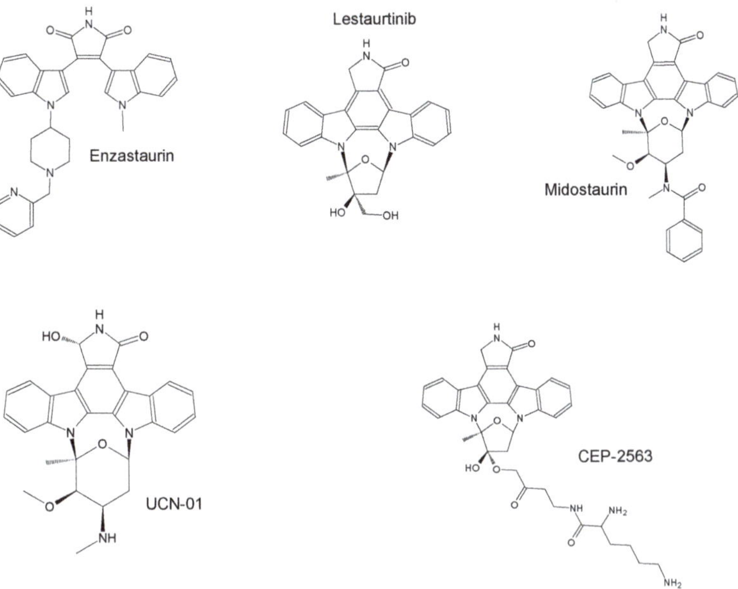

Figure 7. Chemical structures of the staurosporine derivatives CEP-2563 [16], enzastaurin, lestaurtinib, midostaurin, and UCN-01 [91].

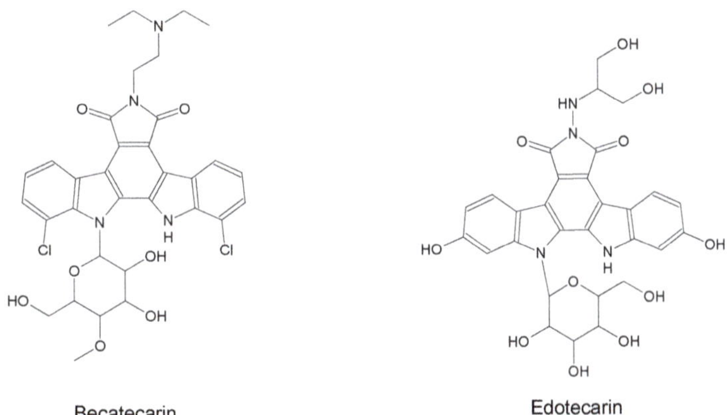

Figure 8. Chemical structures of the rebeccamycin derivatives becatecarin and edotecarin [92].

The chromatographic bioanalytical methods can be divided in different steps (Figure 9), starting with the collection of a biological sample. Next, the drug, or analyte, needs to be extracted from the biological matrix, and the sample is cleaned up to such an extent that all other (endogenous) compounds extracted from the sample do not interfere with analysis. Thereafter, the sample is injected onto a chromatographic system, and the mobile phase together with the stationary phase enable the separation of the analyte of interest from the other components in the sample. Finally, the analyte is detected by a detector. Derivatization might be necessary prior to sample cleanup if an ultraviolet (UV) detector or a fluorescence detector (FLD) is used. The next paragraphs describe these individual steps of the drug analysis.

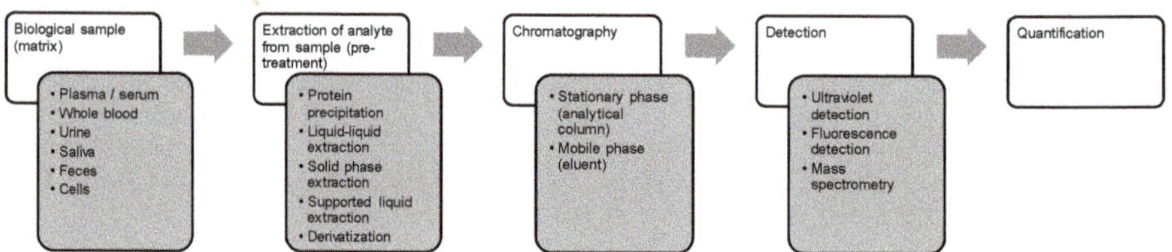

Figure 9. General set up of drug analysis.

3.2. Matrices

3.2.1. Plasma and Serum

Most bioanalytical assays have been developed for the plasma and serum matrix. Of the 23 manuscripts included, 21 cover concentration determinations in plasma or serum. Generally, this is done to assess plasma pharmacokinetic characteristics and to link this to clinical outcomes (antitumor efficacy and/or toxicities). Ideally, quantification of the drug is done at its site of action, but this is (usually) not possible. Instead, blood samples are taken, plasma or serum is obtained by centrifugation, and the drug concentrations are taken as a substitute for the concentrations at the site of action. It is relatively simple to draw a blood sample, and although invasive, the burden of multiple venipunctures can be reduced by placing a peripheral intravenous catheter. This way, a complete pharmacokinetic profile could technically be constructed with a single vein puncture.

Because the only difference between plasma and serum is the presence of coagulation factors, the concentrations in plasma and serum are usually regarded as equivalent. Serum samples are, however, cleaner compared to plasma because of the removal of clotting factors. This could be beneficial, because coagulants can contaminate the analysis instrumentation (i.e., the analytical column). The downside is that it takes some time for the blood to coagulate, so sample processing is relatively more tedious compared to obtaining plasma. On the other hand, a serum sample is slightly more concentrated than a plasma sample, because the coagulation factors have been removed from the sample. No bioanalytical methods were found for lestaurtinib and enzastaurin, but pharmacokinetic studies have been published describing serum pharmacokinetics of these compounds [93–95], indicating that assays exist but are not in the public domain as far as we know.

3.2.2. Whole Blood

Whole blood samples are less easy to process than plasma samples due to their viscosity and need to be handled more carefully than other sample types. For instance, to homogenize and thus distribute compounds evenly, a whole blood sample should only be shaken very gently and not too vigorously. An advantage of whole blood can be the presence of hemoglobin in the blood, which might have a stabilizing effect, making the analyte present less prone to photo-oxidation compared with plasma [96]. However, hemoglobin could also interfere with analysis. Whole blood analyses are considered when a drug is bound to red blood cells, such as plitidepsin [73–75]. Plitidepsin concentrations in whole blood are approximately fourfold higher than its plasma levels, so red blood cells are an important distribution compartment in this case [97,98]. Quantifying the drug in red blood cells, therefore, has added value. A bioanalytical method has been developed to quantify eribulin mesylate in whole blood as well, but there were no significant differences in the plasma and whole blood concentrations measured [64,99].

3.2.3. Urine

Drug concentrations are frequently measured in urine to get insight into drug and metabolite excretory pathways. Urine collection is fairly easy: it is non-invasive, and urine quantities are large. Drug levels can often be measured for longer periods of time after administration compared with blood [100]. It also adds information in terms of pharmacokinetics. By dividing the amount of drug excreted unchanged by the dose given, the fraction of the amount entering the general circulation that is excreted unchanged (f_e) can be calculated. This provides a quantitative measure of the contribution of renal excretion to overall drug elimination [101]. The drawbacks of urine analysis, however, include the large variety in urine composition between patients but also between intervals/timepoints. Furthermore, in order to obtain complete excretion data, urine collection needs to be complete, and this is even more challenging in a clinical setting when the patients need to collect samples at home. Another potential problem with urine analysis is non-selective adsorption of the analyte to the container wall. Urine is an aqueous fluid, which means that hydrophobic analytes might prefer to adhere to the container wall than to be in the aqueous solution. This specific type of nonselective adsorption has been described for plitidepsin only [75], despite the fact that this problem would seem to be more prevalent due to the hydrophobic nature of most of the compounds discussed here. A reason for this is that not much urine analysis has been executed in the first place. Eribulin is less hydrophobic than, for instance, plitidepsin, and for the UCN-01 assay, 0.5% polyoxyethylene (20) sorbitan monolaurate was added to the tubes to prevent adsorption from taking place. Beumer et al. also briefly mentioned the prevention of the adsorption of trabectedin to the container wall by adding bovine serum albumin to the tubes before urine collection [90]. Mostly, urine drug analysis is done to support the monitoring of renal excretion of the drug in (pre-)clinical studies [70,73,102] and more specifically to support human mass balance studies, in which excretory pathways are investigated [64,75,98,99,103].

3.2.4. Saliva

It is not uncommon to measure drug concentrations in saliva. It might be preferred over blood sampling, because it is non-invasive and collection is fairly easy, causing much less stress and discomfort [104]. Saliva analysis could, in fact, even be used for therapeutic drug monitoring (TDM); however, if concentrations are linked to plasma concentrations, then the risk for overestimation owing to preferential partitioning into the saliva is high [105], and sample volumes are usually small [100]. Moreover, oral contamination might occur and intra- and inter-patient variability in pH can impact quantification [104]. The only publication included here that quantifies drug levels in saliva is for UCN-01 analysis. In this case, it was believed that saliva concentrations reflected free plasma concentrations, and this unbound fraction is responsible for the pharmacological effect [82,100,106]. Thus, the method was developed not only for plasma but for saliva as well. Dees et al. investigated the relationship between saliva concentrations of UCN-01 and hypotension. Hypotension was the most severe adverse event that occurred during the clinical trial with the drug [59].

3.2.5. Feces

Quantification of drugs in feces could be considered for bioanalysis as part of a mass balance study where excretory pathways of a drug and its metabolites are investigated. However, the major concern is that feces composition varies greatly, and this may have an effect on the sample preparation recovery [107]. One manuscript included in the review demonstrates drug quantification in feces. The method was validated to support a human mass balance clinical trial in which the excretion of eribulin was investigated and hence included quantification in feces [64]. The mass balance study had shown that the majority of eribulin and eribulin-metabolites was excreted via the feces [99], meaning that the ability to quantify eribulin mesylate levels reliably had substantial added value. Potential problems were overcome by diluting the feces in water, in order to obtain samples with similar compositions [99].

3.2.6. Cells

Occasionally, drug concentrations are measured inside tissues or cells. The obvious advantage here is that drug concentrations are measured directly in the tumor and that drug distribution in the tumor can be determined, whereas plasma measurements are usually a surrogate for drug determination at the site of action. The drawbacks are that cells need to be isolated, counted and lysed, and tissues need to be homogenized even before sample pre-treatment. One publication describes the quantification of a drug in cells. In this case, it determined trabectedin concentrations in the liver cells and tumor cells originating from xenograft mice to investigate the tissue distribution of trabectedin [80]. The authors believed that the available data on plasma concentrations did not predict accurately the drug levels achievable in tumor or tissues, hence the reason for the development of this method.

3.3. Sample Preparation and Recovery

Before analysis, samples are usually pre-treated in order to extract the analyte from the matrix abundant with endogenous, interfering compounds. Moreover, because marine-derived drugs are administered at low doses, the analyte most likely needs to be concentrated as well. Sample preparation can be highly specific to extract one analyte of interest. However, when multi-drug assays are to be developed, the sample cleanup should not be too specific, otherwise some of the analytes of interest might be lost. Below is an overview of the sample pre-treatment methods applied to extract marine-derived drugs from biological samples.

3.3.1. Protein Precipitation

The sample pre-treatment method generally employed most often is protein precipitation (PP). Plasma and serum especially contain many proteins, which should be removed from the sample before analysis, because they can interfere with or contaminate analysis, also clogging the analytical column [108]. Precipitation of proteins is usually done with organic solvents, such as acetonitrile or methanol (or a mixture of the two), and often contains an acid, such as formic acid (FA) or trichloroacetic acid (TCA). In the included methods, the recovery of the analyte after PP is consistently high, with recoveries over 80% [69,78,79,82,83]. Recoveries above 100% have been reported and could occur for two reasons. First of all, it could be due to ion enhancement in the source of the mass spectrometer (MS) [64]. Secondly, it could be due to adsorption of the analyte to the equipment/materials in a sample free of matrix [78].

3.3.2. Liquid–Liquid Extraction

Liquid–liquid extraction (LLE) can occur when two immiscible solvents are used, resulting in two phases, an organic and an aqueous phase. The separation of the analyte from other (endogenous) compounds in the sample is based on the different distributions of these compounds in the two phases. The advantages of LLE include the low cost and easy application. A downside, however, is the fact that large volumes of organic solvents are often used. LLE is usually performed with ether (diisopropyl ether, *tert*-butyl methyl ether, or diethyl ether), ethyl acetate, chloroform, and *n*-butanol, possibly with pH modifications, and is therefore suitable for extraction of lipophilic compounds [109]. The next step is to evaporate the organic phase, because these solvents are not compatible with LC-MS/MS, at least when large injection volumes are used (this is generally not a problem if only 1 μL is injected onto the analytical column). This method has been applied for the lipophilic drugs elisidepsin, eribulin, midostaurin, plitidepsin, and zalypsis. In these cases, the analyte is removed from the aqueous plasma layer and partitioned into the organic layer. Whenever reported, the sample pre-treatment recovery was moderate to high (57–115%) with three exceptions: LLE with a mixture of ethyl acetate, methanol, and ethanol resulted in an extraction recovery of 33–45.3% for eribulin, whereas this was more than 60% for another method published using the same extraction solvent. The authors claim this to be due to matrix effects [63].

3.3.3. Solid Phase Extraction

Solid phase extraction (SPE) remains a popular cleanup method as well. The advantage is that it often results in very clean samples, because the SPE columns are very efficient at retaining endogenous, interfering substances. Despite this major advantage, SPE remains relatively laborious and costly, especially compared with PP and LLE. During SPE, the analyte is retained on a sorbent, after which matrix interferences are removed by a washing step. Then the analyte is eluted from the SPE column using an elution solvent. This sample pre-treatment method has been applied in four cases included in this review [71,74,76,77]. C18 cartridges and cyano cartridges were used for plitidepsin and trabectedin, respectively. Cleanup of samples containing trabectedin using SPE resulted in high recoveries (>87%) [76,77]. It resulted in lower recoveries for plitidepsin (>55%) [71,74]. However, it was found to be reproducible.

3.3.4. Supported Liquid Extraction

Supported liquid extraction (SLE) is a relatively new technique but analogous to LLE. The difference between the two is that the two immiscible solvents are not mixed together, but instead, they are immobilized on an inert support, and the organic phase is used to extract the analyte from the SLE column [110]. It has been used only for the sample cleanup of lurbinectedin [65]. Despite its presumed ability to have thorough cleanup of the samples, getting rid of most endogenous compounds, extraction recovery was rather low and very variable (data on file). In this case, a stable isotopically labeled (SIL) internal standard is recommended to ensure that the variable recoveries are sufficiently corrected for. Though, the signal-to-noise ratio was quite high, with values around 7 for the lower limit of quantification (LLOQ), meaning that SLE was able to thoroughly clean up these biological samples.

It is well known that phospholipids, proteins that are highly abundant in plasma, cause major ion suppression effects in mass spectrometric detection and thereby influence analysis [111]. LLE and the analogous SLE are efficient in extracting analytes from phospholipids, resulting in clean samples. During SPE, on the other hand, efficient phospholipid removal is not a guarantee. Instead, specific types of cartridges need to be used in order to separate the analyte from the phospholipids. These include, for instance, strong cation exchange sorbents or hybrid SPE [111].

3.3.5. Derivatization

Some sample pre-treatment methods involve a derivatization step in order to enable ultraviolet (UV) detection or fluorescence detection. The obvious drawback is that this extra step makes the method more laborious and complicated. Two methods have been described to quantify plitidepsin levels in biological fluids, by first derivatizing plitidepsin using trans-4-hydrazino-2-stilbazole to enable fluorescence detection [71,74]. With the advent of liquid chromatography coupled with mass spectrometry detectors, the use of this type of analysis has become less prevalent.

3.4. Analytical Methods

3.4.1. Chromatography

As it was a criterion for inclusion, all methods make use of high performance liquid chromatography (HPLC). During HPLC analysis, a sample is brought onto an analytical column, the stationary phase, and eluted from it using a suitable eluent, the mobile phase.

All manuscripts included in this review describe the use of reversed-phase (RP) liquid chromatography (LC). On RP columns, polar compounds elute before non-polar compounds. The most widespread used RP columns have regular C18 stationary phases, most notably Polaris C18, µBondapak RP-18, Hypersil-5 ODS, and the Symmetry C18 columns. A few assays report the use of other types of RP columns. For plitidepsin analysis, a Zorbax Bonus-RP column has been applied, which has C14 instead of C18 alkyl chains, with an amide linkage in between [72]. For lurbinectedin, an ACE pentafluorophenyl (PFP) column has been reported, which has a PFP group attached to the C18 alkyl chain [65]. RP analytical columns with phenyl chemistries have been utilized for the analysis of UCN-01 [82,83]. Unfortunately, justification has not always been given for those choices.

Because most compounds included in the review are hydrophobic, they elute from the analytical column with relatively high concentrations of organic mobile phases. The most widespread mobile phases used include water as the aqueous phase and acetonitrile or methanol as the organic phase. Modifiers include ammonium acetate, ammonium formate, formic acid, triethylamine, tetrahydrofuran, trifluoroacetic acid, and phosphoric acid. A few authors justify their choice of mobile phase.

For instance, Celli et al. have explained that the use of formic acid as the modifier has the advantage that the formation of sodium and potassium adducts of plitidepsin are reduced, and higher MS sensitivity is thereby obtained [70]. Moreover, trifluoroacetic acid could inhibit the ionization of the analyte and should therefore not be added to the mobile phase. The addition of ammonium acetate or ammonium formate would produce ammonium adducts and lower the sensitivity [70,73]. However, our group has reported that ammonium acetate added to the mobile phase could significantly improve peak shape in the plitidepsin assay [75]. Despite other groups reporting the disadvantage of the ammonium adduct formation for sensitive MS detection, this was not a problem in the latest assay, because sensitivity was more than sufficient [75].

Rosing et al. have also experimented with various modifiers in the assays for trabectedin quantification [77]. They found that ammonium acetate and acetic acid reduced MS electrospray sensitivities greatly. However, small amounts (5 mM) of ammonium acetate were added, because buffer capacity was needed to separate trabectedin from its internal standard. On the other hand, formic acid needed to be added as well, because this greatly improved sensitivity [77].

An isocratic system, where the mobile phase composition does not change throughout the analytical run, has been reported for midostaurin, plitidepsin, trabectedin, and UCN-01, for instance, by van Gijn et al. [66] and Illmer et al. [68] for the quantification of midostaurin. The first group quantified midostaurin and three potential metabolites [66]. When the method was adapted to include a fourth potential metabolite, isocratic elution seemed to be inappropriate [67]. Although isocratic elution should in theory lead to enhanced separation of analytes, gradient elution was applied, and it showed to be successful in separating the parent drug and all four metabolites. Moreover, triethylamine had a positive effect on the chromatography overall [67]. Kurata et al. stated that the addition of triethylamine to the isocratic mobile phase reduced peak tailing of UCN-01 [81].

It is surprising that none of the publications describe ultra-high-performance liquid chromatography (UHPLC). This type of chromatography has been available since 2004, and it has improved sensitivity, selectivity and speed owing to the smaller particle size (2 μm) [112].

3.4.2. Detection

UV Detection

UV detection is not commonly used for the analysis of marine-derived compounds; in only two reports is analyte detection by UV described [76,82]. The absorbance wavelengths for trabectedin and UCN-01 are 210 nm and 295 nm, respectively [77,82]. The lowest LLOQ achieved using UV detection was 1 ng/mL for trabectedin with sample volumes of 500 μL [76].

Fluorescence Detection

Fluorescence detection has been described for the detection of midostaurin, plitidepsin, and UCN-01 [66–68,71,74,81–83]. A compound needs to possess a chromophore to enable fluorescence detection, hence some analytes lacking this need to be derivatized before enabling the detection by fluorescence [71,74]. The lowest LLOQ achieved using fluorescence detection was 0.2 ng/mL with sample volumes of 50 μL (UCN-01) and 100 μL (midostaurin) [67,81].

Mass Spectrometry

Most recent articles make use of mass spectrometry (MS) detection, often consisting of a triple quadrupole detector after electrospray ionization (ESI) or atmospheric pressure chemical ionization (APCI). The reason for this shift is that MS is in general more sensitive than UV, and fluorescence detection has shown improved selectivity, and it has become more readily available.

The marine-derived drugs are commonly administered at very low doses, meaning that lower limits of quantification (LLOQs) in the nanograms and even pictograms per milliliter range are often required. This shift towards MS is seen for a number of compounds included in this review.

For instance, Rosing et al. have developed a method to quantify trabectedin in a linear range of 1–50 ng/mL [76]. An HPLC-UV method was validated first. However, after the first clinical trials in humans, the assay seemed to be lacking in sensitivity, hence a more sensitive HPLC-MS/MS method was developed [77]. A similar trend was seen for midostaurin [66–69]. The recent method for midostaurin quantification was developed for high-throughput analysis and, more specifically, for TDM-purposes [69]. Interestingly, no metabolites were taken into account for this assay, whereas they were included in the older assays. Some of these metabolites have been shown to be active, so it might have been appropriate to include the metabolites in the assay meant for TDM [113].

As just mentioned, MS provides superior sensitivity, hence it should be used in case of possible sensitivity problems. Sensitivity issues can arise for numerous reasons. For instance, Rosing et al. have emphasized the challenges in terms of sensitivity that occurs for those drugs that are administered at prolonged infusion schedules [77]. Zalypsis, elisidepsin, plitidepsin, and trabectedin have all been administered for over 24 h [50,58,114,115], and UCN-01 has even been infused to cancer patients for over 72 h [116]. This fact, together with the low dose and high volumes of distribution require the need for sensitive methods. Moreover, sensitivity issues have been described for plitidepsin besides the lengthy infusion times. Plitidepsin exists in two conformers (*cis* and *trans*), and the conformation equilibrium reaction is slow enough at room temperature for the two conformers to be separated as two chromatographic peaks. This is a well-known phenomenon for proline-containing peptides. In the case of plitidepsin, it is not desirable to distinguish between the two conformers, because they do not differ in clinical activity [117]. Moreover, lack of sensitivity can occur, because the MS signal is distributed over two chromatographic peaks, thereby reducing the sensitivity of the assay. This means that accurate quantification is especially challenging at low concentrations, and the assay needs to be sufficiently sensitive because of the low dose administered, long infusion times, and high volumes of distribution.

Interestingly, for the isoquinolines, the most pronounced ion observed in the mass spectrum corresponds to a loss of water from the molecule, a phenomenon that has not been described for any of the other compounds included in the review [65,77–80,84]. Allowing quantification based on the parent mass could therefore induce sensitivity problems as well.

The lowest LLOQ achieved using MS detection was 0.01 ng/mL with sample volumes of 200 µL (zalypsis) and 500 µL (trabectedin) [77,84], whereas the lowest LLOQ achieved by UV detection was 1 ng/mL trabectedin with a sample volume of 500 µL [76].

Still, there are a few disadvantages. First of all, MS is costlier than UV and still not available in every laboratory, and secondly, matrix effects could seriously hamper analysis. Correct quantification is based on the ionization of the analyte. If co-eluting endogenous compounds in the matrix interferes with ionization, thereby causing ion suppression or ion enhancement, accurate quantification cannot be guaranteed [118].

3.5. Quantification

3.5.1. Analytical Range

Assay sensitivity is very important, especially for those drugs that are administered at low doses and show large volumes of distribution. The analytical range should be appropriate for the therapeutic window or target concentrations to be reached, but the range should not be too broad either, because this could lead to non-linearity, which could impact the accuracy of the method. Ideally, at least 90% of the study samples should fall within the analytical range. Re-analysis is not preferred, especially if dilution integrity has not been tested for during method validation. Furthermore, samples with concentrations above the upper limit of quantification (ULOQ) can lead to carry-over problems, meaning that concentrations in the subsequent samples might be overestimated. Occasionally, adjustments to methods are necessary to accommodate for samples in clinical trials. For instance, an assay was developed to quantify trabectedin in human plasma. In a phase I trial with prolonged infusion of the drug, it appeared that the measured concentrations were lower than the anticipated analytical range for which the method was validated [76]. Hence, a new, more sensitive method was developed [77]. The analytical range was thus decreased from 1–50 ng/mL to 0.01–0.25 ng/mL.

The importance of the analytical range is also demonstrated for elisidepsin, eribulin mesylate, and UCN-01. An assay to quantify elisidepsin had a range of 0.05–100 ng/mL [62]. A phase I clinical trial published in 2016 reported C_{max} values of ~100 ng/mL, and samples could be measured up to 336 h, covering the complete pharmacokinetic profile [119]. The published method is therefore suitable for the analysis of samples obtained after dosing patients as reported in the phase I clinical study. Perhaps a few samples could be above the ULOQ and would have to be diluted prior to analysis. Dilution integrity was, however, not tested during validation [62].

Eribulin mesylate methods have been developed and validated for plasma, whole blood, urine, and feces, with analytical ranges of 0.2–100 ng/mL (plasma), 0.5–100 ng/mL (whole blood and urine), and 0.1–25 µg/mL (feces homogenates) [64]. The drug has been registered at a dosage of 1.4 mg/m^2 (equals 1.23 eribulin freebase), and the most recent pharmacokinetic data published report plasma C_{max} values ranging from 109 ng/mL (normal liver function) to 236 ng/mL (impaired liver function), meaning that not all samples could be measured by the published methods without the need for dilution [120]. The methods published did test for dilution integrity and appear appropriate [63,64].

An assay was developed for quantification of UCN-01 with an ULOQ of 100 ng/mL. The consequence was that samples collected in clinical trials [121–123] needed to be diluted 300 times prior to injection to ensure that concentrations fell within the calibration range [81,82]. A method published later had extended the analytical range up to 20 µg/mL, which means that even C_{max} samples from the phase I trials could be measured without dilution [82].

3.5.2. Internal Standard

It is customary to use an internal standard to correct for variability in the sample pre-treatment, as well as sample analysis and detection (including matrix effects, such as ion suppression or enhancement when MS detection is used), thereby improving accuracy and precision of quantitation [124,125]. Preferably, an internal standard has similar physiochemical properties as the analyte of interest. Typically, two types are commonly used: structural analogues, both for UV and MS detection, and stable isotopically labeled (SIL) internal standards for MS methods; 21 out of 23 manuscripts report the use of an internal standard, of which only eight MS methods out of 14 make use of a SIL internal standard. It is highly recommended to use SIL internal standards for bioanalytical methods to compensate for matrix effects. Occasionally, the internal standard is a compound that is not even remotely similar to the analyte, which was for instance the case when propyl-*p*-hydroxybenzoate was used as the internal standard in an assay to quantify trabectedin in plasma and *N*-phenyl-1-naphthylamine in a midostaurin assay [68,76]. Availability and costs are probably the most determining factors in choosing the internal standard. Nevertheless, it seems as if the internal standards used in these assays fulfilled their purpose, with excellent accuracy and precision values reported. Two publications do not report the use of an internal standard at all [71,74]. In the first, it was mentioned that an appropriate internal standard was not available, but nevertheless, the accuracy and precision of the assay was acceptable. Interferences were investigated, and it was concluded that there were none that could influence quantification of plitidepsin. Interestingly, peak height was used for the quantification of plitidepsin in the second method, whereas the peak area was used in the first [71,74].

The first study discussed in this review to use a SIL internal standard was in 2004 for the quantification of trabectedin in plasma [78]. Our group has investigated whether the use of a SIL internal standard was always superior over other types of internal standards [124]. We have looked into various assays that were modified in order to substitute the structural analogue with a SIL internal standard, and we have demonstrated this for a marine-derived anticancer compound kahalalide F as well [126]. We found that a structural analogue often leads to insufficient accuracy and precision and concluded that a SIL internal standard usually corrected best for variation compared with a structural analogue, because its structure is more similar and hence the compound behaves more similarly in terms of recovery, ionization, and stability. The latter issue was also demonstrated for the kahalalide F assay: putative stability of the analyte was significantly prolonged using the SIL internal standard [126].

When a SIL internal standard is used, generally ^{13}C or ^{15}N isotopes are preferred over ^{2}H-labeled internal standards, because these isotopes are more stable. Moreover, it has been shown that ^{2}H-labeled internal standards do not always co-elute with the analyte of interest. This is due to slight differences in physiochemical properties, because ^{2}H has stronger binding capacities than does hydrogen [124]. Co-elution is important, because it ensures similar matrix effects for internal standard and analyte and hence improved correction for variation [124]. Furthermore, differences in physiochemical properties can induce differences in extraction recovery due to exchange of ^{2}H atoms with hydrogen atoms.

If, still a structural analogue is used, it should not resemble or correspond to any metabolites that could have been formed in vivo [124]. For instance, Rosing et al. (1998) used ET-729, demethylated trabectedin, as an internal standard, but a later mass balance and metabolite profiling study has detected this compound in feces [127]. It could not be confirmed if the product was indeed a metabolite or a degradation product, but regardless, its use as an internal standard is generally undesirable. This was addressed, and it was concluded that no ET-729 was present in patient samples after trabectedin administration [77]. The internal standard used in bioanalytical methods to quantify eribulin also used a structural analogue, where the NH_2-group had been replaced by an OH group [63,64]. Technically, this could also result from in vivo biotransformations, but a mass balance study has revealed that it is not formed, hence the use of this internal standard was most likely appropriate [99]. The obvious advantage of using structural analogues is that they are often readily available and therefore much cheaper than a SIL internal standard, which requires dedicated synthesis.

Finally, it is worth mentioning that the importance of the internal standard's resemblance to the analyte depends on the sample pre-treatment as well. A simple dilution or PP method might require less resemblance of the internal standard to the analyte than a liquid–liquid extraction (LLE) or a solid-phase extraction (SPE) method [96]. However, as mentioned before, the internal standard needs to correct for matrix effects as well, which is a problem separate from extraction recovery. If it has been proven that matrix effect is not observed, a structural analogue might well be appropriate.

3.6. Metabolites

In some cases, quantification methods include drug metabolites as well, especially if metabolites have been known to be active and/or toxic. Midostaurin is biotransformed into active metabolites [113], and metabolites have been included in bioanalytical assays [66–68]. In these cases, one internal standard was used to correct for variation in analytical procedures for all analytes. As mentioned above, a recent method for midostaurin quantification was validated for high-throughput analysis for TDM, but none of the metabolites were quantified in this assay, whereas they were included in the older assays [69]. Some of these metabolites (CGP52421 and CGP62221) have been shown to be active, so it might have been interesting to include the metabolites in the assay [113]. Inclusion and quantification of metabolites in the assay might be valuable in metabolite profiling studies. However, this is only possible if the identities are already known, because reference standards should be available. When pre-clinical studies have demonstrated the formation of a particular major metabolite, it might be useful to include it in the validated assay. Additionally, an appropriate internal standard may be sought for the individual metabolites and parent compound. Often, the internal standard that is used to ensure accurate quantification of the parent drug is also used for the metabolite. Clearly, physiochemical properties are in this case not identical. For MS methods, matrix effects could be most effectively compensated for the quantification of the parent drug and its metabolites if the analytes co-elute with the SIL internal standard; however, the method should be sufficiently selective if the analytes are isomers, or if there is a risk that conjugates are re-converted into the parent drug in the source during ionization.

4. Conclusions

Various analytical methods have been described to quantify a range of anticancer drugs of marine origin, including liquid chromatography with ultraviolet detection (LC-UV), coupled with fluorescence detection (LC-FLD), and in combination with mass spectrometry detection (LC-MS). A total of 35 compounds, of which 5 have been approved, were included in the current review. A total of 19 of those (excluding the ADCs and cytarabine) were finally entered into the PubMed database to search for quantification methods. These were found for eight of the compounds included.

This review has selected many aspects of bioanalysis of the compounds isolated from sea-based organisms. The most used biomatrices include plasma and urine, and saliva, whole blood, feces and cells are occasionally analyzed as well. Sample cleanup is mostly done by PP, LLE, and SPE. The recent advances in this topic include the use of SLE, a method analogous to LLE. Samples are often concentrated during this step in the bioanalytical procedure. Some types of analysis might not be suitable for certain (groups of) drugs. For instance, we have seen that SPE has not been applied to the indocarbazoles and the marcrolides, and PP is most likely not useful for the extraction of depsipeptides from a biological sample. PP and LLE resulted in similar high recoveries of midostaurin, but only PP was applied to UCN-01 samples, also with good recoveries. The isoquinoline alkaloids seem to be extracted well from the matrices using LLE, SPE, and PP.

Chromatographic separation is necessary to isolate the analyte of interest from endogenous interferences. Reversed-phase chromatography seems to be suitable for elution of all analytes discussed herein. C18 columns are clearly the most widespread columns used. Owing to most of the analytes' hydrophobic properties, elution occurs at relatively high concentrations of organic solvents ($\geq$50%), which is advantageous for the sensitivity of MS methods.

Because this particular group of compounds has been the subject of clinical investigation for decades, the advances in bioanalysis are nicely demonstrated. We have seen a shift from LC-UV, a relatively cheap and widely available method, to LC-MS/MS, which has some advantages, most importantly, the superior sensitivity of LC-MS/MS compared with LC-UV. The sensitivity issues concerning the included compounds discussed in the present review include those arising from drugs administered at prolonged infusion times, which was the case for some of the depsipeptides, the isoquinolines, and one of the staurosporine derivatives, the existence of the two conformers that can occur for all proline-containing molecules, such as the depsipeptide plitidepsin, and the fact that the parent mass is not always the most pronounced ion observed in the mass spectrum, which has been described for the isoquinolines zalypsis, trabectedin, and lurbinectedin. Not only does the implementation of MS allow for the quantification of lower concentrations, but also it has the consequence that smaller sample volumes can be used for analysis. Although LLOQs as low as 1 ng/mL have been achieved using UV detection, sample volumes were quite large (500 μL). The superior sensitivity of LC-MS/MS compared with LC-UV also means that it is more suitable for metabolite quantification, because these are often present in circulation at even lower concentrations than the parent compound. Analytical run times are also reduced, due to the superior selectivity of LC-MS/MS. Internal standards are required to compensate for matrix effects resulting in improved accuracy and precision of the method, and the use of a SIL internal standard for MS bioanalysis is highly encouraged.

Continuous efforts are being made to discover, develop, and improve existing marine-derived drugs to treat cancer. Because ecological concerns are limiting new discoveries, more efforts are made in the laboratories, creating derivatives and analogues of the proven effective compounds discussed here. When these newly developed compounds show resemblance to the original compounds, the existing methods could be a basis for the newly validated assays. Depending on the drug properties, suitable cleanup, separation, and detection methods need to be developed and applied. The availability of resources and instrumentation is most likely the most common bottleneck: sometimes one can only work with what has been given.

Overall, this review provides information about the recent developments in bioanalysis of marine-derived anticancer compounds and can serve as a guide towards fast development and validation of new methods to quantify new marine-derived anticancer drugs.

Author Contributions: Conceptualization, J.H.B. and L.v.A.; Methodology, J.H.B. and L.v.A.; Writing—Original Draft Preparation, L.v.A.; Writing—Review & Editing, H.R., J.H.B. and J.H.M.S; Supervision, H.R., J.H.B. and J.H.M.S.

Conflicts of Interest: The authors declare no conflict of interest.

References

1. Indumathy, S.; Dass, C.R. Finding chemo: The search for marine-based pharmaceutical drugs active against cancer. *J. Pharm. Pharmacol.* **2013**, *65*, 1280–1301. [CrossRef] [PubMed]

2. Mudit, M.; El Sayed, K.A. Cancer control potential of marine natural product scaffolds through inhibition of tumor cell migration and invasion. *Drug Discov. Today* **2016**, *21*, 1745–1760. [CrossRef] [PubMed]

3. Hassan, S.S.U.; Anjum, K.; Abbas, S.Q.; Akhter, N.; Shagufta, B.I.; Shah, S.A.A.; Tasneem, U. Emerging biopharmaceuticals from marine actinobacteria. *Environ. Toxicol. Pharmacol.* **2017**, *49*, 34–47. [CrossRef] [PubMed]

4. Newman, D.J.; Cragg, G.M. Advanced preclinical and clinical trials of natural products and related compounds from marine sources. *Curr. Med. Chem.* **2004**, *11*, 1693–1713. [CrossRef] [PubMed]

5. Haefner, B. Drugs from the deep: Marine natural products as drug candidates. *Drug Discov. Today* **2003**, *8*, 536–544. [CrossRef]

6. Gerwick, W.H.; Moore, B.S. Lessons from the Past and Charting the Future of Marine Natural Products Drug Discovery and Chemical Biology. *Chem. Biol.* **2012**, *19*, 85–98. [CrossRef] [PubMed]

7. Suarez-Jimenez, G.M.; Burgos-Hernandez, A.; Ezquerra-Brauer, J.M. Bioactive peptides and depsipeptides with anticancer potential: Sources from marine animals. *Mar. Drugs* **2012**, *10*, 963–986. [CrossRef] [PubMed]

8. Ruiz-Torres, V.; Encinar, J.A.; Herranz-López, M.; Pérez-Sánchez, A.; Galiano, V.; Barrajón-Catalán, E.; Micol, V. An updated review on marine anticancer compounds: The use of virtual screening for the discovery of small-molecule cancer drugs. *Molecules* **2017**, *22*, 1037. [CrossRef] [PubMed]

9. Gomes, N.G.M.; Dasari, R.; Chandra, S.; Kiss, R.; Kornienko, A. Marine invertebrate metabolites with anticancer activities: Solutions to the "supply problem". *Mar. Drugs* **2016**, *14*, 98. [CrossRef] [PubMed]

10. Ye, J.; Zhou, F.; Al-Kareef, A.M.Q.; Wang, H. Anticancer agents from marine sponges. *J. Asian Nat. Prod. Res.* **2015**, *17*, 64–88. [CrossRef] [PubMed]

11. Desbois, A.P. How might we increase success in marine-based drug discovery? *Expert Opin. Drug Discov.* **2014**, *9*, 985–990. [CrossRef] [PubMed]

12. Martins, A.; Vieira, H.; Gaspar, H.; Santos, S. Marketed marine natural products in the pharmaceutical and cosmeceutical industries: Tips for success. *Mar. Drugs* **2014**, *12*, 1066–1101. [CrossRef] [PubMed]

13. Pelay-Gimeno, M.; García-Ramos, Y.; Jesús Martin, M.; Spengler, J.; Molina-Guijarro, J.M.; Munt, S.; Francesch, A.M.; Cuevas, C.; Tulla-Puche, J.; Albericio, F. The first total synthesis of the cyclodepsipeptide pipecolidepsin A. *Nat. Commun.* **2013**, *4*, 1–10. [CrossRef] [PubMed]

14. Mitsiades, C.S.; Ocio, E.M.; Pandiella, A.; Maiso, P.; Gajate, C.; Garayoa, M.; Vilanova, D.; Montero, J.C.; Mitsiades, N.; McMullan, C.J.; et al. Aplidin, a marine organism-derived compound with potent antimyeloma activity in vitro and in vivo. *Cancer Res.* **2008**, *68*, 5216–5225. [CrossRef] [PubMed]

15. Newman, D.J.; Cragg, G.M. Drugs and drug candidates from marine sources: An assessment of the current "state of play". *Planta Med.* **2016**, *82*, 775–789. [CrossRef] [PubMed]

16. Palanisamy, S.K.; Rajendran, N.M.; Marino, A. Natural products diversity of marine ascidians (tunicates; ascidiacea) and successful drugs in clinical development. *Nat. Prod. Bioprospect.* **2017**, *7*, 1–111. [CrossRef] [PubMed]

17. Jansen, R.S.; Rosing, H.; Schellens, J.H.M.; Beijnen, J.H. Mass spectrometry in the quantitative analysis of therapeutic intracellular nucleotide analogs. *Mass Spectrom. Rev.* **2011**, *30*, 321–343. [CrossRef] [PubMed]

18.	Astellas R&D Pipeline (As of January 2017). Available online: https://www.astellas.com/en/ir/library/pdf/3q2017_rd_en.pdf (accessed on 20 November 2017).

19.	European Medicines Agency Summary of Product Characteristics Adcetris. Available online: http://www.ema.europa.eu/docs/en_GB/document_library/EPAR_-_Product_Information/human/002455/WC500135055.pdf (accessed on 27 November 2017).

20.	European Medicines Agency Summary of Product Characteristics DepoCyte. Available online: http://www.ema.europa.eu/docs/en_GB/document_library/EPAR_-_Product_Information/human/000317/WC500035649.pdf (accessed on 27 November 2017).

21.	Seattle Genetics Denintuzumab Mafodotin. Available online: http://www.seattlegenetics.com/pipeline/denintuzumab-mafodotin (accessed on 20 November 2017).

22.	Abbvie Depatuxizumab Mafodotin (ABT-414). Available online: https://www.abbvie.com/our-science/pipeline/depatuxizumab-mafodotin.html (accessed on 20 November 2017).

23.	Seattle Genetics Enfortumab Vedotin. Available online: http://www.seattlegenetics.com/pipeline/enfortumab-vedotin (accessed on 20 November 2017).

24.	European Medicines Agency Summary of Product Characteristics Halaven. Available online: http://www.ema.europa.eu/docs/en_GB/document_library/EPAR_-_Product_Information/human/002084/WC500105112.pdf (accessed on 27 November 2017).

25.	Celldex Therapeutics Glembatumumab Vedotin—Antibody-Drug Conjugate Targeting gpNMB in Metastatic Breast Cancer and Metastatic Melanoma. Available online: http://www.celldex.com/pipeline/cdx-011.php (accessed on 20 November 2017).

26.	GSK Product Pipeline. Available online: https://www.gsk.com/en-gb/investors/product-pipeline/ (accessed on 12 April 2017).

27.	PharmaMar Oncology Pipeline. Available online: https://www.pharmamar.com/science-and-innovation/oncology-pipeline/ (accessed on 20 November 2017).

28.	Triphase Marizomib. Available online: http://triphaseco.com/marizomib/ (accessed on 20 November 2017).

29.	Gescher, A. Staurosporine analogues—Pharmacological toys or useful antitumour agents? *Crit. Rev. Oncol. Hematol.* **2000**, *34*, 127–135. [CrossRef]

30.	European Medicines Agency Summary of Product Characteristics Rydapt. Available online: http://www.ema.europa.eu/docs/en_GB/document_library/EPAR_-_Product_Information/human/004095/WC500237581.pdf (accessed on 27 November 2017).

31.	BeyondSpring Broadening Our Developmental Pipeline. Available online: http://www.beyondspringpharma.com/en/pipeline/ (accessed on 20 November 2017).

32.	Genentech Pipeline. Available online: https://www.gene.com/medical-professionals/pipeline (accessed on 20 November 2017).

33.	Genmab Tisotumab Vedotin. Available online: http://www.genmab.com/product-pipeline/products-in-development/humax-tf-adc (accessed on 20 November 2017).

34.	European Medicines Agency Summary of Product Characteristics Yondelis. Available online: http://www.ema.europa.eu/docs/en_GB/document_library/EPAR_-_Product_Information/human/000773/WC500045832.pdf (accessed on 27 November 2017).

35.	Astellas R&D Pipeline (As of October 2017). Available online: https://www.astellas.com/en/system/files/2q2018_rd_en.pdf (accessed on 20 November 2017).

36.	Merchant, J.; Tutsch, K.; Dresen, A.; Arzoomanian, R.; Alberti, D.; Feierabend, C.; Binger, K.; Marnoccha, R.; Thomas, J.; Cleary, J.; et al. Phase I clinical and pharmacokinetic study of NSC 655649, a rebeccamycin analogue, given in both single-dose and multiple-dose formats. *Clin. Cancer Res.* **2002**, *8*, 2193–2201. [PubMed]

37.	Exelixis Helsinn to Discontinue Becatecarin Trial Program. Available online: https://www.prnewswire.com/news-releases/helsinn-to-discontinue-becatecarin-trial-program-56441722.html (accessed on 26 October 2017).

38.	Stewart, J.A.; Low, J.B.; Roberts, J.D.; Blow, A. A phase I clinical trial of didemnin B. *Cancer* **1991**, *68*, 2550–2554. [CrossRef]

39.	Newman, D.J.; Cragg, G.M. Marine-sourced anti-cancer and cancer pain control agents in clinical and late preclinical development. *Mar. Drugs* **2014**, *12*, 255–278. [CrossRef] [PubMed]

40.	Kitagaki, J.; Shi, G.; Miyauchi, S.; Murakami, S.; Yang, Y. Cyclic depsipeptides as potential cancer therapeutics. *Anti-Cancer Drugs* **2015**, *26*, 259–271. [CrossRef] [PubMed]

41. Williams, R. Discontinued drugs in 2012: Oncology drugs. *Expert Opin. Investig. Drugs* **2013**, *22*, 1627–1644. [CrossRef] [PubMed]

42. Williams, R. Discontinued in 2013: Oncology drugs. *Expert Opin. Investig. Drugs* **2015**, *24*, 95–110. [CrossRef] [PubMed]

43. Almhanna, K.; Miron, M.L.L.; Wright, D.; Gracian, A.C.; Hubner, R.A.; Van Laethem, J.L.; López, C.M.; Alsina, M.; Muñoz, F.L.; Bendell, J.; et al. Phase II study of the antibody-drug conjugate TAK-264 (MLN0264) in patients with metastatic or recurrent adenocarcinoma of the stomach or gastroesophageal junction expressing guanylyl cyclase C. *Investig. New Drugs* **2017**, *35*, 235–241. [CrossRef] [PubMed]

44. InPress Media Group LLC Lifastuzumab Vedotin is Well-tolerated + Improves Objective Response Rate in Phase II Trial in Platinum-resistant Ovarian Cancer. Available online: https://adcreview.com/tag/lifastuzumab-vedotin/ (accessed on 13 April 2018).

45. Martin, M.J.; Rodriguez-Acebes, R.; Garcia-Ramos, Y.; Martinez, V.; Murcia, C.; Digon, I.; Marco, I.; Pelay-Gimeno, M.; Fernández, R.; Reyes, F.; et al. Stellatolides, a new cyclodepsipeptide family from the sponge *Ecionemia acervus*: Isolation, solid-phase total synthesis, and full structural assignment of stellatolide A. *J. Am. Chem. Soc.* **2014**, *136*, 6754–6762. [CrossRef] [PubMed]

46. Tolcher, A.W.; Eckhardt, S.G.; Kuhn, J.; Hammond, L.; Weiss, G.; Rizzo, J.; Aylesworth, C.; Hidalgo, M.; Patnaik, A.; Schwartz, G.; et al. Phase I and pharmacokinetic study of NSC 655649, a rebeccamycin analog with topoisomerase inhibitory properties. *J. Clin. Oncol.* **2001**, *19*, 2937–2947. [CrossRef] [PubMed]

47. Benvenuto, J.A.; Newman, R.A.; Bignami, G.S.; Raybould, T.J.G.; Raber, M.N.; Esparza, L.; Walters, R.S. Phase II clinical and pharmacological study of didemnin B in patients with metastatic breast cancer. *Investig. New Drugs* **1992**, *10*, 113–117. [CrossRef]

48. Yamada, Y.; Tamura, T.; Yamamoto, N.; Shimoyama, T.; Ueda, Y.; Murakami, H.; Kusaba, H.; Kamiya, Y.; Saka, H.; Tanigawara, Y.; et al. Phase I and pharmacokinetic study of edotecarin, a novel topoisomerase I inhibitor, administered once every 3 weeks in patients with solid tumors. *Cancer Chemother. Pharmacol.* **2006**, *58*, 173–182. [CrossRef] [PubMed]

49. Saif, M.W.; Sellers, S.; Diasio, R.B.; Douillard, J.-Y. A phase I dose-escalation study of edotecarin (J-107088) combined with infusional 5-fluorouracil and leucovorin in patients with advanced/metastatic solid tumors. *Anti-Cancer Drugs* **2010**, *21*, 716–723. [CrossRef] [PubMed]

50. Salazar, R.; Jones, R.J.; Oaknin, A.; Crawford, D.; Cuadra, C.; Hopkins, C.; Gil, M.; Coronado, C.; Soto-Matos, A.; Cullell-Young, M.; et al. A phase I and pharmacokinetic study of elisidepsin (PM02734) in patients with advanced solid tumors. *Cancer Chemother. Pharmacol.* **2012**, *70*, 673–681. [CrossRef] [PubMed]

51. Ratain, M.J.; Geary, D.; Undevia, S.D.; Coronado, C.; Alfaro, V.; Iglesias, J.L.; Schilsky, R.L.; Miguel-Lillo, B. First-in-human, phase I study of elisidepsin (PM02734) administered as a 30-min or as a 3-hour intravenous infusion every three weeks in patients with advanced solid tumors. *Investig. New Drugs* **2015**, *33*, 901–910. [CrossRef] [PubMed]

52. Welch, P.A.; Sinha, V.P.; Cleverly, A.L.; Darstein, C.; Flanagan, S.D.; Musib, L.C. Safety, tolerability, QTc evaluation, and pharmacokinetics of single and multiple doses of enzastaurin HCl (LY317615), a protein kinase C-beta inhibitor, in healthy subjects. *J. Clin. Pharmacol.* **2007**, *47*, 1138–1151. [CrossRef] [PubMed]

53. Millward, M.; Price, T.; Townsend, A.; Sweeney, C.; Spencer, A.; Sukumaran, S.; Longenecker, A.; Lee, L.; Lay, A.; Sharma, G.; et al. Phase I clinical trial of the novel proteasome inhibitor marizomib with the histone deacetylase inhibitor vorinostat in patients with melanoma, pancreatic and lung cancer based on in vitro assessments of the combination. *Investig. New Drugs* **2012**, *30*, 2303–2317. [CrossRef] [PubMed]

54. Mita, M.M.; Spear, M.A.; Yee, L.K.; Mita, A.C.; Heath, E.I.; Papadopoulos, K.P.; Federico, K.C.; Reich, S.D.; Romero, O.; Malburg, L.; et al. Phase I first-in-human trial of the vascular disrupting agent plinabulin (NPI-2358) in patients with solid tumors or lymphomas. *Clin. Cancer Res.* **2010**, *16*, 5892–5899. [CrossRef] [PubMed]

55. Nalda-Molina, R.; Valenzuela, B.; Ramon-Lopez, A.; Miguel-Lillo, B.; Soto-Matos, A.; Perez-Ruixo, J.J. Population pharmacokinetics meta-analysis of plitidepsin (Aplidin) in cancer subjects. *Cancer Chemother. Pharmacol.* **2009**, *64*, 97–108. [CrossRef] [PubMed]

56. Jimeno, A.; Sharma, M.R.; Szyldergemajn, S.; Gore, L.; Geary, D.; Diamond, J.R.; Fernandez Teruel, C.; Soto Matos-Pita, A.; Iglesias, J.L.; Cullell-Young, M.; et al. Phase I study of lurbinectedin, a synthetic tetrahydroisoquinoline that inhibits activated transcription, induces DNA single- and double-strand breaks, on a weekly× 2 every-3-week schedule. *Investig. New Drugs* **2017**, 1–7. [CrossRef] [PubMed]

57. Paz-Ares, L.; Forster, M.; Boni, V.; Szyldergemajn, S.; Corral, J.; Turnbull, S.; Cubillo, A.; Teruel, C.F.; Calderero, I.L.; Siguero, M.; et al. Phase I clinical and pharmacokinetic study of PM01183 (a tetrahydroisoquinoline, Lurbinectedin) in combination with gemcitabine in patients with advanced solid tumors. *Investig. New Drugs* **2017**, *35*, 198–206. [CrossRef] [PubMed]

58. Capdevila, J.; Clive, S.; Casado, E.; Michie, C.; Piera, A.; Sicart, E.; Carreras, M.J.; Coronado, C.; Kahatt, C.; Soto Matos-Pita, A.; et al. A phase I pharmacokinetic study of PM00104 (Zalypsis®) administered as a 24-h intravenous infusion every 3 weeks in patients with advanced solid tumors. *Cancer Chemother. Pharmacol.* **2013**, *71*, 1247–1254. [CrossRef] [PubMed]

59. Dees, E.C.; Baker, S.D.; O'Reilly, S.; Rudek, M.A.; Davidson, S.B.; Aylesworth, C.; Elza-Brown, K.; Carducci, M.A.; Donehower, R.C. A phase I and pharmacokinetic study of short infusions of UCN-01 in patients with refractory solid tumors. *Clin. Cancer Res.* **2005**, *11*, 664–671. [PubMed]

60. Jimeno, A.; Rudek, M.A.; Purcell, T.; Laheru, D.A.; Messersmith, W.A.; Dancey, J.; Carducci, M.A.; Baker, S.D.; Hidalgo, M.; Donehower, R.C. Phase I and pharmacokinetic study of UCN-01 in combination with irinotecan in patients with solid tumors. *Cancer Chemother. Pharmacol.* **2008**, *61*, 423–433. [CrossRef] [PubMed]

61. Undevia, S.D.; Vogelzang, N.J.; Mauer, A.M.; Janisch, L.; Mani, S.; Ratain, M.J. Phase I clinical trial of CEP-2563 dihydrochloride, a receptor tyrosine kinase inhibitor, in patients with refractory solid tumors. *Investig. New Drugs* **2004**, *22*, 449–458. [CrossRef] [PubMed]

62. Yin, J.; Avilés, P.; Lee, W.; Ly, C.; Guillen, M.J.; Munt, S.; Cuevas, C.; Faircloth, G. Development of a liquid chromatography/tandem mass spectrometry assay for the quantification of PM02734, a novel antineoplastic agent, in dog plasma. *Rapid Commun. Mass Spectrom.* **2006**, *20*, 2535–2740. [CrossRef] [PubMed]

63. DesJardins, C.; Saxton, P.; Lu, S.X.; Li, X.; Rowbottom, C.; Wong, Y.N. A high-performance liquid chromatography-tandem mass spectrometry method for the clinical combination study of carboplatin and anti-tumor agent eribulin mesylate (E7389) in human plasma. *J. Chromatogr. B Anal. Technol. Biomed. Life Sci.* **2008**, *875*, 373–382. [CrossRef] [PubMed]

64. Dubbelman, A.C.; Rosing, H.; Thijssen, B.; Lucas, L.; Copalu, W.; Wanders, J.; Schellens, J.H.M.; Beijnen, J.H. Validation of high-performance liquid chromatography-tandem mass spectrometry assays for the quantification of eribulin (E7389) in various biological matrices. *J. Chromatogr. B Anal. Technol. Biomed. Life Sci.* **2011**, *879*, 1149–1155. [CrossRef] [PubMed]

65. Pernice, T.; Bishop, A.G.; Guillen, M.J.; Cuevas, C.; Avilés, P. Development of a liquid chromatography/tandem mass spectrometry assay for the quantification of PM01183 (Lurbinectedin), a novel antineoplastic agent, in mouse, rat, dog, Cynomolgus monkey and mini-pig plasma. *J. Pharm. Biomed. Anal.* **2016**, *123*, 37–41. [CrossRef] [PubMed]

66. Van Gijn, R.; van Tellingen, O.; de Clippeleir, J.J.M.; Hillebrand, M.J.X.; Boven, E.; Vermorken, J.B.; ten Bokkel Huinink, W.W.; Schwertz, S.; Graf, P.; Beijnen, J.H. Analytical procedure for the determination of the new antitumour drug N-benzoylstaurosporine and three potential metabolites in human plasma by reversed-phase high-performance liquid chromatography. *J. Chromatogr. B Biomed. Sci. Appl.* **1995**, *667*, 269–276. [CrossRef]

67. Van Gijn, R.; Havik, E.; Boven, E.; Vermorken, J.B.; ten Bokkel Huinink, W.W.; van Tellingen, O.; Beijnen, J.H. High-performance liquid chromatographic analysis of the new four potential metabolites in micro-volumes of plasma. *J. Pharm. Biomed. Anal.* **1995**, *14*, 165–174. [CrossRef]

68. Illmer, T.; Thiede, H.-M.; Thiede, C.; Bornhauser, M.; Schaich, M.; Schleyer, E.; Ehninger, G. A highly sensitive method for the detection of PKC412 (CGP41251) and its metabolites by high-performance liquid chromatography. *J. Pharmacol. Toxicol. Methods* **2007**, *56*, 23–27. [CrossRef] [PubMed]

69. Bourget, P.; Amin, A.; Chandesris, M.-O.; Vidal, F.; Merlette, C.; Hirsch, I.; Cabaret, L.; Carvalhosa, A.; Mogenet, A.; Frenzel, L.; et al. Liquid chromatography–tandem mass spectrometry assay for therapeutic drug monitoring of the tyrosine kinase inhibitor, midostaurin, in plasma from patients with advanced systemic mastocytosis. *J. Chromatogr. B* **2014**, *944*, 175–181. [CrossRef] [PubMed]

70. Celli, N.; Gallardo, A.M.; Rossi, C.; Zucchetti, M.; D'Incalci, M.; Rotilio, D. Analysis of aplidine (dehydrodidemnin B), a new marine-derived depsipeptide, in rat biological fluids by liquid chromatography-tandem mass spectrometry. *J. Chromatogr. B Biomed. Sci. Appl.* **1999**, *731*, 335–343. [CrossRef]

71. Sparidans, R.W.; Kettenes-Van Den Bosch, J.J.; Van Tellingen, O.; Nuijen, B.; Henrar, R.E.C.; Jimeno, J.M.; Faircloth, G.; Floriano, P.; Rinehart, K.L.; Beijnen, J.H. Bioanalysis of aplidine, a new marine antitumoral depsipeptide, in plasma by high-performance liquid chromatography after derivatization with trans-4′-hydrazino-2-stilbazole. *J. Chromatogr. B Biomed. Sci. Appl.* **1999**, *729*, 43–53. [CrossRef]

72. Yin, J.; Avilés, P.; Lee, W.; Ly, C.; Floriano, P.; Ignacio, M.; Faircloth, G. Development of a liquid chromatography/tandem mass spectrometry assay for the quantification of Aplidin, a novel marine-derived antineoplastic agent, in human plasma. *Rapid Commun. Mass Spectrom.* **2003**, *17*, 1909–1914. [CrossRef] [PubMed]

73. Celli, N.; Mariani, B.; Di Carlo, F.; Zucchetti, M.; Lopez-Lazaro, L.; D'Incalci, M.; Rotilio, D. Determination of Aplidin®, a marine-derived anticancer drug, in human plasma, whole blood and urine by liquid chromatography with electrospray ionisation tandem mass spectrometric detection. *J. Pharm. Biomed. Anal.* **2004**, *34*, 619–630. [CrossRef]

74. Sparidans, R.W.; Schellens, J.H.M.; López-Lázaro, L.; Jimeno, J.M.; Beijnen, J.H. Liquid chromatographic assay for the cyclic depsipeptide aplidine, a new marine antitumor drug, in whole blood using derivatization with trans-4′-hydrazino-2-stilbazole. *Biomed. Chromatogr.* **2004**, *18*, 16–20. [CrossRef] [PubMed]

75. Van Andel, L.; Rosing, H.; Fudio, S.; Avilés, P.; Tibben, M.M.; Gebretensae, A.; Schellens, J.H.M.; Beijnen, J.H. Liquid chromatography-tandem mass spectrometry assay to quantify plitidepsin in human plasma, whole blood and urine. *J. Pharm. Biomed. Anal.* **2017**, *145*, 137–143. [CrossRef] [PubMed]

76. Rosing, H.; Hillebrand, M.J.X.; Jimeno, J.M.; Gómez, A.; Floriano, P.; Faircloth, G.; Cameron, L.; Henrar, R.E.C.; Vermorken, J.B.; Bult, A.; et al. Analysis of ecteinascidin 743, a new potent marine-derived anticancer drug, in human plasma by high-performance liquid chromatography in combination with solid-phase extraction. *J. Chromatogr. B Biomed. Appl.* **1998**, *710*, 183–189. [CrossRef]

77. Rosing, H.; Hillebrand, M.J.X.; Jimeno, J.M.; Gómez, A.; Floriano, P.; Faircloth, G.; Henrar, R.E.C.; Vermorken, J.B.; Cvitkovic, E.; Bult, A.; et al. Quantitative determination of Ecteinascidin 743 in human plasma by miniaturized high-performance liquid chromatography coupled with electrospray ionization tandem mass spectrometry. *J. Mass Spectrom.* **1998**, *33*, 1134–1140. [CrossRef]

78. Stokvis, E.; Rosing, H.; López-Lázaro, L.; Beijnen, J.H. Simple and sensitive liquid chromatographic quantitative analysis of the novel marine anticancer drug Yondelis™ (ET-743, trabectedin) in human plasma using column switching and tandem mass spectrometric detection. *J. Mass Spectrom.* **2004**, *39*, 431–436. [CrossRef] [PubMed]

79. Zangarini, M.; Ceriani, L.; Sala, F.; Marangon, E.; Bagnati, R.; D'Incalci, M.; Grosso, F.; Zucchetti, M. Quantification of trabectedin in human plasma: Validation of a high-performance liquid chromatography–mass spectrometry method and its application in a clinical pharmacokinetic study. *J. Pharm. Biomed. Anal. J. Pharm. Biomed.* **2014**, *95*, 107–112. [CrossRef] [PubMed]

80. Ceriani, L.; Ferrari, M.; Zangarini, M.; Licandro, S.A.; Bello, E.; Frapolli, R.; Falcetta, F.; D'Incalci, M.; Libener, R.; Grosso, F.; et al. HPLC–MS/MS method to measure trabectedin in tumors: Preliminary PK study in a mesothelioma xenograft model. *Bioanalysis* **2015**, *7*, 1831–1842. [CrossRef] [PubMed]

81. Kurata, N.; Kuramitsu, T.; Tanii, H.; Fuse, E.; Kuwabara, T.; Kobayashi, H.; Kobayashi, S. Development of a highly sensitive high-performance liquid chromatographic method for measuring an anticancer drug, UCN-01, in human plasma or urine. *J. Chromatogr. B* **1998**, *708*, 223–227. [CrossRef]

82. Bauer, K.S.; Lush, R.M.; Rudek, M.A.; Shih, C.; Sausville, E.; Figg, W.D. A high-performance liquid chromatography method using ultraviolet and fluorescence detection for the quantitation of UCN-01, 7-hydroxystaurosporine, from human plasma and saliva. *Biomed. Chromatogr.* **2000**, *14*, 338–343. [CrossRef]

83. Smith, J.A.; Cortes, J.; Newman, R.A.; Madden, T.L. Development of a simplified, sensitive high-performance liquid chromatographic method using fluorescence detection to determine the concentration of UCN-01 in human plasma. *J. Chromatogr. B Biomed. Sci. Appl.* **2001**, *760*, 247–253. [CrossRef]

84. Yin, J.; Avilés, P.; Lee, W.; Ly, C.; Guillen, M.J.; Munt, S.; Cuevas, C.; Faircloth, G. Development of a liquid chromatography/tandem mass spectrometry assay for the quantification of PM00104, a novel antineoplastic agent, in mouse, rat, dog, and human plasma. *Rapid Commun. Mass Spectrom.* **2005**, *19*, 689–695. [CrossRef] [PubMed]

85. Russo, P.; Del Bufalo, A.; Fini, M. Deep sea as a source of novel-anticancer drugs: Update on discovery and preclinical/clinical evaluaton in a systems medicine perspective. *EXCLI J.* **2015**, *14*, 228–236. [CrossRef] [PubMed]

86. Yamazaki, Y.; Tanaka, K.; Nicholson, B.; Deyanat-Yazdi, G.; Potts, B.; Yoshida, T.; Oda, A.; Kitagawa, T.; Orikasa, S.; Kiso, Y.; et al. Synthesis and structure-activity relationship study of antimicrotubule agents phenylahistin derivatives with a didehydropiperazine-2,5-dione structure. *J. Med. Chem.* **2012**, *55*, 1056–1071. [CrossRef] [PubMed]

87. Bourhill, T.; Narendran, A.; Johnston, R.N. Enzastaurin: A lesson in drug development. *Crit. Rev. Oncol. Hematol.* **2017**, *112*, 72–79. [CrossRef] [PubMed]

88. Hexner, E.; Roboz, G.; Hoffman, R.; Luger, S.; Mascarenhas, J.; Carroll, M.; Clementi, R.; Bensen-Kennedy, D.; Moliterno, A. Open-label study of oral CEP-701 (lestaurtinib) in patients with polycythaemia vera or essential thrombocythaemia with JAK2-V617F mutation. *Br. J. Haematol.* **2014**, *164*, 83–93. [CrossRef] [PubMed]

89. Berg, S.L.; Aleksic, A.; McGuffey, L.; Dauser, R.; Nuchtern, J.; Bernacky, B.; Blaney, S.M. Plasma and cerebrospinal fluid pharmacokinetics of rebeccamycin (NSC 655649) in nonhuman primates. *Cancer Chemother. Pharmacol.* **2004**, *54*, 127–130. [CrossRef] [PubMed]

90. Beumer, J.H.; Rademaker-Lakhai, J.M.; Rosing, H.; Lopez-Lazaro, L.; Beijnen, J.H.; Schellens, J.H.M. Trabectedin (Yondelis, formerly ET-743), a mass balance study in patients with advanced cancer. *Investig. New Drugs* **2005**, *23*, 429–436. [CrossRef] [PubMed]

91. Nakano, H.; Omura, S. Chemical biology of natural indolocarbazole products: 30 Years since the discovery of staurosporine. *J. Antibiot.* **2009**, *62*, 17–26. [CrossRef] [PubMed]

92. Anizon, F.; Moreau, P.; Sancelme, M.; Laine, W.; Bailly, C.; Prudhomme, M. Rebeccamycin analogues bearing amine substituents or other groups on the sugar moiety. *Bioorg. Med. Chem.* **2003**, *11*, 3709–3722. [CrossRef]

93. Marshall, J.L.; Kindler, H.; Deeken, J.; Bhargava, P.; Vogelzang, N.J.; Rizvi, N.; Luhtala, T.; Boylan, S.; Dordal, M.; Robertson, P.; et al. Phase I trial of orally administered CEP-701, a novel neurotrophin receptor-linked tyrosine kinase inhibitor. *Investig. New Drugs* **2005**, *23*, 31–37. [CrossRef] [PubMed]

94. Kreisl, T.N.; Kim, L.; Moore, K.; Duic, P.; Kotliarova, S.; Walling, J.; Musib, L.; Thornton, D.; Albert, P.S.; Fine, H.A. A phase I trial of enzastaurin in patients with recurrent gliomas. *Clin. Cancer Res.* **2009**, *15*, 3617–3623. [CrossRef] [PubMed]

95. Kreisl, T.N.; Kotliarova, S.; Butman, J.A.; Albert, P.S.; Kim, L.; Musib, L.; Thornton, D.; Fine, H.A. A phase I/II trial of enzastaurin in patients with recurrent high-grade gliomas. *Neuro Oncol.* **2010**, *12*, 181–189. [CrossRef] [PubMed]

96. *Handbook of LC-MS Bioanalysis: Best Practices, Experimental Protocols, and Regulations*; Li, W.; Zhang, J.; Tse, F.L.S. (Eds.) John Wiley & Sons Inc.: Hoboken, NJ, USA, 2013; ISBN 9781118671276.

97. Faivre, S.; Chièze, S.; Delbaldo, C.; Ady-Vago, N.; Guzman, C.; Lopez-Lazaro, L.; Lozahic, S.; Jimeno, J.; Pico, F.; Armand, J.P.; et al. Phase I and pharmacokinetic study of aplidine, a new marine cyclodepsipeptide in patients with advanced malignancies. *J. Clin. Oncol.* **2005**, *23*, 7871–7880. [CrossRef] [PubMed]

98. Van Andel, L.; Fudio, S.; Rosing, H.; Munt, S.; Miguel-Lillo, B.; González, I.; Tibben, M.M.; de Vries, N.; de Vries Schultink, A.H.M.; Schellens, J.H.M.; et al. Pharmacokinetics and excretion of 14C–Plitidepsin in patients with advanced cancer. *Investig. New Drugs* **2017**, *35*, 589–598. [CrossRef] [PubMed]

99. Dubbelman, A.-C.; Rosing, H.; Jansen, R.S.; Mergui-Roelvink, M.; Huitema, A.D.R.; Koetz, B.; Lymboura, M.; Reyderman, L.; Lopez-Anaya, A.; Schellens, J.H.M.; et al. Mass balance study of [14C]eribulin in patients with advanced solid tumors. *Drug Metab. Dispos.* **2012**, *40*, 313–321. [CrossRef] [PubMed]

100. Kidwell, D.A.; Holland, J.C.; Athanaselis, S. Testing for drugs of abuse in saliva and sweat. *J. Chromatogr. B Biomed. Appl.* **1998**, *713*, 111–135. [CrossRef]

101. Rowland, M.; Tozer, T.N. *Clinical Pharmacokinetics. Concepts and Applications*, 3rd ed.; Lippincott Williams & Wilkins: Philadelphia, PA, USA, 1995; ISBN 978-0-683-07404-8.

102. Fuse, E.; Tanii, H.; Takai, K.; Asanome, K.; Kurata, N.; Kobayashi, H.; Kuwabara, T.; Kobayashi, S.; Sugiyama, Y. Altered pharmacokinetics of a novel anticancer drug, UCN-01, caused by specific high affinity binding to alpa1-acid glycoprotein in humans. *Cancer Res.* **1999**, *59*, 1054–1060. [PubMed]

103. Nijenhuis, C.M.; Schellens, J.H.M.; Beijnen, J.H. Regulatory aspects of human radiolabeled mass balance studies in oncology: Concise review. *Drug Metab. Rev.* **2016**, *48*, 266–280. [CrossRef] [PubMed]

104. Raju, K.S.R.; Taneja, I.; Singh, S.P. Wahajuddin Utility of noninvasive biomatrices in pharmacokinetic studies. *Biomed. Chromatogr.* **2013**, *27*, 1354–1366. [CrossRef] [PubMed]

105. Rizk, M.L.; Zou, L.; Savic, R.M.; Dooley, K.E. Importance of Drug Pharmacokinetics at the Site of Action. *Clin. Transl. Sci.* **2017**, *10*, 133–142. [CrossRef] [PubMed]

106. Edelman, M.J.; Bauer, K.S.; Wu, S.; Smith, R.; Bisacia, S.; Dancey, J. Phase I and pharmacokinetic study of 7-hydroxystaurosporine and carboplatin in advanced solid tumors. *Clin. Cancer Res.* **2007**, *13*, 2667–2674. [CrossRef] [PubMed]

107. Dubbelman, A.-C.; Rosing, H.; Schellens, J.H.M.; Beijnen, J.H. Bioanalytical aspects of clinical mass balance studies in oncology. *Bioanalysis* **2011**, *3*, 2637–2655. [CrossRef] [PubMed]

108. Liu, G.; Aubry, A.-F. Best Practices in Biological Sample Preparation for LC-MS Bioanalysis. In *Handbook of LC-MS Bioanalysis: Best Practices, Experimental Protocols, and Regulations*; Li, W., Zhang, J., Tse, F.L.S., Eds.; John Wiley & Sons Inc.: Hoboken, NJ, USA, 2013; pp. 165–184.

109. Stokvis, E.; Rosing, H.; Beijnen, J.H. Liquid chromatography-mass spectrometry for the quantitative bioanalysis of anticancer drugs. *Mass Spectrom. Rev.* **2005**, *24*, 887–917. [CrossRef] [PubMed]

110. Danaceau, J.P.; Haynes, K.; Chambers, E.E. *A Comprehensive Comparison of Solid Phase Extraction (SPE) vs. Solid Liquid Extraction (SLE) vs. Liquid Liquid Extraction (LLE) Sample Prep Techniques in Bioanalysis and Forensic Toxicology Analyses*; Waters Corporation: Milford, MA, USA, 2017.

111. Pucci, V.; Di Palma, S.; Alfieri, A.; Bonelli, F.; Monteagudo, E. A novel strategy for reducing phospholipids-based matrix effect in LC-ESI-MS bioanalysis by means of HybridSPE. *J. Pharm. Biomed. Anal.* **2009**, *50*, 867–871. [CrossRef] [PubMed]

112. Nováková, L.; Vlčková, H. A review of current trends and advances in modern bio-analytical methods: Chromatography and sample preparation. *Anal. Chim. Acta* **2009**, *656*, 8–35. [CrossRef] [PubMed]

113. He, H.; Tran, P.; Gu, H.; Tedesco, V.; Zhang, J.; Lin, W.; Gatlik, E.; Klein, K. Midostaurin, a novel protein kinase inhibitor for the treatment of acute nyelogenous leukemia: Insights from human absorption, metabolism, and excretion studies of a BDDCS II Drugs. *Drug Metab. Dispos.* **2017**, *412*, 540–555. [CrossRef] [PubMed]

114. Schöffski, P.; Guillem, V.; Garcia, M.; Rivera, F.; Tabernero, J.; Cullell, M.; Lopez-Martin, J.A.; Pollard, P.; Dumez, H.; Del Muro, X.G.; et al. Phase II randomized study of plitidepsin (aplidin), alone or in association with L-carnitine, in patients with unresectable advanced renal cell carcinoma. *Mar. Drugs* **2009**, *7*, 57–70. [CrossRef] [PubMed]

115. Baruchel, S.; Pappo, A.; Krailo, M.; Baker, K.S.; Wu, B.; Villaluna, D.; Lee-Scott, M.; Adamson, P.C.; Blaney, S.M. A phase 2 trial of trabectedin in children with recurrent rhabdomyosarcoma, Ewing sarcoma and non-rhabdomyosarcoma soft tissue sarcomas: A report from the Children's Oncology Group. *Eur. J. Cancer* **2012**, *48*, 579–585. [CrossRef] [PubMed]

116. Sausville, E.A.; Arbuck, S.G.; Messmann, R.; Headlee, D.; Bauer, K.S.; Lush, R.M.; Murgo, A.; Figg, W.D.; Lahuse, T.; Jaken, S.; et al. Phase I trial of 72-hour continuous infusion UCN-01 in patients with refractory neoplasms. *J. Clin. Oncol.* **2001**, *19*, 2319–2333. [CrossRef] [PubMed]

117. Cárdenas, F.; Thormann, M.; Feliz, M.; Caba, J.M.; Lloyd-Williams, P.; Giralt, E. Conformational analysis of dehydrodidemnin B (aplidine) by NMR spectroscopy and molecular mechanics/dynamics calculations. *J. Org. Chem.* **2001**, *66*, 4580–4584. [CrossRef] [PubMed]

118. Matuszewski, B.K.; Constanzer, M.L.; Chavez-Eng, C.M. Strategies for the assessment of matrix effect in quantitative bioanalytical methods based on HPLC-MS/MS. *Anal. Chem.* **2003**, *75*, 3019–3030. [CrossRef] [PubMed]

119. Petty, R.; Anthoney, A.; Metges, J.P.; Alsina, M.; Gonçalves, A.; Brown, J.; Montagut, C.; Gunzer, K.; Laus, G.; Iglesias Dios, J.L.; et al. Phase Ib/II study of elisidepsin in metastatic or advanced gastroesophageal cancer (IMAGE trial). *Cancer Chemother. Pharmacol.* **2016**, *77*, 819–827. [CrossRef] [PubMed]

120. Tan, A.R.; Sarantopoulos, J.; Lee, L.; Reyderman, L.; He, Y.; Olivo, M.; Goel, S. Pharmacokinetics of eribulin mesylate in cancer patients with normal and impaired renal function. *Cancer Chemother. Pharmacol.* **2015**, *76*, 1051–1061. [CrossRef] [PubMed]

121. Kummar, S.; Gutierrez, M.E.; Gardner, E.R.; Figg, W.D.; Melillo, G.; Dancey, J.; Sausville, E.A.; Conley, B.A.; Murgo, A.J.; Doroshow, J.H. A phase I trial of UCN-01 and prednisone in patients with refractory solid tumors and lymphomas. *Cancer Chemother. Pharmacol.* **2010**, *65*, 383–389. [CrossRef] [PubMed]

122. Gojo, I.; Perl, A.; Luger, S.; Baer, M.R.; Norsworthy, K.J.; Bauer, K.S.; Tidwell, M.; Fleckinger, S.; Carroll, M.; Sausville, E.A. Phase I study of UCN-01 and perifosine in patients with relapsed and refractory acute leukemias and high-risk myelodysplastic syndrome. *Investig. New Drugs* **2013**, *31*, 1217–1227. [CrossRef] [PubMed]

123. Hotte, S.J.; Oza, A.; Winquist, E.W.; Moore, M.; Chen, E.X.; Brown, S.; Pond, G.R.; Dancey, J.E.; Hirte, H.W. Phase I trial of UCN-01 in combination with topotecan in patients with advanced solid cancers: A Princess Margaret Hospital Phase II Consortium study. *Ann. Oncol.* **2006**, *17*, 334–340. [CrossRef] [PubMed]
124. Stokvis, E.; Rosing, H.; Beijnen, J.H. Stable isotopically labeled internal standards in quantitative bioanalysis using liquid chromatography/mass spectrometry: Necessity or not? *Rapid Commun. Mass Spectrom.* **2005**, *19*, 401–407. [CrossRef] [PubMed]
125. Wang, S.; Cyronak, M.; Yang, E. Does a stable isotopically labeled internal standard always correct analyte response? A matrix effect study on a LC/MS/MS method for the determination of carvedilol enantiomers in human plasma. *J. Pharm. Biomed. Anal.* **2007**, *43*, 701–707. [CrossRef] [PubMed]
126. Stokvis, E.; Rosing, H.; López-Lázaro, L.; Schellens, J.H.M.; Beijnen, J.H. Switching from an analogous to a stable isotopically labeled internal standard for the LC-MS/MS quantitation of the novel anticancer drug Kahalalide F significantly improves assay performance. *Biomed. Chromatogr.* **2004**, *18*, 400–402. [CrossRef] [PubMed]
127. Beumer, J.H.; Rademaker-Lakhai, J.M.; Rosing, H.; Hillebrand, M.J.X.; Bosch, T.M.; Lopez-Lazaro, L.; Schellens, J.H.M.; Beijnen, J.H. Metabolism of trabectedin (ET-743, Yondelis) in patients with advanced cancer. *Cancer Chemother. Pharmacol.* **2007**, *59*, 825–837. [CrossRef] [PubMed]

Article

The In Vitro Anti-Tumor Activity of Phycocyanin against Non-Small Cell Lung Cancer Cells

Shuai Hao [1], Yan Yan [1], Shuang Li [1], Lei Zhao [1], Chan Zhang [1], Liyun Liu [2,*] and Chengtao Wang [1,*]

[1] Beijing Advanced Innovation Center for Food Nutrition and Human Health, Beijing Engineering and Technology Research Center of Food Additives, Beijing Technology and Business University, Beijing 100048, China; shmilyhs321@163.com (S.H.); 15128470659@163.com (Y.Y.); lishuangldw@163.com (S.L.); zhaolei@th.btbu.edu.cn (L.Z.); zhangchan@th.btbu.edu.cn (C.Z.)

[2] State Key Laboratory of Infectious Disease Prevention and Control, National Institute for Communicable Disease Control and Prevention, Collaborative Innovation Center for Diagnosis and Treatment of Infectious Disease, Chinese Center for Disease Control and Prevention, Beijing 102206, China

*   Correspondence: liuliyun@icdc.cn (L.L.); ctwangbtbu@163.com (C.W.); Tel.: +86-10-68984857 (C.W.)

Received: 17 April 2018; Accepted: 22 May 2018; Published: 23 May 2018

Abstract: Phycocyanin, a type of functional food colorant, is shown to have a potent anti-cancer property. Non-small cell lung cancer (NSCLC) is one of the most aggressive form of cancers with few effective therapeutic options. Previous studies have demonstrated that phycocyanin exerts a growth inhibitory effect on NSCLC A549 cells. However, its biological function and underlying regulatory mechanism on other cells still remain unknown. Here, we investigated the in vitro function of phycocyanin on three typical NSCLC cell lines, NCI-H1299, NCI-H460, and LTEP-A2, for the first time. The results showed that phycocyanin could significantly induce apoptosis, cell cycle arrest, as well as suppress cell migration, proliferation, and the colony formation ability of NSCLC cells through regulating multiple key genes. Strikingly, phycocyanin was discovered to affect the cell phenotype through regulating the NF-κB signaling of NSCLC cells. Our findings demonstrated the anti-neoplastic function of phycocyanin and provided valuable information for the regulation of phycocyanin in NSCLC cells.

Keywords: phycocyanin; non-small cell lung cancer; proliferation; apoptosis; NF-κB signaling

---

1. Introduction

Extensive research has suggested that many natural products derived from food and food supplements have various health-promoting effects [1]. Recently, marine natural products with pharmacological activity have been shown to have multiple potent biological functions, with less or no toxic side effects [2]. Thus, they have become one of the most important resources of novel lead compounds for critical diseases, which have seen important development and utilization in recent years [3]. Phycocyanin, a marine natural blue photosynthetic pigment protein purified from *Spirulina*, is one of the accepted natural functional foods around the world [4]. Phycocyanin has excellent anti-tumor activity. Studies have shown that phycocyanin plays anti-proliferation and pro-apoptotic effects on different cancer cell lines in vitro, while it has no side effects on normal tissue cells [5,6]. In addition, it also shows good therapeutic values such as antioxidant, anti-inflammatory, immunomodulation, blood vessel-relaxing, and blood liquid-lowering activities, etc. [7–11]. Thus, further investigating on the function and mechanism of phycocyanin has important guiding significance and research value.

Several studies have demonstrated that phycocyanin has bioactivity in kinds of cancers, including breast cancer [6,12], histiocytic tumor [13], ovarian cancer [14], colon cancer [15], prostate cancer [16], melanoma, and lung cancer [17]. Among them, lung cancer is one of the most common health threats in the world, especially given its high mortality rates [18,19]. Human lung cancer is generally classified into two major categories, small cell lung cancer (SCLC) and non-small cell lung cancer (NSCLC). NSCLC is responsible for more than 85% of all lung carcinoma cases, with the characteristics of higher mortality, lower cure rate, and stronger metastasis [20]. Therefore, exploring the anti-cancer function and underlying mechanism of phycocyanin on NSCLC is critical.

The inhibiting effect of phycocyanin on NSCLC has been reported in several studies. Li et al. revealed that phycocyanin could inhibit the growth of NSCLC A549 cells in vivo and in vitro, which also has a synergistic anti-tumor effect with all-trans retinoic acid [21,22]. Bingula et al. reported that phycocyanin and betaine have a synergistic inhibiting effect on the viability of A549 cells [23]. Baudelet et al. discovered that glaucophyte *Cyanophora paradoxa* extracts could significantly inhibit the growth of three cancer cell lines, including A549 cells [17]. However, the abovementioned studies all investigated the function of phycocyanin in one single NSCLC cell line. Moreover, the anti-lung cancer mechanism of phycocyanin remains unclear. Herein, we investigated the growth inhibitory effects and underlying mechanism of phycocyanin in three human NSCLC cell lines, NCI-H1299, LTEP-A2, and NCI-H460. The results laid a solid theoretical foundation for the treatment of NSCLC and the development and utilization of phycocyanin.

2. Results

2.1. Phycocyanin Induces Morphological Changes in NSCLC Cells

To address the relationship between phycocyanin and its effect on non-small cell lung cancer, the morphology of NSCLC cells, H1299, H460, and LTEP-A2, was first studied upon treatment with various doses of phycocyanin. As shown in Figure 1, the normal morphology of H1299 cells was fusiform or triangular. After treatment with 4.8 µM phycocyanin for 72 h, cells appeared in anomalous forms, some of which became needle-shaped. Similarly, the morphology of H460 and LTEP-A2 cells were also abnormally changed by phycocyanin. Furthermore, the number of cells was obviously reduced after phycocyanin treatment. These results suggested that phycocyanin might have a pro-apoptotic effect on NSCLC cells.

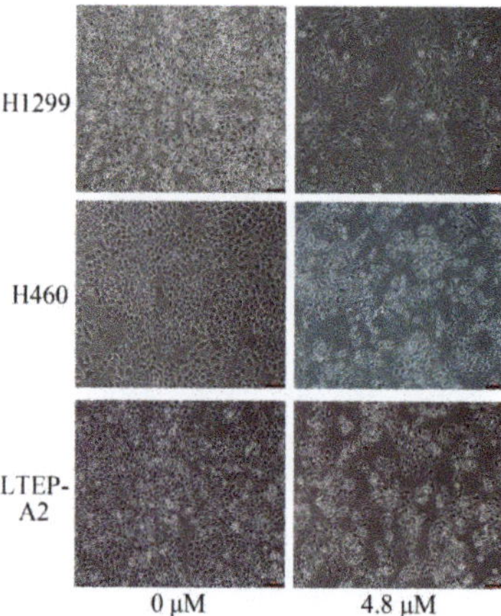

Figure 1. Phycocyanin induces morphological changes in non-small cell lung cancer (NSCLC) cells. H1299, H460, and LTEp-A2 cells were treated with different doses (0 and 4.8 µM) of phycocyanin for 72 h, and photographed under a light microscope (100×). Scale bars represent 100 µm.

2.2. Phycocyanin Induces Apoptosis in NSCLC Cells

As phycocyanin induces morphological changes in NSCLC cells, we next studied the extent of apoptosis in H1299, H460, and LTEP-A2 cells by Annexin V-FITC and 7AAD staining. Figure 2 shows that phycocyanin-treated NSCLC cells demonstrated an induction of apoptosis in comparison to untreated cells. The late apoptotic percentages of H1299 (4.53 ± 0.27%), H460 (2.68 ± 0.37%), and LTEP-A2 cells (4.88 ± 0.55%) increased after incubation with 2.4 μM phycocyanin, as compared to the control groups. In addition, the apoptosis degree of NSCLC cells presented a dose-dependent effect with phycocyanin. A high concentration of phycocyanin (4.8 μM) significantly increased the late apoptotic percentages of H1299 (11.30 ± 0.16%), H460 (3.72 ± 0.98%), and LTEP-A2 cells (14.50 ± 0.68%).

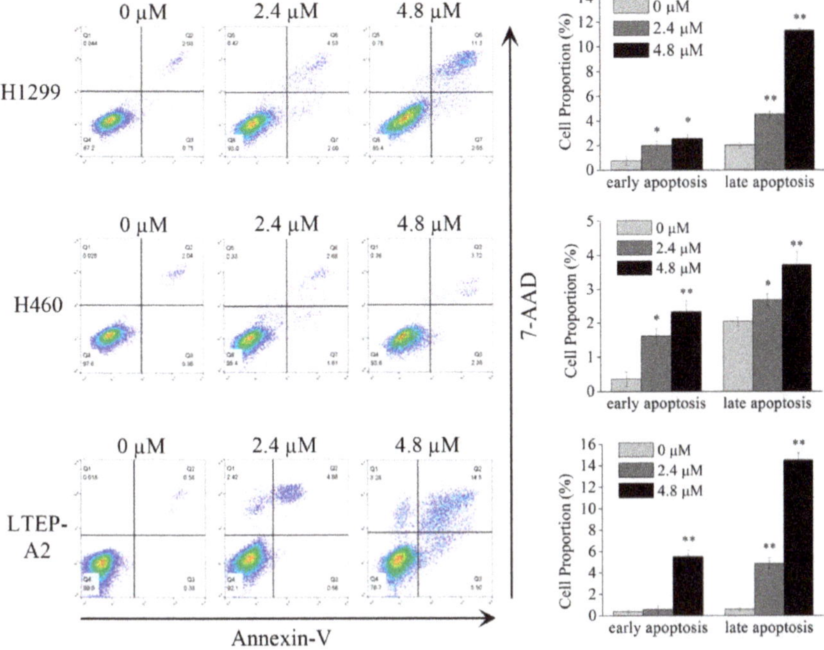

Figure 2. Phycocyanin induces apoptosis in NSCLC cells. H1299, H460, and LTEP-A2 cells were incubated with different concentrations of phycocyanin (0, 2.4, and 4.8 μM) for 48 h and subjected to apoptosis tests. The proportion of early apoptotic and late apoptotic cells were analyzed. Bars represent mean ± SD. *, $p < 0.05$; **, $p < 0.01$.

To gain a deeper insight into the mechanism of apoptosis induced by phycocyanin, we tested the expressions of apoptotic markers using quantitative real-time PCR (qRT-PCR) and Western blot. As shown in Figure 3, phycocyanin could significantly increase the transcriptional levels of pro-apoptotic genes *Bim*, *Bak*, *Bax*, and *Bad*, in addition to reducing the levels of *Bcl-xL* and *Bcl-2*, two anti-apoptotic genes in H1299 and LTEP-A2 cell lines. These results were further supported by Western blot analysis. It was interesting to find that in H460 cells, although the protein level of Bcl-xL was downregulated, its transcriptional level increased after phycocyanin treatment. In addition, the protein level of Bcl-2 showed no obvious alteration while its transcriptional level significantly decreased. These results indicated that a post-transcriptional mechanism might involve the regulation of these genes in H460 cells. Taken together, the above results suggested that phycocyanin could induce apoptosis in NSCLC cells.

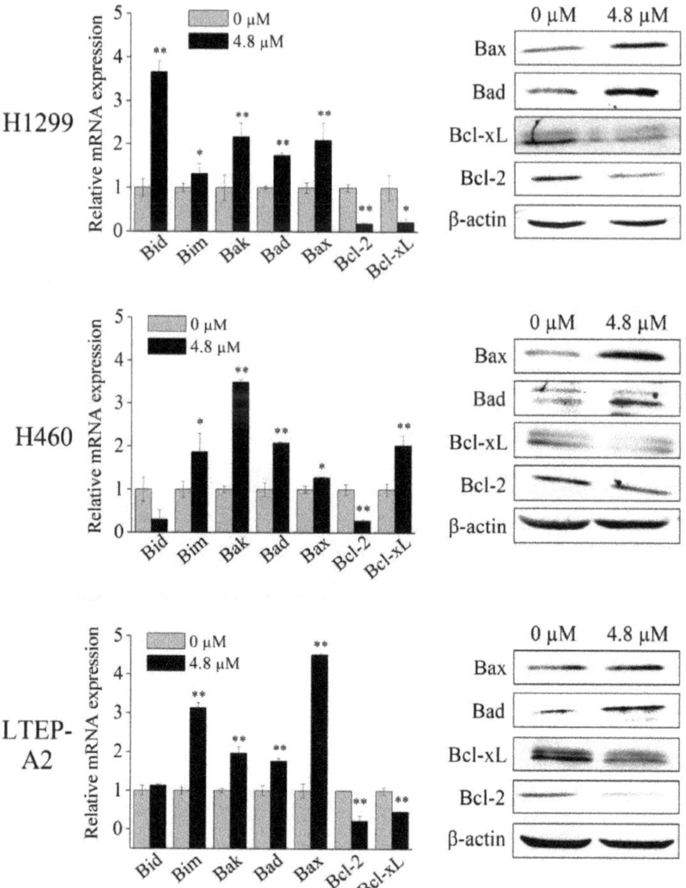

Figure 3. Quantitative real-time PCR (qRT-PCR) and Western blot analysis of apoptotic markers in NSCLC cells after phycocyanin treatment. H1299, H460, and LTEP-A2 cells were incubated with 4.8 μM phycocyanin. qRT-PCR and Western blot were performed at 24 h and 48 h after treatment, respectively. Bars represent mean ± SD. *, $p < 0.05$; **, $p < 0.01$.

2.3. Phycocyanin Displays Anti-Migratory Effect against NSCLC Cells

A wound-healing assay was employed to determine the effect of phycocyanin on cell migration. As shown in Figure 4A, phycocyanin significantly suppressed the migration of H1299, H460, and LTEP-A2 cells in dose- and time-dependent manners (left panel); the migration rates were calculated and are presented in the right panel. After phycocyanin treatment (4.8 μM) for 48 h, the wound closure of H1299 cells clearly decreased from 77.60 ± 0.24% to 37.35 ± 6.24%. Similar results were found in H460 and LTEP-A2 cells. It is worth mentioning that in this study, we cultured cells with medium containing 3% instead of 10% fetal bovine serum (FBS), which eliminated the contribution of proliferation to the phycocyanin-induced inhibition of cell migration. Matrix metalloproteinase-2

(MMP2) and matrix metalloproteinase-9 (MMP9) are gelatinases of the matrix metalloproteinase family, which play a crucial role in cancer cell growth and migration due to their ability to degrade extracellular matrix proteins [24]. In present study, we found that phycocyanin treatment significantly reduced the expression of MMP2 and MMP9 in NSCLC cells (Figure 4B), which was in accordance with the wound-healing analysis. Taken together, these results suggested that phycocyanin displayed inhibitory activity on NSCLC cell migration.

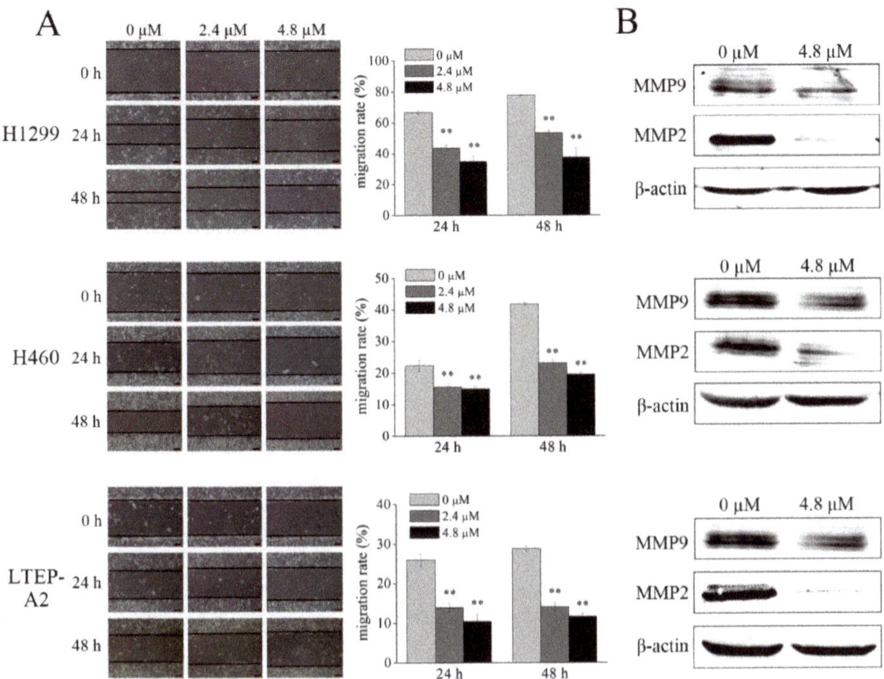

Figure 4. Phycocyanin displays anti-migratory effect against NSCLC cells. (**A**) The wound-healing assay showed representative effects of phycocyanin (0, 2.4, and 4.8 µM) on H1299, H460, and LTEP-A2 cell migration at 24 and 48 h. Quantification of wound closure was shown by histogram. Scale bars represent 100 µm. (**B**) Western blot analysis of MMP2 and MMP9 expression in NSCLC cells after 4.8 µM phycocyanin treatment for 48 h. MMP2, matrix metalloproteinase-2; MMP9, matrix metalloproteinase-9. Bars represent mean ± SD. **, $p < 0.01$.

2.4. Phycocyanin Inhibits Proliferation and Colony Formation Ability of NSCLC Cells

The inhibitory effects of phycocyanin on the viability and proliferation of NSCLC cells were determined. As shown in Figure 5A, compared with control cells, incubation with phycocyanin (1.2, 2.4, and 4.8 µM) dose-dependently inhibited the viability of the three NSCLC cell lines. In addition, MTT assay (Figure 5B) showed that phycocyanin (4.8 µM) could significantly suppress the proliferation of NSCLC cells from the second (H460 and LTEP-A2 cells) or the third day (H1299 cells). These results suggested that phycocyanin exerted anti-proliferation effects on tested cells. To further establish the inhibitory role of phycocyanin on the transforming properties of NSCLC cells, we performed a clonogenic assay. Figure 5C shows that the three types of NSCLC cells could all form well-defined and distinct colonies, only the colonies of H1299 cells were bigger than the other two cell lines. Results showed that phycocyanin-treated cells displayed significant reduction in colony formation when compared to controls, which was indicative of potent inhibition of cell growth and reproductive integrity.

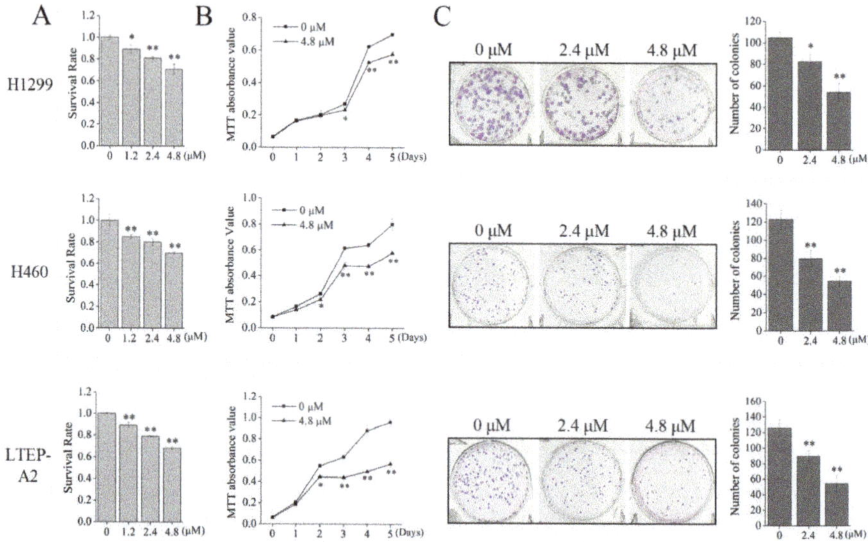

Figure 5. Phycocyanin inhibits the proliferation and colony formation ability of NSCLC cells. (**A**) Cell viability analysis of H1299, H460, and LTEP-A2 after different concentrations (0, 1.2, 2.4, and 4.8 μM) of phycocyanin treatment for 72 h. (**B**) MTT analysis of cell proliferation of H1299, H460, and LTEP-A2 after 4.8 μM phycocyanin treatment. (**C**) Colony formation assay of H1299, H460, and LTEP-A2 cells after treatment of different concentrations (0, 2.4, and 4.8 μM) of phycocyanin for 10–14 days. The quantitative representation of the reduction in number of colonies was shown by histogram. Bars represent mean ± SD. *, $p < 0.05$; **, $p < 0.01$.

2.5. Phycocyanin Induces Cell Cycle Arrest in NSCLC Cells

To elucidate the mechanism of growth inhibition on NSCLC cells, the effects of phycocyanin on cell cycle progression were determined in H1299, H460, and LTEP-A2 cells. Figure 6A showed that phycocyanin caused significant changes in the cell cycle distribution of H1299 and H460 cells. After incubation with 4.8 μM phycocyanin, the proportion of S phase cells reached 40.10 ± 1.06% and 30.60 ± 1.55% in H1299 and H460, respectively, as compared to the control groups (27.20 ± 0.80% and 21.10 ± 0.26% in H1299 and H460 cells, respectively), suggesting that phycocyanin caused S phase arrest in these two cell lines. Interestingly, unlike H1299 and H460 cells, phycocyanin (4.8 μM) induced a significant G1 phase increase (51.79 ± 0.80%) in LTEP-A2 cells compared to the control (49.39 ± 0.38%). Although phycocyanin could suppress the proliferation of NSCLC cells, the present results indicated that the inhibitory mechanism in LTEP-A2 might differ from that in H1299 and H460 cells. To further confirm the above results, we tested the levels of cell cycle regulatory genes involved in S/G2 and G1/S transition. As shown in Figure 6B, after phycocyanin treatment, Cyclin A and CDK2, two key positive regulators in S/G2 checkpoint [25], were significantly downregulated, whereas p21, an important cell cycle suppressor gene [26], was upregulated in H1299 and H460 cells. However, in LTEP-A2 cells, phycocyanin inhibited the transcriptional level of Cyclin E, a G1/S checkpoint regulator [27]. The qRT-PCR results strongly indicated that phycocyanin could trigger G1 phase arrest of LTEP-A2 cells and S phase arrest of H1299 and H460 NSCLC cells.

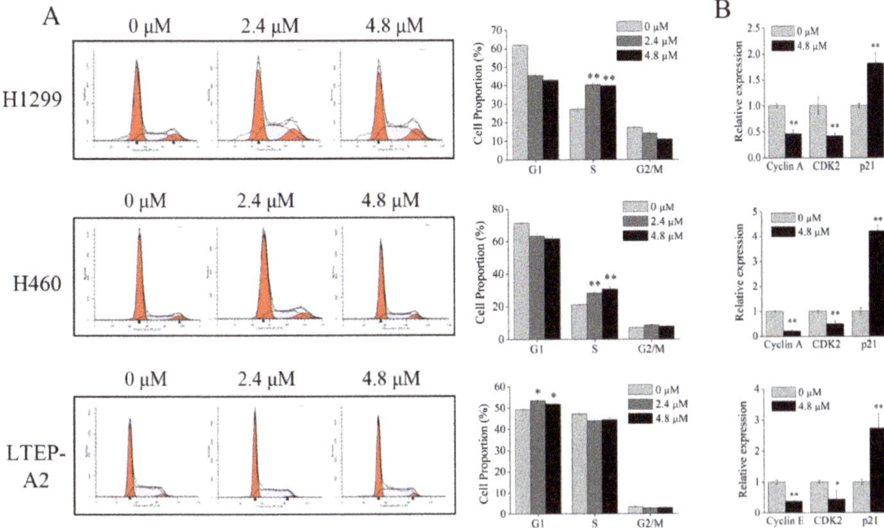

Figure 6. Phycocyanin induces cell cycle arrest in NSCLC cells. (**A**) Cell cycle analysis of NSCLC cells after treated with different concentrations of phycocyanin (0, 2.4, and 4.8 µM) for 48 h. (**B**) qRT-PCR analysis of the transcriptional levels of cell cycle regulatory genes at 24 h after phycocyanin treatment. Glyceraldehyde-3-phosphate dehydrogenase (GAPDH) was used as an internal control for normalization. Bars represent mean ± SD. *, $p < 0.05$; **, $p < 0.01$.

2.6. Phycocyanin Reduces NF-κB Signaling Activity in NSCLC Cells

To investigate the underlying regulatory mechanism of phycocyanin in NSCLC cells, we determined the protein expressions of NF-κB signaling in H1299, H460, and LTEP-A2 cells after 4.8 µM phycocyanin treatment. IKKα and IKKβ served as the catalytic subunits of IκBα, an inhibitory protein of p65. Phosphorylation of IKKα/β could phosphorylate the IκBα protein at Ser32, resulting in the ubiquitin-mediated proteasome-dependent degradation of IκBα, followed by p65 phosphorylation and NF-κB activation [28]. As shown in Figure 7, although the total amounts of IKKα/β, IκBα, and p65 remained stable, their phosphorylation levels (phospho-IKKα/β-Ser176/180, phospho-IκBα-Ser32, and phospho-p65-Ser536) were significantly decreased after phycocyanin treatment, suggesting that the activity of NF-κB signaling was inhibited by phycocyanin. These results demonstrated that phycocyanin could significantly suppress the activity of NF-κB signaling in NSCLC cells. In addition, phycocyanin could also attenuate the phosphorylation levels of AKT in H1299, H460, and LTEP-A2 cells.

To further explore whether NF-κB was related to phenotypic factors of NSCLC cells, we performed a pyrrolidine dithiocarbamate (PDTC, NK-κB inhibitor) addition experiment. As shown in Figure 8, although the total p65 protein amounts remained stable, the phosphorylation levels of p65 were decreased after 10 µM PDTC treatment, indicating that PDTC inhibited the NF-κB pathway activity of H1299, H460, and LTEP-A2 cells. Strikingly, the protein levels of CDK2, MMP2, and Bcl-xL were downregulated, and the amount of Bad was upregulated after PDTC addition. These results strongly suggested that NF-κB had a regulatory effect on the phenotypic factors of NSCLC cells. Taken together, our study demonstrated that phycocyanin could inhibit the migration and proliferation, as well as promote the apoptosis of H1299, H460, and LTEP-A2 cells through regulating the NF-κB pathway. Although the precise mechanism still needs further investigation, the present research illuminated the function of phycocyanin in NSCLC cells and provided solid evidence that warrants further exploration of the anti-cancer mechanism of phycocyanin.

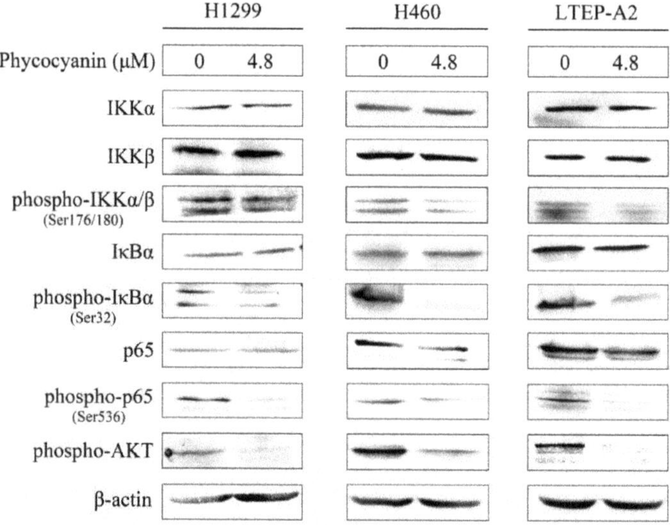

Figure 7. Phycocyanin reduces NF-κB signaling activity in NSCLC cells. The protein expressions of NF-κB signaling and phospho-AKT were analyzed by Western blot in H1299, H460, and LTEP-A2 cells after 0 and 4.8 μM phycocyanin treatment for 48 h.

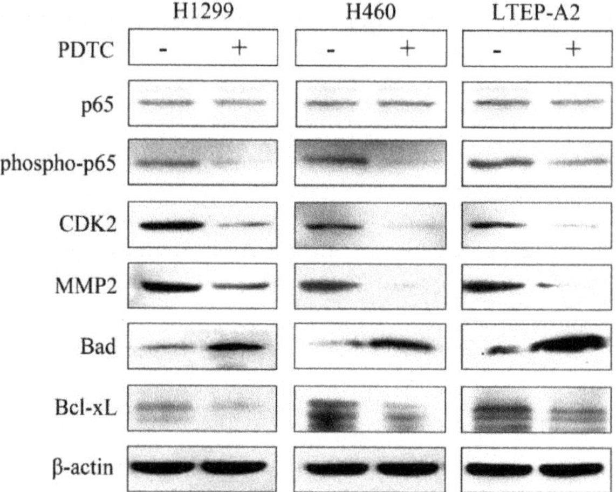

Figure 8. NF-κB signaling has regulatory effect on phenotypic factors of NSCLC cells. The protein expressions of phospho-p65, total p65, CDK2, MMP2, Bad, and Bcl-xL were analyzed by Western blot at 48 h in H1299, H460, and LTEP-A2 cells after 10 μM PDTC treatment for 24 h. PDTC, pyrrolidine dithiocarbamate.

3. Discussion

Over the past few decades, the application of natural products for chemoprevention and therapy has gained great importance [29–31]. More and more studies have demonstrated that pharmacological active marine-derived compounds have potent biological activity with little or no side effects [32–34]. Phycocyanin, a type of phycobiliprotein derived from *Spirulina*, is one of the compounds that have considerable anti-cancer effects on solid malignancies [35,36]. Non-small cell lung cancer, an extremely aggressive form of cancer with few effective therapeutic options, has attracted that attention of many investigators. Previous studies have suggested that phycocyanin exerts an inhibitory effect on A549 cells, a type of NSCLC cell line [21–23]. In this study, we demonstrated that phycocyanin inhibits the growth of H1299, H460, and LTEP-A2 NSCLC cell lines. These results are consistent with previous studies showing that phycocyanin can suppress the proliferation of a variety of tumor cell lines [12,13,37]. To the best of our knowledge, this is the first study to demonstrate the anti-cancer effect of phycocyanin on these NSCLC cell lines, which also highlights the possible mechanism underlying phycocyanin's cytotoxic and anti-metastatic effects. Our study clearly demonstrated that phycocyanin regulated the NF-κB signaling pathway and also altered the expression of proteins involved in cell cycle and cell survival, by which it mediated growth inhibition and apoptosis.

Cell cycle regulation plays an important role in cell proliferation, differentiation, and apoptosis. It has been reported that cell cycle regulation dysfunction is closely related to the development of tumors [38]. Peyressatre et al. found that the deregulated activity of cyclin-dependent kinases (CDKs) contributes to altered cellular proliferation in a wide variety of human cancers [39]. What is interesting is that in the present study, phycocyanin was discovered to induce S phase arrest in H1299 and H460 cell lines, while it caused G1 phase arrest in LTEP-A2 cells (Figure 6), which indicated different regulation mechanisms of phycocyanin in different NSCLC cell lines. In fact, phycocyanin could act as an anti-cancer compound through different mechanisms in different types of tumors. It has been reported that phycocyanin induces G1 cell cycle arrest in colon cancer HT-29 cells [7], breast cancer MDA-MB-231 cells [6], and chronic myelocytic leukemia K562 cells [40]. Meanwhile, it could also block G2/M cell cycle progression in pancreatic cancer PANC-1 cells [41], ovarian cancer SKOV-3 cells [14], and liver cancer HepG2 cells [42]. Thangam et al. reported that phycocyanin could induce G1 phase arrest in lung adenocarcinoma A549 cells [7], which is in agreement with the result of LTEP-A2 cells in our study. It is worth noting that H1299 and LTEP-A2 both belong to lung adenocarcinoma cell lines, but the cell cycles are restrained in different phases. Interestingly, Yao et al. discovered that pseudolaric acid B, a diterpene acid isolated from the root of *Pseudolarix kaempferi*, could inhibit the growth and cause G2/M arrest of A549 cells, but has no effect on H1299 cells because H1299 is a type of p53 null cell line [43]. Therefore, we speculate that p53 might be involved in phycocyanin-induced cell cycle arrest in H1299 (p53 null) and LTEP-A2 (p53 wild type) cells. However, the underlying mechanism still needs further investigation. Taken together, our study provided useful information on the regulation approach of phycocyanin in NSCLC cells.

Cells arrested in mitosis upon the activation of mitotic catastrophe have different fates, including death during mitosis or the entrance into the subsequent cell cycle followed by cell death [44], indicating that cell cycle regulation is closely related to apoptosis. In the present study, phycocyanin was found to induce apoptosis in NSCLC cells. The apoptotic percentage of H460 cells reached $3.72 \pm 0.98\%$ with phycocyanin treatment, which was significantly different than that in control cells, but markedly lower than that in H1299 ($11.3 \pm 0.16\%$) and LTEP-A2 cells ($14.5 \pm 0.68\%$). It is worth noting that Shin et al. discovered that (E)-2-benzylidene-3-(cyclohexylamino)-2,3-dihydro-1H-inden-1-one (BCI, an inhibitor of dual specific phosphatase 1/6 and mitogen-activated protein kinase) significantly inhibits the viability of H1299 cells as compared to H460 cells [45], suggesting that different mechanisms might exist in lung adenocarcinoma (H1299 and LTEP-A2 cells) and undifferentiated large cell lung carcinoma (H460 cells) cell lines. Particularly, the transcription and protein levels of two apoptotic markers (Bcl-2 and Bcl-xL) were not consistently expressed in H460 cells (Figure 3), which further supports the above hypothesis. Interestingly, although the degree of apoptosis in H460 cells is low, it shows the highest expression level of p21 (Figure 6B). It is known that the growth inhibition and the expression of regulation factors might show discrepancy in some cases. Tsui et al. investigated the anti-cancer functions of flavonoids in H460 and A549 cells, and discovered that although H460 cells are more susceptible to flavonoids than A549, p53 level was constitutive and not significantly altered [46]. In the present study, despite the fact that the apoptotic proportion of H460 was low at 48 h after phycocyanin treatment, the proliferation rate dramatically decreased after the third day (Figure 5B). In this case, we speculate that the increased p21 might cause a delay in cell growth and activity inhibition, in spite of the low apoptotic ratio at 48 h.

NF-κB signaling, one of the classical pathways in cells, has been reported to regulate proliferation and apoptosis in various kinds of cancers [47]. Recent studies have demonstrated that the NF-κB pathway is involved in phycocyanin-induced growth inhibition in liver and pancreatic carcinoma cells [41,48]. However, few researchers reported the relationship between NF-κB and phycocyanin in lung cancer; only Bingula et al. discovered that the combined treatment of phycocyanin and betaine could reduce the stimulation of NF-κB expression in A549 cells [23]. In the present study, phycocyanin was first discovered to reduce the activity of NF-κB signaling in H1299, H460, and LTEP-A2 cells (Figure 7), which suggests that phycocyanin could suppress proliferation and induce apoptosis through the inactivation of the NF-κB pathway in NSCLC cell lines.

Cancer metastasis is a complex and multistep event, wherein cancer cells leave the site of the primary tumor and disseminate to distant sites in the body [49]. It is the most destructive stage of cancer progression and the leading cause of cancer-related deaths [50]. More recently, extensive attention has been drawn toward phycocyanin for its potent migration inhibition effect on cancer cells. It has been reported that phycocyanin could suppress the migration of breast cancer and melanoma cells through MAPK signaling [6,51]. Our study showed that phycocyanin exerted a remarkable migration inhibition effect on different NSCLC cells through regulating MMP2 and MMP9 for the first time (Figure 4). Strikingly, NF-κB is reported to play a key role in tumor migration [52] and has a synergistic expression pattern with MMP2 [53], which suggests that MMP2/NF-κB could be involved in phycocyanin-induced migration inhibition regulation in NSCLC cells.

4. Materials and Methods

4.1. Materials and Reagents

Dulbecco's modified Eagle's medium (DMEM) was purchased from Invitrogen (Carlsbad, CA, USA). Phycocyanin (derived from *Spirulina platensis*) standard substance was purchased from Envirologix (Portland, ME, USA). Fetal bovine serum was purchased from Hyclone (Logan, UT, USA). The cell culture consumables were purchased from Corning (Tewksbury, MA, USA). Cell apoptosis analysis kit, RIPA lysis buffer, and protease and phosphatase inhibitors were purchased from Roche (Mannheim, Germany). Propidium iodide, RNase, and skim milk were purchased from Becton Dickinson (Franklin Lakes, NJ, USA). Polyvinylidene difluoride (PVDF) membrane and enhanced chemiluminescence (ECL) kit were purchased from Millipore (Schwalbach, Germany). Antibodies were purchased from Cell Signaling Technology (Danvers, MA, USA). PrimeScript RT Master Mix was purchased from Takara (Dalian, China).

4.2. Cell Line and Culture Conditions

Human NSCLC cell lines NCI-H1299, NCI-H460, and LTEP-A2 were purchased from American Type Cell Collection (ATCC, Manassas, VA, USA). All cell lines were cultured in DMEM media supplemented with 10% heat-inactivated fetal bovine serum (FBS), 0.1 mg/mL of streptomycin, and 100 units/mL of penicillin at 37 °C in a humidified atmosphere with 5% CO_2. Cells were sub-cultured every 3–5 days. Cells between 3–15 passages were used in all experiments.

4.3. Cell Viability Assay

The cell viability was performed by MTT assay. Briefly, cells were seeded at a density of 5000 cells in 100 μL of complete medium per well into 96-well plates. After 12 h of incubation for cell attachment, phycocyanin was added to each well with a final concentration of 0, 1.2, 2.4, or 4.8 μM. Control cells were treated with equal volumes of phosphate buffer solution (PBS). After incubation for 24 h, the culture medium was supplemented with 1 mg/mL MTT for 4 h at 37 °C. The medium was removed and the cells were solubilized with DMSO. The absorbance was then measured at 450 nm and 630 nm. The results were expressed as a survival rate of the absorbance reading of the control cells.

4.4. Cell Proliferation Assay

Cell proliferation was determined by MTT assay. Briefly, cells were seeded at 5000 cells in 100 μL of complete medium per well in quadruplicate in 96-well plates. After 12 h of incubation for cell attachment, phycocyanin was added to each well with a final concentration of 4.8 μM. Each day, 10 μL MTT/well was added to test cells and incubated for 4 h at 37 °C. Then SDS-HCl solution (10% SDS, 0.01 M HCl) was added into each well and incubated for 14 h at 37 °C. The absorbance of formazan was measured at a wavelength of 570 nm using a microplate reader. The assay lasted for 5 or 6 days after treatment. Three independent experiments were performed.

4.5. Colony Formation Assay

Cells in the exponential growth phase were harvested and seeded at about 300 cells per well in six-well plates. After 12 h incubation, cells were treated for another 24 h with 0, 2.4, or 4.8 μM phycocyanin, and then continuously incubated in fresh medium at 37 °C in 5% humidified CO_2. After incubation for 10–14 days, cells were washed with PBS twice, fixed with methanol for 15 min, and stained with 0.5% crystal violet for 15 min at room temperature. The number of colonies was counted for analysis.

4.6. Cell Apoptosis Assay

After being treated with phycocyanin (0, 2.4, and 4.8 μM) for 48 h, cells were harvested and washed twice with cold PBS and then resuspended gently in 500 μL binding buffer. Thereafter, cells were stained in 5 μL Annexin V-FITC/7-aminoactinomycin D (7-AAD) according to the manufacturer's protocol. Stained cells were analyzed by FACSCalibur (Becton Dickinson).

4.7. Cell Cycle Assay

After treatment with phycocyanin for 48 h, cells were harvested and fixed in 1 mL 70% cold ethanol in test tubes and incubated at 4 °C for at least 48 h. Cells were centrifuged at 1500 rpm for 5 min and the cell pellets were resuspended in 500 μL of PI/RNase staining buffer, incubated on ice for 30 min, and washed twice with cold PBS. Cell cycle distribution was measured using FACSCalibur (Becton Dickinson).

4.8. Western Blot Analysis

Proteins were extracted by RIPA lysis buffer (1% NP40, 0.1% SDS, 5 mM EDTA, 0.5% sodium deoxycholate, 1 mM sodium orthovanadate) containing protease and phosphatase inhibitors. Equivalent amounts of proteins were separated by 12% SDS-PAGE, and then electro-transferred onto PVDF membranes. After blocking with 5% skim milk, the membranes were incubated with primary antibodies at 4 °C overnight, followed by incubation with horseradish peroxidase-conjugated secondary antibodies. Signals were detected by an ECL system.

4.9. Quantitative RT-PCR

Total RNA was extracted using Trizol reagent and reverse-transcribed with PrimeScript RT Master Mix. Real-time PCR analysis was performed in an Applied Biosystems Step One-Plus (Waltham, MA, USA) under the following conditions: 95 °C for 30 s, followed by 40 cycles at 95 °C for 5 s, and 60 °C for 40 s. The relative expression of each targeted gene was calculated and normalized using $2^{-\Delta\Delta Ct}$ method relative to glyceraldehyde-3-phosphate dehydrogenase (GAPDH). Each assay was performed in quadruplicate.

4.10. Wound-Healing Assay

Cells in the exponential growth phase were harvested and seeded in six-well plates. After 12 h incubation, cells were treated for another 24 h with 0, 2.4, or 4.8 μM phycocyanin, and then continuously incubated in fresh medium (containing 3% fetal bovine serum) at 37 °C in 5% humidified CO_2. The culture insert provided two cell culture reservoirs that were separated by a thick wall. After removing the culture inserts on the second day, a "wound" was formed between the two cell patches. Photos of the wounds were taken every 12 h. The widths of the wounds were measured at three positions for each replicate using the Leica Application Suite (Leica Microsystems GmbH, Wetzlar, Germany).

4.11. Statistical Analysis

The numerical data were expressed as means ± standard deviation (SD). Two-tailed Student's *t*-test was performed for comparison among the different groups. In addition, $p < 0.05$ (*) or $p < 0.01$ (**) was considered as statistically significant.

5. Conclusions

The present study first demonstrated that phycocyanin, a natural functional food colorant, inhibited migration, proliferation, and promoted apoptosis in three NSCLC cell lines (H1299, H460, and LTEP-A2). Molecular studies further revealed that NF-κB was involved in this process. Phycocyanin could affect the cell phenotype through downregulating the NF-κB pathway. Although the precise mechanism still needs further investigation, the present findings illuminated the function of phycocyanin in NSCLC cells and provided solid evidence to warrant the further exploration of the anti-cancer mechanism of phycocyanin.

Author Contributions: S.H., C.W., and L.L. designed the research study and interpreted the data. Y.Y. and S.L. performed most of the experiments. S.H. wrote the manuscript. L.Z. and C.Z. contributed to the sample collection and performed the statistical analysis. L.Z. edited the manuscript. All authors read and approved the final manuscript.

Funding: This research was funded by the National Natural Science Foundation of China (NSFC), grant number [31701575, 31571801]; Construction of Scientific Research Innovation Service Ability-Basic Scientific Research Operating Expense-Food Feature Project, grant number [PXM2018_014213_000033]; National Key Research and Development Program, grant number [2016YFD0400502]; Beijing Municipal Science and Technology Project, grant number [Z171100002217019]; Beijing Excellent Talents Training Project, grant number [2016000020124G025]; and the Support Project of High-level Teachers in Beijing Municipal Universities in the Period of 13th Five-year Plan, grant number [CIT&TCD201704042, CIT&TCD201804023].

Acknowledgments: This work was supported by the National Natural Science Foundation of China (NSFC, Grant No. 31701575, 31571801), Construction of Scientific Research Innovation Service Ability-Basic Scientific Research Operating Expense-Food Feature Project (PXM2018_014213_000033), National Key Research and Development Program (Grant No. 2016YFD0400502), Beijing Municipal Science and Technology Project (Grant No. Z171100002217019), Beijing Excellent Talents Training Project (Grant No. 2016000020124G025), and the Support Project of High-level Teachers in Beijing Municipal Universities in the Period of 13th Five-year Plan (Grant No. CIT&TCD201704042, CIT&TCD201804023).

Conflicts of Interest: The authors declare no conflict of interest.

References

1. Amin, A.R.; Kucuk, O.; Khuri, F.R.; Shin, D.M. Perspectives for cancer prevention with natural compounds. *J. Clin. Oncol.* **2009**, *27*, 2712–2725. [CrossRef] [PubMed]
2. Jung, I.L. Soluble extract from Moringa oleifera leaves with a new anticancer activity. *PLoS ONE* **2014**, *9*, e95492. [CrossRef] [PubMed]
3. Lobo, V.; Patil, A.; Phatak, A.; Chandra, N. Free radicals, antioxidants and functional foods: Impact on human health. *Pharmacogn. Rev.* **2010**, *4*, 118–126. [CrossRef] [PubMed]
4. De Jesus Raposo, M.F.; de Morais, R.M.; de Morais, A.M. Health applications of bioactive compounds from marine microalgae. *Life Sci.* **2013**, *93*, 479–486. [CrossRef] [PubMed]
5. Liu, Q.; Huang, Y.; Zhang, R.; Cai, T.; Cai, Y. Medical Application of Spirulina platensis Derived C-Phycocyanin. *Evid. Based Complement. Altern. Med.* **2016**, *2016*, 7803846.
6. Ravi, M.; Tentu, S.; Baskar, G.; Rohan Prasad, S.; Raghavan, S.; Jayaprakash, P.; Jeyakanthan, J.; Rayala, S.K.; Venkatraman, G. Molecular mechanism of anti-cancer activity of phycocyanin in triple-negative breast cancer cells. *BMC Cancer* **2015**, *15*, 768. [CrossRef] [PubMed]
7. Thangam, R.; Suresh, V.; Asenath Princy, W.; Rajkumar, M.; Senthilkumar, N.; Gunasekaran, P.; Rengasamy, R.; Anbazhagan, C.; Kaveri, K.; Kannan, S. C-Phycocyanin from Oscillatoria tenuis exhibited an antioxidant and in vitro antiproliferative activity through induction of apoptosis and G0/G1 cell cycle arrest. *Food Chem.* **2013**, *140*, 262–272. [CrossRef] [PubMed]
8. Zhu, C.; Ling, Q.; Cai, Z.; Wang, Y.; Zhang, Y.; Hoffmann, P.R.; Zheng, W.; Zhou, T.; Huang, Z. Selenium-Containing Phycocyanin from Se-Enriched Spirulina platensis Reduces Inflammation in Dextran Sulfate Sodium-Induced Colitis by Inhibiting NF-κB Activation. *J. Agric. Food Chem.* **2016**, *64*, 5060–5070. [CrossRef] [PubMed]
9. Riss, J.; Décordé, K.; Sutra, T.; Delage, M.; Baccou, J.C.; Jouy, N.; Brune, J.P.; Oréal, H.; Cristol, J.P.; Rouanet, J.M. Phycobiliprotein C-phycocyanin from Spirulina platensis is powerfully responsible for reducing oxidative stress and NADPH oxidase expression induced by an atherogenic diet in hamsters. *J. Agric. Food Chem.* **2007**, *55*, 7962–7967. [CrossRef] [PubMed]
10. Zhang, H.; Chen, T.; Jiang, J.; Wong, Y.S.; Yang, F.; Zheng, W. Selenium-containing allophycocyanin purified from selenium-enriched Spirulina platensis attenuates AAPH-induced oxidative stress in human erythrocytes through inhibition of ROS generation. *J. Agric. Food Chem.* **2011**, *59*, 8683–8690. [CrossRef] [PubMed]
11. Nemoto-Kawamura, C.; Hirahashi, T.; Nagai, T.; Yamada, H.; Katoh, T.; Hayashi, O. Phycocyanin enhances secretary IgA antibody response and suppresses allergic IgE antibody response in mice immunized with antigen-entrapped biodegradable microparticles. *J. Nutr. Sci. Vitaminol.* **2004**, *50*, 129–136. [CrossRef] [PubMed]
12. Bharathiraja, S.; Seo, H.; Manivasagan, P.; Santha Moorthy, M.; Park, S.; Oh, J. In Vitro Photodynamic Effect of Phycocyanin against Breast Cancer Cells. *Molecules* **2016**, *21*, 1470. [CrossRef] [PubMed]
13. Pardhasaradhi, B.V.; Ali, A.M.; Kumari, A.L.; Reddanna, P.; Khar, A. Phycocyanin-mediated apoptosis in AK-5 tumor cells involves down-regulation of Bcl-2 and generation of ROS. *Mol. Cancer Ther.* **2003**, *2*, 1165–1170. [PubMed]
14. Ying, J.; Wang, J.; Ji, H.; Lin, C.; Pan, R.; Zhou, L.; Song, Y.; Zhang, E.; Ren, P.; Chen, J.; et al. Transcriptome analysis of phycocyanin inhibitory effects on SKOV-3 cell proliferation. *Gene* **2016**, *585*, 58–64. [CrossRef] [PubMed]
15. Saini, M.K.; Sanyal, S.N. Targeting angiogenic pathway for chemoprevention of experimental colon cancer using C-phycocyanin as cyclooxygenase-2 inhibitor. *Biochem. Cell Biol.* **2014**, *92*, 206–218. [CrossRef] [PubMed]
16. Gantar, M.; Dhandayuthapani, S.; Rathinavelu, A. Phycocyanin induces apoptosis and enhances the effect of topotecan on prostate cell line LNCaP. *J. Med. Food* **2012**, *15*, 1091–1095. [CrossRef] [PubMed]
17. Baudelet, P.H.; Gagez, A.L.; Bérard, J.B.; Juin, C.; Bridiau, N.; Kaas, R.; Thiéry, V.; Cadoret, J.P.; Picot, L. Antiproliferative activity of *Cyanophora paradoxa* pigments in melanoma, breast and lung cancer cells. *Mar. Drugs* **2013**, *11*, 4390–4406. [CrossRef] [PubMed]
18. Jemal, A.; Bray, F.; Center, M.M.; Ferlay, J.; Ward, E.; Forman, D. Global cancer statistics. *CA Cancer J. Clin.* **2011**, *61*, 69–90. [CrossRef] [PubMed]

19. Torre, L.A.; Bray, F.; Siegel, R.L.; Ferlay, J.; Lortet-Tieulent, J.; Jemal, A. Global cancer statistics, 2012. *CA Cancer J. Clin.* **2015**, *65*, 87–108. [CrossRef] [PubMed]

20. Siegel, R.; Naishadham, D.; Jemal, A. Cancer statistics, 2013. *CA Cancer J. Clin.* **2013**, *63*, 11–30. [CrossRef] [PubMed]

21. Li, B.; Gao, M.H.; Lv, C.Y.; Yang, P.; Yin, Q.F. Study of the synergistic effects of all-transretinoic acid and C-phycocyanin on the growth and apoptosis of A549 cells. *Eur. J. Cancer Prev.* **2016**, *25*, 97–101. [CrossRef] [PubMed]

22. Li, B.; Gao, M.H.; Chu, X.M.; Teng, L.; Lv, C.Y.; Yang, P.; Yin, Q.F. The synergistic antitumor effects of all-trans retinoic acid and C-phycocyanin on the lung cancer A549 cells in vitro and in vivo. *Eur. J. Pharmacol.* **2015**, *749*, 107–114. [CrossRef] [PubMed]

23. Bingula, R.; Dupuis, C.; Pichon, C.; Berthon, J.Y.; Filaire, M.; Pigeon, L.; Filaire, E. Study of the Effects of Betaine and/or C-Phycocyanin on the Growth of Lung Cancer A549 Cells In Vitro and In Vivo. *J. Oncol.* **2016**, *2016*, 8162952. [CrossRef] [PubMed]

24. Zhang, C.; Wang, L.; Chen, J.; Liang, J.; Xu, Y.; Li, Z.; Chen, F.; Du, D. Knockdown of Diaph1 expression inhibits migration and decreases the expression of MMP2 and MMP9 in human glioma cells. *Biomed. Pharmacother.* **2017**, *96*, 596–602. [CrossRef] [PubMed]

25. Oakes, V.; Wang, W.; Harrington, B.; Lee, W.J.; Beamish, H.; Chia, K.M.; Pinder, A.; Goto, H.; Inagaki, M.; Pavey, S.; et al. Cyclin A/Cdk2 regulates Cdh1 and claspin during late S/G2 phase of the cell cycle. *Cell Cycle* **2014**, *13*, 3302–3311. [CrossRef] [PubMed]

26. Martín, A.; Odajima, J.; Hunt, S.L.; Dubus, P.; Ortega, S.; Malumbres, M.; Barbacid, M. Cdk2 is dispensable for cell cycle inhibition and tumor suppression mediated by p27Kip1 and p21Cip1. *Cancer Cell* **2005**, *7*, 591–598. [CrossRef] [PubMed]

27. Yue, H.; Yu, J.; Zhao, X.; Song, F.; Feng, X. Expression of P57(kip2) and cyslinE proteins in human pancreatic cancer. *Chin. Med. J.* **2003**, *116*, 944–946. [PubMed]

28. Yamamoto, Y.; Gaynor, R.B. IkappaB kinases: Key regulators of the NF-kappaB pathway. *Trends Biochem. Sci.* **2004**, *29*, 72–79. [CrossRef] [PubMed]

29. Cuzick, J. Preventive therapy for cancer. *Lancet Oncol.* **2017**, *18*, e472–e482. [CrossRef]

30. Goyal, S.; Gupta, N.; Chatterjee, S.; Nimesh, S. Natural Plant Extracts as Potential Therapeutic Agents for the Treatment of Cancer. *Curr. Top. Med. Chem.* **2017**, *17*, 96–106. [CrossRef] [PubMed]

31. Sung, B.; Pandey, M.K.; Aggarwal, B.B. Fisetin, an inhibitor of cyclin-dependent kinase 6, down-regulates nuclear factor-kappaB-regulated cell proliferation, antiapoptotic and metastatic gene products through the suppression of TAK-1 and receptor-interacting protein-regulated IkappaBalpha kinase activation. *Mol. Pharmacol.* **2007**, *71*, 1703–1714. [PubMed]

32. Tan, S.; Yang, B.; Liu, J.; Xun, T.; Liu, Y.; Zhou, X. Penicillixanthone A, a marine-derived dual-coreceptor antagonist as anti-HIV-1 agent. *Nat. Prod. Res.* **2017**, *19*, 1–5. [CrossRef] [PubMed]

33. Castro-Carvalho, B.; Ramos, A.A.; Prata-Sena, M.; Malhao, F.; Moreira, M.; Gargiulo, D.; Dethoup, T.; Buttachon, S.; Kijjoa, A.; Rocha, E. Marine-derived Fungi Extracts Enhance the Cytotoxic Activity of Doxorubicin in Nonsmall Cell Lung Cancer Cells A459. *Pharmacogn. Res.* **2017**, *9*, S92–S98.

34. Sawadogo, W.R.; Boly, R.; Cerella, C.; Teiten, M.H.; Dicato, M.; Diederich, M. A Survey of Marine Natural Compounds and Their Derivatives with Anti-cancer Activity Reported in 2012. *Molecules* **2015**, *20*, 7097–7142. [CrossRef] [PubMed]

35. Bechelli, J.; Coppage, M.; Rosell, K.; Liesveld, J. Cytotoxicity of algae extracts on normal and malignant cells. *Leuk. Res. Treatment* **2011**, *2011*, 373519. [CrossRef] [PubMed]

36. Jiang, L.Q.; Wang, Y.J.; Yin, Q.F.; Liu, G.X.; Liu, H.H.; Huang, Y.J.; Li, B. Phycocyanin: A Potential Drug for Cancer Treatment. *J. Cancer* **2017**, *8*, 3416–3429. [CrossRef] [PubMed]

37. Wang, H.; Liu, Y.; Gao, X.; Carter, C.L.; Liu, Z.R. The recombinant beta subunit of C-phycocyanin inhibits cell proliferation and induces apoptosis. *Cancer Lett.* **2007**, *247*, 150–158. [CrossRef] [PubMed]

38. Ristic, B.; Bhutia, Y.D.; Ganapathy, V. Cell-surface G-protein-coupled receptors for tumor-associated metabolites: A direct link to mitochondrial dysfunction in cancer. *Biochim. Biophys. Acta* **2017**, *1868*, 246–257. [CrossRef] [PubMed]

39. Peyressatre, M.; Prével, C.; Pellerano, M.; Morris, M.C. Targeting cyclin-dependent kinases in human cancers: From small molecules to Peptide inhibitors. *Cancers* **2015**, *7*, 179–237. [CrossRef] [PubMed]

40. Subhashini, J.; Mahipal, S.V.; Reddy, M.C.; Mallikarjuna Reddy, M.; Rachamallu, A.; Reddanna, P. Molecular mechanisms in C-Phycocyanin induced apoptosis in human chronic myeloid leukemia cell line-K562. *Biochem. Pharmacol.* **2004**, *68*, 453–462. [CrossRef] [PubMed]

41. Liao, G.; Gao, B.; Gao, Y.; Yang, X.; Cheng, X.; Ou, Y. Phycocyanin Inhibits Tumorigenic Potential of Pancreatic Cancer Cells: Role of Apoptosis and Autophagy. *Sci. Rep.* **2016**, *6*, 34564. [CrossRef] [PubMed]

42. Wang, C.Y.; Wang, X.; Wang, Y.; Zhou, T.; Bai, Y.; Li, Y.C.; Huang, B. Photosensitization of phycocyanin extracted from Microcystis in human hepatocellular carcinoma cells: Implication of mitochondria-dependent apoptosis. *J. Photochem. Photobiol. B.* **2012**, *117*, 70–79. [CrossRef] [PubMed]

43. Yao, G.D.; Yang, J.; Li, Q.; Zhang, Y.; Qi, M.; Fan, S.M.; Hayashi, T.; Tashiro, S.; Onodera, S.; Ikejima, T. Activation of p53 contributes to pseudolaric acid B-induced senescence in human lung cancer cells in vitro. *Acta Pharmacol. Sin.* **2016**, *37*, 919–929. [CrossRef] [PubMed]

44. Vitale, I.; Galluzzi, L.; Castedo, M.; Kroemer, G. Mitotic catastrophe: A mechanism for avoiding genomic instability. *Nat. Rev. Mol. Cell Biol.* **2011**, *12*, 385–392. [CrossRef] [PubMed]

45. Shin, J.W.; Kwon, S.B.; Bak, Y.; Lee, S.K.; Yoon, D.Y. BCI induces apoptosis via generation of reactive oxygen species and activation of intrinsic mitochondrial pathway in H1299 lung cancer cells. *Sci. China Life Sci.* **2018**. [CrossRef] [PubMed]

46. Tsui, K.C.; Chiang, T.H.; Wang, J.S.; Lin, L.J.; Chao, W.C.; Chen, B.H.; Lu, J.F. Flavonoids from Gynostemma pentaphyllum Exhibit Differential Induction of Cell Cycle Arrest in H460 and A549 Cancer Cells. *Molecules* **2014**, *19*, 17663–17681. [CrossRef] [PubMed]

47. Li, F.; Zhang, J.; Arfuso, F.; Chinnathambi, A.; Zayed, M.E.; Alharbi, S.A.; Kumar, A.P.; Ahn, K.S.; Sethi, G. NF-κB in cancer therapy. *Arch. Toxicol.* **2015**, *89*, 711–731. [CrossRef] [PubMed]

48. Nishanth, R.P.; Ramakrishna, B.S.; Jyotsna, R.G.; Roy, K.R.; Reddy, G.V.; Reddy, P.K.; Reddanna, P. C-Phycocyanin inhibits MDR1 through reactive oxygen species and cyclooxygenase-2 mediated pathways in human hepatocellular carcinoma cell line. *Eur. J. Pharmacol.* **2010**, *649*, 74–83. [CrossRef] [PubMed]

49. Guo, M.; Ding, G.B.; Yang, P.; Zhang, L.; Wu, H.; Li, H.; Li, Z. Migration Suppression of Small Cell Lung Cancer by Polysaccharides from Nostoc commune Vaucher. *J. Agric. Food Chem.* **2016**, *64*, 6277–6285. [CrossRef] [PubMed]

50. Clark, A.G.; Vignjevic, D.M. Modes of cancer cell invasion and the role of the microenvironment. *Curr. Opin. Cell Biol.* **2015**, *36*, 13–22. [CrossRef] [PubMed]

51. Wu, L.C.; Lin, Y.Y.; Yang, S.Y.; Weng, Y.T.; Tsai, Y.T. Antimelanogenic effect of c-phycocyanin through modulation of tyrosinase expression by upregulation of ERK and downregulation of p38 MAPK signaling pathways. *J. Biomed. Sci.* **2011**, *18*, 74. [CrossRef] [PubMed]

52. Dolcet, X.; Llobet, D.; Pallares, J.; Matias-Guiu, X. NF-κB in development and progression of human cancer. *Virchows Arch.* **2005**, *446*, 475–482. [CrossRef] [PubMed]

53. Cheng, W.T.; Wang, N. Correlation between MMP-2 and NF-κB expression of intracranial aneurysm. *Asian Pac. J. Trop. Med.* **2013**, *6*, 570–573. [CrossRef]

Article

Anti-Tumorigenic and Anti-Metastatic Activity of the Sponge-Derived Marine Drugs Aeroplysinin-1 and Isofistularin-3 against Pheochromocytoma In Vitro

Nicole Bechmann [1,*], Hermann Ehrlich [2], Graeme Eisenhofer [1,3], Andre Ehrlich [4], Stephan Meschke [4], Christian G. Ziegler [3] and Stefan R. Bornstein [3,5]

[1] Institute of Clinical Chemistry and Laboratory Medicine, University Hospital Carl Gustav Carus, Technical University Dresden, Fetscherstrasse 74, 01307 Dresden, Germany; Graeme.eisenhofer@uniklinikum-dresden.de

[2] Institute of Experimental Physics, TU Bergakademie Freiberg, Leipziger 23, 09599 Freiberg, Germany; Hermann.Ehrlich@physik.tu-freiberg.de

[3] Department of Medicine III, University Hospital Carl Gustav Carus, Technical University Dresden, Fetscherstrasse 74, 01307 Dresden, Germany; Christian.ziegler@uniklinikum-dresden.de (C.G.Z.); Stefan.bornstein@uniklinikum-dresden.de (S.R.B.)

[4] BromMarin GmbH, Wernerstraße 1, 09599 Freiberg, Germany; Andre.ehrlich@brommarin.de (A.E.); Stephan.meschke@brommarin.de (S.M.)

[5] Center for Regenerative Therapies Dresden, Technical University Dresden, Fetscherstrasse 105, 01307 Dresden, Germany

* Correspondence: Nicole.bechmann@uniklinikum-dresden.de; Tel.: +49-351-458-19687

Received: 24 April 2018; Accepted: 15 May 2018; Published: 20 May 2018

Abstract: Over 10% of pheochromocytoma and paraganglioma (PPGL) patients have malignant disease at their first presentation in the clinic. Development of malignancy and the underlying molecular pathways in PPGLs are poorly understood and efficient treatment strategies are missing. Marine sponges provide a natural source of promising anti-tumorigenic and anti-metastatic agents. We evaluate the anti-tumorigenic and anti-metastatic potential of Aeroplysinin-1 and Isofistularin-3, two secondary metabolites isolated from the marine sponge *Aplysina aerophoba*, on pheochromocytoma cells. Aeroplysinin-1 diminished the number of proliferating cells and reduced spheroid growth significantly. Beside these anti-tumorigenic activity, Aeroplysinin-1 decreased the migration ability of the cells significantly ($p = 0.01$), whereas, the invasion capacity was not affected. Aeroplysinin-1 diminished the high adhesion capacity of the MTT cells to collagen ($p < 0.001$) and, furthermore, reduced the ability to form spheroids significantly. Western Blot and qRT-PCR analysis showed a downregulation of integrin β1 that might explain the lower adhesion and migration capacity after Aeroplysinin-1 treatment. Isofistularin-3 showed only a negligible influence on proliferative and pro-metastatic cell properties. These in vitro investigations show promise for the application of the sponge-derived marine drug, Aeroplysinin-1 as anti-tumorigenic and anti-metastatic agent against PPGLs for the first time.

Keywords: marine sponges; Aeroplysinin; Isofistularin; pheochromocytoma and paraganglioma; metastasis; cancer progression; cell adhesion molecules; integrin β1; hypoxia

---

1. Introduction

Pheochromocytomas and extra-adrenal paragangliomas are rare neural crest-derived tumors with a highly heterogeneous genetic background associated with variable diseases aggressiveness. Pheochromocytoma generally occurs as a benign tumor, but over 10% are malignant at the first presentation or at recurrence usually with metastases at lymph node, bone, liver, and/or lung.

Occurrence of metastatic disease is associated with a 5 years survival of >50% [1,2]. Currently, no methods for the identification, the prediction, or the cure of malignant pheochromocytomas are available. The development of malignancy and the underlying molecular pathways in PPGLs are poorly understood and efficient treatment strategies are missing.

Natural sources of potential anti-tumorigenic and anti-metastatic agents are, for example, provided by diverse marine sponges [3,4]. Lacking an effective external defense mechanism, marine sponges developed chemical defense strategies by the production of secondary metabolites with various bioactivities [5]. A prominent agent is the bioactive (+)-Aeroplysinin-1 isolated from the marine demosponge *Aplysina aerophoba* (Figure 1) [6], which is also the source of chitinous scaffolds for diverse biomedical [7,8] and technological [9] applications. Beside the antibiotic and antiviral activity, Aeroplysinin-1 triggers key molecules of the inflammatory response as especially cyclooxygenase-2 (COX-2), metalloproteinases 1 (MMP-1), and 2 (MMP-2) [10]. Moreover, Aeroplysinin-1 demonstrates an anti-angiogenic activity in vivo and in vitro [11]. An anti-tumorigenic effect of Aeroplysinin-1 has been demonstrated for two human breast cancer cell lines (ZR-75-1 and MCF-7). Sallam et al. [12] showed an inhibitory activity of several dibromotyrosine analogues of Aeroplysinin-1 on human prostate cancer proliferation, migration, and invasion. Treatment with Aeroplysinin-1 blocks the epidermal growth factor (EGF)-dependent proliferation probably due to the inhibition of EGF receptor phosphorylation [13]. Nevertheless, this receptor tyrosine kinase inhibitory activity of Aeroplysinin-1 was controversially discussed in the literature (reviewed in [11]). In conclusion, Aeroplysinin-1 addresses four hallmarks of cancer; proliferation, inflammation, angiogenesis, and metastasis, but the underlying mechanism is mainly unclear.

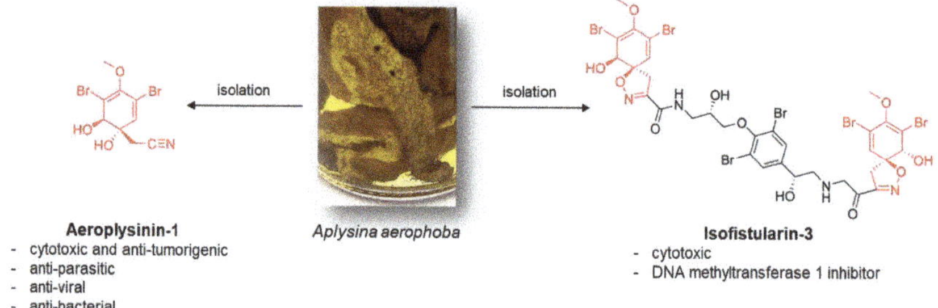

Figure 1. Schematic view: fresh collected 15 cm large *A. aerophoba* demosponge that grow under marine ranching conditions and the chemical structure of its bioactive secondary metabolites Aeropysinin-1 and Isofistularin-3.

A less characterized brominated compound derived from *A. aerophoba* is Isofistularin-3 (Figure 1). Cytotoxic activity of Isofistularin-3 against HeLa cells has been reported (IC$_{50}$ = 8.5 ± 0.2 μM) [14]. Recently, Florean et al. [15] described Isofistularin-3 as a new DNA methyltransferase (DNMT) 1 inhibitor. The agent reduces viability, colony formation as well as in vivo tumor growth in two lymphoma cell lines without affecting the viability of peripheral blood mononuclear cells or zebrafish development. The influence of Isofistularin-3 on cells´ pro-metastatic behavior has not yet been determined.

The absence of effective treatment strategies for malignant pheochromocytoma prompted us to investigate the anti-tumorigenic and anti-metastatic activity of Aeroplysinin-1 and Isofistularin-3 against pheochromocytoma cells in vitro. To the best of our knowledge, data on the impact of Aeroplysinin-1 and Isofistularin-3 pheochromocytoma cells are lacking. The choice to study bromotyrosines of the *A. aerophoba* demosponge origin was motivated by well-developed marine farming of this sponge species and its recognition as a renewable source to isolate large amounts of both bromotyrosines and chitin-based scaffolds [16].

2. Results

2.1. Anti-Proliferative Activity of Aeroplysinin-1 and Isofistularin-3 in Vitro

To investigate the anti-proliferative activity of Aeroplysinin-1 and Isofistularin-3 against PPGLs, three different pheochromocytoma cell lines were used. Aeroplysinin-1 diminished the cell viability (Figure 2A and Figure S1) of all three cell lines in a micromolar concentration (EC_{50} = 10–11 µM). Twenty-four hours of incubation under extrinsic hypoxia (1% oxygen) in the presence of Aeroplysinin-1 resulted in a slight decrease of the effect (EC_{50} = 12–15 µM). Interestingly, 24 h incubation with Isofistularin-3 under normoxic or hypoxic conditions had no influence on the viability of the rat PC12 cells up to a concentration of 100 µM. The viability of mouse pheochromocytoma cell lines, MPC and MTT, was reduced in a high micromolar range ($EC_{50,normoxia}$ = 43–44 µM; $EC_{50,hypoxia}$ = 59–91 µM). Isofistularin-3 had no influence on the number of proliferating cells, whereas, Aeroplysinin-1 decreased the number of proliferating cells in all three cell lines (Figure 2B–D). The pheochromocytoma cell lines are affected by the reduction of oxygen and stopped cell division under hypoxic conditions.

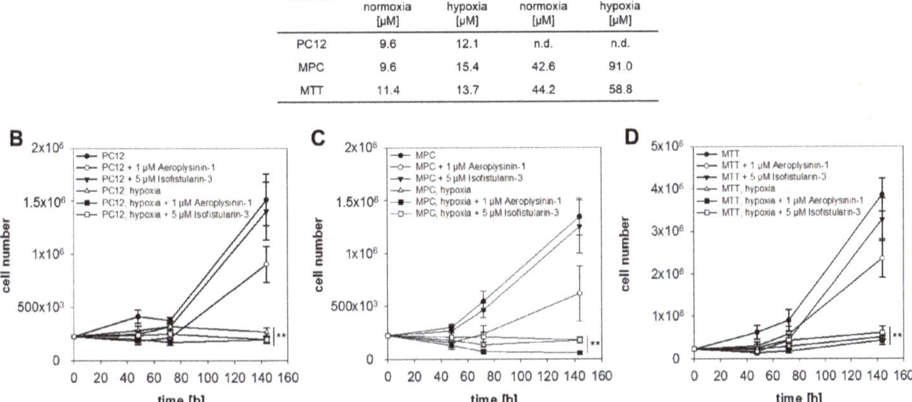

Figure 2. Anti-proliferative activity of Aeroplysinin-1 and Isofistularin-3 on pheochromocytoma cells in monolayer culture. (**A**) Aeroplysinin-1 decreased the viability of all three pheochromocytoma cell lines significantly after 24 h treatment. Isofistularin-3 only affected the viability of the mouse pheochromocytoma cells, whereas, the PC12 rat pheochromocytoma cells was not affected up to a concentration of 100 µM under normoxic and hypoxic conditions. Cultivation under hypoxia increased the necessary effective concentration to reduce the viability to 50% (EC_{50}). (**B**) Furthermore, the effect on the number of proliferating cells was analyzed under normoxic and hypoxic conditions. Treatment with 1 µM Aeroplysinin-1 reduced the number of proliferating cells in all three cell lines in trend. Under hypoxic conditions (1% oxygen), pheochromocytoma cells stopped cell division. Four to five independent experiments were performed (n = 4–14). Average ± standard error of the mean (SEM); Analysis of variance (ANOVA) and Bonferroni *post hoc* test comparison vs. control * $p < 0.05$, ** $p < 0.01$.

Aeroplysinin-1 induced apoptosis, analyzed by caspase (casp)-3 and casp-7 activity assay, in a concentration-dependent manner but predominant at lower concentrations (Figure 3A–C). Furthermore, gene expression analysis demonstrated a reduction of *casp-3* and *casp-7* after treatment with 10 µM Aeroplysinin-1 in the MTT cells (Figure 3D). In contrast to the programmed cell death during apoptosis, necrosis causes traumatic cell death due to an acute cellular injury. Several markers for necrosis (e.g., cyclophilin A and D, RIPK, and BNIP3) were described, but gene expression analysis showed no difference in the expression of these genes after treatment with Aeroplysinin-1 (Figure 3E,F).

Moreover, Belcin 1 (*Becn1*) a marker of cellular stress and a crucial regulator of the crosstalk between apoptosis and autophagy [17], was not affected by the treatment with Aeroplysinin-1 (Figure 3G).

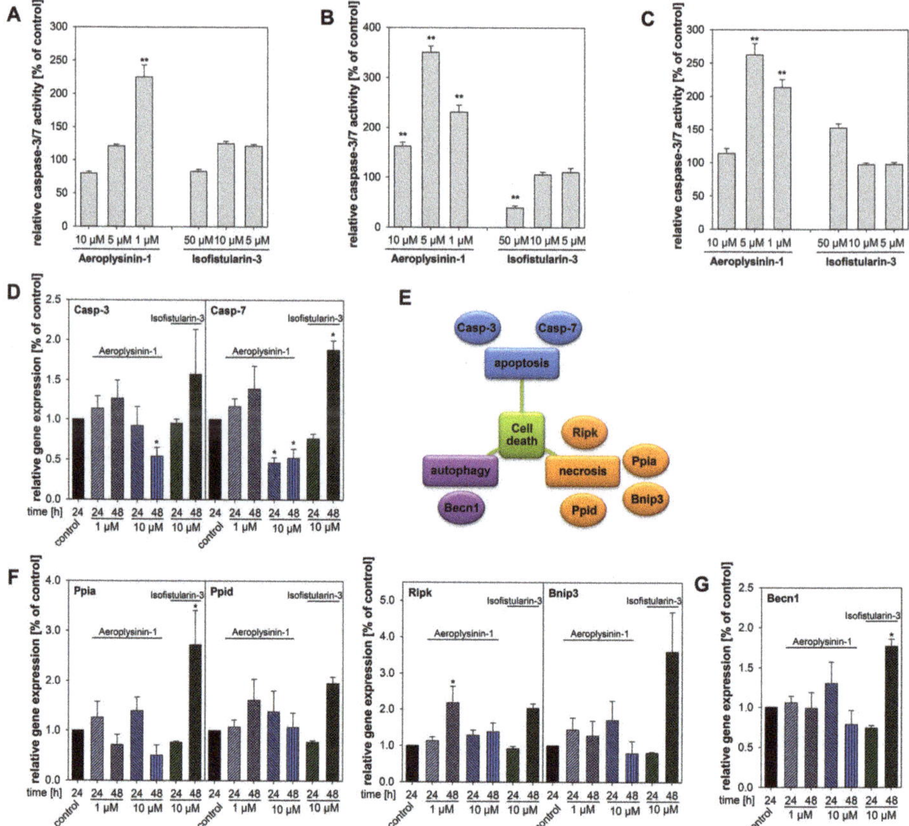

Figure 3. Influence of Aeroplysinin-1 and Isofistularin-3 on cell death related pathways. Induction of apoptosis by the treatment with Aeroplysinin-1 and Isofistularin-3 was analyzed by the measurement of the relative caspase-3 and caspase-7 activity in (**A**) MPC, (**B**) MTT, and (**C**) PC12 cells. Moreover, the impact on the (**D–G**) gene expression levels of the MTT cells was determined by qRT-PCR. (**E**) Beside apoptosis, necrosis and autophagy are also forms of cell death regulated by several genes. (**F**) The analyzed markers for necrosis *Ppia* (cyclophilin A), *Ppid* (cyclophilin D), *Ripk* and *Bnip3* were not affected by the treatment with Aeroplysinin-1. (**G**) Furthermore, no difference of the *Becn1* expression level was detectable after Aeroplysinin-1 treatment. Three to four independent experiments (n = 3–4). Average ± SEM; ANOVA and Bonferroni *post hoc* test comparison vs. control * $p < 0.05$ or ** $p < 0.01$.

Isofistularin-3 reduced caspase activity at higher concentrations (50 μM) (Figure 3A–C), whereas, expression of the caspase genes was induced after 48 h treatment with a concentration of 10 μM. The necrosis-associated genes (Figure 3E,F) were up-regulated after 48 h treatment with Isofistularin-3. Furthermore, *Becn1* as a marker of autophagy was increased after Isofistularin-3 treatment (Figure 3G).

Our investigations in monolayer culture demonstrate the promising effects of Aeroplysinin-1 as an anti-proliferative agent. In the next step, we aimed to determine whether Aeroplysinin-1 is also effective in a more complex model that is closer to the in vivo conditions by the formation of therapeutic resistant hypoxic and necrotic areas. Therefore, 3-dimensional tumor cell spheroids of the MTT cells were generated. Tumor cell spheroids are characterized by a pH, oxygen, and nutrient gradient and allow the screening of different drugs with regard to their potential anti-tumorigenic activity [18]. A single treatment with 5–10 µM Aeroplysinin-1 diminished the spheroid growth significantly over a time period of 12 days (Figure 4A). The medium of the spheroids was replaced every four days. Consequently, four days after treatment there was no Aeroplysinin-1 remaining in the medium. A single treatment with Aeroplysinin-1 resulted in a decelerated spheroid growth, but was not able to inhibit spheroid growth completely. Therefore, we decided to perform a fractionated treatment as usually performed in the clinic. Treatment with Aeroplysinin-1 took place on day four, 8, 11, and 15 after spheroid generation and resulted in a full inhibition of spheroid growth at a concentration of 10 µM (Figure 4B). Moreover, even 5 µM Aeroplysinin-1 significantly decreased spheroid growth over the entire time period. To sum up, Aeroplysinin-1 showed promising anti-proliferative activity in monolayer as well as spheroid culture of pheochromocytoma cells, indicating a potential anti-tumorigenic activity of this sponge-derived drug.

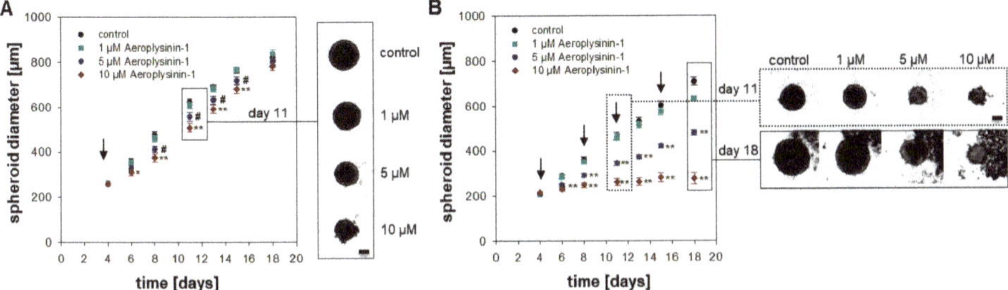

Figure 4. Spheroid growth inhibition by a single or fractionated treatment with Aeroplysinin-1. (**A**) Spheroids were treated with different concentrations Aeroplysinin-1 after completed spheroid formation (day 4) and growth was monitored over a time period of 14 days. Furthermore, (**B**) a fractionated treatment on day four, 8, 11, and 15 took place. Three to four independent experiments (n = 15–48). Average ± SEM; ANOVA and Bonferroni *post hoc* test comparison vs. control * $p < 0.05$, ** $p < 0.01$; or vs. 5 µM Aeropylysinin-1 # $p < 0.05$. Arrows mark the different treatment time points. Scale bar: 200 µm.

2.2. Influence of Aeroplysinin-1 and Isofistularin-3 on Cells' Pro-Metastatic Behavior

Beside the inhibition of tumor growth, the prevention of the development of tumor metastases is a major goal of a potential successful cancer treatment. Formation of metastases is particularly dependent on the migration, invasion, and adhesion capacity of a single cell (pro-metastatic properties). In this context, we asked whether the sponge-derived marine drugs Aeroplysinin-1 and Isofistularin-3 are able to reduce the pro-metastatic behavior of pheochromocytoma cells. To determine the effect of Aeroplysinin-1 and Isofistularin-3 on the migration and invasion capacity of MTT cells Boyden-Chamber assays with or without Matrigel® coating were performed. Aeroplysinin-1 decreased the migration ability of the cells significantly, whereas, the invasion capacity was not affected (Figure 5A,B). In contrast, Isofistularin-3 had no influence on MTT cell migration and invasion (Figure S2). Furthermore, the adhesion to the extracellular matrix proteins collagen and fibronectin was analyzed (Figure 5C,D). MTT cells adhered with a comparable low affinity to fibronectin that was mildly affected by treatment with Aeroplysinin-1. In contrast, Aeroplysinin-1 significantly diminished the high adhesion capacity of the MTT cells to collagen. The adhesion capacity of the pheochromocytoma cell was not affected by Isofistularin-3 (Figure S2). To further investigate the ability of Aeroplysinin-1 to prevent the formation of metastases the effect of this compound on spheroid formation was examined. Therefore, cell suspension was treated with different concentrations of Aeroplysinin-1 before cells were seeded to form spheroids. A concentration of 1 µM Aeroplysinin-1 significantly inhibits the ability of MTT cells to form spheroids (Figure 5E). These data indicate a potential influence of Aeroplysinin-1 on cell adhesion molecules (e.g., integrins, cadherins, and selectins) and, moreover, gives a first hint regarding the potential anti-metastatic effects of this sponge-derived drug.

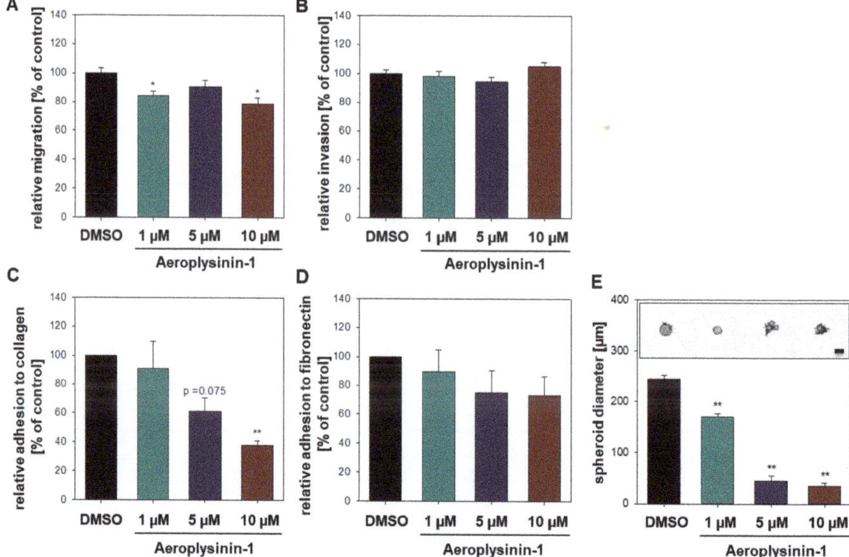

Figure 5. Aeroplysinin-1 influences the pro-metastatic behavior of MTT cells. Impact of different concentrations of Aeroplysinin-1 on (**A**) MTT cell migration and (**B**) invasion were analyzed in Boyden-Chamber assays (with (**B**) or without (**A**) Matrigel coating) after 24 h. Furthermore, the adhesion capacity of MTT cells to (**C**) collagen and (**D**) fibronectin after 24 h treatment with Aeroplysinin-1 was determined. (**E**) Spheroid formation was tracked over four days and the spheroid diameter was analyzed four days after seeding. Three independent experiments ($n = 12$–18). Average ± SEM; ANOVA and Bonferroni *post hoc* test comparison vs. control * $p < 0.05$ or ** $p < 0.01$. Scale bar: 200 µm.

2.3. Impact of Aeroplysinin-1 on Cell Adhesion Molecules

Cell adhesion molecules (CAMs) are the major players for the binding with other cells and the extracellular matrix (ECM). In consideration with the previous results, we hypothesized that Aeroplysinin-1 influences the expression of these CAMs. Integrins are responsible for the cell interaction with the ECM glycoproteins (e.g., collagen, fibronectin, and laminin) and are heterodimeric proteins consisting of alpha and beta subunits. Here we focused on the integrins α1β1 (ligands: collagens, laminins), α2β1 (ligands: collagens, laminins), and α4β1 (ligands: fibronectin, VCAM-1). An up-regulation of ITGB1 is associated with a pro-metastatic behavior, for example in lung [19] and breast cancer [20], as well as renal cell carcinomas [21]. A concentration of 10 µM Aeroplysinin-1 repressed the expression of the beta subunit (Itgb1) in the MTT cells significantly, whereas the expression of *Itga1* and *Itga3* was not affected by Aeroplysinin-1 (Figure 6A–C). MTT cells showed no *Itga4* expression that could explain the relatively low affinity of MTT cells to fibronectin.

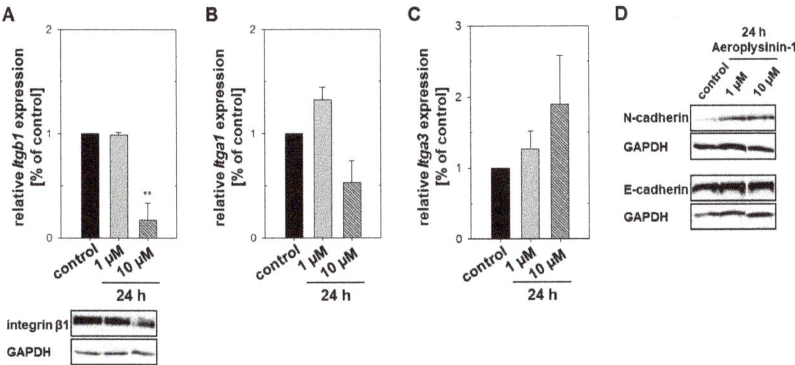

Figure 6. Regulation of cell adhesion molecules in mouse pheochromocytoma cells (MTT) by Aeroplysinin-1. The impact of 24 h treatment with Aeroplysinin-1 on the gene expression was confirmed by qRT-PCR and the protein expression was analyzed by western blot analysis. Three to four independent experiments were performed (n = 3–4) and a representative section of the immunochemical detection is shown. Treatment with 10 µM Aeroplysinin-1 diminished the expression of integrin β1 (**A**) significantly, whereas, the gene expression of *Itga1* (**B**) and *Itga3* (**C**) was not affected. Furthermore, Aeroplysinin-1 had no impact on the protein expression of the calcium-dependent cadherins. Average ± SEM; ANOVA and Bonferroni *post hoc* test comparison vs. control ** $p < 0.01$.

A second type of CAMs is the calcium-dependent cadherins. These transmembrane proteins are important for the formation of adherens junctions, which are crucial for cell-cell interactions. A loss of epithelial(E)-cadherin (CDH1) expression or function diminishes the strength of cellular adhesion within a tissue and results in an increase in cellular motility that correlates with tumor progression and metastasis [22]. Furthermore, the expression of neuronal(N)-cadherin (CDH2), normally expressed from mesenchymal cells, promotes cellular motility and the invasiveness of tumor cells. This switch from E-cadherin to N-cadherin has a functional significance in cancer metastasis [23]. Treatment with Aeroplysinin-1 had no impact on the protein expression of E- and N-cadherin (Figure 6D). Beta-catenin as a subunit of the cadherin protein complex is also involved in the regulation and coordination of cell-cell adhesion and acts as a signal transducer in the Wnt signaling pathway. Park et al. [24] demonstrated that Aeroplysinin-1 inhibits the proliferation of colon cancer cells by promoting β-catenin degradation. Beta-catenin gene expression was significantly reduced after 48 h treatment with Aeroplysinin-1 (data not shown) indicating an impact of Aeroplysinin-1 on the Wnt/β-catenin signaling.

3. Discussion

For the cure of metastatic PPGLs effective treatment strategies are missing. In our present study, we evaluate the impact of two secondary metabolites, Aeroplysinin-1 and Isofistularin-3, isolated from the marine demosponge *A. aerophoba* on the proliferative and pro-metastatic behavior of pheochromocytoma cell lines.

Aeroplysinin-1 diminished the number of proliferating pheochromocytoma cells and their viability in a micromolare range. Furthermore, the MTT cell spheroid growth was significantly repressed by the fractionated treatment with Aeroplysinin-1. The induction of apoptosis only occurred at a lower dose (1–5 µM) and the hypothesis that the effects of higher concentrations are related to necrosis or autophagy could not be confirmed. The impact on cell viability is in line with the findings of other groups on different tumor cell lines (reviewed in [11]). Koulman and coworkers concluded that the formation of free radicals by the semiquinone structure is at least contributing to the cytotoxicity of Aeroplysinin-1 [25]. Another sign in this direction is provided by the work from Stuhldreier et al., who demonstrate that Aeroplysinin-1 stimulates the phosphorylation of histone H2AX (γ-H2AX), a marker for DNA damage, in acute myeloid (NOMO-1) and acute monocytic (THP-1) cells [26]. Moreover, an EGF-dependent anti-proliferative effect was described by the group of Kreuter [13]. The inhibition of receptor tyrosine kinases that are involved in the transduction of mitogenic signals could furthermore be responsible for the effect of Aeroplysinin-1 on cell growth and proliferation [27]. Our data provide a first hint that Aeroplysinin-1 could demonstrate an anti-tumorigenic activity on PPGLs, however, additional in vivo experiments are needed to confirm these findings. Moreover, a specific therapeutic target of Aeroplysinin-1 is largely unknown and that is why the toxicity to the normal tissue must be evaluated carefully to identify unintentional negative side-effects. Anti-angiogenic activity [28,29] and the previously presented data underline the potential anti-tumorigenic activity of Aeroplysinin-1 on the one hand, but indicate an impact on normal blood vessel endothelial cells on the other hand. Kreuter and colleagues demonstrate that Aeroplysinin-1 blocked the proliferation of two breast cancer cell lines (0.25–0.5 µM) significantly; whereas a 10-fold higher concentration did not reveal any cytotoxicity on human fibroblast [13,30].

Cancer cell metastasis is a multistep cascade that is dependent on dynamic changes in the adhesive and migratory ability of tumor cells. Aeroplysinin-1 decreased the migration ability of pheochromocytoma cells significantly, whereas the invasion capacity was not affected. Furthermore, treatment with Aeroplysinin-1 reduced the adhesion capacity of the MTT cells. Quantitative RT-PCR and western blotting confirmed a reduction of the integrin β1 expression after 24 h treatment. The integrin $\alpha v\beta$3 and $\alpha v\beta$5 antagonist cilengitide showed favorable results in clinical phase 2 trials in patients with glioblastoma [31,32], non-small-cell lung [33,34], and prostate [35] cancer (reviewed [36]). Therapeutic strategies targeting integrin β1 have also shown efficacy to reduce tumor growth in preclinical models. Volociximab, a monoclonal antibody blocking the function of integrin $\alpha 5\beta$1 inhibits angiogenesis and decreases tumor growth in vitro [37]. A phase 1 trial in patients with advanced solid malignancies [38] and a phase 2 trial in patients with platinum-resistant advanced epithelial ovarian or primary peritoneal cancer [39] demonstrated that Voleciximab was well tolerated and may improve the clinical outcome. A peptide antagonist of integrin $\alpha 5\beta$1 with anti-angiogenic activity, ATN-161, prolonged stable disease in one-third of the patients with advanced solid tumors [40]. Interestingly, Motuporamine C, a cytotoxic alkaloid isolated from the marine sponge *Xestospongia exigua*, inhibits the invasion of breast and prostate carcinoma and glioma cell lines by a reduction of the integrin ß1 activity [41]. ß1 integrins induce adhesion-dependent activation of the focal adhesion kinase (FAK) and proto-oncogene tyrosine-protein kinase SRC, leading to proliferation, migration, invasion and the survival of tumor cells bound to the ECM [36]. Consequently, a downregulation of β1 integrin via Aeroplysinin-1 offers an excellent therapeutic opportunity to suppress tumor cell migration and invasion, and could furthermore explain the anti-proliferative activity of this sponge-derived secondary metabolite. Further studies in vivo are necessary to confirm our promising in vitro results.

An increased resistance of several tumors against commonly used treatment modalities, such as radiation therapy and chemotherapy, is a major limitation in the therapeutic regime of these tumors. The tumor microenvironment and extracellular matrix play a critical role in the response to different therapeutic options. β1 integrin is described as promising molecular target to enhance radiation therapy [42] and modulate the chemotherapy resistance [43]. A combination of Aeroplysinin-1 with one of the common treatment modalities should be the objective of further studies to evaluate the potential of Aeroplysinin-1 as potential chemo-/radio-sensitizing agent. This could be important especially for PPGLs with an activation in pseudohypoxic pathways, including those with mutations in HIF2α, VHL, PHD and particularly SDHB that carry a higher risk of malignancy [44]. Moreover, stabilization of hypoxia-inducible factors (HIFs) promotes the expression of integrin β1 [45].

In contrast to the auspicious impact of Aeroplysinin-1 on the proliferative and pro-metastatic behavior of the pheochromocytoma cell lines, Isofistularin-3 showed only a negligible activity on these cell properties. In line with the findings from Florean and coworkers, our qRT-PCR results indicate an induction of autophagy- and necrosis-related genes after 48 h treatment [15]. Nevertheless, an impact on the number of proliferating cells was not detected in our pheochromocytoma cell model.

For the first time, our in vitro investigations show promise for the application of Aeroplysinin-1 as anti-tumorigenic and anti-metastatic agent against PPGLs. We suggest that the mechanism is based on the reduction of the integrin β1 expression. The application of Aeroplysinin-1 and other sponge-derived secondary metabolites provides a promising therapeutic strategy especially for the treatment of metastatic PPGLs and other tumor entities with the tendency for metastatic spread.

4. Materials and Methods

BromMarin GmbH (Freiberg, Germany) kindly provided Isofistularin-3 and Aeroplysinin-1 with the purity grade of 99.9%.

4.1. Cell Culture

The mouse pheochromocytoma cells (MPC) generated from heterozygous neurofibromatosis knockout mice and its more aggressive derivate termed MTT as well as the rat pheochromocytoma cell line PC12 were acquired from Arthur Tischler [46–48]. For the cultivation of the PC12 cells RPMI-1640 containing 10% horse serum (HS) and 5% fetal calf serum (FCS) were used. The additional supplementation of this medium with 2 mM Glutamax is necessary for the MPC cell cultivation. The MTT cells were cultivated in Dulbecco's Modified Eagle Medium (DMEM) plus Glutamax supplemented with 10% HS, 5% FCS, and 1 mM sodium pyruvate. In general, cells were cultivated under normoxic conditions in a CO_2 incubator. In order to simulate hypoxic conditions (extrinsic hypoxia) cells were cultivated at reduced oxygen partial pressure ($\leq$1% O_2) in a special incubator equipped with an oxygen-sensor (Gasboy, Labotect, Göttingen, Germany). In all cases, cells were cultured at 37 °C, 5% CO_2, and 95% humidity. MycoAlert Mycoplasma Detection Kit (Lonza, Basel, Switzerland) was used for testing cells to be mycoplasma free. After trypsinization (trypsin/EDTA; 0.05%/0.02%) cells were diluted with complete medium (DMEM plus Glutamax with 10% HS and 5% FCS) and counted by using C-CHIPs (Neubauer improved). All experiments were performed antibiotic free after at least one passage after re-cultivation. The cultivation and the experimental work were performed by using collagen A coated cell culture dishes.

4.2. Viability Assay

To investigate the effect of Isofistularin-3 and Aeroplysinin-1 on cell viability the CellTiter 96® AQueous One Solution Cell Proliferation Assay (Promega, Mannheim, Germany) was used. In analogy to manufacturer´s instructions both cell lines (1.75×10^4) were seeded in 96-well plates and incubated for 24 h with different concentrations of the compounds. After 3 h incubation at 37 °C with CellTiter 96® AQueous One reagent the absorption of the whole plate was measured at 490 nm by Anthos htIII plate reader. The half maximal effective concentration (EC_{50}) was calculated from the dose-response curve by using the dose-response fit model of the SigmaPlot 12.5 software package (SYSTAT Software, San Jose, CA, USA).

4.3. Proliferation Assay

2.25×10^5 cells were seeded in 6-well plates and allowed to attach for 24 h. Cells were treated with Isofistularin-3 or Aeroplysinin-1 (1 μM or 10 μM) and cultivated for 48 h, 72 h, or 144 h under normoxic or hypoxic conditions. After incubation, cells were washed with PBS, trypsinated, and after careful resuspension in medium (total volume: 1 mL) cells were counted by using C-CHIPs (Neubauer improved). Each well was counted in duplicate.

4.4. Apoptosis Assay

To determine the effect of Isofistularin-3 and Aeroplysinin-1 on cell viability the Apo-ONE® Homogeneous Caspase-3/7 Assay (Promega, Germany) was used. In analogy to the manufacturer´s instructions all cell lines (1.75×10^4) were seeded in 96-well plates. Cells were treated with different concentrations of Isofistularin-3 and Aeroplysinin-1 for 24 h under normoxic or hypoxic conditions. After 1 h incubation at room temperature with 80 µL of the supernatant was transferred (duplicate) in a black 96-well plate and the fluorescence was analyzed at $485_{Ex}/530_{Em}$ by LB940 Multilabel Reader Mithras (Berthold Technologies GmbH Co. KG, Bad Wildbad, Germany; Software: MikroWin 2000, Labsis Laborsysteme GmbH, Neunkirchen-Seelscheid, Germany).

4.5. Migration Assay

The influence of Aeroplysinin-1 and Isofistularin-3 on the ability of MTT cells to migrate through 8 µm pores was ascertained by using TC-Inserts (Item No. 83.3931.800; Sarstedt AG & Co. KG, Nümbrecht, Germany). 5×10^6 cells were plated in a cell culture flask (T75) and cultivated for 24 h under normoxia. Medium was removed and DMEM plus Glutmax containing 0.2% bovine serum albumin (BSA) was added, followed by 24 h incubation. Cells were washed with PBS, trypsinized, and different concentrations of Aeroplysinin-1 (10, 5 and 1 µM), or Isofistularin-3 (10 and 5 µM), or DMSO as control was added to the cell suspension containing 1×10^6 cells per milliliter in DMEM containing 0.2% BSA. As chemoattractant complete DMEM + Glutamax (10% HS, 5% FCS, 1 mM sodium pyruvate) was filled in each well of a 12-well plate and the treated single cell suspensions (2×10^5 cells/insert) were added in the upper compartment of the cell culture insert. After 24 h incubation, culture medium was replaced by DMEM + Glutamax (0.2% BSA) with 1 µM calcein (BD™ calcein AM Fluorescent Dye, BD Biosciences, Franklin Lakes, NJ, USA) for 1 h at 37 °C. Afterwards, the lower compartment was washed with PBS and the cells that migrated through the pores were trypsinized and the fluorescence of the calcein-stained cells was measured at $485_{Ex}/528_{Em}$ by VICTOR3 1420 Multilabel Counter (Perkin Elmer, Hong Kong, China).

4.6. Invasion Assay

For invasion experiments the TC-Inserts were coated with basement membrane matrigel (Matrigel (BD Bioscience)/DMEM + Glutamax, 1/3, *v/v*) and the experimental procedure was analog to the migration assay above.

4.7. Adhesion Assay

4×10^5 cells were plated in each well of a 6-well plate (pre-culture). After 24 h cells were treated with different concentrations of Aeroplysinin-1 or Isofistularin-3. DMSO was used as a control. Furthermore, 24-well plates were coated with human fibronectin (5 mg/mL in PBS, Biochrom Ltd., Cambridge, UK) overnight at 4 °C or as previously described with collagen A. After 24 h the fibronectin or collagen coated plates were washed two times with PBS and unspecific binding sites were blocked with PBS containing 2% BSA for 1 h at 37 °C. The 24 h treated cells were washed two times with PBS, detached with trypsin (comment: detachment with EDTA or accutase was not useful for the cells) and resuspended in DMEM + Glutamax containing 0.2% BSA. 2×10^5 cells per well that were seeded in the fibronectin- or collagen-coated wells and allowed to adhere for 30 min. Non-adherent cells were washed away with PBS. The remaining cells were fixed for 5 min in PBS/methanol followed by 10 min in 100% methanol and stained for 15 min with crystal violet. After four washing steps (tap water) cells were dried on air and lysed by using PBS containing 0.5% Triton-X-100 for 30 min under continuous shaking. After transfer in a 96-well plate, absorption at 550 nm (reference 650 nm) was measured by VICTOR3 1420 Multilabel Counter. PBS containing 0.5% Triton-X-100 was used as blank.

4.8. Generation and Cultivation of Tumor Cell Spheroids

5×10^3 cells resupended in complete cell culture medium containing 20% of a 1.2% methylcellulose solution (0.24% (*w/v*), prepared in DMEM + Glutamax) were seeded in nonadherent round-bottom 96-well plates for suspension culture (Greiner, Kremsmünster, Austria). After 3–4 days´ cultivation consumed medium was replaced. To determine the influence of Aeroplysinin-1 or Isofistularin-3 four days' old spheroids were treated with different concentrations of the marine drugs. The treatment of the spheroids takes place one-time and was removed after 4 days by the medium replacement (Figure 4A). In a second experimental setting a fractionated treatment of spheroids was performed to determine if a fractionated treatment is able to reduce the necessary dose (Figure 4B). Therefore, spheroids were treated at day 4, 8, 11, and 15. The size of each spheroid was measured by using an inverse microscope Axiovert 200 M (Software: AxioVision 4.8; Carl Zeiss AG, Oberkoch, Germany). The area (A) of each spheroid was analyzed by using the software package Fiji (ImageJ, 1.51t, National Institutes of Health, Bethesda, MD, USA). The diameter (d) was calculated under acceptance of an approximately spherical form of the spheroids (d = 2 $\times \sqrt{(A/\pi)}$).

4.9. Tumor Cell Spheroid Formation Assay

To analyze the influence of Aeroplysinin-1 or Isofistularin-3 on the formation of the MTT cell spheroids the cell suspension containing 20% of a 1.2% methylcellulose solution was treated with different concentrations of the marine drug. After seeding, formation was tracked over a time period of four days.

4.10. RNA Isolation

RNA from cell pellets was isolated using RNeasy Plus Mini kit (Qiagen, Hilden, Germany) according to the manufacturer's instructions. RNA concentration was analyzed using BopPhotometer (Eppendorf, Hamburg, Germany). For the reverse transcription of 1 µg RNA the iScript RT kit (Bio-Rad, Hercules, CA, USA) was utilized. One microliter of the transcribed cDNA per reaction was used for qRT-PCR analysis.

4.11. Quantitative Real-Time PCR

Mouse *ß-actin, Ripk, Bnip3, Ppia, Ppid, Becn1, Casp3, Casp7, Cdh1, Cdh2, Itga1, Itga3, Itga4, Itgb1* mRNA expression was analyzed using the Quantitec SYBR PCR Master Mix (Qiagen GmbH, Hilden, Germany) and was expressed relative to ß-actin RNA levels as internal control. The sequence of each primer pair and the optimized annealing temperature is described in detail in the supporting information (Table S1). The amplification protocol consisted of a denaturation step at 95 °C for 7 min, followed by 40–45 cycles with a 95 °C denaturation step for 15 s, an annealing step for 20 s, and the extension step at 72 °C for 15 s. The expression of all genes was determined using Bio-Rad CFX 384 Real-Time System (Bio-Rad) and was analyzed using the comparative threshold cycle (CT) method [49].

4.12. SDS-PAGE and Western Blotting

Protein synthesis of integrin β1, E-cadherin, N-cadherin, and GAPDH was determined by Western blot analysis. After incubation with Aeroplysinin-1 under normoxic or hypoxic conditions, cells were washed with PBS, detached with trypsin/EDTA, and resuspended in cold medium. After centrifugation at 4 °C the obtained pellet was washed twice with PBS and stored at −80 °C. Whole cell lysates were prepared on ice using CellLytic™ M (Sigma-Aldrich, St. Louis, MO, USA, C2978) with protease inhibitors (1:100, Sigma-Aldrich; P8340). After 30 min incubation and thoroughly mixing, cell lysate were centrifuged to remove cell debris. The protein concentration of each supernatant was quantified using Bradford assay. Thirty microgram proteins were mixed with four-fold LDS sample buffer (C.B.S. Scientific, San Diego, CA, USA; FB31010) and 5% mercaptoenthanol, denatured at 99 °C for 5 min and separated by 10% SDS-polyacrylamide gels (PAGE). Protein transfer to a polyvinylidine difluoride membrane (0.45 μm; Whatman, Maidstone, UK) was performed by semi-dry electroblotting. Non-specific binding sites on the membrane were blocked by one hour incubation with 5% skimmed milk powder plus 2% bovine serum albumin in TBS-T (blocking solution) at room temperature. Membrane was incubated with primary antibodies anti-integrin beta 1 (1:500; ab179471; abcam plc., Cambridge, UK), anti-E cadherin (1:500, ab76319; abcam plc.), anti-N cadherin (1:500; ab18203; abcam plc.), and anti-GAPDH (1:1000; #2118, Cell Signaling Technology, Frankfurt am Main, Germany) for one hour at room temperature followed by an overnight incubation at 4 °C. After three washing steps in TBS-T, membranes were incubated for one hour at room temperature with peroxidase-conjugated secondary antibody goat anti-rabbit IgG (1:5000; sc-2004; Santa Cruz Biotechnology, Dallas, TX, USA) or goat anti-mouse IgG (1:5000; sc-2005; Santa Cruz Biotechnology). All antibodies were diluted in blocking solution. After washing with TBS-T, proteins were visualized by chemiluminescence under the use of SuperSignal® West Pico and Femto Chemiluminescent Substrate (Thermo Fisher Scientific, Waltham, MA, USA) and detected using G:BOX Chemi-XL1.4 imaging system from Syngene (Cambridge, UK).

4.13. Statistical Analysis

Descriptive data were expressed as average ± SEM. The number of n represents the number of technical and biological replicates within the independent experiments. Statistical analyses were carried out by a one-way analysis of variance with post hoc Bonferroni using SigmaPlot 12.5 (Systat Software GmbH, Erkrath, Germany).

Supplementary Materials: The following are available online at http://www.mdpi.com/1660-3397/16/5/172/s1, Table S1: Primer sequences and the targeted genes, Figure S1: Dose-response curves of three different pheochromocytoma cell lines after treatment with Aeroplysinin-1 or Isofistularin-3 under normoxic and hypoxic conditions, Figure S2: Impact of Isofistularin-3 on the pro-metastatic behavior of MTT cells.

Author Contributions: N.B. conceived and designed the experiments. N.B., H.E., G.E., S.R.B., C.G.Z., A.E., S.M. participated in the data interpretation, wrote the manuscript, and revised the paper.

Funding: This research was funded by the Deutsche Forschungsgemeinschaft (DFG) within the CRC/Transregio 205/1, "The Adrenal: Central Relay in Health and Disease", and the Paradifference Foundation as well as within DFG Project HE 394/3-2.

Acknowledgments: The excellent technical assistance of Linda Friedrich and Isabel Poser is greatly acknowledged. The authors thank Arthur Tischler (MD) for providing the MPC, MTT and PC12 cell lines.

Conflicts of Interest: The authors declare no conflict of interest.

References

1. Eisenhofer, G.; Bornstein, S.R.; Brouwers, F.M.; Cheung, N.-K.V.; Dahia, P.L.; De Krijger, R.R.; Giordano, T.J.; Greene, L.A.; Goldstein, D.S.; Lehnert, H. Malignant pheochromocytoma: Current status and initiatives for future progress. *Endocr. Relat. Cancer* **2004**, *11*, 423–436. [CrossRef] [PubMed]

2. John, H.; Ziegler, W.H.; Hauri, D.; Jaeger, P. Pheochromocytomas: Can malignant potential be predicted? *Urology* **1999**, *53*, 679–683. [CrossRef]

3. Bhatnagar, I.; Kim, S.-K. Marine antitumor drugs: Status, shortfalls and strategies. *Mar. Drugs* **2010**, *8*, 2702–2720. [CrossRef] [PubMed]

4. Dyshlovoy, S.A.; Honecker, F. *Marine Compounds and Cancer: 2017 Updates*; Multidisciplinary Digital Publishing Institute: Basel, 2018.

5. Rodríguez-Nieto, S.; González-Iriarte, M.; Carmona, R.; Muñoz-Chápuli, R.; Medina, M.A.; Quesada, A.R. Antiangiogenic activity of aeroplysinin-1, a brominated compound isolated from a marine sponge. *FASEB J.* **2002**, *16*, 261–263. [CrossRef] [PubMed]

6. Fattorusso, E.; Minale, L.; Sodano, G. Aeroplysinin-1, an antibacterial bromo-compound from the sponge verongia aerophoba. *J. Chem. Soc. Perkin 1* **1972**, *1*, 16–18. [CrossRef] [PubMed]

7. Ehrlich, H.; Ilan, M.; Maldonado, M.; Muricy, G.; Bavestrello, G.; Kljajic, Z.; Carballo, J.; Schiaparelli, S.; Ereskovsky, A.; Schupp, P. Three-dimensional chitin-based scaffolds from verongida sponges (demospongiae: Porifera). Part i. Isolation and identification of chitin. *Int. J. Biol. Macromol.* **2010**, *47*, 132–140. [CrossRef] [PubMed]

8. Mutsenko, V.V.; Bazhenov, V.V.; Rogulska, O.; Tarusin, D.N.; Schütz, K.; Brüggemeier, S.; Gossla, E.; Akkineni, A.R.; Meißner, H.; Lode, A. 3d chitinous scaffolds derived from cultivated marine demosponge aplysina aerophoba for tissue engineering approaches based on human mesenchymal stromal cells. *Int. J. Biol. Macromol.* **2017**, *104*, 1966–1974. [CrossRef] [PubMed]

9. Wysokowski, M.; Motylenko, M.; Bazhenov, V.V.; Stawski, D.; Petrenko, I.; Ehrlich, A.; Behm, T.; Kljajic, Z.; Stelling, A.L.; Jesionowski, T. Poriferan chitin as a template for hydrothermal zirconia deposition. *Front. Mater. Sci.* **2013**, *7*, 248–260. [CrossRef]

10. Martínez-Poveda, B.; García-Vilas, J.A.; Cardenas, C.; Melgarejo, E.; Quesada, A.R.; Medina, M.A. The brominated compound aeroplysinin-1 inhibits proliferation and the expression of key pro-inflammatory molecules in human endothelial and monocyte cells. *PLoS ONE* **2013**, *8*, e55203. [CrossRef] [PubMed]

11. García-Vilas, J.A.; Martínez-Poveda, B.; Quesada, A.R.; Medina, M.Á. Aeroplysinin-1, a sponge-derived multi-targeted bioactive marine drug. *Mar. Drugs* **2016**, *14*, 1. [CrossRef] [PubMed]

12. Sallam, A.A.; Ramasahayam, S.; Meyer, S.A.; Sayed, K.A.E. Design, synthesis, and biological evaluation of dibromotyrosine analogues inspired by marine natural products as inhibitors of human prostate cancer proliferation, invasion, and migration. *Bioorg. Med. Chem.* **2010**, *18*, 7446–7457. [CrossRef] [PubMed]

13. Kreuter, M.-H.; Leake, R.E.; Rinaldi, F.; Müller-Klieser, W.; Maidhof, A.; Müller, W.E.G.; Schröder, H.C. Inhibition of intrinsic protein tyrosine kinase activity of egf-receptor kinase complex from human breast cancer cells by the marine sponge metabolite (+)-aeroplysinin-1. *Comp. Biochem. Physiol. B Biochem.* **1990**, *97*, 151–158. [CrossRef]

14. Teeyapant, R.; Woerdenbag, H.; Kreis, P.; Hacker, J.; Wray, V.; Witte, L.; Proksch, P. Antibiotic and cytotoxic activity of brominated compounds from the marine sponge verongia aerophoba. *Z. Naturforsch. C* **1993**, *48*, 939–945. [PubMed]

15. Florean, C.; Schnekenburger, M.; Lee, J.-Y.; Kim, K.R.; Mazumder, A.; Song, S.; Kim, J.-M.; Grandjenette, C.; Kim, J.-G.; Yoon, A.-Y.; et al. Discovery and characterization of isofistularin-3, a marine brominated alkaloid, as a new DNA demethylating agent inducing cell cycle arrest and sensitization to trail in cancer cells. *Oncotarget* **2016**, *7*, 24027–24049. [CrossRef] [PubMed]

16. Ehrlich, H.; Bazhenov, V.; Meschke, S.; Bürger, M.; Ehrlich, A.; Petovic, S.; Durovic, M. Marine invertebrates of boka kotorska bay unique sources for bioinspired materials science. *Boka Kotorska Bay Environ.* **2016**, *54*, 313–334.

17. Kang, R.; Zeh, H.J.; Lotze, M.T.; Tang, D. The beclin 1 network regulates autophagy and apoptosis. *Cell Death Differ.* **2011**, *18*, 571–580. [CrossRef] [PubMed]

18. Kunz-Schughart, L.A.; Freyer, J.P.; Hofstaedter, F.; Ebner, R. The use of 3-d cultures for high-throughput screening: The multicellular spheroid model. *J. Biomol. Screen.* **2004**, *9*, 273–285. [CrossRef] [PubMed]

19. Wang, X.-M.; Li, J.; Yan, M.-X.; Liu, L.; Jia, D.-S.; Geng, Q.; Lin, H.-C.; He, X.-H.; Li, J.-J.; Yao, M. Integrative analyses identify osteopontin, lamb3 and itgb1 as critical pro-metastatic genes for lung cancer. *PLoS ONE* **2013**, *8*, e55714. [CrossRef] [PubMed]

20. Lowy, C. Tenascin C Interacts with Integrin Receptors to Promote Breast Cancer Metastasis to the Lungs. Ph.D. Dissertation, Ruperto-Carola University of Heidelberg, Heidelberg, Germany, 2017. [CrossRef]

21. Erdem, M.; Erdem, S.; Sanli, O.; Sak, H.; Kilicaslan, I.; Sahin, F.; Telci, D. In Up-regulation of tgm2 with itgb1 and sdc4 is important in the development and metastasis of renal cell carcinoma. *Urol. Oncol.* **2014**, *25*, e13–e25. [CrossRef] [PubMed]

22. Beavon, I.R.G. The e-cadherin-catenin complex in tumour metastasis: Structure, function and regulation. *Eur. J. Cancer* **2000**, *36*, 1607–1620. [CrossRef]

23. Araki, K.; Shimura, T.; Suzuki, H.; Tsutsumi, S.; Wada, W.; Yajima, T.; Kobayahi, T.; Kubo, N.; Kuwano, H. E/N-cadherin switch mediates cancer progression via tgf-β-induced epithelial-to-mesenchymal transition in extrahepatic cholangiocarcinoma. *Br. J. Cancer* **2011**, *105*, 1885. [CrossRef] [PubMed]

24. Park, S.; Kim, J.-H.; Kim, J.E.; Song, G.-Y.; Zhou, W.; Goh, S.-H.; Na, M.; Oh, S. Cytotoxic activity of aeroplysinin-1 against colon cancer cells by promoting β-catenin degradation. *Food Chem.Toxicol.* **2016**, *93*, 66–72. [CrossRef] [PubMed]

25. Koulman, A.; Proksch, P.; Ebel, R.; Beekman, A.C.; van Uden, W.; Konings, A.W.T.; Pedersen, J.A.; Pras, N.; Woerdenbag, H.J. Cytoxicity and mode of action of aeroplysinin-1 and a related dienone from the sponge aplysina aerophoba. *J. Nat. Prod.* **1996**, *59*, 591–594. [CrossRef] [PubMed]

26. Stuhldreier, F.; Kassel, S.; Schumacher, L.; Wesselborg, S.; Proksch, P.; Fritz, G. Pleiotropic effects of spongean alkaloids on mechanisms of cell death, cell cycle progression and DNA damage response (ddr) of acute myeloid leukemia (aml) cells. *Cancer Lett.* **2015**, *361*, 39–48. [CrossRef] [PubMed]

27. Waldmann, H.; Hinterding, K.; Herrlich, P.; Rahmsdorf, H.J.; Knebel, A. Selective inhibition of receptor tyrosine kinases by synthetic analogues of aeroplysinin. *Angew. Chem. Int. Ed.* **1997**, *36*, 1541–1542. [CrossRef]

28. Martínez-Poveda, B.; Rodríguez-Nieto, S.; García-Caballero, M.; Medina, M.-Á.; Quesada, A.R. The antiangiogenic compound aeroplysinin-1 induces apoptosis in endothelial cells by activating the mitochondrial pathway. *Mar. Drugs* **2012**, *10*, 2033–2046. [CrossRef] [PubMed]

29. Córdoba, R.; Tormo, N.S.; Medarde, A.F.; Plumet, J. Antiangiogenic versus cytotoxic activity in analogues of aeroplysinin-1. *Bioorg. Med. Chem.* **2007**, *15*, 5300–5315. [CrossRef] [PubMed]

30. Kreuter, M.; Bernd, A.; Holzmann, H.; Müller-Klieser, W.; Maidhof, A.; Weissmann, N.; Kljajić, Z.; Batel, R.; Schröder, H.; Müller, W. Cytostatic activity of aeroplysinin-1 against lymphoma and epithelioma cells. *Z. Naturforsch. C* **1989**, *44*, 680–688. [PubMed]

31. Reardon, D.A.; Fink, K.L.; Mikkelsen, T.; Cloughesy, T.F.; O'Neill, A.; Plotkin, S.; Glantz, M.; Ravin, P.; Raizer, J.J.; Rich, K.M. Randomized phase ii study of cilengitide, an integrin-targeting arginine-glycine-aspartic acid peptide, in recurrent glioblastoma multiforme. *J. Clin. Oncol.* **2008**, *26*, 5610–5617. [CrossRef] [PubMed]

32. Stupp, R.; Hegi, M.E.; Gorlia, T.; Erridge, S.C.; Perry, J.; Hong, Y.-K.; Aldape, K.D.; Lhermitte, B.; Pietsch, T.; Grujicic, D. Cilengitide combined with standard treatment for patients with newly diagnosed glioblastoma with methylated mgmt promoter (centric eortc 26071–22072 study): A multicentre, randomised, open-label, phase 3 trial. *Lancet Oncol.* **2014**, *15*, 1100–1108. [CrossRef]

33. Manegold, C.; Vansteenkiste, J.; Cardenal, F.; Schuette, W.; Woll, P.J.; Ulsperger, E.; Kerber, A.; Eckmayr, J.; von Pawel, J. Randomized phase ii study of three doses of the integrin inhibitor cilengitide versus docetaxel as second-line treatment for patients with advanced non-small-cell lung cancer. *Investig. New Drugs* **2013**, *31*, 175–182. [CrossRef] [PubMed]

34. Vansteenkiste, J.; Barlesi, F.; Waller, C.; Bennouna, J.; Gridelli, C.; Goekkurt, E.; Verhoeven, D.; Szczesna, A.; Feurer, M.; Milanowski, J. Cilengitide combined with cetuximab and platinum-based chemotherapy as first-line treatment in advanced non-small-cell lung cancer (nsclc) patients: Results of an open-label, randomized, controlled phase ii study (certo). *Ann. Oncol.* **2015**, *26*, 1734–1740. [CrossRef] [PubMed]

35. Beekman, K.W.; Colevas, A.D.; Cooney, K.; DiPaola, R.; Dunn, R.L.; Gross, M.; Keller, E.T.; Pienta, K.J.; Ryan, C.J.; Smith, D. Phase ii evaluations of cilengitide in asymptomatic patients with androgen-independent prostate cancer: Scientific rationale and study design. *Clin. Genitourin. Cancer* **2006**, *4*, 299–302. [CrossRef] [PubMed]

36. Desgrosellier, J.S.; Cheresh, D.A. Integrins in cancer: Biological implications and therapeutic opportunities. *Nat. Rev. Cancer* **2010**, *10*, 9. [CrossRef] [PubMed]

37. Bhaskar, V.; Zhang, D.; Fox, M.; Seto, P.; Wong, M.H.; Wales, P.E.; Powers, D.; Chao, D.T.; DuBridge, R.B.; Ramakrishnan, V. A function blocking anti-mouse integrin α5β1 antibody inhibits angiogenesis and impedes tumor growth in vivo. *J. Transl. Med.* **2007**, *5*, 61. [CrossRef] [PubMed]

38. Ricart, A.D.; Tolcher, A.W.; Liu, G.; Holen, K.; Schwartz, G.; Albertini, M.; Weiss, G.; Yazji, S.; Ng, C.; Wilding, G. Volociximab, a chimeric monoclonal antibody that specifically binds α5β1 integrin: A phase i, pharmacokinetic, and biological correlative study. *Clin. Cancer Res.* **2008**, *14*, 7924–7929. [CrossRef] [PubMed]

39. Bell-McGuinn, K.M.; Matthews, C.M.; Ho, S.N.; Barve, M.; Gilbert, L.; Penson, R.T.; Lengyel, E.; Palaparthy, R.; Gilder, K.; Vassos, A. A phase ii, single-arm study of the anti-α5β1 integrin antibody volociximab as monotherapy in patients with platinum-resistant advanced epithelial ovarian or primary peritoneal cancer. *Gynecol. Oncol.* **2011**, *121*, 273–279. [CrossRef] [PubMed]

40. Cianfrocca, M.; Kimmel, K.; Gallo, J.; Cardoso, T.; Brown, M.; Hudes, G.; Lewis, N.; Weiner, L.; Lam, G.; Brown, S. Phase 1 trial of the antiangiogenic peptide atn-161 (ac-phscn-nh 2), a beta integrin antagonist, in patients with solid tumours. *Br. J. Cancer* **2006**, *94*, 1621. [CrossRef] [PubMed]

41. Roskelley, C.D.; Williams, D.E.; McHardy, L.M.; Leong, K.G.; Troussard, A.; Karsan, A.; Andersen, R.J.; Dedhar, S.; Roberge, M. Inhibition of tumor cell invasion and angiogenesis by motuporamines. *Cancer Res.* **2001**, *61*, 6788–6794. [PubMed]

42. Cordes, N.; Park, C.C. Beta 1 integrin as a molecular therapeutic target. *Int. J. Radiat. Biol.* **2007**, *83*, 753–760. [CrossRef] [PubMed]

43. Zhang, H.; Ozaki, I.; Mizuta, T.; Matsuhashi, S.; Yoshimura, T.; Hisatomi, A.; Tadano, J.; Sakai, T.; Yamamoto, K. B1-integrin protects hepatoma cells from chemotherapy induced apoptosis via a mitogen-activated protein kinase dependent pathway. *Cancer* **2002**, *95*, 896–906. [CrossRef] [PubMed]

44. Qin, N.; De Cubas, A.A.; Garcia-Martin, R.; Richter, S.; Peitzsch, M.; Menschikowski, M.; Lenders, J.W.; Timmers, H.J.; Mannelli, M.; Opocher, G. Opposing effects of HIF1α and HIF2α on chromaffin cell phenotypic features and tumor cell proliferation: Insights from MYC-associated factor X. *Int. J. Cancer* **2014**, *135*, 2054–2064. [CrossRef] [PubMed]

45. Keely, S.; Glover, L.E.; MacManus, C.F.; Campbell, E.L.; Scully, M.M.; Furuta, G.T.; Colgan, S.P. Selective induction of integrin β1 by hypoxia-inducible factor: Implications for wound healing. *FASEB J.* **2009**, *23*, 1338–1346. [CrossRef] [PubMed]

46. Powers, J.F.; Evinger, M.J.; Tsokas, P.; Bedri, S.; Alroy, J.; Shahsavari, M.; Tischler, A.S. Pheochromocytoma cell lines from heterozygous neurofibromatosis knockout mice. *Cell Tissue Res.* **2000**, *302*, 309–320. [CrossRef] [PubMed]

47. Martiniova, L.; Lai, E.W.; Elkahloun, A.G.; Abu-Asab, M.; Wickremasinghe, A.; Solis, D.C.; Perera, S.M.; Huynh, T.-T.; Lubensky, I.A.; Tischler, A.S.; et al. Characterization of an animal model of aggressive metastatic pheochromocytoma linked to a specific gene signature. *Clin. Exp. Metastasis* **2009**, *26*, 239–250. [CrossRef] [PubMed]

48. Tischler, A.S.; Greene, L.A.; Kwan, P.W.; Slayton, V.W. Ultrastructural effects of nerve growth factor on pc 12 pheochromocytoma cells in spinner culture. *Cell Tissue Res.* **1983**, *228*, 641–648. [CrossRef] [PubMed]

49. Rao, X.; Huang, X.; Zhou, Z.; Lin, X. An improvement of the 2^(–delta delta ct) method for quantitative real-time polymerase chain reaction data analysis. *Biostat. Bioinform. Biomath.* **2013**, *3*, 71–85.

 *marine drugs*

Review

Marine Microalgae with Anti-Cancer Properties

Kevin A. Martínez Andrade, Chiara Lauritano, Giovanna Romano and Adrianna Ianora *

Department of Integrative Marine Ecology, Stazione Zoologica Anton Dohrn, 80121 Naples, Italy;
kevin.martinez@szn.it (K.A.M.A.); chiara.lauritano@szn.it (C.L.); romano@szn.it (G.R.)
* Correspondence: adrianna.ianora@szn.it; Tel.: +39-081-583-3246

Received: 23 April 2018; Accepted: 12 May 2018; Published: 15 May 2018

Abstract: Cancer is the leading cause of death globally and finding new therapeutic agents for cancer treatment remains a major challenge in the pursuit for a cure. This paper presents an overview on microalgae with anti-cancer activities. Microalgae are eukaryotic unicellular plants that contribute up to 40% of global primary productivity. They are excellent sources of pigments, lipids, carotenoids, omega-3 fatty acids, polysaccharides, vitamins and other fine chemicals, and there is an increasing demand for their use as nutraceuticals and food supplements. Some microalgae are also reported as having anti-cancer activity. In this review, we report the microalgal species that have shown anti-cancer properties, the cancer cell lines affected by algae and the concentrations of compounds/extracts tested to induce arrest of cell growth. We also report the mediums used for growing microalgae that showed anti-cancer activity and compare the bioactivity of these microalgae with marine anticancer drugs already on the market and in phase III clinical trials. Finally, we discuss why some microalgae can be promising sources of anti-cancer compounds for future development.

Keywords: marine biotechnology; microalgae; anti-cancer

---

1. Introduction

Cancer includes a large group of pathologies related to the unrestrained proliferation of cells in the body [1]. There are more than 200 different types of cancers, and some cancers may eventually spread into other tissues causing metastases that are often lethal. Cancer is the leading cause of death globally, largely due to aging and growth of the world's population. According to the European Cancer Observatory [2], estimates for the four most common types of cancer in the European Union in 2012 were as follows: 342,137 cases of colon cancer, 309,589 cases of lung cancer (including trachea and bronchus cancer), 358,967 cases of breast cancer and 82,075 cases of skin melanoma. Finding more effective methods to treat cancer remains a challenge, and development of new therapeutic agents for cancer treatment is essential for continued progress against the disease. According to Dyshlovoy and Honecker [3] approximately 60% of the drugs used in hematology and oncology have their origin in natural sources, and one third of the most sold are either natural compounds or derivatives thereof. There has also been growing interest in marine bioprospecting, because potent natural compounds (e.g., terpenes, steroids, alkaloids, polyketides, etc.) have already been discovered from marine organisms. Currently there are seven drugs of marine origin on the market, four of which are anticancer drugs. There are also close to 26 marine natural products in clinical trials of which 23 are anti-cancer compounds [4]. Oceans cover nearly 70% of the planet, but remain largely unexplored. To date, more than 28,000 compounds isolated from marine organisms have been reported, and this number is rapidly growing each year [4]. However, despite the number of compounds isolated from marine organisms and the biological activities attributed to many of these, the search for ocean medicines is relatively recent and only in the middle part of the 20th century did scientists begin to systematically probe the oceans for new drugs. Today, the pipeline from the initial demonstration that a molecule may have therapeutic potential to the production of an approved drug involves pre-clinical testing, complex clinical trials in humans, and post-trial regulatory approval by the Food and Drug Administration (FDA). For drugs, this process can take 10 to 15 years (Figure 1) and costs millions of dollars [5], with less than 12% of the potential drugs receiving final approval [6].

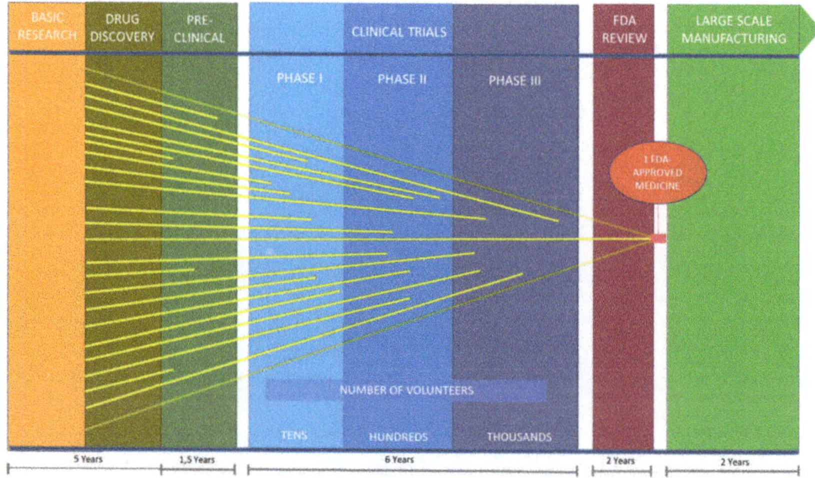

Figure 1. Time estimates for research and development of new Food and Drug Administration (FDA) approved drugs.

Several factors, such as difficulties in harvesting organisms, low quantities of active compounds in extracts, finding adequate procedures for isolation and purification, possible toxicity of the compounds and sustainable production of compounds may slow down the entire pipeline. Notwithstanding these difficulties, the discovery of new ocean medicines is one of the most promising new directions of marine science today. Novel initiatives with marine organisms aimed at enabling environmentally-friendly approaches to drug discovery have been tackled in several European Union 7th Framework Programme (EU FP7) and European Union Horizon 2020 projects (EU H2020), such as Biologically Active Molecules of Marine Based Origin (BAMMBO), Bluegenics, European Marine Biological Research Infrastructure Cluster (EMBRIC), Genetic Improvement of Algae for Value Added Products (GIAVAP), Marine Microorganisms Cultivation Methods for Improving their Biotechnological Applications (MaCuMBA) and PharmaSea. This has led to several technological advancements in culturing micro- and macro-organisms, increased sampling efforts in diverse and often extreme habitats that have led to the discovery of species that are new for science, massive sequencing of genomes and transcriptomes allowing for the identification of new metabolic pathways and/or assignment of potential functions to unknown genes. Here, we discuss the marine microalgae which have shown anti-cancer activity. This is the first review on this subject because recent studies have indicated that microalgae may represent a reservoir for new bioactive compounds that can act as anti-cancer drugs.

2. Marine Microalgae

Microalgae are eukaryotic plants that contribute up to 40% of global productivity [7]. They are at the base of aquatic food webs, have short generation times (doubling time = 5–8 h for some species) and have colonized almost all biotopes, from temperate to extreme environments (e.g., cold environments and hydrothermal vents). Their advantage in marine drug discovery is their metabolic plasticity, which can trigger the production of several compounds with possible applications in various biotechnology sectors (i.e., food, energy, health, environment and biomaterials) [8,9]. They can be easily cultivated in photo-bioreactors (e.g., in 100,000 L bioreactors) to obtain a huge biomass and represent a renewable and still poorly-explored resource for drug discovery. They use solar energy and fix CO_2 which contributes to the mitigation of greenhouse gas effects and the removal of nitrogen and phosphorous derivatives which can be pollutants depending on their concentration [10]. Table 1 reports the microalgal species that have shown anti-cancer properties, the cancer cell lines affected by microalgae and the concentrations that have been tested to induce the arrest of cell growth.

Table 1. Active microalgal species, active fraction/compounds tested and cell lines against which these have proven to be effective (CV stands for cell viability).

Microalgae	Fraction/Compound	Target Cells	Active Concentration	Reference
Thalassiosira rotula, Skeletonema costatum and *Pseudonitzschia delicatissima.* Commercial source, not from microalgae	Polyunsaturated aldehydes (PUAs)	Colon adenocarcinoma (Caco-2) Lung adenocarcinoma (A549) Colon adenocarcinoma (COLO 205)	11 to 17 µg/mL (arrest of cell growth) 0.22 to 1.5 µg/mL (CV of 80% to 0% depending on the conditions)	[11] [12]
Chlorella ellipsoidea	Carotenoid extract	Colon carcinoma (HCT-116)	40 µg/mL (IC$_{50}$)	[13]
Synedra acus	Chrysolaminaran (polysaccharide)	Colorectal adenocarcinoma (HT-29 and DLD-1)	54.5 and 47.7 µg/mL (IC$_{50}$ for HT-29 and DLD-1)	[14]
Dunaliella tertiolecta	Violaxanthin (carotenoid already identified in *C. ellipsoidea*)	Breast adenocarcinoma (MCF-7)	40 µg/mL (to observe cytostatic activity)	[15]
Cocconeis scutellum	Eicosapentaenoic acid (EPA)	Breast carcinoma (BT20)	Not clarified	[16]
Chaetoseros sp., Cylinrotheca closterium, Odontella aurita and *Phaeodactylum tricornutum*	Fucoxanthin (carotenoid)	Promyelocytic leukemia (HL-60), Caco-2, colon adenocarcinoma (HT-29), DLD-1 and prostate cancer (PC-3, DU145 and LNCaP)	29.78 µg/mL (CV of 17.3% for HL-60) 10.01 µg/mL (CV of 14.8%, 29.4% and 50.8% for Caco-2, DLD-1 and HT-29) 13.18 µg/mL (CV of 14.9%, 5.0% and 9.8% for PC-3, DU145 and LNCaP)	[17]
Chaetoceros calcitrans	EtOH extract AcOEt extract	MCF-7 Breast adenocarcinoma (MDA-MB-231)	3.00 µg/mL (IC$_{50}$) 60 µg/mL (IC$_{50}$)	[18] [19]
Amphidinium carterae	CH$_3$Cl fraction Hexane fraction AcOEt fraction	HL-60 HL60, Skin melanoma (B16F10), A549	50 µg/mL (CV of 40%) 25–50 µg/mL (CV between 50% and 90%)	[20]
Eleven strains of benthic diatoms *Ostreopsis ovata Amphidinium operculatum*	MeOH extract	HL-60	50 µg/mL (CV of 48% for *O. ovata* and 58% for *A. operculatumi*)	[21]
Navicula incerta	Stigmasterol (phytosterol)	Liver hepatocellular carcinoma (HepG2)	8.25 µg/mL (CV of 54%)	[22]
Phaeodactylum tricornutum	Nonyl-8-acetoxy-6-methyloctanoate (NAMO, fatty alcohol ester) Monogalactosyl glycerols [1]	HL-60 Mouse epithelial cell lines (W2, D3)	22.3 µg/mL (IC$_{50}$) 40-50 µg/mL (concentration necessary to induce apoptosis)	[23] [24]
Skeletonema costatum Skeletonema marinoi	Hydrophobic fraction and PUAs Hydrophobic fraction	Caco-2 (A2058 not affected) Skin melanoma (A2058)	11 to 17 µg/mL (PUAs) 50 µg/mL (CV of 60%)	[11] [8]
Canadian marine microalgal pool	Aqueous extract	A549, lung carcinoma (H460), prostate carcinoma (PC-3, DU145), stomach carcinoma (N87), MCF-7, pancreas adenocarcinoma (BxPC-3) and osteosarcoma (MNNG)	5000 µg/mL (CV between 30% and 80% depending on the cell line)	[25]
Chlorella sorokiniana	Aqueous extract	A549 and lung adenocarcinoma (CL1-5)	0.0156 to 1 µg/mL (CV reduced down to 20% progressively)	[26]

[1] (2*S*)-1-*O*-5,8,11,14,17-eicosapentaenoyl-2-*O*-6,9,12-hexadecatrienoyl-3-*O*-[β-D-galactopyranosyl]-glycerol and (2*S*)-1-*O*-3,6,9,12,15-octadecapentaenoyl-2-*O*-6,9,12,15-octadecatetraenoyl-3- *O*-β-D-galactopyranosyl-sn-glycerol.

As reported in previous studies, the bioactivity of microalgae may differ for different clones and can vary depending on the culturing conditions (e.g., nutrient availability, temperature, light intensity) [8,27] and growth phase [28]. For example, Ingebrigtsen et al. [27] demonstrated that the bioactivity of various marine microalgae extracts (i.e., the diatoms *Attheya longicornis*, *Chaetoceros socialis*, *Chaetoceros furcellatus*, *Skeletonema marinoi* and *Porosira glacialis*) with anti-cancer activity against melanoma A2058 cells changed when they were cultured under different light and temperature conditions. Lauritano et al. [8] also showed that microalgal bioactivity can vary depending on the nutrient concentrations used for their cultivation. These authors showed that the diatom *Skeletonema marinoi* had anti-cancer activity exclusively when cultured under nitrogen starvation conditions. Considering the importance of culturing conditions (Table 2), we report the mediums used for growing marine microalgae that showed anti-cancer activity (e.g., Conway's medium, Guillard's F/2 medium or variations of both mediums) and, where available, the sampling locations and harvesting times.

Table 2. Active microalgal species, sources, culturing conditions and references.

Microalgae	Source	Culturing Conditions	Harvesting Time	Reference
Synedra acus	Lake Baikal	Culture medium consisting of (mg/L) $Ca(NO_3)_2 \cdot 4H_2O$ (20), KH_2PO_4 (2), $MgSO_4$ (12), $NaHCO_3$ (30), Na_2EDTA (2.2), H_3BO_3 (2.4), $MnCl_2 \cdot 4H_2O$ (1.3), $(NH_4)_6Mo_7O_{24} \cdot 4H_2O$ (1), $Na_2SiO_3 \cdot 9H_2O$ (25), $FeCl_3$ (1.6), cyanocobalamine (0.04), thiamine (0.04), and biotin (0.04). 12 °C and 250–300 $\mu mol \cdot m^{-2} \cdot s^{-1}$ light intensity.	Not provided	[14]
Dunaliella tertiolecta	DT strain CCMP364 (NCMA, USA)	Conway medium. 20 °C, 180 $\mu mol \cdot m^{-2} \cdot s^{-1}$ light intensity.	Late exponential phase	[15]
Cocconeis scutellum	Mediterranean Sea, Stazione Zoologica A. Dohrn	Guillard's F/2 medium. 18 °C, 140 $\mu mol \cdot m^{-2} \cdot s^{-1}$ light intensity and 12 h:12 h photoperiod.	Not provided	[16]
Chaetoceros calcitrans	Strain UPMAAHU10 University Putra Malaysi	Conway medium. 24 °C, 120 $\mu mol \cdot m^{-2} \cdot s^{-1}$ light intensity, automatic oscillating shaker at 110 rpm and harvested at stationary phase (6–7 days). Conway medium. Conditions not provided.	Stationary phase Not provided	[18] [19]
Amphidinium carterae	Korea Marine Microalgae Culture Center	Conway medium. 20 °C, 34 $\mu mol \cdot m^{-2} \cdot s^{-1}$ light intensity and 24 h:0 h photoperiod.	Days 8–10	[20]
Eleven strains of benthic dinoflagellates	Coast of Jeju Island (Korea)	Daigo IMK medium (Nihon Pharmaceutical Co., Ltd.) and Guillard's F/2 medium. 20 °C, 180 $\mu mol \cdot m^{-2} \cdot s^{-1}$ light intensity and 12 h:12 h photoperiod.	Exponential phase.	[21]
Navicula incerta	Korea Marine Microalgae Culture Center.	Guillard's F/2 medium. Conditions not provided.	Not provided	[22]
Phaeodactylum tricornutum	Korea Marine Microalgae Culture Center Provasoli-Guillard National Center	Conway medium. 20 °C, 34 $\mu mol \cdot m^{-2} \cdot s^{-1}$ light intensity and 24 h:0 h photoperiod. Guillard's F/2 medium. 18 °C and 100 $\mu mol \cdot m^{-2} \cdot s^{-1}$ light intensity.	Days 8–10 Not provided	[23] [24]
Skeletonema marinoi FE6 (1997) FE60 (2005)	Adriatic Sea (Mediterranean Sea)	Guillard's F/2 medium. 19 °C, 100 $\mu mol \cdot m^{-2} \cdot s^{-1}$ light intensity and 12 h:12 h photoperiod.	Late stationary phase	[8]

3. Active Fractions from Marine Microalgae

3.1. Carotenoid Extract from Chlorella Ellipsoidea

Chlorella species are widely known as being a good commercial source of carotenoids such as lutein, β-carotene, zeaxanthin and astaxanthin. Kwang et al. 2008 [13] tested the anti-proliferative effect of the carotenoids extracted from the green algae *C. ellipsoidea* and *C. vulgaris* on a human colon carcinoma cell line (HCT116). Briefly, a freeze-dried *Chlorella* powder was extracted with ethanol, treated with a solution of KOH for saponification and further partitioned with hexane. This hexane phase, rich in carotenoids, was analyzed using HPLC-ESI-MS in order to identify the major carotenoid composition. They found that the carotenoid extract of *C. ellipsoidea* was mainly composed of violaxanthin and, in lower ratios, by two other xanthophylls (antheraxanthin and zeaxanthin). The extract from *C. vulgaris* was composed mainly of lutein. Anti-cancer activity was measured using the MTT assay after 24 h of exposure with the microalgal extracts. The half maximal inhibitory concentration (IC_{50}) value was 40.73 ± 3.71 µg/mL for *C. ellipsoidea* and 40.31 ± 4.43 µg/mL for *C. vulgaris*—much higher than the IC_{50} of pure lutein (21.02 ± 0.85 µg/mL). In order to understand if apoptosis is linked to the observed anti-proliferative effect, the authors also performed an annexin V-fluorescein assay to check phosphatidylserine translocation (index of apoptosis). The apoptotic effect was confirmed after treatment with both *C. ellipsoidea* and *C. vulgaris* extracts. They reported that apoptosis was 2.5-fold higher in the case of the *C. ellipsoidea* extract. The carotenoid extract was not tested on normal human cell lines.

3.2. Ethanol and Ethyl Acetate Extracts from Chaetoceros Calcitrans

Nigjeh et al. [18] tested the ethanolic extract from the planktonic diatom *Chaetoceros calcitrans* on breast adenocarcinoma (MCF-7), breast epithelial (MCF-10A) and peripheral blood mononuclear cells (PMBC). The ethanolic extract was obtained after the homogenization of the microalgal biomass with absolute ethanol and further filtration of the supernatant using filter cotton and a 0.2 µm filtration unit. The results were compared with the effects of tamoxifen, an already known drug used for the treatment of breast cancer. Cell viability was performed using the MTT assay on MCF-7, MCF-10A and PMBC cells. The results were expressed as IC_{50} values obtained by screening different concentrations (0 to 30 µg/mL) of *C. calcitrans* extract for 24 and 72 h. The IC_{50} values of the ethanolic extract screened on MCF-7 were 3.00 ± 0.65 µg/mL for 24 h and 2.69 ± 0.24 µg/mL for 72 h, while the IC_{50} values of MCF-10A cells were 12.00 ± 0.59 µg/mL for 24 h and 3.30 ± 0.36 µg/mL for 72 h. *C. calcitrans* extract did not display any cytotoxicity on PMBC cells even at the highest concentrations; the activity was specific to the cancer cells. The IC_{50} of tamoxifen on MCF-7 was 12.00 ± 0.52 µg/mL for 24 h and 9.00 ± 0.40 µg/mL for 72 h. The comparison with tamoxifen shows that the anti-cancer activity of *C. calcitrans* extract is very interesting. In addition, annexin V/propidium iodide analyses were performed and the results indicated apoptosis induction in MCF-7 cells after treatment with the extract. The authors also observed an increase in the proapoptotic protein Bax, and the caspases 3 and 7 transcripts.

Goh et al. [19] analyzed the effects of the extracts from *C. calcitrans* against a wide range of cancer cell lines. In particular, they studied the cytotoxicity of four crude solvent extracts (hexane, dichloromethane, ethyl acetate and methanol) in the following cancer cell lines: human breast adenocarcinoma (MDA-MB-231), MCF-7, mouse breast carcinoma (4T1), liver hepatocellular carcinoma (HepG2), cervix epithelial carcinoma (HeLa), human prostate carcinoma (PC-3), human lung adenocarcinoma (A549), human colon adenocarcinoma (HT-29), and human ovarian adenocarcinoma (Caov3). A mouse embryo fibroblast (3T3) cell line was used to measure cytotoxicity against non-tumorigenic cells. A freeze-dried powder of *C. calcitrans* was shaken in hexane, dichloromethane, ethyl acetate or methanol (MeOH) for 24 h and filtered through cotton to obtain the different extracts. The MTT assay was carried out after 72 h of treatment with the microalgal extracts and doxorubicin was used as a control (60 µg/mL of doxorubicin). The authors observed that crude ethyl acetate extract from *C. calcitrans* had cytotoxic properties in the MDA-MB-231 cancer cell line with IC_{50} of 60 µg/mL. An assay on a non-tumorigenic fibroblast cell line also revealed that the cytotoxic effect was specific against cancer cells; the extract did not have cytotoxic effects on the 3T3 cell line.

3.3. Organic Fractions from Amphidinium Carterae

Samarakoon et al. [20] tested the anti-proliferative activity of various fractions from the dinoflagellate *Amphidinium carterae* extract on different cancer cell lines: HL-60 (Human promyelocytic leukemia cells), B16F10 (mouse melanoma tumor cells), and A549 (adenocarcinomic human alveolar basal epithelial cells). Cytotoxicity assays were also carried out using the mouse monocyte macrophage cell line (RAW 264.7). Freeze-dried biomass from the cultured marine microalgae was grounded into fine powder, extracted with methanol (80%) and homogenized by sonication at 25 °C for 90 min. The crude methanol extract was concentrated by evaporating the solvent under reduced pressure using a rotary evaporator and further partitioned. Analytical grade n-hexane, chloroform, ethyl acetate, and water were used in solvent-solvent partition chromatography in order to obtain the fractions to be tested. Cell growth inhibition was measured with the MTT assay. *A. carterae* chloroform fraction was the most active and reduced HL-60 cell viability by about 50% after 24 h exposure at a concentration of 50 μg/mL. No tests were performed on normal human cell lines.

3.4. Methanolic Extracts from Amphidinium Carterae, Prorocentrum Rhathymum, Symbiodinium sp., Coolia Malayensis, Ostreopsis Ovata, Amphidinium operculatum and Heterocapsa psammophila

Shah et al. [21] cultivated up to eleven different strains of benthic dinoflagellates (*Amphidinium carterae*, *Prorocentrum rhathymum*, *Symbiodinium sp.*, *Coolia malayensis* strain 1, *Ostreopsis ovata* strain 1, *Ostreopsis ovata* strain 2, *Coolia malayensis* strain 2, *Amphidinium operculatum* strain 1, *Heterocapsa psammophila*, *Coolia malayensis* strain 3 and *Amphidinium operculatum* strain 2) isolated from the coast of Jeju Island (Korea) in 2011, to screen on RAW 264.7 (murine macrophage cell line) and HL-60 (human promyelocytic leukemia cell line) cells. They specified the specific sampling location and the microalgal growth phase tested (Table 2). To obtain the methanolic extracts, the freeze-dried biomass from the cultured marine microalgae was ground into fine powder, extracted with methanol (80%) and homogenized by sonication at 25 °C for 90 min. An MTT assay was carried out to study cell viability after extract exposure for 24 h at 37 °C. In this case, only *Ostreopsis ovata* 1 and *Amphidinium operculatum* 1 significantly inhibited the growth of HL-60 cancer cells (reducing cell viability between 40 and 60% compared to the control, at a concentration of 50 μg/mL). No tests were performed on normal human cell lines.

3.5. Hydrophobic Fraction from Skeletonema Marinoi

Lauritano et al. [8] studied the effects of 32 species of microalgae identified by microscopy and 18S sequencing. The microalgal biomass from the cultured marine microalgae was extracted with a ratio of acetone:water (1:1) and further fractionated using Amberlite RXAD16N resin with acetone as the resin eluent to obtain an hydrophobic fraction. The authors found that hydrophobic fractions from *Skeletonema marinoi*, *Alexandrium minutum*, *Alexandrium tamutum* and *Alexandrium andersoni* were active against a melanoma cancer cell line (A2058) at a concentration of 100 μg/mL. Further testing on a normal lung fibroblast (MRC-5) cell line showed that *Alexandrium* species were toxic. Two different strains of *Skeletonema marinoi* (FE6 and FE60 strains from the Adriatic Sea) were tested on an A2058 cell line. The colorimetric MTS (3-(4,5-dimethylthiazol-2-yl)-5-(3-carboxymethoxyphenyl)-2-(4-sulfophenyl)-2*H*-tetrazolium, a tetrazolium dye used for the quantification of viable cells) assay was performed to check the cytotoxicity in both normal and cancer cell lines. The results showed that only the FE60 strain was active against A2058 and only when cultured under nitrogen-starvation conditions. FE60 reduced the cell viability to 60% when screened at 50 μg/mL and to 10% when screened at 100 μg/mL.

3.6. Aqueous Extract from a Canadian Marine Microalgal Pool

Somasekharan et al. [25] studied the anti-proliferative effect of a raw marine microalgal material from Canada (dried powder) on eight different cancer cell lines, together with the anti-colony forming activity in the same lines. The dried microalgal powder was suspended in distilled water at a concentration of 30 mg/mL and then sonicated with short bursts. The sonicated suspension was then passed through a 25-gauge needle to release the cytosolic contents, followed by syringe filtration through 0.2 μm filters. This aqueous extract from the raw material was tested on A549, H460 (lung adenocarcinoma cell lines), PC-3, DU145 (prostate cancer cell lines), N87 (stomach cancer cell line), MCF7 (breast cancer cell line), BxPC-3 (pancreas cancer cell line) and MNNG (bone cancer cell line). The cells were incubated with the extract for 72 h at 0 (control), 1, 2 and 5 mg/mL. Cell viability was determined using the MTT assay. The extract did not show any significant activity at 1–2 mg/mL except for on the MNNG cell line (50% reduction in cell viability at 2 mg/mL). At 5 mg/mL proliferation of almost all the cell lines were significantly inhibited. Tests on normal human cell lines were not performed. The authors performed a crystal violet test at 0.5–5 mg/mL to check the anti-colony activity of the extract. This test indicated that the extract successfully inhibited the colony forming ability of all cancer cells tested even at the lowest concentration (0.5 mg/mL).

3.7. Aqueous Extract from Chlorella Sorokiniana

Chlorella sp. biomass is widely used as a dietary supplement in many countries and is mostly produced in Asia (https://www.marketresearchfuture.com/reports/chlorella-market-4413). There are also some patents related to its use as a dietary supplement (e.g., US 2005/0196389 A1 [29]). Lin et al. [26] studied the effects of hot water extracts from the diatom *Chlorella sorokiniana* (marine strain) on lung adenocarcinoma cell lines (A549 and CL1-5). The extracts were obtained by reflux extraction of the dried biomass with distilled water for 1 h and further filtration with N0.5 filter paper. To determine the cytotoxicity, the authors performed the MTT assay at a concentration range of 15.625 to 1000 ng/mL; the results indicated a dose-dependent reduction in cell viability on both cancer cell lines. The cytotoxicity on normal human cells was not tested. They also studied the mechanism of action of *C. sorokiniana* extract using annexin V/propidium iodide staining (flow cytometry analysis) to confirm a possible cell cycle arrest/apoptotic process. Cell cycle arrest was not observed even after 24 h of exposure, but an increment in the number of cells in sub-G1 phase was observed, which is a phenomenon that typically indicates apoptosis. Protein expression (Western blot analysis) of the cleaved and activated forms of caspase 9, caspase 3 and Poly (ADP-ribose) polymerase (PARP) increased in both cell lines after exposure to the microalgal extract after 24 h. The activation of caspase 9 and caspase 3 suggested that the main pathway involved in apoptosis was the mitochondrial pathway. In addition, the ratio of Bax/Bcl-2 (pro/antiapoptotic proteins) increased after 24 h of treatment which is another sign of apoptosis.

4. Active Compounds from Marine Microalgae

4.1. Polyunsaturated Aldehides (PUAs)

Miralto et al. [11] isolated three polyunsaturated aldehydes (PUAs, Figure 2) from the marine diatoms *Thalassiosira rotula, S. costatum* and *P. delicatissima*. They found that 2-*trans*-4-*cis*-7-*cis*-decatrienal, 2-*trans*-4-*trans*-7-*cis*-decatrienal and 2-*trans*-4-*trans*-decadienal had anti-proliferative activity on the human colon adenocarcinoma cell line (Caco-2). To check the anti-proliferative activity, they used different concentrations of PUAs between 0 and 20 µg/mL after 48 h of incubation. Concentrations of 11–17 µg/mL were enough to reduced cell viability to almost 0%. In addition, a TUNEL assay was performed to check DNA fragmentation and to verify that apoptosis had occurred.

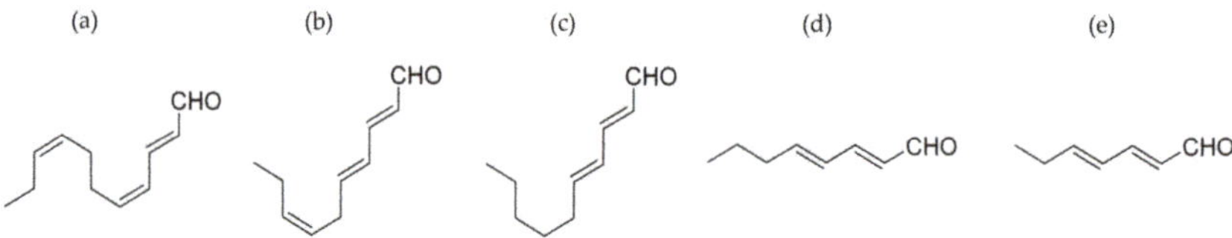

Figure 2. Polyunsaturated aldehydes. From left to right: 2-*trans*-4-*cis*-7-*cis*-decatrienal (**a**); 2-*trans*-4-*trans*-7-*cis*-decatrienal (**b**); 2-*trans*-4-*trans*-decadienal (**c**); 2-*trans*-4-*trans*-octadienal (**d**) and 2-*trans*-4-*trans*-heptadienal (**e**).

Sansone et al. [12] tested the effect of the commercially-available PUAs 2-*trans*-4-*trans*-decadienal (DD), 2-*trans*-4-*trans*-octadienal (OD) and 2-*trans*-4-*trans*-heptadienal (HD) (Figure 2) on the adenocarcinoma cell lines, lung A549 and colon COLO 205. The authors tested these three polyunsaturated aldehydes at different exposure times (i.e., 48 and 72 h) and concentrations (i.e., 2, 5 and 10 µM). For quantities of 2, 5 and 10 µM DD decreases in cell viability of 70%, 50% and 18% respectively, were induced in A549 cells after 24 h. For the COLO 205 cell line, cell viability decreased to 80%, 44% and 26% with 2, 5 and 10 µM DD, respectively. In the case of OD, 10 µM of the compound decreased cell viability to 35% after 72 h in A549 cells. At 2, 5 and 10 µM, OD also reduced cell viability to 60%, 60% and 41% in COLO 205 cells, respectively, after 72 h. At 10 µM, HD reduced cell viability to 10% in A549 cells after 48 h and 0% after 72 h, while the same concentration tested on COLO 205 cells reduced cell viability to 40% after 48 h and 28% after 72 h. The authors also tested the three PUAs on a normal lung/brunch epithelial BEAS-2B cell line to check cytotoxicity on normal cells. None of the PUAs were toxic.

4.2. Chrysolaminaran Polysaccharide

Kusaikin et al. [14] isolated one polysaccharide of the chrysolaminaran family (Figure 3) from the diatom *Synedra acus*. These storage polysaccharides are well known to be common water soluble biopolymers synthetized by diatoms [30]. The anti-tumor activity of the chrysolaminaran extracted from *S. acus* was studied on HTC-116 and DLD-1 human colon cancer cell lines. The MTS method was carried out to determine cell viability. Cancer cells were treated with 25, 50 and 100 µg/mL of chrysolaminaran for up to 72 h. The inhibition trend in the different experiments was irregular, but the IC_{50} values were determined for each cell line: 54.5 µg/mL for HCT-116 and 47.7 µg/mL for DLD-1. The authors did not find any toxicity on the HTC-116 and DLD-1 cell lines at concentrations above 200 mg/mL and, considering that most anti-tumor drugs are toxic at these concentrations, this is a very promising property. The degree of cytotoxicity on normal human cells was not tested.

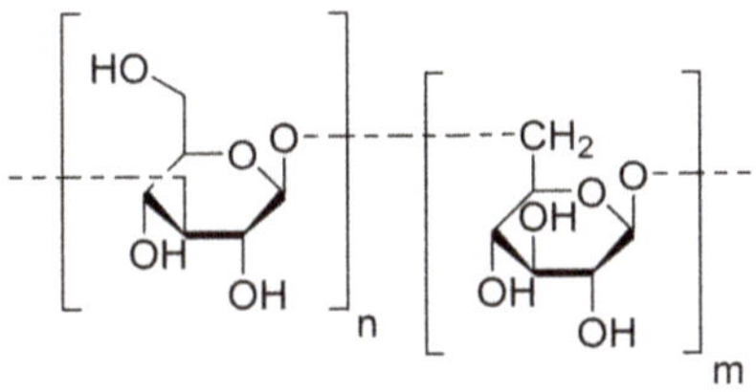

Figure 3. Chrysolaminaran monomer.

4.3. Violaxanthin

Pasquet et al. [15] performed an anti-cancer screening of extracts from the green algae *Dunaliella tertiolecta* on four different cancer cell lines: MCF-7, MDA-MB-231, A549 and LNCaP. They prepared different extracts using a wide range of solvents in terms of polarity, including dichloromethane, ethanol and ultrapure water. The MTT assay was used to evaluate the cells' viability. The dichloromethane extract showed significant activity against MCF-7 cancer cells. RP-HPLC analysis and fractionation was used to obtain one subfraction of the dichloromethane extract that was also screened for 72 h at concentrations between 0.1 µg/mL and 40 µg/mL. The subfraction was then identified as violaxanthin (Figure 4) at a rate of 95%. In addition to these results, the DNA of non-treated and treated cells was extracted and analysed using standard electrophoresis. Despite indications of early apoptosis (phosphatidylserines translocation detected using annexin-V-Alexa 568 fluorochrome), the violaxanthin subfraction did not cause any DNA fragmentation. Cytotoxicity tests were not performed in normal human cell lines.

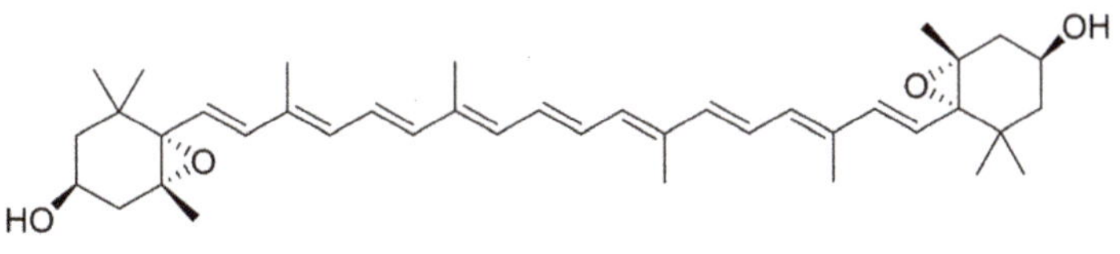

Figure 4. Violaxanthin.

4.4. Eicosapentaenoic Acid (EPA)

Nappo et al. [16] screened the extracts from the marine diatom *Cocconeis scutellum* on the following cells lines: BT20 (human breast cancer), MB-MDA468 (human breast cancer), LNCaP (human prostate adenocarcinoma cells), COR (Epstein-Barr Virus-transformed B cells isolated from human tonsils), JVM2 (lymphoblast immortalized with Epstein-Barr virus) and BRG-M (Burkitt's lymphoma cells). The results of the screening were not entirely published but the authors determined that *C. scutellum* extract was more effective on the BT20 cell line. The degree of cytotoxicity normal human cell lines was not studied. The fractionation of diethyl ether extract (the most active) from *C. scutellum* produced three fractions with differentiated activities. Fractions 1–2 did not induce any significant reduction in cell viability compared to the control but fraction 3 reduced the viability to 56.2%. DNA fragmentation was evaluated with the annexin V/propidium iodide staining methods. The analysis of the composition of these fractions indicated that fraction 1 contained glycerides (77.2% of total ion current) and fatty acids (2.4%), fraction 2 contained fatty acids (66.7%), monoglycerides (11.0%) and sterols (3.2%), and finally, fraction 3 contained fatty acids (81.7%) and 4-methylcholesterol (2.3%). The authors concluded that the fatty acid subfractions were responsible for this activity, specifically, eicosapentaenoic acid (EPA, Figure 5). This conclusion was reached because EPA was the only product in the fraction that has been reported to induce apoptosis [31]. The activation of caspases 8 and 3 was also confirmed by Western blot analysis. The authors concluded that it is not yet clear whether EPA is the only factor involved in the apoptosis of BT20 cells or if there is a synergic association among different compounds in the same fraction.

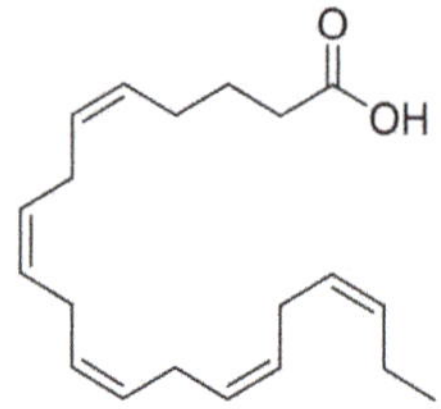

Figure 5. Eicosapentaenoic acid.

4.5. Fucoxanthin

Fucoxanthin (Figure 6) is one of the most studied compounds that can be found in marine micro- and macroalgae. It is a pigment from the family of the xanthophylls and a major carotenoid in brown algae [32]. Kadekaru et al. [33] evaluated fucoxanthin toxicity by providing oral doses (10 mg/kg and 50 mg/kg) to rats for a period of 28 days. Fucoxanthin did not show any obvious toxicity and is hence considered safe as a pharmaceutical ingredient. Ishikawa et al. [34] performed a similar analysis in mice using a metabolite of fucoxanthin, fucoxanthinol. In this case, they used a higher dose (200 mg/kg) for 28 days, but it did not show any toxicity either.

Hosokawa et al. [35] showed that fucoxanthin had strong anti-proliferative activity against HL-60 cells and could also induce apoptosis. Cells treated with 11.3 and 45.2 µM of fucoxanthin showed viabilities of 46.0% and 17.3%, respectively, after 24 h. Cell viability was determined by the dye exclusion test using trypan blue. Hosokawa et al. 2004 also tested the cell viability on three human colon cancer cell lines (Caco-2, DLD-1 and HT-29) treated with fucoxanthin. Caco-2, DLD-1 and HT-29 cells showed a dose-time dependent trend. The Caco-2 cell line was more affected than the other two cell lines (analyzed by WST-1 assay). Normal human cells were not tested.

Kotake-Nara et al. [36] examined 15 types of carotenoids (Neoxanthin, fucoxanthin, phytofluene, lycopene, phytoene, canthaxanthin, β-cryptoxanthin, zeaxanthin, β-carotene, α-carotene, γ-carotene, astaxanthin, capsanthin, lutein and violaxanthin) on three different prostate cancer cell lines (PC-3, DU145 and LNCaP) and found that fucoxanthin was one of the most active anti-cancer compounds.

The percentages of viable cells after 72 h when fucoxanthin was added at 20 µM were 14.9% for PC-3, 5.0% for DU145 and 9.8% for LNCaP, respectively (determined by MTT assay). Normal human cells were not tested.

Peng et al. [17] summarized the studies related to fucoxanthin, the microalgae that are known to produce it (i.e., *Chaetoseros* sp., *Cylinrotheca closterium*, *Odontella aurita* and *Phaeodactylum tricornutum*) and its role as a bioactive compound (e.g., as an antioxidant, anti-inflamatory, anti-cancer, anti-diabetic, skin protective agent, bone protective agent, etc.).

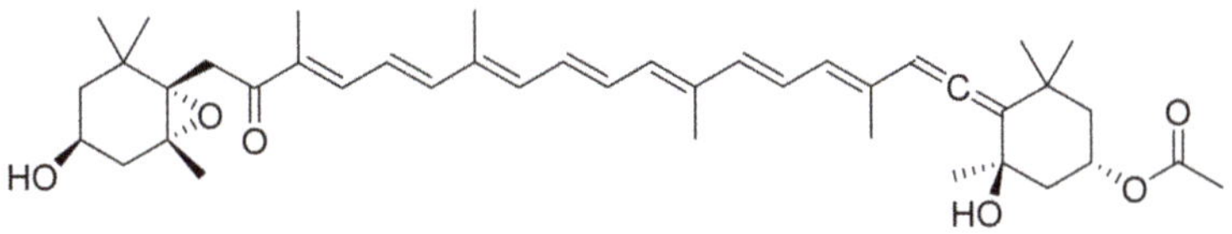

Figure 6. Fucoxanthin.

4.6. Stigmasterol

Kim et al. [22] isolated Stigmasterol (Figure 7) from *Navicula incerta* extracts using chromatography techniques such as silica gel open column chromatography and preparative thin layer chromatography (PTLC). They screened the anti-proliferative effect of the isolated stigmasterol at 5, 10 and 20 µM on HepG2 (human liver cancer cell line). Cytotoxicity values of 40%, 43% and 54% were found, respectively, which indicated a dose-dependent trend. The cytotoxic effects on normal human cells were not studied. The phytosterol-like structures with double bonds in the C-5 and C-22 positions, like stigmasterol, have been shown to induce apoptosis [37]. In this case, apoptosis was studied by controlling morphological changes, fluorescence-activated cell sorting, apoptosis pathways analysis, gene expression levels and also, with flow cytometric measurement of cell cycle arrest. All these assays indicated that stigmasterol has a huge apoptosis induction capability, probably via an apoptosis signaling pathway in the mitochondria.

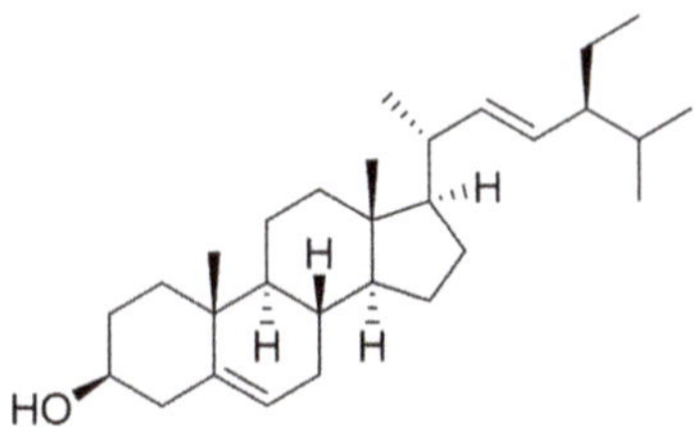

Figure 7. Stigmasterol.

4.7. NAMO (Nonyl 8-acetoxy-6-methyloctanoate)

Samarakoon et al. [23] tested the anti-cancer activity of 8-acetoxy-6-methyloctanoate (NAMO, Figure 8) obtained from *Phaeodactylum tricornutum* against three different cell lines: human promyelocytic leukemia cell line (HL-60), a human lung carcinoma cell line (A549) and a mouse melanoma cell line (B16F10). NAMO was screened at 25 and 50 µg/mL for 48 h. NAMO was only active against HL-60 cells at both concentrations tested. The highest growth inhibitory activity of about 70% on HL-60 cells was observed at a concentration of 50 µg/mL NAMO. The cytotoxic effects on normal human cells were not studied. Regarding its mechanism of action, NAMO induced DNA damage and increased apoptotic body formation. Cell cycle arrest and the accumulation of cells in the sub-G1 phase were observed to occur proportionally to the concentration of NAMO. The authors also observed activation of the pro-apoptotic protein Bax, suppression of the anti-apoptotic protein Bcl-x, and an increase in the expression of both caspase-3 and p53 proteins.

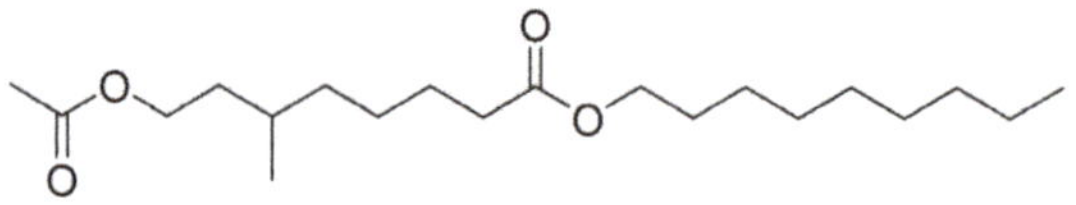

Figure 8. Nonyl 8-acetoxy-6-methyloctanoate (NAMO).

4.8. Monogalactosyl Glycerols

Andrianasolo et al. [24] isolated two different monogalactosyl glycerols (Figures 9 and 10) from *Phaeodactylum tricornutum* and tested them against immortal mouse epithelial cells (W2 and D3). The W2 cell line is a wild type, while D3 cells have the apoptosis function disabled through gene deletion (this assay is one of the approaches for the study of apoptosis and its role in cancer and oncogenesis). The minimum values required for apoptosis induction with this test were a death rate of 20% on the W2 cell line and, at a growth rate of at least 10% on the D3 cell line. For compound **1** (Figure 9) (52 µM) the W2 death rate was 18% ± 1% and the D3 growth rate was 10% ± 1%. For compound **2** (Figure 10) (64 µM), the W2 death rate was 18% ± 1% and the D3 growth rate was 14% ± 1%. The results confirmed that the isolated compounds have specific apoptosis activity against the W2 cell line.

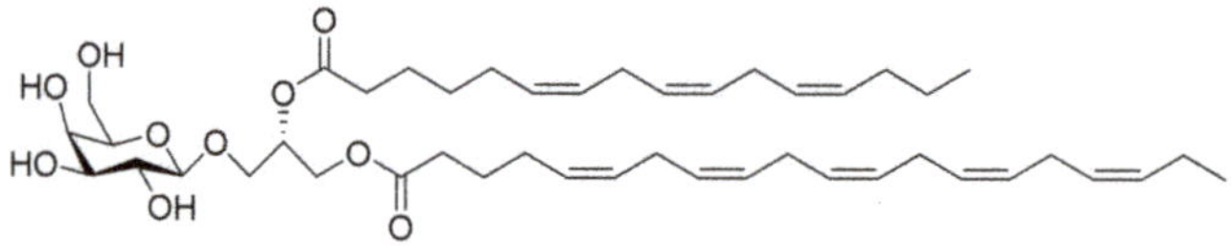

Figure 9. Monogalactosyl Glycerol (Compound **1**): (2S)-1-O-5,8,11,14,17-eicosapentaenoyl-2-O-6,9,12-hexadecatrienoyl-3-O-[β-D-galactopyranosyl]-glycerol.

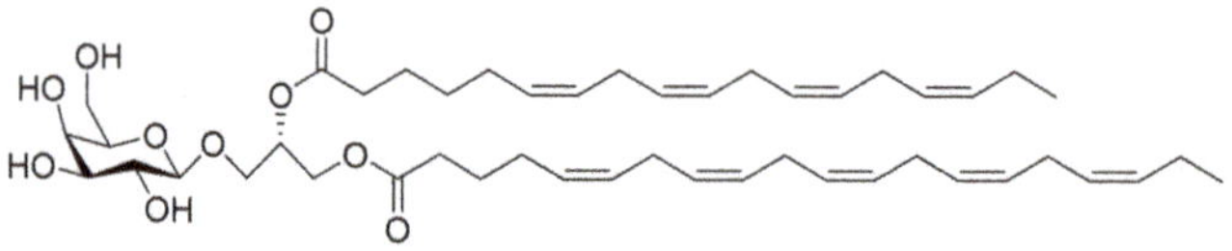

Figure 10. Monogalactosyl Glycerol (Compound **2**): (2S)-1-O-3,6,9,12,15-octadecapentaenoyl-2-O-6,9,12,15-octadecatetraenoyl-3-O-β-D-galactopyranosyl-sn-glycerol.

5. Active Compounds from Other Marine Organisms

To better understand the potential of microalgal species as important sources of anti-cancer compounds, we compared their bioactivity with marine anti-cancer drugs already on the market. Tables 3 and 4 report the anti-cancer compounds already on the market and in phase III clinical trials, the marine organisms from which the compounds were isolated, the active compounds, the target cancer cell lines and the active concentrations that arrest the growth of cancer cells.

Table 3. Active marine-derived compounds available on the market. The table reports the producing marine organisms, the active compounds, the target cancer cell lines, the active concentrations and references.

Marine Organism	Compound	Target	Active Concentration	Reference
Ecteinascidia turbinata	Ecteinascidin/ Trabectedin (alkaloid)	MFC7 A549	0.6 ng/mL (IC_{70}) 5.6 ng/mL (IC_{70})	[38]
Dolabella auricularia/Symploca sp. VP642	Brentuximab vedotin (antibody-drug conjugate)	Non-Hodgkin's lymphoma cells (Karpas 299)	2.5 ng/mL (IC_{50})	[39]
Halichondria okadai	Eribulin mesylate (macrolide)	DLD-1 LNCaP HL-60	6.934 ng/mL (IC_{50}) 0.365 ng/mL (IC_{50}) 0.657 ng/mL (IC_{50})	[40]
Cryptotheca crypta	Cytarabine (nucleoside)	Acute Myeloid Leukemia (AML) cells	272 ng/mL (IC_{50})	[41]

Table 4. Active marine-derived compounds in phase III clinical trials. The table reports the producing marine organisms, the active compounds, the target cancer cell lines, the active concentrations and references.

Marine Organism	Compound	Target	Active Concentration	Reference
Aspergillus sp. CNC139	Plinabulin (diketopiperazine)	Multiple myeloma cells (MM.1S, MM.1R, RPMI8226, and INA-6)	2.7 to 3.375 ng/mL (IC_{50})	[42]
Aplidium albicans	Plitidepsin (depsipeptide)	MCF-7	55.5 ng/mL (IC_{50})	[43]
Halichondria okadai	Lurbinectedin (alkaloid)	Ovarian cancer cells (RMG1, RMG2, KOC7C, HAC2, A2780, HeyA8 and SKOV-3)	0.78 to 2.34 ng/mL (IC_{50})	[44]

These studies give an overview on the active concentrations for each compound and delimit a concentration range (0.365–272 ng/mL IC_{50}) that can be used to understand whether a compound is active enough to be considered as a drug candidate. The sources of the active compounds in all of these cases were multicellular organisms which implies that there are difficulties in harvesting biomass and obtaining low amounts of secondary metabolites and a huge environmental impact due to incorrect harvesting [4]. Most of the compounds in Tables 3 and 4 are small-medium size molecules, but of particular interest is brentuximab vedotin, an antibody-drug conjugate where the compound monomethyl auristatin E is linked to a monoclonal antibody (mAb) that recognizes a specific marker expression in cancer cells and directs monomethyl auristatin E to the targeted cancer cell. The reason why it is used as a complex and not as a pure compound is because it is too potent (100–1000 times more potent than doxorubicin, a common drug used for chemotherapy) and less specific to cancer cells [45].

6. Discussion

Anti-cancer compounds from marine microalgae have been poorly investigated. Most studies have been done on microalgal extracts or fractions obtained using low resolution methods such as liquid-liquid partitioning or solid phase extractions. It is uncommon to see dereplication methodologies, fractionations based on high throughput techniques (e.g., HPLC or gas chromatography) or a complete structural elucidation of the compounds that have been found.

Despite the low availability of data so far, the studies performed on *Chlorella sorokiniana* and *Chaetoceros calcitrans* show interesting activities compared to commercially available marine anti-cancer drugs [18,26]. Most of the marine pharmaceuticals on the market are active at the level of 0.6–7 ng/mL (Table 3) while the fractions from *Chlorella sorokiniana* and *Chaetoceros calcitrans* display significant activity at 1–3 µg/mL (Table 1). Even if fractions and pure compounds cannot be directly compared in terms of activity, the anti-cancer activities of *C. sorokiniana* and *C. calcitrans* extracts seem very promising and appear preferential for further investigation and purification of the active molecules. Considering that 0.0001% of the crude aqueous ethanol extract of the ascidian *Ecteinascidia turbinate* [46] leads to the isolation of trabectedin ET-743 and the development of the anti-cancer drug Yondelis, it would not be difficult to find a compound from *C. sorokiniana* and *C. calcitrans* with an activity as high as that of the marketed drugs. These data highlight that microalgae can be a promising source of anti-cancer compounds. Regarding the other fractions/extracts obtained from marine microalgae, it cannot be excluded that other compounds are present with clinical or biotechnological potential, but further studies are necessary to demonstrate this possibility.

Several compounds isolated from microalgae mentioned in this review have been studied not only as possible anti-cancer compounds, but also for other biotechnological applications. Compounds such as polyunsaturated aldehydes [11,12], eicosapentaenoic acid (EPA) [16], fucoxanthin [17], violaxanthin [15], stigmasterol [22] or chrysolaminaran [14] (Table 1) may have a potential role as high-value products (e.g., nutraceuticals, cosmeceuticals, additives, fuel precursors and biomaterials) or as possible future new drugs. For example, the polyunsaturated aldehydes have shown anti-proliferative activities on Caco-2 [11], A549 and COLO 205 cancer cell lines [12], but also anti-bacterial activities [47,48], making them possible new candidates for drug development. The polyunsaturated fatty acid, EPA, has been widely studied as a nutraceutical or dietary supplement, with beneficial effects on fetal development, prevention of cardiovascular diseases and even, the improvement of cognitive functioning in patients with Alzheimer's disease [49].

The carotenoids, fucoxanthin and violaxanthin, have been used as precursors of vitamins in food and animal feed, additives, cosmetics, food coloring agents and biomaterials [50]. In particular, fucoxanthin has been shown to possess potential anti-inflammatory, antioxidant, anti-obesity, anti-diabetic, anti-tumorigenic and cardioprotective activities [51]. On the market, there are several fucoxanthin-based products used as dietary supplements such as "Solaray Fucoxanthin Special Formula Vegetarian Capsules" or "BRI NUTRITION ® Fucoxanthin Capsules".

Phytosterols such as stigmasterol have been receiving increasing attention because of their capacity to reduce blood cholesterol concentrations, prevent cardiovascular disorders and because of their health benefits in general [52]. Stigmasterol has been studied not only for its anti-cancer activity [22] but also for its antioxidant activity [53]. On the market, the phytosterol, β-sitosterol, is sold as a nutraceutical under brand names like "NOW ® Foods Beta-Sitosterol Plant Sterols Softgels". β-sitosterol has been widely studied for several biological activities such as anti-inflammatory, antioxidant, anti-diabetic or anti-cancer activities [54].

Finally, microalgal polysaccharides have the capacity to modulate the immune system and inflammatory reactions, making them interesting candidates for cosmetic additives, food ingredients and natural therapeutic agents [50]. Chrysolaminaran, for example, is a storage polysaccharide that is well known to be a common water soluble biopolymer synthetized by diatoms. Water soluble storage carbohydrates are more accessible, energetically and biophysically, than insoluble starch [30] so they can also be useful for microalgal energy applications.

There is sufficient evidence confirming the key roles of culturing conditions (e.g., light intensity, photoperiod, nutrient availability or harvesting time) in the modification of microalgal bioactivity. Studies such as those by Lauritano et al. [8], Ingebrigten et al. [27] and Ribalet et al. [28] have shown how light intensity, temperature, nutrient concentration and harvesting time have strong impacts on the activity of different microalgal strains. The use of stressful conditions (such as nutrient starvation, light or temperature variation) seems to play a key role in the production of active metabolites. Unfortunately, very few studies on anti-cancer activities from microalgae have reported the culturing conditions and the growth phases at which the species were tested in detail (Table 2). This makes stressful culturing conditions still an unexploited tool with a high potential for the bioprospecting of novel bioactive metabolites from marine microalgae.

Author Contributions: K.A.M.A., C.L. and A.I. conceived and analyzed the data; K.A.M.A., C.L., A.I. and G.R. wrote the paper.

Mar. Drugs **2018**, *16*, 165

Funding: This research received no external funding.

Acknowledgments: Kevin A. Martínez Andrade has been supported by a fellowship Marie Skłodowska-Curie Innovative Training Networks PhD (Project Marpipe MSCA-ITN-ETN Proposal number: 721421). We thank Flora Palumbo for graphics.

Conflicts of Interest: The authors declare no conflict of interest. The founding sponsors had no role in the writing of the manuscript.

References

1. What Is Cancer. Available online: https://www.cancer.gov/about-cancer/understanding/what-is-cancer (accessed on 12 April 2018).
2. European Cancer Observatory. Available online: http://eco.iarc.fr/ (accessed on 12 April 2018).
3. Dyshlovoy, S.A.; Honecker, F. Marine compounds and cancer: Where do we stand? *Mar. Drugs* **2015**, *13*, 5657–5665. [CrossRef] [PubMed]
4. Jaspars, M.; De Pascale, D.; Andersen, J.H.; Reyes, F.; Crawford, A.D.; Ianora, A. The marine biodiscovery pipeline and ocean medicines of tomorrow. *J. Mar. Biol. Assoc. UK* **2016**, *96*, 151–158. [CrossRef]
5. Van Norman, G.A. Drugs, Devices, and the FDA: Part 1: An Overview of Approval Processes for Drugs. *JACC Basic Transl. Sci.* **2016**, *1*, 170–179. [CrossRef]
6. Biopharmaceutical Research & Development. Available online: http://phrma-docs.phrma.org/sites/default/files/pdf/rd_brochure_022307.pdf (accessed on 12 April 2018).
7. Moreno-Garrido, I. Microalgae immobilization: Current techniques and uses. *Bioresour. Technol.* **2008**, *99*, 3949–3964. [CrossRef] [PubMed]
8. Lauritano, C.; Andersen, J.H.; Hansen, E.; Albrigtsen, M.; Escalera, L.; Esposito, F.; Helland, K.; Hanssen, K.Ø.; Romano, G.; Ianora, A. Bioactivity screening of microalgae for antioxidant, anti-inflammatory, anticancer, anti-diabetes and antibacterial activities. *Front. Mar. Sci.* **2016**, *3*, 1–12. [CrossRef]
9. Romano, G.; Costantini, M.; Sansone, C.; Lauritano, C.; Ruocco, N.; Ianora, A. Marine microorganisms as a promising and sustainable source of bioactive molecules. *Mar. Environ. Res.* **2017**, *128*, 58–69. [CrossRef] [PubMed]
10. De Morais, M.G.; Vaz, B.D.S.; De Morais, E.G.; Costa, J.A.V. Biologically active metabolites synthesized by microalgae. *Biomed. Res. Int.* **2015**, *2015*. [CrossRef] [PubMed]
11. Miralto, A.; Barone, G.; Romano, G.; Poulet, S.A.; Ianora, A.; Russo, G.L.; Buttino, I.; Mazzarella, G.; Laabir, M.; Cabrini, M.; et al. The insidious effect of diatoms on copepod reproduction. *Nature* **1999**, *402*, 173–176. [CrossRef]
12. Sansone, C.; Braca, A.; Ercolesi, E.; Romano, G.; Palumbo, A.; Casotti, R.; Francone, M.; Ianora, A. Diatom-derived polyunsaturated aldehydes activate cell death in human cancer cell lines but not normal cells. *PLoS ONE* **2014**, *9*. [CrossRef] [PubMed]
13. Kwang, H.C.; Song, Y.I.K.; Lee, D.U. Antiproliferative effects of carotenoids extracted from *Chlorella ellipsoidea* and *Chlorella vulgaris* on human colon cancer cells. *J. Agric. Food Chem.* **2008**, *56*, 10521–10526. [CrossRef]
14. Kusaikin, M.I.; Ermakova, S.P.; Shevchenko, N.M.; Isakov, V.V.; Gorshkov, A.G.; Vereshchagin, A.L.; Grachev, M.A.; Zvyagintseva, T.N. Structural characteristics and antitumor activity of a new chrysolaminaran from the diatom alga *Synedra acus*. *Chem. Nat. Compd.* **2010**, *46*, 1–4. [CrossRef]
15. Pasquet, V.; Morisset, P.; Ihammouine, S.; Chepied, A.; Aumailley, L.; Berard, J.B.; Serive, B.; Kaas, R.; Lanneluc, I.; Thiery, V.; et al. Antiproliferative activity of violaxanthin isolated from bioguided fractionation of *Dunaliella tertiolecta* extracts. *Mar. Drugs* **2011**, *9*, 819–831. [CrossRef] [PubMed]
16. Nappo, M.; Berkov, S.; Massucco, C.; Di Maria, V.; Bastida, J.; Codina, C.; Avila, C.; Messina, P.; Zupo, V.; Zupo, S. Apoptotic activity of the marine diatom *Cocconeis scutellum* and eicosapentaenoic acid in BT20 cells. *Pharm. Biol.* **2012**, *50*, 529–535. [CrossRef] [PubMed]
17. Peng, J.; Yuan, J.P.; Wu, C.F.; Wang, J.H. Fucoxanthin, a marine carotenoid present in brown seaweeds and diatoms: Metabolism and bioactivities relevant to human health. *Mar. Drugs* **2011**, *9*, 1806–1828. [CrossRef] [PubMed]
18. Nigjeh, S.E.; Yusoff, F.; Banu, N.; Alitheen, M.; Rasoli, M.; Keong, Y.S.; Rahman, A. Cytotoxic effect of ethanol extract of microalga, *Chaetoceros calcitrans*, and its mechanisms in inducing apoptosis in human breast cancer cell line. *Biomed. Res. Int.* **2013**, *2013*, 1–9. [CrossRef] [PubMed]

19. Goh, S.H.; Alitheen, N.B.; Yusoff, F.M.; Yap, S.K.; Loh, S.P. Crude ethyl acetate extract of marine microalga, *Chaetoceros calcitrans*, induces Apoptosis in MDA-MB-231 breast cancer cells. *Pharmacogn. Mag.* **2014**, *10*, 1–8. [CrossRef] [PubMed]

20. Samarakoon, K.W.; Ko, J.Y.; Shah, M.M.R.; Lee, J.H.; Kang, M.C.; O-Nam, K.; Lee, J.B.; Jeon, Y.J. In vitro studies of anti-inflammatory and anticancer activities of organic solvent extracts from cultured marine microalgae. *Algae* **2013**, *28*, 111–119. [CrossRef]

21. Shah, M.R.; Kalpa, W.S.; Ju-Young, K.; Lakmal, H.H.C.; Ji-Hyeok, L.; So-Jeong, A.; You-Jin, J.; Joon-Baek, L. Potentiality of benthic dinoflagellate cultures and screening of their bioactivities in Jeju Island, Korea. *Afr. J. Biotechnol.* **2014**, *13*, 792–805. [CrossRef]

22. Kim, Y.-S.; Li, X.-F.; Kang, K.-H.; Ryu, B.; Kim, S.-K. Stigmasterol isolated from marine microalgae *Navicula incerta* induces apoptosis in human hepatoma HepG2 cells. *BMB Rep.* **2014**, *47*, 433–438. [CrossRef] [PubMed]

23. Samarakoon, K.W.; Ko, J.Y.; Lee, J.H.; Kwon, O.N.; Kim, S.W.; Jeon, Y.J. Apoptotic anticancer activity of a novel fatty alcohol ester isolated from cultured marine diatom, *Phaeodactylum tricornutum*. *J. Funct. Foods* **2014**, *6*, 231–240. [CrossRef]

24. Andrianasolo, E.H.; Haramaty, L.; Vardi, A.; White, E.; Lutz, R.; Falkowski, P. Apoptosis-inducing galactolipids from a cultured marine diatom, *Phaeodactylum tricornutum*. *J. Nat. Prod.* **2008**, *4883*, 2–6. [CrossRef] [PubMed]

25. Somasekharan, S.P.; El-Naggar, A.; Sorensen, P.H.; Wang, Y.; Cheng, H. An aqueous extract of marine microalgae exhibits antimetastatic activity through preferential killing of suspended cancer cells and anticolony forming activity. *Evid. Based Complement. Altern. Med.* **2016**, *2016*, 9730654. [CrossRef] [PubMed]

26. Lin, P.-Y.; Tsai, C.-T.; Chuang, W.-L.; Chao, Y.-H.; Pan, I.-H.; Chen, Y.-K.; Lin, C.-C.; Wang, B.-Y. *Chlorella sorokiniana* induces mitochondrial-mediated apoptosis in human non-small cell lung cancer cells and inhibits xenograft tumor growth in vivo. *BMC Complement. Altern. Med.* **2017**, *17*, 88. [CrossRef] [PubMed]

27. Ingebrigtsen, R.A.; Hansen, E.; Andersen, J.H.; Eilertsen, H.C. Light and temperature effects on bioactivity in diatoms. *J. Appl. Phycol.* **2016**, *28*, 939–950. [CrossRef] [PubMed]

28. Ribalet, F.; Wichard, T.; Pohnert, G.; Ianora, A.; Miralto, A.; Casotti, R. Age and nutrient limitation enhance polyunsaturated aldehyde production in marine diatoms. *Phytochemistry* **2007**, *68*, 2059–2067. [CrossRef] [PubMed]

29. Dockery, N.; Higashida, K.; Verdes, R.P.; Mooneyham, T.P. *Chlorella* Containing Nutritional Supplement Having Improved Digestability. Patent Application US 10/795,804, 8 September 2005.

30. Hildebrand, M.; Manandhar-Shrestha, K.; Abbriano, R. Effects of chrysolaminarin synthase knockdown in the diatom *Thalassiosira pseudonana*: Implications of reduced carbohydrate storage relative to green algae. *Algal Res.* **2017**, *23*, 66–77. [CrossRef]

31. Chajès, V.; Sattler, W.; Stranzl, A.; Kostner, G.M. Influence of n-3 fatty acids on the growth of human breast cancer cells in vitro: Relationship to peroxides and vitamin-E. *Breast Cancer Res. Treat.* **1995**, *34*, 199–212. [CrossRef] [PubMed]

32. Kong, Z.; Kao, N.; Hu, J.; Wu, C. Fucoxanthin-rich brown algae extract decreases inflammation and attenuates colitis-associated colon cancer in mice. *J. Food Nutr. Res.* **2016**, *4*, 137–147. [CrossRef]

33. Kadekaru, T.; Toyama, H.; Yasumoto, T. Safety evaluation of Fucoxanthin purified from *Undaria pinnatifida*. *Nippon Shokuhin Kagaku Kogaku Kaishi* **2008**, *55*, 304–308. [CrossRef]

34. Ishikawa, C.; Tafuku, S.; Kadekaru, T.; Sawada, S.; Tomita, M.; Okudaira, T.; Nakazato, T.; Toda, T.; Uchihara, J.N.; Taira, N.; et al. Antiadult T-cell leukemia effects of brown algae fucoxanthin and its deacetylated product, fucoxanthinol. *Int. J. Cancer* **2008**, *123*, 2702–2712. [CrossRef] [PubMed]

35. Hosokawa, M.; Wanezaki, S.; Miyauchi, K.; Kurihara, H.; Kohno, H.; Kawabata, J.; Odashima, S.; Takahashi, K. Apoptosis-inducing effect of fucoxanthin on human leukemia cell line HL-60. *Food Sci. Technol. Res.* **1999**, *5*, 243–246. [CrossRef]

36. Kotake-nara, E.; Kushiro, M.; Zhang, H.; Sugawara, T.; Miyashita, K.; Nagao, A. Carotenoids affect proliferation of human prostate cancer cells. *J. Nutr.* **2001**, *131*, 3303–3306. [CrossRef] [PubMed]

37. Ryu, B.; Li, Y.; Qian, Z.J.; Kim, M.M.; Kim, S.K. Differentiation of human osteosarcoma cells by isolated phlorotannins is subtly linked to COX-2, iNOS, MMPs, and MAPK signaling: Implication for chronic articular disease. *Chem. Biol. Interact.* **2009**, *179*, 192–201. [CrossRef] [PubMed]

38. Ghielmini, M.; Colli, E.; Erba, E.; Bergamaschi, D.; Pampallona, S.; Jimeno, J.; Faircloth, G.; Sessa, C. In vitro schedule-dependency of myelotoxicity and cytotoxicity of Ecteinascidin 743 (ET-743). *Ann. Oncol.* **1998**, *9*, 989–993. [CrossRef] [PubMed]

39. Francisco, J.A.; Cerveny, C.G.; Meyer, D.L.; Mixan, B.J.; Klussman, K.; Chace, D.F.; Rejniak, S.X.; Gordon, K.A.; DeBlanc, R.; Toki, B.E.; et al. cAC10-vcMMAE, an anti-CD30-monomethyl auristatin E conjugate with potent and selective antitumor activity. *Blood* **2003**, *102*, 1458–1465. [CrossRef] [PubMed]

40. Towle, M.J.; Salvato, K.A.; Budrow, J.; Wels, B.F.; Kuznetsov, G.; Aalfs, K.K.; Welsh, S.; Zheng, W.; Seletsky, B.M.; Palme, M.H.; et al. In vitro and in vivo anticancer activities of synthetic macrocyclic ketone analogues of halichondrin B. *Cancer Res.* **2001**, *61*, 1013–1021. [PubMed]

41. Desai, U.; Shah, K.; Mirza, S.; Panchal, D.; Parikh, S.; Rawal, R. Enhancement of the cytotoxic effects of Cytarabine in synergism with Hesperidine and Silibinin in Acute Myeloid Leukemia: An in-vitro approach. *J. Cancer Res. Ther.* **2015**, *11*, 352–357. [CrossRef] [PubMed]

42. Singh, A.V.; Bandi, M.; Raje, N.; Richardson, P.; Palladino, M.A.; Anderson, K.C.; Singh, A.V.; Bandi, M.; Raje, N.; Richardson, P.; et al. A novel vascular disrupting agent plinabulin triggers JNK-mediated apoptosis and inhibits angiogenesis in multiple myeloma cells A novel vascular disrupting agent plinabulin triggers JNK-mediated apoptosis and inhibits angiogenesis in multiple myeloma cells. *Blood* **2011**, *117*, 5692–5700. [CrossRef] [PubMed]

43. Gómez-Fabre, P.M.; De Pedro, E.; Medina, M.A.; Núñez De Castro, I.; Márquez, J. Polyamine contents of human breast cancer cells treated with the cytotoxic agents chlorpheniramine and dehydrodidemnin B. *Cancer Lett.* **1997**, *113*, 141–144. [CrossRef]

44. Takahashi, R.; Mabuchi, S.; Kawano, M.; Sasano, T.; Matsumoto, Y.; Kuroda, H.; Kozasa, K.; Hashimoto, K.; Sawada, K.; Kimura, T. Preclinical investigations of PM01183 (Lurbinectedin) as a single agent or in combination with other anticancer agents for clear cell carcinoma of the ovary. *PLoS ONE* **2016**, *11*, e0151050. [CrossRef] [PubMed]

45. Monomethyl Auristatin E (MMAE). Available online: https://adcreview.com/adc-university/adcs-101/cytotoxic-agents/monomethyl-auristatin-e-mmae/ (accessed on 12 April 2018).

46. Cuevas, C.; Francesch, A. Development of Yondelis® (trabectedin, ET-743). A semisynthetic process solves the supply problem. *Nat. Prod. Rep.* **2009**, *26*, 322. [CrossRef] [PubMed]

47. Amaro, H.; Guedes, A.; Malcata, F. Antimicrobial activities of microalgae: An invited review. *Sci. Microb. Pathog. Commun. Curr. Res. Technol. Adv.* **2011**, 1272–1280. [CrossRef]

48. Paul, C.; Reunamo, A.; Lindehoff, E.; Bergkvist, J.; Mausz, M.A.; Larsson, H.; Richter, H.; Wängberg, S.Å.; Leskinen, P.; Bamstedt, U.; et al. Diatom derived polyunsaturated aldehydes do not structure the planktonic microbial community in a mesocosm study. *Mar. Drugs* **2012**, *10*, 775–792. [CrossRef] [PubMed]

49. Swanson, D.; Block, R.; Mousa, S.A. Omega-3 Fatty Acids EPA and DHA: Health benefits throughout life. *Adv. Nutr.* **2012**, *3*, 1–7. [CrossRef] [PubMed]

50. Chew, K.W.; Yap, J.Y.; Show, P.L.; Suan, N.H.; Juan, J.C.; Ling, T.C.; Lee, D.J.; Chang, J.S. Microalgae biorefinery: High value products perspectives. *Bioresour. Technol.* **2017**, *229*, 53–62. [CrossRef] [PubMed]

51. Zhang, H.; Tang, Y.; Zhang, Y.; Zhang, S.; Qu, J.; Wang, X.; Kong, R.; Han, C.; Liu, Z. Fucoxanthin: A promising medicinal and nutritional ingredient. *Evid. Based Complement. Altern. Med.* **2015**, *2015*, 723515. [CrossRef] [PubMed]

52. Luo, X.; Su, P.; Zhang, W. Advances in microalgae-derived phytosterols for functional food and pharmaceutical applications. *Mar. Drugs* **2015**, *13*, 4231–4254. [CrossRef] [PubMed]

53. Panda, S.; Jafri, M.; Kar, A.; Meheta, B.K. Thyroid inhibitory, antiperoxidative and hypoglycemic effects of stigmasterol isolated from Butea monosperma. *Fitoterapia* **2009**, *80*, 123–126. [CrossRef] [PubMed]

54. Saeidnia, S. The Story of Beta-sitosterol—A Review. *Eur. J. Med. Plants* **2014**, *4*, 590–609. [CrossRef]

Review

Investigation of the Anti-Prostate Cancer Properties of Marine-Derived Compounds

Meiqi Fan [1,†], Amit Kumar Nath [1,†], Yujiao Tang [1,2], Young-Jin Choi [1], Trishna Debnath [3], Eun-Ju Choi [4] and Eun-Kyung Kim [1,*]

1   Division of Food Bioscience, College of Biomedical and Health Sciences, Konkuk University, Chungju 27478, Korea; fanmeiqi@kku.ac.kr (M.F.); chinmoyamit@gmail.com (A.K.N.); yuanxi00@126.com (Y.T.); choijang11@kku.ac.kr (Y.-J.C.)
2   School of Bio-Science and Food Engineering, Changchun University of Science and Technology, Changchun 130-600, China
3   Department of Food Science and Biotechnology, Dongguk University, Goyang 10326, Korea; trishna_rahul@yahoo.com
4   Department of Physical Education, College of Education, Daegu Catholic University, Gyeongsan 38430, Korea; cej0915@cu.ac.kr
*   Correspondence: eunkyungkim@kku.ac.kr
†   These authors contributed equally to this work.

Received: 8 April 2018; Accepted: 9 May 2018; Published: 12 May 2018

Abstract: This review focuses on marine compounds with anti-prostate cancer properties. Marine species are unique and have great potential for the discovery of anticancer drugs. Marine sources are taxonomically diverse and include bacteria, cyanobacteria, fungi, algae, and mangroves. Marine-derived compounds, including nucleotides, amides, quinones, polyethers, and peptides are biologically active compounds isolated from marine organisms such as sponges, ascidians, gorgonians, soft corals, and bryozoans, including those mentioned above. Several compound classes such as macrolides and alkaloids include drugs with anti-cancer mechanisms, such as antioxidants, anti-angiogenics, antiproliferatives, and apoptosis-inducing drugs. Despite the diversity of marine species, most marine-derived bioactive compounds have not yet been evaluated. Our objective is to explore marine compounds to identify new treatment strategies for prostate cancer. This review discusses chemically and pharmacologically diverse marine natural compounds and their sources in the context of prostate cancer drug treatment.

Keywords: prostate cancer; antioxidant; anti-proliferative; apoptosis; natural marine compounds

1. Introduction

Throughout the history of humanity, marine sources have played an important role as a source of natural medicinal products. The ocean covers almost 70% of the earth's surface and contains varied environmental conditions. Half of the previously described novel marine natural compounds have been shown to be biologically active [1]. The marine ecological system is unique. Therefore, marine organisms must survive and adapt to these harsh environmental conditions. Researchers have become increasingly interested in marine sources in the development of potential anticancer drugs [2].

Cancer is the most devastating disease in recent years. According to a report on prostate cancer from Research Fund International, 1.1 million prostate cancer cases were reported in 2012, corresponding to 8% of all new cancer patients and 15% of male cancer patients. Prostate cancer is one of the most common malignant diseases in men but is primarily seen in developed countries. Prostate cancer is the second-most common cause of death among all male cancer patients [3].

Prostate cancer can spread to other parts of the body, especially the bones and lymph nodes [4]. Prostate cancer is an androgen-dependent carcinoma that is mediated through androgen receptor

(AR)-regulated genes [5]. The testes produce the most androgens, with a small amount of androgen produced through the conversion of adrenal steroids [6]. In the early stages of prostate cancer, cancerous cells can metastasize through the lymphatic fluid to other organs. Prostate cancer progresses through the activation of growth factors that promote critical signaling of cascades [7].

Marine compounds have shown potential for the treatment of prostate cancer. In this review, we discuss the various species and compounds and their specific effects on prostate cancer. Among the diverse number of invertebrates, sponges have the most significant medicinal importance. Screening of cyanobacteria, fungi, sponges, algae, and tunicates has yielded a large number of anticancer compounds; these marine-derived chemical compounds include alkaloids, macrolides, terpenoids, among other compounds.

2. Molecular Targets for Anti-Prostate Cancer Compounds

Prostate cancer is a disease of the prostate characterized by disordered cell growth and proliferation. Mutations of the oncogenes are responsible for prostate cancer. These oncogenes encode altered proteins that result in increased cell growth and proliferation [8]. The proteins encoded by oncogenes include growth factors; receptor tyrosine kinases such as epidermal growth factor receptor (EGFR) and vascular endothelial growth factor receptor (VEGFR); tyrosine kinases including Src family of protein tyrosine kinase (Src) and Bruton's tyrosine kinase (BTK); serine/threonine kinase families such as rapidly accelerated fibrosarcoma (RAF), cyclin dependent kinase (CDK), and checkpoint kinase (CHK); guanosine triphosphate hydrolase enzymes (GTPases) such as the rat sarcoma (Ras) protein; and various transcription factors including myelocytomatosis oncogene (Myc). Mutation or overexpression of oncogenes leads to aberrant and excessive cell proliferation [9].

Androgens stimulate the growth of prostate cancer cells. Higher levels of androgens might contribute to prostate cancer risk in some men. P53 is a tumor suppressor gene responsible for the control of cell growth and proliferation as well as diminishing AR-mediated signaling in prostate cancer cell lines [10]. Insulin-like growth factor-1 (IGF-1) is associated with the risk of prostate cancer [11]. E-cadherin is an important biomarker for prostate cancer diagnosis and is involved in cell-cell adhesion, which is linked to invasion and metastasis. E-cadherin binds to β-catenin and forms a protein complex to prevent adhesion and migration.

Matrix metalloproteinases (MMP)-1 and MMP-2 play a key role in the metastasis, invasion, and angiogenesis of prostate cancer. MMP-1 and MMP-2 are novel molecular biomarkers and tissue inhibitors of matrix metalloproteinases (TIMPs) responsible for the regulation of angiogenesis and are potentially invasive and metastatic [12].

In addition, overexpression of caveolin-1 (CAV-1), zinc-dependent mammalian histone deacetylase (HDAC), 3-phosphoinositide-dependent protein kinase-1 (PDPK1), PG receptor EP2 and FP, prostaglandin-degrading enzyme (15-PGDH), and prostaglandin-endoperoxide synthase protein cyclooxygenase-2 (COX-2) also trigger the development of prostate cancer [13].

3. Bioactive Products with Potential for Prostate Cancer Treatment

Marine bioactive compounds and their biological activity towards prostate cancer are summarized in Table 1.

Table 1. Marine sources and compounds with potential for anti-prostate cancer drug development.

Source Group	Compounds	Structure	Biological Activity on Prostate Cancer	References
Bacteria	Kahalalide F		Cytotoxicity (IC$_{50}$: 0.07 µM in PC-3 cells; 0.28 µM in DU-145 cells)	[14]
			50% of PSA decline for ≥4 weeks at 80 µg/kg/day in clinical trial	[15]
Marine fungi	Demethoxyfumitremorgin C		Inhibition of proliferation (50% inhibition at 100 µM in PC-3 cells)	[16]
	Apochalasin V		Cytotoxicity (IC$_{50}$: 30.4 µM in PC-3 cells)	[17]
Marine sponges	Rhizochalin		Cytotoxicity (IC$_{50}$: 16.55 µM in PC-3 cells, IC$_{50}$: 10.75 µM in DU-145 cells, IC$_{50}$: 7.88 µM in LNCaP cells, IC$_{50}$: 7.37 µM in 22Rv1 cells, IC$_{50}$: 5.81 µM in VCaP cells)	[18]
	Rhizochalinin		Cytotoxicity (IC$_{50}$: 1.14 µM in PC-3 cells, IC$_{50}$: 1.05 µM in DU-145 cells, IC$_{50}$: 1.69 µM in LNCaP cells, IC$_{50}$: 0.87 µM in 22Rv1 cells, IC$_{50}$: 0.42 µM in VCaP cells)	[19]
	latrunculin A		Inhibition of invasion (23% inhibition at 100 nM in PC-3 cells)	[20]

Table 1. *Cont.*

Source Group	Compounds	Structure	Biological Activity on Prostate Cancer	References
	Halichondramide		Cytotoxicity (IC$_{50}$: 0.81 µM in PC-3 cells)	[21]
	Spongistatin 1		Inhibition of proliferation (50% inhibition at 500 pmol in LNCaP cells)	[22]
	Furospinosulin-1		Inhibition of proliferation (60% inhibition at 100 µM in DU-145 cells)	[23]
	Sodwanone and Yardenone		Inhibition of HIF-1α expression at 15 µM in PC-3 cells	[24]
	Niphatenone B		Inhibition of proliferation (90% inhibition at 250 µM in LNCaP cells)	[25]
	Agelasine B		Cytotoxicity (IC$_{50}$: 0.04 µg/mL in DU-145 cells)	[26]
Cyanobacteria	Cryptophycin 52		Apoptosis (40% at 250 µg/mL in LNCaP cells)	[27]

Table 1. *Cont.*

Source Group	Compounds	Structure	Biological Activity on Prostate Cancer	References
	Lagunamide C	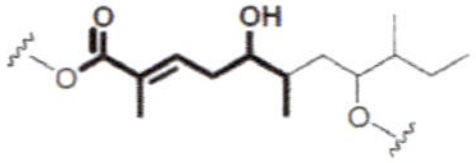	Cytotoxicity (IC$_{50}$: 2.6 nM in PC-3 cells)	[28]
	Dolastatins	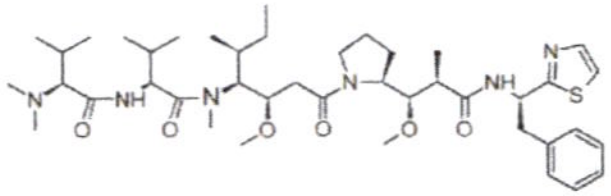	Cell cycle arrest (G2/M arrest in DU-145 cells)	[29]
	C-phycocyanin (C-PC)	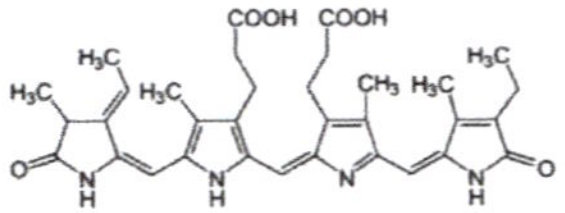	Inhibition of proliferation (30% inhibition at 500 µg/mL in LNCaP cells)	[30]
	Iejimalide B	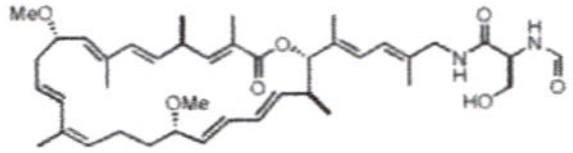	Cell cycle arrest (G0 / G1 arrest in LNCaP cells)	[31]
Rhodophyta	Bromophycolide D	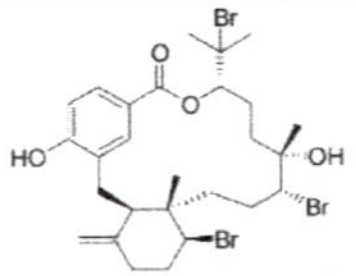	Cytotoxicity (IC$_{50}$: 9.0 µM in PC-3 cells)	[32]
Chlorophyta	14-keto-stypodiol diacetate (SDA)	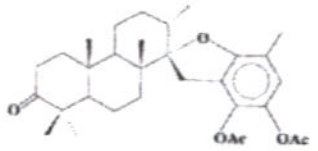	Cytotoxicity (IC$_{50}$: 2.7 µM in DU145 cells)	[33]
	Astaxanthin	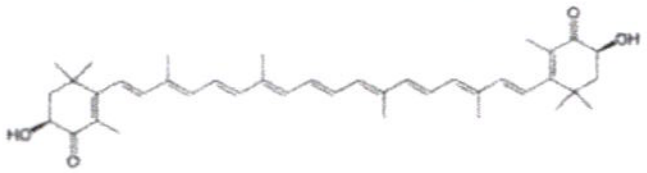	Inhibition of proliferation (38% inhibition at 0.01 µg/mL in LNCaP cells)	[34]

Table 1. *Cont.*

Source Group	Compounds	Structure	Biological Activity on Prostate Cancer	References
Phaeophyta	Fucoidan		Apoptosis (15.2% at 10 μg/mL, 29.8% at 50 μg/mL, 39.3% at 100 μg/mL, and 45.1% at 200 μg/mL in PC3 cells)	[35,36]
Marine diatoms	Fucoxanthin		Inhibition of proliferation (50% inhibition at 2.5 μM in LNCaP cells)	[37,38]
	Fucoxanthin, Fucoxanthinol, and Amarouciaxanthin A		Cytotoxicity (IC$_{50}$: 2.0–4.6 μM in PC-3 cells)	[39]
Corals	Pachycladins A–E	Pachycladins A Pachycladins B Pachycladins C Pachycladins D Pachycladins E	Inhibition of invasion (87% inhibition at 50 μM in PC-3 cells)	[40]

Table 1. *Cont.*

Source Group	Compounds	Structure	Biological Activity on Prostate Cancer	References
	Metabolite 1 from *Sarcophyton ehrenbergi*, synthetic enantiomer (*R*)-1	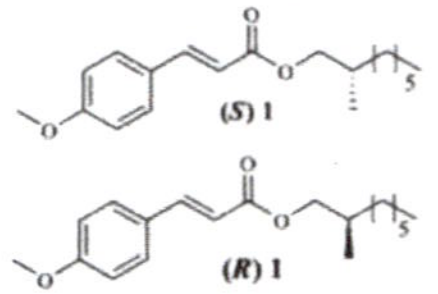	Cytotoxicity ((*S*)-1 (IC$_{50}$: 161 mM in DU-145 cells); (*R*)-1 (IC$_{50}$: 77.2 in DU145 cells)	[41]
Holothurians	Frondoside A	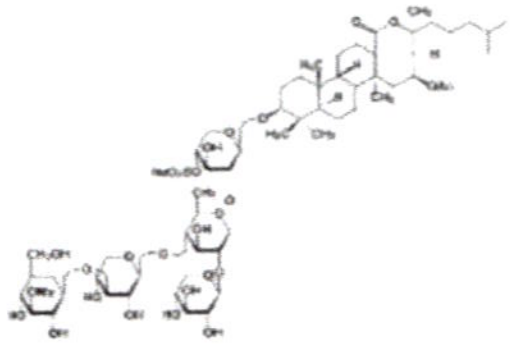	Cell cycle arrest (G2/M-phase at 0.5 μM in PC-3 cells)	[42]
	12-methyltetradecanoic acid	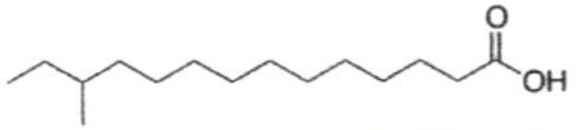	Cytotoxicity (IC$_{50}$: 35.48 μg/mL in DU-145 cells, IC$_{50}$: 20.45 μg/mL in PC-3 cells)	[43]

3.1. Marine Bacteria

Marine microorganisms offer a unique source for potential anticancer drugs. Scientists are interested in marine microorganisms for the development of these drugs. Marine microorganisms have yielded novel anti-inflammatory agents such as pseudopterosins, topsentins, scytonemin, manoalide, topsentins, and scytonemin [44]. These agents show cytotoxic activity against cancer cell lines like PC-3 (prostate cancer cells).

Microbes associated with the mollusk *Elysia rufescens* synthesize kahalalide F (KF), a compound with both in vitro and in vivo antitumor activity in various solid tumor models. In vitro antiproliferative studies have found activity in certain prostate cancer cell lines (PC-3, DU-145), but none against the hormone-sensitive LNCaP line [14]. Therefore, KF exhibits antitumor activity against solid prostate tumors []. In clinical trials, adult patients with advanced or metastatic androgen-refractory prostate cancer have received intravenous administration of KF. One patient had a partial response at a dose of 80 µg per kg per day, showing a prostate-specific antigen decline of at least 50% for $\geq$4 weeks. Five patients showed stable disease. KF can be safely administered as a one-hour infusion for five consecutive days at a dose of 560 µg per kg per day once every three weeks [15].

3.2. Marine Fungi

Marine fungi provide a rich profile of biologically active metabolites. Despite the interests toward studying of biologically active metabolites, such studies remain scarce. The effect of marine fungal metabolite 1386A from the South China Sea on the proliferation of androgen-independent cells has been reported in DU-145 cells. The half-maximal inhibitory concentrations (IC_{50}) of 1386A incubated with DU-145 cells for 24, 48, and 72 h were 25.31, 8.62, and 4.79 µmol/L respectively [45]. This activity may be useful in diseases including prostate cancer as a therapeutic or food additive [45].

Demethoxyfumitremorgin C, a secondary metabolite of the marine fungus, Aspergillus fumigatus, had been reported to inhibited the cell viability on PC-3 cells [16].

An investigation into new bioactive metabolites of marine gut fungi revealed aspochalasins isolated from the gut of the marine isopod *Ligia oceanica*. Aspochalasins are a subgroup of cytochalasans consisting of a macrocyclic ring, isoindolone moiety, and a 2-methyl-propyl side-chain. Aspochalasins showed various bioactivities such as cytotoxicity [17], anti-HIV [46], and TNF-alpha [47] and melanogenesis inhibitors [48]. Cytotoxicity against the prostate cancer PC-3 cell line was assayed using the MTT method. Apochalasin V showed moderate activity at IC_{50} values of 30.4 µM, respectively [17].

3.3. Marine Sponges

Marine sponges are an abundant source of alkaloids. For example, rhizochalin is a bioactive substance initially isolated from the marine sponge *Rhizochalina incrustata*. Rhizochalin exhibited anticancer properties in human castration-resistant prostate cancer cells, induced apoptosis, and G2/M cell cycle arrest, and inhibited autophagy [18]. Rhizochalinin (Rhiz) is a sphingolipid-like semi-synthetic compound hydrolytically derived from rhizochalin. Rhiz had cytotoxic effects on all human prostate cancer cell lines (PC-3, DU145, LNCaP, 22Rv1, VCaP) at low micromolar concentrations. In general, aglycones are more cytotoxic than glycosides [19]. Functional analyses confirmed an anti-migratory effect of Rhiz in PC-3 cells. Additionally, a predicted ERK1/2 activation was confirmed by Western blot analysis, and the prosurvival effects in Rhiz-treated prostate cancer cells indicated a potential mechanism of resistance [49].

In addition, two prominent alkaloids, heliclonadiamines (HCA) from ethanol extracts of the marine sponge *Haliclona* spp. show strong cytotoxic effects against PC-3 cells, with 50% viability at 100 µM [50]. Overexpression of phosphatase of regenerating liver-3 (PRL-3) in these cells was suppressed by treatment with HCA. HCA activates E-cadherin and downregulates highly overexpressed N-cadherin.

The macrolide compound latrunculin A isolated from the Red Sea sponge *Negombata magnifica* exhibits anti-invasive activity against PC-3 cells [20]. Halichondramide is a trisoxazole-containing macrolide extracted from *Chondrosia corticata* that modulates prostate cancer-related biomarkers such as E-cadherin, N-cadherin, MMP2, and MMP9 at both transcriptional and transitional levels [21].

Epithelial to mesenchymal transition (EMT) biomarkers indicate the metastatic characteristics of prostate cancer [51]. Spongistatin 1 is a macrocyclic lactone derived from the marine sponge *Spongia* sp., that has been shown to induces apoptosis and caspase independent cell death in DU-145 cells [52]. In addition, Spongistatin 1 also upregulates BIM, pro-apoptotic BCL-2 family member BIM, by acting on both the microtubular complex and the antiapoptotic MCL-1. BIM is an important genetic factor that plays a role in upregulating caspase-independent apoptotic signaling pathways executed by mitochondria in prostate cancer [22]. Marine-sponge-derived furanosesterterpene furospinosulin-1 has selective antiproliferative activity against DU-145 cells under hypoxic conditions [23].

Sodwanone and yardenones derived from *Axinella* sp. inhibited PC-3 cells by deactivating HIF-1 [24]. The glycerol ether niphatenone B is a natural product that leads to the development of castration-recurrent prostate cancer that has been isolated from crude methanolic extracts of *Niphates digitalis*. It induces proliferation of LNCaP cells but not PC-3 cells. Consequently, there is no functional AR support against target-specific anti-proliferation. Niphatenone B prominently binds the activation function-1 region of the AR N-terminus domain (NTD) [25]. Finally, agelasine B has been isolated from the marine sponge *Agelas clathrodes*. This compound has been shown to inhibit the viability of PC-3 cells. It significantly reduces the Ca^{2+} concentration in these cells and induces the fragmentation of DNA [26].

3.4. Marine Algae

3.4.1. Cyanobacteria

Cyanobacterium (marine blue algae) are a diverse group of prokaryotic organisms. Cryptophycin 52 is a naturally macrocyclic anticancer compound isolated from the marine cyanobacteria *Nostoc* spp. [27].

A new cyclic depsipeptide, lagunamide C was isolated from the marine cyanobacterium, *L. majuscula*, collected from Pulau Hantu Besar, Singapore. Lagunamide C was tested against PC3 cells, with an IC_{50} of 2.6 nM. It also possesses significant antimalarial properties [28].

Cytotoxic peptides like dolastatins isolated from *Dolabella auricularia* and their synthetic analogs dolastatin 10 in symploca and its non-cyanobacterial analog dolastin are responsible for cell cycle arrest in the G2/M phase [29].

Marine cyanobacteria-derived compounds can induce the alteration of caspases and activate the pathway to induced cell death. Caspase-3 is the most well-known caspase in the apoptosis of prostate cancer cell lines. C-phycocyanin (C-PC) s isolated from the *Limnothrix* sp. cyanobacterium has previously been shown to have anticancer properties. We found that only 10% of a typical dose of the topotecan (TPT) anticancer drug combined with C-PC killed cancer cells at a higher rate than that of TPT being used alone at full dose. We also detected an increased magnitude of the increased activities of caspase-9 and caspase-3 when these two compounds were used in combination [30].

The BCL-2 protein family acts as an important regulator of apoptosis in prostate cancer. Cryptophycin 52 promotes BCL-2 and BCL-xL phosphorylation in several prostate cancer cell lines including PC-3, LNCaP, and DU-145 [27].

Iejimalide B, a marine macrolide, was first extracted from the tunicate *Eudistoma cf. rigida*. Iejimalide B is active in both LNCaP and PC-3 cell lines in the nanomolar range, but the effects on the two cell lines differed significantly. One experiment showed that iejimalide B doses below 30 nM induced cell cycle arrest in G0/G1 and cell death at doses at and above 50 nM in LNCaP cells, but neither of these doses induced apoptosis in PC-3 cells after 72 h [31].

3.4.2. Chlorophytes

Chlorophyta is a group of green photosynthetic algae. Most seaweeds are classified as marine chlorophytes and are an important source of vitamins and minerals; they are also promising for their activities against prostate cancer [53]. There has been some research on green algae, which have isolated several potential anticancer compounds. For example, 14-keto-stypodiol diacetate (SDA) extracted from the seaweed *Stypopodium flabelliforme* inhibits cell growth and tumor invasion in DU-145 cells. The studies suggest that this novel derivative from a marine natural product induces the mitotic arrest of tumor cells, an effect that could be associated with alterations in the normal microtubule assembly process. In addition, a salient feature of this compound is that it affects protease secretion and in vitro invasive capacity, both properties of cells from metastases. The secretion of plasminogen activator (u-PA) and the capacity of DU-145 cells to migrate through a Matrigel-coated membrane was significantly inhibited in the presence of micromolar concentrations of SDA. [33]. *Haematococcus pluvialis* is a rich source of carotenoid astaxanthin that is an efficient promoter of antioxidants and apoptosis by inhibiting NF-kappa B, which subsequently inhibits growth in prostate cancer cell lines [34,54].

KF is a significant bioactive compound isolated from *Elysia rufescens*; the actual source of kahalalide is believed to be *Bryopsis* sp. The compound triggered oncogenesis in a prostate cancer cell line. Therefore, KF induced lysosomal and cell membrane permeability and induced apoptosis by inhibiting the PI3K/AKT pathways [55].

3.4.3. Rhodophyta

Rhodophyta, known as red algae, is primarily found in the sea. However, there is not much evidence on the use of red algae extracts as a drug in prostate cancer treatment. Bromophycolides C-I has been isolated from *Callophycus serratus* and has cytotoxic activity against a wide range of cancer cells. Among them, the effect of Bromophycolides D is the most significant. [32].

3.4.4. Phaeophyta

Phaeophyta (brown algae) produce complex diterpenoids and metabolites of mixed terpenoid-aromatic origins. Previous research showed that many of these compounds could be potent antibiotic, antifungal, antiviral, or anticancer agents [56]. They can be isolated from *Cladosiphon novaecaledoniae*, *Undaria pinnatifida*, and other species of brown algae [57]. Fucoidan inhibited PC-3 cells and activated intrinsic and extrinsic apoptosis [35]. This apoptosis was followed by extracellular signal-regulated kinase mitogen-activated protein kinase (ERK1/2 MAPK) and p38 MAPK inactivation. Furthermore, it inactivated phosphatidylinositol 3-kinase (PI3K)/Akt. In addition, p21Cip1/Waf was upregulated following the application of fucoidan. Therefore, fucoidan downregulates E2F-1 cell-cycle-related proteins and upregulates the Wnt/β-catenin signaling pathway. GSK-3β protein activation decreased the β-catenin level and c-MYC and cyclin D1 expression in PC-3 cells [35].

Transforming growth factor β (TGFβ) and receptors (TGFRs) play an important role in the EMT of cancer cells. In one study, fucoidan prominently reversed TGFR-induced EMT morphological changes [36]. Therefore, fucoidan upregulates epithelial markers and downregulates mesenchymal markers as well as decreasing the expression of the transcriptional repressors snail, slug, and twist in prostate cancer cells.

3.5. Marine Diatoms

To date, few natural bioactive products have been derived from diatoms despite the abundance of diatoms. Fucoxanthin is an important marine compound in prostate cancer treatment that was isolated from *Sargassum* sp. Fucoxanthin inhibits the growth of LNCaP cells [37]. A growth inhibitory effect was shown by the induction of GADD45A and G1 cell cycle arrest. Fucoxanthin is a highly conjugated natural compound relatively safe for use as an antitumor compound in prostate cancer [38].

Ingested fucoxanthin was reportedly deacetylated in the intestinal lumen and transported via blood in White Leghorn which was fed the brown seaweed *F. serratus*; thus, fucoxanthinol was present as one of the main carotenoids in the egg yolks [58]. Asai et al. investigated the biotransformation of fucoxanthinol in ICR mice, reporting an unknown metabolite which was previously found in the marine tunicate *Amaroucium pliciferum* that was identified as amarouciaxanthin A. Both fucoxanthinol and amarouciaxanthin A reduced the viability of PC-3 cells, with 50% inhibitory concentrations of fucoxanthin, fucoxanthinol, and amarouciaxanthin A of 3.0, 2.0, and 4.6 μM [39], respectively. However, there are few studies on this topic.

3.6. Marine Diatom Metabolites

Many novel and physiologically active natural organic compounds have been isolated from soft corals, and further research on these complexes is essential for both the development of marine drugs and the search for new drugs. Five new eunicellin diterpenes, pachycladins A–E (1–5), were isolated from the Red Sea soft coral *Cladiella pachyclados*. Some of the new metabolites exhibited significant anti-invasive activity in PC-3 cells [40]. New metabolite 1 has been isolated from the marine soft coral *Sarcophyton ehrenbergi* along with the known diterpenoids 2 and 3 and cholesterol 4. All of these compounds showed moderate anticancer activity. (S)-1 showed modest activity against DU145 cells in the 106–161 mM range. Its synthetic enantiomer (R)-1 demonstrated better cytotoxicity against DU145 cells, with an IC_{50} of 77.2 ± 2.5 mM. The naturally obtained membrane 3 compound exhibited good potency against DU145 cells, with an IC_{50} of 75.0 ± 3.8 mM [41].

3.7. Holothurians

Holothurians (sea cucumbers) are marine invertebrates that have been used in traditional Asian medicine for centuries [59]. Triterpene glycoside frondoside A (FrA) was initially isolated from an extract of the edible sea cucumber *Cucumaria frondosa*. The FrA compound showed high efficacy and low toxicity in human prostate cancer cells, including cell lines with resistance to standard therapies. Its unique combination of properties includes the simultaneous induction of apoptosis coupled with cell cycle arrest and inhibition of pro-survival autophagy, as well as potential immune modulatory effects [42].

12-MTA inhibited prostate cancer cell lines. PI staining showed that 12-MTA caused PC-3 cell death through the induction of apoptosis, in which caspase-3 may play a role. At relevant biological concentrations, 12-MTA can selectively inhibit the formation of 5-hydroxyeicosatetraenoic acid (5-HETE), a metabolite of 5-lipoxygenase. This agent may be a novel adjunctive therapy for selected malignancies including prostate cancer [43].

4. Conclusions

The resistance of marine compounds to prostate cancer has been recognized by academics at home and abroad. The study of the anti-prostate cancer effects of marine compounds is exploring their direct effects on tumor cells by boosting host immune function.

In conclusion, marine compounds have significant potential as anticancer drug compounds. In recent years, the incidence of prostate cancer has been increasing. From an application standpoint, it is difficult to rely on the compounds extracted from land-grown animals and plants to meet clinical needs. Not only is it difficult to collect large numbers of terrestrial animals and plants as it cannot be multiplied, but since some are endangered species, there is a problem of resource competition. Marine sources can be beneficial to prostate cancer research. However, there has been little research on this topic. The organisms studied in the assessment of the anticancer aspects of marine compounds represent only a tiny fraction of the millions of marine creatures. Marine flora exists in large quantities in nature, and many anticancer bioactive compounds have been isolated from them.

This review discussed a number of marine-derived compounds that are related to prostate cancer. Full elucidation of the anti-cancer mechanisms of these compounds, including the clear structure-activity relationship between these compounds and dose-effect may simplify the extraction process. Reasonable pharmacological screening and clinical observation of marine compounds offer promising leads for the development of anticancer drugs or anticancer adjuvants. Marine compounds can play their due role in improving the quality of life of patients with prostate cancer. Further research on these compounds is required for the development of new anti-prostate cancer drugs.

Author Contributions: M.F. and A.K.N. wrote the manuscript. Y.-J.C., T.D. participated in literature analysis and manuscript editing. Y.T. and E.-J.C. contributed to the conception of the study. E.-K.K. designed the main structure of the manuscript, and review and edit the manuscript. All authors reviewed and approved the final version of the manuscript.

Acknowledgments: This research was supported by Basic Science Research Program through the National Research Foundation of Korea (NRF) funded by the Ministry of Education, Science and Technology (NRF-2017R1D1A1B03036247) and the Korea Institute of Planning and Evaluation for Technology in Food, Agriculture, Forestry and Fisheries (IPET) through the Agri-Bio Industry Technology Development Program (316027-5 & 116163-2) funded by the Ministry of Agriculture, Food and Rural Affairs (MAFRA).

Conflicts of Interest: The authors declare no conflict of interest.

References

1. Kiuru, P.; D'Auria, M.V.; Muller, C.D.; Tammela, P.; Vuorela, H.; Yli-Kauhaluoma, J. Exploring marine resources for bioactive compounds. *Planta Med.* **2014**, *80*, 1234–1246. [CrossRef] [PubMed]
2. Müller, M.; Mentel, M.; van Hellemond, J.J.; Henze, K.; Woehle, C.; Gould, S.B.; Yu, R.Y.; van der Giezen, M.; Tielens, A.G.; Martin, W.F. Biochemistry and evolution of anaerobic energy metabolism in eukaryotes. *Microbiol. Mol. Biol. Rev.* **2012**, *76*, 444–495. [CrossRef] [PubMed]
3. Torre, L.A.; Bray, F.; Siegel, R.L.; Ferlay, J.; Lortet-Tieulent, J.; Jemal, A. Global cancer statistics, 2012. *CA Cancer J. Clin.* **2015**, *65*, 87–108. [CrossRef] [PubMed]
4. Mao, Q.Q.; Lin, Y.W.; Chen, H.; Yang, K.; Kong, D.B.; Jiang, H. Monitoring of prostate cancer growth and metastasis using a PSA luciferase report plasmid in a mouse model. *Asian. Pac. J. Trop. Biomed.* **2014**, *7*, 879–883. [CrossRef]
5. Lonergan, P.E.; Tindall, D.J. Androgen receptor signaling in prostate cancer development and progression. *J. Carcinog.* **2011**, *10*, 20. [PubMed]
6. Wilson, J.D. The role of androgens in male gender role behavior. *Endocr. Rev.* **1999**, *20*, 726–737. [CrossRef] [PubMed]
7. Dasgupta, S.; Srinidhi, S.; Vishwanatha, J.K. Oncogenic activation in prostate cancer progression and metastasis: Molecular insights and future challenges. *J. Carcinog.* **2012**, *11*, 4. [CrossRef] [PubMed]
8. Wieduwilt, M.J.; Moasser, M.M. The epidermal growth factor receptor family: biology driving targeted therapeutics. *Cell Mol. Life Sci.* **2008**, *65*, 1566–1584. [CrossRef] [PubMed]
9. Katz, M.; Amit, I.; Yarden, Y. Regulation of MAPKs by growth factors and receptor tyrosine kinases. *Biochim. Biophys. Acta* **2007**, *1773*, 1161–1176. [CrossRef] [PubMed]
10. Rodier, F.; Campisi, J.; Bhaumik, D. Two faces of p53: aging and tumor suppression. *Nucleic Acids Res.* **2007**, *35*, 7475–7484. [CrossRef] [PubMed]
11. Rowlands, M.A.; Gunnell, D.; Harris, R.; Vatten, L.J.; Holly, J.M.; Martin, R.M. Circulating insulin-like growth factor peptides and prostate cancer risk: A systematic review and meta-analysis. *Int. J. Cancer* **2009**, *124*, 2416–2429. [CrossRef] [PubMed]
12. Shang, X.; Lin, X.; Alvarez, E.; Manorek, G.; Howell, S.B. Tight junction proteins claudin-3 and claudin-4 control tumor growth and metastases. *Neoplasia* **2009**, *14*, 974–985. [CrossRef]
13. Vezza, R.; Rokach, J.; FitzGerald, G.A. Prostaglandin F2α receptor-dependent regulation of prostaglandin transport. *Mol. Pharmacol.* **2001**, *59*, 1506–1513. [CrossRef] [PubMed]
14. Suárez, Y.; González, L.; Cuadrado, A.; Berciano, M.; Lafarga, M.; Muñoz, A. Kahalalide F, a new marine-derived compound, induces oncosis in human prostate and breast cancer cells. *Mol. Cancer Ther.* **2003**, *2*, 863–872. [PubMed]

15. Rademaker-Lakhai, J.M.; Horenblas, S.; Meinhardt, W.; Stokvis, E.; de Reijke, T.M.; Jimeno, J.M.; Lopez-Lazaro, L.; Martin, J.A.L.; Beijnen, J.H.; Schellens, J.H. Phase I clinical and pharmacokinetic study of kahalalide F in patients with advanced androgen refractory prostate cancer. *Clin. Cancer Res.* **2005**, *11*, 1854–1862. [PubMed]

16. Kim, Y.S.; Kim, S.K.; Park, S.J. Apoptotic effect of demethoxyfumitremorgin C from marine fungus *Aspergillus fumigatus* on PC3 human prostate cancer cells. *Chem.-Biol. Interact.* **2017**, *269*, 18–24. [CrossRef] [PubMed]

17. Liu, Y.; Zhao, S.; Ding, W.; Wang, P.; Yang, X.; Xu, J. Methylthio-aspochalasins from a marine-derived fungus aspergillus sp. *Mar. Drugs* **2014**, *12*, 5124–5131. [CrossRef] [PubMed]

18. Dyshlovoy, S.A.; Otte, K.; Tabakmakher, K.M.; Hauschild, J.; Makarieva, T.N.; Shubina, LK.; Fedorov, S.N.; Bokemeyer, C.; Stonik, V.A.; von Amsberg, G. Synthesis and anticancer activity of the derivatives of marine compound rhizochalin in castration resistant prostate cancer. *Oncotarget* **2018**, *9*, 16962–16973. [CrossRef] [PubMed]

19. Dyshlovoy, S.A.; Otte, K.; Alsdorf, W.H.; Hauschild, J.; Lange, T.; Venz, S.; Schumacher, U.; Schröder-Schwarz, J.; Makarieva, T.N.; Guzii, A.G.; et al. Marine compound rhizochalinin shows high in vitro and in vivo efficacy in castration resistant prostate cancer. *Oncotarget* **2016**, *7*, 69703. [CrossRef] [PubMed]

20. Sayed, K.A.; Khanfar, M.A.; Shallal, H.M.; Muralidharan, A.; Awate, B.; Youssef, D.T.; Liu, Y.; Zhou, Y.D.; Nagle, D.G.; Shah, G. Latrunculin A and its C-17-O-carbamates inhibit prostate tumor cell invasion and HIF-1 activation in breast tumor cells. *J. Nat. Prod.* **2008**, *71*, 396–402. [CrossRef] [PubMed]

21. Shin, Y.; Kim, G.D.; Jeon, J.E.; Shin, J.; Lee, S.K. Antimetastatic effect of halichondramide, a trisoxazole macrolide from the marine sponge *Chondrosia corticata*, on human prostate cancer cells via modulation of epithelial-to-mesenchymal transition. *Mar. Drugs* **2013**, *11*, 2472–2485. [CrossRef] [PubMed]

22. Schneiders, U.M.; Schyschka, L.; Rudy, A.; Vollmar, A.M. BH3-only proteins Mcl-1 and Bim as well as endonuclease G are targeted in spongistatin 1–induced apoptosis in breast cancer cells. *Mol. Cancer Ther.* **2009**, *8*, 2914–2925. [CrossRef] [PubMed]

23. Arai, M.; Kawachi, T.; Setiawan, A.; Kobayashi, M. Hypoxia-selective growth inhibition of cancer cells by furospinosulin-1, a furanosesterterpene isolated from an Indonesian marine sponge. *Chem. Med. Chem.* **2010**, *5*, 1919–1926. [CrossRef] [PubMed]

24. Dai, J.; Fishback, J.A.; Zhou, Y.D.; Nagle, D.G. Sodwanone and yardenone triterpenes from a South African species of the marine sponge Axinella inhibit hypoxia-inducible factor-1 (HIF-1) activation in both breast and prostate tumor cells. *J. Nat. Prod.* **2006**, *69*, 1715–1720. [CrossRef] [PubMed]

25. Meimetis, L.G.; Williams, D.E.; Mawji, N.R.; Banuelos, C.A.; Lal, A.A.; Park, J.J.; Tien, A.H.; Fernandez, J.G.; de Voogd, N.J.; Sadar, M.D.; et al. Niphatenones, glycerol ethers from the sponge *Niphates digitalis* block androgen receptor transcriptional activity in prostate cancer cells: structure elucidation, synthesis, and biological activity. *J. Med. Chem.* **2012**, *55*, 503–514. [CrossRef] [PubMed]

26. Gordaliza, M. Terpenyl-purines from the sea. *Mar. Drugs* **2009**, *7*, 833–849. [CrossRef] [PubMed]

27. Drew, L.; Fine, R.L.; Do, T.N.; Douglas, G.P.; Petrylak, D.P. The novel antimicrotubule agent cryptophycin 52 (LY355703) induces apoptosis via multiple pathways in human prostate cancer cells. *Clin. Cancer. Res.* **2002**, *8*, 3922–3932. [PubMed]

28. Tripathi, A.; Puddick, J.; Prinsep, M.R.; Rottmann, M.; Chan, K.P.; Chen, D.Y.K.; Tan, L.T. Lagunamide C, a cytotoxic cyclodepsipeptide from the marine cyanobacterium Lyngbya majuscula. *Phytochemistry* **2011**, *72*, 2369–2375. [CrossRef] [PubMed]

29. Costa, M.; Costa-Rodrigues, J.; Fernandes, M.H.; Barros, P.; Vasconcelos, V.; Martins, R. Marine cyanobacteria compounds with anticancer properties: A review on the implication of apoptosis. *Mar. Drugs* **2012**, *10*, 2181–2207. [CrossRef] [PubMed]

30. Gantar, M.; Dhandayuthapani, S.; Rathinavelu, A. Phycocyanin induces apoptosis and enhances the effect of topotecan on prostate cell line LNCaP. *J. Med. Food.* **2012**, *15*, 1091–1095. [CrossRef] [PubMed]

31. Wang, W.L.W.; McHenry, P.; Jeffrey, R.; Schweitzer, D.; Helquist, P.; Tenniswood, M. Effects of Iejimalide B, a marine macrolide, on growth and apoptosis in prostate cancer cell lines. *J. Cell. Biochem.* **2008**, *105*, 998–1007. [CrossRef] [PubMed]

32. Kubanek, J.; Prusak, A.C.; Snell, T.W.; Giese, R.A.; Fairchild, C.R.; Aalbersberg, W.; Hay, M.E. Bromophycolides C− I from the Fijian Red Alga *Callophycus s erratus*. *J. Nat. Prod.* **2006**, *69*, 731–735. [CrossRef] [PubMed]

33. Farooqi, A.A.; Butt, G.; Razzaq, Z. Algae extracts and methyl jasmonate anti-cancer activities in prostate cancer: choreographers of 'the dance macabre'. *Cancer Cell Int.* **2012**, *12*, 50. [CrossRef] [PubMed]

34. Anderson, M.L. A preliminary investigation of the enzymatic inhibition of 5α-reductase and growth of prostatic carcinoma cell line LNCap-FGC by natural astaxanthin and saw palmetto lipid extract in vitro. *J. Herb. Pharmacother.* **2005**, *5*, 17–26. [CrossRef] [PubMed]

35. Boo, H.J.; Hong, J.Y.; Kim, S.C.; Kang, J.I.; Kim, M.K.; Kim, E.J.; Hyun, J.W.; Koh, Y.S.; Yoo, E.S.; Kwon, J.M.; et al. The anticancer effect of fucoidan in PC-3 prostate cancer cells. *Mar. Drugs* **2013**, *11*, 2982–2999. [CrossRef] [PubMed]

36. Hsu, H.Y.; Lin, T.Y.; Hwang, P.A.; Tseng, L.M.; Chen, R.H.; Tsao, S.M.; Hsu, J. Fucoidan induces changes in the epithelial to mesenchymal transition and decreases metastasis by enhancing ubiquitin-dependent TGFβ receptor degradation in breast cancer. *Carcinogenesis* **2012**, *34*, 874–884. [CrossRef] [PubMed]

37. Martin, L.J. Fucoxanthin and its metabolite fucoxanthinol in cancer prevention and treatment. *Mar. Drugs* **2015**, *13*, 4784–4798. [CrossRef] [PubMed]

38. Satomi, Y. Fucoxanthin induces GADD45A expression and G1 arrest with SAPK/JNK activation in LNCap human prostate cancer cells. *Anticancer Res.* **2012**, *32*, 807–813. [PubMed]

39. Asai, A.; Sugawara, T.; Ono, H.; Nagao, A. Biotransformation of fucoxanthinol into amarouciaxanthin A in mice and HepG2 cells: formation and cytotoxicity of fucoxanthin metabolites. *Drug Metab. Dispos.* **2004**, *32*, 205–211. [CrossRef] [PubMed]

40. Hassan, H.M.; Khanfar, M.A.; Elnagar, A.Y.; Mohammed, R.; Shaala, L.A.; Youssef, D.T.; Hifnawy, M.S.; El Sayed, K.A. Pachycladins A-E, prostate cancer invasion and migration inhibitory eunicellin-based diterpenoids from the Red Sea soft coral *Cladiella pachyclados*. *J. Nat. Prod.* **2010**, *73*, 848–853. [CrossRef] [PubMed]

41. Bhujanga, R.C.; Babu, D.C.; Bharadwaj, T.V.; Srikanth, D.; Vardhan, K.S.; Raju, T.V.; Bunce, R.A.; Venkateswarlu, Y. Isolation, structural assignment and synthesis of (SE)-2-methyloctyl 3-(4-methoxyphenyl) propenoate from the marine soft coral *Sarcophyton ehrenbergi*. *Nat. Prod. Res.* **2015**, *29*, 70–76. [CrossRef] [PubMed]

42. Dyshlovoy, S.A.; Menchinskaya, E.S.; Venz, S.; Rast, S.; Amann, K.; Hauschild, J.; Alsdorf, W. The marine triterpene glycoside frondoside A exhibits activity in vitro and in vivo in prostate cancer. *Int. J. Cancer* **2016**, *138*, 2450–2465. [CrossRef] [PubMed]

43. Yang, P.; Collin, P.; Madden, T.; Chan, D.; Sweeney-Gotsch, B.; McConkey, D.; Newman, R.A. Inhibition of proliferation of PC3 cells by the branched-chain fatty acid, 12-methyltetradecanoic acid, is associated with inhibition of 5-lipoxygenase. *Prostate* **2003**, *55*, 281–291. [CrossRef] [PubMed]

44. Lindequist, U. Marine-derived pharmaceuticals–challenges and opportunities. *Biomol. Ther.* **2016**, *24*, 561–571. [CrossRef] [PubMed]

45. Lin, W.; Fang, L.K.; Liu, J.W.; Cheng, W.Q.; Yun, M. Effect of marine fungal metabolites from the south china sea on prostate cancer cell line du-145. *J. Intern. Med.* **2008**, *35*(10), 562–563.

46. Rochfort, S.; Ford, J.; Ovenden, S.; Wan, S.S.; George, S.; Wildman, H.; Tait, R.M.; Meurer-Grimes, B.; Cox, S.; Coates, J.; et al. A novel aspochalasin with HIV-1 integrase inhibitory activity from *Aspergillus flavipes*. *J. Antibiot. (Tokyo)* **2005**, *58*, 279–283. [CrossRef] [PubMed]

47. Liu, J.; Hu, Z.; Huang, H.; Zheng, Z.; Xu, Q. Aspochalasin U, a moderate TNF-alpha inhibitor from *Aspergillus* sp. *J. Antibiot. (Tokyo)* **2012**, *65*, 49–52. [CrossRef] [PubMed]

48. Choo, S.J.; Yun, B.S.; Ryoo, I.J.; Kim, Y.H.; Bae, K.H.; Yoo, I.D. Aspochalasin I, a melanogenesis inhibitor from *Aspergillus* sp. *J. Microbiol. Biotechnol.* **2009**, *19*, 368–371. [CrossRef] [PubMed]

49. Dyshlovoy, S.A.; Otte, K.; Venz, S.; Hauschild, J.; Junker, H.; Makarieva, T.N.; Balabanov, S.; Alsdorf, W.H.; Madanchi, R.; Honecker, F.; et al. Proteomic-based investigations on the mode of action of the marine anticancer compound rhizochalinin. *Proteomics* **2017**, *17*, 170048. [CrossRef] [PubMed]

50. Sima, P.; Vetvicka, V. Bioactive substances with anti-neoplastic efficacy from marine invertebrates: Porifera and Coelenterata. *World. J. Clin. Oncol.* **2011**, *2*, 355–361. [CrossRef] [PubMed]

51. Ko, L.J.; Prives, C. p53: puzzle and paradigm. *Genes Dev.* **1996**, *10*, 1054–1072. [CrossRef] [PubMed]

52. Scanlon, C.S.; Van Tubergen, E.A.; Inglehart, R.C.; D'Silva, N.J. Biomarkers of epithelial-mesenchymal transition in squamous cell carcinoma. *J. Dent. Res.* **2013**, *92*, 114–121. [CrossRef] [PubMed]

53. Kalimuthu, S.; Venkatesan, J.; Kim, S.K. Marine derived bioactive compounds for breast and prostate cancer treatment: A review. *Curr. Bioact. Compd.* **2014**, *10*, 62–74. [CrossRef]

54. Ambati, R.R.; Phang, S.M.; Ravi, S.; Aswathanarayana, R.G. Astaxanthin: sources, extraction, stability, biological activities and its commercial applications—a review. *Mar. Drugs* **2014**, *12*, 128–152. [CrossRef] [PubMed]

55. Sithranga, B.N.; Kathiresan, K. Anticancer drugs from marine flora: an overview. *J. Oncol.* **2010**, *2010*, 1–18. [CrossRef] [PubMed]

56. Mayer, A.M.S.; Hamann, M.T. Marine pharmacology in 1999: compounds with antibacterial, anticoagulant, antifungal, anthelmintic, anti-inflammatory, antiplatelet, antiprotozoal and antiviral activities affecting the cardiovascular, endocrine, immune and nervous systems, and other miscellaneous mechanisms of action. *Comp. Biochem. Physiol. C Toxicol. Pharmacol.* **2002**, *132*, 315–339. [PubMed]

57. Peng, J.; Yuan, J.P.; Wu, C.F.; Wang, J.H. Fucoxanthin, a marine carotenoid present in brown seaweeds and diatoms: metabolism and bioactivities relevant to human health. *Mar. Drugs* **2011**, *9*, 1806–1828. [CrossRef] [PubMed]

58. Strand, A.; Herstad, O.; Liaaen-Jensen, S. Fucoxanthin metabolites in egg yolks of laying hens1. *Comp. Biochem. Phys. A.* **1998**, *119*, 963–974. [CrossRef]

59. Aminin, D.; Menchinskaya, E.; Pisliagin, E.; Silchenko, A.S.; Avilov, S.A.; Kalinin, V.I. Anticancer activity of sea cucumber triterpene glycosides. *Mar. Drugs* **2015**, *13*, 1202–1223. [CrossRef] [PubMed]

 *marine drugs*

Article

A Low Molecular Weight Protein from the Sea Anemone *Anemonia viridis* with an Anti-Angiogenic Activity

Erwann P. Loret [1,*], José Luis [2], Christopher Nuccio [3], Claude Villard [2], Pascal Mansuelle [4], Régine Lebrun [4] and Pierre Henri Villard [1]

[1] Aix-Marseille University (AMU), Université d'Avignon, Centre National de la Recherche Scientifique (CNRS), Institut de la Recherche et du Développement (IRD), Institut Méditerranéen de Biologie et d'Ecologie. CNRS UMR 7263 IRD 237 Faculté de Pharmacie, 27 Bd Jean Moulin, 13385 Marseille, France; pierre.villard@univ-amu.fr

[2] AMU, CNRS, Institut de Neurophysio Pathologie, 13385 Marseille, France; jose.luis@univ-amu.fr (J.L.); claude.villard@univ-amu.fr (C.V.)

[3] AMU, Institut National de la Santé Et de la Recherche Scientifique, 13385 Marseille, France; christopher.nuccio@univ-amu.fr

[4] AMU, CNRS Formation de Recherche 3479, Institut de Microbiologie de la Méditerranée, Plateforme Protéomique, 31 Chemin Joseph Aiguier, 13402 Marseille, France; pmansuelle@imm.cnrs.fr (P.M.); rlebrun@imm.cnrs.fr (R.L.)

* Correspondence: erwann.loret@univ-amu.fr; Tel.: +33-625-072-362

Received: 16 March 2018; Accepted: 12 April 2018; Published: 19 April 2018

Abstract: Sea anemones are a remarkable source of active principles due to a decentralized venom system. New blood vessel growth or angiogenesis is a very promising target against cancer, but the few available antiangiogenic compounds have limited efficacy. In this study, a protein fraction, purified from tentacles of *Anemonia viridis*, was able to limit endothelial cells proliferation and angiogenesis at low concentration (14 nM). Protein sequences were determined with Edman degradation and mass spectrometry in source decay and revealed homologies with Blood Depressing Substance (BDS) sea anemones. The presence of a two-turn alpha helix observed with circular dichroism and a trypsin activity inhibition suggested that the active principle could be a Kunitz-type inhibitor, which may interact with an integrin due to an Arginine Glycin Aspartate (RGD) motif. Molecular modeling showed that this RGD motif was well exposed to solvent. This active principle could improve antiangiogenic therapy from existing antiangiogenic compounds binding on the Vascular Endothelial Growth Factor (VEGF).

Keywords: sea anemone; drug discovery; cancer; antiangiogenic; endothelial cells; RGD motif; kunitz type inhibitor

---

1. Introduction

Sea anemones have been understudied as a source of new pharmacological tools or therapeutic leads [1]. Sea anemones belong to *Cnidaria* that also includes corals, jellyfish and sponges. They have toxic peptides to incapacitate and immobilize prey and to defend from potential predators [2]. Their toxin arsenal is complex, targeting a variety of pharmacological targets such as ionic channels, inflammatory receptors [3] or pore forming protein in cellular membranes [4]. Anti-hyperglycemic and anti-diabetic activities were also observed from sea anemone extract [5].

Active principles from venomous species such as snakes or scorpions have a centralized venom system and have a structural homogeneity, which is not the case for sea anemones. Scorpion active principles for instance have a scaffold characterized by three beta sheets, an alpha helix and four disulfide bridges [6]. In sea anemones, the venom system is decentralized in all parts of the animal body showing a higher diversity in size and scaffold of protein active principles. The active principles binding on receptors such as ionic channels are short size proteins between 3000 and 5000 Da cross linked with three disulfide bridges [2]. They have different structural scaffolds regarding their pharmacological targets. Active principles binding on ionic channel are characterized by three beta strands [7], while those binding on enzymes such as the family of Kunitz-type inhibitors have two beta strands and an alpha helix [8]. Surprisingly, the sea anemone toxin Bg1, with a scaffold completely different, can compete at nanomolar concentrations with the Aah II scorpion toxin on the same pharmacological site on sodium channel [9]. However, the superposition of the 3D structure of these two proteins shows that the lateral chains of four basic residues are in exactly the same positions [9].

Cancer treatments are mainly based on anti-mitotic compounds that have side effects because they block the division of both cancer cells and healthy cells. Cancer cells need to be vascularized by endothelial cells, in a process called angiogenesis, to grow as a tumor and then to spread as metastases inducing the patient death. It was proposed almost 50 years ago to block angiogenesis to fight against cancer because angiogenesis is no longer important after embryogenesis [10]. There are very few antiangiogenic compounds compared to antimitotic compounds and it has only been a decade since antiangiogenic compounds began being tested in clinical trials. Amazingly, antiangiogenic compounds showed a limited efficacy because they all bind on the Vascular Endothelial Growth Factor (VEGF) or the VEGF receptor and tumor cells can trigger different biological ways to have angiogenesis [11]. Among antiangiogenic compounds binding on VEGF, the most used in clinical trial is Bevacizumab (known as Avastin), which is a monoclonal antibody [11]. Bevacizumab is costly due to its size and difficulties to have germ free production as recombinant protein. Furthermore, resistance to Bevacizumab is observed due to the upregulation of other redundant angiogenic factors different from VEGF [12]. Compounds binding on VEGF receptors that are not proteins have toxic effects that limit their use. Resistance towards VEGF-centered antiangiogenic therapy represents a substantial clinical challenge [10]. There is therefore a need to have compounds blocking angiogenesis that do not bind on VEGF or VEGF receptor, are easy to produce and have no long-term toxicity.

Short size synthetic proteins (less than 50 residues) are suitable compounds for this goal. *Anemonia viridis* (also called *Anemonia sulcata*) is well studied and proven to be a remarkable source of low molecular proteins with different pharmacological binding sites on ionic channel receptors [7]. Seven short size proteins (42–49 residues) binding on ionic channels were purified and characterized from *Anemonia viridis* [7]. Among them, AS2 is a 47-residue-long protein (4934 Da) having only beta structures [13]. There is also BDS-1 that is a blood depressing protein of 43 residues (4708 Da) binding on a specific potassium channel [14]. The main secondary structure of BDS-1 is a triple-stranded antiparallel beta-sheet without alpha helix [7]. Recently, a partially purified extract of *Anemonia viridis* was reported to affect the growth and viability of selected tumor cell lines [15].

In this study, two cellular antiangiogenic screening tests were used with Human Microvascular Endothelial Cells (HMEC) having the Epithelial Growth Factor instead of VEGF [16]. The first one measured the proliferation of HMEC and the second one the formation of HMEC capillary network. These screening tests made it possible to identify from *Anemonia viridis* a protein fraction able to limit HMEC proliferation and angiogenesis at low concentration (IC$_{50}$ 14 nM). Trypsin inhibition and the presence of a two turns alpha helix revealed by circular dichroism suggest that the active principle is a kunitz type inhibitor.

2. Results

2.1. Purification from Full Sea Anemone Body

To preserve the integrity of potential protein active principles, a new purification protocol was established. This protocol described in the experimental procedure was tested with both the full animal body and tentacles in preliminary studies using two animal bodies. HPLC analysis after precipitation, centrifugation and filtrations steps revealed that tentacles had the highest content in short size proteins compared to the full body (data not shown). Tentacles were therefore chosen to obtain pure proteins with semi preparative HPLC. Before HPLC, it was necessary to carry out different steps of extractions and filtrations. To avoid protein enzyme digestions or degradations, the tentacles from 11 *Anemonia viridis* were cut and immediately frozen at $-20\,^{\circ}$C. After 24 h at $-20\,^{\circ}$C, the tentacles were lyophilized for three days. Then, the dry tentacles (20.8 g) were mashed. A first precipitation and a centrifugation were carried out with a hydrophobic solvent to eliminate lipids and membrane proteins. After evaporation for 24 h, the weight of the pellet was 19.2 g. The pellet was suspended in water 0.1% TFA and a centrifugation was carried out to eliminate the calcic skeleton. A gel filtration (cut off 30 KDa) and a 0.22 μm filtration made it possible to select low molecular weight compounds. A UV absorption spectrum of this solution (data not shown) revealed two major bands at 265 nm and 330 nm due to alkaloid pigments that are low molecular weight compounds abundant in sea anemones [17]. After lyophilization, only 0.984 g remained from the initial 20.8 g dry tentacles (4.7%). Semi preparative HPLC purification was carried out on this 0.984 g and made it possible to obtain 15 fractions. The first fraction contained essentially alkaloid pigments. The total amount of proteins found in the other fractions was only 30 mg. The low molecular weight proteins that are the targets of this purification corresponded to 0.14% of the dry tentacles in the beginning of this purification protocol.

HPLC analysis before semi preparative HPLC purification showed a major peak at 4 min (Figure 1B) and the spectral analysis in (Figure 1A) revealed a major absorbance at 265 nm and 330 nm that were dominant in the spectral analysis carried out before semi preparative HPLC. This was not a protein spectrum. Protein spectra are characterized by a first absorbance at 280 nm due to aromatic residues and then a major absorbance due to peptidic bonds that begins at 210 nm [16]. The elution time showed that these compounds were not retained in C8 column and confirmed that they were alkaloid pigments with aromatic cycles [18].

Semi prepartive HPLC made it possible to collect nine main protein fractions. Mass spectroscopy analysis revealed that the molecular weight of these proteins was going from 4690 to 6700 Da. They displayed typical protein UV spectra (data not shown). The most abundant fraction eluted at 7 min (Figure 1) and had an absorbance spectrum with a bump at 290 nm typical of tryptophan (W) that were highly conserved in active principles binding on ionic channel identified in *Anemonia viridis* but the molecular weight of this protein (6620 Da) was not compatible with known *Anemonia viridis* active principles that had sizes <50 amino acid residues [6]. The other proteins eluting from 7 min to 15 min displayed absorbance spectra without W contribution characterized by a shift of the maximum absorbance from 280 to 278 nm (data not shown). Other fractions had W contribution, particularly the 31-min fraction. It was possible to recognize and isolate AS2 in this fraction due to its molecular weight and its N-terminal sequence, which was a 47-residue-long protein having three W in its sequence [12].

2.2. HMEC Proliferation Inhibition (n = 2)

A screening was carried out with the nine protein fractions and the alkaloid pigments as control to measure the capacity of these proteins to inhibit HMEC proliferation. This experiment was performed twice with fractions from the second purification. The test was carried out with a 10 μM concentration for the nine protein fractions and the alkaloid pigments. It was important in this test that living cells remained detectable to differentiate between cytotoxicity and proliferation inhibition. Proliferation inhibition was characterized by a decrease of absorbance (Figure 2). The absorbance was proportional to the mitochondrial activity. Only three protein fractions were able to limit HMEC proliferation up or below to 50% at 7 min, 12 min and 28 min (Figure 2). No absorbance due to cellular death was observed with the alkaloid pigments (Figure 2, 4 min). Cytotoxicity of *Anemonia viridis* alkaloid pigments was previously reported [17]. Trace of alkaloid pigments was observed in analytical HPLC of 7 min and 12 min proteins (data not shown) and the decrease of absorbance >50% could be due to cytotoxicity and not to an inhibition of proliferation due to an antiangiogenic activity. Alkaloid pigments were present in fraction 17 min, 22 min and 26 min in the second purification and may explain the high variation between the two experiments.

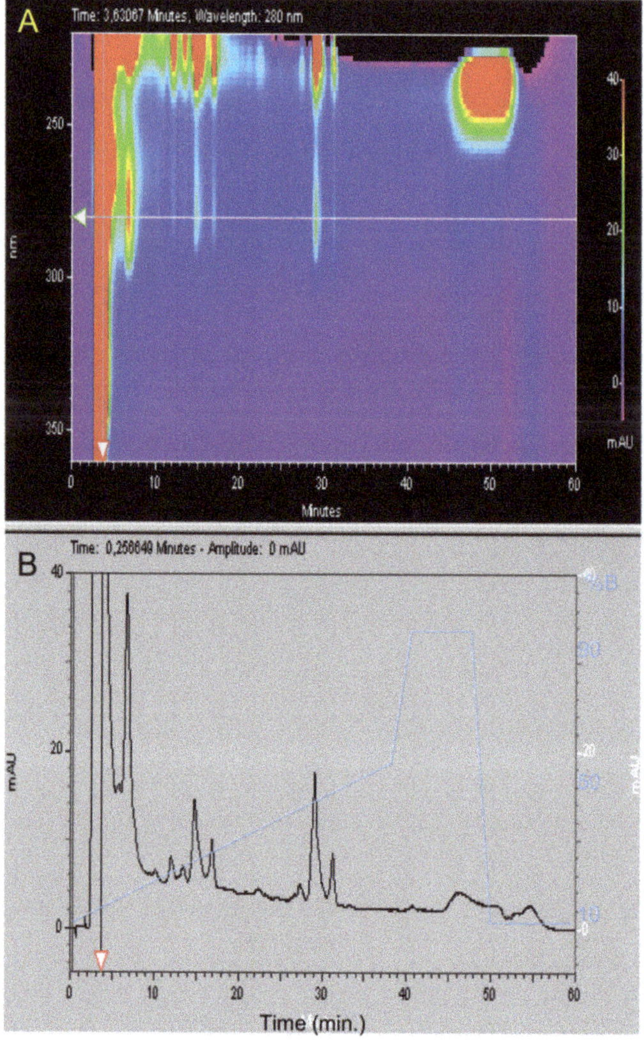

Figure 1. 3D Analytical HPLC of *Anemonia viridis* tentacles after phase extractions, centrifugation and filtrations and before semi preparative HPLC. The UV absorbance of the effluent was measured with a diode ray detector. (**A**) A 3D plot displaying elution time from 0 to 60 min (*x* axis) regarding wavelengths from 220 to 330 nm (*y* axis). The *z* axis corresponds to different peak intensities and is represented with a color scale going from 0 (purple) to 40 mAU (red). The horizontal arrow shows the 280 nm wavelength. (**B**) Elution time regarding absorbance at 280 nm from 0 to 40 mAU and the acetonitrile (%B) gradient in blue. The vertical arrow shows the retention time of the major peak at 4 min that is out scaled with an absorbance at 500 mAU.

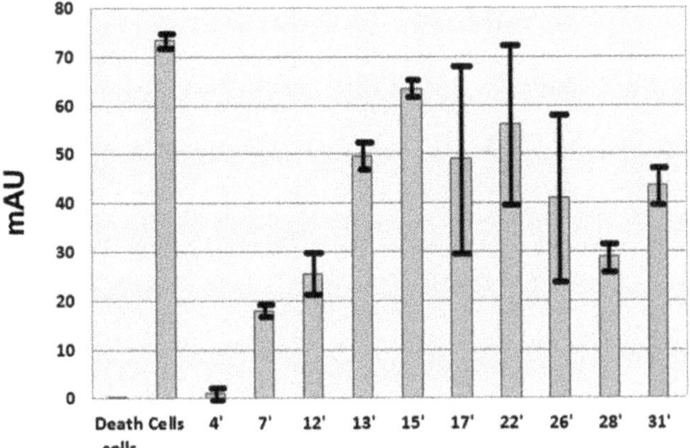

Figure 2. HMEC proliferation inhibition with 50 µg/L of the alkaloid fraction (4 min) and the nine protein fractions purified from semipreparative HPLC. This experiment was carried out twice with two different purifications. Alkaloid pigments were present in fraction 17 min, 22 min and 26 min in the second purification and may explain the differences between the two experiments.

2.3. HMEC Tubulogenesis Assay (n = 3)

A first screening at 10 µM made with the fractions obtained from SP HPLC showed that the capillary network formation was inhibited only with the 28-min fraction and not with the 7 min- and 12-min fractions where the remaining living cells were able to display a network (data not shown). To measure a dose effect of the 28-min fraction on HMEC tubulogenesis, seven dilutions from 10 µM to 3.2 nM (dilution factor 1/5) were tested (Figure 3). A capillary network with interconnected lines similar to control is observed only at 3.2 nM (Figure S1 in Supplementary Materials). To determine the half effect, a dose–response curve was established by measuring the capillary length with METAMORPH 7.6 software for each dilution (Figure 3).

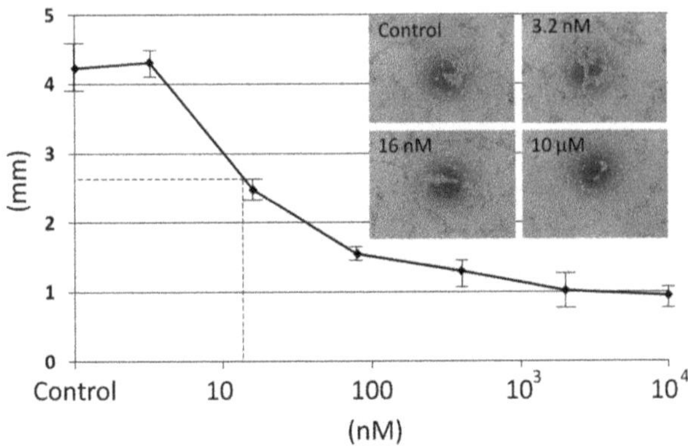

Figure 3. HMEC tubulogenesis assay with the 28-min fraction with concentrations going from 10 µM to 3.2 nM in a triplicate experiment (*n* = 2). A network with interconnected lines similar to control was observed only at 3.2 nM. The dose–response curve was established by measuring capillary length with the METAMORPH software. The six dilutions were represented with the control in a decimal logarithm scale regarding the capillary length measured in mm. The half effect corresponding to a capillary length of 2.65 mm was observed at 14 nM.

2.4. Primary Structure Analyses

The Edman N-terminal sequencing of the 28-min fraction made it possible to identify the first 31 amino acid residues and revealed a sequence homology with BDS sea anemones [7]. The sequence was identical to BDS-1 excepted on position 18 with a phenylalanine (F) instead of Leucine (L) in BDS-1. Although the 28-min fraction gave only one peak in analytical HPLC, the sequencing revealed the presence of contaminants with a proline (P) instead of S7 and glycine (G) instead of Lysine (K) 9. The mutation S/P 7 was observed in BDS2 [7].

A first Matrix-Assisted Laser Desorption Ionization Mass Spectrometry (MALDI MS) at low resolution gave a molecular weight of 4730 Da for the 28-min fraction. However, high resolution MALDI MS revealed that the 28-min fraction was most heterogeneous and contained up to ten proteins (Figure 4). The three main proteins had average molecular weights corresponding to 4692, 4708 and 4742 Da. To further investigate on the monoisotopic mass of the protein content, the 28-min fraction was also analyzed with high resolution Orbitrap Mass Spectrometry (Figure S2 in Supplementary Materials).

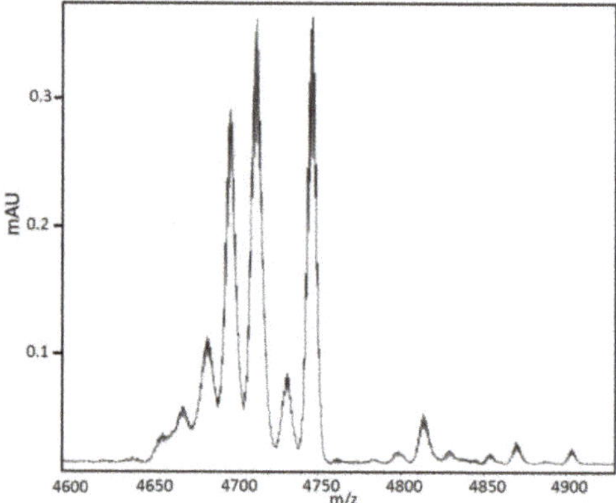

Figure 4. High resolution MALDI MS of the 28-min fraction. Each peak was decomposed in different isotopic forms. The isotopic average masses of the three main peaks were, respectively, 4692, 4708 and 4742 Da. Their mono isotopic masses were, respectively, 4689, 4705 and 4739 Da.

It was possible to determine the sequences of these three main proteins in the 28-min fraction with both MALDI MS using In Source Decay (ISD) technology and HPLC-MS/MS analyses of the digested fractions after Trypsin treatment (Figures S3–S5 in Supplementary Materials). ISD confirmed the sequence homology between BDS1 and BDS5 from the N-terminal Alanine to F18 for BDS5 and L18 for BDS1. The experiment revealed a third chain (BDS16) with a F18 as BDS5 but with a difference of 50 Da compared to BDS5. A bottom up approach using HPLC-MS/MS was performed using a digest of the fraction (Figure S4). The major peaks validated ISD results and identified a peptide with minus 50 Da regarding the masses of the N-terminus of BDS5 and BDS1. This peptide corresponding to the N-terminus of BDS16 was sequenced with ISD (Figure S5) to provide the full sequence of BDS16. The three proteins identified in the 28-min fraction were therefore BDS-1 and two others proteins named BDS-5 and BDS-16.

```
      1 10 20 30 40

BDS-1   AAPCFCSGKPGRGDLWILRGTCPGGYGYTSNCYKWPNICCYPH 4708 Da

BDS-5   ------------------F----------------------- 4742 Da

BDS-16  ----S-P----------F----------------------- 4692 Da
```

Enzymatic digestion was carried out on the 28-min fraction with a Trypsin/Lys C mixture. The disulfide bridges were not altered to allow the possibility of identifying their location. Amazingly, in regular condition corresponding to 1 h incubation at 37 °C, the HPLC peak of the 28-min fraction remained almost unaltered (Figure 5A). It was necessary to wait three days in similar conditions to observe the complete digestion of the 28-min fraction. Resistance to Trypsin digestion was not reported when BDS-1 was sequenced [14]. Sequencing of the peptides purified after complete digestion of the 28-min fraction revealed that the peaks between 22 min and 24 min were the same sequence ^{13}**GDLWIFR**^{19} corresponding to the central part of BDS-5. The peptide corresponding to this sequence had probably different conformational states that may explain the different elution times observed between 22 min and 24 min. The Edman sequencing either for the N-terminus or in peptides obtained from enzymatic digestion did not confirm in the 28-min fraction, the presence of BDS-1 characterized by L18. Sequencing of the peak eluting at 20′ revealed the presence of three peptides corresponding to expected sequences, the C terminus with ^{35}**WPNICCYPH**^{43}, the N-terminus (^{1}**AAPCFCSGK**^{9}) and ^{20}**GTCPGGYGYTSNCYK**^{34}. The gap between ^{31}**NCYK**^{34} was determined by ISD. It was not possible to identify a peptide corresponding to the ^{10}**PGR**^{12} sequence but this part of the BDS-5 sequence was previously identified in the N terminal sequencing of the 28-min fraction. Another digestion of the non-reduced 28-min fraction was performed at different times to control from 20 h to 70 h (Figure S6 in Supplementary Materials). The three peptides produced four different peptides, two for the peptide 13–19 of BDS-5 and BDS-1 and two others with an average mass of 3842 and 3892 attesting that all the other peptides were crosslinked by disulfide bridges. It was possible to deduce that the three disulfide bridges of BDS-5 were similar to BDS-1 and corresponded to ^{4}C–C^{39}, ^{6}C–C^{32} and ^{22}C–C^{40}.

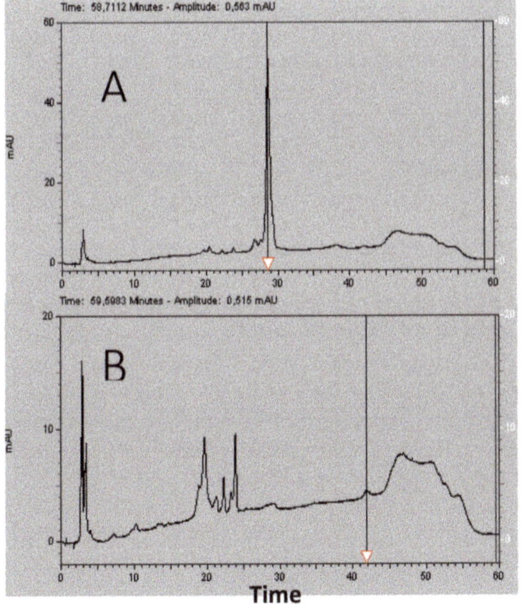

Figure 5. Enzymatic digestion of the 28-min fraction with a Trypsin/Lys C mixture at 37 °C for: one hour (**A**); and three days (**B**). The 28-min fraction displayed only one peak and remained almost unchanged after one-hour digestion, while it should have completely disappeared. This experiment suggested that the 28-min fraction contained a Trypsin inhibitor. The Trypsin/Lys C mixture eluted at 42 min. Another digestion experiment with only Trypsin gave the same results with 72 h necessary to have a complete digestion (Figure S6 in Supplementary Materials).

2.5. Secondary and Tertiary Structure Analyses

A far UV study (178–260 nm) of AS2 and the 28-min fraction was carried out (Figure 6). Absorbance spectra were similar and characterized by absorption at 190 nm of the π–π^* transition of the peptide bonds amid chromophore. A bump due to n–π^* transition was observed at 210 nm [17]. The absorption spectra revealed that the protein concentrations were similar and the light transmission was correct due to the linearity of the signal up to 180 nm (Figure 6A). CD spectra revealed a major difference between AS2 and the 28-min fraction, thus a difference in the content in secondary structures (Figure 7B). AS2 had a CD spectra characteristic of beta sheet and beta turns with no alpha helix. This type of folding is typical of *Anemonia viridis* proteins binding on ionic channel such as AS2 or BDS-1 [7]. The shift of this negative band at 196 nM showed that this was a well-structured protein characterized by beta structures in accordance with a former Laser Raman structural study made with AS2 [13]. The CD spectrum of the 28-min fraction was characteristic of an alpha helix with the splitting of the π–π^* transition in two bands, a positive one at 190 nm and a negative one at 205 nm, and a negative band at 215 nm due to n–π^* transition [17]. The low intensity of the CD bands and a bump at 200 nm in the positive band revealed contributions of beta structures [17]. The CD spectrum of BDS-1 had no positive band and was typical of CD spectra with only beta structures [7]. The alpha helix CD signal was due probably to BDS-5 and might correspond to a two-turn alpha helix [17].

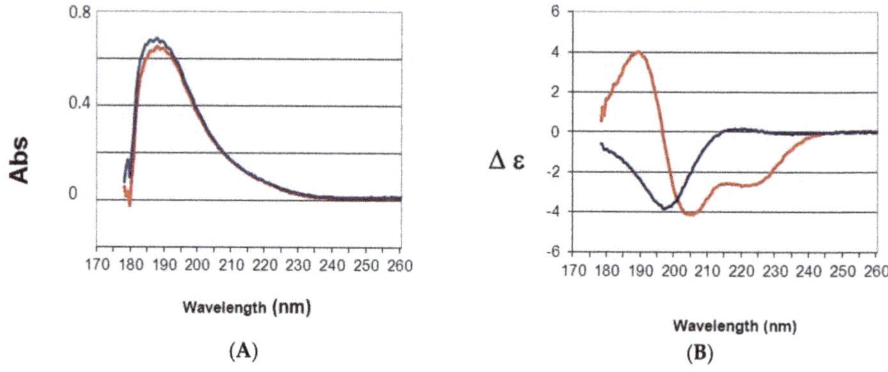

Figure 6. Far UV (178–260 nm) absorbance (**A**); and circular dichroism (**B**) spectra of AS2 (Blue) and the 28-min fraction (red). CD spectra are presented as $\Delta\varepsilon$ per amide rather than in ΔA unit to have a measure independent of mass and concentration [17].

Based on CD data, the BDS-5 model was built from the crystal structure of SHPI-1 in a complex with Elastase [18]. SHPI-1 was purified from the *Stichodactyla helianthus* sea anemone [18]. This structure model was chosen because it was a low molecular weight protein (55 residues) cross linked with three disulfide bridges and the content in secondary structures was compatible with CD data, which was not the case for BDS-1 [7]. The atomic coordinates of alpha carbons of SHPI-1 N-terminus (residues 1–15) and C-terminus of SHPI-1 (residues 32–55) were used as template for BDS-5. Then, atomic coordinates of lateral chains of BDS-5 residues were assigned without overlap to avoid energy bumps. Ten loops were generated to connect BDS-5 residues 15–20. The lowest root means square deviation loop regarding optimum dihedral angles was used to assign atomic coordinates to the last five missing BDS-5 residues to complete the final structure.

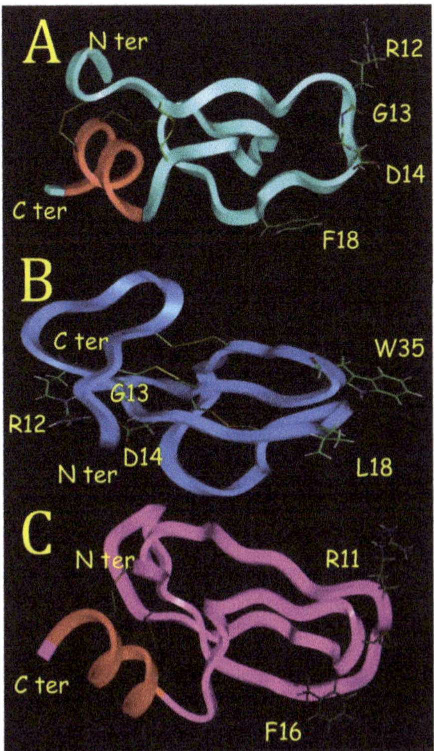

Figure 7. (**A**) The BDS-5 model 3D structure determined with molecular modeling from atomic coordinates of SHPI-1 [18] in (**C**). It was also possible to build a BDS-5 model from BDS-1 structure (**B**) since only one mutation F/L 18 existed between these two proteins but this model was not compatible with CD data. Although one disulfide bridge (colored in yellow) is different in SHPI-1 structure, molecular modeling showed that it is possible to have a BDS-5 model structure compatible with a Kunitz-type structure characterized by a C-terminal short alpha helix (colored in red).

A first energy minimization and dynamic procedure was carried out with only one disulfide bridge (^{4}C-C^{39}) to connect the N and C terminus of BDS-5 with the four other cysteines being free. Structural changes due to dynamic procedure and the constraint related to this ^{4}C-C^{39} disulfide bridge made it possible to have the SH group at the right position to constitute two other disulfide bridges, ^{6}C-C^{32} and ^{22}C-C^{40}. A final energy minimization procedure made it possible to obtain a final BDS-5 model 3D structure with a low van der Waals energy (Figure 7). Kunitz-type structures are low molecular weight proteins characterized by a short alpha helix, a two-beta strand sheet and three disulfide bridges as the SHPI-1 sea anemone protein [18]. Molecular modeling showed that it was possible to have a BDS-5 model 3D structure compatible with a Kunitz-type structure with disulfide bridges similar to BDS-1. Attempt to create a BDS-5 model using a first disulfide bridge constraint with ^{4}C-C^{40} was not possible because it induced the collapse of the C terminal alpha helix to constitute the two other disulfide bridges. It is important to note that another BDS-5 model was possible from BDS-1 NMR structure [7]. In that case, the BDS-5 model was similar to BDS-1 with the F18 lateral chain located at the L18 position (Figure 7B). In the two BDS-5 models, F18 appeared located similarly (data not shown). It is interesting to note that F16 in the SHPI-1 crystal structure had also a similar position (Figure 7A,C). R11 and F18 in SHPI-1 were the residues involved in the inhibition of Elastase [18], and the BDS-5 3D structure model showed that R12 was located similarly to R11 in SHPI-1 (Figure 7A,C).

The presence of a C terminal alpha helix induced a major change in the location of the ^{12}**RGD**^{14} motif for the BDS-5 structure model (Figure 7A) compared to BDS-1 NMR structure (Figure 7B). Contrary to BDS-1, the ^{12}**RGD**^{14} motif in the BDS-5 structure model was well exposed to solvent and on the opposite side (Figure 7A).

3. Discussion

The purpose of this study was to identify and purify low molecular weight proteins from *Anemonia viridis* susceptible to have antiangiogenic activity not acting on the VEGF pathway. The purification protocol was set not to purify all compounds in *Anemonia viridis* having an antiangiogenic activity but only low molecular weight proteins. The reason is that synthetic proteins less than 50 residues can now be produced at low cost and have the great advantage to make possible a sterile production, which is not the case for monoclonal antibodies such as Avastin requiring a biological production. The high cost of Avastin is not related to the production as recombinant protein but to the purification process to have a germ free pharmaceutical production. A low molecular weight compound (<1000 Da) would be certainly less expensive to produce than a synthetic protein of 43 residues corresponding to BDS-5. However, a new chemical family of active principles requires now very expensive toxicological studies to have a Drug Master File suitable for clinical studies. Moreover, actual preclinical toxicity studies required for clinical studies are not sufficient to guarantee an absence of long term toxicities. These long-term toxicities are often due to accumulation in tissues of chemical compounds that cannot be or are insufficiently degraded. Therefore, a synthetic protein (with a molecular weight <5000 Da) represents a good compromise between cost of production and long-term safety issues.

This study shows that sea anemones are certainly a very interesting source of low molecular weight protein active principles, which have been understudied [1]. Although most of these proteins have three disulfide bridges and share sequence homologies, point mutations can totally change their structures and their pharmacological properties [1]. This is well outlined in this study if we compare BDS-5 with BDS-1. With its C-terminal alpha helix, BDS-5 appears to have a 3D structure different from sea anemone protein binding on ionic channels such as BDS-1. Proteins binding on ionic channel identified in sea anemones or in scorpions are characterized by a hydrophobic surface, which helps for a correct positioning of basic residues located on the opposite side. These basic residues create ionic bonds with acidic residues located in their ionic channel binding sites [6]. This is illustrated with BDS-1 that have a hydrophobic surface, with L18 and W35 (Figure 7B), and on the opposite side R12, which is one of the basic residues interacting with the binding site on potassium channel [7].

Although BDS-5 had never been purified and characterized, its sequence was previously known from a search for amino acid sequence motifs in sea anemone polypeptides, where 14 BDS-like sequences were identified in the *Anemonia viridis* genome [19]. This search was made in a cDNA library built from an Expressed Sequence Tag (EST) analysis performed on *Anemonia viridis* revealing 14,504 unique protein sequences in its genome [20]. Another EST analysis made it possible to discover BDS-15 [21]. Interestingly this study showed that BDS-1, BDS-3, BDS-4, BDS-5 and BDS-6 were the most represented transcripts among BDS like proteins [22]. The F18 mutation is found in half of the BDS like sequence. BDS-16 purified and sequenced in this study was never referenced before. Amazingly, the RGD motif is highly conserved in BDS sequences but no activity on integrins has been reported with BDS-1 or BDS-2. This is probably due to the N terminus in BDS-1 and BDS-2, which is masking a part of the RGD motif (Figure 7B).

Kunitz-Type Inhibitors (KTI) have a conserved scaffold (a chain of 40–60 residues stabilized by three disulfide bridges, a two-beta strand sheet and a short C terminal alpha helix) and the most known is the Bovine Pancreatic Trypsin Inhibitor [7]. Another property of KTI is to have an active site with amphoteric characteristic as described in the crystal structure of a Caribbean Sea anemone SHPI-1 in a complex with an Elastase enzyme used in this study [18]. The amphoteric characteristic is due to hydrophobic and hydrophilic poles made of F16 and R11 for SHPI-1 (Figure 7C), and F16 and R12 for BDS-5 (Figure 7A), respectively. A KTI from a tick was able to inhibit Trypsin, Elastase and interfere with blood vessel formation [22]. The RGD motif is recognized by Integrins that are membrane proteins playing a major role in cancer progression and metastasis [23]. BDS-5 appears to be a KTI with a RGD motif that could block angiogenesis in binding on integrins.

4. Materials and Methods

4.1. Phase Extractions and Filtrations from Crude Anemonia viridis

Living animals were collected in the south bay at Marseille in a harbor called "Base nautique du Prado" where shallow and calm water made the development of an important colony of at least 1000 *Anemonia viridis* possible. No other sea anemone species were observed in this spot. They are spread on little stones and they were collected alive fixed on their stone. The very same day the anemones ($n = 10$) with an average size of 10 cm were transported in sea water to the lab at the faculty of Pharmacy and the tentacles were immediately cut and frozen at $-20\ ^\circ$C. Tentacles were then lyophilized (Jouan, Nante, France) for 3 days. The dried tentacles were mashed and suspended in tert-butyl-methyl-Ether (Merck, Lyon, France). After a first centrifugation at 4500 rpm for 20 min (Hettich, Vlotho, Germany), the organic liquid phase containing hydrophobic compounds was eliminated and the dry pellet was suspended in water containing 0.1% Trifluoroatic acid (TFA) (Carlo Erba, Barcelona, Spain). A second centrifugation at 4500 rpm for 20 min was carried out and the pellet was eliminated. The liquid phase containing hydrophilic compounds was then centrifuged in a 30 kDa Centrifugal Filter Unit (Millipore, Burlington, MA, USA) at 4500 rpm for 20 min and then lyophilized. The liquid extract containing compounds <30,000 Da was filtered at 0.22 μm with a filter (Millipore, Guyancourt, France) fixed on a syringe and then lyophilized. UV spectrum of the filtered solution was carried out from 240 to 340 nm with a Beckman DU640.

4.2. High Liquid Performance Chromatography (HPLC)

After filtration at 0.22 μm, the lyophilized material was suspended in water 0.1% TFA at 50 mg/mL. Purification was carried out using a Beckman HPLC System Gold 125 (Brea, CA, USA) apparatus with a Merck Lichrospher C8 reverse phase column (15 × 100 mm). Buffer A was water supplemented with 0.1% (v/v) TFA and buffer B was acetonitrile (Merck, Lyon, France) supplemented with 0.1% (v/v) TFA. Gradient was buffer B from 15% to 35% in 40 min, then 5 min at 90% B and a return to initial condition for 10 min at 15% B with a 2 mL/min flow rate. Fractions were collected manually and their absorptions were measured with a Beckman DU 640B before being frozen at $-20\ ^\circ$C and lyophilized. HPLC analysis was carried out using a Merck Lichrospher RP-C8 column (4.6 × 100 mm) with similar buffers but a 0.8 mL/min flow rate using a linear gradient from 10% to 50% B in 40 min, then 50% to 90% B in 2 min, 5 min at 90% B, a return to initial condition in two min at 10% B and a final stationary phase of 10 min at 10% B. The total duration of the analytical run was 60 min. The UV absorbance of the effluent was measured from 220 to 330 nm with a diode ray Beckman 168 Gold detector and is expressed in milli Absorbance Unit (mAU).

4.3. VEGF Free HMEC Proliferation Assay

HMEC-1 cells were cultured in MCDB-131 medium containing 10% heat inactivated fetal bovine serum, 2 mmol/L glutamine, 100 units/mL penicillin and streptomycin, 1 mg/mL hydrocortisone and 10 ng/mL Epithelial Growth Factor. HMEC-1 cells were seeded on 0.1% gelatin-coated flasks and cultured at 37 $^\circ$C with 5% CO_2. HMEC-1 cells were transferred in 96-well plates at 5000 cells/well in culture medium. Twenty-four hours after seeding, cells were treated with extracts or the control vehicle solution. After 72 h of treatment, cells were exposed to 0.5 mg/mL of diMethyl Thiazol Tetrazolim (MTT) bromide for 3 h at 37 $^\circ$C in culture medium. Cells were then washed with Phosphate Buffered Saline (PBS), formazan crystals were solubilized with 100 μL DMSO and absorbance was measured at

600 nm. MTT is degraded by a mitochondrial enzyme in living cells and forms a purple precipitate in mitochondria that is measured at 600 nm.

4.4. Tubulogenesis Assays

HMEC-1 cells were cultured as described in the proliferation inhibition assay. HMEC-1 capillary network formation on Matrigel were performed on Matrigel™ (Gibco, Beziers, France) added to 96-well plates and allowed to solidify for 30 min at 37 °C. HMEC-1 cells (10,000 cells/well) were added atop the Matrigel™, treated with fractions or a control vehicle solution and incubated for 5 h at 37 °C. Capillary-like structures formed in the gel were photographed and their length was quantified using the METAMORPH 7.6 software [15].

4.5. Mass Determination and Sequencing

Matrix-Assisted Laser Desorption Ionization–Mass Spectrometry (MALDI-MS) spectra were obtained on an UltraflXtreme apparatus (Bruker, Bremen, Germany) operating in positive linear mode with delayed extraction. The samples were co-crystallized with a 10 mg/mL solution of Hydroxy α-Cyano-4-Cinnamic Acid (HCCA) on the MALDI-MS target by the dry droplet method. MALDI-MS spectra were acquired with an accelerating potential of 20 KV and a laser power set to the minimum level necessary to get an efficient signal. Spectrum mass calibration was based on external calibration using an appropriate peptides standard mixture (Peptide Calibration Standard, Brüker, Bremen, Germany). Sequencing was carried out with the Edman degradation method on a PPSQ31 B Shimadzu sequencer. Enzymatic digestion was carried out on 50 μg of the 28-min fraction and 1 μg a Trypsin/Lys C mixture (Promega, Madison, WI, USA) in ammonium bicarbonate 100 mM pH 9 at 37 °C. Different digestions were carried out from one hour to three days. For Nano-HPLC-MS/MS mass spectrometry, the 28-min fraction was reduced with DTT (Sigma-Aldrich, St. Louis, MO, USA), and alkylated with iodoacetamide (Sigma-Aldrich) and incubated overnight at 37 °C in a solution of 12.5 ng/μL of trypsin (Sequencing grade, Roche, Basel, Switzerland) in 25 mM NH4HCO3. The protein digests were sequenced by Nano-LC–MS/MS (Dionex RSLC coupled to a hybrid Q orbitrap mass spectrometer equipped with a Nano-ESI source; Q exactive ThermoFisher scientific, Waltham, WA, USA) in the data-dependent acquisition mode (method top 10). Data were matched to the Uniprot protein database. MALDI MS In Source Decay (ISD) was performed using an Ultraflextreme TOF/TOF mass spectrometer controlled by the FlexControl 3.3 software (BrukerDaltonics, Bremen Germany). The laser was increased (20%) to fragment protein in the source of the mass spectrometer using a specific matrix enabling the generation of hydrogen radicals breaking the peptide backbone producing C ions and z ions from 1000 to 5000 Da generating tag masses that could be used to search proteins in databases. Spectra were acquired in positive reflectron ion mode with 2000 laser shots accumulated and the laser power was set with an increase of 20% of fluensce with a frequency of 1000 Hz. The mass spectrometer parameters were set according manufacturer's settings for optimal acquisition performance. Sequences were analyzed on Flex Analysis 3.0 software, (BrukerDaltonics, Bremen, Germany).

4.6. Far UV Absorption and Circular Dichroism (CD) Spectra

Spectra were measured with a 100 μm path length from 260 to 178 nm at 20 °C on a JASCO J-810 spectropolarimeter. Data were collected at 0.5 nm intervals using a step auto response procedure (JASCO, Oklahoma City, OK, USA) and the buffer background was subtracted. Protein concentrations were 1 mg/mL in 20 mM pH 7 phosphate buffer. Absorption spectra were measured to verify both concentration and light transmission. CD spectra are presented as Δε per amide rather than in ΔA unit to have a measure independent of mass and concentration according to the Beer Lambert law. Furthermore, the signal was divided by the number of amino acid residues of the protein minus one to have Δε per amide [17]. The instrument was calibrated with (+)-10-camphorsulfonic acid, a ratio of 2:1 was found between the positive CD band at 290.5 nm and the negative band at 192.5 nm.

4.7. Molecular Modelling

BDS-5 model was built with the Insight II software from MSI Technologies, Inc. (San Diego, CA, USA). The model was optimized with the Consistent Valence Force Field (CVFF) in term of the internal energies, using the van der Waals energy to monitor each step of the model. Model of BDS-5 was built with the InsightII Homology pulldown from the crystal structure of SHPI 1 in a complex with Elastase [19] available in the Brookhaven data bank (PDB code 1SHP). Minimization was performed with steepest descent and conjugate gradient algorithms. Dynamic was performed at 300 K for 1.1 ps using 1000 steps.

5. Conclusions

In conclusion, BDS-5 could be a member of a new family of antiangiogenic compounds, although the mechanism through which it acts has yet to be demonstrated. Further investigations are necessary to demonstrate that BDS-5 is acting without binding to VEGF or to the VEGF receptor. BDS-5 appears to be structurally related to Kunitz-type inhibitor and its antiangiogenic activity might be due to an interaction with an Integrin due to a RGD motif well exposed to the solvent. BDS-5 could improve cancer treatment in complementing existing antiangiogenic compounds binding on the VEGF in antiangiogenic combinatory therapies. Complementary pharmacological characterization and structural study with 2D NMR are planned with BDS-5. Chemical synthesis of BDS-5 might open the road to a clinical development in cancer therapy, particularly for children, for whom anti-mitotic compounds have devastating side effects.

Supplementary Materials: The following are available online at http://www.mdpi.com/1660-3397/16/4/134/s1, Figure S1: HMEC tubulogenesis assay with the 28′ fraction with concentrations going from 10 µM to 3.2 nM in a triplicate experiment. A network with interconnected lines similar to control was observed only at 3.2 nM. Figure S2: High resolution nano-HPLC MS of fraction 28′. The numbers in square indicate the charge state of the peptides, the insert shows the isotopic envelopes of the quintupled charged main species. Figure S3: Maldi ISD of fraction 28. A: annotation of the ions obtained by Maldi ISD of the three main species of the fraction. B: The insert shows the common n-terminal part of BDS1 and BDS5. Figure S4: MALDI spectra of digest of the 28′ fraction after reduction before Nano-HPLC ms/ms analysis on Orbitrap Q-exactive: The peaks 1 to 4 match with the sequence of BDS-5 (table in insert), A is specific to BDS1 with a Leu/PHE mutation, B match with the BDS16 sequence and was de Novo sequenced by hplc-ms/ms as shown below. Figure S5: MS/MS spectrum of the peptide of the N-terminal part of BDS-16 with two mutations (S5 et P7 instead of F5 et S7 for BDS1 and BDS5). The spectrum is annotated by the y-ions generated by the HCD fragmentation from the Q-exactive Orbitrap. Figure S6: Digestion of the non-reduced fraction from 30 to 70 h. A: 30 H, B: 40 H C: 70 H, D: Reduction of the digest 70 H.

Acknowledgments: Erwann P. Loret thanks Thierry le Taux from the Base Nautique du Prado and students from Biology Technical Schools for their help. Erwann P. Loret thanks Florence Sabatier, Gerard Pepe and Frederic Zuberer for fruitful discussions. Erwann P. Loret thanks Ian Mc Bitter for improving the English.

Author Contributions: E.P.L. designed the study, collected sea anemones, performed purifications and circular dichroism and wrote the manuscript. J.L. performed cellular test. C.V. and C.N. performed mass spectroscopy. P.M. and R.L. performed sequencing. All authors discussed and approved the paper.

Conflicts of Interest: The authors declare no conflict of interest.

Sources of Funding: This research was supported by CNRS.

References

1. Prentis, P.J.; Pavasovic, A.; Norton, R.S. Sea Anemones: Quiet Achievers in the Field of Peptide Toxins. *Toxins* **2018**, *10*, 36. [CrossRef] [PubMed]
2. Catterall, W.A.; Beress, L. Sea anemone toxin and scorpion toxin share a common receptor site associated with the action potential sodium ionophore. *J. Biol. Chem.* **1978**, *253*, 7393–7396. [PubMed]
3. Logashina, Y.A.; Mosharova, I.V.; Korolkova, Y.V.; Shelukhina, I.V.; Dyachenko, I.A.; Palikov, V.A.; Palikova, Y.A.; Murashev, A.N.; Kozlov, S.A.; Stensvåg, K.; et al. Peptide from Sea Anemone *Metridium senile* Affects Transient Receptor Potential Ankyrin-repeat 1 (TRPA1) Function and Produces Analgesic Effect. *J. Biol. Chem.* **2017**, *292*, 2992–3004. [CrossRef] [PubMed]

4. Macrander, J.; Daly, M. Evolution of the Cytolytic Pore-Forming Proteins (Actinoporins) in Sea Anemones. *Toxins* **2016**, *8*, 1–16. [CrossRef] [PubMed]

5. Lauritano, C.; Ianora, A. Marine Organisms with Anti-Diabetes Properties. *Mar. Drugs* **2016**, *14*, 220. [CrossRef] [PubMed]

6. Loret, E.P.; Hammock, B. Structure and Neurotoxicity of venom. In *The Biology of Scorpion*; Philip, B., Gary, P., Eds.; Oxford University Press: Oxford, UK, 2001; pp. 204–234.

7. Diochot, S.; Loret, E.; Bruhn, T.; Beress, L.; Lazdunski, M. APETx1, a new toxin from the sea anemone *Anthopleura elegantissima*, blocks voltage-gated human ether-a-go-go-related gene potassium channels. *Mol. Pharmacol.* **2003**, *64*, 59–69. [CrossRef] [PubMed]

8. Mourão, C.B.; Schwartz, E.F. Protease inhibitors from marine venomous animals and their counterparts in terrestrial venomous animals. *Mar. Drugs* **2013**, *11*, 2069–2112. [CrossRef] [PubMed]

9. Loret, E.; Menendez, R.; Mansuelle, P.; Sampieri, F.; Rochat, H. Positvely Charged Amino Acid Residues Located Similarly in Sea Anemone and Scorpion Active principles. *J. Biol. Chem.* **1994**, *269*, 16785–16788. [PubMed]

10. Folkman, J.; Merler, E.; Abernathy, C.; Williams, G. Isolation of a tumor factor responsible for angiogenesis. *J. Exp. Med.* **1971**, *133*, 275–288. [CrossRef] [PubMed]

11. Carmeliet, P.; Jain, R.K. Principles and mechanisms of vessel normalization for cancer and other angiogenic diseases. *Nat. Rev. Drug Discov.* **2011**, *473*, 417–427. [CrossRef] [PubMed]

12. Kong, D.H.; Kim, M.R.; Jang, J.H.; Na, H.J.; Lee, S. A Review of Anti-Angiogenic Targets for Monoclonal Antibody Cancer Therapy. *Int. J. Mol. Sci.* **2017**, *18*, 1786–1801. [CrossRef] [PubMed]

13. Prescott, B.; Thomas, G.J., Jr.; Béress, L.; Wunderer, G.; Tu, A.T. Structural properties of toxin II of sea anemone (*Anemone sulcata*) determined by laser Raman spectroscopy. *FEBS Lett.* **1976**, *64*, 144–147. [CrossRef]

14. Diochot, S.; Schweitz, H.; Beress, L.; Lazdunski, M. Sea anemone peptides with a specific blocking activity against the fast inactivating potassium channel Kv3.4. *J. Biol. Chem.* **1998**, *273*, 6744–6749. [CrossRef] [PubMed]

15. Bulati, M.; Longo, A.; Masullo, T.; Vlah, S.; Bennici, C.; Bonura, A.; Salamone, M.; Tagliavia, M.; Nicosia, A.; Mazzola, S.; et al. Partially Purified Extracts of Sea Anemone *Anemonia viridis* Affect the Growth and Viability of Selected Tumour Cell Lines. *Immunobiology* **2016**, *221*, 1374–1377.

16. Pilorget, A.; Demeule, M.; Barakat, S.; Marvaldi, J.; Luis, J.; Béliveau, R. Modulation of P-glycoprotein function by sphingosine kinase-1 in brain endothelial cells. *J. Neurochem.* **2007**, *100*, 1203–1210. [CrossRef] [PubMed]

17. Johnson, W.C., Jr. Circular dichroism and its empirical application to biopolymers. *Methods Biochem. Anal.* **1985**, *31*, 161–163.

18. García-Fernández, R.; Perbandt, M.; Rehders, D.; Ziegelmüller, P.; Piganeau, N.; Hahn, U.; Betzel, C.; Chávez Mde, L.; Redecke, L. Three-dimensional Structure of a Kunitz-type Inhibitor in Complex with an Elastase-like Enzyme. *J. Biol. Chem.* **2015**, *290*, 14154–14165. [CrossRef] [PubMed]

19. Kotslov, S.; Grishin, E. The mining of toxin-like polypeptides from EST database by single residue distribution analysis. *BMC Genom.* **2011**, *12*, 88–100.

20. Sabourault, C.; Ganot, P.; Deleury, E.; Allemand, D.; Furla, P. Comprehensive EST analysis of the symbiotic sea anemone, *Anemonia viridis*. *BMC Genom.* **2009**, *10*, 333–345. [CrossRef] [PubMed]

21. Nicosia, A.; Maggio, T.; Mazzola, S.; Cuttitta, A. Evidence of accelerated evolution and ectodermal-specific expression of presumptive BDS toxin cDNAs from *Anemonia viridis*. *Mar. Drugs* **2013**, *11*, 4213–4231. [CrossRef] [PubMed]

22. Soares, T.S.; Oliveira, F.; Torquato, R.J.; Sasaki, S.D.; Araujo, M.S.; Paschoalin, T.; Tanaka, A.S. Kunitz type inhibitor from *Rhipicephalus microplus* able to interfere in vessel formation. *Vet. Parasitol.* **2016**, *219*, 44–52. [CrossRef] [PubMed]

23. Nieberler, M.; Reuning, U.; Reichart, F.; Notni, J.; Wester, H.J.; Schwaiger, M.; Weinmüller, M.; Räder, A.; Steiger, K.; Kessler, H. Exploring the Role of RGD-Recognizing Integrins in Cancer. *Cancers* **2017**, *9*, 116. [CrossRef] [PubMed]

 *marine drugs*

Review

A Potential Adjuvant Agent of Chemotherapy: Sepia Ink Polysaccharides

Fangping Li, Ping Luo and Huazhong Liu *

College of Chemistry & Environment, Guangdong Ocean University, Zhanjiang 524088, China;
15709482571@163.com (F.L.); luopingna@163.com (P.L.)
* Correspondence: liuhz@gdou.edu.cn; Tel.: +86-759-238-3300

Received: 18 January 2018; Accepted: 25 March 2018; Published: 28 March 2018

Abstract: Sepia ink polysaccharide (SIP) isolated from squid and cuttlefish ink is a kind of acid mucopolysaccharide that has been identified in three types of primary structures from squid (*Illex argentinus* and *Ommastrephes bartrami*), cuttlefish *Sepiella maindroni,* and cuttlefish *Sepia esculenta* ink. Although SIP has been proved to be multifaceted, most of the reported evidence has illuminated its chemopreventive and antineoplastic activities. As a natural product playing a role in cancer treatment, SIP may be used as chemotherapeutic ancillary agent or functional food. Based on the current findings on SIP, we have summarized four topics in this review, including: chemopreventive, antineoplastic, chemosensitive, and procoagulant and anticoagulant activities, which are correlative closely with the actions of anticancer agents on cancer patients, such as anticancer, toxicity and thrombogenesis, with the latter two actions being common causes of death in cancer cases exposed to chemotherapeutic agents.

Keywords: Sepia ink polysaccharides; chemoprevention; antitumour; chemosensitization; anticoagulation

---

1. Introduction

Sepia ink, a black suspension of melanin granules, is a traditional Chinese medicine listed in the *Compendium of Materia Medica* compiled by Shizhen Li, a famous doctor at the time of the Ming Dynasty, and has been used in Asia for millennia [1]. The ancient medicine book records the treatment efficacies on heart pain and haemostasis, especially gynaecological haemostasis [1]. Based on the plentiful findings in the latest two decades, the dark ink has been proved useful and to be a kind of multifunctional bioactive marine substance as antioxidant [1–8], anti-inflammatory [9,10], anti-ulcerogenic [10,11], anti-retroviral [12], anti-hypertensive [13], antimicrobial [14–17], and anti-radiation reagent [18], and to have anticancer properties [9,17,19–21], as well as haematopoietic [1,18], immunoregulatory [1,4,7], procoagulant [22] and chemoprophylactic activities [1,5–8]. Sepia ink is a mixture secreted from two glands: the ink gland in the ink sac, and a mucus-producing gland that is a poorly understood funnel organ [23]. The ink contains melanin, proteins, peptidoglycans, amino acids, lipid, metals, tetrodotoxin, etc. [1,23]. The peptidoglycans are composed of sepia ink polysaccharide (SIP) and oligopeptide (SIO) [23]. SIPs derived from ink of different cuttlefishes and squids have distinct primary structures. To date, only one kind of SIO has been characterized, and this is a tripeptide consisting of glutamine (Gln), proline (Pro) and lysine (Lys), the peptide chain is *N*-Gln-Pro-Lys-*C* derived from *Sepia esculenta* ink [21].

The biological functions of melanin, proteins, amino acids, SIO, and metals in the ink have been investigated by various researchers [23]. According to the published work, in contrast, SIP has undoubtedly attracted more attention. The polysaccharide is a glycosaminoglycan that can be absorbed rapidly by the host gastrointestinal tract, and its content in serum can reach a peak at 1 h after gavage [24]. Reports have outlined the activities of this marine polysaccharide. In this paper, the biological properties based on chemopreventive, antineoplastic and chemosensitive effects, the molecular mechanisms involved, and the molecular characteristics of SIPs have been summarized.

2. Molecular Characteristics of SIPs

As shown in Table 1, the polysaccharide from the ink of the squid *Illex argentinus* was the first known SIP that was reported by Takaya et al. in 1994 [25]. The fucose-rich glycosaminoglycan is composed by equimolar ratios of glucuronic acid (GlcA), N-acetylgalactosamine (GalNAc) and fucose (Fuc). Its primary structure was initially determined to have a linear repeating structure of $(-6GalNAc\alpha1-3GlcA\beta1-3Fuc\alpha1-)_n$ [25], but this was amended to $(-3GlcA\beta1-4(GalNAc\alpha1-3)Fuc\alpha1-)_n$ by the discoverers themselves in their next work [26]. The main chain of the polysaccharide was a repeating unit of di-saccharide, GlcA-Fuc, branched at Fuc H-3 by GalNAc [26]. Interestingly, the SIP's primary structure is identical to another SIP derived from squid *Ommastrephes bartrami* ink that was reported by Chen et al. [27].

Apart from the two kinds of SIPs from squids, SIPs from cuttlefish have been reported in recent years. A heteropolysaccharide was isolated from cuttlefish *Sepiella maindroni* ink using enzymolysis, anion-exchange, and gel-permeation chromatography [28]. This SIP contained GlcA, mannose (Man), GalNAc, and Fuc in a molar ratio of 1:1:2:2. Its primary structure was determined to comprise a main chain composed of a repeating unit of $(-Fuc-Fuc-GalNAc-Man-GalNAc-)_n$ and a branch of GlcA at Man H-3; the structural characteristic was $(-4Fuc\beta1-4Fuc\beta1-4GalNAc\alpha1-6(GlcA\alpha1-3)Man\alpha1-4GalNAc\alpha1-)_n$, which differentiates it from squid ink polysaccharides.

Recently, a novel SIP was isolated from the ink of cuttlefish *Sepia esculenta* in our laboratory [29]. This polysaccharide has a unique primary structure mainly composed of galactosamine (GalN) and arabinose (Ara) in an approximate molar ratio of 1:1. The two monosaccharides account for 81.72% of the total monosaccharide mass. This SIP also contains small amounts of Fuc (9.00%), xylose (Xyl, 4.32%), Man (0.09%), glucosamine (GlcN, 1.35%), GlcA (1.98%), and galacturonic acid (GalA, 1.53%). The detailed molecular structure of this SIP will be revealed in a future report.

Table 1. Molecular characteristics of SIPs.

Species	Monosaccharides (Molar Ratio)	Primary Structure	Sulphate (Molar Ratio: Sulphate/Monsaccharides)	Literature
Illex argentinus *Ommastrephes bartrami*	GlcA, GalNAc, Fuc (1:1:1)	$(-3GlcA\beta1-4(GalNAc\alpha1-3)Fuc\alpha1-)_n$	1/3 no	[25,26] [27]
Sepiella maindroni	Fuc, GalNAc, GlcA, Man (2:2:1:1)	$(-4Fuc\beta1-4Fuc\beta1-4GalNAc\alpha1-6$ $(GlcA\alpha1-3)Man\alpha1-4GalNAc\alpha1-)_n$	unknown	[28]
Sepia esculenta	GalN, Ara, Fuc (5:5:1)	unknown	unknown	[29]

To date, a great number of bioactive polysaccharides have been characterized from marine organisms, including marine animals, plants, and microorganisms. Most of these reports are focused on polysaccharides from marine animals and plants. The published marine plant origin carbohydrates mainly include marine algae saccharides, such as alginates, carrageenans, and fucoidans. Chitosans, hyaluronans and chondroitin sulphates are the main polysaccharides of marine animals, and have been studied for several decades. Table 2 indicates that these polysaccharides, which have unique molecular characteristics, possess biological activities and action mechanisms [30–32]. Obviously, SIP is a different polysaccharide compared with other marine origin polysaccharides. Based on structure and function observations, SIP possesses specific activities that differ from the well-studied marine polysaccharides. Therefore, the following section reviews research progress of SIP properties in recent years.

Table 2. Molecular characteristics and properties of some marine polysaccharides [30–32].

Species	Polysaccharides	Monosaccharides	Properties
Marine plants	alginate	L-guluronate, D-mannuronate	antibacterial, tissue regeneration
	carrageenan	D-galactose, D/L-galactose	anticoagulant, antitumour, immunomodulatory, antihyperlipidemic, antioxidant, antibacterial, antifungal, antiviral
	fucoidan	L-fucose	antitumour, anticoagulant, anti-adhensive,'antiviral
Marine animals	chitosan	D-glucosamine	antimicrobial, antitumour, anti-inflammatory
	chondroitin sulphate	glucuronic, N-acetyl-galactosamine	Improving function and elasticity of the articular cartilage, hemostasis and anti-inflammation, regulation of cell development, cell adhesion, cell proliferation, cell differentiation, anticoagulation
	hyaluronan	N-acetyl-D-glucosamine, D-glucuronic acid	tissue regeneration, cell prolifernation, cell differentiation, cell migration

3. Biological Activities of SIPs

SIP has been confirmed to have chemoprevention, antitumour, chemosensitization and anticoagulation activities. This section summarizes the properties listed in Table 3.

Table 3. Biological activities of SIPs.

Species	Sulfation	Properties	Targets	Literature
Illex argentinus	no	no antitumouractivity	Meth-A	[33]
Ommastrephes bartrami	no	chemoprevention	intestinal tract (mice)	[34–40]
	yes	antitumour	HepG2	[41]
		anticoagulant	blood (in vitro experiment)	[42]
Sepiella maindroni	yes	antitumour	SKOV-3, KB, HT-29, S180, B16F10	[43–47]
Sepia esculenta	no	chemoprevention	testis, ovary, spleen, kidney, liver, lung, heart, bone marrow (mice)	[29,48–56]
		antitumour	B16F10, MDA-MB-231	[57,58]

3.1. Chemoprevention

3.1.1. Protection of the Reproductive System

With increasing incidence and mortality rates, cancer is the leading cause of death in China and is a major public health problem. Because of its large population, China's cancer cases constitute almost 22% of global new cancer cases and close to 27% of global cancer deaths [59]. Furthermore, cancers are becoming more likely to be found in younger patients, resulting in increasing numbers of cancer patients of childbearing age. Since chemotherapy is still a major therapeutic method for cancer, anticancer agents exerting toxic effects on the reproductive system in patients of childbearing age is almost inevitable, and can potentially lead to damage and consequent infertility. Therefore, screening substances with chemopreventive properties in order to attenuate the negative effects of chemotherapeutic drugs is urgent for the treatment of the increasing number of cancer cases of childbearing age.

SIP has been verified to have chemoprophylactic actions on the reproductive system [29,48–53]. When male mice exposed to cyclophosphamide were administered SIP, the abnormal rates of their sperm declined, and the foetal abnormalities in female mice mated with them also declined, with total foetal count and average foetal count increasing [48].

The toxicity mechanisms of cyclophosphamide are complicated. Drug-induced oxidative stress and DNA strand breakage are two critical causes [60–64]. Cyclophosphamide-exposed mice/rats showed disruption of testicular antioxidant capacity and histopathologic changes through suppression of the nuclear factor erythroid 2 related factor 2 (Nrf2)/antioxidant response element (ARE) signalling pathway [48–52,60–62]; however, testicular functional disorders of the chemotherapeutic model animals were prevented by SIP via activation of the antioxidant signalling pathway [48–52]. In addition, SIP can prevent animals from cyclophosphamide-mediated mutation in vivo and H_2O_2/UV-induced DNA strand break in vitro [28,65]. The testicular cells of cyclophosphamide-exposed mice, including spermatogonia, Sertoli cells, and Leydig cells, were protected by SIP via repression of cyclophosphamide-induced autophagy-associated cell death and apoptosis; the mechanisms involved p38 mitogen-activated protein kinase and phosphoinositide 3-kinase (PI3K)/Akt signalling pathways [51–53]. Similarly, for cyclophosphamide-mediated ovarian failure, SIP also successfully inhibited follicle deletion and granule cell disruption by repressing cyclophosphamide-induced autophagy-associated cell death and apoptosis via regulation of p38 mitogen-activated protein kinase and PI3K/Akt pathways, resulting in functional rescue of the ovaries of cyclophosphamide-exposed mice [29]. These data show that SIP can prevent mice from reproductive system damage caused by cyclophosphamide-associated toxicities.

3.1.2. Protection of Intestinal Tract

Cyclophosphamide-mediated augmentation of intestinal pathogenic bacterial counts and intestinal permeability was found to have a negative effect on cancer patients. Intestinal imbalance and consequent infections were consequences of the immune system disruption resulting from chemotherapy [66], and can be partly attributed to a decrease in immunoglobulin A (IgA) production due to the anticancer agent-induced reduction of IgA-producing cells [67]. Tang et al. found that, in cyclophosphamide-exposed mice, SIP could recover the balance of intestinal microflora by blocking the anticancer agent-mediated reduction of the quantity of probiotic Bifidobacterium [34]. Intestinal microbiota promote development and regulation of the acquired mucosal immune system [68,69]. As an important element of the intestinal mucosal immune system, under exposure to a chemotherapeutic agent, IgA-producing cell reduction is responsible for the imbalance in intestinal microflora. In one study, SIP promoted the expression of IL-6, IL-10, and TNF-α and up-regulated the expression of IgA J chain gene in IgA-producing cells and pIgR gene in epithelial cells. Meanwhile, SIP increased the expression of unfolded protein response effecters XBP-1s and Bip to accelerate IgA secretion. As a result, the IgA content in the intestinal tract of mice exposed to SIP was elevated [35]. A further investigation with high-throughput sequencing analysis revealed that SIPs altered the imbalance of the gut microbial ecology caused by cyclophosphamide; amounts of *Ruminococcus, Bilophila, Oscillospira, Dorea* and, especially, *Mucispirillum* were reduced, resulting in the repression of early disruption of the colonic surface mucus layer and an increase in the risk of inflammatory disorders [36].

The intestinal epithelium is a vital barrier contributing to preventing infection and to innate immunity, and maintaining its integrity is necessary for normal intestinal function and a healthy body. However, maintaining the integrity of the intestinal mucosa is almost impossible under exposure to chemotherapeutic drugs. Chemotherapy-induced disruption of intestinal barrier function has been confirmed [68,69]. The goblet cells, Paneth cells, and epithelial junctions (tight junctions and adherent junctions) are responsible for the integrity of the barrier, but these three important elements can be destroyed by the chemotherapeutic drug cyclophosphamide [37–40]. As a type of major epithelial cell in the small intestine, goblet cells provide first-line protection for the host against possible pathogens, which is an important part of the innate mucosal immune system. SIPs are capable of increasing quantities of goblet cells in mice to express more mucins, such as Cyto18, avoiding pathogens penetrating or colonising in the intestinal mucosa, and rescuing mucosal immunity of cyclophosphamide-treated mice [37]. Additionally, Paneth cells, another intestinal epithelial cell contributing to innate immunity by secreting antimicrobial proteins onto the mucosal surface, can be promoted by SIP to express antimicrobial proteins, including lysozyme, angiogenin-4, defensin alpha 5, and type-2 secretory phospholipase. The mechanisms depend on the relatively highly developed endoplasmic reticulum structure, not on increases in the quantity of endoplasmic reticulum, which is associated with the SIP-activated, IRE-1 mediated, XBP-1s pathway [38]. In addition, chemotherapy damages epithelial junctions and destroys the intestinal barrier, resulting in disruption of the innate immune system and consequent infections, and chemotherapeutic mucositis. Zuo et al. discovered that SIP effectively improves expression of occludin, zonulae occluden 1/2/3, claudin, cingulin, and E-cadherin genes, stabilizing tight junctions and adherent junctions, which was helpful for protecting immune function of intestinal mucosa in mice exposed to chemotherapeutic drugs [39].

Histopathological observation showed that mice treated with SIP have longer small intestinal villi, deeper crypts, and a larger ratio of villus height/crypt depth compared with cyclophosphamide-treated mice [40]. Moreover, SIP-treated mice have stronger antioxidant capacity in the intestinal tissue when compared with model mice [40].

3.1.3. Protection of Other Tissues/Organs

Apart from the reproductive system and intestinal tract, chemoprevention of some other organs/tissues by SIP was also investigated in our laboratory, including liver, kidney, heart, spleen, lung, and bone marrow. SIP repressed cyclophosphamide-induced alterations of biochemistry indicators in the serum and tissues/organs of model animals, such as relative masses of liver and spleen, activities of glutamic-pyruvic transamine, glutamic-oxalacetic transaminease, catalase, lactic dehydrogenase, and creatine kinase in serum, antioxidant capacity of liver and heart, serum urea nitrogen content, peripheral blood profile including quantities of erythrocytes, leukocytes, and platelets, and haemoglobin content, as well as the DNA content in bone marrow [54–56].

3.2. Antitumour Activities

Initially, SIP's antitumour activity was investigated by Takaya et al. in 1994 [33]. Their results showed that SIP had no inhibitory activity on Meth-A fibrosarcoma cells transplanted into BALB/c mice, but peptidoglycan had. Therefore, the researchers deduced that the antitumour activity was attributable to the complex of SIP and other components (for example, peptide or melanin) [33]. However, in the last decade, many reports have verified the antitumour effects of SIP on several types of tumours.

Unmodified naturally occurring SIPs were found to be able to inhibit both melanoma cell B16F10 [57] and human adenocarcinoma cell line MDA-MB-231 [58]. Growth and proliferation of B16F10 were repressed, tyrosinase activity and melanin production were also effectively reduced. Similarly, proliferation and migration of MDA-MB-231 were significantly blocked [58]. It is well known that polysaccharide sulphates have stronger antitumour properties. Now, several studies have also reported anticancer activities of sulphated SIP (S-SIP).

A S-SIP isolated by Chen et al. inhibited invasion and migration, but not proliferation, of human hepatocellular liver carcinoma cell line HepG2, and inhibited angiogenesis in a chick embryo chorioallantoic membrane model [41].

Another S-SIP decreased MMP-2 expression of SKOV-3 and human umbilical vein vascular endothelial cells ECV304, leading to inhibition of SKOV-3 cell penetration and ECV304 cell migration [43]. The SIP inhibits proliferation, migration, invasion, and MMP-2 expression of human epidermoid carcinoma cell line KB by inhibiting the EGFR/Akt/p38 MAPK/MMP-2 signalling pathway [44]. Furthermore, the sulphated SIP combines with the cell membrane of human ovarian cancer cell line SKOV-3, human colorectal adenocarcinoma cell line HT-29, and mouse fibroblast cell line L929 cells. In SKOV-3 cells, SIP binds to epidermal growth factor receptor (EGFR) and inhibits EGF-induced expression and activation of EGFR as well as cell migration. Consequently, the SIP suppressed EGFR-mediated p38/MAPK and PI3K/Akt/mTOR signalling pathways to inhibit migration, invasion, and MMP-2 expression of SKOV-3 cells [45].

Additionally, in vivo data showed that the derivative SIP repressed tumour growth and enhanced immune function in S180-bearing mice, also induced SKOV-3 cells apoptosis in vivo and in vitro [46].

Further investigation indicated that sulphated SIP decreased melanoma cell B16F10 pulmonary metastasis in mice models, and down-regulated expression of the intercellular adhesion molecule 1 (ICAM-1) and basic fibroblast growth factor (bFGF) in lung metastasis nodules. Furthermore, neovascularisation was suppressed in chick chorioallantoic membrane exposed to S-SIP [46]. In vitro experiments exhibited an expression reduction of ICAM-1 and bFGF in SKOV-3 and EA.hy926 cells, respectively [46]. These results suggested that S-SIP down-regulated the expression of ICAM-1 and bFGF to inhibit tumour adhesion and angiogenesis. Consequently, invasion and migration of tumour cells were restrained.

3.3. Chemosensitization

Combination treatment is frequently used in cancer treatment to reduce drug resistance, alleviate adverse effects, and enhance anticancer efficacy.

Currently, only two papers have reported chemosensitization by SIP. Zong et al. found that sulphated SIP increased the killing effects of cyclophosphamide on tumours and reduced the toxicity of the chemotherapy drug on the thymus in S180-bearing mice [47]. Our previous work indicated that SIP enhanced inhibition of proliferation and migration of MDA-MB-231 cells by cisplatin [58].

3.4. Anticoagulant and Procoagulant Activities

It has now been confirmed that chemotherapy induces hypercoagulability of blood, and consequent thrombus formation is a critical cause of cancer death, so a potential anticancer agent should possess anticoagulation properties.

Ancient Chinese medicine used sepia ink as a coagulant drug for internal haemorrhage, especially as a coagulant for gynaecology. Modern medicine, directly or indirectly, has noted the coagulant property of the ink [22,70]. Although there is no direct evidence, to date, indicating the coagulant activity of SIP, a report of the haemostatic effects of a SIP-chitosan hybrid haemostatic sponge implies that natural SIP might possess procoagulant activity [71]; however, the confusion should be removed as early as possible by future investigation.

In contrast, a paper showed anticoagulant activity of SIP. Chen et al. prepared a derivative SIP that was sulphated chemically in pyridine-sulphur-trioxide complex in a dimethyl sulphoxide system or triethylamine pyridine-sulphur-trioxide complex in a dimethyl sulphoxide system. The sulphation mainly occurred at the 4,6-positions of GalNAc, the active primary structure of the sulphated SIP was identified to be $(\text{-GlcA}\beta1\text{-}4(4,6\text{-SO}_4\text{-GalNAc}\alpha1\text{-}3)\text{Fuc}\alpha1\text{-})_n$. The sulphated SIP in vitro increased partial thromboplastin time and prothrombin time, suggesting that the derivative SIP could play an anticoagulation role by inhibiting endogenous and exogenous blood coagulation processes. The sulphated SIP effectively suppressed activities of clotting factors, FIIa and FXa, mediated by antithrombin III or heparin cofactor II [42].

Other reports have shown that the sulphated group in polysaccharides has an important function with regard to anticoagulant activity; sulphation can promote anticoagulation of a non-sulphated group polysaccharide, and the degree of sulphation is positively correlated to anticoagulation [42,72]. Recently reported natural SIPs include acid mucopolysaccharides with or without a small quantity of sulphated groups. The low sulphated group content of polysaccharides has been deduced to have procoagulant activity, but an experiment has shown that sulphation-modified SIP exhibited anticoagulant activity [42], which implies that sulphated SIP is more suitable for developing an ancillary antitumour drug for cancer treatment.

4. Conclusions

This review summarized chemopreventive, antineoplastic, chemosensitive, and procoagulant/anticoagulant properties of SIPs, as well as their molecular characteristics. Various SIPs with distinct primary structures from different sepia inks share similar biological actions. The number of sulphated groups is crucial to the coagulant actions of SIP, with low sulphated group content leading to

Mar. Drugs **2018**, *16*, 106

procoagulation activities and high sulphated group content leading to anticoagulation activities. The sulphated SIP may be an important bioactive marine substance, which could be developed as a clinical antitumour agent or chemotherapy-supplementary functional food for application in the clinical treatment of cancer.

Acknowledgments: We thank past and present members of our laboratory for their hard work on SIP and its biological activities, appreciate professor Yanqun Li for revising this paper, and are very grateful for support from the National Natural Science Foundation of China (Grant No. 31171667) and the Natural Science Foundation of Guangdong Province, China (Grant No. 2016A030313753).

Author Contributions: Fangping Li and Ping Luo conceived and designed the review. Fangping Li collected data, and Ping Luo analysed the data. Fangping Li, Ping Luo and Huazhong Liu wrote this paper.

Conflicts of Interest: The author declares no conflict of interest.

References

1. Zhong, J.P.; Wang, G.; Shang, J.H.; Pan, J.Q.; Li, K.; Huang, Y.; Liu, H.Z. Protective effects of squid ink extract towards hemopoietic injuries induced by cyclophosphamide. *Mar. Drugs* **2009**, *7*, 9–18. [CrossRef] [PubMed]

2. Saleh, H.; Soliman, A.M.; Mohamed, A.S.; Marie, M.A.S. Antioxidant effect of *Sepia* ink extract on extrahepatic cholestasis induced by bile duct ligation in rats. *Biomed. Environ. Sci.* **2015**, *28*, 582–594. [PubMed]

3. Vate, N.K.; Benjakul, S.; Agustini, T.W. Application of melanin-free ink as a new antioxidative gel enhancer in sardine surimi gel. *J. Sci. Food Agric.* **2015**, *95*, 2201–2207. [CrossRef] [PubMed]

4. Liu, H.Z.; Luo, P.; Chen, S.H.; Shang, J.H. Effects of squid ink on growth performance, antioxidant functions and immunity in growing broiler chickens. *Asian Austral. J. Anim. Sci.* **2011**, *24*, 1752–1756. [CrossRef]

5. Liu, H.Z.; Wang, G.; Guo, Y.Z.; Pan, J.Q.; Huang, Y.; Zhong, J.P.; Li, K. *Sepia* ink extract attenuates renal injury caused by cyclophosphamide-induced oxidative stress. *Chin. J. Nephrol.* **2009**, *25*, 804–805.

6. Wang, G.; Guo, Y.Z.; Guan, S.B.; Zhong, J.P.; Pan, J.Q.; Huang, Y.; Liu, H.Z. Protective effects of sepia ink extract on cyclophosphamide-induced pulmonary fibrosis in mice. *Chin. J. Mar. Drugs* **2009**, *28*, 36–40.

7. Wang, G.; Pan, J.Q.; Zhong, J.P.; Li, K.; Huang, Y.; Wu, J.L.; Liu, H.Z. Protective effects of sepia ink extract against cyclophosphamide-induced oxidative damage in mice spleen. *Food Sci.* **2009**, *30*, 219–222.

8. Wang, G.; Liu, H.Z.; Wu, J.L.; Cao, Q.W.; Chen, Y.P.; Yang, C.L.; Zhong, J.P. Study of sepia ink extract on protection from oxidative damage of cardiac muscle and brain tissue in mice. *Chin. J. Mod. Appl. Pharm.* **2010**, *27*, 95–99.

9. Fahmy, S.R.; Soliman, A.M. In vitro antioxidant, analgesic and cytotoxic activities of *Sepia officinalis* ink and *Coelatura aegyptiaca* extracts. *Afr. J. Pharm. Pharmacol.* **2013**, *7*, 1512–1522. [CrossRef]

10. Mimura, T.; Itoh, S.; Tsujikawa, K.; Nakajima, H.; Satake, M.; Kohama, Y.; Okabe, M. Studies on biological activities of melanin from marine animals. V. Anti-inflammatory activity of low-molecular-weight melanoprotein from squid (Fr. SM II). *Chem. Pharm. Bull.* **1987**, *35*, 1144–1150. [CrossRef] [PubMed]

11. Mimura, T.; Maeda, K.; Terada, T.; Oda, Y.; Morishita, K.; Aonuma, S. Studies on biological activities of melanin from marine animals. III. Inhibitory effect of SM II (low molecular weight melanoprotein from squid) on phenylbutazone-induced ulceration in gastric mucosa in rats, and its mechanism of action. *Chem. Pharm. Bull.* **1985**, *33*, 2052–2060. [CrossRef] [PubMed]

12. Rajaganapathi, J.; Thyagarajan, S.P.; Patterson Edward, J.K. Study on cephalopod's ink for anti-retroviral activity. *Indian J. Exp. Biol.* **2000**, *38*, 519–520. [PubMed]

13. Kim, S.Y.; Kim, S.H.; Song, K.B. Characterization of a partial purification and angiotensin-converting enzyme inhibitor from squid ink. *Agric. Chem. Biotechnol.* **2003**, *46*, 122–123.

14. Fahmy, S.R.; Soliman, A.; Ali, E.M. Antifungal and antihepatotoxic effects of sepia ink extract against oxidative stress as a risk factor of invasive pulmonary aspergillosis in neutropenic mice. *Afr. J. Tradit. Complement. Altern. Med.* **2014**, *11*, 148–159. [CrossRef] [PubMed]

15. Petkovic, M.V. Determination of the antimicrobial activity of purified melanin from the ink Octopus mimus Gould, 1852 (Cephalopoda: Octopodidae). *Lat. Am. J. Aquat. Res.* **2013**, *41*, 584–587.

16. Shi, L.S.; Liu, H.Z.; Zhong, J.P.; Pan, J.Q. Fresh-keeping effects of melanin-free extract from squid ink on yellowfin sea bream (*Sparus latus*) during cold storage. *J. Aquat. Food Prod. Technol.* **2015**, *24*, 199–212. [CrossRef]

17. Kumar, P.; Kannan, M.; ArunPrasanna, V.; Vaseeharan, B.; Vijayakumar, S. Proteomics analysis of crude squid ink isolated from *Sepia esculenta* for their antimicrobial, antibiofilm and cytotoxic properties. *Microb. Pathog.* **2018**, *116*, 345–350. [CrossRef] [PubMed]

18. Lei, M.; Wang, J.F.; Wang, Y.M.; Pang, L.; Wang, Y.; Xu, W.; Xue, C.H. Study of the radio-protective effect of cuttlefish ink on hemopoietic injury. *Asia Pac. J. Clin. Nutr.* **2007**, *16* (Suppl. 1), 239–243. [PubMed]

19. Sasaki, J.; Ishita, K.; Takaya, Y.; Uchisawa, H.; Matsue, H. Anti-tumor activity of squid ink. *J. Nutr. Sci. Vitaminol.* **1997**, *43*, 455–461. [CrossRef] [PubMed]

20. Ding, G.F.; Huang, F.F.; Yang, Z.S.; Yu, D.; Yang, Y.F. Anticancer activity of an oligopeptide isolated from hydrolysates of sepia ink. *Chin. J. Nat. Med.* **2011**, *9*, 51–55.

21. Huang, F.; Yang, Z.; Yu, D.; Wang, J.; Li, R.; Ding, G. Sepia ink oligopeptide induces apoptosis in prostate cancer cell lines via caspase-3 activation and elevation of Bax/Bcl-2 ratio. *Mar. Drugs* **2012**, *10*, 2153–2165. [CrossRef] [PubMed]

22. Xie, G.L.; Lv, C.L.; Hong, M.B. Experimental study of the effective mechanisms of sepia ink in promoting coagulation of blood. *J. China Med. Univ.* **1994**, *23*, 530–531.

23. Derby, C.D. Cephalopod ink: Production, chemistry, functions and applications. *Mar. Drugs* **2014**, *12*, 2700–2730. [CrossRef] [PubMed]

24. Zuo, T.; Zhang, N.; Zhang, Q.; Shi, H.; Lu, S.; Xue, C.; Tang, Q.J. Transportation of squid ink polysaccharide SIP through intestinal epithelial cells and its utilization in the gastrointestinal tract. *J. Funct. Foods* **2016**, *22*, 408–416. [CrossRef]

25. Takaya, Y.; Uchisawa, H.; Hanamatsu, K.; Narumi, F.; Okuzaki, B.; Matsue, H. Novel fucose-rich glycosaminoglycans from squid ink bearing repeating unit of trisaccharide structure (-6GalNAcα1-3GlcAβ1-3Fucα1-)$_n$. *Biochem. Biophy. Res. Commun.* **1994**, *198*, 560–567. [CrossRef] [PubMed]

26. Takaya, Y.; Uchisawa, H.; Narumi, F.; Matsue, H. Illexins A, B and C from squid ink should have a branched structure. *Biochem. Biophy. Res. Commun.* **1996**, *226*, 335–338. [CrossRef] [PubMed]

27. Chen, S.G.; Xu, J.; Xue, C.H.; Dong, P.; Sheng, W.J.; Yu, G.L.; Chai, W.G. Sequence determination of a non-sulfated glycosaminoglycan-like polysaccharide from melanin-free ink of the squid *Ommastrephes bartrami* by negative-ion electrospray tandem mass spectrometry and NMR spectroscopy. *Glycoconj. J.* **2008**, *5*, 481–492. [CrossRef] [PubMed]

28. Liu, C.H.; Li, X.D.; Li, Y.H.; Feng, Y.; Zhou, S.; Wang, F.S. Structural characterization and antimutagenic activity of a novel polysaccharide isolated from *Sepiella maindroni* ink. *Food Chem.* **2008**, *110*, 807–813. [CrossRef] [PubMed]

29. Liu, H.Z.; Tao, Y.X.; Luo, P.; Deng, C.M.; Gu, Y.P.; Yang, L.; Zhong, J.P. Preventive effects of a novel polysaccharide from *Sepia esculenta* ink on ovarian failure and its action mechanisms in cyclophosphamide-treated mice. *J. Agric. Food Chem.* **2016**, *64*, 5759–5766. [CrossRef] [PubMed]

30. Cardoso, M.J.; Costa, R.R.; Mano, J.F. Marine origin polysaccharides in drug delivery systems. *Mar. Drugs* **2016**, *14*, 34. [CrossRef] [PubMed]

31. De Jesus Raposo, M.F.; de Morais, A.M.B.; de Morais, R.M.S.C. Marine polysaccharide from algae with potential biomedical applications. *Mar. Drugs* **2015**, *13*, 2967–3028. [CrossRef] [PubMed]

32. Vazqez, J.A.; Rodriguez-Amado, I.; Montemayor, M.I.; Fraguas, J.; del Pilar Gonzalez, M.; Murado, M.A. Chondronitin sulfate, hyaluronic acid and chitin/chitosan production using marine waste sources: Characteristics, applications and eco-friendly processes: A review. *Mar. Drugs* **2013**, *11*, 747–774. [CrossRef] [PubMed]

33. Takaya, Y.; Uchisawa, H.; Matsue, H.; Okuzaki, B.I.; Narumi, F.; Sasaki, J.I.; Ishida, K. An investigation of the antitumor peptidoglycan fraction from squid ink. *Biol. Pharm. Bull.* **1994**, *17*, 846–849. [CrossRef] [PubMed]

34. Tang, Q.; Zuo, T.; Lu, S.; Wu, J.; Wang, J.; Zheng, R.; Chen, S.; Xue, C. Dietary squid ink polysaccharides ameliorated the intestinal microflora dysfunction in mice undergoing chemotherapy. *Food Funct.* **2014**, *5*, 2529–2535. [CrossRef] [PubMed]

35. Zuo, T.; Cao, L.; Sun, X.; Li, X.; Wu, J.; Lu, S.; Xue, C.; Tang, Q. Dietary squid ink polysaccharide could enhance SIgA secretion in chemotherapeutic mice. *Food Funct.* **2014**, *5*, 3189–3196. [CrossRef] [PubMed]

36. Lu, S.; Zuo, T.; Zhang, N.; Shi, H.; Liu, F.; Wu, J.; Wang, Y.; Xue, C.; Tang, Q. High throughput sequencing analysis reveals amelioration of intestinal dysbiosis by squid ink polysaccharide. *J. Funct. Foods* **2016**, *20*, 506–515. [CrossRef]

37. Zuo, T.; Cao, L.; Xue, C.; Tang, Q.J. Dietary squid ink polysaccharide induces goblet cells to protect small intestinal from chemotherapy induced injury. *Food Funct.* **2015**, *6*, 981–986. [CrossRef] [PubMed]

38. Zuo, T.; He, X.; Cao, L.; Xue, C.; Tang, Q.J. The dietary polysaccharide from *Ommastrephes bartrami* prevents chemotherapeutic mucositis by promoting the gene expression of antimicrobial peptides in Paneth cells. *J. Funct. Foods* **2015**, *12*, 530–539. [CrossRef]

39. Zuo, T.; Cao, L.; Li, X.; Zhang, Q.; Xue, C.; Tang, Q. The squid ink polysaccharides protect tight junctions and adherens junctions from chemotherapeutic injury in the small intestinal epithelium of mice. *Nutr. Cancer* **2015**, *67*, 364–371. [CrossRef] [PubMed]

40. Cao, L.; Zuo, T.; Li, X.M.; Yu, Q.; Cao, B.B.; Wang, J.F. Protective effects of polysaccharides from the ink of *Ommastrephes bartrami* on the intestinal mucosa of the mice with chemotherapy and its mechanism. *Chin. Pharmacol. Bull.* **2013**, *29*, 1558–1562.

41. Chen, S.G.; Wang, J.F.; Xue, C.H.; Li, H.; Sun, B.B.; Xue, Y.; Chai, W.G. Sulfation of a squid ink polysaccharide and its inhibitory effect on tumor cell metastasis. *Carbohyd. Polym.* **2010**, *81*, 560–566. [CrossRef]

42. Chen, S.G.; Li, Z.G.; Wang, Y.M.; Li, G.Y.; Sun, B.B.; Xue, C.H. Sulfation of a squid ink polysaccharide and its anticoagulant activities. *Chem. J. Chin. Univ.* **2010**, *31*, 2407–2412.

43. Wang, S.; Cheng, Y.; Wang, F.; Sun, L.; Liu, C.; Chen, G.; Li, Y.; Ward, S.G.; Qu, X. Inhibition activity of sulfated polysaccharide of *Sepiella maindroni* ink on matrix metalloproteinase (MMP)-2. *Biomed. Pharmacother.* **2008**, *62*, 297–302. [CrossRef] [PubMed]

44. Jiang, W.; Tian, W.; Ijaz, M.; Wang, F. Inhibition of EGF-induced migration and invasion by sulfated polysaccharide of *Sepiella maindroni* ink via the suppression of EGFR/Akt/p38 MAPK/MMP-2 signaling pathway in KB cells. *Biomed. Pharmacother.* **2017**, *95*, 95–102. [CrossRef] [PubMed]

45. Jiang, W.; Cheng, Y.; Zhao, N.; Li, L.; Shi, Y.; Zong, A.; Wang, F. Sulfated polysaccharide of *Sepiella Maindroni* ink inhibits the migration, invasion and matrix metalloproteinase-2 expression through suppressing EGFR-mediated p38/MAPK and PI3K/Akt/mTOR signaling pathways in SKOV-3 cells. *Int. J. Bio. Macromol.* **2018**, *107*, 349–362. [CrossRef] [PubMed]

46. Zong, A.Z.; Zhao, T.; Zhang, Y.; Song, X.L.; Shi, Y.K.; Cao, H.Z.; Liu, C.H.; Cheng, Y.N.; Qu, X.J.; Cao, J.C.; et al. Anti-metastatic and anti-angiogenic activities of sulfated polysaccharide of *Sepiella maindroni* ink. *Carbohyd. Polym.* **2013**, *91*, 403–409. [CrossRef] [PubMed]

47. Zong, A.Z.; Liu, Y.H.; Zhang, Y.; Song, X.; Shi, Y.K.; Cao, H.Z.; Liu, C.H.; Cheng, Y.; Jiang, W.J.; Du, F.L.; et al. Anti-tumor activity and the mechanism of SIP-S: A sulfated polysaccharide with anti-metastatic effect. *Carbohyd. Polym.* **2015**, *129*, 50–54. [CrossRef] [PubMed]

48. Luo, P.; Liu, H.Z.; Le, X.Y.; Du, H.; Kang, X.H. Squid ink polysaccharide prevents chemotherapy induced injury in the testes of reproducing mice. *Pak. J. Pharm. Sci.* **2018**, *31*, 25–29. [PubMed]

49. Le, X.Y.; Luo, P.; Gu, Y.P.; Tao, Y.X.; Liu, H.Z. Interventional effects of squid ink polysaccharide on cyclophosphamide-associated testicular damage in mice. *Bratislava Med. J.* **2015**, *5*, 334–339. [CrossRef]

50. Le, X.Y.; Luo, P.; Gu, Y.P.; Tao, Y.X.; Liu, H.Z. Squid ink polysaccharide reduces cyclophosphamide-induced testicular damage via Nrf2/ARE activation pathway in mice. *Iran. J. Basic Med. Sci.* **2015**, *18*, 827–831. [PubMed]

51. Gu, Y.P.; Yang, X.M.; Duan, Z.H.; Luo, P.; Shang, J.H.; Xiao, W.; Tao, Y.X.; Zhang, D.Y.; Zhang, Y.B.; Liu, H.Z. Inhibition of chemotherapy-induced apoptosis of testicular cells by squid ink polysaccharide. *Exp. Ther. Med.* **2017**, *14*, 5889–5895. [CrossRef] [PubMed]

52. Gu, Y.P.; Yang, X.M.; Duan, Z.H.; Shang, J.H.; Luo, P.; Xiao, W.; Zhang, D.Y.; Liu, H.Z. Squid ink polysaccharide prevents autophagy and oxidative stress affected by cyclophosphamide in Leydig cells of mice: A pilot study. *Iran. J. Basic Med. Sci.* **2017**, *20*, 1194–1199. [PubMed]

53. Gu, Y.P.; Yang, X.M.; Luo, P.; Li, Y.Q.; Tao, Y.X.; Duan, Z.H.; Xiao, W.; Zhang, D.Y.; Liu, H.Z. Inhibition of acrolein-induced autophagy and apoptosis by a glycosaminoglycan from *Sepia esculenta* ink in mouse Leydig cells. *Carbohyd. Polym.* **2017**, *163*, 270–279. [CrossRef] [PubMed]

54. Liu, H.Z.; Wang, G.; Wu, J.L.; Shi, L.S.; Zhong, J.P.; Pan, J.Q. Amelioration of sepia ink polysaccharides on internal organs injured by cyclophosphamide in mice. *Chin. J. Mod. Appl. Pharm.* **2012**, *29*, 89–93.

55. Wu, J.L.; Wang, G.; Shi, L.S.; Zhong, J.P.; Pan, J.Q.; Liu, H.Z. Sepia ink polysaccharide preparation attenuates cyclophosphamide toxicity in mice. *Chin. J. Mar. Drugs* **2012**, *31*, 49–51.

56. Wang, G.; Zhong, J.P.; Liu, H.Z. Amelioration of sepia ink polysaccharides on functional inhibition of bone marrow by cyclophosphamide in rats. *Sci. Technol. Food Ind.* **2012**, *33*, 365–367.

57. Cao, S.Q.; Liu, L.; Zhu, M.F.; Chen, Q.P.; Qi, X.Y.; Yang, H. Preparation of polysaccharide from sepia ink and its effects on B16F10 cells. *J. Nucl. Agric. Sci.* **2017**, *31*, 906–912.

58. Liu, H.Z.; Xiao, W.; Gu, Y.P.; Tao, Y.X.; Zhang, D.Y.; Du, H.; Shang, J.H. Polysaccharide from *Sepia esculenta* ink and cisplatin inhibit synergistically proliferation and metastasis of triple-negative breast cancer MDA-MB-231 cells. *Iran. J. Basic Med. Sci.* **2016**, *19*, 1292–1298. [PubMed]

59. Chen, W.; Zheng, R.; Baade, P.D.; Zhang, S.; Zeng, H.; Bray, F.; Jemal, A.; Yu, X.Q.; He, J. Cancer statistics in China, 2015. *CA Cancer J. Clin.* **2016**, *66*, 115–132. [CrossRef] [PubMed]

60. Senthilkumar, S.; Yogeeta, S.K.; Subashini, R.; Devaki, T. Attenuation of cyclophosphamide induced toxicity by squalene in experimental rats. *Chem. Biol. Interact.* **2006**, *3*, 252–260. [CrossRef] [PubMed]

61. Selvakumar, E.; Prahalathan, C.; Sudharsan, P.T.; Varalakshmi, P. Protective effect of lipoic acid on cyclophosphamide-induced testicular toxicity. *Clin. Chim. Acta* **2006**, *7*, 114–119. [CrossRef] [PubMed]

62. Motawi, T.M.K.; Sadik, N.A.H.; Refaat, A. Cytoprotective effects of DL-alpha-lipoic acid or squalene on cyclophosphamide-induced oxidative injury: An experimental study on rat myocardium, testicles and urinary bladder. *Food Chem. Toxicol.* **2010**, *48*, 2326–2336. [CrossRef] [PubMed]

63. Aguilar-Mahecha, A.; Hales, B.F.; Robaire, B. Effects of acute and chronic cyclophosphamide treatment on meiotic progression and the induction of DNA double-strand breaks in rat spermatocytes. *Biol. Reprod.* **2005**, *72*, 1297–1304. [CrossRef] [PubMed]

64. Elangovan, N.; Chiou, T.J.; Tzeng, W.F.; Chu, S.T. Cyclophosphamide treatment causes impairment of sperm and its fertilizing ability in mice. *Toxicology* **2006**, *222*, 60–70. [CrossRef] [PubMed]

65. Luo, P.; Liu, H.Z. Antioxidant ability of squid ink polysaccharides as well as their protective effects on DNA damage in vitro. *Afr. J. Pharm. Pharmacol.* **2013**, *7*, 1382–1388. [CrossRef]

66. Yang, J.; Liu, K.X.; Qu, J.M.; Wang, X.D. The changes induced by cyclophosphamide in intestinal barrier and microflora in mice. *Eur. J. Pharmacol.* **2013**, *714*, 120–124. [CrossRef] [PubMed]

67. Meng, Y.; Huang, Y.K. Chemotherapeutics impairment on intestinal mucosal barrier and the prevention and treatment. *Med. Recapitul.* **2012**, *18*, 1325–1327.

68. Round, J.L.; Mazmanian, S.K. The gut microbiota shapes intestinal immune responses during health and disease. *Nat. Rev. Immunol.* **2009**, *9*, 313–323. [CrossRef] [PubMed]

69. Macpherson, A.J.; Harris, N.L. Interactions between commensal intestinal bacteria and the immune system. *Nat. Rev. Immunol.* **2004**, *4*, 478–485. [CrossRef] [PubMed]

70. Zhang, W.; Sun, Y.L.; Chen, D.H. Effects of chitin and sepia ink hybrid hemostatic sponge on the blood parameters of mice. *Mar. Drugs* **2014**, *12*, 2269–2281. [CrossRef] [PubMed]

71. Huang, N.; Lin, J.; Li, S.; Deng, Y.; Kong, S.; Hong, P.; Yang, P.; Liao, M.; Hu, Z. Preparation and evaluation of squid ink polysaccharide-chitosan as a wound-healing sponge. *Mat. Sci. Eng. C Mater.* **2018**, *82*, 354–362. [CrossRef] [PubMed]

72. Franz, S.A. Structure-activity relationship of antithrombotic polysaccharide derivatives. *Int. J. Bio. Macromol.* **1995**, *17*, 311–314. [CrossRef]

 *marine drugs*

Article

Effects of the Combination of Gliotoxin and Adriamycin on the Adriamycin-Resistant Non-Small-Cell Lung Cancer A549 Cell Line

Le Van Manh Hung [1,†], Yeon Woo Song [2,†] and Somi Kim Cho [2,3,*]

[1] School of Biomaterial Science and Technology, College of Applied Life Sciences, Jeju National University, Jeju 63243, Korea; manhhung.levan@gmail.com
[2] Faculty of Biotechnology, College of Applied Life Sciences, Jeju National University, Jeju 63243, Korea; syw1212@naver.com
[3] Subtropical/Tropical Organism Gene Bank, Jeju National University, Jeju 63243, Korea
*   Correspondence: somikim@jejunu.ac.kr; Tel.: +82-010-8660-1842
†   These authors contributed equally to this work.

Received: 7 February 2018; Accepted: 24 March 2018; Published: 27 March 2018

Abstract: Acquired drug resistance constitutes an enormous hurdle in cancer treatment, and the search for effective compounds against resistant cancer is still advancing. Marine organisms are a promising natural resource for the discovery and development of anticancer agents. In this study, we examined whether gliotoxin (GTX), a secondary metabolite isolated from marine-derived *Aspergillus fumigatus*, inhibits the growth of adriamycin (ADR)-resistant non-small-cell lung cancer (NSCLC) cell lines A549/ADR. We investigated the effects of GTX on A549/ADR cell viability with the 3-(4,5-dimethylthiazol-2-yl)-2,5-diphenyltetrazolium bromide (MTT) assay and the induction of apoptosis in A549/ADR cells treated with GTX via fluorescence-activated cell sorting analysis, Hoechst staining, annexin V/propidium iodide staining, tetraethylbenzimidazolylcarbocyanine iodide (JC-1) staining, and western blotting. We found that GTX induced apoptosis in A549/ADR cells through the mitochondria-dependent pathway by disrupting mitochondrial membrane potential and activating p53, thereby increasing the expression levels of p21, p53 upregulated modulator of apoptosis (PUMA), Bax, cleaved poly (ADP-ribose) polymerase (PARP), and cleaved caspase-9. More importantly, we discovered that GTX works in conjunction with ADR to exert combinational effects on A549/ADR cells. In conclusion, our results suggest that GTX may have promising effects on ADR-resistant NSCLC cells by inducing mitochondria-dependent apoptosis and through the combined effects of sequential treatment with ADR.

Keywords: gliotoxin; NSCLC; adriamycin resistance; apoptosis

---

1. Introduction

Non-small-cell lung cancer (NSCLC), which accounts for 85% of lung cancer cases, is one of the world's leading causes of death, with an incidence of 1.8 million cases and 1.5 million deaths in 2012 [1,2]. Approximately 50% of patients diagnosed with NSCLC present with advanced disease (stage III or IV) that is not responsive to curative treatment. Patients with advanced NSCLC survive for only 9–12 months [3]. One explanation for these discouraging statistics is the development of drug resistance. Chemotherapy resistance has long constituted a major hurdle in the treatment of cancer, especially in NSCLC. This leads to increases in the drug dosage, which in turn increase the cytotoxicity and undesirable effects to normal cells/tissues. Chemotherapeutic resistance may occur as a consequence of several factors, including increased efflux capability of membrane P-glycoproteins such as multidrug resistance-related protein 1 (MRP1), multidrug resistance 1 (MDR1), or lung resistance protein (LRP);

altered drug-metabolizing enzymes decreasing drug activation and increasing drug degradation; conjugation of the drug with increased glutathione; subcellular redistribution; drug interactions; elevated DNA repair; and failure of cells to undergo apoptosis [4]. Among these factors, apoptosis resistance is one of the most well studied, and it may be caused by the altered expression of genes involved in the apoptotic pathway [5]. Adriamycin (ADR), a topoisomerase II inhibitor belonging to the anthracycline family, is commonly used in cancer chemotherapy [6,7]. However, the use of ADR still has several limitations, such as acquired drug resistance, chronic cardiomyopathy, and congestive heart failure [8]. Moreover, the response rate of NSCLC to ADR is only 30–50% [9]. Thus, there is an urgent need to identify and develop an alternative approach to enhance the therapeutic outcome or increase the efficacy of this drug.

Developing anticancer drugs from natural sources is one of the most important strategies in the production of novel cancer therapies. Due to their extensive biodiversity, marine organisms represent a tremendous library of bioactive compounds. Marine-originated compounds have gained great interest as anticancer agents [10]. For example, Tuberatolide B isolated from the Korean marine algae *Sargassum macrocarpum* inhibits the proliferation of breast, lung, colon, prostate, and cervical cancer cells by disrupting the signal transducer and activator of transcription 3 (STAT3) signaling pathway [11]. Kempopeptin C isolated from a marine cyanobacterium from the Florida Keys inhibits the growth of invasive breast cancer by acting as a serine protease inhibitor [12]. Sinulariolide isolated from the soft coral *Sinularia flexibilis* decreases the epithelial–mesenchymal transition (EMT) in human bladder cancer cells [13]. Gliotoxin (GTX) is a secondary metabolite isolated from the marine-derived fungus *Aspergillus fumigatus*. It is an epipolythiodioxopiperazine characterized by a disulfide bridge across a piperazine ring (Please see Section 2.2) [14]. GTX has shown multiple biological effects, including antifungal [15], antiviral [16], anti-inflammatory [17], and antitumor activities in different cell lines [18,19]. In terms of its anticancer activity, GTX has shown great potential in targeting neurogenic locus notch homolog protein 2 (NOTCH2) [20,21], the Wnt/β-catenin pathway [18], farnesyltransferase, and geranylgeranyltransferase I [22]. However, the effects of GTX on chemoresistant cancer cells remain to be clarified.

In this study, we investigated the activities of GTX on the induction of apoptosis in ADR-resistant A549 cells using cell viability assays, detection of the formation of apoptotic bodies, cell cycle assay, mitochondrial membrane analysis, and western blotting. We also found that pretreatment with GTX for 12 h potentiated the effects of ADR, thereby reducing the effective concentration of ADR. Taken together, GTX shows promise as a potential therapeutic agent against chemoresistant NSCLC.

2. Results

2.1. Establishment of an ADR-Resistant Cell Line

Over 3 months, we established a cell line with acquired ADR resistance, generated from the A549 NSCLC cell line, named A549/ADR. The two cell lines were observed under a microscope and significant morphological differences were detected between the A549/ADR cell line and the original A549 line. The A549/ADR cells acquired an oval shape distinctly different from the epithelial-like shape of the parent cells (Figure 1a). The viability of A549 and A549/ADR cells was investigated using the 3-(4,5-dimethylthiazol-2-yl)-2,5-diphenyltetrazolium bromide (MTT) assay, which is based on the transformation of yellow tetrazolium salt MTT to purple formazan crystals by metabolically active cells. The IC$_{50}$ values of ADR in A549 and A549/ADR cells were 0.55 and 1.40 μM at 48 h time point, respectively, for a resistance index of 2.55 for A549/ADR cells. Furthermore, 0.5 μM ADR did not decrease the proliferation of A549/ADR cells, whereas the viability of A549 cells gradually decreased after 48 h (Figure 1b).

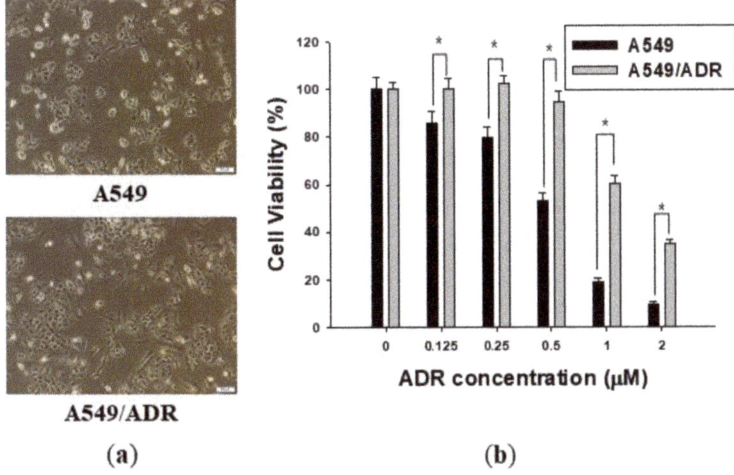

Figure 1. Characterization of the adriamycin (ADR)-resistant A549 cell line. (**a**) The morphologies of A549 and A549/ADR cells were observed under a microscope; (**b**) Effects of ADR on A549 and A549/ADR cells for 48 h. Cell viability was determined by the MTT assay. The results of independent experiments were averaged and are presented as percentage cell viability. Values represent means $\pm$ standard deviation (SD) ($n = 3$) (* $p < 0.05$).

2.2. GTX Overcame ADR Resistance in A549 NSCLC Cells

Whereas ADR had strong cytotoxic effects on A549 cells but relatively weak effects on A549/ADR cells, GTX was effective in inhibiting the proliferation of both the parental A549 and resistant A549/ADR cells, with IC_{50} values of 0.40 and 0.24 µM, respectively. Interestingly, treatment of cells with GTX 0.5 and 1 µM significantly reduced the viability of A549/ADR cells than the viability of A549 cells. Apparently, there was no significant resistance against GTX compared to ADR. Moreover, GTX was more effective in inhibiting the proliferation of both cell lines than ADR (IC_{50} 0.40 and 0.24 vs. 0.55 and 1.40 µM) (Figure 2b).

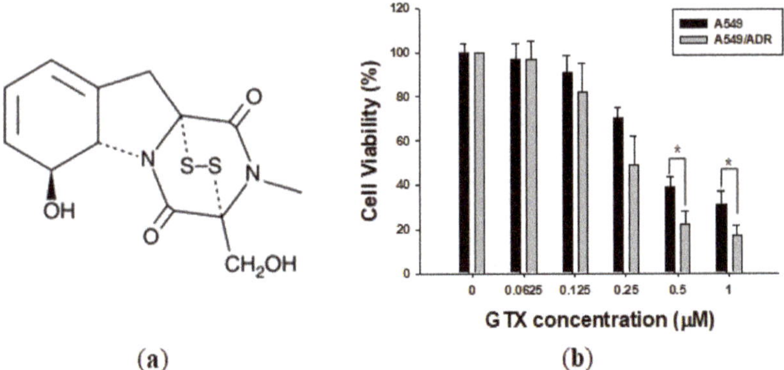

Figure 2. Gliotoxin (GTX) treatment reduces A549/ADR cell viability. (**a**) Chemical structure of GTX; (**b**) Effects of GTX on A549 and A549/ADR cells for 48 h. Cell viability was determined by the MTT assay. Results of independent experiments were averaged and are presented as percentage cell viability. Values represent means $\pm$ standard deviation (SD) ($n = 3$) (* $p < 0.05$).

2.3. GTX Induced Apoptosis in A549/ADR Cells

2.3.1. GTX Induced Cell Cycle Arrest in A549/ADR Cells

Propidium iodide (PI) staining and flow cytometry analysis were performed to investigate the cell cycle distribution of A549/ADR cells treated with 0.0625, 0.125, 0.25, and 0.5 µM GTX for 24 h (Figure 3a). Compared with the control sample, there was a dose-dependent increase of the sub-G1 population, from 1.37 to 52.49%, coupled with a decrease in the G1 population, from 65.41 to 28.44% (Figure 3a). This indicates that GTX-induced cell death of A549/ADR cells was mediated by sub-G1 cell cycle arrest and apoptosis.

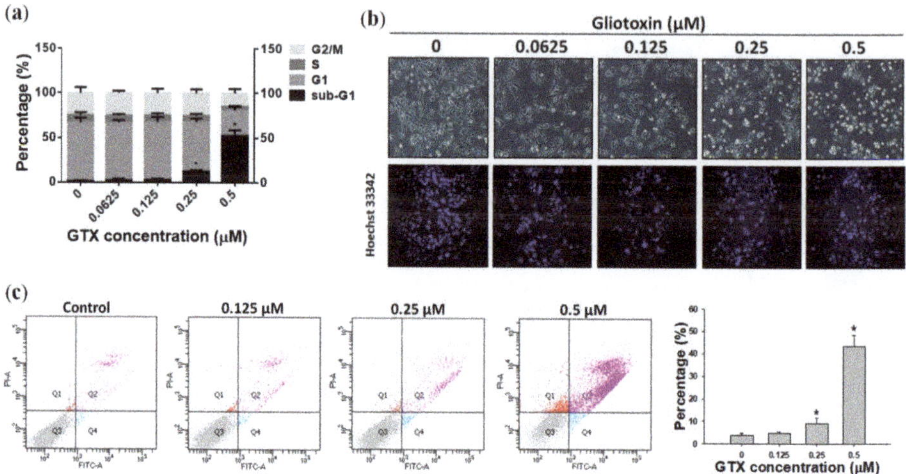

Figure 3. GTX treatment induces apoptosis in A549/ADR cells. (**a**) Cell cycle analysis of A549/ADR cells treated with GTX. Cells were seeded in 60-mm dishes and treated with different concentrations of GTX (0, 0.0625, 0.125, 0.25, and 0.5 µM) for 24 h. Cells were then stained with propidium iodide (PI) solution and analyzed by flow cytometry; (**b**) Cells were treated with increasing doses of GTX. After 24 h, apoptotic cells were detected by staining with Hoechst 33342 and observed under a fluorescence microscope; (**c**) Annexin V/PI staining analysis by flow cytometry. After cells were treated with 0, 0.125, 0.25, and 0.5 µM GTX for 24 h, they were stained with PI and annexin V-fluorescein isothiocyanate (FITC) together with binding buffer for 15 min before analysis. Values represent means $\pm$ standard deviation (SD) ($n = 3$) ($^*\,p < 0.05$).

2.3.2. Hoechst 33342 Staining of A549/ADR Cells Treated with GTX

Chromatin condensation and apoptotic body formation, two characteristics of apoptosis, were investigated by Hoechst 33342 staining assay. Hoechst 33342 is a cell-permeable DNA stain that can be absorbed by both viable and dead cells. Viable cells with intact DNA show weak fluorescence signals, whereas cells undergoing apoptosis with condensed chromatin exhibit stronger fluorescence when observed under a fluorescence microscope. In this experiment, A549/ADR cells were treated with four concentrations of GTX for 24 h. As shown in Figure 3b, the number of A549/ADR cells with intense fluorescence signals increased in a dose-dependent manner, which indicates that apoptosis was the major cell death mechanism induced by GTX treatment.

2.3.3. Annexin V/PI Staining

To continue to assess the lethality of GTX, A549/ADR cells were subjected to flow cytometry analysis after treatment with 0.125, 0.25, and 0.5 µM GTX for 24 h, and double stained with annexin

V-fluorescein isothiocyanate (FITC) and PI solution. Detecting apoptosis with annexin V is based on the location of the membrane phospholipid phosphatidylserine (PS). In healthy cells, PS is located on the cytoplasmic side of the plasma membrane. However, in the early stages of apoptosis, PS translocates to the outer side of the membrane and can be detected by fluorescence-bound annexin V. The results are illustrated as a quadrant model with PI signal on the y-axis and annexin V intensity on the x-axis. The lower left quadrant shows the viable cells (negative for both PI and annexin V). The lower right quadrant shows the early apoptotic cells (PI negative, annexin V positive). The upper left quadrant shows necrotic cells (PI positive, annexin V negative), and the upper right quadrant shows the late apoptotic population (PI positive, annexin V positive). The results of quadrant statistical analysis showed that the number of annexin V-positive cells increased with increasing GTX dose. The apoptotic population in untreated cells was 3.73%, whereas it increased to 9.37% in 0.25 μM GTX-treated samples, and 43.37% in 0.5 μM GTX-treated samples (Figure 3c). Taken together with the previous experiments, these results indicate that GTX caused apoptosis in ADR-resistant A549 cells.

2.4. Disruption of Mitochondrial Membrane Potential by GTX in A549/ADR Cells

Mitochondrial membrane potential ($\Delta\Psi$m) is an important indicator of cell condition, and depletion of $\Delta\Psi$m is evidence of the early induction of apoptosis. Figure 4 shows that treatment with GTX (0–0.5 μM) increased the depolarization of mitochondrial membranes in A549/ADR cells, as indicated by the reduction in the level of $\Delta\Psi$m (Figure 4). These data suggest that apoptosis caused by GTX treatment was induced through the intrinsic pathway.

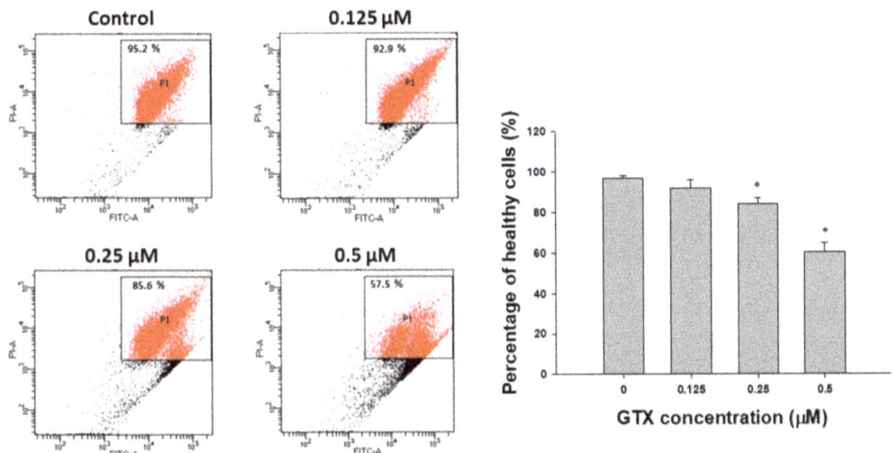

Figure 4. GTX attenuates mitochondrial membrane potential in A549/ADR cells. Cells were treated with three concentrations of GTX for 24 h, stained with tetraethylbenzimidazolylcarbocyanine iodide (JC-1)-FITC together with binding buffer, and analyzed by flow cytometry. Values represent means ± SD ($n = 3$) (* $p < 0.05$).

2.5. Effects of GTX on the Expression of Apoptosis-Related Proteins

Western blot analysis was used to examine the expression of total p53, phosphorylated p53 (p-p53), p21, p53 upregulated modulator of apoptosis (PUMA), Bax, poly (ADP-ribose) polymerase (PARP), cleaved PARP, cleaved caspase-9, cleaved caspase-8 and Bid in A549/ADR cells after GTX treatment. β-actin was used as the loading control. As shown in Figure 5, GTX treatment significantly increased the expression of p53 and p-p53, and consequently, its downstream targets, including p21, PUMA, and Bax. Furthermore, the levels of cleaved PARP and cleaved caspase-9, two important apoptotic markers, were dose-dependently increased by GTX treatment (Figure 5).

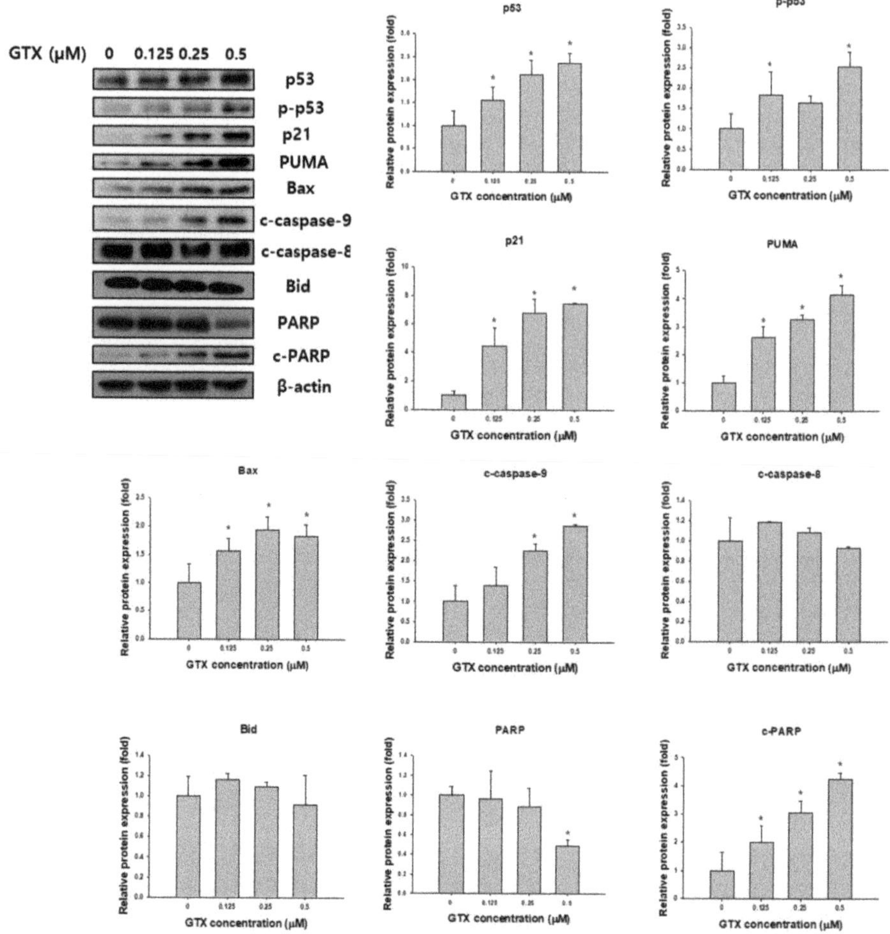

Figure 5. GTX induces A549/ADR cell death via the mitochondria-dependent pathway. Cells were treated with 0, 0.125, 0.25, and 0.5 µM GTX for 24 h and cell lysates were subjected to western blotting for various apoptosis-related markers including p53, phosphorylated p53 (p-p53), p21, PUMA, Bax, PARP, cleaved PARP (c-PARP), cleaved caspase-9 (c-caspase-9), cleaved caspase-8 and Bid. β-actin was used as a loading control. The graphs are the densitometric quantification of the western blotting result (* $p < 0.05$).

2.6. GTX Potentiates the Effects of ADR on A549/ADR Cells

To determine the effects of the two drugs on ADR-resistant A549 cells, we treated these cells with GTX, ADR, or both (24 h GTX pretreatment) for 24 or 48 h. The MTT assay, cell cycle assay, and western blot analysis were performed to determine the ability of GTX to potentiate the effects of ADR in the ADR-resistant cell line. As shown in Figure 6a, the MTT assay indicated that A549/ADR cell viability was more severely decreased by the combinatorial treatment than with the single treatments. The CI values calculated by using Calcusyn (Biosoft, UK) were ranged between 0.8 and 1.3, which is comparable to the additive effect ranging from 0.9 to 1.1. Moreover, to study the cell cycle distribution after different treatments, we pretreated GTX for 24 h and then ADR for another 24 h, then we stained

A549/ADR cells with PI solution and analyzed them with flow cytometry. The sub-G1 population was elevated from 11.44% for 0.125 μM and 50.02% for 0.25 μM GTX treatment alone to 29.09% and 61.16% for the combination treatment (Figure 6b). After we determined that GTX in combination with ADR potentiated the cytotoxicity in A549/ADR cells, we conducted western blot analysis to investigate changes in cleaved PARP and PARP, with β-actin as a loading control. We found that the level of PARP decreased in the GTX alone treatment and the expression of cleaved PARP in the GTX+ADR treatment group were higher than those in the individual treatments (Figure 6c). Moreover, the level of Bax was increased while the expression of Bcl-2 remained constant, this led to the increase of the Bax/Bcl-2 ratio, which plays an important role in apoptosis induction.

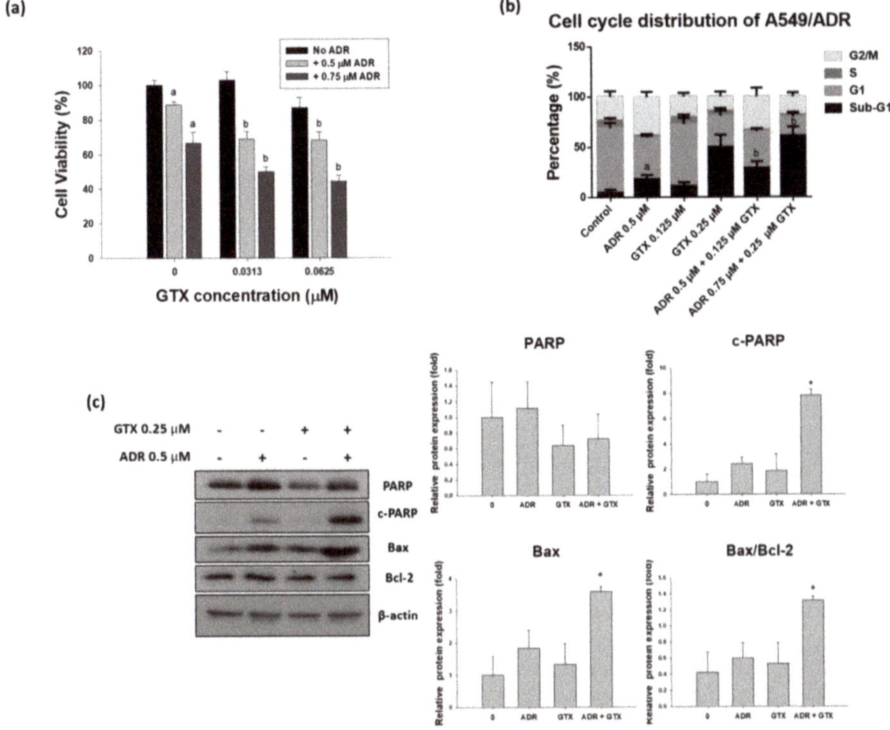

Figure 6. Pre-treatment with GTX for 24 h potentiates the effects of ADR in A549/ADR cells. A549/ADR cells were pretreated with GTX for 24 h and subsequently treated with ADR. The effects of the combination treatment were investigated by MTT assay (**a**); cell cycle analysis (**b**); and western blotting (**c**), the bar graphs represent the densitometric quantification of the western blotting result (* $p < 0.05$). Columns with different letters are significantly different between samples of the same ADR dose treatment ($p < 0.05$).

3. Discussion

Cancer raises many treatment challenges owing to the variable effectiveness of the primary therapies, surgery, radiation, and chemotherapy. Acquired drug resistance is considered a major barrier to a positive treatment outcome. ADR has been widely used to treat various kinds of solid tumors, including NSCLC. However, its efficacy is limited, as cancer cells often become resistant after a prolonged period of drug exposure. Natural resources have been used to develop antitumor agents [23]. In addition to terrestrial plants and microorganisms, marine species have emerged as a source of biologically active compounds for various purposes including anticancer effects. Among candidate marine-derived compounds, GTX is a fungal secondary metabolite isolated from the marine-derived *Aspergillus fumigatus*. Not much is known about the effects of marine products on drug resistant cancer cells. Previous studies have investigated the cytotoxicity of GTX in various cancer cell lines, however, the effects of GTX on chemoresistant cancer cells remains to be elucidated. In the present study, we examined the potential of GTX against drug resistant NSCLC A549/ADR cells and explored its underlying mechanism of action.

After testing a wide array of natural compounds in search of the most potent chemicals against both typical NSCLC A549 cells and ADR-resistant A549/ADR cells, we identified GTX as a compound that overcame ADR resistance and significantly inhibited both cell lines. Moreover, we reasoned that compounds with a lower IC_{50} are of interest for further study, as we can minimize the generic toxicity linked to ion transport. Although ADR had relatively low effectiveness in A549/ADR cells compared with A549 cells, GTX inhibited both cell lines in a similar manner. Interestingly, A549/ADR cell line was even more sensitive to GTX than A549 at high concentrations. This phenomenon could be explained by the fact that our A549/ADR has a higher expression level of multidrug resistance-associated protein 1 (MRP1) compared to the A549 cell line (data not shown). Previously, it has been reported that MRP1-overexpressing cells are more sensitive to a selective subset of chemical compounds due to collateral sensitivity [24,25]. In addition, De Groot et al. indicated that ADR-resistant small-cell lung cancer cells overexpressing MRP1 was more sensitive to indomethacin compared to the parental cell line [26]. In this study, the IC_{50} of GTX in A549 cells was 0.46 μM, which was more effective than in HeLa cervical cancer cells or SW1353 chondrosarcoma cells (both $IC_{50} > 90$ μM) [19], and equally potent in six breast cancer cell lines (IC_{50} ranging from 38 to 985 nM) [19]. Furthermore, GTX was also more effective than ADR (IC_{50} values were 0.40 and 0.24 compared with 0.55 and 1.40 μM, respectively). As the concentrations of IC_{50} or higher of ADR or GTX themselves exert dramatic effect on A549/ADR cells, we decided to choose the concentrations that are lower than the IC_{50} for our study on the combinational effect of ADR and GTX on A549/ADR cells.

Apoptosis is an ordered and well-arranged biological phenomenon that occurs under both physiological and pathological conditions. Apoptosis has been extensively studied for years in the cancer field. It is widely recognized that the induction of apoptosis is one of the most essential criteria in the development of antitumor agents [27]. In this study, we found that GTX inhibited the proliferation of A549/ADR cells by promoting mitochondria-mediated apoptosis. First, apoptosis induction by GTX was evidenced by the accumulation of populations of sub-G1 cells, the formation of apoptotic bodies, and increased numbers of annexin V-positive A549/ADR cells. Beside the apoptotic population, the percentage of necrotic cells were 0.3 ± 0.2, 0.5 ± 0.3, 0.9 ± 0.2, 4.5 ± 0.9 for GTX treatment at 0, 0.125, 0.25, 0.5 μM, respectively. In fact, apoptosis can be accompanied by necrosis, or necrosis can occur as the secondary phase after the apoptotic program has been completed [28]. To determine the type of apoptosis, we next used JC-1 staining to investigate whether the mitochondrial membrane potential was disrupted by GTX treatment. We observed a shift in fluorescence signal from red (~590 nm) to green (~529 nm), indicating the depolarization of the mitochondrial membrane as a result of mitochondria-dependent apoptosis. Additionally, we further explored the apoptotic pathway using western blotting. p53, which plays a crucial role in regulating the apoptotic pathway, was wild-type in both cell lines. As indicated by western blotting, p53 was upregulated, and the downstream targets of p53 were subsequently upregulated, including p21 (a cell cycle regulator), PUMA, and Bax (two pro-apoptotic proteins belonging to the Bcl-2 family). Moreover, the protein levels of proteolytically cleaved caspase-9 and cleaved PARP increased, clearly indicating the onset of apoptotic cell death. Furthermore, to exclude the possibility the extrinsic pathway is involved in GTX-induced apoptosis, we examined the expression level of cleaved caspase-8 and Bid. We observed that the expressions of both cleaved caspase-8 and Bid were not significantly altered by GTX treatment, therefore GTX induced apoptosis in A549/ADR via the intrinsic pathway.

A combination of natural compounds and approved chemotherapeutic agents to increase their efficacy has gained enormous attention in recent years. Several phytochemicals have been shown to elevate the outcome of ADR treatment, for example, schisandrin B reversed ADR resistance in MCF-7 breast cancer cells [29], neferine improved the effects of ADR on A549 lung adenocarcinoma cells [30], and resveratrol and didox increased the effects of ADR on HCT116 colorectal cancer cells [31]. Our findings suggest that GTX enhanced the cytotoxicity of ADR in an ADR-resistant A549 cell line. First, the MTT assay revealed that combined treatment resulted in a higher proportion of cell death. Then, by using a cell cycle distribution analysis, we demonstrated that sequential treatment with GTX and ADR elevated the sub-G1 population. Finally, western blot results showed that the combination of GTX and ADR enhanced the expression of significant protein in the apoptotic pathway, such as cleaved PARP and the Bax/Bcl-2 ratio. Moreover, the ability of GTX to reduce PARP protein level, which plays an important role in DNA damage repair pathway [32], could serve as a mechanism to potentiate the effect of DNA-damage-inducing agent ADR.

4. Materials and Methods

4.1. Cell Line, Reagents, and Chemicals

F-12K, bovine serum albumin, trypsin/ethylenediaminetetraacetic acid, fetal bovine serum, and $100\times$ antibiotic-antimycotic were purchased from Invitrogen (Carlsbad, CA, USA). Adriamycin ($\geq$ 98% purity), gliotoxin ($\geq$ 98% purity), hoechst 33342, dimethyl sulfoxide, MTT, PI, RNase A, and anti-β-actin antibodies were purchased from Sigma-Aldrich (St. Louis, MO, USA). Anti-cleaved PARP, -PARP, -Bax, -PUMA, -cleaved caspase-9, -p21, -p53, and -p-p53 (serine 15) antibodies were purchased from Cell Signaling Technology, Inc. (Beverly, MA, USA). Polyvinylidene fluoride membranes for western blotting were purchased from Millipore (Billerica, MA, USA). Mitochondrial membrane potential detection JC-1 kit and an annexin V-FITC apoptosis detection kit were purchased from BD Biosciences (San Jose, CA, USA).

4.2. Cell Culture

A549 human NSCLC and ADR-resistant NSCLC A549/ADR cells were cultured in F-12K medium supplemented with 10% heat-inactivated fetal bovine serum (FBS), 1% antibiotics at 37 °C in a CO_2 incubator. All cells were gradually passaged when they reached 90% confluence. All cultures were maintained at 37 °C in a humidified incubator with 5% CO_2. A549/ADR cells were obtained by using the dose-escalating method. Briefly, after examining the cell viability of A549 cells treated with different concentrations of ADR for 48 h, we chose 0.03 μM as our starting concentration for our procedure. A549 cells were maintained in medium containing ADR for two weeks with each concentration. ADR doses were gradually increased until it reached 0.5 μM, this process lasted for approximately three months. After that, we enriched three colonies that remained on the dish and performed MTT assay as well as western blotting for the MRP1 protein expression and chose the one with the highest cell viability and MRP1 expression as our ADR-resistant cell line.

4.3. Cell Viability Assay

The cell viability assay was performed as previously described [33]. Briefly, exponential-phase cells were seeded to 96-well plates (2.5×10^4 cells/mL). After 24 h, cells were incubated in with various concentration of gliotoxin for 48 h. Then MTT agent was added into each well and incubated for 4 h at 37 °C. The MTT results were read immediately at 570 nm with a Sunrise microplate reader (Tecan, Salzburg, Austria). The percentage of viable cells was calculated based on the following formula: mean value of (control group–treated group)/control group $\times$ 100%. All results were assessed in triplicate at each concentration.

4.4. Flow Cytometry

PI and JC-1 staining for investigating cell cycle distribution and mitochondrial membrane potential were performed as previously described [33]. Briefly, A549/ADR cells were plated in 60 mm plates (15×10^4 cells/plate) and treated with gliotoxin (0–0.5 μM) for 24 h. Then cells were harvested, washed with phosphate-buffered saline (PBS), and fixed in 70% ethanol. Before analysis, cells were treated with the solution containing 2 mM EDTA-PBS, RNase A (25 ng/mL) and propidium iodide (40 μg/mL). In terms of JC-1 staining assay, we harvested the GTX treated cells, then cells were washed with $1\times$ assay buffer, stained with JC-1 for 15 min at 37 °C and washed again with $1\times$ assay buffer. All analyses were performed using a FACSCalibur flow cytometer (BD Biosciences). Data were analyzed with CellQuest software (BD Biosciences). Each experiment was repeated at least three times.

4.5. Western Blot Analysis

Western blotting was performed as previously described [33]. Briefly, A549/ADR cells were plated in 60 mm dishes (15×10^4 cells/plate). 24 h after being treated, cells were harvested and lysed in lysis buffer (20 nM Tris-HCl, 150 mM NaCl, 1 mM Na_2EDTA, 1 mM EGTA, 1% NP-40, 2.5 mM sodium pyrophosphate, 1 mM β-glycerophosphate, 1 mM Na_3VO_4). Protein concentrations were measured and normalized using a BCA Protein Assay kit. Lysates were then separated by 10–15% sodium dodecyl sulfate-polyacrylamide gel electrophoresis and transferred onto a PVDF membrane using glycine transfer buffer. Membranes were then blocked for non-specific bindings by 5% skim milk solution. After that, membranes were incubated with primary antibodies overnight at 4 °C, followed by an additional 40 min incubation with secondary antibodies. The resultant membranes were analyzed using a BS ECL Plus kit (Biosesang Inc., Seongnam, Korea).

4.6. Statistical Analysis

Group comparisons were performed using Student's *t*-test and one-way analysis of variance with Statistical Package for the Social Sciences software (SPSS v. 20.0, IBM Corp., Armonk, NY, USA). $p < 0.05$ was considered statistically significant.

Acknowledgments: The authors acknowledge Jeong Yong Moon for the excellent technical support. This research was part of the "2018 Jeju Sea Grant" project funded by the Ministry of Oceans and Fisheries, South Korea.

Author Contributions: S.K.C. conceived and designed the experiments, analyzed the data, and wrote the manuscript; L.V.M.H. and Y.W.S. performed the experiments and analyzed the data, and wrote the draft of the article.

Conflicts of Interest: The authors declare no conflict of interest.

References

1. Sher, T.; Dy, G.K.; Adjei, A.A. Small Cell Lung Cancer. *Mayo Clin. Proc.* **2008**, *83*, 355–367. [CrossRef] [PubMed]
2. Ferlay, J.; Soerjomataram, I.; Dikshit, R.; Eser, S.; Mathers, C.; Rebelo, M.; Parkin, D.M.; Forman, D.; Bray, F. Cancer incidence and mortality worldwide: Sources, methods and major patterns in globocan 2012. *Int. J. Cancer* **2015**, *136*. [CrossRef] [PubMed]
3. Paz-Ares, L.; de Marinis, F.; Dediu, M.; Thomas, M.; Pujol, J.-L.; Bidoli, P.; Molinier, O.; Sahoo, T.P.; Laack, E.; Reck, M. Maintenance therapy with pemetrexed plus best supportive care versus placebo plus best supportive care after induction therapy with pemetrexed plus cisplatin for advanced non-squamous non-small-cell lung cancer (paramount): A double-blind, phase 3, randomised controlled trial. *Lancet Oncol.* **2012**, *13*, 247–255. [PubMed]
4. Tsvetkova, E.; Goss, G. Drug resistance and its significance for treatment decisions in non-small-cell lung cancer. *Curr. Oncol.* **2012**, *19*, S45–S51. [PubMed]
5. Igney, F.H.; Krammer, P.H. Death and anti-death: Tumour resistance to apoptosis. *Nat. Rev. Cancer* **2002**, *2*, 277–288. [CrossRef] [PubMed]

6.   Arcamone, F.; Cassinelli, G.; Fantini, G.; Grein, A.; Orezzi, P.; Pol, C.; Spalla, C. Adriamycin, 14-hydroxydaimomycin, a new antitumor antibiotic from s. Peucetius var. Caesius. *Biotechnol. Bioeng.* **1969**, *11*, 1101–1110. [CrossRef] [PubMed]

7.   Cortés-Funes, H.; Coronado, C. Role of anthracyclines in the era of targeted therapy. *Cardiovasc. Toxicol.* **2007**, *7*, 56–60. [CrossRef] [PubMed]

8.   Minotti, G.; Menna, P.; Salvatorelli, E.; Cairo, G.; Gianni, L. Anthracyclines: Molecular advances and pharmacologic developments in antitumor activity and cardiotoxicity. *Pharmacol. Rev.* **2004**, *56*, 185–229. [CrossRef] [PubMed]

9.   Mi, J.; Zhang, X.; Rabbani, Z.N.; Liu, Y.; Reddy, S.K.; Su, Z.; Salahuddin, F.K.; Viles, K.; Giangrande, P.H.; Dewhirst, M.W. Rna aptamer-targeted inhibition of nf-κb suppresses non-small cell lung cancer resistance to doxorubicin. *Mol. Ther.* **2008**, *16*, 66–73. [CrossRef] [PubMed]

10.   Simmons, T.L.; Andrianasolo, E.; McPhail, K.; Flatt, P.; Gerwick, W.H. Marine natural products as anticancer drugs. *Mol. Cancer Ther.* **2005**, *4*, 333–342. [PubMed]

11.   Choi, Y.K.; Kim, J.; Lee, K.M.; Choi, Y.-J.; Ye, B.-R.; Kim, M.-S.; Ko, S.-G.; Lee, S.-H.; Kang, D.-H.; Heo, S.-J. Tuberatolide B suppresses cancer progression by promoting ros-mediated inhibition of stat3 signaling. *Mar. Drugs* **2017**, *15*, 55. [CrossRef] [PubMed]

12.   Al-Awadhi, F.H.; Salvador, L.A.; Law, B.K.; Paul, V.J.; Luesch, H. Kempopeptin c, a novel marine-derived serine protease inhibitor targeting invasive breast cancer. *Mar. Drugs* **2017**, *15*, 290. [CrossRef] [PubMed]

13.   Cheng, T.-C.; Din, Z.-H.; Su, J.-H.; Wu, Y.-J.; Liu, C.-I. Sinulariolide suppresses cell migration and invasion by inhibiting matrix metalloproteinase-2/-9 and urokinase through the pi3k/akt/mtor signaling pathway in human bladder cancer cells. *Mar. Drugs* **2017**, *15*, 238. [CrossRef] [PubMed]

14.   Gardiner, D.M.; Waring, P.; Howlett, B.J. The epipolythiodioxopiperazine (etp) class of fungal toxins: Distribution, mode of action, functions and biosynthesis. *Microbiology* **2005**, *151*, 1021–1032. [CrossRef] [PubMed]

15.   Reilly, H.C.; Schatz, A.; Waksman, S.A. Antifungal properties of antibiotic substances. *J. Bacterial.* **1945**, *49*, 585.

16.   McDougall, J. Antiviral action of gliotoxin. *Arch. Virol.* **1969**, *27*, 255–267. [CrossRef]

17.   López-Franco, O.; Suzuki, Y.; Sanjuán, G.; Blanco, J.; Hernández-Vargas, P.; Yo, Y.; Kopp, J.; Egido, J.; Gómez-Guerrero, C. Nuclear factor-κb inhibitors as potential novel anti-inflammatory agents for the treatment of immune glomerulonephritis. *Am. J. Pathol.* **2002**, *161*, 1497–1505. [CrossRef]

18.   Chen, J.; Wang, C.; Lan, W.; Huang, C.; Lin, M.; Wang, Z.; Liang, W.; Iwamoto, A.; Yang, X.; Liu, H. Gliotoxin inhibits proliferation and induces apoptosis in colorectal cancer cells. *Mar. Drugs* **2015**, *13*, 6259–6273. [CrossRef] [PubMed]

19.   Nguyen, V.-T.; Lee, J.S.; Qian, Z.-J.; Li, Y.-X.; Kim, K.-N.; Heo, S.-J.; Jeon, Y.-J.; Park, W.S.; Choi, I.-W.; Je, J.-Y. Gliotoxin isolated from marine fungus aspergillus sp. Induces apoptosis of human cervical cancer and chondrosarcoma cells. *Mar. Drugs* **2013**, *12*, 69–87. [CrossRef] [PubMed]

20.   Hubmann, R.; Hilgarth, M.; Schnabl, S.; Ponath, E.; Reiter, M.; Demirtas, D.; Sieghart, W.; Valent, P.; Zielinski, C.; Jäger, U. Gliotoxin is a potent notch2 transactivation inhibitor and efficiently induces apoptosis in chronic lymphocytic leukaemia (CLL) cells. *Br. J. Haematol.* **2013**, *160*, 618–629. [CrossRef] [PubMed]

21.   Hubmann, R.; Sieghart, W.; Schnabl, S.; Araghi, M.; Hilgarth, M.; Reiter, M.; Demirtas, D.; Valent, P.; Zielinski, C.; Jäger, U. Gliotoxin targets nuclear notch2 in human solid tumor derived cell lines in vitro and inhibits melanoma growth in xenograft mouse model. *Front. Pharmacol.* **2017**, *8*, 319. [CrossRef] [PubMed]

22.   Vigushin, D.; Mirsaidi, N.; Brooke, G.; Sun, C.; Pace, P.; Inman, L.; Moody, C.; Coombes, R. Gliotoxin is a dual inhibitor of farnesyltransferase and geranylgeranyltransferase i with antitumor activity against breast cancer in vivo. *Med. Oncol.* **2004**, *21*, 21–30. [CrossRef]

23.   Li, J.W.-H.; Vederas, J.C. Drug discovery and natural products: End of an era or an endless frontier? *Science* **2009**, *325*, 161–165. [CrossRef] [PubMed]

24.   Lorendeau, D.; Dury, L.; Genoux-Bastide, E.; Lecerf-Schmidt, F.; Simoes-Pires, C.; Carrupt, P.-A.; Terreux, R.; Magnard, S.; Di Pietro, A.; Boumendjel, A. Collateral sensitivity of resistant mrp1-overexpressing cells to flavonoids and derivatives through gsh efflux. *Biochem. Pharmacol.* **2014**, *90*, 235–245. [CrossRef] [PubMed]

25.   Lorendeau, D.; Dury, L.; Nasr, R.; Boumendjel, A.; Teodori, E.; Gutschow, M.; Falson, P.; Di Pietro, A.; Baubichon-Cortay, H. Mrp1-dependent collateral sensitivity of multidrug-resistant cancer cells: Identifying selective modulators inducing cellular glutathione depletion. *Curr. Med. Chem.* **2017**, *24*, 1186–1213. [CrossRef] [PubMed]

26. De Groot, D.; Van Der Deen, M.; Le, T.; Regeling, A.; De Jong, S.; De Vries, E. Indomethacin induces apoptosis via a mrp1-dependent mechanism in doxorubicin-resistant small-cell lung cancer cells overexpressing mrp1. *Br. J. Cancer* **2007**, *97*, 1077–1083. [CrossRef] [PubMed]

27. Fischer, U.; Schulze-Osthoff, K. Apoptosis-based therapies and drug targets. *Cell Death Differ.* **2005**, *12*, 942–961. [CrossRef] [PubMed]

28. Silva, M.T. Secondary necrosis: The natural outcome of the complete apoptotic program. *FEBS Lett.* **2010**, *584*, 4491–4499. [CrossRef] [PubMed]

29. Wang, S.; Wang, A.; Shao, M.; Lin, L.; Li, P.; Wang, Y. Schisandrin b reverses doxorubicin resistance through inhibiting p-glycoprotein and promoting proteasome-mediated degradation of survivin. *Sci. Rep.* **2017**, *7*, 8419. [CrossRef] [PubMed]

30. Poornima, P.; Kumar, V.B.; Weng, C.F.; Padma, V.V. Doxorubicin induced apoptosis was potentiated by neferine in human lung adenocarcima, a549 cells. *Food Chem. Toxicol.* **2014**, *68*, 87–98. [CrossRef] [PubMed]

31. Khaleel, S.A.; Al-Abd, A.M.; Ali, A.A.; Abdel-Naim, A.B. Didox and resveratrol sensitize colorectal cancer cells to doxorubicin via activating apoptosis and ameliorating p-glycoprotein activity. *Sci. Rep.* **2016**, *6*, 36855. [CrossRef] [PubMed]

32. Javle, M.; Curtin, N. The role of parp in DNA repair and its therapeutic exploitation. *Br. J. Cancer* **2011**, *105*, 1114–1122. [CrossRef] [PubMed]

33. Kim, H.; Moon, J.Y.; Ahn, K.S.; Cho, S.K. Quercetin induces mitochondrial mediated apoptosis and protective autophagy in human glioblastoma u373mg cells. *Oxid. Med. Cell. Longev.* **2013**, *2013*, 596496. [CrossRef] [PubMed]

 *marine drugs*

Article

Antiproliferative Activity of Glycosaminoglycan-Like Polysaccharides Derived from Marine Molluscs

Abdullah Faisal Aldairi, Olanrewaju Dorcas Ogundipe and David Alexander Pye *

School of Environment and Life Sciences, Cockcroft Building, University of Salford, Manchester M5 4WT, UK;
A.Aldairi@edu.salford.ac.uk (A.F.A.); dorcasogundipe14@gmail.com (O.D.O.)
* Correspondence: d.pye@salford.ac.uk; Tel.: +44-(0)161-295-470

Received: 10 January 2018; Accepted: 12 February 2018; Published: 15 February 2018

Abstract: Despite the increasing availability of new classes of cancer treatment, such as immune- and targeted therapies, there remains a need for the development of new antiproliferative/cytotoxic drugs with improved pharmacological profiles that can also overcome drug resistant forms of cancer. In this study, we have identified, and characterised, a novel marine polysaccharide with the potential to be developed as an anticancer agent. Sulphated polysaccharides isolated from the common cockle (*Cerastoderma edule*) were shown to have antiproliferative activity on chronic myelogenous leukaemia and relapsed acute lymphoblastic leukaemia cell lines. Disaccharide and monosaccharide analysis of these marine polysaccharides confirmed the presence of glycosaminoglycan-like structures that were enriched in ion-exchange purified fractions containing antiproliferative activity. The antiproliferative activity of these glycosaminoglycan-like marine polysaccharides was shown to be susceptible to heparinase but not chondrotinase ABC digestion. This pattern of enzymatic and antiproliferative activity has not previously been seen, with either marine or mammalian glycosaminoglycans. As such, our findings suggest we have identified a new type of marine derived heparan sulphate/heparin-like polysaccharide with potent anticancer properties.

Keywords: marine mollusc; glycosaminoglycans; antiproliferative; anticancer; heparan sulphate

---

1. Introduction

Glycosaminoglycans (GAGs) are a complex family of polysaccharides found in both vertebrates and invertebrates. They bind to many proteins and mediate a diverse range of biological functions, including both cellular and physiological events [1]. These molecules are widely exploited as therapeutics, for example, hyaluronic acid (HA) and chondroitin sulphate (CS) are used to treat osteoarthritis, and of course heparin has been used as an anticoagulant and antithrombotic drug for more than 70 years [2]. However, no native mammalian GAGs have been shown to have direct cytotoxic or antiproliferative effects on cancer cells. There are five families of mammalian GAGs, each based on their repeating disaccharide units. These are heparan sulphate (HS) including the related molecule heparin, CS, dermatan sulphate (DS), keratan sulphate and the unsulphated HA. Heparan sulphate and CS/DS GAG chains are initially synthesised as an alternating backbone of *N*-acetylated amino sugar, either *N*-acetylglucosamine (GlcNAc) in HS or *N*-acetylgalactosamine (GalNAc) in CS/DS and glucuronic acid (GlcA), or its C5 epimer iduronic acid (IdoA). Epimerisation takes place to differing extents in HS and heparin and in the conversion of CS into DS. Further modifications occur by *N*-sulphation of the amino sugar (GlcNS) in HS/heparin and *O*-sulphation at various positions on the sugar units, commonly 2-*O*- and 6-*O*-positions [3]. The extent of both epimerisation and sulphation defines the interactions that take place between GAGs and their protein receptor partners [4]. Mammalian GAGs, in particular HS, are very heterogeneous in structure and are organised into subdomains that contain regions of high sulphation [5], which are important for interactions with proteins [6].

Marine invertebrates are a rich source of potential therapeutics and marine carbohydrates have been shown to have complex structures and biological activities [7]. Marine polysaccharides differ considerably from their mammalian counterparts, in terms of their structure and biological activities. As a result, they provide important new opportunities as carbohydrate based therapeutics [8]. The principle marine glycans are comprised of non-sulphated chitin, chitosan, sulphated GAGs, sulphated fucans and sulphated galactans. Marine polysaccharides from seaweed and algae have been shown to have anticancer activities but these are largely related to polysulphated fucans that are not directly members of the GAG family [8]. Sulphated fucans or fucoidans from marine sources are comprised of homogeneous unbranched polymers of (1–3) or (1–4) linked L-fucose units, which can be substituted with sulphate groups at positions C-2, C-4 and rarely at C-3. They may also be heterogeneous branched polymers with variable content of uronic acids and neutral sugars [9]. Single fucose units, or short fuco-oligosaccharide branches, can also be present on the polysaccharide backbone, usually at the C-4 position. Structural features within Fucoidans have been linked to many beneficial effects including anticancer [10–12], antiviral [13], anti-inflammatory [14] and antithrombotic activities [15]. GAG-like polysaccharides have been isolated from many marine species, including molluscs and in some instances, HS structures were found to be similar to heparin. Marine DS and CS have backbone structures comparable to mammalian GAGs but with differing sulphation patterns. Fucosylated CS and CS/DS hybrid chains have also been identified from marine sources [16,17]. Marine GAGs have not previously been linked with antiproliferative or cytotoxic effects on cancer cells.

In this study, we have shown that marine derived GAG-like polysaccharides from the common cockle (*Cerastoderma edule*) have antiproliferative activity on two cancer cell lines. Enzyme depolymerisation and structural analysis suggest that novel HS-like structures are responsible for the antiproliferative activity of the marine derived polysaccharides.

2. Results

2.1. Polysaccharide Isolation and Anticancer Activity

Numerous studies have portrayed the potential medicinal properties of marine polysaccharides, mainly from seaweed, and some have even demonstrated cytotoxic effects using both in vitro and in vivo models [18]. In this study, we show for the first time that marine polysaccharides isolated from molluscs, with structural similarities to mammalian GAGs, have in vitro antiproliferative effects on cancer cell lines.

Marine polysaccharides were isolated from common cockles, following a typical cetylpyridinium chloride extraction procedure [19]. The cockle polysaccharides were assessed against two leukaemia cell lines (K562 and MOLT4) for their ability to suppress cancer cell growth. The cockle polysaccharides showed considerable inhibition of cell proliferation with these cell lines, as determined by 3-(4,5-dimethylthiazole-2-yl)-2,5-diphenyltetrazolium bromide (MTT) assay (Figure 1).

Some batch-to-batch variability in IC_{50} values was seen with different isolations of cockle polysaccharides, presumably because of the complex heterogeneous mixture of polysaccharide chains present. Cisplatin was used in all assays, as a control for comparison of different preparations. Typical IC_{50} values were around 9 µg/mL and 1 µg/mL, respectively, for cell lines K562 and MOLT4.

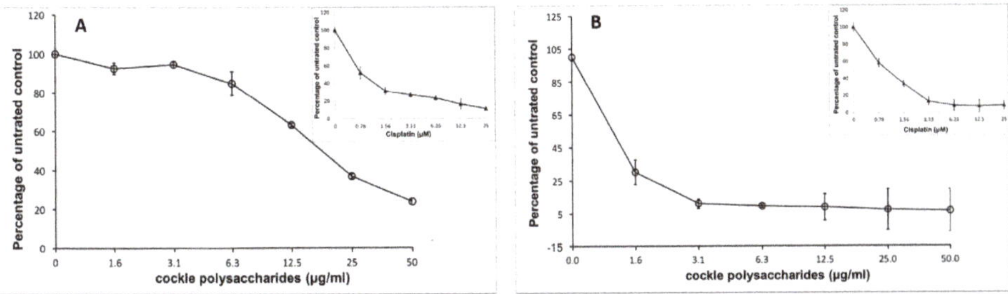

Figure 1. Antiproliferative activity of cockle polysaccharides on cancer cell lines. Two cancer cell lines K562 (**A**) and MOLT4 (**B**) were treated with increasing doses of cockle polysaccharides and cell viability was determined by MTT assay, as detailed in the "Materials and Methods" section. Inserts show effects of cisplatin treatment on viable cell number. Cells were cultured under standard conditions and maintained at 37 °C in a humidified 5% CO_2 atmosphere. Cell viability is expressed as a percentage relative to untreated control cells. All experiments were conducted in triplicate and the IC_{50} values were calculated using non-linear regression analysis (GraphPad Prism 5.0).

2.2. Annexin V Apoptosis Detection and Cell Cycle Analysis

Cell Cycle analysis of MOLT-4 cells (Figure 2A,C) showed an increase in cell numbers in both G1 and G2/M phases after 24 h exposure to cockle polysaccharides, accompanied by a lowering of cells in S-phase. These data, and the presence of cells in sub G1, indicate a complex mechanism of cell death, following treatment of MOLT-4 cells with cockle polysaccharides, potentially mediated via apoptosis. Flow cytometry and Annexin V/propidium iodide (PI) staining, was used to further investigate the mechanism of cell death following cockle polysaccharide treatment (Figure 2B,D). Leukaemia cell line MOLT-4 showed small but significant increases in late apoptotic (Annexin V^+/PI^+) cells when treated with cockle polysaccharide, However, early apoptotic (Annexin V^+/PI^-) showed the biggest change in cell populations after a 24 h incubation. Necrotic (Annexin V^-/PI^+) cells showed little change over untreated controls. The results shown in Figure 2 clearly implicate apoptosis as the most likely cause of the cytotoxicity seen following treatment by cockle polysaccharides. However, primary necrosis cannot be ruled out. Data from K562 cells were not obtained, as their inherent tendency to aggregate proved problematic in the flow cytometry studies.

2.3. Effect of Enzymatic Degradation on Cockle Polysaccharide Anticancer Activities

The extraction procedure used, whilst designed to enrich the content of sulphated GAG or GAG-like structures, will also bring contamination from other marine polysaccharides. These non-GAG polysaccharides may contribute solely, or in part, to the observed antiproliferative activity of the cockle polysaccharides. Enzymatic treatment of the cockle polysaccharides by heparinases I, II, III and chondroitinase ABC was used to investigate any link between typical GAG-like structures and the observed antiproliferative activity. Figure 3 demonstrates that there is no appreciable change in antiproliferative activity when cockle polysaccharides are incubated with chondroitinase ABC. Heparinase treatment (individually and in combination) however, did lead to a significant loss of biological activity, with an approximate increase in IC_{50} values for combined heparinases I, II, III of around 4- and 8-fold for K562 and MOLT4 cell lines respectively. There was little difference between the individual heparinases enzymes, however in replicates heparinase II, was consistently more effective at reducing the antiproliferative effects of the cockle polysaccharides. Despite the significant loss in activity, inhibition was not entirely destroyed by heparinase treatment, even when used in combination. This suggest that a proportion of resistant disaccharide linkages may exist and that resistant fragments are of sufficient size to still function as an antiproliferative. Taken together, the data suggest that

HS/heparin like GAG structures are key contributors to the antiproliferative activity of the cockle polysaccharides, with CS-like GAGs contributing little to the anticancer affect.

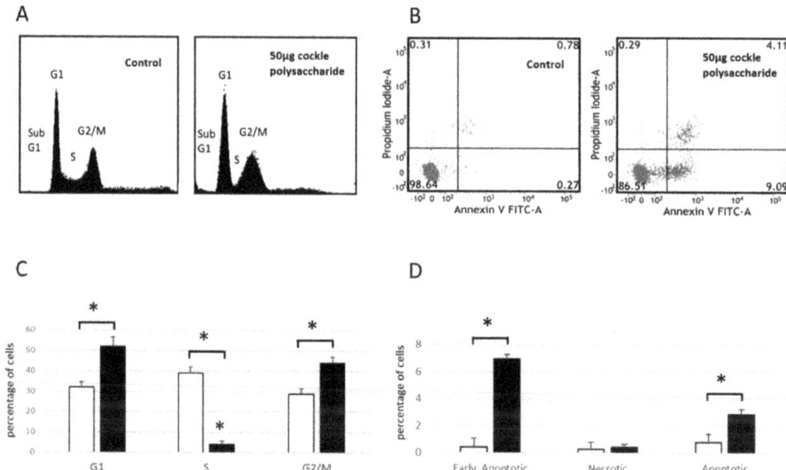

Figure 2. Cell cycle analysis and Annexin V apoptosis assay. MOLT4 cells were treated with 50 µg of cockle polysaccharides for 24 h then stained using Annexin V conjugated Fluorescein isothiocyanate (Annexin V-FITC) and/or PI. (**A**) Flow cytometry cell cycle analysis of PI stained cells with or without cockle polysaccharide treatment. A single representative experiment is shown. (**B**) Flow cytometry scatter plot of Annexin V-FITC/PI stained cells with or without cockle polysaccharide treatment. A single representative experiment is shown. (**C**) Quantitative cell cycle analysis as determined by PI staining and flow cytometry. The percentage of cells in each phase is shown for control (□) and cockle polysaccharide treated (■) cells. Results are presented as the mean ± SD of three independent experiments. Statistical significance was determined using the two-tailed Student's *t*-test. $p < 0.05$ was considered statistically significant (*). (**D**) Quantitative analysis of apoptosis as determined by Annexin V-FITC/PI staining and flow cytometry. The percentage of cells in each quadrant is shown for control (□) and cockle polysaccharide treated (■) cells. Results are presented as the mean ± SD of three independent experiments. Statistical significance was determined using the two-tailed Student's *t*-test. $p < 0.05$ was considered statistically significant (*).

2.4. Comparisons of Marine GAG Antiproliferative Activity with Mammalian GAGs

The potent antiproliferative effect of cockle polysaccharides was surprising, as the anticancer activity appeared to be linked to the presence of mammalian HS/heparin-like structures, which to our knowledge have never shown direct antiproliferative effects on cancer cell lines in vitro. Further investigation confirmed that mammalian derived GAGs (porcine mucosal) had no antiproliferative activity in our MTT assay system. Figure 4 shows that no measurable IC_{50} could be determined for mammalian HS, with either cancer cell line, yet the cockle polysaccharides continued to show significant antiproliferative activity. In fact, at low concentrations mammalian HS showed slight stimulatory activity on cell growth. The results suggest that, if HS-like-GAG chains are involved in the antiproliferative activity of cockle polysaccharides, then they must contain unique structural features not found in typical mammalian GAGs.

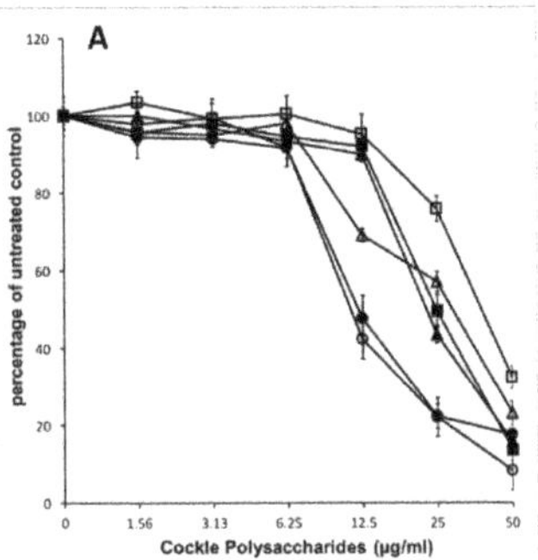 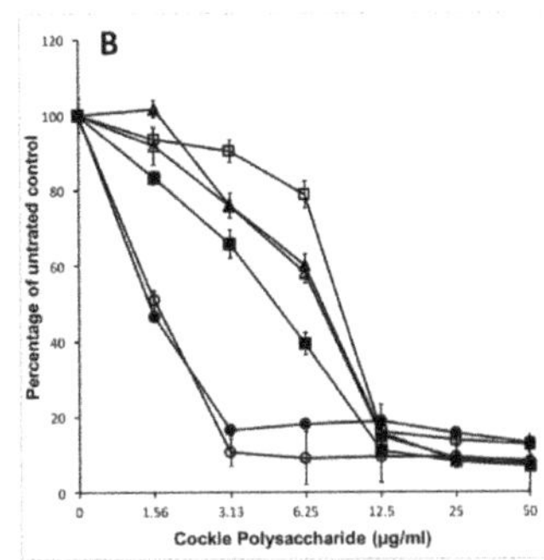

Figure 3. Effect of heparinase (I, II, and III) and chondroitinase ABC enzymatic degradation on cockle polysaccharide antiproliferative activity. Sensitivity of the cockle polysaccharide antiproliferative activity to enzymatic degradation was determined by MTT assay. Antiproliferative activity of cockle polysaccharides on K562 cells (**A**) and MOLT4 cells (**B**), with and without heparinase or chondrotinase ABC digestion. Intact cockle polysaccharides (○), heparinase I treated (△), heparinase II treated (▲), heparinase III treated (■), heparinase I, II, and III treated (□) and chondroitinase ABC (●). The data are presented as the percentage of viable cells following treatment with cockle polysaccharides, relative to untreated control. All experiments were conducted in triplicate and the results are shown as the mean ± the SD. Cells were cultured in suspension and maintained at 37 °C in humidified 5% CO_2 atmosphere.

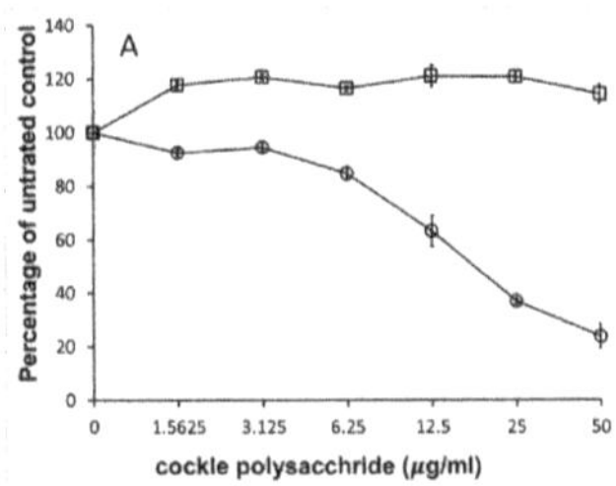 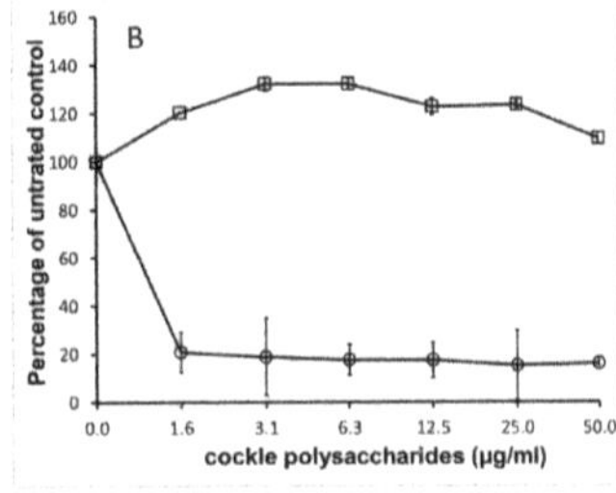

Figure 4. Comparison of mammalian GAGs and cockle polysaccharides antiproliferative activity. Differences in biological activities of mammalian GAGs and cockle polysaccharides on K562 (**A**) and MOLT4 (**B**) cell lines was assessed by MTT assay, cockle polysaccharides (○), mammalian HS (□). Data are presented as the percentage of viable cells following treatment with cockle polysaccharides, relative to untreated control. All experiments were conducted in triplicate and the results are shown as the mean ± the SD. Cells were cultured in suspension and maintained at 37 °C in humidified 5% CO_2 atmosphere.

2.5. Disaccharide Analysis of GAGs

Disaccharide analysis was used solely to confirm the presence of typical unsaturated HS and CS/DS disaccharides, following treatment of the cockle polysaccharides with heparinase I, II and III or chondroitinase ABC and was not part of any detailed structure/activity study.

Initial analysis of the HS/heparin disaccharide compositions, produced by combined heparinase I, II, III digestion, showed that susceptible chains within the cockle polysaccharide preparations did contain the major disaccharide species found in mammalian HS and heparin. Care must be taken in attributing total HS disaccharide compositions to these chains, as activity data (Figure 3) suggest that the combined heparinase I, II, and III digests are possibly incomplete. The major differences between disaccharides derived from heparinase digestion of mammalian HS and the cockle polysaccharide (Table 1) were the low levels of unsulphated disaccharide ΔHexA-GlcNAc and the high levels of the disulphated disaccharide ΔHexA-GlcNS(6S) released from the cockle polysaccharides. Other disaccharides were obtained at similar levels to those seen with many typical mammalian HS types. The heparinases failed to generate high quantities of the trisulphated disaccharide ΔHexA(2S)-GlcNS(6S) from the crude cockle polysaccharides, indicating that the HS-like cockle polysaccharides chains are more akin to mammalian HS than heparin. CS/DS disaccharide analysis was carried out, despite the fact that chondrotinase ABC treatment of cockle polysaccharides failed to show any significant loss of antiproliferative activity. The principle disaccharide components produced by chondroitinase ABC digestion were ΔHexA-GalNAc(4S), ΔHexA-GalNAc(6S) and ΔHexA-GalNAc(4S)(6S). The results clearly show that CS/DS-like chains are also present in the cockle polysaccharide preparations.

Disaccharide analysis data support the initial suggestion that the reduction in antiproliferative activity, following treatment of cockle polysaccharides with heparinases, was due to their breakdown. Resulting in the release of unsaturated disaccharides commonly found in mammalian HS/heparin. The lack of activity seen with the mammalian GAGs in the MTT assay, and the presence of heparinase resistant fragments within the cockle polysaccharides, suggests that the active HS-like-chains differ considerably from their mammalian counterparts.

Table 1. Disaccharide analysis of crude unfractionated cockle polysaccharides. 1: Data are presented as a percentage of the moles of CS/DS and HS unsaturated disaccharides produced by chondroitinase ABC and heparinase I, II, and III digestion.

HS Disaccharides	Disaccharides Produced (%) [1]
ΔHexA-GlcNAc	26.6
ΔHexA(2S)-GlcNAc	0.0
ΔHexA(2S)-GlcNH$_2$	0.0
ΔHexA-GlcNAc(6S)	4.1
ΔHexA(2S)-GlcNAc(6S)	4.0
ΔHexA-GlcNS	25.5
ΔHexA(2S)-GlcNS	9.5
ΔHexA-GlcNS(6S)	24.7
ΔHexA-GlcNH$_2$(6S)	0.0
ΔHexA(2S)-GlcNS(6S)	5.6
ΔHexA(2S)-GlcNH$_2$(6S)	0.0
CS/DS Disaccharides	**Disaccharides Produced (%) [1]**
ΔHexA-GalNAc	3.2
ΔHexA-GalNAc(4S)	33.5
ΔHexA-GalNAc(6S)	17.2
ΔHexA(2S)-GalNAc(4S)	0.0
ΔHexA(2S)-GalNAc(6S)	0.7
ΔHexA-GalNAc(4S)(6S)	45.4
ΔHexA(2S)-GalNAc(4S)(6S)	0.0

Abbreviations: ΔHexA, 4,5 unsaturated uronic acid; GlcNAc, *N*-acetylglucosamine; GlcNS, *N*-sulphated glucosamine; GlcNH$_2$, glucosamine; GalNAc, *N*-acetylgalactosamine; 2S, 2-*O*-sulphate; 4S, 4-*O*-sulphate; 6S, 6-*O*-sulphate.

2.6. Ion-Exchange Chromatography

The complexity of the cockle polysaccharide mixtures limited our ability to identity particular structural features responsible for their antiproliferative effects. Clearly, the cockle polysaccharides contain both HS and CS/DS-like chains and probably other non-GAG chains. Consequently, ion-exchange chromatography was performed to evaluate further the structure/activity features of the components contained within the unfractionated cockle polysaccharides preparations. The elutant was monitored at 280 nM to detect the peptide fragments still attached to the polysaccharide chains following the initial protease digestion. A typical elution profile is shown in Figure 5, with six peaks generally occurring. The absorbance measured under each peak was broadly comparable to the dry weights of polysaccharides obtained after desalting and lyophilisation.

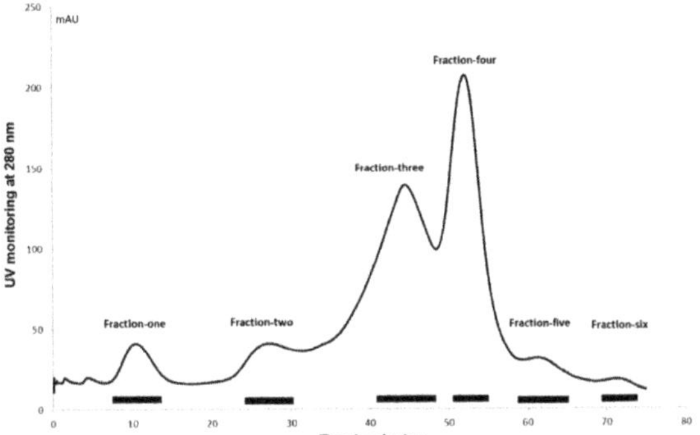

Figure 5. Ion-exchange chromatography of cockle polysaccharides. Cockle polysaccharides were applied to a DEAE-Sepharose column and eluted using a 0–1.5 M NaCl gradient over 70 min. Peaks were pooled as indicated by the bars shown, lyophilised and Fractions 1–6 stored at $-20\,^{\circ}\text{C}$ for further analysis.

2.7. MTT Assay of Ion-Exchange Purified Cockle Polysaccharide Fractions

Antiproliferative activity of ion-exchange purified cockle polysaccharides against K562 and MOLT4 cells (Figure 6) showed that most of the antiproliferative activity was seen with Fraction 5, with negligable antiproliferative activity observed with Fractions 1–3. The elution profile from the ion-exchange column suggests that the low levels of activity observed with Fractions 4 and 6 are probably due to cross contamination from Fraction 5. The results indicate that the antiproliferative cockle glycans are a relatively minor component of the unfractionated mixtures (approximately 1–2% of the total dry mass applied to the column).

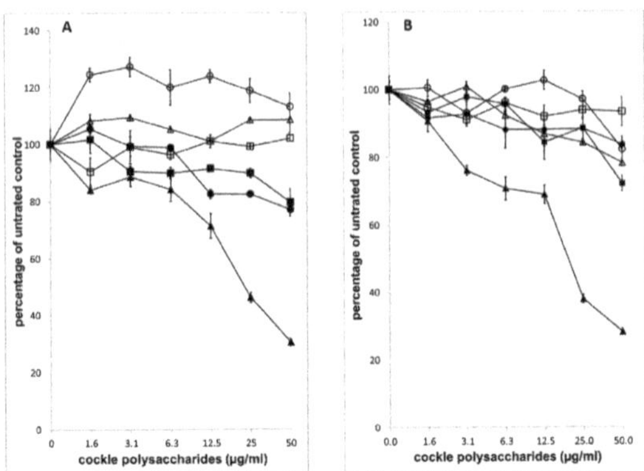

Figure 6. Antiproliferative properties of the ion-exchange purified cockle polysaccharide fractions. Measurement of antiproliferative activity of ion-exchanged purified fractions on K562 (**A**) and MOLT4 (**B**) cells was achieved by MTT assay. Fraction 1 (○), Fraction 2 (□), Fraction 3 (Δ), Fraction 4, (■), Fraction 5 (▲) and Fraction 6 (●). The data are presented as the percentage of viable cells following treatment with cockle polysaccharides, relative to untreated control. All experiments were conducted in triplicate and the results are shown as the mean $\pm$ the SD. Cells were cultured in suspension and maintained at 37 $^{\circ}$C in humidified 5% CO_2 atmosphere.

2.8. Disaccharide Analysis of Ion-Exchange Purified Polysaccharides

Disaccharide analysis of the ion-exchange purified fractions was used to identify further the nature of the different glycan chain types within the cockle polysaccharide preparations and to confirm the role of HS-like-GAG structures in the antiproliferative activity. Table 2 shows the relative distribution of unsaturated disaccharides and the amounts produced when each fraction was treated exhaustively with a combination of heparinases I, II and III. The total weight of disaccharides produced from each microgram of sample digested showed that only Fraction 5 produced a significant amount of unsaturated disaccharides. Fractions 2, 3, 4 and 6 produced approximately 30-, 150-, 30-, 6- and 5-fold less disaccharide than Fraction 5, respectively, suggesting that the early eluting fraction contain mainly non HS-like chains. The composition of Fractions 5 and 6 showed the presence of all the major disaccharides found in mammalian HS/heparin. The enrichment of the trisulphated disaccharide ΔHexA(2S)-GlcNS(6S) and the disulphated disaccharides ΔHexA(2S)-GlcNS with an approximate four-fold and three-fold increase, respectively, in comparison to the unfractionated cockle polysaccharides (Table 1), likely contributes significantly to the later elution of these chains from the ion-exchange column. Fraction 1 appears to generate a higher proportion of sulphated disaccharides than any other fraction; however, the quantity, in μg, of disaccharides produced is very small in comparison to the total material digested. In fact, there is little correlation between the increase in the number of sulphates per disaccharide produced from heparinase digestion and the elution position of Fractions 1–4. This is likely linked to the limited digestion of the glycan chains contained within them.

Table 2. HS Disaccharide analysis of ion-exchange purified cockle polysaccharide fractions. Data are presented as a percentage of the moles of unsaturated disaccharides produced by heparinase I, II, III digestion of the ion-exchange fractions (F1–F6).

HS Disaccharides	F1 (%)	F2 (%)	F3 (%)	F4 (%)	F5 (%)	F6 (%)
ΔHexA-GlcNAc	4.6	28.0	84.4	31.9	9.8	6.8
ΔHexA(2S)-GlcNAc	0.0	0.0	0.0	0.0	0.9	0.4
ΔHexA(2S)-GlcNH$_2$	0.0	0.0	0.0	0.1	0.2	0.0
ΔHexA-GlcNAc(6S)	0.2	2.4	1.6	5.3	4.9	3.3
ΔHexA(2S)-GlcNAc(6S)	30.4	10.9	0.0	0.0	3.1	4.6
ΔHexA-GlcNS	0.7	8.6	12.9	28.4	10.4	5.9
ΔHexA(2S)-GlcNS	16.6	25.2	0.5	12.8	24.4	34.0
ΔHexA-GlcNS(6S)	4.9	10.9	0.6	15.9	19.4	16.4
ΔHexA-GlcNH$_2$(6S)	0.0	0.0	0.0	0.1	0.2	0.0
ΔHexA(2S)-GlcNS(6S)	42.7	14.0	0.0	5.7	26.3	28.6
ΔHexA(2S)-GlcNH$_2$(6S)	0.0	0.0	0.0	0.0	0.3	0.1
Unsulphated	4.6	28.0	84.4	31.9	9.8	6.8
N-SO$_3$	64.8	58.8	14.0	62.6	80.5	84.8
2-O-SO$_3$	89.6	50.2	0.6	18.5	55.3	67.6
6-O-SO$_3$	78.1	38.2	2.2	27.0	54.2	53.0
Total GAG disaccharides produced by heparinase digestion (μg)	0.01	0.002	0.01	0.05	0.29	0.06
Average sulphate per disaccharide	2.33	1.47	0.17	1.08	1.9	2.05

Abbreviations: ΔHexA, 4,5 unsaturated uronic acid; GlcNAc, *N*-acetylglucosamine; GlcNS, *N*-sulphated glucosamine; GlcNH$_2$, glucosamine; GalNAc, 2S, 2-O-sulphate; 6S, 6-O-sulphate

Disaccharide analysis of each fraction, following chondroitinase ABC digestion (Table 3), also showed early eluting fractions lacked correlation between the sulphate content of the resultant disaccharides and the fractions elution position. Again, this is likely linked to the incomplete nature of the digests, as confirmed by the quantity, in μg, of disaccharides released by digestion. The later eluting fractions produced disaccharides compositions close to those seen for the unfractionated cockle polysaccharides (Table 1). The later eluting Fractions 5 and 6 contain CS-like-GAG chains, however Figure 3 suggests that these chains do not contribute to the antiproliferative activity of Fraction 5.

Table 3. CS/DS Disaccharide analysis of ion-exchange purified cockle polysaccharide fractions. Data are presented as a percentage of the moles of unsaturated disaccharides produced by chondroitinase ABC digestion of the ion-exchange fractions (F1–F6).

CS/DS Disaccharides	F1 (%)	F2 (%)	F3 (%)	F4 (%)	F5 (%)	F6 (%)
ΔHexA-GalNAc	70.6	72.4	86.4	8.8	3.5	3.6
ΔHexA-GalNAc(4S)	23.1	26.7	13.6	89.9	56.8	35.0
ΔHexA-GalNAc(6S)	3.8	0.2	0.0	1.4	3.7	4.3
ΔHexA(2S)-GalNAc(4S)	0.0	0.0	0.0	0.0	0.0	0.0
ΔHexA(2S)-GalNAc(6S)	0.0	0.0	0.0	0.0	1.8	3.9
ΔHexA-GalNAc(4S)(6S)	2.5	0.8	0.0	0.0	34.2	53.2
ΔHexA(2S)-GalNAc(4S)(6S)	0.0	0.0	0.0	0.0	0.0	0.0
Unsulfated	70.6	72.4	86.4	8.8	3.5	3.6
2-O-SO$_3$	0.0	0.0	0.0	0.0	1.8	3.9
4-O-SO$_3$	25.6	27.5	13.6	89.9	91.0	88.2
6-O-SO$_3$	6.3	1.0	0.0	1.4	39.7	61.5
Total GAG disaccharides produced by ABC lyase digestion (µg)	0.002	0.006	0.006	0.026	0.285	0.086
Average sulfate per disaccharide	0.32	0.28	0.14	0.91	1.31	1.51

Abbreviations: ΔHexA, 4,5 unsaturated uronic acid; GalNAc, *N*-acetylgalactosamine; 2S, 2-O-sulphate; 4S, 4-O-sulphate; 6S, 6-O-sulphate

Taken together, the data imply that the majority of heparinases and chondrotinase ABC susceptible linkages, and therefore GAG-like structures, are present in the later eluting fractions from the ion-exchange separation. The production of a significant quantity of HS disaccharides, from the highly active Fraction 5, confirms the link between HS-GAG-like structures and the antiproliferative activity of the cockle polysaccharides. However, the residual activity of the heparinase resistant chains must also be taken into account when judging any structure/activity relationships.

2.9. Monosaccharide Analysis of Ion-Exchange Purified Polysaccharides

The lack of information regarding the identity of the uronic acid moieties, using the method of disaccharide analysis employed in this study, means that we cannot identity potentially important structural features within the GAG-like chains. We are also missing structural detail on the content of non-GAG-like components of the cockle polysaccharide preparations. These issues can be clarified by total acid hydrolysis and monosaccharide analysis of both the unfractionated cockle polysaccharides and the ion-exchanged purified fractions. Analysis of the unfractionated cockle polysaccharide preparations (Table 4) showed the presence of significant quantities of the monosaccharides, GalNH$_2$, GlcNH$_2$, and GlcA, which are typically found following acid hydrolysis of mammalian and marine GAG chains. Interestingly, there was no detectable IdoA component. Fucose made up about 10% of the monosaccharides produced and could be derived from fucoidan-like or fucosylated CS glycans. The neutral sugars glucose (Glc) and galactose (Gal) were by far the most predominant monosaccharaides present in the unfractionated cockle polysaccharide preparations and represent components typically found in marine *N*- and *O*-linked glycans. Mannose represented less than 4% of the monosaccharides released by hydrolysis, and was predominantly found in Fractions 2 and 6. The data support the presence of GAG-like chains, as was indicated by both the susceptibility of cockle polysaccharides to heparinase/chondroitinase ABC digestion, and the production of a range of unsaturated disaccharides, typically found in mammalian GAG chains. Analysis of ion-exchange purified fractions showed enrichment of the GAG related monosaccharides GalNH$_2$, GlcNH$_2$, and GlcA in the later eluting fractions (Fractions 3–6) and critically the antiproliferative Fraction 5, which also contained the highest amount of GlcA.

Table 4. HPAEC-PAD analysis of monosaccharides derived from unfractionated (crude) cockle polysaccharide and ion-exchange purified fractions (F1–F6). Samples (50 µg) were degraded to monosaccharides by treatment with TFA prior to high performance anion exchange chromatography with pulsed amperometric detection (HPAEC-PAD) analysis. The peaks observed were identified by comparison to the elution position of known monosaccharide standards. Data are presented as a percentage of the moles of monosaccharide produced by acid hydrolysis.

Monosaccharide	Crude (%)	F1 (%)	F2 (%)	F3 (%)	F4 (%)	F5 (%)	F6 (%)
Fuc	11.1	5.1	5.8	3.3	3.6	14.2	2.1
GalNH$_2$	16.7	3	4.0	11.7	35.6	21.8	9.5
GlcNH$_2$	9.9	3.9	2.1	11.9	10.4	14.5	3.6
Gal	19.1	2.3	3.0	11.0	14.4	20.1	8.8
Glc	35.2	73.5	57.4	52.9	30.0	10.4	50.4
Man	3.7	5.0	28.1	7.4	2.7	4.9	23.5
GlcA	4.3	1.0	0.7	1.9	3.3	14.1	2.1
IdoA	0.0	0.0	0.0	0.0	0.0	0.0	0.0

Abbreviations: Fuc, fucose; GalNH$_2$, galactosamine; GlcNH$_2$, glucosamine; Gal, galactose; Glc, glucose; Man, mannose; GlcA, glucuronic acid; IdoA, iduronic acid :

Overall, the data indicate the presence of a diverse range of glycan chains, including GAG-like structures, within the unfractionated cockle polysaccharide preparations. The components found in the active Fraction 5 are clearly GAG-like in nature. The lack of IdoA is an interesting observation, yet this does not, on its own, explain the unique activity of the cockle derived HS-like GAG chains. The complexity of the cockle polysaccharide preparations, even within the ion-exchange purified fractions, and the limitations of the compositional analysis tools used in this study make it difficult to determine the exact structural differences between the antiproliferative cockle HS-like GAG chains and their inactive mammalian counterparts. We also need to consider the nature of the antiproliferative activity that remains following extensive heparinase digestion.

3. Discussion

We have shown for the first time that marine derived polysaccharides, with proven anticancer activity, are susceptible to the effects of heparinase digestion. The results, in conjunction with the lack of antiproliferative activity of mammalian GAGs, suggest that atypical GAG structures are present in marine polysaccharides from mollusc sources. These unique structures can be directly linked, at least in part, to the antiproliferative activity seen on human cancer cell lines treated with cockle polysaccharides.

Anti-metastatic and antiproliferative activity of marine derived polysaccharides has previously been observed from a variety of sources, such as algae, ascidians, euechinoidea, molluscs and bacteria [20]. The active molecules identified to date are classes of sulphated polysaccharides, such as sulphated fucans and sulphated galactans, which are defined as non-GAG glycans. They differ from typical mammalian GAG family members in several ways, including their relative homogeneity, in terms of their monosaccharide compositions and distribution. Their most dramatic structural differences lie in the type of monosaccharide building blocks that make up these chains. Sulphated fucans are polymers of fucopyranso units, which may contain variable amounts of neutral monosaccharides and uronic acids and are predominantly classed as fucoidans [9,21], whereas sulphated galactans are almost exclusively composed of Gal units [22]. None of the glycosidic linkages within these marine polysaccharides has any reported susceptibility to the enzymes used in this study. Hence, the reduction in antiproliferative activity we observed following heparinase digestion of the cockle polysaccharides suggests that the anticancer activity of the intact chains is linked to the presence of archetypal mammalian GAG disaccharide structures. Clearly, this is divergent with the literature that to date has exclusively linked anticancer activity of marine polysaccharides to non-GAG glycans.

Some heterogeneity has been reported for some of the sulphated fucans [23], and galactans [24]. However, none of the variations was connected to the presence of mammalian like GAG disaccharides.

We have not yet considered previously discovered marine polysaccharides that are closely related to traditional mammalian GAG structures, such as fucosylated CS, dermatan sulphate-like polysaccharides, GlcA containing HS and highly sulphated hybrid heparin/HS chains. Despite their widespread presence, no antiproliferative activity, on cancer cell lines, has previously been attributed to these polysaccharides classes. Fucosylated CS from *holothuroidea* has been reported to be insensitive to chondroitinase ABC digestion, without prior removal of fucose residues [25] and highly sulphated dermatan-like polysaccharides from tunicates have been shown to be sensitive to chondroitinase ABC digestion [26,27]. In our study, unsaturated disaccharides were produced by chondroitinase ABC digestion of cockle polysaccharides, although this had no effect on the antiproliferative activity. We have not confirmed the presence of fucosylated CS or dermatan like polysaccharides in the active cockle polysaccharide fractions. As such, we cannot rule out a contribution by these types of structure to the observed antiproliferative effects. However, the absence of IdoA in the monosaccharide analysis suggest that dermatan-like structures are unlikely to be present in the cockle polysaccharide preparations.

The heparin/HS type of polysaccharide chains identified from marine sources include a report of a unique hybrid heparin/HS polysaccharide found in the head of shrimp. These marine heparin/HS hybrid GAGs were extensively degraded by heparinase enzymes [28]. The GlcA containing HS-like GAGs derived from the mollusc *Nodipecten nodosus* represents another heparinase sensitive structure that may be linked to the antiproliferative activity of the cockle polysaccharides [29]. Although no antiproliferative activity has been reported, to date, with either of these types of glycan chains, the cockle polysaccharides did loose significant antiproliferative activity, following heparinase degradation (Figure 3). This observation might be linked to the presence of similar hybrid heparin/HS or GlcA-rich HS structures in the cockle polysaccharide preparations. However, appreciable antiproliferative activity did remain, even after extensive heparinase treatment of the unfractionated cockle polysaccharides. Suggesting that other atypical GAG structures are potentially involved in the antiproliferative activity. A reductionist assessment of the structural features within these heparinase resistant sequences represents an attractive approach that may well lead to a more thorough understanding of precise structural features required to facilitate that anticancer activity seen in this study.

Disaccharide analysis was employed to identify compositional differences between the building blocks of known marine/mammalian GAGs and the antiproliferative chains within cockle-derived polysaccharides. Our data showed that heparinase I, II, III treatment of cockle polysaccharides liberated all of the most abundant unsaturated disaccharide types typically observed in mammalian GAGs. The rare 3-*O*-sulphated disaccharides were not detected in any of the cockle polysaccharide samples, using the methods described here. The most significant feature of the analysis appeared to be related to the overall levels of *N*-sulphated disaccharides produced by heparinase digestion, which accounted for approximately two thirds of the disaccharide units obtained from the unfractionated cockle polysaccharide preparations. The antiproliferative ion-exchange purified, Fraction 5, yielded in access of 80% *N*-sulphated disaccharides, as opposed to the much lower levels, typically around 50%, produced by combined heparinase digestion of mammalian HS. For example, porcine mucosal HS contains approximately 40% *N*-sulphated disaccharides [30]. Mammalian heparins typically have higher *N*-sulphate levels, averaging around 85% of the disaccharide units [31]. The other notable differences are in the levels of the disaccharides ΔHexA(2S)-GlcNS and ΔHexA-GlcNS(6S) generated by combined heparinase digestion of some of the ion-exchange purified fractions, making up 24.4% and 19.4% of the disaccharides liberated from the active Fraction 5, respectively, as opposed to the percentage of these disaccharide in porcine mucosal heparin (0.9% and 14.6%, respectively), porcine mucosal HS (10.4% and 3.9%, respectively), bovine kidney (2.9% and 5.1%, respectively) and rat liver HS (14.5% and 6.2%, respectively) [30,31]. Despite Fraction 5 showing significant antiproliferative activity in comparison to the inactive porcine mucosal HS, it seems unlikely that the increased presence of these disaccharides is solely responsible for the differences in antiproliferative activity observed in this study. In fact, the analysis of the disaccharides liberated from the unfractionated cockle polysaccharides, and the active Fraction 5, by heparinase gave no definitive clues to the differences in biological activity seen between the marine sourced cockle polysaccharides and mammalian GAGs. Overall, the lack of antiproliferative activity of mammalian GAGs suggest that a radically different type of HS-like structure may be present in the cockle polysaccharides.

Analysis of the monosaccharide composition of the cockle polysaccharides, again failed to give a conclusive insight into the specific glycan structures that might be facilitating the antiproliferative activity of these glycan chains. The monosaccharide analysis did confirm that GAG-like structures where enriched in the unfractionated cockle polysaccharides and the active Fraction 5. However, this compositional detail and the complete absence of IdoA in the cockle polysaccharides does not yet represent sufficient evidence to conclude the exact nature of the antiproliferative glycan structures.

It might be suggested that the antiproliferative effects observed in this study, simply represent a non-specific structure/activity relationship linked to the very high sulphate content of the cockle polysaccharides and the active Fraction 5. To the authors knowledge no studies have demonstrated any direct antiproliferative effects of mammalian heparins on cancer cells, using in vitro MTT assays, despite their high sulphation levels. A report has been published, in which a clinical preparation of low molecular weight heparin showed slight inhibition of a human lung adenocarcinoma A549 cell line by MTT assay. However, 50% inhibition was never achieved at the concentrations used [32]. In addition, the low molecular weight heparins used failed to induce apoptosis, unlike the cockle polysaccharides (Figure 2). Intact high molecular weight porcine mucosal heparins, used in our study, failed to show any inhibition of cancer cell growth (results not shown). In addition, the activity profile for the ion-exchange purified fractions showed that the more heavily sulphated Fraction 6 had significantly less antiproliferative activity than the earlier eluting Fraction 5. However, both fractions produced HS/heparin like disaccharides following heparinase treatment and Fraction 6 had a higher overall sulphate content (Table 2). Taken together, these results suggest that the high levels of sulphation found in these cockle polysaccharides are not the principle reason for their antiproliferative activity. Hence, it is likely that specific sequences within the polysaccharides are mediating their biological activities in a distinct manner.

Marine polysaccharides continue to attract great interest as molecules with the potential for therapeutic development. In this study, we have identified antiproliferative and apoptotic activity in sulphated polysaccharides isolated from a marine mollusc. These novel GAG-like chains differ significantly from their mammalian counterparts, in terms of structure and anticancer activity, and have the potential to be developed into new a new class of cancer therapeutics.

4. Materials and Methods

4.1. Extraction of Sulphated Polysaccharides

Polysaccharides were extracted from common cockle (*Cerastoderma edule*), obtained from the Irish Sea, British Isles, using a standard protocol [19]. Shells were removed and the entire soft body tissue was defatted by incubation with acetone for 72 h. Defatted tissue was left to dry for 24–48 h then ground to a fine powder. The powder (4 g) was suspended in 40 mL of 0.05 M of sodium carbonate pH 9.2, and 2 mL of Alcalase enzyme (Merck, Millipore, Watford, UK) added. Samples were then incubated at 60 °C for 48 h with constant agitation at 200 rpm. The mixture was then cooled at 4 °C and trichloroacetic acid (Sigma-Aldrich, Gillingham, UK) added to a concentration of 5% (*w/v*) and left for 10 min. Precipitated peptides were removed by centrifugation (5000× *g* for 10 min). The supernatant was retained and three volumes of 5% (*w/v*) potassium acetate in ethanol was added to one volume of supernatant and the mixed solution left overnight at 4 °C. Precipitated cockle polysaccharide chains were recovered by centrifugation (5000× *g* for 30 min) and the pellet washed with absolute alcohol.

The recovered precipitate (1 g) was then dissolved in 40 mL of 0.2 M NaCl solution and centrifuged (5000× g for 30 min) to remove any insoluble material. Cetylpyridinium chloride (Sigma-Aldrich) (0.5 mL of a 5% (w/v) solution) was added to the supernatant and the precipitate formed, recovered by centrifugation (8000× g for 30 min). The precipitate was subsequently dissolved in 10 mL of 2.5 M NaCl solution, followed by the addition of 5 volumes of ethanol. The precipitated cockle polysaccharide chains were recovered by centrifugation (8000× g for 30 min) before being dialysed against water for 72 h. The dialysate was lyophilized to obtain a white powder containing approximately 2.0 mg of cockle polysaccharides.

4.2. Maintenance of Cell Lines

Human cancer cell lines K562 (Chronic myelogenous leukaemia) and MOLT-4 (acute T lymphoblastic leukaemia) were grown in RPMI-1640 medium containing 1 g/L glucose (Lonza Group Ltd., Basel, Switzerland), supplemented with 10% (v/v) inactivated FBS (Labtech International Ltd., Heathfield, UK), 2 mM L-glutamine (Labtech International Ltd.), 100 Units/mL penicillin and 100 ug/mL streptomycin (Labtech International Ltd.). All cell lines were maintained in 25 cm^2 flasks under a humidified atmosphere of 95% air and 5% CO_2 at 37 °C.

4.3. Cell Viability Assay

Cell viability was assessed using an MTT (3-(4,5 dimethylthiazol2-yl)-2,5 diphenyltetrazolium bromide) (Sigma-Aldrich) assay. Cells were seeded at a density of 5×10^4 cells/well and cultured overnight in 96-well plates containing 100 μL of culture medium prior to treatment with polysaccharide samples. Treatment was conducted for 96 h following addition of various concentrations of cockle polysaccharides and mammalian GAGs (Celsus, Cincinnati, OH, USA) (0–50 μg/mL), with cisplatin (0–25 mM) as a positive control. At the end of the incubation period, 50 μL MTT solution (5 mg/mL in PBS) was added to each well and incubated for 3 h at 37 °C. Next, 200 μL of DMSO was added to each well and the plates agitated to dissolve the formamazan crystal product. The amount of MTT converted to formazan is indicative of the number of viable cells. The plates were gently agitated until the colour reaction was uniform and the absorbance was measured at 570 nm using a multi-well plate reader. The cell viability effects from exposure of cells to each concentration of crude cockle polysaccharides, commercial mammalian GAGs and cisplatinum (Sigma-Aldrich) were analysed as percentages of the untreated control cell absorbance. The average cell survival obtained from triplicate determinations, at each concentration, was plotted as a dose–response curve. The IC_{50} values were calculated using nonlinear regression analysis (GraphPad Prism 5.0, La Jolla, CA, USA)

4.4. Cell Cycle and Annexin V Apoptosis Assay

Cells were seeded into sterile 6-well plates at 5×10^5/mL, and incubated with or without cockle polysaccharides at 37 °C. Cells were harvested at designated time intervals, centrifuged at 2000× g for 5 min, washed twice with cold PBS and resuspended in 1 × binding buffer (0.1 M HEPES (Sigma-Aldrich), 1.4 M NaCl and 25 mM $CaCl_2$) at a concentration of 1×10^6 cells/mL. Cells (100 μL) were then transferred into 5 mL culture tube and stained with Annexin V-FITC (5 μL) and propidium iodide (PI) (Sigma-Aldrich) (10 μL). The stained cells were gently vortex for few seconds and incubated in the dark for 15 min at room temperature. Binding buffer (400 μL) was added to each tube prior analysis by flow cytometer (BD FACSVerse, Franklin Lakes, NJ, USA). Prior to flow cytometry cell cycle analysis, cells (100 μL) were centrifuged, washed in PBS three times and incubated at room temperature with 50 μL of Ribonuclease A (RNase) (50 mg/mL) before staining with a solution containing 50 μg/mL propidium iodide (PI) for 30 min in the dark.

4.5. Enzymatic Digestion

Polysaccharide samples were digested by heparinase or chondroitinase ABC prior to MTT assay, as follows. Enzyme buffer comprised of 10 mM calcium acetate and 50 mM sodium acetate pH 7.0 containing 30 mIU of heparinase I, II, III (Grampian Enzymes, Aberdeen, UK) separately or in combination was added to solutions of 100 µg of cockle polysaccharides in enzyme buffer. Chondroitinase ABC (Grampian Enzymes, Aberdeen, UK) (30 mIU) in enzyme buffer was added to 100 µg cockle polysaccharides and incubated for 24 h at 37 °C. The enzyme digests were monitored spectrophotometrically at 232 nM and the reactions terminated by heating at 100 °C for 5 min.

4.6. Ion-Exchange Chromatography

Crude cockle polysaccharide preparations were fractionated by anion exchange chromatography, using an FPLC system (Pharmacia, Stockholm, Sweden). Samples were applied to an ion-exchange column (16 × 200 mm), packed with 10 mL of DEAE-Sepharose (GE Healthcare, Little shallot, UK). Cockle polysaccharides were eluted using a linear 0–1.5 M NaCl gradient in 50 mM sodium phosphate buffer pH 7.0 over 70 min at a flow rate of 1 mL/min. Absorbance was monitored at 280 nm, 1 mL fractions were collected. and pooled as indicated.

Pooled fractions, corresponding to the peaks in the elution profile were dialysed extensively against water using 14 kDa molecular weight cut-off tubing (Scientific Laboratory Supplies, Nottingham, UK). After dialysis, peaks were lyophilised, and stored at −20 °C.

4.7. HPAEC-PAD Monosaccharide Analysis

HPAEC-PAD was preformed to determine monosaccharide composition, using a Dionex ICS-3000 system (Sunnyvale, CA, USA) [33]. Crude cockle polysaccharides (50 µg) or pooled peaks from the ion-exchange chromatography separation, were dissolved in 200 µL of MilliQ water. The mixture was then hydrolysed using 1 mL of trifluoroacetic acid at 100 °C for 6 h to ensure complete hydrolysis of polysaccharides to monosaccharides. Monosaccharide samples were next centrifuged at 2000 rpm for 2 min. Acid was removed using a dry nitrogen flush after addition of 50 µL of 50% (*v/v*) aqueous isopropyl alcohol. Monosaccharide samples (20 µg) were applied to a Dionex CarboPac PA1 column, 4 mm × 250 mm, using a 4 µm and 4 mm × 50 guard column at a flow rate 1 mL/min. Monosaccharides were eluted using a gradient formed from three solvents: water as solvent A, 100 mM NaOH with 5 mM NaOAc as solvent B, and 100 mM NaOH with 250 mM NaOAc as solvent C. Peaks were detected using pulsed amperometric detector with standard quad waveform.

4.8. Disaccharide Analysis

The cockle polysaccharide samples were fractionated by ion-exchange chromatography prior to disaccharide analysis as described previously.

Crude cockle polysaccharide or ion-exchange fractionated polysaccharides were incubated with a mixture of 10 mIU each of heparinases I, II, and III in lyase buffer (100 mM sodium acetate pH 7.0, containing 0.1 mM calcium acetate). Samples were incubated overnight at 37 °C and the reaction terminated by heating to 100 °C for 5 min. The samples were passed through a 10 K MWCO spin filter then dried before further analysis. GC-MS and isotopic aniline tagging (GRIL-Glycan Reductive Isotope Labelling) was used for composition analysis and mass detection of disaccharide yield from heparinase treated cockle polysaccharides [34]. 15 µL of $^{12}C_6$ labelled Aniline and 15 µL of 1 M sodium cyanoborohydride, freshly prepared in DMSO:acetic acid (7:3, *v/v*), was added to 8 pmol of dried heparinase derived disaccharides. Reactions were carried out at 37 °C for 16 h, and then dried in a centrifugal evaporator. The dried samples were prepared for LCQ-MS analysis by resuspending in running buffer (8 mM acetic acid, 5 mM dibutylamine (DBA)) followed by centrifugation at 14,000× g for 7 min. The supernatant (5 µL) was spiked with an 8 pmol solution of unsaturated disaccharide standards tagged with $^{13}C_6$ aniline (2 µL) and the sample made up to 10 µL with running buffer. Aniline isotopic and non-isotopic disaccharides were separated on a C_{18} reversed-phase column (0.46 cm × 25 cm, Vydac). The solvent system used to elute the samples was 100% buffer A (8 mM acetic acid, 5 mM DBA) for 10 min, 17% buffer B (70%, methanol 8 mM acetic acid 5 mM DBA) for 15 min, 32% buffer B for 15 min, 40% buffer B for 15 min, 60% buffer B for 15 min, 100% buffer B for 10 min and 100% buffer A. Ions of interest were detected in negative ion mode and the capillary temperature and spray voltage were kept at 140 °C and 4.75 kV, UV detection was at 232 nm.

Acknowledgments: We would like to acknowledge Umm Al-Qura University and the ministry of higher education of Saudi Arabia for funding Abdullah Faisal Aldairi PhD studentship. The Authors would also like to thank the University of Salford for technical and administrative support.

Author Contributions: Experimental work was carried out by Abdullah Faisal Aldairi and Olanrewaju Dorcas Ogundipe and monosaccharide and disaccharide analysis was performed by the UC San Diego GlycoAnalytics Core. Experiments were conceived by David Alexander Pye, Abdullah Faisal Aldairi, and Olanrewaju Dorcas Ogundipe. The manuscript was written by David Alexander Pye.

Conflicts of Interest: The authors declare no conflict of interests.

References

1. Hileman, R.E.; Fromm, J.R.; Weiler, J.M.; Linhardt, R.J. Glycosaminoglycan-protein interactions: Definition of consensus sites in glycosaminoglycan binding proteins. *Bioessays* **1998**, *20*, 156–167. [CrossRef]
2. Volpi, N. Therapeutic applications of glycosaminoglycans. *Curr. Med. Chem.* **2006**, *13*, 1799–1810. [CrossRef] [PubMed]
3. Zhang, L. Glycosaminoglycan (gag) biosynthesis and gag-binding proteins. *Prog. Mol. Biol. Transl. Sci.* **2010**, *93*, 1–17. [PubMed]
4. Lindahl, U.; Kjellén, L. Pathophysiology of heparan sulphate: Many diseases, few drugs. *J. Intern. Med.* **2013**, *273*, 555–571. [CrossRef] [PubMed]
5. Gallagher, J.T.; Turnbull, J.E.; Lyon, M. Heparan sulphate proteoglycans: Molecular organisation of membrane—Associated species and an approach to polysaccharide sequence analysis. *Adv. Exp. Med. Biol.* **1992**, *313*, 49–57. [PubMed]
6. Meneghetti, M.C.; Hughes, A.J.; Rudd, T.R.; Nader, H.B.; Powell, A.K.; Yates, E.A.; Lima, M.A. Heparan sulfate and heparin interactions with proteins. *J. R. Soc. Interface* **2015**, *12*, 0589. [CrossRef] [PubMed]
7. Valcarcel, J.; Novoa-Carballal, R.; Pérez-Martín, R.I.; Reis, R.L.; Vázquez, J.A. Glycosaminoglycans from marine sources as therapeutic agents. *Biotechnol. Adv.* **2017**, *35*, 711–725. [CrossRef] [PubMed]
8. Laurienzo, P. Marine polysaccharides in pharmaceutical applications: An overview. *Mar. Drugs* **2010**, *8*, 2435–2465. [CrossRef] [PubMed]
9. Ale, M.T.; Meyer, A.S. Fucoidans from brown seaweeds: An update on structures, extraction techniques and use of enzymes as tools for structural elucidation. *RSC Adv.* **2013**, *3*, 8131–8141. [CrossRef]
10. Alekseyenko, T.V.; Zhanayeva, S.Y.; Venediktova, A.A.; Zvyagintseva, T.N.; Kuznetsova, T.A.; Besednova, N.N.; Korolenko, T.A. Antitumor and antimetastatic activity of fucoidan, a sulfated polysaccharide isolated from the okhotsk sea fucus evanescens brown alga. *Bull. Exp. Biol. Med.* **2007**, *143*, 730–732. [CrossRef] [PubMed]
11. Ale, M.T.; Maruyama, H.; Tamauchi, H.; Mikkelsen, J.D.; Meyer, A.S. Fucoidan from *Sargassum* sp. and *Fucus vesiculosus* reduces cell viability of lung carcinoma and melanoma cells in vitro and activates natural killer cells in mice in vivo. *Int. J. Biol. Macromol.* **2011**, *49*, 331–336. [CrossRef] [PubMed]
12. Han, Y.S.; Lee, J.H.; Lee, S.H. Antitumor effects of fucoidan on human colon cancer cells via activation of akt signaling. *Biomol. Ther.* **2015**, *23*, 225–232. [CrossRef] [PubMed]
13. Makarenkova, I.D.; Deriabin, P.G.; L'Vov, D.K.; Zviagintseva, T.N.; Besednova, N.N. [Antiviral activity of sulfated polysaccharide from the brown algae laminaria japonica against avian influenza a (h5n1) virus infection in the cultured cells]. *Vopr. Virusol.* **2010**, *55*, 41–45. [PubMed]
14. Preobrazhenskaya, M.E.; Berman, A.E.; Mikhailov, V.I.; Ushakova, N.A.; Mazurov, A.V.; Semenov, A.V.; Usov, A.I.; Nifant'ev, N.E.; Bovin, N.V. Fucoidan inhibits leukocyte recruitment in a model peritoneal inflammation in rat and blocks interaction of p-selectin with its carbohydrate ligand. *Biochem. Mol. Biol. Int.* **1997**, *43*, 443–451. [PubMed]
15. Zhu, Z.; Zhang, Q.; Chen, L.; Ren, S.; Xu, P.; Tang, Y.; Luo, D. Higher specificity of the activity of low molecular weight fucoidan for thrombin-induced platelet aggregation. *Thromb. Res.* **2010**, *125*, 419–426. [CrossRef] [PubMed]
16. Myron, P.; Siddiquee, S.; Al Azad, S. Fucosylated chondroitin sulfate diversity in sea cucumbers: A review. *Carbohydr. Polym.* **2014**, *112*, 173–178. [CrossRef] [PubMed]

17. Nandini, C.D.; Mikami, T.; Ohta, M.; Itoh, N.; Akiyama-Nambu, F.; Sugahara, K. Structural and functional characterization of oversulfated chondroitin sulfate/dermatan sulfate hybrid chains from the notochord of hagfish. Neuritogenic and binding activities for growth factors and neurotrophic factors. *J. Biol. Chem.* **2004**, *279*, 50799–50809. [CrossRef] [PubMed]

18. de Jesus Raposo, M.F.; de Morais, A.M.; de Morais, R.M. Marine polysaccharides from algae with potential biomedical applications. *Mar. Drugs* **2015**, *13*, 2967–3028. [CrossRef] [PubMed]

19. Kim, Y.S.; Jo, Y.Y.; Chang, I.M.; Toida, T.; Park, Y.; Linhardt, R.J. A new glycosaminoglycan from the giant african snail *Achatina fulica*. *J. Biol. Chem.* **1996**, *271*, 11750–11755. [CrossRef] [PubMed]

20. Fedorov, S.N.; Ermakova, S.P.; Zvyagintseva, T.N.; Stonik, V.A. Anticancer and cancer preventive properties of marine polysaccharides: Some results and prospects. *Mar. Drugs* **2013**, *11*, 4876–4901. [CrossRef] [PubMed]

21. Li, B.; Lu, F.; Wei, X.; Zhao, R. Fucoidan: Structure and bioactivity. *Molecules* **2008**, *13*, 1671–1695. [CrossRef] [PubMed]

22. Pomin, V.H. Structural and functional insights into sulfated galactans: A systematic review. *Glycoconj. J.* **2010**, *27*, 1–12. [CrossRef] [PubMed]

23. Berteau, O.; Mulloy, B. Sulfated fucans, fresh perspectives: Structures, functions, and biological properties of sulfated fucans and an overview of enzymes active toward this class of polysaccharide. *Glycobiology* **2003**, *13*, 29R–40R. [CrossRef] [PubMed]

24. Pereira, M.G.; Benevides, N.M.; Melo, M.R.; Valente, A.P.; Melo, F.R.; Mourão, P.A. Structure and anticoagulant activity of a sulfated galactan from the red alga, gelidium crinale. Is there a specific structural requirement for the anticoagulant action? *Carbohydr. Res.* **2005**, *340*, 2015–2023. [CrossRef] [PubMed]

25. Pomin, V.H. Holothurian fucosylated chondroitin sulfate. *Mar. Drugs* **2014**, *12*, 232–254. [CrossRef] [PubMed]

26. Kozlowski, E.O.; Lima, P.C.; Vicente, C.P.; Lotufo, T.; Bao, X.; Sugahara, K.; Pavão, M.S. Dermatan sulfate in tunicate phylogeny: Order-specific sulfation pattern and the effect of [→4idoa(2-sulfate)β-1→3galnac(4-sulfate)β-1→] motifs in dermatan sulfate on heparin cofactor ii activity. *BMC Biochem.* **2011**, *12*, 29. [CrossRef] [PubMed]

27. Pavão, M.S.; Aiello, K.R.; Werneck, C.C.; Silva, L.C.; Valente, A.P.; Mulloy, B.; Colwell, N.S.; Tollefsen, D.M.; Mourão, P.A. Highly sulfated dermatan sulfates from ascidians. Structure versus anticoagulant activity of these glycosaminoglycans. *J. Biol. Chem.* **1998**, *273*, 27848–27857. [CrossRef] [PubMed]

28. Brito, A.S.; Cavalcante, R.S.; Palhares, L.C.; Hughes, A.J.; Andrade, G.P.; Yates, E.A.; Nader, H.B.; Lima, M.A.; Chavante, S.F. A non-hemorrhagic hybrid heparin/heparan sulfate with anticoagulant potential. *Carbohydr. Polym.* **2014**, *99*, 372–378. [CrossRef] [PubMed]

29. Gomes, A.M.; Kozlowski, E.O.; Pomin, V.H.; de Barros, C.M.; Zaganeli, J.L.; Pavão, M.S. Unique extracellular matrix heparan sulfate from the bivalve *Nodipecten nodosus* (linnaeus, 1758) safely inhibits arterial thrombosis after photochemically induced endothelial lesion. *J. Biol. Chem.* **2010**, *285*, 7312–7323. [CrossRef] [PubMed]

30. Deakin, J.A.; Lyon, M. A simplified and sensitive fluorescent method for disaccharide analysis of both heparan sulfate and chondroitin/dermatan sulfates from biological samples. *Glycobiology* **2008**, *18*, 483–491. [CrossRef] [PubMed]

31. Saad, O.M.; Ebel, H.; Uchimura, K.; Rosen, S.D.; Bertozzi, C.R.; Leary, J.A. Compositional profiling of heparin/heparan sulfate using mass spectrometry: Assay for specificity of a novel extracellular human endosulfatase. *Glycobiology* **2005**, *15*, 818–826. [CrossRef] [PubMed]

32. Yu, C.J.; Ye, S.J.; Feng, Z.H.; Ou, W.J.; Zhou, X.K.; Li, L.D.; Mao, Y.Q.; Zhu, W.; Wei, Y.Q. Effect of fraxiparine, a type of low molecular weight heparin, on the invasion and metastasis of lung adenocarcinoma a549 cells. *Oncol. Lett.* **2010**, *1*, 755–760. [CrossRef] [PubMed]

33. Hardy, M.R.; Townsend, R.R.; Lee, Y.C. Monosaccharide analysis of glycoconjugates by anion exchange chromatography with pulsed amperometric detection. *Anal. Biochem.* **1988**, *170*, 54–62. [CrossRef]

34. Lawrence, R.; Olson, S.K.; Steele, R.E.; Wang, L.; Warrior, R.; Cummings, R.D.; Esko, J.D. Evolutionary differences in glycosaminoglycan fine structure detected by quantitative glycan reductive isotope labeling. *J. Biol. Chem.* **2008**, *283*, 33674–33684. [CrossRef] [PubMed]

 *marine drugs*

Article

A Novel Bromophenol Derivative BOS-102 Induces Cell Cycle Arrest and Apoptosis in Human A549 Lung Cancer Cells via ROS-Mediated PI3K/Akt and the MAPK Signaling Pathway

Chuan-Long Guo [1,2,3,†], Li-Jun Wang [1,2,†], Yue Zhao [1,2], Hua Liu [1,2,3], Xiang-Qian Li [1,2], Bo Jiang [1,2], Jiao Luo [1,2,3], Shu-Ju Guo [1,2], Ning Wu [1,2] and Da-Yong Shi [1,2,3,*]

[1] Key Laboratory of Experimental Marine Biology, Institute of Oceanology, Chinese Academy of Sciences, Qingdao 266071, China; gcl_cpu@126.com (C.-L.G.); wanglijun@qdio.ac.cn (L.-J.W.); zhaoyue19931104@163.com (Y.Z.); baihualin55100@163.com (H.L.); lnu101@163.com (X.-Q.L.); jiangbo@qdio.ac.cn (B.J.); luojiao2012@163.com (J.L.); guoshuju@qdio.ac.cn (S.-J.G); wuning@qdio.ac.cn (N.W.)
[2] Laboratory for Marine Drugs and Bioproducts, Qingdao National Laboratory for Marine Science and Technology, Qingdao 266071, China
[3] University of Chinese Academy of Sciences, Beijing 10049, China
* Correspondence: shidayong@qdio.ac.cn; Tel.: +86-532-8289-8719; Fax: +86-532-8289-8741
† These authors contributed equally to this work.

Received: 5 December 2017; Accepted: 23 January 2018; Published: 25 January 2018

Abstract: Bromophenol is a type of natural marine product. It has excellent biological activities, especially anticancer activities. In our study of searching for potent anticancer drugs, a novel bromophenol derivative containing indolin-2-one moiety, 3-(4-(3-([1,4′-bipiperidin]-1′-yl)propoxy)-3-bromo-5-methoxybenzylidene)-N-(4-bromophenyl)-2- oxoindoline-5-sulfonamide (**BOS-102**) was synthesized, which showed excellent anticancer activities on human lung cancer cell lines. A study of the mechanisms indicated that **BOS-102** could significantly block cell proliferation in human A549 lung cancer cells and effectively induce G0/G1 cell cycle arrest via targeting cyclin D1 and cyclin-dependent kinase 4 (CDK4). **BOS-102** could also induce apoptosis, including activating caspase-3 and poly (ADP-ribose) polymerase (PARP), increasing the Bax/Bcl-2 ratio, enhancing reactive oxygen species (ROS) generation, decreasing mitochondrial membrane potential (MMP, $\Delta\Psi_m$), and leading cytochrome c release from mitochondria. Further research revealed that **BOS-102** deactivated the PI3K/Akt pathway and activated the mitogen-activated protein kinase (MAPK) signaling pathway resulting in apoptosis and cell cycle arrest, which indicated that **BOS-102** has the potential to develop into an anticancer drug.

Keywords: bromophenol; molecular mechanisms; apoptosis; cell cycle; PI3K/Akt; p38/ERK; ROS; human lung cancer

---

1. Introduction

In China, cancer morbidity and mortality are increasing year by year. It has become the leading cause of death and seriously affected the quality of life of people with cancer [1]. Lung cancer is the cancer with the highest mortality rate in the world [1–3]. Among lung cancers, approximately 85% of cases are non-small cell lung cancer (NSCLC). Thus, it is highly desirable to develop safe and effective drugs to treat NSCLC and to improve the quality of life of patients with lung cancer.

Apoptosis can occur by one of two representative pathways: the mitochondrial pathway and the death receptor-mediated pathway [4]. In the mitochondrial pathway, ROS plays an important role [5]. Generated ROS directly activates the mitochondrial permeability transition and results in mitochondrial membrane potential (MMP, $\Delta\Psi_m$) loss, causing the release of cytochrome c. Then the released cytochrome c activates caspases, and the activated caspases eventually lead to apoptosis [6–8]. In addition, ROS could also induce other various signaling pathways, such as the PI3K/Akt signaling pathway, which plays an important role in cell proliferation and survival [9]. When cells are subject to external stimuli, such as anticancer drug stimulation, the PI3K/Akt signaling pathway may be down-regulated, and finally induce cell apoptosis. The PI3K/Akt pathway also plays an important role in cell cycle regulation; the mechanism of this pathway is the regulation of CDK4 and cyclin D1, which could block cell cycle in G1 phase [10].

ROS is a second messenger in cells, and plays an important role in cell proliferation or apoptosis [11]. It has been demonstrated that ROS generation in cells could activate mitogen-activated protein kinase (MAPK) signaling pathways including p38 MAPK and ERK1/2 [12]. MAPK pathways are key mediators of eukaryotic transcription and they control gene expression via phosphorylation [13].

Bromophenols, an important kind of natural marine product, extracted from a variety of marine organisms, exhibit excellent biological activities, such as antitumor and antibacterial activities [14–16]. However, the natural bromophenol in the marine organisms is in low content, and its separation is difficult. Additionally, there are more or less deficiencies in its efficacy, stability, biological toxicity, and bioavailability. These are extremely limited for its further research and development [15].

Synthesis of new bromophenol compounds, or modifications to the structure of natural bromophenols, can effectively improve the biological activity of bromophenol compounds. Based on this foundation, a series of bromophenol compounds with potent anticancer activities were designed and synthesized in our previous works [14]. These compounds showed anticancer activity on human cancer cell lines, such as the human lung cancer cell line, human hepatoma cell lines, the human cervical cancer cell line, and the human colon cancer cell line. Among of them, a novel bromophenol derivative containing indolin-2-one moiety, 3-(4-(3-([1,4′-bipiperidin]-1′-yl)propoxy)-3-bromo-5-methoxybenzylidene)-*N*-(4-bromophenyl)-2-oxoindoline-5-sulfonamide (**BOS-102, Figure** 1) showed excellent anticancer activities on human lung cancer cell lines [17]. In order to study whether compound **BOS-102** has the potential to develop into an anticancer drug, more experiments regarding the molecular mechanisms, including cell apoptosis analysis, cell cycle analysis, ROS generation analysis, mitochondrial membrane potential analysis, and potential signaling pathways (PI3K/Akt and MAPK) analysis in A549 lung cancer cells were explored in this study.

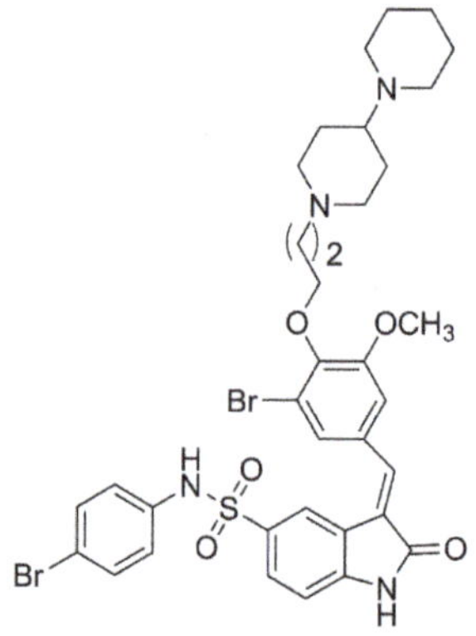

Figure 1. The structure of **BOS-102**.

2. Results

2.1. *BOS-102 Inhibits Cell Proliferation*

We examined the effect of **BOS-102** on cell viability using MTT assay. The results suggested that **BOS-102** could induce cytotoxicity of several tumor cell lines, including human lung cancer cell line A549, human hepatoma cell line HepG2, human primary glioblastoma cell line U87 MG, and human pancreatic cell line PANC-1. **BOS-102** could also induce cytotoxicity on normal cells, such as human umbilical vein endothelial cells (HUVECs). As shown in Figure 2A, **BOS-102** could inhibit cell proliferation of cancer cell lines, including A549 (IC_{50} = 4.29 ± 0.79 μM), HepG2 (IC_{50} = 13.87 ± 1.40 μM), U87 MG (IC_{50} = 23.98 ± 8.80 μM), PANC-1 (IC_{50} = 12.48 ± 1.66 μM), and HUVECs (15.43 ± 1.07 μM). In our study, A549 cells were much more sensitive. Thus, A549 cells were used in the subsequent experiments.

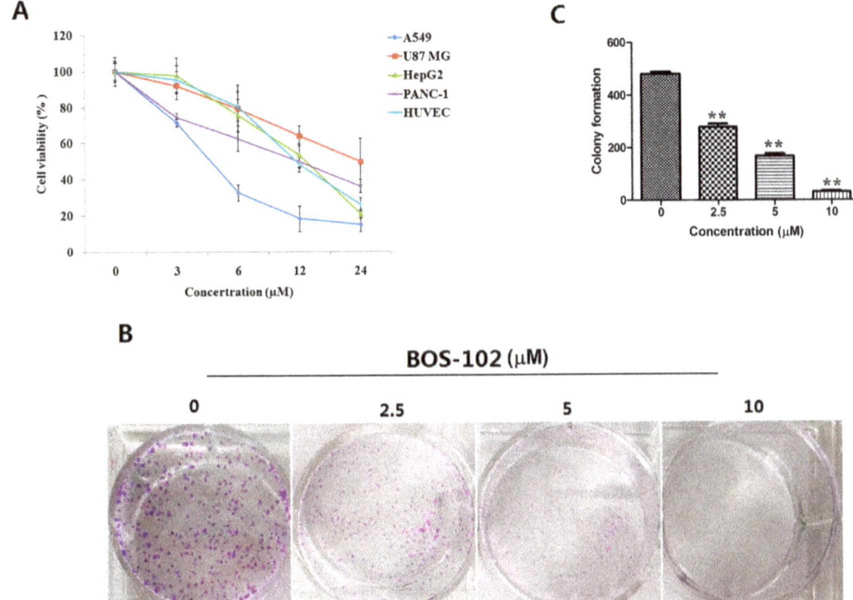

Figure 2. **BOS-102** inhibits the viability, migration, and colony formation in human cancer cell lines. (**A**) The inhibitory effect of **BOS-102** on the cell proliferation of A549, HepG2, U87 MG, and PANC-1 cells. Cells were treated with various concentrations of **BOS-102** (0, 3, 6, 12, 24 μM) for 48 h. After incubation, cell viability was evaluated by MTT assay; (**B,C**) A549 cells were treated with **BOS-102** (0, 2.5, 5, 10 μM) for 10 day and colony formation was determined by staining with crystal violet. The data represent mean values (±SD) obtained from three separate experiments. ** $p < 0.01$ vs. control group.

2.2. *BOS-102 Inhibits Colony Formation*

A colony formation assay was performed to study the effect of **BOS-102** on cell colony formation. A549 cells were seeded in six-well plates at a density of 500 cells per well. After 24 h, cells were treated with **BOS-102** (0, 2.5, 5, 10 μM), and incubated for 10 days. In our study, the results showed that **BOS-102** can significantly inhibit the colony formation of A549 cells (Figure 2B,C).

2.3. *BOS-102 Induces A549 Apoptosis*

To evaluate effect of **BOS-102** on the induction of apoptosis, A549 cells were treated with **BOS-102** (0, 2.5, 5, 10 μM) for 48 h. After stained with Annexin V/PI, cells were analyzed by flow cytometry. As shown in Figure 3A,B, **BOS-102** induced apoptosis in A549 cells in a concentration-dependent manner. Compared with treatment of **BOS-102** at 2.5 μM, the percentage of apoptotic cells was increased from $16.2 \pm 2.5\%$ to $79.2 \pm 4.5\%$ after treatment with **BOS-102** at 10 μM (Figure 3A,B). Moreover, Z-VAD-FMK (the pan-caspase inhibitor) was used in our study. The results showed that Z-VAD-FMK could inhibit **BOS-102**-induced apoptosis (Figure 3D) and **BOS-102**-induced cytotoxicity in A549 cells (Figure 3E).

Apoptosis often causes cell morphological changes, such as nuclear apoptotic bodies [18]. It is interesting to investigate the effect of **BOS-102** apoptosis induction by Hoechst 33258 staining in the A549 cell line. A549 cells were treated with **BOS-102** (0, 2.5, 5, 10 μM) for 48 h. As shown in Figure 3C, after staining with Hoechst 33258, cell nuclear condensation, chromosome condensation, and apoptotic bodies were observed in **BOS-102**-treated cells.

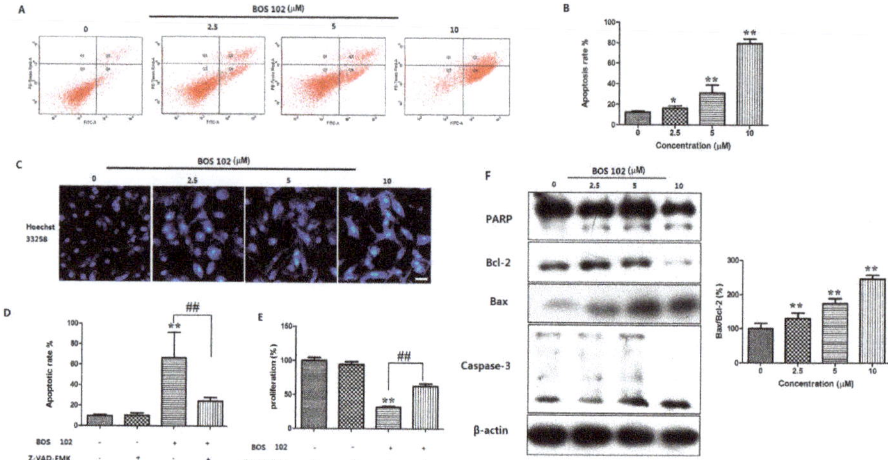

Figure 3. **BOS-102** induces intrinsic apoptosis in A549 cells. (**A,B**) FACS analysis via Annexin V/PI staining was used to identify apoptosis induced by **BOS-102**. A549 cells were treated with various concentrations of **BOS-102** (0, 2.5, 5, 10 μM) for 48 h; (**C**) A549 cells were treated with **BOS-102** (0, 2.5, 5, 10 μM) for 48 h. Hoechst 33258 staining was used to detected the apoptosis and photographed using fluorescence microscopy (Bar = 50 μm); (**D**) A549 cells were treated with 5 μM **BOS-102** alone or in combination with Z-VAD-FMK (10 μM) for 48 h. The percentages of apoptotic cells were determined by flow cytometr (FACS) analysis via Annexin V/PI staining; (**E**) A549 cells were treated with 5 μM **BOS-102** alone or in combination with Z-VAD-FMK (10 μM) for 48 h, cell viability was evaluated by MTT assay; and (**F**) Western blot analysis of apoptosis-related proteins, including PARP, Bcl-2, Bax, and Caspase-3. β-actin was used to normalize the protein content. The data represent mean values ($\pm$SD) obtained from three separate experiments. * $p < 0.05$, ** $p < 0.01$ vs. control group, [##] $p < 0.01$ vs. 102(+)/Z-VAD-FMK(−) group.

2.4. *Effect of BOS-102 on the Expression of Apoptosis-Related Proteins*

When apoptosis occurred, the expression of apoptosis related proteins, such as Bax, Bcl-2, caspase-3, and PARP may change. Western blot was used to detect the expression of these proteins. After treatment with **BOS-102** for 48 h, the expression of Bax was increased while the Bcl-2 was decreased (Figure 3F). Furthermore, caspase-3 and PARP were also activated after **BOS-102** treatment

(Figure 3F). Our results indicated that **BOS-102** induced apoptosis on A549 cells probably through the mitochondrial-mediated apoptotic pathway.

2.5. *BOS-102 Induces G0/G1 Cell Cycle Arrest and Down-Regulates Cyclin D1 and CDK4 in A549 Cells*

To investigate the effects of **BOS-102** on cell cycle distribution, A549 cells were treated with **BOS-102** (0, 2.5, 5, 10 μM) for 48 h and analyzed by flow cytometry. The results showed that the G0/G1 phase was increased in a dose-dependent manner after **BOS-102** treatment. (Figure 4A,B). Treatment with **BOS-12** for 48h caused a remarkable dose-dependent accumulation of cells in G0/G1 phase; from 46.06% (0 μM) to 74.37% (10 μM), these findings denoted that **BOS-102** could induce G0/G1 cell cycle arrest.

To explore the mechanism for **BOS-102**-induced cell cycle arrest, Western blot analysis was used to examine the effect of **BOS-102** treatment on the expression levels of cell cycle-related proteins, such as cyclin D1 and CDK4. In our study, **BOS-102** treatment caused a significant decrease in the protein levels of cyclin D1 and CDK4 (Figure 4C).

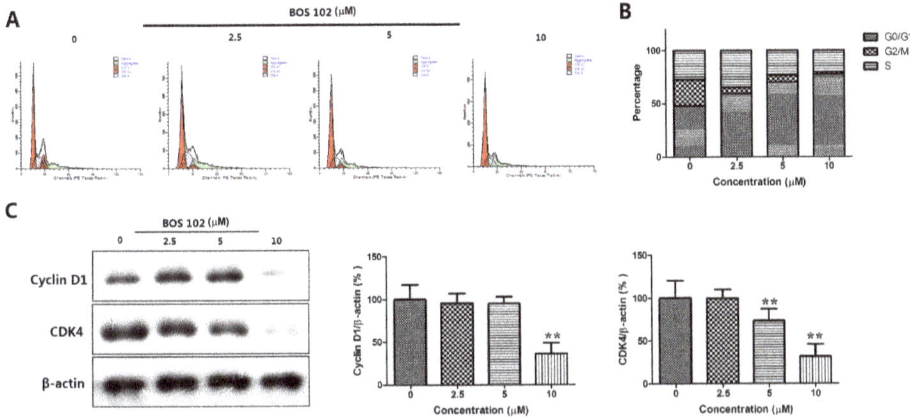

Figure 4. **BOS-102** induces G0/G1 cell cycle arrest. (**A,B**) Cell cycle distribution was monitored by FACS. A549 cells were treated with various concentrations of **BOS-102** (0, 2.5, 5, 10 μM) for 48 h. Cells were harvested and fixed in 70% ethanol overnight, then cells were stained with PI and analysis by FACS; and (**C**) Western blot analysis of cell cycle-related proteins, including Cyclin D1 and CDK4. β-actin was used to normalize protein content. The data represent mean values (±SD) obtained from three separate experiments. ** $p < 0.01$ vs. control group.

2.6. *BOS-102 Induces Mitochondrial Dysfunctionin in A549 Cells*

The maintenance of MMP ($\Delta\Psi_m$) is significant for mitochondrial integrity and bioenergetic function [19]. The decline of MMP is a marker of apoptosis. We used JC-1 to detect the decline of MMP of A549 cells after **BOS-102** treatment. In this study, the results indicated that **BOS-102** significantly reduced the MMP in A549 cells compared with the control group (Figure 5E,F). Additionally, we detected the expression of cytochrome c in mitochondria and cytoplasm using Western blot analysis. The results indicated that **BOS-102** decreased cytochrome c in mitochondria, while increased it in cytoplasm (Figure 5G). The results indicated that **BOS-102** induced apoptosis in A549 cells through the mitochondrial apoptotic pathway.

2.7. *BOS-102 Induces ROS Generation in A549 Cells*

ROS generation is considered as one of the key mediators of apoptotic signaling. We used DCFH-DA to detect ROS in A549 cells. In our study, a rapid production of ROS was occurred after the treatment of **BOS-102** (0, 2.5, 5, 10 μM) for 48 h. The result showed that DCFH-DA fluorescence intensity was increased in a dose-dependent manner. Compared with control group, the content of ROS was increased to 185.84 ± 24.91%, 211.50 ± 7.69%, and 233.36 ± 18.52% (Figure 5A–C). Next, A549 cells were treated with 5 μM **BOS-102** combined with/without 5 mM *N*-acetyl cysteine (NAC) for 48 h, respectively. Interestingly, we observed that the ROS inhibitor NAC blocked ROS generation in A549 cells (Figure 5D). We also observed that NAC blocked **BOS-102**-induced apoptosis in A549 cells (data not shown). These results indicated that ROS generation is important in **BOS-102**-induced apoptosis.

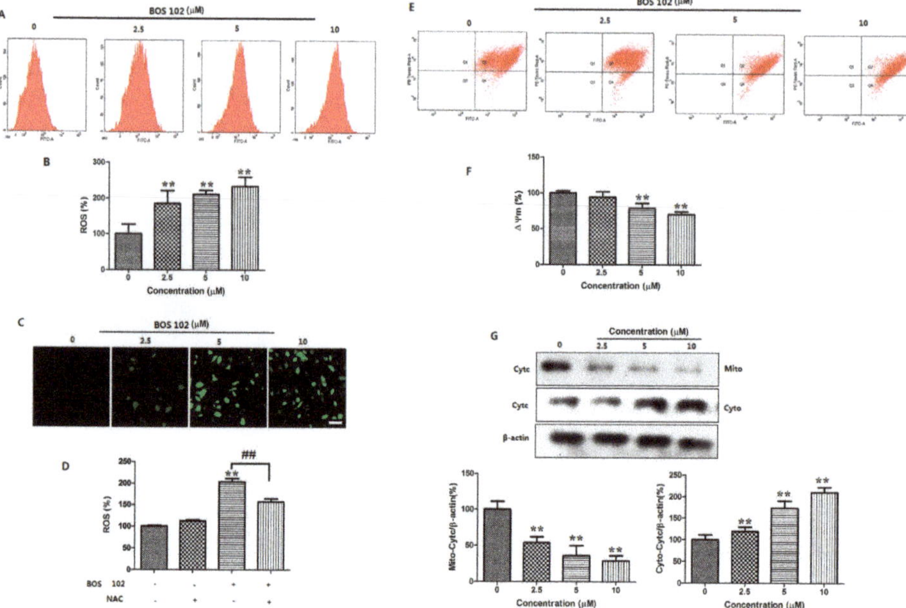

Figure 5. **BOS-102** induces ROS generation and loss of MMP in A549 cells. (**A–C**) A549 cells were treated with various concentrations of **BOS-102** (0, 2.5, 5, 10 μM) for 48 h, the medium was discarded, and cells were incubated at 37 °C in the dark for 20 min with culture medium containing DCFDA. Cells were harvested and analysis using FACS and fluorescence microscopy (Bar = 50 μm); (**D**) A549 cells were treated with 5 μM **BOS-102** alone or in combination with NAC (5 mM) for 48 h. ROS generation was detected by FACS after staining with DCFDA; (**E,F**) A549 cells were treated with various concentrations of **BOS-102** (0, 2.5, 5, 10 μM) for 48 h. The cells were collected and incubated with JC-1 for 20 min at 37 °C, cells were then washed twice with PBS and the values of MMP were analyzed by FACS; (**G**) Immunoblotting of mito-cytochrome c and cyto-cytochrome c. The data represent mean values (±SD) obtained from three separate experiments. ** $p < 0.01$ vs. control group. ## $p < 0.01$ vs. 102(+)/NAC(−) group.

2.8. *BOS-102 Suppresses the PI3K/Akt Signaling Pathway*

When apoptosis occurs, the PI3K/Akt signaling pathway plays an important role, which is also an important modulator in ROS-related cell apoptosis. In our study, we used Western blot analysis to detect the expression levels of PI3K/Akt after treatment of **BOS-102**. As shown in Figure 6A, the phosphorylation of Akt and PI3K were decreased in a dose-dependent manner after **BOS-102** treatment. Furthermore, pretreatment of antioxidant NAC, a ROS inhibitor, could efficiently reverse the **BOS-102**-induced Akt phosphorylation (Figure 6B). These results indicated that **BOS-102** induced cell apoptosis by inhibiting the PI3K/Akt signaling pathway.

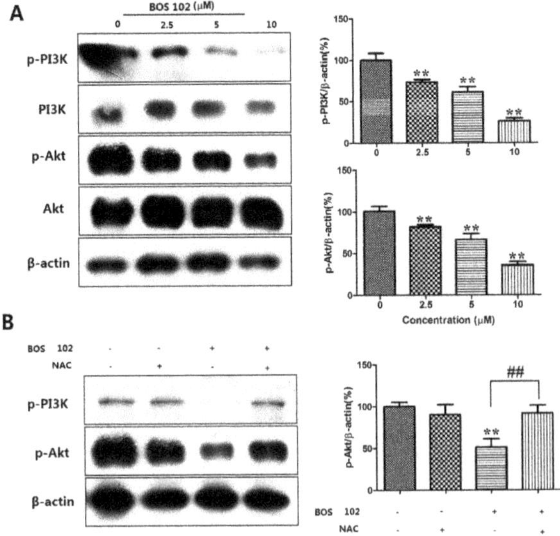

Figure 6. Effects of **BOS-102** on PI3K/Akt pathway. (**A**) A549 cells were treated with various concentrations of **BOS-102** (0, 2.5, 5, 10 μM) for 48 h and then the expressions and phosphor of PI3K and Akt were assessed by western blot analysis; and (**B**) the effect of antioxidant NAC on **BOS-102**-induced Akt activation. All data were representative of three independent experiments. ** $p < 0.01$ vs. control group. ## $p < 0.01$ vs. 102(+)/NAC(−) group.

2.9. *BOS-102 Activates the P38/ERK Signaling Pathway*

It is well known that the MAPK signaling pathway plays a critical role in regulating cell apoptosis. The effect of **BOS-102** on this pathway was detected using Western blot. As shown in Figure 7, at various concentrations after application of **BOS-102**, the phosphorylation of ERK and p38 were significantly increased. These results indicated that the ERK/p38 signaling pathway was involved in **BOS-102**-induced apoptosis.

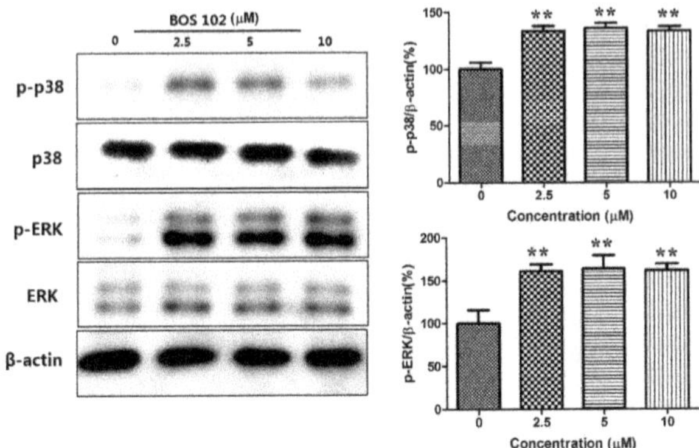

Figure 7. Effects of **BOS-102** on the MAPK pathway. A549 cells were treated with various concentrations of **BOS-102** (0, 2.5, 5, 10 μM) for 48 h and then the expressions and phosphor of p38 and ERK were assessed by Western blot analysis. All data were representative of three independent experiments. ** $p < 0.01$ vs. control group.

3. Discussion

Non-small cell lung cancer is the most common form of lung cancer globally. Currently, surgery and radiotherapy are available therapeutic approaches for lung cancer patients, but both of them cause severe pain and side-effects [2]. It is highly desirable to develop safe and effective drugs to treat lung cancer and to reduce the pain of patients with lung cancer. In the current study, we reported that **BOS-102** could inhibit the proliferation of several cancer cell lines, including human lung cancer cell line A549 (IC_{50}: 4.29 ± 0.79 µM), human hepatoma cell line HepG2 (IC_{50}: 13.87 ± 1.40 µM), human primary glioblastoma cell line U87 MG (IC_{50}: 23.98 ± 8.80 µM) and human pancreatic cell line PANC-1 (IC_{50}: 12.48 ± 1.66 µM). In addition, **BOS-102** could inhibit the proliferation of normal cells such as human umbilical vein endothelial cells (HUVECs). In the following experiment, we demonstrated that **BOS-102** significantly inhibited colony formation in A549 cells (Figure 2B,C), which was largely due to the effect of **BOS-102** induced G0/G1 cell cycle arrest and apoptosis. Furthermore, we investigated the molecular mechanisms of anti-cancer effect of **BOS-102**, which was likely mediated through various signaling pathways.

As we all know, there are two main methods for cell growth inhibition: cell cycle arrest and apoptosis. In mammalian cells, cell-cycle progression is controlled by a series of cyclin-dependent kinase (CDK)–cyclin complexes. CDK family members, such as CDK2, CDK4, and CDK6, play a key role in the cell-cycle control of tumor cells [20]. Our results showed that the proportion in the G1 phase was significantly increased from 46.06% to 74.37% after treatment of **BOS-102** for 48 h (Figure 4A,B), denoting that **BOS-102** induced G0/G1 arrest in A549 cells. Furthermore, the mechanisms of **BOS-102**-induced cell cycle arrest were investigated using Western blot analysis. Our results showed that the expression of cyclin D1 and CDK4 was decreased after treatment of **BOS-102** in A549 cells (Figure 4C). As we all know, activation of cyclin D1 is important when cells were transiting through G1 into the S phase, then the activated cyclin D1 binds to CDK4 and, finally, induces cells from G1 to the S phase [21,22]. In our study, the expression of cyclin D1 and CDK4 were decreased after treatment of **BOS-102,** indicating that **BOS-102** could exert its anti-proliferative effect via inducing G0/G1 cell cycle arrest, and the molecular mechanisms may be through the modulation of cyclin D1 and CDK4.

Apoptosis is a normal physiological phenomenon that can be observed in various tissues and cells. When apoptosis happens, various morphological changes will appear, such as cell surface changes, nuclear pyknosis, DNA fragmentation, and chromosome condensation [23]. Morphological changes in A549 cells were analyzed by Hoechst 33258 staining (Figure 3C). The phosphatidylserine externalization (cell surface changes) in A549 cells was detected by flow cytometry after staining with Annexin V-FITC/PI, and our results showed that **BOS-102** induced A549 apoptosis in a concentration-dependent manner (Figure 3A,B). In order to detect whether caspase was involved in **BOS-102**-induced apoptosis, we used the pan-caspase inhibitor, Z-VAD-FMK, which can inhibit caspase processing and apoptosis induction in tumor cells in vitro. In our study, Z-VAD-FMK inhibited **BOS-102**-induced A549 apoptosis and cell hypoproliferation (Figure 3D). These experiments indicated that **BOS-102** could induce apoptosis in A549 cells via a caspase-dependent pathway.

It is well known that apoptosis can be activated by various stressors, among which ROS is the major cause of toxicity. It has been reported that ROS acts as a second messenger in the signal process [7,24]. ROS can induce apoptosis via a variety of mechanisms. For instance, increased ROS can active the intrinsic pathway by stimulating the depolarization of MMP [25]. The depolarization of MMP can activate the intrinsic apoptosis pathway and be tightly regulated bythe Bcl-2 family, such as Bcl-2 (anti-apoptotic protein) and Bax (pro-apoptotic protein) [26]. In the present study, an increase of ROS generation was observed in the **BOS-102**-treated A549 cells (Figure 5A–C). Furthermore, the ROS generation ability of **BOS-102** could be inhibited by NAC, a specific inhibitor of ROS (Figure 5D). Our results also clearly indicated that MMP was collapsed after treatment of **BOS-102** (Figure 5E,F). As we all know, the loss of MMP leads to cytochrome c release, and the released cytochrome c can activate caspase-3 and PARP. In this study, after **BOS-102** treatment, the expression of Bax was increased while Bcl-2 was decreased (Figure 3E). Additionally, cytochrome c was released to cytoplasm (Figure 5G), resulting in caspase and PARP activation (Figure 3F) and cell apoptosis (Figure 3A,B). In general, our data demonstrated that **BOS-102** could induce A549 cells apoptosis via the ROS-mediated mitochondria pathway.

The PI3K/Akt pathway plays an important role in regulating cell survival and death [27]. Therefore, the suppression of the PI3K/Akt signaling pathway may be an effective approach to the treatment of human lung cancer. Phosphorylated Akt can promote the apoptosis-related protein Bcl-2 and cell cycle-related protein cyclin D. Our study found that **BOS-102** treatment could induce decreased phosphorylation of PI3K and Akt (Figure 6A). In our previous study, our results suggested that ROS generation played an important role in **BOS-102**-induced apoptosis, whether the PI3K/Akt pathway was also connected with ROS generation. In the following study, pretreatment of NAC could efficiently reverse the **BOS-102**-induced PI3K and Akt phosphorylation (Figure 6B), suggesting the PI3K/Akt pathway might also be implicated in **BOS-102**-induced apoptosis, and this process was via excessive generation of ROS.

Mitogen-activated protein kinase (MAPK) signaling pathways also play an important role in the development of cancer. An increase in intracellular ROS can activate MAPK signaling pathways, including p38 MAPK and ERK1/2 [12]. It has been demonstrated that p38 and ERK play a vital role in cell survival and apoptosis [13]. On the one hand, some literatures reported that ERK activation was beneficial to cell proliferation and survival [28]. On the other hand, there is also some literature reported that ERK activation could induce cell apoptosis [4,13,29]. In our study, we found that the phosphorylation of p38 and ERK were increased after treatment of **BOS-102** (Figure 7), indicating that the MAPK signaling pathway was also related to **BOS-102**-induced apoptosis in A549 cells.

4. Materials and Methods

4.1. Reagents

BOS-102 was synthesized by our lab according to our previous publish [17], with a purity of 98%. The purity of compound **BOS-102** was measured by HPLC (Shimadzu, Kyoto, Japan), carried out on a Shimadzu LC-20A system (Shimadzu, Kyoto, Japan) equipped with a Shimadzu InertSustain C-18 reverse phase column (4.6 mm × 250 mm × 5 μm, Shimadzu, Kyoto, Japan) and SPD-20A detector (Shimadzu, Kyoto, Japan). In this study, **BOS-102** was dissolved in dimethyl sulfoxide (DMSO) and shored in −20 °C for less than one month before use. The vehicle (DMSO) was used as a control group, and the concentration of DMSO used in the experiments was less than 0.1%. The Dulbecco's modified Eagle's medium (DMEM) and Eagle's Minimum Essential Medium (MEM) were obtained from Hyclone (Logan, UT, USA). Fetal bovine serum (FBS) was obtained from ExCell Bio (Shanghai, China). Medium 200 and Low Serum Growth Supplement (LSGS) were obtained from Thermo Fisher Scientific (Waltham, MA, USA). The ROS assay kit, JC-1 assay kit, apoptosis assay kit, and Hoechst 33258 staining kit and *N*-acetyl cysteine (NAC), the pan-caspase inhibitor (Z-VAD-FMK) were obtained from Beyotime (Nanjing, China). Antibodies against cyclin D1, CDK4, Caspase-3, Bcl-2, Bax, PARP, phosphorylation-Akt, Akt, phosphorylation-p38, p38, phosphorylation-ERK1/2, ERK1/2, cytochrome c, and β-actin were purchased from Cell Signaling Technology (Beverly, MA, USA). Phosphorylation-PI3K and PI3K were purchased from Abcam (Cambridge, UK). The anti-mouse IgG and anti-rabbit secondary antibodies raised from goat were obtained from Abcam (Cambridge, UK).

4.2. Cell Culture

Human lung cancer cell line A549 cells and human pancreatic cell line PANC-1 cells were cultured in Dulbecco's Modified Eagle's medium (DMEM) supplemented with 10% FBS, 100 U/mL penicillin, and 100 U/mL streptomycin. Human hepatoma cell line HepG2 cells and human primary glioblastoma cell line U87 MG cells were cultured in Eagle's Minimum Essential Medium (MEM) supplemented with 10% FBS, 100 U/mL penicillin and 100 μg/mL streptomycin. Human umbilical vein endothelial cells (HUVECs) were cultured in Medium 200-supplemented LSGS. Cells were cultured at 37 °C in humidified CO_2 (5%).

4.3. In Vitro Cytotoxicity Test

The cytotoxicity of **BOS-102** was tested with standard 3-(4,5-dimethyl-2-thiazolyl)-2,5-diphenyl tetrazolium bromide (MTT) testing. Briefly, cells (A549, HepG2, U87 MG, PANC-1, and HUVECs) were seeded in 96-well plates (3×10^3 cells/well for A549, U87 MG, and PANC-1 cells, 5×10^3 cells/well for HepG2 and HUVEC cells) and allowed to settle 24 h, then cells were treated with varying concentrations of **BOS-102** (0, 3, 6, 12, 24 µM). After 48 h, MTT (5 mg/mL) was added in the plates, and incubated at 37 °C for 4 h. After removing the supernatant, the transformed crystals were dissolved in 150 µL DMSO and measured using a microplate reader (BioTek, Winooski, VT, USA) at 490 nm. Experiments were performed in triplicate on six wells for each measurement.

4.4. Colony Forming Assay

A549 cells were seeded in 6-well plates at a density of 500 cells per well. After 24 h, cell were treated with **BOS-102** (0, 2.5, 5, 10 µM). Then cells were incubated for 10 days to allow for colony formation. After staining with crystal violet, colonies containing more than 50 cells were counted and evaluated [30].

4.5. Assays for Apoptosis

Cell apoptosis analysis was detected using Annexin V/PI staining assay. Briefly, cells (1×10^5 cells/well) were seeded in a six-well plate and incubated for 24 h. Then, cells were treated with **BOS-102** (0, 2.5, 5, 10 µM) for 48 h. Cells were harvested, washed with cold PBS, and stained with Annexin V-FITC (Annexin V-Fluorescein isothiocyanate) and PI (propidine iodide). Then cells were analyzed in three different experiments using flow cytometry (Becton Dickinson, Franklin Lakes, NJ, USA).

4.6. Flow Cytometric Analysis of Cell Cycle

A549 cells were seeded into 6-well plates at a density of 5×10^5 cells per well and treated with **BOS-102** (0, 2.5, 5, 10 µM) for 48 h. Cells were harvested, fixed in cold 75% ethanol at −20 °C overnight. Cells were washed with PBS, re-suspended with cold PBS containing 20 µg/mL RNaseA and 50 µg/mL PI. After incubation at room temperature in the dark for 30 min. Cells were analyzed in three different experiments using flow cytometry (Becton Dickinson, Franklin Lakes, NJ, USA).

4.7. Morphological Analysis of Apoptosis

Cell morphology of apoptosis cells was evaluated by Hoechst 33258 staining. A549 cells were treated with **BOS-102** (0, 2.5, 5, 10 µM) for 48 h. Then cells were washed with PBS and stained with Hoechst 33258. After washing twice with PBS, cells were visualized on a fluorescence microscope (Olympus, Tokyo, Japan).

4.8. ROS Determination

ROS production was measured after staining with 2′,7′-dichlorohydrofluorescein (DCFH-DA). Briefly, A549 cells (1×10^6) were seed in six-well plates and treated with **BOS-102** (0, 2.5, 5, 10 µM). After 48 h incubation, cells were stained with 10 µM DCFH-DA for 20 min at 37 °C in the dark. Then cells were washed with PBS and harvested. Fluorescence was detected on a flow cytometer (Becton Dickinson, Franklin Lakes, NJ, USA) and a fluorescence microscope (Olympus, Tokyo, Japan).

4.9. Analysis of the MMP

MMP was assessed using flow cytometry after staining with JC-1. Briefly, A549 cells (1×10^6) were seed in six-well plates and treated with **BOS-102** (0, 2.5, 5, 10 µM) for 48 h. Cells were harvested and incubated with JC-1 at 37 °C for 20 min. After two washes with PBS, cells were analyzed using flow cytometry (Becton Dickinson, Franklin Lakes, NJ, USA).

4.10. Western Blot Analysis

A549 cells were treated with various concentrations of **BOS-102** (0, 2.5, 5, 10 µM) for 48 h. Cells were lysed in RIPA lysis buffer on ice for 15 min. The protein concentrations of the cell lysates were determined with a BCA protein kit (Beyotime, Nanjing, China). Proteins were mixed with loading buffer containing 5% 2-mercaptoethanol and then heated for 5 min at 95 °C. Protein lysates were subsequently loaded into each lane of a 12% SDS–PAGE and transferred onto PVDF membranes. Membranes were blocked in 5% (*w/v*) non-fat milk for 1 h. Then, the membranes were probed with a primary antibody at 4 °C overnight, and incubated with HRP-conjugated secondary antibody. Finally, proteins were visualized using a BeyoECL Plus enhanced chemiluminescence system (Beyotime, Nanjing, China). The protein level was normalized to β-actin. Protein bands were quantified with Image J (Image 2×, NIH, Bethesda, MD, USA).

4.11. Statistical Analysis

Statistical analysis was performed using GraphPad Prism 5.0 (San Diego, CA, USA). The data were presented as mean ± SD. Statistical comparisons were performed using one-way analysis of variance. Value of $p < 0.05$ was considered statistically significant.

5. Conclusions

In summary, our study extensively evaluated the anti-proliferative effect of **BOS-102** in human lung cancer cells, which was mediated through the arresting cell cycle in the G0/G1 phase and inducing cell apoptosis. Furthermore, our results provided evidence that **BOS-102** induced apoptosis in A549 cells through ROS-mediated inhibition of PI3K/Akt and activation of p38/ERK signaling pathways (Figure 8). Overall, our study indicated that the novel bromophenol derivative **BOS-102** has the potential to develop into an anticancer drug.

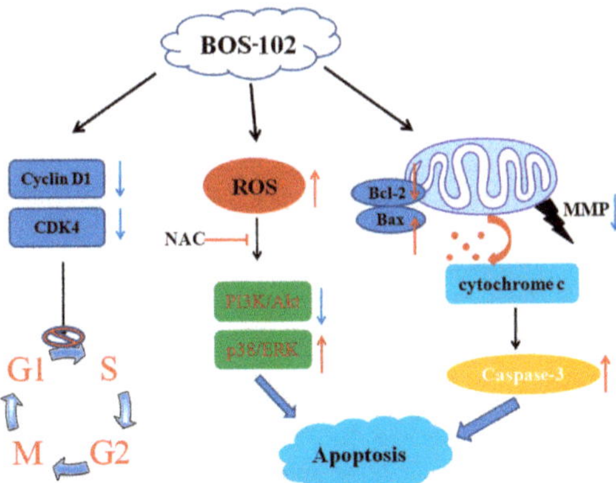

Figure 8. The proposed molecular mechanisms of cell cycle arrest and apoptosis induced by **BOS-102** in A549 cells.

Acknowledgments: This work was supported by the National Natural Science Foundation of China (Nos. 81773586 and 81703354); the Key Research and Development Project of Shandong province (2016GSF201193, 2016ZDJS07A13, 2016GSF115002, and 2016GSF115009); the Key Research Program of Frontier Sciences, CAS (QYZDB-SSW-DQC014); the Project of Discovery, Evaluation, and Transformation of Active Natural Compounds, Strategic Biological Resources Service Network Programme of Chinese Academy of Sciences (ZSTH-026), and the NSFC-Shandong Joint Fund for Marine Science Rearch Centers (U1606403); the Scientific and Technological Innovation Project financially supported by the Qingdao National Laboratory for Marine Science and Technology (No. 2015ASKJ02); Aoshan Talents Program Supported by Qingdao National Laboratory for Marine Science and Technology (No. 2015ASTP); and the National Program for the Support of Top-notch Young Professionals; and the Taishan scholar Youth Project of Shandong province.

Author Contributions: Chuan-Long Guo, Li-Jun Wang, and Da-Yong Shi contributed to the study concept and design, and the manuscript preparation. Chuan-Long Guo and Li-Jun Wang performed the experimental studies and analyzed the data. Yue Zhao, Hua Liu, Xiang-Qian Li, Bo Jiang, Jiao Luo, Shu-Ju Guo, and Ning Wu contributed in critical reading and discussion on the manuscript. All the authors approved the final version.

Conflicts of Interest: The authors declare no conflict of interest.

References

1. Chen, W.; Zheng, R.; Baade, P.D.; Zhang, S.; Zeng, H.; Bray, F.; Jemal, A.; Yu, X.Q.; He, J. Cancer statistics in China, 2015. *CA Cancer J. Clin.* **2016**, *66*, 115–132. [CrossRef] [PubMed]

2. Rami-Porta, R.; Asamura, H.; Travis, W.D.; Rusch, V.W. Lung cancer—Major changes in the American Joint Committee on Cancer eighth edition cancer staging manual. *CA Cancer J. Clin.* **2017**, *67*, 138–155. [CrossRef] [PubMed]

3. Siegel, R.L.; Miller, K.D.; Jemal, A. Cancer Statistics, 2017. *CA Cancer J. Clin.* **2017**, *67*, 7–30. [CrossRef] [PubMed]

4. Zhang, L.; Wang, H.; Xu, J.; Zhu, J.; Ding, K. Inhibition of cathepsin S induces autophagy and apoptosis in human glioblastoma cell lines through ROS-mediated PI3K/AKT/mTOR/p70S6K and JNK signaling pathways. *Toxicol. Lett.* **2014**, *228*, 248–259. [CrossRef] [PubMed]

5. Fleury, C.; Mignotte, B.; Vayssiere, J.L. Mitochondrial reactive oxygen species in cell death signaling. *Biochimie* **2002**, *84*, 131–141. [CrossRef]

6. Hainaut, P.; Plymoth, A. Targeting the hallmarks of cancer: Towards a rational approach to next-generation cancer therapy. *Curr. Opin. Oncol.* **2013**, *25*, 50–51. [CrossRef] [PubMed]

7. Kim, W.; Youn, H.; Kang, C.; Youn, B. Inflammation-induced radioresistance is mediated by ROS-dependent inactivation of protein phosphatase 1 in non-small cell lung cancer cells. *Apoptosis* **2015**, *20*, 1242–1252. [CrossRef] [PubMed]

8. Li, Z.; Qin, B.; Qi, X.; Mao, J.; Wu, D. Isoalantolactone induces apoptosis in human breast cancer cells via ROS-mediated mitochondrial pathway and downregulation of SIRT1. *Arch. Pharm. Res.* **2016**, *39*, 1441–1453. [CrossRef] [PubMed]

9. Zhao, Y.; Wang, X.; Sun, Y.; Zhou, Y.; Yin, Y.; Ding, Y.; Li, Z.; Guo, Q.; Lu, N. LYG-202 exerts antitumor effect on PI3K/Akt signaling pathway in human breast cancer cells. *Apoptosis* **2015**, *20*, 1253–1269. [CrossRef] [PubMed]

10. Romashkova, J.A.; Makarov, S.S. NF-kappa B is a target of AKT in anti-apoptotic PDGF signalling. *Nature* **1999**, *401*, 86–90. [CrossRef] [PubMed]

11. Chen, K.; Chu, B.Z.; Liu, F.; Li, B.; Gao, C.M.; Li, L.L.; Sun, Q.S.; Shen, Z.F.; Jiang, Y.Y. New benzimidazole acridine derivative induces human colon cancer cell apoptosis in vitro via the ROS-JNK signaling pathway. *Acta Pharmacol. Sin.* **2015**, *36*, 1074–1084. [CrossRef] [PubMed]

12. Sauer, H.; Wartenberg, M.; Hescheler, J. Reactive oxygen species as intracellular messengers during cell growth and differentiation. *Cell. Physiol. Biochem.* **2001**, *11*, 173–186. [CrossRef] [PubMed]

13. Ki, Y.W.; Park, J.H.; Lee, J.E.; Shin, I.C.; Koh, H.C. JNK and p38 MAPK regulate oxidative stress and the inflammatory response in chlorpyrifos-induced apoptosis. *Toxicol. Lett.* **2013**, *218*, 235–245. [CrossRef] [PubMed]

14. Wang, L.J.; Wang, S.Y.; Jiang, B.; Wu, N.; Li, X.Q.; Wang, B.C.; Luo, J.; Yang, M.; Jin, S.H.; Shi, D.Y. Design, synthesis and biological evaluation of novel bromophenol derivatives incorporating indolin-2-one moiety as potential anticancer agents. *Mar. Drugs* **2015**, *13*, 806–823. [CrossRef] [PubMed]

15. Liu, M.; Hansen, P.E.; Lin, X.K. Bromophenols in Marine Algae and Their Bioactivities. *Mar. Drugs* **2011**, *9*, 1273–1292. [CrossRef] [PubMed]

16. Oztaskin, N.; Cetinkaya, Y.; Taslimi, P.; Goksu, S.; Gulcin, I. Antioxidant and acetylcholinesterase inhibition properties of novel bromophenol derivatives. *Bioorg. Chem.* **2015**, *60*, 49–57. [CrossRef] [PubMed]

17. Wang, L.-J.; Guo, C.-L.; Li, X.-Q.; Wang, S.-Y.; Jiang, B.; Zhao, Y.; Luo, J.; Xu, K.; Liu, H.; Guo, S.-J.; et al. Discovery of Novel Bromophenol Hybrids as Potential Anticancer Agents through the Ros-Mediated Apoptotic Pathway: Design, Synthesis and Biological Evaluation. *Mar. Drugs* **2017**, *15*, 343. [CrossRef] [PubMed]

18. Wu, N.; Luo, J.; Jiang, B.; Wang, L.; Wang, S.; Wang, C.; Fu, C.; Li, J.; Shi, D. Marine bromophenol bis (2,3-dibromo-4,5-dihydroxy-phenyl)-methane inhibits the proliferation, migration, and invasion of hepatocellular carcinoma cells via modulating beta1-integrin/FAK signaling. *Mar. Drugs* **2015**, *13*, 1010–1025. [CrossRef] [PubMed]

19. Park, S.H.; Kim, J.H.; Chi, G.Y.; Kim, G.Y.; Chang, Y.C.; Moon, S.K.; Nam, S.W.; Kim, W.J.; Yoo, Y.H.; Choi, Y.H. Induction of apoptosis and autophagy by sodium selenite in A549 human lung carcinoma cells through generation of reactive oxygen species. *Toxicol. Lett.* **2012**, *212*, 252–261. [CrossRef] [PubMed]

20. Treite, F.; Kohler, L.; Mosch, B.; Pietzsch, J. Cell cycle regulating kinase Cdk4 as a potential target for tumour visualisation in vivo. *Eur. J. Nucl. Med. Mol. Imaging* **2008**, *35*, S325.

21. Tsai, J.H.; Hsu, L.S.; Huang, H.C.; Lin, C.L.; Pan, M.H.; Hong, H.M.; Chen, W.J. 1-(2-Hydroxy-5-methylphenyl)-3-phenyl-1,3-propanedione Induces G1 Cell Cycle Arrest and Autophagy in HeLa Cervical Cancer Cells. *Int. J. Mol. Sci.* **2016**, *17*, 1274. [CrossRef] [PubMed]

22. Yin, H.P.; Guo, C.L.; Wang, Y.; Liu, D.; Lv, Y.B.; Lv, F.X.; Lu, Z.X. Fengycin inhibits the growth of the human lung cancer cell line 95D through reactive oxygen species production and mitochondria-dependent apoptosis. *Anti-Cancer Drugs* **2013**, *24*, 587–598. [CrossRef] [PubMed]

23. Hanahan, D.; Weinberg, R.A. Hallmarks of Cancer: The Next Generation. *Cell* **2011**, *144*, 646–674. [CrossRef] [PubMed]

24. Li, P.; Zhao, Q.L.; Wu, L.H.; Jawaid, P.; Jiao, Y.F.; Kadowaki, M.; Kondo, T. Isofraxidin, a potent reactive oxygen species (ROS) scavenger, protects human leukemia cells from radiation-induced apoptosis via ROS/mitochondria pathway in p53-independent manner. *Apoptosis* **2014**, *19*, 1043–1053. [CrossRef] [PubMed]

25. Lee, J.H.; Won, Y.S.; Park, K.H.; Lee, M.K.; Tachibana, H.; Yamada, K.; Seo, K.I. Celastrol inhibits growth and induces apoptotic cell death in melanoma cells via the activation ROS-dependent mitochondrial pathway and the suppression of PI3K/AKT signaling. *Apoptosis* **2012**, *17*, 1275–1286. [CrossRef] [PubMed]

26. Zhang, Z.; Zheng, Y.; Zhu, R.; Zhu, Y.; Yao, W.; Liu, W.; Gao, X. The ERK/eIF4F/Bcl-XL pathway mediates SGP-2 induced osteosarcoma cells apoptosis in vitro and in vivo. *Cancer Lett.* **2014**, *352*, 203–213. [CrossRef] [PubMed]

27. Madhunapantula, S.V.; Mosca, P.J.; Robertson, G.P. The Akt signaling pathway: An emerging therapeutic target in malignant melanoma. *Cancer Biol. Ther.* **2011**, *12*, 1032–1049. [CrossRef] [PubMed]

28. Xia, Z.G.; Dickens, M.; Raingeaud, J.; Davis, R.J.; Greenberg, M.E. Opposing Effects of Erk and Jnk-P38 Map Kinases on Apoptosis. *Science* **1995**, *270*, 1326–1331. [CrossRef] [PubMed]

29. Lee, W.J.; Hsiao, M.; Chang, J.L.; Yang, S.F.; Tseng, T.H.; Cheng, C.W.; Chow, J.M.; Lin, K.H.; Lin, Y.W.; Liu, C.C.; et al. Quercetin induces mitochondrial-derived apoptosis via reactive oxygen species-mediated ERK activation in HL-60 leukemia cells and xenograft. *Arch. Toxicol.* **2015**, *89*, 1103–1117. [CrossRef] [PubMed]

30. Wen, Q.Y.; Wang, W.Y.; Luo, J.D.; Chu, S.Z.; Chen, L.J.; Xu, L.N.; Zang, H.J.; Alnemah, M.M.; Ma, J.; Fan, S.Q. CGP57380 enhances efficacy of RAD001 in non-small cell lung cancer through abrogating mTOR inhibition-induced phosphorylation of eIF4E and activating mitochondrial apoptotic pathway. *Oncotarget* **2016**, *7*, 27787–27801. [CrossRef] [PubMed]

Review

Sponges: A Reservoir of Genes Implicated in Human Cancer

Helena Ćetković [1,*,†], Mirna Halasz [1,†] and Maja Herak Bosnar [2,†]

[1] Laboratory for Molecular Genetics, Division of Molecular Biology, Ruđer Bošković Institute, 10000 Zagreb, Croatia; mimesek@irb.hr
[2] Laboratory for Protein Dynamics, Division of Molecular Medicine, Ruđer Bošković Institute, 10000 Zagreb, Croatia; mherak@irb.hr
*   Correspondence: cetkovic@irb.hr; Tel.: +385-(0)1-4561-115
†   These authors contributed equally to this work.

Received: 31 October 2017; Accepted: 4 January 2018; Published: 10 January 2018

Abstract: Recently, it was shown that the majority of genes linked to human diseases, such as cancer genes, evolved in two major evolutionary transitions—the emergence of unicellular organisms and the transition to multicellularity. Therefore, it has been widely accepted that the majority of disease-related genes has already been present in species distantly related to humans. An original way of studying human diseases relies on analyzing genes and proteins that cause a certain disease using model organisms that belong to the evolutionary level at which these genes have emerged. This kind of approach is supported by the simplicity of the genome/proteome, body plan, and physiology of such model organisms. It has been established for quite some time that sponges are an ideal model system for such studies, having a vast variety of genes known to be engaged in sophisticated processes and signalling pathways associated with higher animals. Sponges are considered to be the simplest multicellular animals and have changed little during evolution. Therefore, they provide an insight into the metazoan ancestor genome/proteome features. This review compiles current knowledge of cancer-related genes/proteins in marine sponges.

Keywords: porifera/sponge; evolution; cancer; cancer genes; molecular oncology

---

1. Introduction

While cancer evolution is still not fully understood, it is clear that the appearance of multicellularity is directly related to the origin of this widespread disease. Cancer can be described as a consequence of errors within the multicellular system causing the proliferation of "selfish" cell lines [1]. It is most likely that cancer appeared in parallel with multicellularity and the development of true tissues and organs, and that the genes responsible for cellular cooperation and multicellularity evolved together. When malfunctioning, these genes may cause cancer [2]. Intensive research in the field of molecular oncology over the recent decades expanded the initial division of cancer genes to oncogenes and tumor suppressor genes, by involving genes connected to angiogenesis, differentiation, adhesion, as well as to invasion processes, inflammation etc. Although little is known about cancer in invertebrates, especially non-bilaterians and unicellular relatives of animals, the available literature sources report on tumorous formations in some phyla, for example flatworms, molluscs, and insects [1]. Even some of the simplest "early branching" non-bilaterian animals, such as cnidarians, may develop tumors [3]. Comparative genomics studies have implied that a number of genes linked to human diseases were already present in species distantly related to humans [4] and similar has been suggested for cancer genes [5]. It seems that most of cancer-related protein domains have appeared during the two main evolutionary peaks, the emergence of unicellular eukaryotes and the transition to multicellular metazoans [5]. Since no systematic studies were thus far performed to reveal the occurrence of tumors in invertebrates, the current knowledge about cancer and status of cancer-related genes in simple animals is scarce. Studying cancer-related genes at the level at which they have first emerged, in a simpler model system, may provide an insight into their basic cellular functions.

The sponges (phylum Porifera) are the simplest and likely the most ancient group of animals that have separated from other metazoans more than 800 million years ago [6]. They are sessile as adults and lack true tissues and organs as well as any recognizable sensory or nerve structures. Despite their simple morphology, sponges genomes are complex [7] and many of their genes bear a striking similarity to vertebrate homologs [8]. Should we postulate that tumors had developed together with the emergence of tissues and organs, it would be likely for sponges to be unable to form tumors. Their genomes nevertheless, contain the majority of genes whose human homologs are implicated in cancer [7]. Research of cancer associated genes in simple animals with no tissues and blood vessels, such as sponges, could help in understanding the more complex signalling circuitry of their homologs in higher animals, such as vertebrates. Although comparative genomics studies have shown that many genes linked to cancer diseases in humans were already present in marine sponges [5,8], a precise understanding of their status and especially their biochemical and biological functions in sponges is more than limited.

Herein, we review the current knowledge on cancer-related genes/proteins in marine sponges (Figure 1) as well as the available experimental data about their biochemical characteristics and functions. Although genes/proteins are roughly divided according to their most prominent function, most of them are known to be involved in several cancer-causing processes.

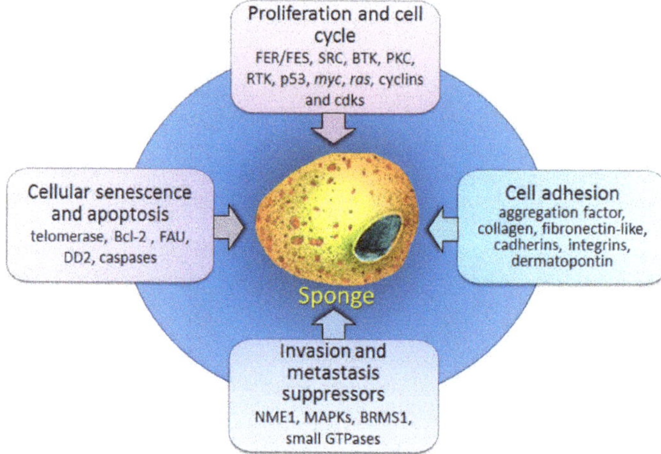

Figure 1. Schematic illustration of described sponge homologs of human cancer-related genes/proteins and cancer-associated processes.

2. Proliferation and Cell Cycle

The fundamental feature of cancer growth involves continuous, unsustained proliferation which eventually leads to tumor mass formation. Cell division in multicellular organisms is normally tightly controlled by numerous mechanisms which maintain a sophisticated balance between positive and negative regulators of cell proliferation. The controlled progression through the cell cycle ensures the well-being of an organism as a whole. Numerous cellular key players control these events, including receptors on the cell surface, various signalling molecules that conduct signals from the membrane to the nucleus, as well as transcription factors and repressors which enable or impede transcription of particular genes [9]. Numerous genes/proteins linked to proliferation and cell cycle progression have been described in sponges (Porifera). First and foremost are protein kinases, which, among various signalling pathways, also control cell proliferation [10].

2.1. Protein Kinases

The protein kinase family encompasses enzymes that catalyse transfer of the phosphate group from a high energy molecule such as adenosine triphosphate (ATP) to a specific substrate. Protein kinases play a significant role in cell-life regulation: in signal transmission pathways, cell differentiation control, proliferation, metabolism, DNA damage repair, cell motility, response to external stimuli, and apoptosis. Mutations, or other genetic alterations causing changes in their activity, often result in malignant transformation.

Protein kinase C. Protein kinase C (PKC) is a class of serine/threonine kinases which represents one of the most studied signaling kinases [11]. Although the expression of PKCs is changed in multiple cancer types, their diverse biological functions make it difficult to define the relationship between these changes and the initiation of a disease. Analysis of the gene coding for PKC in the sponge *Geodia cydonium* showed that it contains 13 exons and 12 introns and that exons encode the regulatory and catalytic domains typical for the metazoan PKCs. Further analysis of the promoter activity revealed that this, phylogenetically oldest, PKC gene contains a promoter functional in the heterologous mammalian cell system [12]. It was shown that sponge aggregation factor (AF) functions not only as a cell adhesion molecule, but also as a mitogenic agent [13]. In this context, PKC is involved in the AF induced transmembrane signalling. The activation of PKC leads to phosphorylation of many nuclear components, including the topoisomerase II, which subsequently activates the DNA replication process [14]. Two PKCs, *GCPKC1* and *GCPKC2*, have been sequenced from *G. cydonium* [15]. A comparison of the complete structures of the sponge PKCs, with those of higher metazoans, but also of protozoan, plant, and bacterial Ser/Thr kinases, revealed that the animal kinase domains display homologies with those from plants, protozoa, and bacteria. This implies that the Ser/Thr kinase domain has an universal common ancestor. However, the overall structure of the metazoan PKCs differs from non-metazoans, which suggests their distinct functions [15].

2.2. Protein Tyrosine Kinases

Protein tyrosine kinases (PTKs) specifically phosphorylate tyrosines on their target proteins. According to their cellular localization PTKs are divided in two major categories: receptor PTKs or transmembrane proteins (RTKs) and non-receptor or cytoplasmic PTKs. They are almost exclusively found in Metazoa and many have been described in sponges [16–21].

The receptor tyrosine kinase. The phylogenetic analysis of the protein-serine/threonine kinases (PS/TKs) from three sponges, the demosponges *G. cydonium* and *Suberites domuncula* and the calcareous sponge *Sycon raphanus*, suggests a common ancestry of PTKs with the PS/TK superfamily, from which the *G. cydonium* RTKs have diverged first [22]. The analysis of the *G. cydonium* RTK gene revealed that it contains introns outside of its TK domain, unlike the introns in higher animals that are inserted into the TK region [23]. Since the RTK from *G. cydonium* has been identified as the phylogenetically oldest member of PTKs [24], it was assumed that introns within the TK domains of genes from higher animals were inserted after the sponge taxa have branched off from all other metazoans [23].

The FES/FER non-receptor tyrosine kinases. Two different forms of Feline Sarcoma and FES Related proteins (FES/FER) exist in mammals [25] and both can be activated by a number of extracellular signals [26]. FER/FES non-RTKs are engaged in cytoskeletal rearrangements, as well as in cell-matrix and cell-cell interactions, while genetic analyses implicate their involvement in the regulation of inflammation and innate immunity [27]. The implication of FES and FER in human pathology still remains to be fully elucidated, but their high oncogenic potential has been implied in several recent studies [28,29]. The analysis of cDNA from the sponge *S. raphanus* disclosed a protein highly similar in its primary structure and organization of domains with tyrosine kinases (TKs) from the FES/FER family of non-RTKs [18]. The protein from *S. raphanus* was named FES/FER_SR since it exhibited high homology to the mammalian FES/FER proteins. Phylogenetic analysis revealed that FES/FER_SR from *S. raphanus* is the most ancient known member of the FES/FER family of non-RTKs [18]. Their role in organisms without tissues and organs, such as sponges, is not yet clear.

The SRC non-receptor tyrosine kinases. SRC (Rous sarcoma oncogene cellular homolog) is a non-receptor PTK that has been implicated in the development of malignant tumors in humans [30]. SRC is involved in many signaling pathways, such as gene transcription, cell cycle progression, cell adhesion, apoptosis, transformation, and migration. An extensive analysis of SRC PTKs from the marine sponge *S. domuncula* revealed nine different cDNAs encoding three different SRC proteins, SRC1SD, SRC2SD, and SRC3SD [16]. Considering that the N-termini of each of these proteins are unique, it can be speculated that the three sponge SRC kinases have different biological functions. Furthermore, it seems that all three SRCs from *S. domuncula* display the highest homology with SRCA PTKs from higher Metazoa. Phylogenetic analysis showed that SRC-related Src Homology 2 and 3 (SH2 and SH3) domains from different species are more conserved than SH2 and SH3 domains within different *S. domuncula* proteins [16]. The described data suggests that *src* genes are very ancient, that they already existed in the common ancestor of all Metazoa and that the ancestral *src* gene was already a multidomain protein composed of SH3, SH2 and tyrosine kinase (TK) domains.

Bruton's tyrosine kinase. Bruton's tyrosine kinase (BTK) belongs to the TEC (Tec Protein Tyrosine Kinase) family of PTKs and is required for maturation of the b-lineage lymphoid cells. Mutations in the BTK gene were shown to cause X-linked agammaglobulinemia (XLA) in humans and X-linked immunodeficiency (Xid) in mice [31,32]. Poriferan BTK-like protein (BTKSD) has been described from the marine sponge *S. domuncula*. This 700-aa-long protein contains all typical TEC family domains and shows the highest homology with the human BTK protein [17]. Phylogenetic analysis of PTKs from the TEC family revealed that the BTK/TEC genes have an ancient origin and have remained highly conserved during evolution [33]. These genes have probably been present in the common ancestor of all metazoans before gene duplications and the separation of sponges from other animal lineages. The structural analysis of the sponge and human promoter implies similar regulation of both genes; the kinase activities of the sponge BTKSD enzyme and its human homolog are in the same range [33]. Therefore, it is reasonable to presume that the ancestral-type protein was structurally and functionally similar to the multifunctional enzyme present in higher animals. Given that the BTK function in mammals is associates with the modulation of actin polymerization, a similar role of BTK-like protein in sponges is also expected. Modulation of actin polymerization by TEC family kinases probably has a significant role in cellular processes such as proliferation, motility, and adhesion [34]. Further, it was shown that PTKs of the TEC family possess a key role in production of the lipopolysaccharide (LPS)-induced tumor necrosis factor involved in the cellular response to bacterial infections [35]. Since *S. domuncula* is very efficient in fighting bacterial infections [36], it is possible that the role of BTK in sponges also includes antimicrobial defence.

Cyclins, cyclin-dependent kinases, and other key proliferation molecules. Cyclin-dependent kinases (CDKs) are a group of serine/threonine kinases, critical in the regulation of the cell cycle [37]. A comprehensive evolutionary study of cyclins and CDKs in metazoans and their unicellular relatives has been published by Lihuan Cao et al. [38]. It seems that the sponge genome lacks three cyclin subfamilies, cyclin A, cyclin D, and cyclin O, while the CDK4/6 subfamily is also missing from its genome. Although it has already been present in the placozoan *Trichoplax adhaerens*, the authors speculate that the establishment of the CDK4/6-cyclin D complex may have been fully functional in cell cycle control during eumetazoan emergence.

Finally, it seems that p53 (tumor protein p53), the guardian of the genome and putative tumor suppressor gene, is also present in the sponge genome as well as the powerful MYC (myelocytomatosis oncogene cellular homolog) and RAS (Rat sarcoma) oncogenes [7,39]. It has even been reported that the choanoflagellate *Monosiga brevicollis* already contains a subset of transcription factors, such as p53 and MYC, previously thought to be specific for metazoans [40]. The evolutionary history of p53 gene family as well as MYC [41] has been studied, but not specifically in the sponge genome [42]. The RAS-like family of proteins will be discussed further in the text.

3. Players in Cellular Senescence and Apoptosis

It is widely accepted that normal cells are able to pass only through a limited number of divisions. The unlimited proliferation of cancer cells has been associated with impairment of two major cellular cancer defence mechanisms: senescence and crisis/apoptosis. Senescence is normally a consequence of shortening of telomeres, multiple tandem hexanucleotide repeats that protect the ends of chromosomes from end-to-end fusion. The progressive shortening of telomeres can be stopped by telomerase, which can add repetitive nucleotide sequences to the ends of chromosomes [43]. In eukaryotic cells the telomere activity is usually restricted to germ and stem cells. In all other normal cells after a limited number of divisions the cell enters a senescent, but viable state. If the cell breaks this barrier, the crisis occurs which immediately redirects the cell to engage in the apoptotic program. The rare event of cellular immortalization of potential cancer cells occurs by the knock-down of both of these barriers. This is usually due to unlocking the transcriptional program which leads to dedifferentiation followed by upregulating of telomerase expression [9].

As expected, telomeres are an integral part of sponge chromosomes. The telomere sequence has been analyzed in three different sponge species and it consists exclusively of (TTAGGG) *n*-type sequences that are widely distributed in many multicellular animals belonging to the various phyla and classes [44]. The telomerase activity was determined in two different sponges: *S. domuncula* and *G. cydonium* [45]. A quantitative analysis revealed that tissues from both sponges contained telomerase activity of approximately 30% and 20%, respectively, of the activity in the positive reference cells. In higher animals the high telomerase activity is strictly limited to germ cells and metastatic tumors, while stem cells, needed for tissue renewal and injury repair, have intrinsically low levels of telomerase [46]. It is known that sponge tissue has a very low germ cell number, so this finding may indicate that all of the sponge cells have some, low telomerase level and, therefore, maintain permanently their proliferation potential. The isolated sponge cell in cell culture, however, stops proliferating [47] and loses the telomerase activity, opposed to the human tumor cells, which in culture conditions obtain both proliferation and telomerase activity. All these findings indicate that sponges have already developed the senescence/apoptotic program.

Apoptosis is a form of programmed cell death which can be found in multicellular metazoans and is a necessary prerequisite for the optimal functioning of these organisms. Apoptosis is a highly controlled mechanism which serves to eliminate redundant, infected, necrotic, or simply unwanted cells during development, senescence, morphogenesis, organogenesis, and defence, or other physiological and pathophysiological events [48]. In multicellular organisms apoptosis serves as a major barrier against uncontrolled proliferation, which leads to cancer. The apoptotic process can generally be divided into two pathways. The extrinsic pathway is triggered upon ligation of tumor necrosis factors (TNFs) transmitting the apoptotic signal into the cell. The intrinsic pathway is stress-dependent and controlled by members of the Bcl-2 protein family. Both pathways culminate with the activation of proteolytic enzymes, caspases, the executors of programmed cell death.

It seems that the sponge genome possesses components of both pathways. Experimental evidence proves that marine sponges, like *G. cydonium* or *S. domuncula*, undergo the process of apoptosis during formation of asexual reproduction bodies or in stress response [49]. The poriferan pro-apoptotic molecule DD2 (death domain 2), a death domain containing molecule, was identified by screening *G. cydonium* library with degenerate primers directed towards a conserved region of human FAS receptor (fas cell surface death receptor). Interestingly, this protein has two death domains which have not been found in any known protein involved in the TNF signalling pathway.

Moreover, evolutionary ancient members of the intrinsic, Bcl-2 (B-cell lymphoma 2) family members have been identified in sponge [50]. The sponge *G. cydonium* expresses a pro-survival Bcl-2 homolog BHP2 that impairs apoptosis in sponge tissue induced by heat shock or tributyltin exposure [51]. Recently, Caria et al. [52] performed structural and binding studies on BHP2 and BAK-2 (Bcl-2 Antagonist/Killer) from *G. cydonium* and the freshwater sponge *Lubomirskia baicalensis*. The results of their findings suggest that the molecular machinery and mechanisms for executing Bcl-2-mediated apoptosis are evolutionary ancient and resemble their mammalian versions.

Furthermore, FAU (Finkel-Biskis-Reilly murine sarcoma virus (FBR-MuSV) ubiquitously expressed), recently identified as a pro-apoptotic regulatory gene mediated by Bcl-G (a pro-apoptotic member of the Bcl-2 family), has also been identified in the sponge genome [53]. FAU gene encodes a fusion protein composed of an ubiquitin-like protein (FUBI) at its N terminus and ribosomal protein S30 at the C terminus. The protein is then post-translationally processed into two separate proteins, FUBI and S30 ribosomal protein. *FAU* has been identified as a pro-apoptotic regulatory gene, whose expression has been downregulated in human ovarian, prostate, and breast cancer [54–56]. A study of the FAU protein from the marine sponge *S. domuncula* showed that sponge FAU protein displays a pro-apoptotic activity, as it increases apoptosis in human HEK293T cells (human embryonic kidney cells) as well as its human homolog and was shown to be more similar to its human homolog than to the one from *Caenorhabditis elegans* or *Drosophila melanogaster*. These findings implicate that the common metazoan ancestor probably possessed the FAU protein that was structurally and functionally similar to its recent version and has changed very little during evolution [57]. Additionally, the number, phases, and position of its introns were not significantly changed from sponges to humans. Ribosomal proteins (RPs) are evolutionary conserved components of ribosomes crucial for the correct ribosome assembly. Many ribosomal protein genes (RPGs) have additional extraribosomal functions, being involved in replication, regulation of cell growth, apoptosis, and cancer development [58]. RPs have shown to be important tools for studying intron evolution, since the analysis of RPG introns in sponges revealed the same rate of conservancy between two different sponge species compared to the conservancy between sponges and humans [59,60]. Since S30 ribosomal protein is produced from the fusion protein transcribed from the pro-apoptotic regulatory gene FAU, its extraribosomal function may be connected to the regulation of apoptosis [57].

As mentioned above, the apoptotic process is eventually executed by specific proteases, caspases. According to Wiens et al. [49], altogether three caspase-like proteins have been found in Porifera. The smaller number of caspase family genes, compared to vertebrates, probably implies the simplicity of this proteolytic cascade in sponges, but undoubtedly suggests that this process was already present and engaged in first simple multicellular animals.

4. Cell Anchoring Molecules

Adhesion of cells is a primary prerequisite for tissues architecture. Cell within tissues adhere to each other, as well as to the extracellular matrix (ECM); these connections are collectively called the anchoring junctions. Several types of anchoring junctions have been identified, depending on the type of adhesion, and they are in general responsible not only for anchoring cells within a certain tissue, but also for their migration. Anchoring junctions are multiprotein complexes in which cell-adhesion molecules (CAMs) play a crucial role. There are five principal classes of CAMs: cadherins, immunoglobulin (Ig) superfamily, selectins, mucins, and integrins. Integrins mediate cell-matrix interactions whereas other types of CAMs participate in cell–cell adhesion [61]. Cell adhesion is essential in all aspects of vertebrate cell growth, migration, and differentiation. CAMs are involved in a wide variety of cellular functions including signal transduction, cellular communication and recognition, embryogenesis, inflammatory and immune responses, and apoptosis [62], therefore, they play a significant role in cancer progression and metastasis. For example, tumor invasion and metastasis frequently coincide with the loss of E-cadherin-mediated cell–cell adhesion [63].

During the last century sponges have been repeatedly used as a model system for studying cell adhesion [26,64]. The first extracellular particle named aggregation factor (AF), which promotes the species-specific aggregation of sponge cells, was isolated from two sponges in 1973 [65,66]. The *G. cydonium* AF was shown to bind the plasma membrane-bound aggregation receptor (AR), after which the RAS gene is expressed [67]. The RAS gene product further binds to the anti-aggregation receptor (aAR), which has been identified as a lectin receptor [68,69]. After the first cell–cell adhesion molecule, galectin [70], and the first cell–matrix adhesion receptor, integrin [71,72], were identified in sponge *G. cydonium*, it became clear that sponges contain proteins involved in adhesion, which also show homology to those in humans [73]. The extracellular matrix (ECM) represents a platform for cell adhesion via integrin receptors, signal transduction, cell development, and cell division. Sponge mesohyl, a space between the external pinacoderm and the internal choanoderm, is composed of collagen, fibronectin-like molecules, and dermatopontin [74], which are also found in the ECM of higher metazoans. Therefore, it seems that the sponge mesohyl could serve as a primitive ECM in sponges.

Cadherins. Cadherins (CDHs) are a major component of adherent junctions, typically present in vertebrate epithelial tissues. Adherent junctions are composed primarily of type I cadherins, transmembrane glycoproteins that form Ca^{+2}-dependent homotypic complexes. Loss of cadherins enables cancer cells to detach from the original tissue and trigger the metastatic cascade. It is widely accepted that the key molecule in metastasis onset is E-cadherin, specifically. Cadherins (CDHs) and cadherin-related proteins are found in the genomes of all sequenced metazoans, including diverse bilaterians, cnidarians, sponges, and also in choanoflagellates, the closest unicellular relatives of animals [75]. CDH1 and CDH2 (type I) are found to be involved in metastasis suppression. Nichols et al. [76] have described the incomplete type I CDH1 from the draft genome of sponge *Oscarella carmela*.

Integrins. Integrins form a large family of cell adhesion receptors involved in cell adhesion, migration, and signalling [77]. Integrins are heterodimers composed of one α and one β subunit. Eighteen different integrin α subunits and eight β subunits have been described [78].

All animals express integrins, indicating that this family evolved early in the history of metazoans. Both α and β subunits have been found throughout the invertebrates species, including sponges [71,72,79].

5. Metastasis Suppressor Genes

Metastasis is a process in which tumor cells detach from the original tissue, travel to nearby or distant sites in the body and form secondary tumors. Metastasis suppressors inhibit the tumor cell ability to metastasize, having little or no effect on the primary tumor growth. Metastasis suppressors are involved in one or several steps of the metastatic cascade, which includes the following: detachment of the cell from the main tissue, breaking through the basal lamina and invasion of the surrounding tissue, entrance to the nearby blood or lymphatic vessels, surviving the transit through the lymphatic or blood system and extravasation from blood/lymphatic vessels into the distant tissue, where they can grow into macroscopic tumors [80]. The ongoing discoveries of genes/proteins that are directly involved in the metastatic cascade seem to be an important step forward in our understanding of this process.

NME1. The human NME1 (non-metastatic 1) is the first identified and the most studied metastasis suppressor [81], present in all three domains of life: Bacteria, Archaea, and Eukarya [82]. The NME1 protein possesses the nucleoside diphosphate kinase (NDPK) activity, but also has some additional functions such as histidine kinase activity, transcription factor activity, etc. [83,84]. A single gene coding for the human NME1 homolog was found in the genome of the marine sponge *Amphimedon queenslandica* [7] and *S. domuncula* cDNAs [8,85]. These sponge homologs are similar in primary structure and possess conserved residues essential for NDPK activity. Biochemical and functional characterizations of non-bilaterian homologs of the human NME1 protein are still scarce. Our group has shown that sponge genes NMEGp1 are intron-rich and that these introns are relatively short. The analysis of sponge NMEGp1 promoters showed that some of the motifs crucial for human promoter activity are also present in sponges. Furthermore, the analysis of protein activity showed that the sponge *S. domuncula* NMEGp1Sd protein possesses a hexameric form and has a similar level of kinase activity as the human NME1 protein. The sponge homolog interacts with its human ortholog/homolog in human cultured cells and shows the same subcellular localization pattern as human NME1. Also, if expressed in human tumor cells, the sponge homolog significantly reduces its migration potential. Based on the evidence presented, we have concluded that the sponge NME protein can replace its human homolog in at least some of its biological functions which are usually associated with "higher" metazoans. We also presumed that the biochemical function of NME1, which is responsible for metastasis suppression in human, is present in the sponge, and was possibly also present in the ancestor of all Metazoa [86,87]. On the contrary, our studies on another member of the NME family in sponges, NME6, indicates that this ancient gene/protein probably changed its biochemical/biological function during the course of evolution. It is not yet elucidated whether this protein is also linked to tumor formation or progression [88].

Small GTPases. RAS-like superfamily of GTPases (small GTPases) are a family of enzymes that bind and/or hydrolyze GTP. The state of the bound nucleotide acts as a switch between the active and inactive state of the enzyme. The RAS superfamily is divided into five subfamilies: RAS, RHO, RAN (Ras-related nuclear protein), RAB (Ras genes from rat brain), and ARF (ADP ribosylation Factor) [89]. Small GTPases play an important part in numerous cellular processes but it is generally accepted that the RAS family is responsible for cell proliferation, RHO for cell morphology, RAN for nuclear transport, and RAB and ARF families for vesicle transport. Therefore, it is clear that the RAS-like superfamily is implicated in several aspects of cancer occurrence and progression [90]. The analysis of the EST database from the marine sponge *S. domuncula* has revealed cDNA sequences encoding 50 different RAS-like small GTPases. Forty-four sponge proteins from the RAS family are described here: six proteins from the RAS subfamily, five from RHO, six from ARF, one RAN, and 26 RABs or RAB-like proteins. Small GTPases from sponge are more related in their primary structure to the orthologues from vertebrates than to those from other invertebrates. These findings imply that diversification of genes encoding RAS-like small GTPases happened after the duplications that occurred very early in the evolution of Metazoa [39].

6. Cancer Associated Genes in Marine Sponges-Final Remarks and Future Challenges

Apart from humans, tumors have mostly been described in other vertebrates, especially in farm animals and pets, as well as in other animals kept in captivity [1]. Furthermore, they have also been reported in invertebrate deuterostomes [1], protostomes [80,91], and even in simple non-bilaterian animals [3,92,93], although these findings should be taken with caution since laboratory breeding and culture conditions may be far from those found in their natural habitats [3]. Therefore, it is doubtful whether these organisms would develop tumors also in a natural environment. Indirect evidence of the presence of tumors in simple non-bilaterian animals also comes from the fact that marine invertebrates, especially sponges, produce bioactive compounds [94], some of which have an antitumor activity on human cells in culture [95–97]. The role of these compounds in sponge biology is not fully understood. Since, sponges are sessile and lack capabilities for physical defences, they are often targets of marine predators. Therefore, it is expected that sponges have developed a range of defensive chemicals in order to disable and repel predators. They also use their defensive substances to keep plants and other animals' offsprings from settling onto their surface [98,99]. Although it is widely accepted that sponges use these compounds for ecological purposes it cannot be excluded that at least some of these substances, are used in physiological processes, e.g., as a primitive form of immunological defence, in the regulation of cell count or for various metabolic purposes. Some of the compounds they produce are directly engaged in specific signalling pathways. However, it should be taken into consideration that more than one pathway is usually affected by each of those compounds.

About a hundred novel compounds that exhibit significant inhibitory activity towards a range of protein kinases have been detected so far [100]. Some of these act as inhibitors of protein kinase C (PKC): xestocyclamine A [101] (Z)-Axinohydantoin), debromo-Z-axinohydantoin, hymenialdisine, debromohymenialdisine [102], frondosins A–E [103], BRS1 (C30 BIS-Amino,

BIS-Hydroxy polyunsaturated lipid) [104], sesquiterpenoid quinones, nakijiquinones A–D [105,106], cytotoxic sesterpenes spongianolides A–E [107], lasonolide A [108–110], azetidine compound penazetidine A [111], and numerous others. Several compounds that display inhibitory activity towards CDKs have also been isolated from marine sponges, such as hymenialdisine [112,113], microxine, variolin B [114,115], fascaplysin [116], and konbu'acidin A [117]. It has also been shown that marine sponges possess tyrosine protein kinase (TPK) inhibitors: the penta-, hexa- and hepta-prenylhydroquinone 4-sulfates [118], melemeleone [119], and polyketide inhibitors [120]. Furthermore, (+)-aeroplysinin-1 from sponge displays a strong antitumor effect by blocking the proliferation of EGFR-dependent human breast cancer cell lines MCF-7 and ZR-75-1 [121], while curcuphenol displays a SRC protein kinase inhibition and curcudiola focal adhesion kinase (FAK) inhibition [122].

So far, as much as 60 compounds isolated from marine sponges have been shown to induce anticancer activity through apoptosis. For the majority of them, the proapoptotic mechanism has not yet been fully elucidated [123]. The pre-clinical study on Renieramycin M reveals that this substance induces apoptosis in cancer cell lines through the p53-dependent pathway and consequently inhibits cancer progression and metastasis [96]. Renieramycin M, a tetrahydroisoquinoline, is the first anticancer drug to be approved by the European Union and by the Food and Drug Administration (FDA) for the treatment of advanced soft tissue sarcomas [124]. Selected compounds for which cellular and molecular mechanisms were explored in depth include: aaptamines [125], psammaplysene A (PsA) [126], stellettin B [127–129], candidaspongiolide [130], heteronemin [131–133], and dideoxypetrosynol A [134]. Many compounds produced by marine sponges display antiproliferative activity targeting various signaling pathways of the cell-cycle progression such as aragusterol A [135], (19Z)-hlichondramide [136], smenospongine [137], and crambescidin 800 [138]. Some can also act as anti-inflammatory molecules targeting TLR (Toll-like receptor) signalling pathways [139], or as inhibitors of matrix proteinases [140]. Further, it has been shown that macrolide halichondramide (HCA) has an antimetastatic effect on human prostate cancer cells PC3, inhibiting their migration and invasion through the modulation of the crucial cadherin switches [141]. The specific role of these substances in sponge ecology and metabolism will hopefully be disclosed by future studies.

Based on the data collected so far, we can only speculate on the presence of cancer during the early animal evolution. It is presumed that tumors in invertebrates do not exhibit a malignant phenotype, while the progress in malignancy probably runs in parallel with the development of the immune system [142], as well as the development of a highly effective vasculatory system in vertebrates [143].

Despite the lack of true organs and tissues in sponges, they obviously possess a number of genes/proteins related to tumor onset and progression. Their genomes contain many gene homologs of those involved in human cancer development. Furthermore, the majority of sponge proteins transcribed from these genes possess the same domain organization as their homologs from higher metazoans, yet, their function within the sponge cells remains to be elucidated. Tumors in sponges have not been identified so far and, because of its simple structure, the formation of tumors within the sponge body is highly unlikely. Nevertheless, a number of studies indicated that sponge proteins may reflect at least some of the biochemical characteristics related to their homologs in higher animals, particularly humans. We know today that several other physiological processes are linked to tumor emergence, such as angiogenesis, inflammation, or cellular energetics [9]. Evidence shows that at least some of the components involved in these processes are present in sponges, but no systematic research is done in this context [7,144].

Sponges represent an important model, not only for studying ancestral metazoan homologs and their features before the diversification and specialization of these genes in higher animals, but also for understanding the basic physiological function of cancer genes in simpler animals. These findings could help to elucidate the more complex interactions of their homologs in humans and consequently explain possible reasons for their oncogenic potential.

Acknowledgments: This work has been fully supported by the Croatian Science Foundation under projects IP-2014-09-6400 and IP-2016-06-4021.

Author Contributions: Helena Ćetković drafted the initial version of the manuscript and oversaw the work and writing process. Helena Ćetković, Maja Herak Bosnar, and Mirna Halasz wrote, edited, and reviewed the manuscript.

Conflicts of Interest: The authors declare no conflict of interest.

References

1. Aktipis, C.A.; Boddy, A.M.; Jansen, G.; Hibner, U.; Hochberg, M.E.; Maley, C.C.; Wilkinson, G.S. Cancer across the tree of life: Cooperation and cheating in multicellularity. *Philos. Trans. R. Soc. B* **2015**, *370*, 20140219. [CrossRef] [PubMed]
2. Davies, P.C.W.; Lineweaver, C.H. Cancer tumors as metazoa 1.0: Tapping genes of ancient ancestors. *Phys. Biol.* **2011**, *8*, 015001. [CrossRef] [PubMed]
3. Domazet-Lošo, T.; Klimovich, A.; Anokhin, B.; Anton-Erxleben, F.; Hamm, M.J.; Lange, C.; Bosch, T.C.G. Naturally occurring tumours in the basal metazoan hydra. *Nat. Commun.* **2014**, *5*, 4222. [CrossRef] [PubMed]
4. Domazet-Lošo, T.; Tautz, D. An ancient evolutionary origin of genes associated with human genetic diseases. *Mol. Biol. Evol.* **2008**, *25*, 2699–2707. [CrossRef] [PubMed]
5. Domazet-Lošo, T.; Tautz, D. Phylostratigraphic tracking of cancer genes suggests a link to the emergence of multicellularity in metazoa. *BMC Biol.* **2010**, *8*, 66. [CrossRef] [PubMed]
6. Love, G.D.; Grosjean, E.; Stalvies, C.; Fike, D.A.; Grotzinger, J.P.; Bradley, A.S.; Kelly, A.E.; Bhatia, M.; Meredith, W.; Snape, C.E.; et al. Fossil steroids record the appearance of demospongiae during the cryogenian period. *Nature* **2009**, *457*, 718–721. [CrossRef] [PubMed]
7. Srivastava, M.; Simakov, O.; Chapman, J.; Fahey, B.; Gauthier, M.E.; Mitros, T.; Richards, G.S.; Conaco, C.; Dacre, M.; Hellsten, U.; et al. The Amphimedon queenslandica genome and the evolution of animal complexity. *Nature* **2010**, *466*, 720–726. [CrossRef] [PubMed]
8. Harcet, M.; Roller, M.; Ćetković, H.; Perina, D.; Wiens, M.; Müller, W.E.; Vlahovicek, K. Demosponge EST sequencing reveals a complex genetic toolkit of the simplest metazoans. *Mol. Biol. Evol.* **2010**, *27*, 2747–2756. [CrossRef] [PubMed]
9. Hanahan, D.; Weinberg, R.A. Hallmarks of cancer: The next generation. *Cell* **2011**, *144*, 646–674. [CrossRef] [PubMed]
10. Hunter, T. A 1001 protein-kinases. *Cell* **1987**, *50*, 823–829. [CrossRef]
11. Garg, R.; Benedetti, L.G.; Abera, M.B.; Wang, H.; Abba, M.; Kazanietz, M.G. Protein kinase c and cancer: What we know and what we do not. *Oncogene* **2014**, *33*, 5225–5237. [CrossRef] [PubMed]
12. Seack, J.; Kruse, M.; Muller, I.M.; Müller, W.E. Promoter and exon-intron structure of the protein kinase C gene from the marine sponge *Geodia cydonium*: Evolutionary considerations and promoter activity. *Biochim. Biophys. Acta* **1999**, *1444*, 241–253. [CrossRef]
13. Müller, W.E.; Rottmann, M.; Diehl-Seifert, B.; Kurelec, B.; Uhlenbruck, G.; Schröder, H.C. Role of the aggregation factor in the regulation of phosphoinositide metabolism in sponges—Possible consequences on calcium efflux and on mitogenesis. *J. Biol. Chem.* **1987**, *262*, 9850–9858. [PubMed]
14. Rottmann, M.; Schröder, H.C.; Gramzow, M.; Renneisen, K.; Kurelec, B.; Dorn, A.; Friese, U.; Müller, W.E. Specific phosphorylation of proteins in pore complex-laminae from the sponge *Geodia cydonium* by the homologous aggregation factor and phorbol ester. Role of protein kinase C in the phosphorylation of DNA topoisomerase II. *EMBO J.* **1987**, *6*, 3939–3944. [PubMed]
15. Kruse, M.; Gamulin, V.; Cetkovic, H.; Pancer, Z.; Müller, I.M.; Müller, W.E.G. Molecular evolution of the metazoan protein kinase C multigene family. *J. Mol. Evol.* **1996**, *43*, 374–383. [CrossRef] [PubMed]
16. Cetkovic, H.; Grebenjuk, V.A.; Müller, W.E.G.; Gamulin, V. Src proteins/src genes: From sponges to mammals. *Gene* **2004**, *342*, 251–261. [CrossRef] [PubMed]
17. Cetkovic, H.; Müller, W.E.G.; Gamulin, V. Bruton tyrosine kinase-like protein, BtkSD, is present in the marine sponge *Suberites domuncula*. *Genomics* **2004**, *83*, 743–745. [CrossRef] [PubMed]
18. Cetkovic, H.; Müller, I.M.; Müller, W.E.; Gamulin, V. Characterization and phylogenetic analysis of a cDNA encoding the Fes/FER related, non-receptor protein-tyrosine kinase in the marine sponge *Sycon raphanus*. *Gene* **1998**, *216*, 77–84. [CrossRef]

19. Suga, H.; Katoh, K.; Miyata, T. Sponge homologs of vertebrate protein tyrosine kinases and frequent domain shufflings in the early evolution of animals before the parazoan-eumetazoan split. *Gene* **2001**, *280*, 195–201. [CrossRef]

20. Suga, H.; Koyanagi, M.; Hoshiyama, D.; Ono, K.; Iwabe, N.; Kuma, K.; Miyata, T. Extensive gene duplication in the early evolution of animals before the parazoan-eumetazoan split demonstrated by G proteins and protein tyrosine kinases from sponge and hydra. *J. Mol. Evol.* **1999**, *48*, 646–653. [CrossRef] [PubMed]

21. Perovic-Ottstadt, S.; Cetkovic, H.; Gamulin, V.; Schröder, H.C.; Kropf, K.; Moss, C.; Korzhev, M.; Diehl-Seifert, B.; Müller, I.M.; Müller, W.E. Molecular markers for germ cell differentiation in the demosponge *Suberites domuncula*. *Int. J. Dev. Biol.* **2004**, *48*, 293–305. [CrossRef] [PubMed]

22. Kruse, M.; Müller, I.M.; Müller, W.E. Early evolution of metazoan serine/threonine and tyrosine kinases: Identification of selected kinases in marine sponges. *Mol. Biol. Evol.* **1997**, *14*, 1326–1334. [CrossRef] [PubMed]

23. Gamulin, V.; Skorokhod, A.; Kavsan, V.; Müller, I.M.; Müller, W.E. Experimental indication in favor of the introns-late theory: The receptor tyrosine kinase gene from the sponge *Geodia cydonium*. *J. Mol. Evol.* **1997**, *44*, 242–252. [CrossRef] [PubMed]

24. Müller, W.E.; Schacke, H. Characterization of the receptor protein-tyrosine kinase gene from the marine sponge *Geodia cydonium*. *Prog. Mol. Subcell. Biol.* **1996**, *17*, 183–208. [PubMed]

25. Fischman, K.; Edman, J.C.; Shackleford, G.M.; Turner, J.A.; Rutter, W.J.; Nir, U. A murine fer testis-specific transcript (ferT) encodes a truncated Fer protein. *Mol. Cell. Biol.* **1990**, *10*, 146–153. [CrossRef] [PubMed]

26. Wilson, H.V. A new method by which sponges may be artificially reared. *Science* **1907**, *25*, 912–915. [CrossRef] [PubMed]

27. Hackenmiller, R.; Kim, J.; Feldman, R.A.; Simon, M.C. Abnormal stat activation, hematopoietic homeostasis, and innate immunity in c-fes-/- mice. *Immunity* **2000**, *13*, 397–407. [CrossRef]

28. Ahn, J.; Truesdell, P.; Meens, J.; Kadish, C.; Yang, X.; Boag, A.H.; Craig, A.W. Fer protein-tyrosine kinase promotes lung adenocarcinoma cell invasion and tumor metastasis. *Mol. Cancer Res.* **2013**, *11*, 952–963. [CrossRef] [PubMed]

29. Oneyama, C.; Yoshikawa, Y.; Ninomiya, Y.; Iino, T.; Tsukita, S.; Okada, M. Fer tyrosine kinase oligomer mediates and amplifies Src-induced tumor progression. *Oncogene* **2016**, *35*, 501–512. [CrossRef] [PubMed]

30. Biscardi, J.S.; Ishizawar, R.C.; Silva, C.M.; Parsons, S.J. Tyrosine kinase signalling in breast cancer—Epidermal growth factor receptor and c-Src interactions in breast cancer. *Breast Cancer Res.* **2000**, *2*, 203–210. [CrossRef] [PubMed]

31. Rawlings, D.J.; Saffran, D.C.; Tsukada, S.; Largaespada, D.A.; Grimaldi, J.C.; Cohen, L.; Mohr, R.N.; Bazan, J.F.; Howard, M.; Copeland, N.G.; et al. Mutation of unique region of bruton's tyrosine kinase in immunodeficient xid mice. *Science* **1993**, *261*, 358–361. [CrossRef] [PubMed]

32. Conley, M.E.; Dobbs, A.K.; Farmer, D.M.; Kilic, S.; Paris, K.; Grigoriadou, S.; Coustan-Smith, E.; Howard, V.; Campana, D. Primary B cell immunodeficiencies: Comparisons and contrasts. *Annu. Rev. Immunol.* **2009**, *27*, 199–227. [CrossRef] [PubMed]

33. Perina, D.; Mikoc, A.; Harcet, M.; Imesek, M.; Sladojevic, D.; Brcko, A.; Cetkovic, H. Characterization of bruton's tyrosine kinase gene and protein from marine sponge *Suberites domuncula*. *Croat. Chem. Acta* **2012**, *85*, 223–229. [CrossRef]

34. Finkelstein, L.D.; Schwartzberg, P.L. Tec kinases: Shaping T-cell activation through actin. *Trends Cell Biol.* **2004**, *14*, 443–451. [CrossRef] [PubMed]

35. Horwood, N.J.; Mahon, T.; McDaid, J.P.; Campbell, J.; Mano, H.; Brennan, F.M.; Webster, D.; Foxwell, B.M.J. Bruton's tyrosine kinase is required for lipopolysaccharide-induced tumor necrosis factor alpha production. *J. Exp. Med.* **2003**, *197*, 1603–1611. [CrossRef] [PubMed]

36. Böhm, M.; Hentschel, U.; Friedrich, A.B.; Fieseler, L.; Steffen, R.; Gamulin, V.; Müller, I.M.; Müller, W.E.G. Molecular response of the sponge *Suberites domuncula* to bacterial infection. *Mar. Biol.* **2001**, *139*, 1037–1045.

37. Sharma, P.S.; Sharma, R.; Tyagi, R. Inhibitors of cyclin dependent kinases: Useful targets for cancer treatment. *Curr. Cancer Drug Target* **2008**, *8*, 53–75. [CrossRef]

38. Cao, L.; Chen, F.; Yang, X.; Xu, W.; Xie, J.; Yu, L. Phylogenetic analysis of CDK and cyclin proteins in premetazoan lineages. *BMC Evol. Biol.* **2014**, *14*. [CrossRef] [PubMed]

39. Cetkovic, H.; Mikoc, A.; Müller, W.E.G.; Gamulin, V. Ras-like small GTpases form a large family of proteins in the marine sponge *Suberites domuncula*. *J. Mol. Evol.* **2007**, *64*, 332–341. [CrossRef] [PubMed]

40. King, N.; Westbrook, M.J.; Young, S.L.; Kuo, A.; Abedin, M.; Chapman, J.; Fairclough, S.; Hellsten, U.; Isogai, Y.; Letunic, I.; et al. The genome of the choanoflagellate *Monosiga brevicollis* and the origin of metazoans. *Nature* **2008**, *451*, 783–788. [CrossRef] [PubMed]

41. Young, S.L.; Diolaiti, D.; Conacci-Sorrell, M.; Ruiz-Trillo, I.; Eisenman, R.N.; King, N. Premetazoan ancestry of the Myc-Max network. *Mol. Biol. Evol.* **2011**, *28*, 2961–2971. [CrossRef] [PubMed]

42. Belyi, V.A.; Ak, P.; Markert, E.; Wang, H.; Hu, W.; Puzio-Kuter, A.; Levine, A.J. The origins and evolution of the p53 family of genes. *Cold Spring Harb. Perspect. Biol.* **2010**, *2*, a001198. [CrossRef] [PubMed]

43. Morin, G.B. The human telomere terminal transferase enzyme is a ribonucleoprotein that synthesizes TTAGGG repeats. *Cell* **1989**, *59*, 521–529. [CrossRef]

44. Sakai, M.; Okumura, S.I.; Onuma, K.; Senbokuya, H.; Yamamori, K. Identification of a telomere sequence type in three sponge species (Porifera) by fluorescence in situ hybridization analysis. *Fish. Sci.* **2007**, *73*, 77–80. [CrossRef]

45. Koziol, C.; Borojevic, R.; Steffen, R.; Muller, W.E.G. Sponges (Porifera) model systems to study the shift from immortal to senescent somatic cells: The telomerase activity in somatic cells. *Mech. Ageing Dev.* **1998**, *100*, 107–120. [CrossRef]

46. Hiyama, E.; Hiyama, K. Telomere and telomerase in stem cells. *Br. J. Cancer* **2007**, *96*, 1020–1024. [CrossRef] [PubMed]

47. Gramzow, M.; Schröder, H.C.; Fritsche, U.; Kurelec, B.; Robitzki, A.; Zimmermann, H.; Friese, K.; Kreuter, M.H.; Müller, W.E. Role of phospholipase a2 in the stimulation of sponge cell proliferation by homologous lectin. *Cell* **1989**, *59*, 939–948. [CrossRef]

48. Wiens, M.; Müller, W.E.G. Cell death in Porifera: Molecular players in the game of apoptotic cell death in living fossils. *Can. J. Zool.* **2006**, *84*, 307–321. [CrossRef]

49. Wiens, M.; Krasko, A.; Perovic, S.; Müller, W.E.G. Caspase-mediated apoptosis in sponges: Cloning and function of the phylogenetic oldest apoptotic proteases from Metazoa. *Biochim. Biophys. Acta* **2003**, *1593*, 179–189. [CrossRef]

50. Wiens, M.; Krasko, A.; Müller, C.I.; Müller, W.E.G. Molecular evolution of apoptotic pathways: Cloning of key domains from sponges (Bcl-2 homology domains and death domains) and their phylogenetic relationships. *J. Mol. Evol.* **2000**, *50*, 520–531. [CrossRef] [PubMed]

51. Wiens, M.; Krasko, A.; Blumbach, B.; Müller, I.M.; Müller, W.E.G. Increased expression of the potential proapoptotic molecule DD2 and increased synthesis of leukotriene B-4 during allograft rejection in a marine sponge. *Cell Death Differ.* **2000**, *7*, 461–469. [CrossRef] [PubMed]

52. Caria, S.; Hinds, M.G.; Kvansakul, M. Structural insight into an evolutionarily ancient programmed cell death regulator—The crystal structure of marine sponge BHP2 bound to LB-Bak-2. *Cell Death Dis.* **2017**, *8*, e2543. [CrossRef] [PubMed]

53. Pickard, M.R.; Mourtada-Maarabouni, M.; Williams, G.T. Candidate tumour suppressor Fau regulates apoptosis in human cells: An essential role for Bcl-G. *Biochim. Biophys. Acta* **2011**, *1812*, 1146–1153. [CrossRef] [PubMed]

54. Pickard, M.R.; Green, A.R.; Ellis, I.O.; Caldas, C.; Hedge, V.L.; Mourtada-Maarabouni, M.; Williams, G.T. Dysregulated expression of Fau and MELK is associated with poor prognosis in breast cancer. *Breast Cancer Res.* **2009**, *11*, R60. [CrossRef] [PubMed]

55. Pickard, M.R.; Edwards, S.E.; Cooper, C.S.; Williams, G.T. Apoptosis regulators Fau and Bcl-G are down-regulated in prostate cancer. *Prostate* **2010**, *70*, 1513–1523. [CrossRef] [PubMed]

56. Moss, E.L.; Mourtada-Maarabouni, M.; Pickard, M.R.; Redman, C.W.; Williams, G.T. FAU regulates carboplatin resistance in ovarian cancer. *Genes Chromosomes Cancer* **2010**, *49*, 70–77. [CrossRef] [PubMed]

57. Perina, D.; Korolija, M.; Hadzija, M.P.; Grbesa, I.; Beluzic, R.; Imesek, M.; Morrow, C.; Marjanovic, M.P.; Bakran-Petricioli, T.; Mikoc, A.; et al. Functional and structural characterization of FAU gene/protein from marine sponge *Suberites domuncula*. *Mar. Drugs* **2015**, *13*, 4179–4196. [CrossRef] [PubMed]

58. Zhou, X.; Liao, W.J.; Liao, J.M.; Liao, P.; Lu, H. Ribosomal proteins: Functions beyond the ribosome. *J. Mol. Cell Biol.* **2015**, *7*, 92–104. [CrossRef] [PubMed]

59. Perina, D.; Korolija, M.; Mikoc, A.; Roller, M.; Plese, B.; Imesek, M.; Morrow, C.; Batel, R.; Cetkovic, H. Structural and functional characterization of ribosomal protein gene introns in sponges. *PLoS ONE* **2012**, *7*, e42523. [CrossRef] [PubMed]

60. Perina, D.; Cetkovic, H.; Harcet, M.; Premzl, M.; Lukic-Bilela, L.; Müller, W.E.; Gamulin, V. The complete set of ribosomal proteins from the marine sponge *Suberites domuncula*. *Gene* **2006**, *366*, 275–284. [CrossRef] [PubMed]

61. Lock, J.G.; Wehrle-Haller, B.; Stromblad, S. Cell-matrix adhesion complexes: Master control machinery of cell migration. *Semin. Cancer Biol.* **2008**, *18*, 65–76. [CrossRef] [PubMed]

62. Cohen, O.; Feinstein, E.; Kimchi, A. Dap-kinase is a Ca^{2+} calmodulin-dependent, cytoskeletal-associated protein kinase, with cell death-inducing functions that depend on its catalytic activity. *EMBO J.* **1997**, *16*, 998–1008. [CrossRef] [PubMed]

63. Cavallaro, U.; Christofori, G. Cell adhesion in tumor invasion and metastasis: Loss of the glue is not enough. *Biochim. Biophys. Acta* **2001**, *1552*, 39–45. [CrossRef]

64. Müller, W.E.G. Cell-membranes in sponges. *Int. Rev. Cytol.* **1982**, *77*, 129–181.

65. Müller, W.E.G.; Zahn, R.K. Purification and characterization of a species-specific aggregation factor in sponges. *Exp. Cell Res.* **1973**, *80*, 95–104. [CrossRef]

66. Henkart, P.; Humphreys, S.; Humphreys, T. Characterization of sponge aggregation factor. A unique proteoglycan complex. *Biochemistry* **1973**, *12*, 3045–3050. [CrossRef] [PubMed]

67. Schröder, H.C.; Kuchino, Y.; Gramzow, M.; Kurelec, B.; Friese, U.; Uhlenbruck, G.; Müller, W.E. Induction of Ras gene expression by homologous aggregation factor in cells from the sponge *Geodia cydonium*. *J. Biol. Chem.* **1988**, *263*, 16334–16340. [PubMed]

68. Müller, W.E.; Kurelec, B.; Zahn, R.K.; Muller, I.; Vaith, P.; Uhlenbruck, G. Aggregation of sponge cells. Function of a lectin in its homologous biological system. *J. Biol. Chem.* **1979**, *254*, 7479–7481. [PubMed]

69. Hanisch, F.G.; Saur, A.; Müller, W.E.G.; Conrad, J.; Uhlenbruck, G. Further characterization of a lectin and its invivo receptor from *Geodia cydonium*. *Biochim. Biophys. Acta* **1984**, *801*, 388–395. [CrossRef]

70. Pfeifer, K.; Frank, W.; Schröder, H.C.; Gamulin, V.; Rinkevich, B.; Batel, R.; Müller, I.M.; Müller, W.E. Cloning of the polyubiquitin cDNA from the marine sponge *Geodia cydonium* and its preferential expression during reaggregation of cells. *J. Cell Sci.* **1993**, *106*, 545–553. [PubMed]

71. Pancer, Z.; Kruse, M.; Müller, I.; Müller, W.E. On the origin of metazoan adhesion receptors: Cloning of integrin alpha subunit from the sponge *Geodia cydonium*. *Mol. Biol. Evol.* **1997**, *14*, 391–398. [CrossRef] [PubMed]

72. Wimmer, W.; Perovic, S.; Kruse, M.; Schröder, H.C.; Krasko, A.; Batel, R.; Müller, W.E. Origin of the integrin-mediated signal transduction. Functional studies with cell cultures from the sponge *Suberites domuncula*. *Eur. J. Biochem.* **1999**, *260*, 156–165. [CrossRef] [PubMed]

73. Müller, W.E. Molecular phylogeny of Metazoa (animals): Monophyletic origin. *Naturwissenschaften* **1995**, *82*, 321–329. [CrossRef] [PubMed]

74. Schutze, J.; Skorokhod, A.; Müller, I.M.; Müller, W.E. Molecular evolution of the metazoan extracellular matrix: Cloning and expression of structural proteins from the demosponges *Suberites domuncula* and *Geodia cydonium*. *J. Mol. Evol.* **2001**, *53*, 402–415. [CrossRef] [PubMed]

75. Hulpiau, P.; Van Roy, F. Molecular evolution of the cadherin superfamily. *Int. J. Biochem. Cell Biol.* **2009**, *41*, 349–369. [CrossRef] [PubMed]

76. Nichols, S.A.; Roberts, B.W.; Richter, D.J.; Fairclough, S.R.; King, N. Origin of metazoan cadherin diversity and the antiquity of the classical cadherin/beta-catenin complex. *Proc. Natl. Acad. Sci. USA* **2012**, *109*, 13046–13051. [CrossRef] [PubMed]

77. Bokel, C.; Brown, N.H. Integrins in development: Moving on, responding to, and sticking to the extracellular matrix. *Dev. Cell* **2002**, *3*, 311–321. [CrossRef]

78. Johnson, M.S.; Lu, N.; Denessiouk, K.; Heino, J.; Gullberg, D. Integrins during evolution: Evolutionary trees and model organisms. *Biochim. Biophys. Acta* **2009**, *1788*, 779–789. [CrossRef] [PubMed]

79. Brower, D.L.; Brower, S.M.; Hayward, D.C.; Ball, E.E. Molecular evolution of integrins: Genes encoding integrin beta subunits from a coral and a sponge. *Proc. Natl. Acad. Sci. USA* **1997**, *94*, 9182–9187. [CrossRef] [PubMed]

80. Stephan, F. Spontaneous tumors in the planarian Dugesia tigrina. *Comptes Rendus des Seances de la Societe de Biologie et de Ses Filiales* **1962**, *156*, 920–922. [PubMed]

81. Steeg, P.S.; Bevilacqua, G.; Kopper, L.; Thorgeirsson, U.P.; Talmadge, J.E.; Liotta, L.A.; Sobel, M.E. Evidence for a novel gene associated with low tumor metastatic potential. *J. Natl. Cancer Inst.* **1988**, *80*, 200–204. [CrossRef] [PubMed]

82. Bilitou, A.; Watson, J.; Gartner, A.; Ohnuma, S. The NM23 family in development. *Mol. Cell. Biochem.* **2009**, *329*, 17–33. [CrossRef] [PubMed]

83. Hartsough, M.T.; Morrison, D.K.; Salerno, M.; Palmieri, D.; Ouatas, T.; Mair, M.; Patrick, J.; Steeg, P.S. Nm23-H1 metastasis suppressor phosphorylation of kinase suppressor of Ras via a histidine protein kinase pathway. *J. Biol. Chem.* **2002**, *277*, 32389–32399. [CrossRef] [PubMed]

84. Postel, E.H. Multiple biochemical activities of NM23/NDP kinase in gene regulation. *J. Bioenerg. Biomembr.* **2003**, *35*, 31–40. [CrossRef] [PubMed]

85. Harcet, M.; Lukic-Bilela, L.; Cetkovic, H.; Müller, W.E.G.; Gamulin, V. Identification and analysis of cDNAs encoding two nucleoside diphosphate kinases (NDPK/NM23) from the marine sponge *Suberites domuncula*. *Croat. Chem Acta* **2005**, *78*, 343–348.

86. Perina, D.; Bosnar, M.H.; Bago, R.; Mikoc, A.; Harcet, M.; Dezeljin, M.; Cetkovic, H. Sponge non-metastatic group I Nme gene/protein—Structure and function is conserved from sponges to humans. *BMC Evol. Biol.* **2011**, *11*, 87. [CrossRef] [PubMed]

87. Cetkovic, H.; Perina, D.; Harcet, M.; Mikoc, A.; Bosnar, M.H. Nme family of proteins—Clues from simple animals. *Naunyn Schmiedebergs Arch. Pharmacol.* **2015**, *388*, 133–142. [CrossRef] [PubMed]

88. Perina, D.; Bosnar, M.H.; Mikoc, A.; Müller, W.E.G.; Cetkovic, H. Characterization of Nme6-like gene/protein from marine sponge *Suberites domuncula*. *Naunyn Schmiedebergs Arch. Pharmacol.* **2011**, *384*, 451–460. [CrossRef] [PubMed]

89. Wennerberg, K.; Rossman, K.L.; Der, C.J. The Ras superfamily at a glance. *J. Cell Sci.* **2005**, *118*, 843–846. [CrossRef] [PubMed]

90. Rojas, A.M.; Fuentes, G.; Rausell, A.; Valencia, A. The Ras protein superfamily: Evolutionary tree and role of conserved amino acids. *J. Cell Biol.* **2012**, *196*, 189–201. [CrossRef]

91. Schaeffer, D.J. Planarians as a model system for in vivo tumorigenesis studies. *Ecotoxicol. Environ. Saf.* **1993**, *25*, 1–18. [CrossRef] [PubMed]

92. Kaczmarsky, L.T. Coral disease dynamics in the central Philippines. *Dis. Aquat. Organ.* **2006**, *69*, 9–21. [CrossRef] [PubMed]

93. Peters, E.C.; Halas, J.C.; Mccarty, H.B. Calicoblastic neoplasms in acropora-palmata, with a review of reports on anomalies of growth and form in corals. *J. Natl. Cancer Inst.* **1986**, *76*, 895–912. [PubMed]

94. Andersen, R.J. Sponging off nature for new drug leads. *Biochem. Pharmacol.* **2017**, *139*, 3–14. [CrossRef] [PubMed]

95. Ruiz-Torres, V.; Encinar, J.A.; Herranz-Lopez, M.; Perez-Sanchez, A.; Galiano, V.; Barrajon-Catalan, E.; Micol, V. An updated review on marine anticancer compounds: The use of virtual screening for the discovery of small-molecule cancer drugs. *Molecules* **2017**, *22*, 1037. [CrossRef] [PubMed]

96. Halim, H.; Chunhacha, P.; Suwanborirux, K.; Chanvorachote, P. Anticancer and antimetastatic activities of renieramycin M, a marine tetrahydroisoquinoline alkaloid, in human non-small cell lung cancer cells. *Anticancer Res.* **2011**, *31*, 193–201. [PubMed]

97. Sharma, S.; Guru, S.K.; Manda, S.; Kumar, A.; Mintoo, M.J.; Prasad, V.D.; Sharma, P.R.; Mondhe, D.M.; Bharate, S.B.; Bhushan, S. A marine sponge alkaloid derivative 4-chloro fascaplysin inhibits tumor growth and VEGF mediated angiogenesis by disrupting PI3K/Akt/mTOR signaling cascade. *Chem. Biol. Interact.* **2017**, *275*, 47–60. [CrossRef] [PubMed]

98. Ye, J.J.; Zhou, F.; Al-Kareef, A.M.Q.; Wang, H. Anticancer agents from marine sponges. *J. Asian Nat. Prod. Res.* **2015**, *17*, 64–88. [CrossRef] [PubMed]

99. Anjum, K.; Abbas, S.Q.; Shah, S.A.A.; Akhter, N.; Batool, S.; Ul Hassan, S.S. Marine sponges as a drug treasure. *Biomol. Ther.* **2016**, *24*, 347–362. [CrossRef] [PubMed]

100. Skropeta, D.; Pastro, N.; Zivanovic, A. Kinase inhibitors from marine sponges. *Mar. Drugs* **2011**, *9*, 2131–2154. [CrossRef] [PubMed]

101. Rodriguez, J.; Peters, B.M.; Kurz, L.; Schatzman, R.C.; Mccarley, D.; Lou, L.; Crews, P. An alkaloid protein-kinase-C inhibitor, xestocyclamine-A, from the marine sponge *Xestospongia* sp. *J. Am. Chem. Soc.* **1993**, *115*, 10436–10437. [CrossRef]

102. Patil, A.D.; Freyer, A.J.; Killmer, L.; Hofmann, G.; Randall, K. Z-axinohydantoin and debromo-Z-axinohydantoin from the sponge *Stylotella aurantium*: Inhibitors of protein kinase C. *Nat. Prod. Lett.* **1997**, *9*, 201–207. [CrossRef]

103. Patil, A.D.; Freyer, A.J.; Killmer, L.; Offen, P.; Carte, B.; Jurewicz, A.J.; Johnson, R.K. Frondosins, five new sesquiterpene hydroquinone derivatives with novel skeletons from the sponge Dysidea frondosa: Inhibitors of interleukin-8 receptors. *Tetrahedron* **1997**, *53*, 5047–5060. [CrossRef]

104. Willis, R.H.; DeVries, D.J. BRS1, A C30 BIS-Amino, BIS-Hydroxy polyunsaturated lipid from an Australian calcareous sponge that inhibits protein kinase C. *Toxicon* **1997**, *35*, 1125–1129. [CrossRef]

105. Shigemori, H.; Madono, T.; Sasaki, T.; Mikami, Y.; Kobayashi, J. Nakijiquinone-A and nakijiquinone-B, new antifungal sesquiterpenoid quinones with an amino-acid residue from an Okinawan marine sponge. *Tetrahedron* **1994**, *50*, 8347–8354. [CrossRef]

106. Kobayashi, J.; Madono, T.; Shigemori, H. Nakijiquinone-C and nakijiquinone-D, new sesquiterpenoid quinones with a hydroxy amino-acid residue from a marine sponge inhibiting c-erbB-2 kinase. *Tetrahedron* **1995**, *51*, 10867–10874. [CrossRef]

107. He, H.Y.; Kulanthaivel, P.; Baker, B.J. New cytotoxic sesterterpenes from the marine sponge *Spongia* sp. *Tetrahedron Lett.* **1994**, *35*, 7189–7192. [CrossRef]

108. Longley, R.E.; Harmody, D. A rapid colorimetric microassay to detect agonists antagonists of protein-kinase-C based on adherence of EL-4.IL-2 cells. *J. Antibiot.* **1991**, *44*, 93–102. [CrossRef] [PubMed]

109. Horton, P.A.; Koehn, F.E.; Longley, R.E.; Mcconnell, O.J. Lasonolide-A, a new cytotoxic macrolide from the marine sponge *Forcepia* sp. *J. Am. Chem. Soc.* **1994**, *116*, 6015–6016. [CrossRef]

110. Isbrucker, R.A.; Guzman, E.A.; Pitts, T.P.; Wright, A.E. Early effects of lasonolide A on pancreatic cancer cells. *J. Pharmacol. Exp. Ther.* **2009**, *331*, 733–739. [CrossRef] [PubMed]

111. Alvi, K.A.; Jaspars, M.; Crews, P.; Strulovici, B.; Oto, E. Penazetidine-A, an alkaloid inhibitor of protein-kinase-C. *Bioorg. Med. Chem. Lett.* **1994**, *4*, 2447–2450. [CrossRef]

112. Cimino, G.; Derosa, S.; Destefano, S.; Mazzarella, L.; Puliti, R.; Sodano, G. Isolation and X-ray crystal-structure of a novel bromo-compound from 2 marine sponges. *Tetrahedron Lett.* **1982**, *23*, 767–768. [CrossRef]

113. Meijer, L.; Thunnissen, A.M.W.H.; White, A.W.; Garnier, M.; Nikolic, M.; Tsai, L.H.; Walter, J.; Cleverley, K.E.; Salinas, P.C.; Wu, Y.Z.; et al. Inhibition of cyclin-dependent kinases, GSK-3 beta and CK1 by hymenialdisine, a marine sponge constituent. *Chem. Biol.* **2000**, *7*, 51–63. [CrossRef]

114. Killday, K.B.; Yarwood, D.; Sills, M.A.; Murphy, P.T.; Hooper, J.N.; Wright, A.E. Microxine, a new cdc2 kinase inhibitor from the Australian marine sponge *Microxina* species. *J. Nat. Prod.* **2001**, *64*, 525–526. [CrossRef] [PubMed]

115. Walker, S.R.; Carter, E.J.; Huff, B.C.; Morris, J.C. Variolins and related alkaloids. *Chem. Rev.* **2009**, *109*, 3080–3098. [CrossRef] [PubMed]

116. Soni, R.; Muller, L.; Furet, P.; Schoepfer, J.; Stephan, C.; Zumstein-Mecker, S.; Fretz, H.; Chaudhuri, B. Inhibition of cyclin-dependent kinase 4 (Cdk4) by fascaplysin, a marine natural product. *Biochem. Biophys. Res. Commun.* **2000**, *275*, 877–884. [CrossRef] [PubMed]

117. Kobayashi, J.; Suzuki, M.; Tsuda, M. Konbu'acidin a, a new bromopyrrole alkaloid with cdk4 inhibitory activity from *Hymeniacidon* sponge. *Tetrahedron* **1997**, *53*, 15681–15684. [CrossRef]

118. Bifulco, G.; Bruno, I.; Minale, L.; Riccio, R.; Debitus, C.; Bourdy, G.; Vassas, A. Bioactive prenylhydroquinone sulfates and a novel c-31 furanoterpene alcohol sulfate from the marine sponge, *Ircinia* sp. *J. Nat. Prod.* **1995**, *58*, 1444–1449. [CrossRef]

119. Alvi, K.A.; Diaz, M.C.; Crews, P.; Slate, D.L.; Lee, R.H.; Moretti, R. Evaluation of new sesquiterpene quinones from 2 Dysidea sponge species as inhibitors of protein tyrosine kinase. *J. Org. Chem.* **1992**, *57*, 6604–6607. [CrossRef]

120. Lee, R.H.; Slate, D.L.; Moretti, R.; Alvi, K.A.; Crews, P. Marine sponge polyketide inhibitors of protein tyrosine kinase. *Biochem. Biophys. Res. Commun.* **1992**, *184*, 765–772. [CrossRef]

121. Kreuter, M.H.; Leake, R.E.; Rinaldi, F.; Mullerklieser, W.; Maidhof, A.; Müller, W.E.G.; Schröder, H.C. Inhibition of intrinsic protein tyrosine kinase-activity of EGF-receptor kinase complex from human breast-cancer cells by the marine sponge metabolite (+)-aeroplysinin-1. *Comp. Biochem. Phys. B* **1990**, *97*, 151–158. [CrossRef]

122. Hertiani, T.; Edrada-Ebel, R.A.; Kubbutat, M.H.G.; van Soest, R.W.M.; Proksch, P. Protein kinase inhibitors from Indonesian sponge *Axynissa* sp. *Indones. J. Pharm.* **2008**, *19*, 78–85.

123. Calcabrini, C.; Catanzaro, E.; Bishayee, A.; Turrini, E.; Fimognari, C. Marine sponge natural products with anticancer potential: An updated review. *Mar. Drugs* **2017**, *15*, 310. [CrossRef] [PubMed]

124. Gordon, E.M.; Sankhala, K.K.; Chawla, N.; Chawla, S.P. Trabectedin for soft tissue sarcoma: Current status and future perspectives. *Adv. Ther.* **2016**, *33*, 1055–1071. [CrossRef] [PubMed]

125. Dyshlovoy, S.A.; Fedorov, S.N.; Shubina, L.K.; Kuzmich, A.S.; Bokemeyer, C.; Keller-von Amsberg, G.; Honecker, F. Aaptamines from the marine sponge *Aaptos* sp. display anticancer activities in human cancer cell lines and modulate AP-1-, NF-kappa b-, and p53-dependent transcriptional activity in mouse JB6 CL41 cells. *BioMed Res. Int.* **2014**, *2014*, 469309. [CrossRef] [PubMed]

126. Berry, E.; Hardt, J.L.; Clardy, J.; Lurain, J.R.; Kim, J.J. Induction of apoptosis in endometrial cancer cells by psammaplysene a involves foxo1. *Gynecol. Oncol.* **2009**, *112*, 331–336. [CrossRef] [PubMed]

127. Chen, Y.L.; Zhou, Q.X.; Zhang, L.; Zhong, Y.X.; Fan, G.W.; Zhang, Z.; Wang, R.; Jin, M.H.; Qiu, Y.L.; Kong, D.X. Stellettin B induces apoptosis in human chronic myeloid leukemia cells via targeting PI3K and Stat5. *Oncotarget* **2017**, *8*, 28906–28921. [CrossRef] [PubMed]

128. Wang, R.; Zhang, Q.; Peng, X.; Zhou, C.; Zhong, Y.X.; Chen, X.; Qiu, Y.L.; Jin, M.H.; Gong, M.; Kong, D.X. Stellettin B induces G1 arrest, apoptosis and autophagy in human non-small cell lung cancer A549 cells via blocking PI3K/Akt/mTOR pathway. *Sci. Rep.* **2016**, *6*, 27071. [CrossRef] [PubMed]

129. Tang, S.A.; Zhou, Q.X.; Guo, W.Z.; Qiu, Y.L.; Wang, R.; Jin, M.H.; Zhang, W.J.; Li, K.; Yamori, T.; Dan, S.G.; et al. In vitro antitumor activity of stellettin B, a triterpene from marine sponge Jaspis stellifera, on human glioblastoma cancer SF295 cells. *Mar. Drugs* **2014**, *12*, 4200–4213. [CrossRef] [PubMed]

130. Trisciuoglio, D.; Uranchimeg, B.; Cardellina, J.H.; Meragelman, T.L.; Matsunaga, S.; Fusetani, N.; Del Bufalo, D.; Shoemaker, R.H.; Melillo, G. Induction of apoptosis in human cancer cells by candidaspongiolide, a novel sponge polyketide. *J. Natl. Cancer Inst.* **2008**, *100*, 1233–1246. [CrossRef] [PubMed]

131. Schumacher, M.; Cerella, C.; Eifes, S.; Chateauvieux, S.; Morceau, F.; Jaspars, M.; Dicato, M.; Diederich, M. Heteronemin, a spongean sesterterpene, inhibits TNF alpha-induced NF-kappa B activation through proteasome inhibition and induces apoptotic cell death. *Biochem. Pharm.* **2010**, *79*, 610–622. [CrossRef]

132. Wu, J.C.; Wang, C.T.; Hung, H.C.; Wu, W.J.; Wu, D.C.; Chang, M.C.; Sung, P.J.; Chou, Y.W.; Wen, Z.H.; Tai, M.H. Heteronemin is a novel c-Met/STAT3 inhibitor against advanced prostate cancer cells. *Prostate* **2016**, *76*, 1469–1483. [CrossRef] [PubMed]

133. Huang, H.H.; Kuo, S.M.; Wu, Y.J.; Su, J.H. Improvement and enhancement of antibladder carcinoma cell effects of heteronemin by the nanosized hyaluronan aggregation. *Int. J. Nanomed.* **2016**, *11*, 1237–1251. [CrossRef] [PubMed]

134. Choi, H.J.; Bae, S.J.; Kim, N.D.; Jung, J.H.; Choi, Y.H. Induction of apoptosis by dideoxypetrosynol A, a polyacetylene from the sponge *Petrosia* sp., in human skin melanoma cells. *Int. J. Mol. Med.* **2004**, *14*, 1091–1096. [CrossRef] [PubMed]

135. Fukuoka, K.; Yamagishi, T.; Ichihara, T.; Nakaike, S.; Iguchi, K.; Yamada, Y.; Fukumoto, H.; Yoneda, T.; Samata, K.; Ikeya, H.; et al. Mechanism of action of aragusterol a (YTA0040), a potent anti-tumor marine steroid targeting the G(1) phase of the cell cycle. *Int. J. Cancer* **2000**, *88*, 810–819. [CrossRef]

136. Bae, S.Y.; Song, J.; Shin, Y.; Kim, W.K.; Oh, J.; Choi, T.J.; Jeong, E.J.; Park, S.H.; Jang, E.J.; Kang, J.I.; et al. Anti-proliferative effect of (19Z)-halichondramide from the sponge *Chondrosia corticata* via G2/M cell cycle arrest and suppression of mTOR signaling in human lung cancer cells. *Cancer Res.* **2014**, *74*. [CrossRef]

137. Aoki, S.; Kong, D.; Matsui, K.; Kobayashi, M. Smenospongine, a spongean sesquiterpene aminoquinone, induces erythroid differentiation in k562 cells. *Anti-Cancer Drug* **2004**, *15*, 363–369. [CrossRef]

138. Aoki, S.; Kong, D.X.; Matsui, K.; Kobayashi, M. Erythroid differentiation in K562 chronic myelogenous cells induced by crambescidin 800, a pentacyclic guanidine alkaloid. *Anticancer Res.* **2004**, *24*, 2325–2330. [PubMed]

139. Fung, S.Y.; Sofiyev, V.; Schneiderman, J.; Hirschfeld, A.F.; Victor, R.E.; Woods, K.; Piotrowski, J.S.; Deshpande, R.; Li, S.C.; de Voogd, N.J.; et al. Unbiased screening of marine sponge extracts for antiinflammatory agents combined with chemical genomics identifies girolline as an inhibitor of protein synthesis. *ACS Chem. Biol.* **2014**, *9*, 247–257. [CrossRef] [PubMed]

140. Zhang, C.; Kim, S.K. Matrix metalloproteinase inhibitors (MMPIs) from marine natural products: The current situation and future prospects. *Mar. Drugs* **2009**, *7*, 71–84. [CrossRef] [PubMed]

141. Shin, Y.; Kim, G.D.; Jeon, J.E.; Shin, J.; Lee, S.K. Antimetastatic effect of halichondramide, a trisoxazole macrolide from the marine sponge *Chondrosia corticata*, on human prostate cancer cells via modulation of epithelial-to-mesenchymal transition. *Mar. Drugs* **2013**, *11*, 2472–2485. [CrossRef] [PubMed]

142. Robert, J. Comparative study of tumorigenesis and tumor immunity in invertebrates and nonmammalian vertebrates. *Dev. Comp. Immunol.* **2010**, *34*, 915–925. [CrossRef] [PubMed]
143. Monahan-Earley, R.; Dvorak, A.M.; Aird, W.C. Evolutionary origins of the blood vascular system and endothelium. *J. Thromb. Haemost.* **2013**, *11*, 46–66. [CrossRef] [PubMed]
144. Müller, W.E.; Blumbach, B.; Müller, I.M. Evolution of the innate and adaptive immune systems: Relationships between potential immune molecules in the lowest metazoan phylum (Porifera) and those in vertebrates. *Transplantation* **1999**, *68*, 1215–1227. [CrossRef] [PubMed]